Earth Systems

Earth Systems: Processes and Issues is the ideal textbook for introductory courses in Earth systems science and environmental science. Integrating the principles of the natural sciences, engineering, and economics as they pertain to the global environment, it explains the complex couplings and feedback mechanisms linking the geosphere, biosphere, hydrosphere, and atmosphere.

An impressive group of internationally respected researchers and lecturers have collaborated to produce this fully integrated environmental textbook. They bring together a vast wealth of teaching experience to cover the necessary breadth of the physical and life sciences, melded with environmental economics and legislation. This book's truly unique contribution lies in the interweaving of these topics with engineering environmental systems, regional case studies, and the economic implications of environmental policy decisions.

The textbook has been designed for the wide range of courses at the first-year university level that touch upon environmental issues: for example, Earth and atmospheric science, oceanography, biologic science, geography, civil engineering, environmental law, political science, and environmental economics. Each chapter includes a reading list of some of the most important references, and follow-on questions will encourage students to explore the subject further.

This text will favorably influence the future development of environmental studies and Earth systems science.

The authors are: Susan E. Alexander; Kathryn Arbeit; Julie K. Bartley; Carol L. Boggs; Peter G. Brewer; Robert Chatfield; Nona Chiariello; Paul R. Ehrlich; Marco T. Einaudi; W. G. Ernst; Christopher B. Field; W. S. Fyfe; Lawrence H. Goulder; James C. Ingle, Jr.; Mark J. Johnsson; Isaac R. Kaplan; Donald Kennedy; Keith Loague; Pamela A. Matson; Patricia A. Maurice; Rosamond L. Naylor; Christopher Place; Terry L. Root; Joan Roughgarden; Stephen H. Schneider; Edward A. G. Schuur; Barton H. Thompson, Jr.; and Jane Woodward.

W. G. Ernst is currently a professor in the Department of Geological and Environmental Sciences at Stanford University. He is the author of 6 books and research memoirs, editor of 11 research volumes, and author of more than 175 scientific papers.

Earth Systems

PROCESSES AND ISSUES

Edited by
W. G. ERNST

 **CAMBRIDGE**
UNIVERSITY PRESS

PUBLISHED BY THE PRESS SYNDICATE OF THE UNIVERSITY OF CAMBRIDGE
The Pitt Building, Trumpington Street, Cambridge, United Kingdom

CAMBRIDGE UNIVERSITY PRESS
The Edinburgh Building, Cambridge CB2 2RU, UK   http: //www.cup.cam.ac.uk
40 West 20th Street, New York, NY 10011-4211, USA   http: //www.cup.org
10 Stamford Road, Oakleigh, Melbourne 3166, Australia
Ruiz de Alarcón 13, 28014 Madrid, Spain

First published 2000

Printed in the United States of America

Typefaces Stone Serif 8.5/12 pt., Eurostile Demibold, and Optima   *System* DeskTopPro$_{/UX}$® [BV]

A catalog record for this book is available from the British Library.

Library of Congress Cataloging-in-Publication Data
Earth systems : processes and issues / edited by W.G. Ernst.
p.   cm.
ISBN 0 521 47323 3 (hardbound)
1. Earth sciences.   I. Ernst, W. G. (Wallace Gary), 1931– .
QE28.E137   1999
550–dc21   98-37539
                                                        CIP

Source for title page photograph: Courtesy of United States Geological Survey.

ISBN 0 521 47323 3 hardback
ISBN 0 521 47895 2 paperback

Contents

Preface

This introductory text integrates principles of the physical sciences, engineering, and economics as they pertain to the global environment. The complex couplings and feedback mechanisms linking the geosphere, biosphere, hydrosphere, and atmosphere are analyzed by more than a score of authors who carry out nationally visible investigations in the fundamental disciplines, and who collaborate across field boundaries in both their research and teaching.

The subject matter has been addressed over the past five years (and is continuing) for an elementary course at Stanford entitled *Introduction to Earth Systems*. The philosophy of presentation is problem-focused, not discipline-focused. Topics have been developed by a team of lecturers who have attempted to provide a seamless integration spanning the traditional subjects of the natural sciences, engineering, and the social sciences. These lecturers include most of the contributors to this text.

This book is unusual in terms of the breadth of its physical and life sciences content; its truly unique aspects, however, lie in the interweaving of these topics with engineering environmental systems, case studies, and the economic implications of environmental public policy decisions. To achieve a relatively continuous gradation and smooth passage through the remarkably broad range of subjects, the editor and his editorial colleagues at Cambridge University Press, Catherine Flack, Tigger Posey, and Lisa Albers, have provided cross-referencing and uniformity of style and content level among the various chapters. In addition, contributors have had access to preliminary drafts of the entire manuscript for review, enabling them to cross-link subjects and areas, thereby enriching both their own contributions and those of the other authors.

The rationale for this fruitful synthesis, involving the melding of disparate disciplines, involves a nontraditional approach to the quantitative understanding of interactive global environmental phenomena and more effective Earth systems problem solving. The target audience is college/university undergraduates. *Earth Systems: Processes and Issues* is an introductory text that assumes a readership familiar with the principles of elementary chemistry and college mathematics. It attempts to provide a high-level but predominantly qualitative rather than quantitative treatment of global environmental subjects. Detailed sections and topical examples are placed in boxes so that interested students can gain in-depth understanding and the more general reader will not be deterred by the complexities. Individual chapters include an introductory box: For chapters on scientific and technologic subjects, this introduction serves as a brief preview of the societal implications; for chapters on policy issues, the introductory statement summarizes the scientific and/or technologic foundation of the problem. Short lists of pertinent references for further reading and questions posed at the end of each chapter encourage the interested reader to explore the subject in greater detail. A glossary is included at the end of the text.

What is abundantly clear from even a casual reading of this book is that the Earth's biosphere is increasingly under siege. A burgeoning human population, driven by hunger, procreativity, and an intense urge to survive, and aided by ever more invasive, efficient technologies, is altering the interconnected planetary ecosystems through a wide range of processes – both obvious and subtle. Impacts include pollution of the global atmosphere, the oceans, lakes, rivers, and groundwaters, the loss of fertile topsoils, habitats, and biodiversity. Increasing desertification and deforestation, and the exhaustion of fisheries, fossil energy deposits, and mineral resources are some of the oncoming results. In aggregate, the thirty-three chapters of *Earth Systems: Processes and Issues* constitute a worldwide wake-up call to constituencies and policy makers of all socioeconomic sectors and political persuasions. The need for conservation, human population management, and wise utilization of the environment requires action at both local and global levels. As this text demonstrates, current human practices are borrowing substantially from the future sustainable carrying capacity of the planet. The time for effective action is short, indeed – and failure to act will only exacerbate the ongoing environmental degradation. The generation for which this book is intended will decide the fate of the Earth's biosphere, one way or the other.

W. G. Ernst, editor, Stanford University, 1998

Contributors

Susan E. Alexander
Dept. of Earth Systems Science and Policy
California State University, Monterey Bay
Seaside, CA 93955

Kathryn Arbeit
Dept. of Petroleum Engineering
Stanford University
Stanford, CA 94305-2220

Julie K. Bartley
Geology Department
State University of West Georgia
Carrollton, GA 30118

Carol L. Boggs
Dept. of Biological Sciences
Stanford University
Stanford, CA 94305-5020

Peter G. Brewer
Monterey Bay Aquarium Research Institute
P. O. Box 628
Moss Landing, CA 95039

Robert Chatfield
Earth Science Division
NASA Ames Research Center; MS 245-5
Moffett Field, CA 94035-1000

Nona Chiariello
Dept. of Plant Biology
Carnegie Institution of Washington
Stanford, CA 94305

Paul R. Ehrlich
Dept. of Biological Sciences
Stanford University
Stanford, CA 94305-5020

Marco T. Einaudi
Dept. Geological and Environmental Sciences
Stanford University
Stanford, CA 94305-2115

W. G. Ernst
Dept. Geological and Environmental Sciences
Stanford University
Stanford, CA 94305-2115

Christopher B. Field
Dept. of Plant Biology
Carnegie Institution of Washington
Stanford, CA 94305

W. S. Fyfe
Dept. of Earth Sciences
University of Western Ontario
London, Ontario N6A 5B7; Canada

Lawrence H. Goulder
Dept. of Economics
Institute for International Studies
Stanford University
Stanford, CA 94305-6072

James C. Ingle, Jr.
Dept. Geological and Environmental Sciences
Stanford University
Stanford, CA 94305-2115

Mark J. Johnsson
Dept. of Geology
Bryn Mawr College
Bryn Mawr, PA 19010

Isaac R. Kaplan
Dept. of Earth and Space Sciences
University of California
Los Angeles, CA 90095-1567

Donald Kennedy
Dept. of Biological Sciences
Stanford University
Stanford, CA 94305-5020

Keith Loague
Dept. Geological and Environmental Sciences
Stanford University
Stanford, CA 94305-2115

Pamela A. Matson
Dept. of Geological and Environmental Sciences
Stanford University
Stanford, CA 94305-2115

Patricia A. Maurice
Dept. of Geology
Kent State University
Kent, OH 44242

Rosamond L. Naylor
Institute for International Studies
Stanford University
Stanford, CA 94305-6055

Christopher Place
Dept. of Petroleum Engineering
Stanford University
Stanford, CA 94305-2220

Terry L. Root
Dept. of Biological Sciences
University of Michigan
Ann Arbor, MI 48109

Joan Roughgarden
Dept. of Biological Sciences
Stanford University
Stanford, CA 94305-5020

Stephen H. Schneider
Dept. of Biological Sciences
Stanford University
Stanford, CA 94305-5020

Edward A. G. Schuur
College of Natural History
University of California
Berkeley, CA 94720

Barton H. Thompson, Jr.
Stanford Law School
Stanford, CA 94305-8610

Jane Woodward
Dept. of Petroleum Engineering
Stanford University
Stanford, CA 94305-2220

INTRODUCTION

THE EARTH AS A SYSTEM

1 Why Study Earth Systems Science?

STEPHEN H. SCHNEIDER

THE GREENHOUSE EFFECT: FACT OR MEDIA HYPE?

Is the so-called greenhouse effect a fact or a controversial hypothesis? As a climatologist, I am reminded of a headline I saw in the *New York Times* in early 1989: "U.S. Data Since 1895 Fail to Show Warming Trend." I must have had fifty phone calls to my office the day after that story came out, asking, "What happened to global warming?" One week later, after a new global average set of thermometer readings were put together, the very same *New York Times* reporter wrote another front-page article, this time stating "Global Warming for 1988 Was Found to Set a Record." Taken together, these two stories caused a lot of confusion. How could there be record warmth globally when the lower forty-eight states didn't warm much? Was there a greenhouse effect or not? What was going on? In fact, the reconciliation of the two stories was quite simple. It is important to place these headlines in perspective, recognizing that the lower forty-eight states constitute only 2 percent of the Earth's surface area. There is not a very high probability of getting the correct temperature of the whole globe by looking at just 2 percent of it. In fact, if you had looked at temperatures for Alaska or for central Eurasia in that same period and tried to make a statement about global temperatures based upon those data, you would have thought that the Earth had warmed up 1.5°C in that same period. Meanwhile, the North Atlantic region cooled 0.5°C or so in the same period.

The conclusion we can draw from a comparison of timely news articles like these is that the global warming problem, to take one example, is indeed global. Often, but not always, that means that what happens in our own backyard in the time frame of our recent experience may be irrelevant to the problems of the following century. A core lesson in Earth systems science teaches that the word "global" means that the experiences we have in our neighborhood (geographic or intellectual) may be instructive about a single component of global issues, but that we can't automatically extrapolate local experience to learn how the interconnected global systems work, let alone make a credible forecast of global changes over the long term. To make such sweeping statements with any authority, we need to look across various scales and disciplines at interconnected systems. Then we need to validate our concepts of global systems, often by going back in time or to local scales to check our global ideas.

The greenhouse effect is a scientific fact. Controversy over this issue arises primarily in discussions of whether humans will make a significant impact and what to do about it. Without good science as a basis for answering important questions such as what can happen, what are the potential consequences, and how likely are these outcomes, we cannot hope to answer authoritatively or confidently the question of what to do. This book gives the student a solid introduction to several crucial scientific disciplines so that he or she may know what questions to ask of the various disciplinarians in order to find both good data and a good solution to today's complex environmental problems.

Earth systems science tries to find solutions for real, global environmental problems at the times and places that they exist. These topics cannot be addressed comprehensively by looking through the limited lens of only one of the traditional disciplines established in academia, such as biology, chemistry, engineering, or economics. We certainly can't solve most global problems without the detailed information that those disciplines provide, but the study of Earth systems science suggests that we also need to find appropriate ways to *integrate* high-quality disciplinary work from several fields. Although scholars from various disciplines may study the Earth locally – in a tax district, a volcano, a thunderstorm, a patch of forest, or a test tube –

Earth systems scientists put the accent on "systems," the multiscale interactions of all these small-scale phenomena.

This introductory chapter is designed to give the reader a quick sketch of the excitement and urgency of this global-scale, systems-oriented approach to environmental science, technology, and policy problems. Our challenge is to be creative in doing something both new and necessary: to put together sets of expertise from various academic disciplines in original ways that will improve our understanding of both nature and humanity. Some will express concern over this approach, feeling that without in-depth content in each disciplinary subcomponent, our systems analyses will be shallow. Without the context of real problems, however, disciplinary specialists will lack the information necessary to solve pressing issues. From the perspective of Earth systems scientists, it is not sensible to debate whether it is worse to lose *context* by approaching real problems in depth, from the narrow purview of one area, or to lose *content* by integrating information across disciplines without studying any of the interrelated subfields in adequate depth. Both context and content are necessary. We need to blend them to a considerable extent, using context to help guide the selection of appropriate content areas. Although practical considerations will, of course, prevent budding Earth systems scientists from studying all relevant fields in tremendous depth, this text gives a solid foundation in the disciplinary sciences necessary to enable the student to engage in future interdisciplinary environmental pursuits, and to choose which content areas to explore more fully in the future.

THE SCALE OF EARTH PROCESSES

At what spatial scale do you think of the Earth's atmosphere as functioning? Think about one of the famous photographs taken from space by astronauts: You can probably visualize white clouds swirling around the blue globe, with the spiral patterns of storms standing out at 1000-kilometer scales. However, if your vantage point were from an airplane during a turbulent flight, you might think of atmospheric action taking place on a scale measurable in tens of meters. A balloonist who is able to see individual rain droplets or snowflakes drift by might conclude that atmospheric action takes place at the microscale of millimeters. Of course, these observations are all "correct," but knowledge of cloud microphysics in great detail does not by itself provide the context for understanding the large-scale atmospheric dynamics visible from space. As mathematical ecologist Simon Levin once put it, the world looks very different depending on the size of the window we are looking through.

Nature has amazing richness across the range of spatial and temporal scales at which processes and their interactions occur. You know from your own experience that winds blow and oceans move, but those aren't the only natural forces that are dynamic. Our "solid" Earth is not solid, if we define "solid" to mean forever immoveable in space and time. In fact, the Earth itself moves about in response to natural forces (see Chap. 6). The drift of continents, as we'll learn later, can have a major influence on both climate and life. Except for local phenonema such as earthquakes, landslides, and mountain glaciers, the time frame for major continent-scale Earth motions is thousands to millions of years. How the "solid" Earth interacts with air, water, and life is essential for understanding the Earth as a system, as knowledge of how and why the Earth system changes over geologic time allows us to calibrate our tools needed to forecast global changes.

Studying these phenomena at all relevant scales is no small task. In order to gain a good working knowledge of the Earth and its processes, we need to understand the interaction not only between systems but also between and among the various scales of activity of the many systems. Will a change in a small-scale biological community, such as the extinction of a species of termite, have any effect upon nutrient cycling, upon emissions of greenhouse gases from the soil, and ultimately upon global-scale weather patterns? At what point will nitrogen fertilizer used in agriculture create sufficient amounts of nitrous oxide emissions to warm the climate or deplete stratospheric ozone? At a global scale, nature exists nearly in a state of balance. Parts are constantly changing, while the whole continues to function as if in near equilibrium. If humans push too many parts out of balance, what will happen to the whole? How much resilience is there in each part at various scales? These are the kinds of questions that Earth systems science must address.

LOCAL VARIABILITY AND GLOBAL CHANGE

Earth systems science focuses on an issue called "global change," a phrase invented by people who study the Earth as a system to refer to the changes on a global scale (or regional changes that are repeated around the globe) that occur to those Earth systems (which could be physical, biologic, and/or social) that are interconnected and that humans have some component in forcing. Why then, you might ask, study continental drift as part of global change if humans are not able to influence the course of continental drift? If we don't understand how drifting continents affect the gases in the atmosphere, the climate, or biologic evolution, then we're not going to have the background knowledge necessary to forecast so-called global change, even though global change is driven in part by human disturbances such as deforestation and air pollution. In this textbook we explore traditional disciplines such as geology, atmospheric science, biology, technology, chemistry, agronomy, and economics. We also explore how humans are disturbing various components of the system. In the chapters that follow, we consider a number of questions:

- How does the entire system work?
- How does it work as a coupled set of subsystems?
- How are humans disturbing the system?
- What have we learned from how the system works that can help us forecast how human disturbances might play themselves out?
- What could – or should – we do about the information we collect?

Several years ago, I traveled to the picturesque town of Argentiere in the French Alps, a trip that demonstrated the dramatic changes that can occur in a short time, geologically speaking. I went there to see a famous glacier that was located far above the town. I took photographs of the glacier, framed against a local church steeple. It is a stunning sight, made all the more impressive when compared with an 1855 etching that pictures the glacier on the very outskirts of the town, as if it were about to devour the town. The more recent photographs show the glacier at some distance from the town, quite a distance up the mountains in the background. What accounts for this dramatic retreat in the century after 1855? One hundred and fifty years ago, the global climate was colder than it is today – about 1°C colder. The warming trend over the past century and a half is correlated with a major response in that glacier. A one-degree change may sound trivial, but if it is a sustained change, then it can have an identifiable impact, particularly on sensitive indicators such as mountain glaciers – most of which have been retreating during the twentieth century.

Small changes can add up to create large ones. For example, a number of years ago a satellite photograph of Israel was taken with a near-infrared wavelength device that showed the boundary between the Israeli Negev, the Egyptian Sinai, and the Gaza Strip, an unnatural, political boundary. Why, then, does it appear in photographs as a physical boundary? The line was visible because there were animal herds grazing more heavily on one side of the border fence, and the vegetation and soils there had been deeply disturbed. That changed the reflectivity of the surface to sunlight, which, in turn, alters the amount of sunlight absorbed, which, in turn, is the primary driving force behind the weather. The climate of the Earth – the natural climate – works from a balance between the amount of absorbed solar energy and the amount of outgoing, so-called infrared radiative energy. The key is, if we can "see" a political boundary as a physical line, then the Sun can "see" it, too. If the Sun can see it, that changes the amount of solar energy absorbed. If humans can change the amount of solar energy absorbed, they can affect the climate.

These simple examples demonstrate that the repeated patterns of local and regional changes that are taking place are significant. The sum of thousands of local to regional changes in the land surface can disrupt transcontinental migration patterns of birds and might have some influence on the overall climate at a larger scale, as well. The climate changes then influence agricultural activity, water supplies, and ecosystems locally.

Another example of local landscape damage is located in the American Great Plains region. A typical aerial view of this region includes perfectly round, dark green circles that mark irrigated fields. This kind of center-pivot irrigation typically uses fossil groundwater at a much faster rate than it can be replenished in the underlying aquifers (i.e., underground natural reservoirs). It is a practice that raises socioeconomic problems – whether it is fair to future generations, for example, for today's farmer to be using up a resource at a nonrenewable rate. The relevant aspect of this example is that we're changing the water balance of the system at the same time that we're changing the brightness of the system (which can affect local rainfall); we're also changing the local habitat for migratory birds that fly between Canada and Mexico – the neotropical migrants – and for waterfowl. Those birds are used to certain kinds of wetlands and certain other forms of habitat at their nesting sites and between there and their wintering grounds. However, when human activities dramatically change habitats, some species thrive and others are endangered. This, then, can create conditions in which extinctions of sensitive species are more likely to take place.

One of the most serious problems of global change connected to Earth systems science is the combined effect of habitat fragmentation and climate change. When climate changes, individual species adjust if they can, as they have in the past. Typically, they move with changing climate; for example, when the last ice age ended some 10,000 years ago, spruce trees moved from their ice age locations in the U.S. mid-Atlantic region to their current location as the northerly Boreal forests of Canada. What would happen if climate changed comparably today and the affected plant and animal species had to move again? Could flora and fauna successfully migrate across freeways, agricultural zones, and cities? The combination of habitat fragmentation and climate change makes it much more difficult for natural communities to adjust. This, in turn, sets up a potentially enormous management problem. Do we have to set aside nature reserves in interconnected areas and not simply isolated reserves and parks? If so, whose farms or houses or fields do we take away in order to create these reserves? How do we deal with risks to wildlife from highways? Do we spend money to create bypasses or elevated sections so that migration routes can be maintained? How much is it worth to protect the survival of a species or a habitat? Although these are essentially value choices, good science is necessary to help answer how such biologic conservation practices can take place in the most economically efficient way. Global change science involves looking at these kinds of questions. To answer them, we go to the various academic disciplines to ask, "What knowledge do you have?" In particular, we ask, "What can happen?" and "What are the odds it might happen?" The Earth systems scientist tries to integrate the information from many disciplines in order to address real problems.

THE BALANCING ACT: WEIGHING LOCAL AND GLOBAL NEEDS

In this discussion about environmental protection, we begin with global-scale causes of environmental degradation. This degradation is most often ascribed to increasing numbers of people striving for higher standards of living and using technologies or practices that often pollute or fragment the landscape. However, when one abandons the global or even the national perspective and looks instead at local environmental problems, these three multiplicative macroscale causes – population times per capita affluence times technology used – may not be easily seen. Corrupt officials, unaccountable industries, poverty, lack of appropriate labor force, or simple ignorance of less environmentally deleterious alternatives stand out as prime causes of local environmental degradation. These problems intersect at large scales with the demands for increasing use of land and resources from burgeoning populations seeking to improve their living standards and willing to use the cheapest available technologies toward that goal.

There is an equity issue involved in these dilemmas: Desire for economic progress today may create environmental problems for later generations or downstream neighbors, neither of which participate in the immediate decision making. We need to find solutions that do not treat nature as a nonrenewable resource for the benefit of a few today at the expense of many later – the problem known as "intergenerational equity." Also, some nations are economically better off than others. The desire for more equality often motivates low-cost development plans (burning unclean coal, for example) that can threaten massive environmental disruptions (global warming or health-damaging smog). This sort of "environment – development" tradeoff issue will lead to major debates in the decades ahead.

On the East Coast of the United States, from Boston to New York to Washington, the amount of heat being released from all the energy uses that take place is approximately 1 percent of the incident energy from the Sun. Did you ever hear a weather forecast for Manhattan that sounded something like: "Tonight it is going to be twenty-five degrees Fahrenheit in the city, and twelve in the suburbs"? Ever wonder why it is so much warmer in the city? The answer is that there is literally a "sun" on at night, heating the city – or, at least, the energy equivalent of a winter's sunny day. The so-called urban heat island effect tells us that if we release energy comparable to a few percent or more of that which arrives from the Sun, we're going to change the climate locally. That effect is important even though, at a global scale, the total amount of heat generated by human activities is a tiny percentage of the Earth's heat budget. The key is that the combination of energy use and all other human modifications to the land, water, and air is already regionally significant – and also very inequitably distributed – and rapidly is becoming global in scope.

If we look at the Earth over the past 30,000 years, we find that up until approximately 15,000 years ago ice sheets several kilometers thick covered most of Canada, and it was only 6000 years ago that the last remnants of ice disappeared over Hudson's Bay. What happened when all that ice that was over land melted? Sea level rose by more than 100 meters, the globally averaged climate warmed up approximately 5°C, whole habitats were reconfigured, and species became extinct – all this was a natural change. Although there were regional and short-term changes that were rapid, the sustained, globally averaged rates at which nature caused ice ages to melt into the warm 10,000-year period of relatively stable climate that saw human civilization develop was on the order of 1°C per 1000 years.

If we go back approximately 150,000 years, we find a comparable cycle of temperature changes, as well as changes in concentrations of methane gas (CH$_4$) and carbon dioxide

(CO$_2$) in the atmosphere. In Antarctica 125,000 years ago, that continent was approximately 2°C warmer on average than at present. Then, the temperature dropped (Fig. 1.1), fluctuated, and finally became extremely cold some 30,000 years ago. The last ice age peaked approximately 20,000 years ago. It took more than 10,000 years for the ice age to end; since then we've been in a 10,000-year so-called interglacial, the Holocene Epoch, during which temperatures have been within a degree or two of present temperatures (see Chap. 3). During this time, two gases changed their atmospheric concentrations in fairly close correlation to the temperature changes. These gases are very important climatically because they trap heat near the Earth's surface and are partly responsible for the so-called greenhouse effect. There is a strong correlation between methane gas (which is produced in nature by the anerobic decomposition of organic matter) and carbon dioxide: an approximate factor of 2 difference between the ice age and the interglacial methane, (CH$_4$), and a difference of approximately 30 to 40 percent CO$_2$ in the ice age and the interglacial. Simply put, lower concentrations of these so-called greenhouses gases occur when it is cold, and higher concentrations occur when it is warm. These fluctuations over time were all the work of nature.

Figure 1.1 shows that carbon dioxide concentrations of

Figure 1.1. Air bubbles trapped in ancient polar ice sheets can be analyzed to determine the changing composition of the atmosphere over hundreds of thousands of years. Such analyses at Vostok in Antarctica show that carbon dioxide (CO$_2$) concentrations were approximately 25 to 30 percent lower in glacial times than an interglacial periods over the past 160,000 years. Local temperatures (in Antarctica) at the extreme glacial times (approximately 20,000 and 150,000 years ago) were approximately 10°C (18°F) colder than at interglacial times. kyr BP = kiloyear (1000 years) before present; ppm = parts per million. (*Source*: Adapted with permission from J. M. Barnola, D. Raynaud, Y. S. Korotkevich, and C. Lorius, Vostok ice core provides 160,000-year record of atmospheric CO$_2$, *Nature*, Copyright © 1987.)

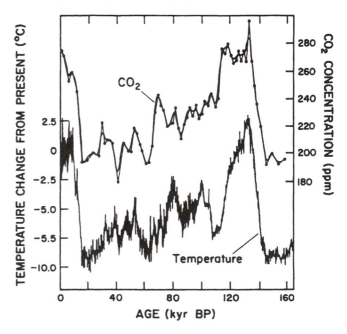

approximately 280 parts per million have remained very stable for the 10,000 years of our current interglacial era. However, as of 1998 that concentration level is at least 370 parts per million, an *un*naturally high number – the result of more people on the planet, demanding higher standards of living, engaging in agriculture, using fossil fuels for energy, developing the land, and cutting down trees. The fact of this global change in atmospheric composition of carbon dioxide is not controversial; is well understood by everyone who has studied the evidence, and humans are almost certainly responsible for it. What is controversial is the consideration of questions such as "What precisely is it going to do to the environment?" and "What can we do about it?"

Many components of Earth systems science are well understood. For example, we can divide the amount of incoming solar energy into percentages and track how much of it reflects from clouds or passes through to the Earth's surface. Likewise, we can track infrared radiative heat from upward emission to downward reradiation by so-called greenhouse gases such as water vapor, carbon dioxide, or methane, all in an effort to explain how the greenhouse effect works (see Chap. 14). That is not controversial, either. What is controversial is how much of the extra heat from such global changes (e.g., the 30 percent increase in carbon dioxide since the Industrial Revolution) will be available to raise surface temperatures directly versus how much will result in increased evaporation, which in turn might change cloudiness, which might reflect away extra sunlight or trap extra infrared radiation. The changes in evaporation or cloudiness, in turn, can "feedback" on the amount of energy retained by the climatic system and either accelerate or retard the initial warming from the global carbon dioxide change that would have occurred in the absence of such feedback.

FEEDBACK MECHANISMS AND THE HUMAN ELEMENT

Humans are what physiologists call homeostatic systems. We contain stabilizing, *negative feedback mechanisms*: If we get too hot, we sweat to cool down; if we get too cold we shiver, which is a mechanical way of generating heat. There are lots of feedback processes in the dynamic climatic system as well, some of them stabilizing, and some of them destabilizing. For example, if the Earth warms, some snow and ice will melt. Bright, white, reflective areas are then replaced by green trees, brown fields, or blue oceans, all of which are darker than the snowfields, so they absorb more sunlight. Hence, if the Earth warms and some snow melts, it absorbs more of the incoming sunlight and this feedback process accelerates the warming. That is a *positive feedback*. However, if more water evaporates and makes wider cloud cover that reflects more sunlight back to space, that is a negative feedback. In addition to the climatalogic feedback responses, we have to deal quantitatively with the biologic system. Trees absorb carbon dioxide from the air through photosynthesis – a potential negative feedback on global warming. However, soil

bacteria, which decompose dead organic matter into carbon dioxide or methane gas, work faster when it is warm – a potential positive feedback on global warming. We need also to understand the history of biologic evolution in order to identify rates of speciation (forming new species) and extinction that are natural, and causes that are natural. Just as we need to understand the geologic backdrop of ice ages coming and going in order to see how the climate works and thus be able to forecast climate change credibly in the future, we need to have some sense of how biologic evolution works in order to see how land and habitat fragmentation, chemical and pesticide release, climate change, species competition, and the synergisms of all these might impact on ecosystems and how specific species will fare in the future.

Into this mix must be added human science. There are approximately 6 billion people in the world today, with 1 billion living on the margins of nutritional deprivation, and many tens of millions who die every year from preventable illnesses because they are malnourished. These people demand and deserve improved standards of living. However, how they achieve those improved standards of living is critical to the environmental future. Will they develop the way more developed countries did it, using the cheapest available means with little regard for nature until society becomes affluent? Or will currently poorer countries develop and grow economically by using better technology and better organization in less environmentally destructive ways? The answers are going to have a dramatic impact on the future nature of the environment, and on what global change portends. Development is inevitable. The open questions are: "What kinds of development?" "Who pays?" "What is the distribution of resources and the distribution of consequences, be they economic, social, or environmental?" To study this problem we've got to look at the human dimension, which is driven by values, feelings, history, tradition, and, power.

I recall Indonesia in the late 1970s. On a particularly steamy afternoon when I was driven around Jogjakarta by a sweating pedicab driver, I asked him what his dream was, and a one-word answer quickly came back: "Toyota." On my return trip two decades later, nearly all the pedicabs I had seen earlier have now been replaced by Toyotas and the like. The tremendous increase in the number of cars worldwide has resulted in smog-choked cities and millions more tons of carbon dioxide being generated. To some – like most people in now developed nations who take their cars for granted – their proliferation is the price of progress. To others, the increase symbolizes quality of life. At present, most people are, reluctantly or enthusiastically, unwilling to trade cars for less polluting modes of transportation, regardless of the potentially beneficial consequences for the atmosphere. The inevitable consequences of population growth and development by business-as-usual technologies seem to be negative environmental side effects.

The open question isn't "Should we protect the environment *or* encourage the economy?" The better question is, simply, "How can we develop in environmentally sustainable ways?'' The answer is that there *are* ways to try to do both, but they are not necessarily the traditional ways of development. To convince people to accept these nontraditional ways can be a ''hard sell.'' It takes hard economics and hard science to describe what is commonly called ''environmentally sustainable development.'' Carbon dioxide is one of the principal greenhouse gases, and it results primarily from burning fossil fuels – coal, oil, and natural gas. Considering the total amount of emissions of carbon dioxide country by country around the world, we find the greatest emissions in the former Soviet Union, China, India, the United States, and Brazil. In India or China, data that might elicit comments such as, ''We're not the problem – the over-affluent are the problem.'' That is because these countries focus on the per capita emissions of carbon dioxide per country, in which developing countries rank low.

To improve the standard of living in many developing countries, those countries need the services provided by more energy. How they acquire it, and what systems they use, is dramatically important. One concept that needs to be considered is ''planetary bargaining,'' trading off rapid population growth or inefficiency or corruption in underdeveloped countries for moderated growth of affluence in developed countries, with the latter largely financing both – at least initially.

One example of planetary bargaining applies Asia and its staple rice crop. Flooded rice paddies produce lots of methane gas. There may be alternative ''dry'' paddy agricultural techniques that produce less methane gas, but the question is, ''Will anybody use those techniques?'' What are the trade-offs in ways to produce the food that is needed to feed people as populations increase, without producing as much extra methane gas? Who will pay to get the process started? These questions form the basis for bargaining between developed and developing countries.

Another example of planetary bargaining concerns deforestation. A tree is made up largely of carbon, which comes from carbon dioxide in the air through photosynthesis. When we cut trees down and burn them, we dump back into the atmosphere all the carbon that it took trees thirty years to take out of the air – in the space of thirty minutes. There is about as much carbon in all the trees on Earth as in the air. However, if we add carbon dioxide to the air, aren't trees going to photosynthesize more and grow faster? Yes, probably, because more carbon dioxide would mean more such ''fertilizer.'' However, other factors are involved. After all, adding carbon dioxide is likely to warm the climate, so overall there could be a negative feedback: Trees are going to take some of the extra carbon dioxide out of the air, holding back some of the warming potential. Is this the end of the feedback story? The amount of carbon in the soils – in dead leaves, dead roots, dead organic matter – is about twice that in trees. Getting carbon from the soil back into the air is the job of trillions upon trillions of microbes that decompose organic material. The rate at which they produce either carbon dioxide or methane depends upon the temperature of the soils. Consequently, if we remove the trees, the soil become hotter, and organic matter is going to decompose faster. Likewise, if carbon dioxide and methane increase global warming, we might experience a positive ''biogeochemical feedback.''

Based on a host of variables – how many people there will be in the world, how they use land, what kinds of energy systems they'll use, what their standards of living will be, and what the feedback mechanisms are in the physical and biologic systems – Earth systems scientists attempt to project what the climate will be like in the future, often using mathematic models.

FUTURE POSSIBILITIES: THE COMMON LANGUAGE OF COST/BENEFIT ANALYSIS

Global temperatures for the past 100 years exhibit a warming trend of about 0.5°C. This trend is consistent with the 30 percent increase in carbon dioxide and 150 percent increase in methane since preindustrial times, but it is not a large enough temperature trend to rule out the possibility that it is simply an unusual natural warming that occurs perhaps one century out of ten. If we are extremely lucky and the negative feedbacks dominate, or there is a breakthrough in the costs and use of solar energy, or if we are unlucky economically and experience a world economic depression that means little growth in energy use of any kind, the lowest number that most current assessments project for future warming over the next 100 years is approximately another 1°C. That increase may sound trivial; however, the sustained natural global average rate of change between the end of the last ice age and our present interglacial is approximately 1°C per 1000 years. Even the most conservative anthropogenic global change estimate is projected to be approximately five times faster than the natural rate. If the positive feedbacks dominate and the economy booms so that energy use triples, then a warming rate fifty times faster than the sustained natural global averaged rate would be expected, as shown in Figure 1.2.

To see what this warming trend could mean in terms of an impact on the environment, consider the case of a certain cool-climate-adapted squirrel living in a restricted habitat in the American Southwest, restricted perhaps to the uppermost regions of a range of mountains. Even a 0.5°C warming could ''lift'' the physiologic needs of its habitat a few hundred meters higher. However, if the squirrel's habitat already is at the top of the peak, then for that species of squirrel, a ''trivial'' change of 0.5°C might mean death – extinction of the entire species, perhaps. If we were unlucky and a change in temperature of 5°C in a century were to occur, that would likely be ecologically catastropic for a large fraction of species on Earth – particularly when combined with habitat fragmentation. It would rearrange species ranges and ecosystems everywhere.

Consider a wildlife reserve, which is an area similar to a national park that is designed to preserve particular species

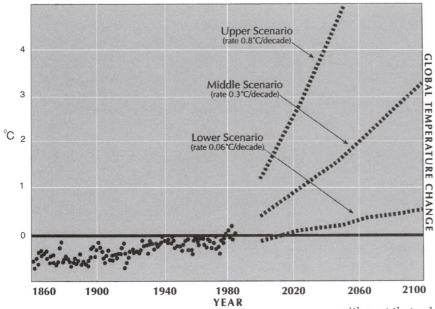

Figure 1.2. Three scenarios for global temperature change to 2100 derived from uncertainties in future trace gas projections combined with those of the biotic and climatic response projections. Sustained global changes beyond 2°C (3.6°F), unprecedented during recent geologic history, represent climatic changes at a pace tens of times faster than the natural average rates of change. (*Source:* Adapted from Stephen Schneider, Degrees of certainty, *Research and Exploration*, 1993.)

and habitats. If we look at range limits of certain trees or birds in a given reserve, they are often there because they cannot exist in other areas (e.g., where it is too hot, or too wet) (see Chap. 21). If the climate were to change suddenly (i.e., within less than 100 years), by +3°C, a new range limit would be established for each species. In a naturally occurring change, a tree would "march" its way poleward, with birds spreading its seeds, or by other slower dispersal mechanisms, depending on particulars, if the climate change were slow enough. What happens if the climate change is occurring at 10 to 100 times the sustained natural rate at which present habitats have evolved? Many trees may be stranded, and many species might even go extinct. What happens if they do have the capacity to move but, as mentioned earlier, encounter farms and interstate highways in their "march?"

The middle range of projected climate change (Fig. 1.2) – several degrees Celsius in a century – is the "best guess" of the majority of the knowledgeable scientific community. There is massive debate about these projections. The literature from groups that are worried about social intervention in economic activities argues that there is so much uncertainty that we should not do anything to slow down our "progress." The literature from environmental advocates often asks what sane person could take a planetary-scale risk with the Earth even if there is much uncertainty.

Economists typically ask, "How much is it going to cost to cut out, say, 20 percent of carbon dioxide emissions?"

They might estimate such costs by considering a tax on carbon in fuels. That would, in turn, show up at the gas pump and in heating bills and in the prices of energy-intensive products. Price increases also have a disproportionate impact on poor versus rich people.

The relevant questions for Earth systems scientists are: What is the price tag economically and politically to reduce environmental risks, and how much is the environment really "worth?" These questions are consistent with those asked by most policy makers in the worlds of business and government. It is essential for us to understand the nature of these arguments before we can expect to replace standard priorities with a set that values nature more highly. We need to understand in some depth the philosophy behind so-called cost/benefit analysis and how it is done (see Chaps. 26 and 27). Only then can we judge whether a modification may be cost effective or morally preferable.

CONCLUSION

The issues introduced in this chapter involve the kinds of interdisciplinary questions so fundamental to Earth system studies. Complex, thorny subjects increasingly will become part of the technology, education, science, and policy agendas of the twenty-first century, including such mind bogglers as: What are the synergisms between habitat fragmentation, chemical dumping, new population pressures, development strategies, introduction of new species, technological choice, and climate change – all happening simultaneously?

The problems of our time are not those of our grandparents or even our parents. With increasing population and information-sharing technology, the realm of human knowledge grows exponentially in fantastically short periods of time. From the masses of highly skilled, detail-oriented disciplinarians, we must reconnect to solve complex, interdisciplinary, real-world problems. Our grandparents' mission was to specialize, to track details within the details, to demystify the minutiae in order to understand the macroscale phenomena of the world. *Our* generation's mission includes pursuit of ever more specialized fields of knowledge, and also to synthesize that newfound knowledge. We must train people such as Earth systems scientists to communicate among disciplines so that disciplinarians may learn from and build on one another's work in the context of real-world problem solving. Good examples of historic success with interdisciplinary knowledge include the great naturalists of the nineteeth century (e.g., Darwin applied his background in geology to his biologic observations, which was essential to his grasping the fact that physical barriers can drive speciation). In this age of specialization, we need Renaissance individuals once again – we must not lose sight of the forest for the trees, even while we catalogue the DNA of each species.

Earth systems science is designed to look at how these and other related real-world problems connect. It will take decades for Earth systems scientists to answer only a few of the many detailed questions that need to be addressed to each of the subdisciplines that constitute Earth systems science. However, human pressures on the Earth's systems are already documented. Policies to alter some of the activities forcing global changes cannot wait decades to reduce uncertainties without taking many risks. These difficult questions must be dealt with now. Earth systems science attempts to bring the relevant content of physical, biologic, and social science disciplines to bear in the context of the real, interdisciplinary problems of environment and development. It aims to analyze causes and assess solutions, recognizing that any solutions will not be perfect or certain, but that on balance they will be better than either wild guesses or actions based upon narrow, specialized views. Finally, this integrated approach necessitates grappling with problems at the scale at which they exist, not ignoring or postponing them, or pretending that they are not the stuff of real scientists or technologists. It is an exciting journey, and we welcome students to this lifelong adventure in learning and doing.

QUESTIONS

1. Think of a familiar environmental problem. Now try to imagine the way a roomful of "disciplinarians" or advocates would speak on the subject. What would an economist have to say? A biologist? A groundwater hydrologist? How about a policy maker, a coal miner, or an environmentalist? Do you think that people from different disciplines or advocacy positions could communicate effectively with one another about the problem? What barriers exist? Who would facilitate communication among them?

2. A positive/negative feedback question: Think of a dynamic system (such as driving a car, or the flow rate of traffic) and list as many positive and negative (stabilizing and destabilizing) feedback mechanisms as you can in three minutes.

3. How would you mediate a dispute between a coal mining company and an environmental organization trying to slow down carbon dioxide emissions? Can you find any "win–win" solutions? How about a similar dispute between a less developed country dependent on coal and a small island state whose existence is threatened by a rising sea level from global warming?

FURTHER READING

Bruce, J., Hoesung Lee and Haites, E. 1996. *Climate change 1995 – economic and social dimensions of climate change: contribution of working group III to the Second Assessment Report of the Intergovernmental Panel on Climate Change.* New York: Cambridge University Press.

Houghton, J. J., Meiro Filho, L. G., Callander, B. A., Harris, N., Kattenberg, A., and Maskell K. (eds.) 1996. *Climate change 1995 – the science of climate change: contribution of working group I to the Second Assessment Report of the Intergovernmental Panel on Climate Change.* New York: Cambridge University Press.

Levin, S. A. 1992. The problem of pattern and scale in ecology. *Ecology* 73:1943–1967.

Peter, R., and Lovejoy, T. 1992. *Global warming and biological diversity.* New Haven, CT: Yale University Press.

Root, T. L., and Schneider, S. H. 1995. Ecology and climate: research strategies and implications. *Science* 269:334–341.

Schneider, S. H. 1997. *Laboratory Earth: the planetary gamble we can't afford to lose.* New York: Basic Books.

Watson, R. T., Zinyowera, M. C., and Moss, R. H. (eds.) 1996. *Climate change 1995 – impacts, adaptations and mitigation of climate change – scientific-technical analyses: contribution of working group II to the Second Assessment Report of the Intergovernmental Panel on Climate Change.* New York: Cambridge University Press.

2 Physical Geography

DONALD KENNEDY

EARTH SYSTEMS SCIENCE – SPACE AGE GEOGRAPHY

The subject called "physical geography" was once taught to general students in many universities. Now it is almost an endangered species; few institutions still have geography departments, and the average contemporary undergraduate – even those studying scientific subjects – would be unable to say what is encompassed by the subject.

For those concerned with the vital matter of environmental quality, physical geography is a core discipline. It attends to the structure and character of the local and the global habitat: landforms, temperature, soils, and climate. It examines the way in which those physical factors determine the pattern of occupancy by living systems – that is, it seeks explanations for the spatial distribution of species of organisms, and of the development, through their interactions, of ecosystems. Finally, it attempts to explain how humans have settled on the land and have used it. Dressed up in a more modern name, physical geography *is* Earth systems science.

The agenda is a monumental one. It would be daunting to many students – except that the subject matter is both fascinating and familiar. Why is corn grown in Iowa and wheat in Kansas? How is it that San Francisco is foggy for the same reason London is? Why did the fifteenth-century voyagers of discovery sail westward from Europe nearly to South America before turning east to round the tip of Africa? What will the weather be like tomorrow? These very different questions all depend on a knowledge of how the Earth spins, how it is heated by the Sun, and how the resulting water and wind movements affect the conditions for life. To describe what headway geologists, geographers, climatologists, oceanographers, ecologists, anthropologists, and sociologists have made with this problem would require many texts, not just a single chapter. However the core of what we know about what lives where and why is not difficult to describe, particularly if we confine ourselves to a portion of the globe. It is a fascinating story!

In this chapter, we begin with an account of some basic forces: geologic history, wind, ocean currents – and how they influence climate and vegetation on a small and relatively isolated island. Many of the general concepts illuminated by this microcosm of the planet are returned to, in much greater detail, in the chapters that follow. As a case study, this story of a mid-oceanic island provides an opportunity to talk in an introductory fashion about geologic history, weather, and rainfall, and to demonstrate their influence on vegetation patterns and, eventually, on human occupancy. Islands have special properties with respect to the origin and diversification of living things, and thus also serve as an appropriate introduction to the subject of biologic diversity.

The chapter then turns to a much more complex system, the physical geography of a large continental landmass. The principles worked out in island systems can be applied on this larger scale, and can be used to explain the origin and distribution of the major plant communities: tundra, coniferous forest, deciduous forest, grasslands, and deserts. In turn, these patterns determine the patterns of human occupation, and the ways in which humans use the land for agricultural or other purposes. Unfortunately, these uses have often ignored important limits in the physical and biologic systems on which they depend; the chapter ends with illustrations of these difficulties as part of a growing environmental crisis.

AN ISLAND ILLUSTRATION

In an effort to limit the scope of this account of physical geography we will focus on the North American continent, with some reference to parts of the United States that are extracontinental. We begin with one of those parts – a western outpost of the United States that offers an instructive glimpse of the physical environment at work under the most simplified circumstances available in nature. The place is the Hawaiian island of Kauai; the simplicity comes about because it is a young island in the middle of a very large ocean.

Only a few million years ago, long after the shifting of the continents had laid out the basic geometries of land and water as we know them now, the Pacific plate was moving

W. G. Ernst (ed.), *Earth Systems: Processes and Issues*. Printed in the United States of America. Copyright © 2000 Cambridge University Press. All rights reserved.

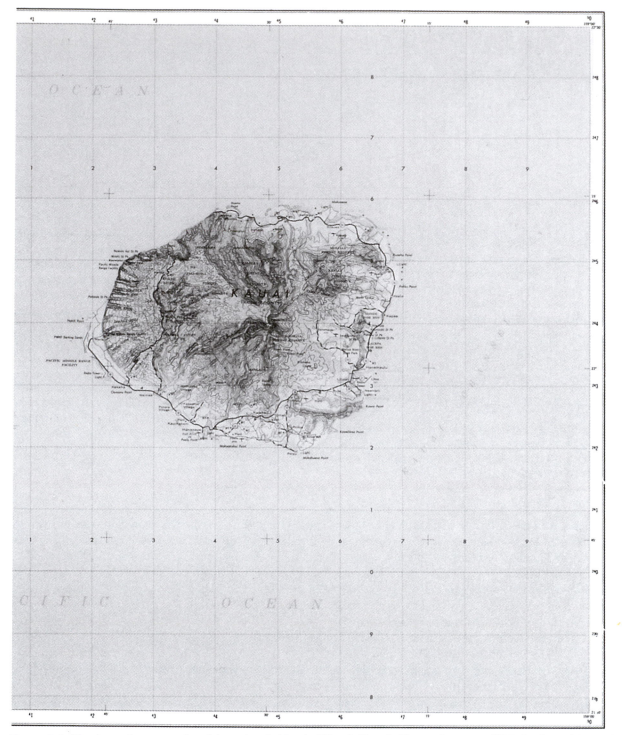

Figure 2.1. The dramatic topography of the volcanic island of Kauai, Hawaii. (*Source:* Courtesy of U.S. Geological Survey.)

north-westward over a hot spot in the vicinity of the present Hawaiian Islands (see Chap. 6 and Fig. 25.2). At intervals, hot magma (molten rock) from the Earth's interior broke through, raising volcanoes from the sea floor – a process described more fully in Chapter 5. The youngest of these volcanic islands, still highly active, is the easternmost – the big island of Hawaii. The most ancient, some 5 million years old, is Kauai, at the western end of the chain.

Early in Kauai's above-the-sea existence, repeated episodes of volcanic activity occurred, each depositing thick layers of volcanic ash as well as lava flows. The peaks of these volcanoes were substantially higher than today's present Mount Waialeale, the 1600-meter ridge that marks the southwestern rim of an old crater and is the highest point on the island. (Kauai's topography is shown in Fig. 2.1.) Plants with long-distance dispersal mechanisms arrived from far away, and clothed the slopes; some of the pioneer species evolved into the unique forest flora that now makes the higher reaches of the island a tropical paradise.

Rich forests require abundant water. Where does Kauai's come from? The answer requires some reference to the heat budget of the Earth, the circulation patterns of the oceans, and the complex result that we call the weather. The brief account given here will preview some of the material presented in Chapter 10, as it applies to the rather special circumstance of Kauai.

ABOUT WEATHER AND RAINFALL

As the Sun warms the Earth, it impinges much more directly on the Equator than on the poles, so that air in the equatorial regions is heated more than that at the poles. Being warmer, it has a lower density and it rises, being replaced by colder and denser air drawn from higher latitudes. This establishes a circulation pattern, shown in Figure 2.2: The rising equatorial air turns poleward, creating a high-altitude current that is exactly balanced by a return surface flow of colder air toward the equator. At approximately 30° N and S latitudes, the poleward-flowing air – having cooled and lost much of its moisture – sinks again and warms. These latitudes, accordingly, are dry, and include many of the world's deserts. At 60° N and S, the air rises again, only to sink at the poles.

Were the Earth a stationary object, our description would be complete. Each hemisphere would have its own giant circulation system, powered by the heat engine of the sun. However the Earth spins, which supplies a deflection to the moving currents of air (the Coriolis effect – much greater at high latitudes than near the Equator). This creates a complex pattern of prevailing surface winds – the northeast and southeast tradewinds at low latitudes, and the westerlies at high latitudes. The navigators of sailing vessels came to know these well: They account for the indirect routes that became standard during the Age of Sail. The pioneering voyage of the Portuguese navigator Vasco da Gama in 1497 laid out the basic template for the *volta*: A vessel leaving Southern Europe

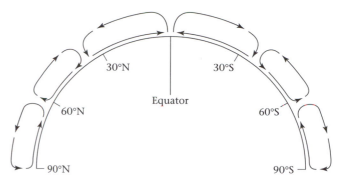

Figure 2.2. Atmospheric circulation pattern of the Earth, resulting from the variable heat received from the Sun at different latitudes of the Earth.

for the tip of Africa would sail southward at first, to the Cape Verde Islands, but would then follow prevailing winds toward the southwest, through the tropics and nearly to the east coast of Brazil. Then it would ride the prevailing winds of the Southern Hemisphere to the South African coast. It was a remarkable discovery (Fig. 2.3).

Wind patterns drive the surface circulation of the oceans, as well; as a result, the major ocean basins have circulation patterns that are clockwise in the Northern Hemisphere and counterclockwise in the Southern Hemisphere. Because cold water is thus brought down the west coasts of the northern continents, fog is a regular feature of both; it comes about because of the temperature contrast between cold seas and land-warmed air.

Regional "weather" is also powerfully affected by shifts in these dynamic, broad-scale circulation patterns, which in turn lead to the formation of domains of colder and warmer air with "fronts" between them. Collision between air masses of different temperatures along such fronts produces complex dynamic changes. When colder, denser air comes to overlie warmer, lighter air, for example, the latter forces its way upward buoyantly, producing winds and often precipitation. Frontal systems interact to produce *cyclones*, zones or "cells" of weather that have low-pressure centers, with counterclockwise rotation in the Northern Hemisphere. These are moved around by prevailing airflows; in the continental United States, they are driven along storm tracks by the west-to-east "jet stream" winds (see Chap. 14).

The distribution and movement of air masses with different properties results in a complex pattern of winds. Satellite mapping has made it possible to reconstruct the global distribution of wind directions and velocities. Of course, these vary from season to season, but the main features are surprisingly constant. Figure 2.4 shows the global mean surface wind field in July.

As the figure shows, at the latitude of Hawaii the prevailing winds in the North Pacific blow from the northeast, and are commonly called the Northeast Trades. These winds arrive at Kauai after traveling thousands of miles across the open ocean. The air is relatively warm and contains a great deal of water vapor. The tradewinds encounter an unusual

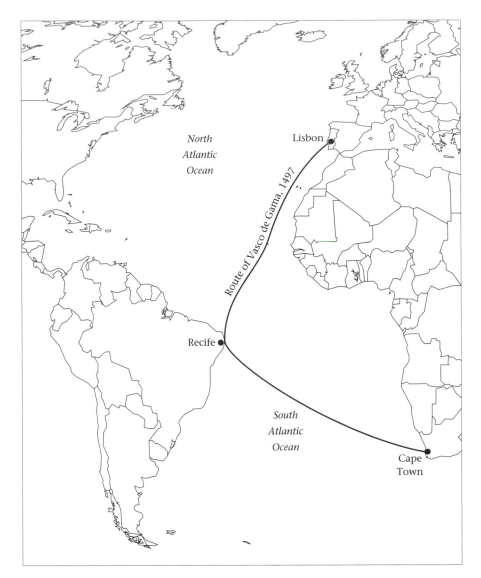

Figure 2.3. Vasco de Gama's route from Lisbon to the tip of Africa on his 1497 voyage.

topography as they reach the island's mountains. Waialeale is the rim of an old volcanic crater, the northeast side of which has been eroded. Thus the island presents the wind with a profile like a half-teacup, its open side directly facing northeast. The warm airstream is thus directed straight upward, forming a vertical plume that, according to experienced helicopter pilots, provides the most powerful sustained lifting force they have ever seen. In its rapid rise, the warm, moisture-laden air cools (see *The Cooling of Rising Air Masses*) as it expands against the cold air around it; when the temperature falls below the dew point, the moisture condenses and falls as rain.

The results of these processes on Kauai are truly spectacular. The windward slopes of Waialeale and the Alakai Swamp are among the wettest places on Earth: Some localities receive well over 1000 centimeters of rain per year. The resulting water flow has created some impressive changes in the landscape – changes made over relatively short geologic time because the soft, weathered volcanic residues erode so easily. The Waimea River, fed by the Waialeale torrent, has taken

less than 5 million years – an instant in geologic time – to cut a spectacular canyon one-third the depth of the Grand Canyon. Waterfalls and steep erosion gullies dot the windward cliffs, making the Na Pali Coast one of the most spectacular meetings of land and water to be found anywhere on Earth.

On the leeward side of the mountains, rainfall is dramatically reduced. The southern coast west of the town of Hanapepe receives 20 centimeters of rain per year or less. These extraordinary differences in local *microclimate* illustrate the truth of the old saying that "Mountains make their own weather." Indeed, early navigators understood that principle well. They knew that even in a trackless, unfamiliar part of the world's warm oceans when one saw a distant, isolated mass of cloud, an island was likely to sit beneath it.

Whenever mountains cause prevailing winds or moisture-containing, moving air masses to rise abruptly, cooling-induced precipitation is the likely result; this phenomenon is termed *orographic rainfall*. Ranchers in the western United States sometimes speak of mountains "squeezing the water"

July mean surface wind   ~ 5 m/s

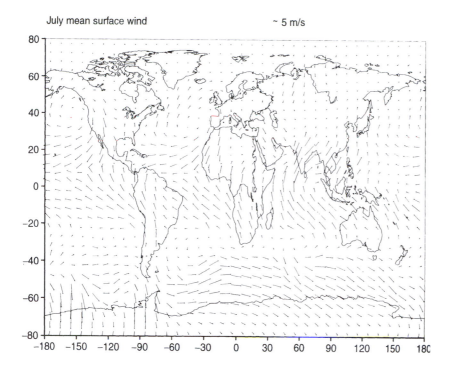

Figure 2.4. Ocean/atmosphere circulation. m/sec = meters per second. (*Source:* Grant R. Bigg, *The oceans and climate*, Cambridge University Press, 1996. Data from A. H. Oort, *Global atmospheric circulation statistics 1958–73*, National Oceanic and Atmospheric Administration, 1983.)

out of storms moving eastward over the continent, and of the dry areas to the east of major mountain ranges as lying in a "rain shadow." The desert-like climate in Oregon east of the Cascades is a familiar North American example. Nowhere on Earth are rain shadows more pronounced than on the island of Kauai.

CLOTHING AN ISLAND

A new environment represents a new opportunity for life. However, the life must get there first, from a nearby island or from a continental landmass. The very first arrivals on volcanic islands are apt to be tiny, wind-borne insects, spiders, and other arthropods, but the establishment of fully interdependent, living communities – ecosystems – depends on the arrival of colonizing plants. Colonizing plants are those whose dispersal mechanisms (for example, very light, fluffy seeds capable of long-distance transport by winds) and physiologic tolerance for an initially nutrient-poor environment make them natural pioneers. Once some vegetation is established, secondary opportunities are created for other plants, and for a variety of animals. The insects, mites, "balloon spiders" and other tiny animals capable of long-distance flights encounter a more hospitable environment, and there will be opportunities for larger, accidental arrivals. Some birds, perhaps, will be blown ashore from a distant mainland by violent storms. "Rafting" on logs or mats of vegetation is another mode of transport for animals. For an island group as isolated as Hawaii, these colonizing events seem highly improbable. However, over tens of thousands of years, the improbable becomes the likely.

By such processes, Kauai gradually became fully clothed with vegetation and developed its own fauna of insects, birds, and a few reptiles and mammals. However, like many young and remote islands, its flora and fauna are special. The early colonization events occurred primarily by unlikely chance; new arrivals encountered an environment with many opportunities for specialization. An ecologist would say that there are many unfilled niches – occupations for species that are free of competition. For example, if there are trees with bark and wood-boring insects but no woodpeckers, the insects represent a resource for which there is no compe-

THE COOLING OF RISING AIR MASSES

The temperature decrease does not occur by conduction – that is, the rising moist air is not cooled by losing heat to the colder air around it. That process would be much too slow. Rather, it undergoes *adiabatic cooling* – that is, the temperature drops because the air mass expands due to the reduced pressure.

The interaction between rising air masses and the air surrounding them is complex. The temperature of still, dry air decreases in altitude by approximately 6.4° C per 1000 meters; this is called the *lapse rate*. If a parcel of air is rising within such a larger air mass, it will have its own rate of adiabatic cooling, depending on whether it is moist or dry. If it is dry, its cooling rate is apt to be faster than the surrounding air, so it will soon reach a height at which it is colder and therefore denser than the air around it. At this point, it will stop rising. A moist air mass will have a slower adiabatic rate, because the condensation of water as the temperature falls below the dew point releases heat, thus retarding the cooling process. Rising columns of moist air are thus apt to keep on rising, often high enough to produce "thunderhead" clouds.

tition. The process of evolution will lead to the natural selection of varieties that can occupy such unfilled niches. As this process proceeds, the biologic diversity of the island increases – there are more species. Yet each will have arisen from a single species, one of the early generalist species involved in the early colonization. The term for this process of diversification, starting with a stem species or group, is "adaptive radiation." In Hawaii this process, acting over relatively short spans of geologic time, has produced whole arrays of species. For example, the islands were colonized at some early time by fruitflies of the genus *Drosophila* – close relatives of the little flies commonly used in genetics laboratories. Because many of the ecologic niches occupied by insects were vacant, these Drosophilids underwent an evolutionary explosion. There are *Drosophila* the size of horseflies, and others smaller than mosquitoes. Some feed on fruit, others on nectar or fungi. By the techniques of modern molecular genetics, in which sequences of DNA can be compared to infer relationship, all these Drosophilids are as closely related to one another as are two nearly identical mainland species. However, they have assumed an almost bizarre array of forms, habits, and life-styles.

Such evolutionary responses to new opportunity occur very rapidly. In the Galapagos Islands, visited by Charles Darwin on the HMS *Beagle* early in the nineteenth century, one can view an early stage of this process. A dozen or so species of finches, many quite similar to one another, are clearly the descendants of a single mainland species. In a matter of a few hundred thousand years, "Darwin's Finches" have developed and exploited a variety of niches. There are several sizes of seedeaters, and even a finch that uses cactus spines as tools to probe for insects beneath the bark of trees (Fig. 2.5). In recent research, it has been shown that significant evolutionary changes in bill size can occur in a few generations following a season of unusual rainfall that results in thicker, more resistant seed coats.

Figure 2.5. Originating from one species of finch, Darwin's Finches of the Galapagos Islands have evolved in different ways to exploit a variety of niches in their environment. (*Source:* Photograph courtesy of Peter Grant, Princeton, NJ.)

Adaptive radiation, when it occurs in isolated groups of islands such as Hawaii or the Galapagos, gives rise to a host of species that are unique to that place, or *endemic*. The same thing happens with mountaintops in a desert, or places that are left as isolated refuges when glacial ice sheets predominantly cover an area. Evolutionary diversification requires geographic isolation, and the geographic properties of islands or desert mountaintops or glacial refuges favor it by providing the conditions for isolation.

The role of isolation can be illustrated by a hypothetic example. Suppose that a species located on a single island occupies quite different habitats on the two ends, such that natural selection is working in different directions. (We might imagine that on the windward side of the island the tendency is for larger, stronger fliers to survive better than smaller individuals; whereas on the leeward side, there would be no such advantage.) It is unlikely that the single species will become two separate, reproductively independent species even though the forms at the two ends of the distribution are quite different from one another (see *The Concept of Species*). That is because there is reproductive continuity between the two populations; we say that genes flow from one end of the distribution to the other. However, if several individuals of, say, the windward type moved to another island, 50 or a 100 miles away, they would be permitted to continue on their evolutionary pathway independent of the gene pool on the parental island with which they were formerly in communication. Eventually, they would accumulate enough differences so that even when the geographically separated species was reunited with birds from the other part of its range, the two could not interbreed.

Thus does geography act to determine the very patterns of evolution that give rise to the ecosystems that characterize the land. Groups of geologically young islands located at a great distance from the nearest mainland and separated from one another by enough space to make transfer of organisms possible but rare are ideal species-generating machines. As already mentioned, there are others: mountaintops intermittently isolated from one another by glaciation, or serving as wet spots in an otherwise dry desert. Still others are the result of strange and unexpected geologic events. A recently discovered Romanian cave has been sealed off from the rest of the world by volcanic deposits for several million years. Its subterranean waters, exposed to exploration by a mining operation, revealed a remarkable lightless, oxygen-free ecosystem whose primary energy source is the oxidation of hydrogen sulfide by bacteria. Forty-seven species of invertebrates were found in this "island under the rock"; fully three-quarters of them were entirely new to science.

In Hawaii the most spectacular case of adaptive radiation is found among a group of birds – the Drepanididae, or honeycreepers. Fifteen or so species remain of a much larger number that originally evolved on the islands from a single ancestral stock of finch-like birds. They form a remarkable array. The majority are nectar-feeders, particularly on the blossoms of *ohia* and other endemic flowering tree species.

THE CONCEPT OF SPECIES

The idea of a species can be defined somewhat more formally. Biologists find much to argue about here; there are problems in drawing lines between species that become separate from one another in time, just as there are in defining species that look different but may occasionally hybridize along a common geographic boundary. The biologic species concept defines a species as a population whose members interbreed freely under natural conditions, and do not interbreed significantly with the members of other species populations. Hybridizing in zoos does not threaten this distinction – lions and tigers are perfectly good species despite tiglons and ligers. In fact, lions and tigers occupied overlapping ranges as recently as the nineteenth century, and there is no evidence that they interbred. Different ecologic preferences and social organization would function to keep them apart. Neither does the occasional naturally occurring hybrid: Some species simply have imperfect sterility barriers. However, tough calls involve species that have distinct ranges but zones of hybridization between them. Bullock's Oriole in the western United States and the Baltimore Oriole in the east are distinguishable but hybridize so freely where they overlap that ornithologists decided to lump them under the less colorful name of Northern Oriole, leaving the Baltimore variety an inhabitant only of the baseball diamond; however, ornithologists later reversed that decision.

However, others have become seedeaters, and still others have adopted leaf-gleaning habits. Three remarkable species have, like one of the Galapagos finches, occupied the niche filled elsewhere by woodpeckers. However, they have done it in a different way – one that, we may assume, required more evolutionary time. Two have evolved long, curved upper mandibles, and another an entire bill, that are used to probe the bark of trees for insects.

A number of these splendid bird species are now extinct on the Hawaiian Islands, and others are nearing extinction. Habitat destruction and the introduction of alien or exotic species – rats and pigs, which destroy nests and eggs, and mosquitoes, which act as distributing agents or *vectors* for various avian diseases including malaria, are largely responsible. European settlement following Captain Cook's landfall at Kauai in 1789 featured widespread destruction of Kauai's lowland forests to make room for sugar plantations – and later, of course, for the more lucrative assembly of beach resorts and golf courses. Indeed, the average tourist visiting any of the Hawaiian Islands will almost surely never see a native plant or animal; near the beaches, one finds carefully tended exotic ornamental vegetation and a variety of birds descended from cage escapees, but no indigenous flora or fauna.

It is natural enough, at first glance, to assume that the transformation of these islands is entirely a European work. Indeed, until fairly recently, experts blamed the devastation of the native biota of Hawaii on the colonists who settled the islands after Captain Cook arrived. Careful examination of "semifossilized" remains found in lava tubes, however, indicates that the earliest Polynesian settlers accomplished a great deal of biodiversity reduction long before Cook first landed (see *Population of the Pacific Islands*). The beautiful colored cloaks worn by the legendary "native" Hawaiian kings were made from Drepanid feathers, and large populations of feral pigs predate European arrival. Once there were four species of endemic "geese";* Cook found only one. Similar reductions occurred among other bird groups, and it is estimated that over 50 percent of the Hawaiian bird extinctions took place at the hands of Polynesian settlers.

The surviving native ecosystems of Hawaii are found only at relatively high altitudes – on Kauai, at 1000 meters or higher. It is possible for a visitor to drive to a lookout beyond Kokee State Park and be in the midst of a beautiful *ohia* forest within minutes; a two-hour hike will allow the adventurous traveler to reach the Alakai Swamp. This high wetland forest, maintained by the drenching rainfall pattern described earlier, is protected against encroachment by its isolation and by its elevation. The forest bird species that have been decimated by avian malaria at lower elevations enjoy a climate-derived benefit here: Seasonal cold weather prevents the breeding of the introduced mosquito species found on Kauai at the altitude of the Alakai Swamp, although they sustain populations in places that are only a few hundred feet lower. That brings us back to climate and geography, for this may be an example – surely but one among many – of an ecosystem that is poised to endure great damage from a slight elevation in global temperature. A degree or two warmer at the critical point in the season could create the opportunity

POPULATION OF THE PACIFIC ISLANDS

The history of Polynesian settlement of remote Pacific islands is full of high drama and tragedy. The remarkable canoe voyages, undertaken in exploratory waves beginning as early as 3600 years ago, proceeded eastward from the regions north of New Guinea and resulted in the occupation of nearly all the major Pacific island groups. There are success stories, such as the stable, sustainable kingdom created on Tonga. However, there are countless disasters as well. The populations on at least a dozen islands outran their resource bases, and became extinct. On Easter Island, the people who created the famous statues used a dense tropical forest to build canoes for fishing, and planted crops on reclaimed land. The population density rose explosively, and eventually the remaining forest was cleared and the land birds driven to extinction by hunting. Erosion washed away the agricultural land, and there were no trees with which to construct fishing canoes. When Easter Island was eventually "discovered" by Europeans, it was a civilization in collapse, with cannibalism rampant.

---

* They were not, strictly speaking, geese. These large, flightless birds are actually more closely related to a group of ducks.

Figure 2.6. A shaded relief map of the United States, showing the landforms and topography. (*Source:* Gail P. Thelain and Richard J. Pike, *Landforms of the conterminous United States – a digital shaded-relief portrayal*, U.S. Geological Survey, 1991.)

for mosquitoes to breed in the swamp, and bring their diseases to decimate one of the richest and most extraordinary endemic faunas in the world.

PHYSICAL GEOGRAPHY OF NORTH AMERICA

The rather special case of Kauai shows how geography may influence every aspect of a habitat. Its youth, the geologic origin of its soils, its great distance from the mainland, its orientation to the tradewinds, the annual temperature regime – all these have shaped not only the physical character of the island and the organisms that inhabit it, but also the historic patterns of human occupancy.

In the same way, we can begin to understand how such influences have played upon the character of the continental United States. However, on this much larger and older landmass, the challenge is far more daunting. The North American continent not only has a much longer geologic history; it is also more complex than the remote young island we took for our first example. The continent has drifted across the globe over the past 200 million years (see Chap. 6), carrying its ever-changing cargo of living things with it; it has also been subject to occasional geologic additions from other sources. It has received catastrophic meteorite impacts at least five times since the origin of complex living creatures, and even during relatively recent times its climate has been drastically modified by the advance and retreat of the great glacial ice sheets of the Pleistocene epoch (see Chap. 3).

The basic patterns of wind and weather, together with the topography of the continental United States, nevertheless tell us much about the character of our ecosystems and even about the human settlement of the landscape. As the dramatic map presented as Figure 2.6 shows, the physiography of the United States is dominated by major mountain ranges that run north and south. Beginning at the Pacific and proceeding eastward, they are: the Coast Ranges, the Sierra Nevada, the Rockies, and the Appalachians.

Because regional weather (circulating systems around centers of low or high pressure) generally moves from west to east across these north-south ranges, the resulting pattern of orographic rainfall produces a sequence of alternating moist and arid regions – superimposed on a general tendency toward drier climates in the west and moister in the east. Figure 2.8A gives the regional distribution of precipitation on the continent. The Coast Ranges receive rainfall (especially in winter) concentrated on the windward slopes and peaks, and on their east side lies the Central Valley of California. The rain shadow of the Coast Ranges means that rainfall here is low, approximately 25 centimeters or less per year. The peaks of the Sierra Nevada receive much more abundant precipitation; being much higher than the Coast Range mountains of the same latitude, they receive a large second dose of orographic precipitation – most of which falls in the form of snow during the winter months. East of the Sierra is the even drier Great Basin – known to geologists as the Basin and Range. Here, rainfall may be only 5 to 10 centimeters per year in the drier lowlands. The Rockies receive more precipitation, especially on the higher western slopes; that precipitation is also more seasonally distributed, because in the summer months moist air intrusion from the Gulf of Mexico may become part of the pattern. The Rockies, too, cast a long rain shadow. As one moves east from the Front Range at the latitude of central Colorado, the annual precipitation gradually increases – from less than 25 centimeters per year to approximately 50, encountered at the line indicated in Figure 2.8A. That line has special significance: It marks the approximate boundary between what has been called "short-grass" prairie, to the west – an area first given over to grazing land and dry-land wheat agriculture – and the "tall-grass" prairie, to the east. Here wheat and, farther east, corn can be grown. We then intersect the remains of the original Great Lakes forest, now largely cleared for agriculture and urban development, and finally, the Appalachians and the Atlantic coastal plain.

The differences we have been discussing involve the interaction between climate and topography. Of course, there are other features as well. The geologic origin of soils (Chap. 8) partly determines their nutritional value to plants. In addition, temperature varies with both latitude and altitude, and is a critical environmental variable.

ECOSYSTEMS OF NORTH AMERICA: BIOMES

It has long been noted that ecosystems, characterized in terms of their dominant vegetation, tend to occupy particular geographic regions. Early North American biogeographers defined the regions occupied by these characteristic plant communities; the regions are called biomes. Although the picture is actually somewhat more complicated than originally thought, the basic classification for North America is still useful. In the extreme north, a zone of tundra, featuring low-growing perennial plants ranging up to the size of dwarf willows, overlies "permafrost" – permanently frozen ground. At somewhat lower latitudes there is a band of boreal forest, or taiga, consisting of large, coniferous trees sufficiently dense that they permit only limited growth of other plants at ground level. In the United States, taiga is found in the extreme northeast and northwest, and elsewhere at higher altitudes. The Great Lakes region and much of eastern North America were originally covered by deciduous forest – a mixture of oak, beech, maple, hickory, and other hardwoods. In the summer, when the leaves of these trees are out, the decid-

uous forest of North America is characterized by a filmy, almost translucent quality; the leafy canopy admits enough light for a relatively rich understory of shrubs and flowering herbs. A broad region of the central part of the country is native grassland, now the richest and most productive agricultural region of the United States. Within this prairie zone, the 50-centimeter rainfall line, referred to earlier, divides "short-grass prairie," largely now devoted to wheat or grazing in the drier portions, and "tall-grass prairie," now primarily the corn belt. Regions of the Great Basin, the Central Valley of California, and the southwest receive 25 centimeters or less of rain per year, and develop deserts of various forms.

This broad-brush, sketchy description of North American biomes should leave the reader who is familiar with the ecologic diversity of the continent somewhat dissatisfied. Where is the oak-pine forest of the southeastern Piedmont? The fog-shrouded redwoods of the northern California coast? The montaine fir belts of the Cascades? The central Florida scrub? Obviously, local conditions have produced a huge variety of special, local ecosystems, and these may constitute glaring departures from the very general specifications laid out in biome descriptions. Even in the more biologically impoverished deserts of the southwestern United States, it is possible to recognize dozens of local types.

In addition, altitude can replace latitude in determining the distribution of natural communities. Indeed, one of the early North American biogeographers, C. H. Merriam, developed a classification of "life zones" based on altitude. Because temperature varies inversely with elevation, one can replicate the biologic experience of traveling northward by climbing a mountain in the tropics. Air temperature decreases with altitude by approximately 6°C for each 1000 meters of elevation – the equivalent of going north approximately 800 kilometers. Winter snows occur regularly on the upper slopes of Mauna Kea and Mauna Loa on the big island of Hawaii; people actually ski there. In ascending a mountain from, for example, an Arizona desert, one may pass through an area of grassland, a deciduous forest, a coniferous forest, and finally a zone that looks a bit like tundra (Fig. 2.7). However, were one to make the climb for the purpose of exploration, one would notice important local variations. Canyons, relatively well-watered, would support local assemblies of deciduous trees even amid surrounding desert. Farther up, north-facing, cool slopes would support coniferous forests whereas south-facing ones would be covered by an assemblage of drought-resistant shrubs. Drainage and soil type, too, would produce special local conditions leading to differences in the character of plant and animal communities.

Implicit in the notion of biomes is the idea that there is a particular plant community – indeed, an entire ecosystem – that exists as a stable entity in a particular place. Such entities were called climax communities by early ecologists, who observed that on newly available ground a succession of vegetation types developed over many years, culminating in a community that appeared to be stable.

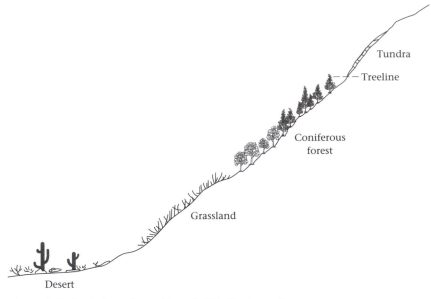

Figure 2.7. Depiction of a section of altitudinal zonation on a southwestern US mountain.

It now appears, however, that the stability of such communities may have been overestimated. Many carefully studied "climax" communities have been shown to be more locally variable and less temporally fixed than was thought at first. Also, over longer time periods, climatic change has altered the vegetation type that can exist in a particular place. We know from the analysis of fossil pollen recovered from lake sediments that at the height of recent glacial advances, coniferous forests grew where magnolias, tulip trees, and other subtropical species are now found. The latest evidence on the history of global climate change indicates that major temperature alterations occurred on a time scale of centuries or even fractions of centuries. Therefore, it is difficult to believe that – at least in the temperate zones – stable "climax" communities have existed for extended periods of time, that is, intervals of geologic scale.

In fact, the disintegration of the notion of a stable climax community for each place is but one example of a change that has swept over all thinking about geologic and biologic systems. First our notions about the stability of the Earth's landforms were revised by the discovery of plate tectonics and continental drift. The idea that the evolution of new species and the extinction of old ones are slow processes was exploded by revelations about the impact of such catastrophic events as meteorite collisions and episodes of volcanism. Finally, our belief in the relative stability of climates over time has been shaken by measurements indicating that in the fairly recent past, significant changes in global climate occurred over time spans of the order of decades rather than centuries or millennia. In short, in our view of the natural world, statics are everywhere being converted into dynamics, and our respect for instability and change is growing.

HUMAN OCCUPANCY

We suggested earlier that patterns of human occupancy of North America have been heavily conditioned by its physical and biologic environment. Human occupancy, in turn, has profoundly altered the ecosystems themselves.

One tends to think of this as a modern event, but it began in Paleolithic times. Just as the first human colonists of Hawaii changed their biologic environment, so – much earlier – did the first invaders to cross the Bering land bridge into North America (or travel by water) between 12,000 and 20,000 years ago. At this time, a magnificent array of large mammals inhabited North America: mammoths, camels, saber-toothed tigers, several kinds of antelopes, dire wolves, tapirs, and native horses. Hunters who arrived were well equipped, though, and the ecologic consequences were disastrous. Three-quarters of the genera of large mammals that existed then are now extinct. Climate change then in progress may have contributed to this massive extinction. Perhaps it did, by placing the megafauna under stress that made it more vulnerable to predation by the new arrivals. Significantly, however, massive extinction events of the same kind followed the human occupancy of Australia, and more recently of Madagascar and New Zealand (See *Human Reshaping of the Environment*).

Settlement in North America had an impact on the environment in even more dramatic ways prior to the arrival of European colonists at the end of the fifteenth century. Earlier estimates of the population of native North Americans have been revised radically upward: It is now believed that there were perhaps 10 million inhabitants in precontact times. The evidence suggests a well-developed civilization: lively trade, even over considerable distances; settled agriculture; herding of animals; and social infrastructures of substantial complexity. These activities, beginning with the earliest extinctions, has already modified the North American continent. European settlement, beginning in the seventeenth century in the northeast, wrought even more significant changes.

PATTERNS OF HUMAN SETTLEMENT. As the descendants of the first New England colonists moved westward, they encountered a continental interior very different from the land on which they had settled. They wrought changes in the biosphere that were even more dramatic than their accomplishments in New England. They cleared much of the Great Lakes forest and planted it to corn, and extended that "corn belt" onto the rich soils of the eastern Great Plains – the old "tallgrass prairie." There, however, they encountered

HUMAN RESHAPING OF THE ENVIRONMENT: AN ILLUSTRATION

A classic example of human participation in large-scale ecologic change – and of the reciprocal effects of ecologic succession on human activity – is found in the history of New England pastures. Settlers of central and northern New England found a mixed mature deciduous forest consisting of beech, maple, and oak with a few coniferous species – hemlock and white pine. They cleared it painstakingly, creating miles and miles of stone walls that the contemporary explorer of any second-growth woodland is likely to stumble over.

Declining agricultural opportunity and the Westward Movement began to empty the farmhouses of New England late in the eighteenth century. As the pastures were abandoned, an unexpected phenomenon occurred. Within dissemination range of nearly every vacated field, there was a big pine tree, whose seeds were carried or dispersed into the newly untenanted opening. The pine seedlings were tolerant of full Sun, whereas those of the deciduous species needed shade. As a result the old fields developed healthy, fast-growing stands of almost even-aged pine. Later in the nineteenth century a burgeoning lumber industry grew up; the trees were cut off and converted into flooring and furniture. The lumbermen assumed that the fields would "come back to pine," but they did not. The lumbering operation and the soil conditions left in its aftermath were more encouraging to second-growth hardwood species, especially ash and gray birch. These do not yield commercially satisfactory lumber, and have been left to become rural woodlots. At the height of New England agriculture, but a small fraction of the land surface remained in forest. Today – despite the proliferation of suburbs and parking lots – the forest area has more than doubled. However, it is a very different forest from the one that clothed New England prior to European contact. What we have now – though technically forest, and composed of indigenous species – is both less lovely and less productive than the mixed forest first encountered by settlers from the Old World.

the first harsh reality of settlement in interior North America: water – or, rather, the lack of it. The limit of corn cultivation approximately coincides with the line separating rainfall of 60 to 100 centimeters per year from a western zone in which the rainfall is less than 50 centimeters per year. Still moving westward, the pioneers planted wheat on the better-watered portions of the old "short-grass prairie," and in the drier parts they allowed livestock to graze on the open range. Some of the land was desired for both purposes, and naturally conflicts arose between the homesteaders, who practiced settled (and fenced) agriculture, and the cattlemen, who initially depended on open, unfenced range. This is the raw material for hundreds of Western dramas, featuring confrontations and the inevitable Hollywood-style shoot-out between cowhands and "sod-busters."

The western half of the United States has been properly described as an "oasis society." Between the Pacific Coast and central Nebraska and Kansas, only the higher elevations are well-watered. The rest is primarily scrubland and desert. Without the agency of human intervention, it cannot support productive plant communities.

Yet we now know that tens of millions of Americans have found a way to live there, through the most remarkable kind of intervention. The Central Valley of California, essentially an arid though locally swampy ancient lake bed, has been made to bloom by the importation of water from higher elevations and by the local pumping of water from underground aquifers. It now supports a rich mixed agriculture: fruit and nut trees, cotton, and even rice – all of it heavily subsidized by artificially low prices for agricultural water.

In the six-state High Plains region – including dry parts of West Texas, Oklahoma, eastern Colorado, eastern New Mexico, Kansas, and Nebraska – a vigorous farming economy occupying nearly 15 million acres now exists, nourished by "fossil water." It is obvious to the airline passenger from 35,000 feet: Huge, intensely green circles stand out from the surrounding dry rangeland. They are irrigated by center-pivot irrigation, deep-bore wells that withdraw the water from a limited underground source called the Ogallala Aquifer (see Chap. 30). As one flies over the western slope of the central and southern Rockies, one sees strips of green valleys as well as more circles. These are supplied with irrigation water obtained by diversion from the Colorado River and its tributaries.

Irrigation has thus changed the face of the American West. But for how long? The original settlers – courageous men and women who dared to try to survive in tough country – brought with them a fund of optimism. Of their efforts to homestead in dry country, they said: "Rain follows the plow." When they discovered the Ogallala Aquifer, they spoke of a "vast, inexhaustible underground lake." In 1922 they allocated withdrawal rights from the Colorado River on the basis of stream flows from a period during the previous century – unaware that it represented the wettest interval in 100 years. Small wonder that the downstream users have been water-starved; indeed, the water finally arriving in Mexico was so salty that it became unusable for agriculture several years ago. In order to conform to the terms of an international agreement, the government had to build a plant to desalinate the water! And small wonder that the Ogallala Aquifer has been drawn down so far that it is no longer economical to pump irrigation water in parts of West Texas. Indeed, the entire six-state High Plains region will, within the next twenty years or less, have to convert from irrigated agriculture unless massive projects are built to transfer water from a distant and as yet unidentified river basin.

Weather, soils, topography, and history have reshaped the North American continent. They have determined the evo-

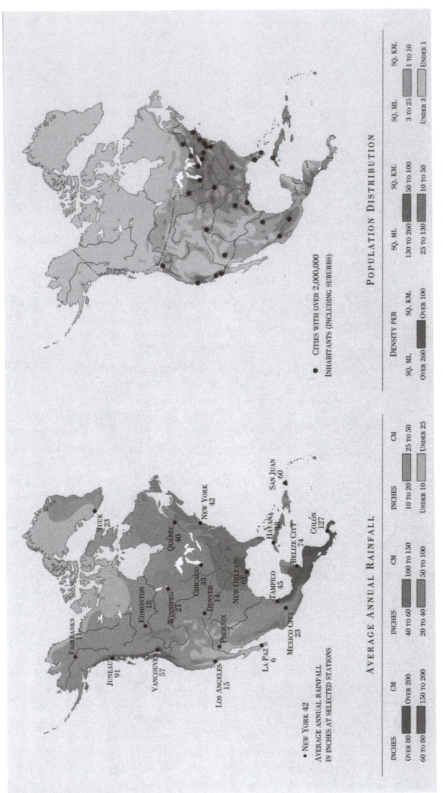

Figure 2.8. (A) Average annual rainfall of the United States; (B) population distribution of the United States. Comparison of these two maps shows the close correlation between settlement and rainfall. cm = centimeters; sq. km. = square kilometers; sq. mi. = square miles. (*Source: The New Comparative World Atlas*, Hammond Inc., © 1997.)

lutionary development of living ecosystems on its surface, and in turn the patterns of human occupancy. No more dramatic illustration of this fundamental fact in our geography can be found than the correspondence between rainfall and population density, shown in Figure 2.8.

The human occupants of North America are changing the continent itself – not only by extinguishing its fauna and replacing its natural ecosystems with settled agriculture, but also by making extraordinary demands on its nonrenewable resources and modifying its landscape. We have even changed the weather, by creating urban hot spots and zones of higher humidity where irrigation is widespread.

No species in the history of life on Earth has achieved such a capacity to alter the physical environment. As we see later in this book, that capacity extends even to the alteration of Earth's atmosphere (Chaps. 13 and 14). Yet we are also, as far as we know, the only species capable of measuring the availability of vital resources and then modifying our behavior to ensure their continued availability. Why we have not yet done so is a troubling paradox; one would surely think that our species could learn to live more lightly on the land.

QUESTIONS

1. Two closely related species of willow occur in very restricted areas in central Alaska. Their centers of distribution are only 400 miles apart, and the extremes of their distributions barely fail to overlap. Although the two species look quite similar, experiments have shown that they are not interfertile. Construct a historic hypothesis to explain these facts.

2. In mid-continent North America during January and February, the following weather sequence is common. There is increasing evening cloudiness with rising temperatures and a falling barometer. In the middle of the night, rain begins and continues through mid-morning on the next day. The rain then changes to snow, which accumulates to a depth of several inches by evening. Barometric pressure is rising, and the snowfall tapers off and stops before midnight. The temperature falls during the night. Explain this sequence, and give the daily television weather forecast for the eight o'clock news the next morning.

3. Indonesia has the highest species diversity per unit land area in the world. Why?

4. What are some ''ideal'' properties for a plant that would equip it to be a successful early colonist on newly formed volcanic islands? What would be some ideal properties for an insect? A mammal?

5. Explain why geographic isolation is an important condition for the formation of new species.

6. What feature of climate or weather accounts for the fact that the relatively isolated mountains of southeastern Arizona, although surrounded by dry desert, are forested on their upper slopes?

7. Suppose that global warming raises the average temperature over interior North America by several degrees Celsius over the space of a century. Describe the sequence of events you might observe (with an extra lifetime to do it!) at a point (A) in central Iowa, and (B) at an altitude of 3000 meters in the Colorado Rockies.

8. It is sometimes said that the western United States is an ''oasis society.'' Explain in terms of natural features, including weather. Describe how human activity has changed the ''oasis'' character of the region.

FURTHER READING

Christopherson, R. W. 1994. *Geosystems: an introduction to physical geography.* New York: Macmillan.

Stegner, W. 1953. *Beyond the hundredth meridian: John Wesley Powell and the second opening of the West.* New York: Penguin.

Cronon, W. 1983. *Changes in the land: Indians, colonists, and the ecology of New England.* New York: Hill and Wang.

Wilson, E. O. 1992. *The diversity of life.* New York: Norton.

3 Time Scales: Geologic, Biologic, and Political

W. G. ERNST

RATES OF CHANGE AND ENVIRONMENTAL PROCESSES

Geologic processes are vast and unstoppable; however, except for a few spectacular examples, such as meteorite impacts, avalanches, and tsunamis (tidal waves), most are infinitely slow – so slow as to go unnoticed by any but careful scientific observers. For instance, we now know that the great lithospheric plates that form the outer rind of the solid Earth are moving with respect to one another, yet sensitive, state-of-the-art measurements are required to detect the differential slip between them. In human terms, the mountains and ocean basins seem to be immutable features; however, geologists have demonstrated that the solid Earth changes gradually, but continuously, with the passage of time. The course of biologic evolution is analogous: While changes in the terrestrial biota take place at a somewhat quickened pace relative to most geologic processes, they are nonetheless imperceptible with respect to our personal time frames. Climate changes are a bit more sprightly, yet the dispersal of biologic species takes place over hundreds of thousands of years.

Humans have recently become a major force of nature, and have wrought profound changes on the interconnected Earth systems over the course of civilization – say, over the past ten millennia. The rate of environmental alteration is accelerating rapidly, and most of the changes, such as loss of topsoil, decrease in biodiversity, and global warming, appear to be profoundly damaging to the integrated web of life. Is there a remedy to some of these adverse impacts, and will we have enough time to accomplish a far-reaching, sustained reversal? Do we even need to take action? No one knows for certain, but past experience in the public policy arena provides little cause for optimism. Perhaps by studying the interconnected Earth systems, humans will develop a better appreciation for how the planet works, and we may be able to preserve more fully the environmental quality of the planet while being sustained by its natural resources. Should we fail in our attempt to maintain worldwide ecologic values, the gradual impoverishment of life – human and otherwise – will be an inevitable consequence.

Look up the word "time" in a dictionary, and read what you find. Does it make sense to you? I can barely comprehend what the dictioner is driving at, let alone fully understand the definition. Yet I know what time is, as do you, and so do all cognizant beings. The concept is innate, intuitive, and reinforced by common experience, but difficult to define other than obliquely as, for example, "a succession of events, duration, or continuum of moments." This phrase is not a very succinct or illuminating expression of a universally experienced phenomenon.

Space and motion somehow are tied up with an understanding of time, even though our unaided abilities to detect both rapid and slow movements are severely limited. Instrumentation and repeated observation have allowed scientists to document differential motion across geologic fault zones of a few centimeters per year, as well as the escape of gases at several hundred kilometers per hour from a volcanic vent. In flight, a bumblebee's wings beat too fast for us to see the motion, and the flowers it flits among grow far too slowly for the change to be perceived, yet we know both processes are occurring over the course of time.

If we evoke the concept of motion, velocity (v), distance (d), and time (t) are related by the equation $v = d/t$. This provides us with a means of measuring time in certain situations. For instance, we can determine the speed of light by sending a pulsed, energetic beam of light (e.g. employing a laser source) to a mirror sited at a previously measured distant location such as the Moon, and measuring the time delay on receipt of the reflected beam (the distance can be determined by triangulation, employing two ground stations and utilizing the well-known principles of trigonometry). The velocity of light is constant (in a vacuum), hence time may be derived quantitatively in terms of this parameter.

However, employing $v = d/t$, we would be unable to evaluate time for a purely static universe. For instance, we may sit, motionless, pondering the nature of time for as long as we wish, and nothing need change – including our understanding. However, time has passed. So, clearly, movement is not an invariant or necessary attribute of the passage of time.

Actually, for most phenomena of interest, our world is in a state of complex motion, locally approaching a sort of dynamic equilibrium in which different portions of the system interact with one another. Sunlight falling on the Earth is balanced by the energy radiating back out into space. The evaporation of water from the surface of the planet is offset by an equal amount of precipitation. Over time, the growth of plants and animals is matched by death and decay of an equivalent biomass. Constant change at the microlevel is compensated for elsewhere in the system, hence net change is imperceptible. Over time, of course, change is gradual but inexorable as the system evolves. In the study of Earth systems, we recognize three different, intergradational time scales and rates of change, namely *geologic, biologic,* and *political,* characterizing the interactive Earth systems, geosphere, biosphere, and anthrosphere. A fourth, the daily and seasonal rhythms of the Earth, is well known to all of us. A much less common fifth frame of reference, the virtually instantaneous, or *cataclysmic,* typifies catastrophes such as planetary collision, but these are exceedingly rare events – at least, we hope so. Generally speaking, geologic time is calculated in millions or even billions of years, biologic time in thousands to millions of years, and political time in months to years. Of course, rare classes of geologic events, such as meteorite impacts and individual earthquakes, are nearly instantaneous, lasting only seconds, whereas some other actions in the political arena, such as tax and national debt reduction, may never happen at all.

GEOLOGIC TIME

The enormity of geologic time far exceeds human experience, so is almost beyond comprehension. The solar system that includes our Earth, for instance, is a relative newcomer to the cosmos. The universe as we know it is believed to have originated during the Big Bang, at a time when all matter was tightly clumped together approximately 15 billion years ago (see Chap. 4). We may wonder what the configuration of mass and energy was before the Big Bang, but no one has a clue as to the situation prior to this "time zero." In any case, at the instant of the Big Bang, the combination of matter with antimatter generated repulsive nuclear forces that explosively blew this concentrated mass apart. The surviving matter has been expanding ever since, apart from the local self-gravitative mass accumulations we know as stars, galaxies, clusters of galaxies, superclusters, interstellar dust, planets, meteorites, and comets.

Condensation of the solar nebula to form the Sun occurred approximately 4.5 to 4.6 billion years ago. The accretionary formation of the Earth and other planets from *planetesimals* (meteorites, asteroids, and comets) took place concurrently. The time of origin of our solar system has been determined through the investigation of meteorites and lunar samples; reflecting the relatively vigorous reworking of

Earth materials, terrestrial rocks more ancient than 3.9 billion years old have not yet been recognized. By what means were such ages obtained, and how reliable are they? The answers to these questions are to be found through the measurement of radioactive isotopes in minerals and rocks. As will now be explained, such systems behave as "atomic clocks," and the derived ages are both precise and – in favorable cases – quite accurate. Scientific investigations involve the analysis of radioactive parent and radiogenic daughter isotopes. These methods have been remarkably successful in unraveling the history of the Earth's rocks and constituent minerals. Examples include muscovite and zircon, minerals formed in granitic rocks (for a description of rocks and minerals, see Chap. 5).

Within its specific atomic arrangement, or crystal structure, every mineral contains certain elements in major proportions, others as minor or trace constituents; still other elements are, for all practical purposes, entirely lacking. For instance, potassium-bearing mica – muscovite, $KAl_2Si_3AlO_{10}(OH)_2$ – contains essential amounts of potassium (K), but totally lacks argon (Ar), because of the extreme rarity in condensed planetary matter of this noble gas. Furthermore, a positively charged cation such as K+ is strongly preferred in the appropriate cation structural site over a neutral, nonbonding inert gas atom such as argon. For those needing a reminder about the systematics, the Periodic Table of the Elements is presented as Table 3.1. Helpful English-metric conversion factors are listed in Table 3.2.

Regardless of the local concentration of potassium, its isotopes (isotopes of an element possess the same number of protons, but different numbers of neutrons) occur in similar proportions everywhere on the Earth, and apparently, in the same proportions throughout the solar system. One of these isotopes, ^{40}K (the superscript indicates atomic mass, a quantity specifying a particular isotope of the element) is radioactive. In approximately 1.5 billion years, half of the ^{40}K originally present will have decayed to daughter products (including argon as well as well as calcium) at a measured and well-known rate proportional to the amount of the parent isotope. Accordingly, the length of time required for half of the original material to be transformed is termed the *halflife.* * Then, in the next 1.5 billion years, half of the surviving ^{40}K will have decayed, and so on. This gradual, exponential decay and the concept of half-lives are illustrated in Figure 3.1. The daughter particles are radiogenic argon (^{40}Ar), produced by electron capture (proton plus electron yields neutron), and calcium (^{40}Ca), produced by electron emission (neutron yields proton plus electron). These daughter parti-

---

* In quantitative terms, radioactive transmutation may be defined in terms of the ratio of parent isotopes now present, N, to those incorporated at time of crystallization of the host mineral, N_0, and the decay constant, λ (the rate of transformation of parent isotope to daughter isotope), thus: $N/N_0 = (1 - \lambda)^t$. If we define the decay constant in terms of one half-life ($\lambda = 0.5$), the exponent t becomes the number of half-lives, hence (because the log of 0.5 is -0.3): log $N/N_0 = -0.3t$.

TABLE 3.1. THE PERIODIC TABLE OF THE ELEMENTS

Period	Group I	Group II	Transition Metals										Group III	Group IV	Group V	Group VI	Group VII	Group VIII
1	1 H Hydrogen																	2 He Helium
2	3 Li Lithium	4 Be Beryllium											5 B Boron	6 C Carbon	7 N Nitrogen	8 O Oxygen	9 F Fluorine	10 Ne Neon
3	11 Na Sodium	12 Mg Magnesium											13 Al Aluminum	14 Si Silicon	15 P Phosphorus	16 S Sulfur	17 Cl Chlorine	18 Ar Argon
4	19 K Potassium	20 Ca Calcium	21 Sc Scandium	22 Ti Titanium	23 V Vanadium	24 Cr Chromium	25 Mn Manganese	26 Fe Iron	27 Co Cobalt	28 Ni Nickel	29 Cu Copper	30 Zn Zinc	31 Ga Gallium	32 Ge Germanium	33 As Arsenic	34 Se Selenium	35 Br Bromine	36 Kr Krypton
5	37 Rb Rubidium	38 Sr Strontium	39 Y Yttrium	40 Zr Zirconium	41 Nb Niobium	42 Mo Molybdenum	43 Tc Technetium	44 Ru Ruthenium	45 Rh Rhodium	46 Pd Palladium	47 Ag Silver	48 Cd Cadmium	49 In Indium	50 Sn Tin	51 Sb Antimony	52 Te Tellurium	53 I Iodine	54 Xe Xenon
6	55 Cs Cesium	56 Ba Barium	57 La Lanthanum	72 Hf Hafnium	73 Ta Tantalum	74 W Tungsten	75 Re Rhenium	76 Os Osmium	77 Ir Iridium	78 Pt Platinum	79 Au Gold	80 Hg Mercury	81 Tl Thallium	82 Pb Lead	83 Bi Bismuth	84 Po Polonium	85 At Astatine	86 Rn Radon
7	87 Fr Francium	88 Ra Radium	89 Ac Actinium															

Metalloids and Nonmetals

Lanthanides (Rare Earth Elements)

58 Ce Cerium	59 Pr Praseodymium	60 Nd Neodymium	61 Pm Promethium	62 Sm Samarium	63 Eu Europium	64 Gd Gadolinium	65 Tb Terbium	66 Dy Dysprosium	67 Ho Holmium	68 Er Erbium	69 Tm Thulium	70 Yb Ytterbium	71 Lu Lutetium

Actinides

90 Th Thorium	91 Pa Protoactinium	92 U Uranium	93 Np Neptunium	94 Pu Plutonium	95 Am Americium	96 Cm Curium	97 Bk Berkelium	98 Cf Californium	99 Es Einsteinium	100 Fm Fermium	101 Md Mendelevium	102 No Nobelium	103 Lw Lawrencium

TABLE 3.2. SELECTED CONVERSION FACTORS

Metric		English	
1 bar	= 0.9869 atmosphere	1 atmosphere	= 14.696 pounds/square inch
	= 10^8 millipascals		
	= 10^5 pascals		
	= 100 kilopascals		= 1.0133 bars
1 kilobar (1000 bars)	= 1000 bars		
	= 100 megapascals		
1 megabar	= 10^6 bars		
	= 100 gigapascals		
Temperature in degrees Celsius (T_C)	= $5/9(T_F - 32)$	Temperature in degrees Fahrenheit (T_F)	= $9/5 T_C + 32$
Temperature in degrees Kelvin (T_K)	= $T_C - 273.16$		
1 centimeter (10 millimeters)	= 0.3937 inch	1 inch	= 2.540 centimeters
			= 25.40 millimeters
1 angstrom (Å)	= 10^{-8} centimeter	1 foot	= 0.3048 meter
1 kilometer	= 0.6214 mile	1 mile (5280 feet)	= 1.609 kilometers
1 gram	= 0.03527 ounce	1 pound	= 0.4536 kilogram
1 kilograms (1000 grams)	= 2.205 pounds	1 ton (2000 pounds)	= 907.03 kilograms
1 metric tonne	= 2205 pounds		

cles, along with radiative energy, are generated at a measured, known, constant rate in fixed proportions from the parental ^{40}K. The decay rate is independent of temperature, pressure, and chemical environment. Because the radiogenic argon and calcium are retained in the host mica in the cation site formerly occupied by the parental potassium, the age of formation of this phase – presumably the time of crystallization or recrystallization of the enclosing rock – can be computed. This is accomplished by measuring the amounts of surviving radioactive parental ^{40}K and accumulated radiogenic daughter ^{40}Ar now present: The higher the ^{40}Ar/^{40}K ratio, the older the mica. It has been assumed that neither potassium nor argon has diffused in or out of the atomic structure since the time of crystallization, and also that the original concentration of ^{40}Ar was zero in the mica.

Another mineral, zircon ($ZrSiO_4$) incorporates trace amounts of uranium (U) during its crystallization but totally rejects lead (Pb), as a consequence of size and charge requirements of cation sites in its atomic structure. Uranium has two radioactive isotopes, ^{238}U and ^{235}U, which decay ultimately to isotopes of lead, ^{206}Pb and ^{207}Pb, respectively. The rates of decay of both uranium isotopes have been measured very accurately. Again, analyzing the proportions among the radioactive parental uranium and radiogenic daughter isotopes of lead allows an accurate assessment of the age at which the host mineral formed, again assuming the initial absence of lead in the host zircon and no import or export of uranium or lead subsequent to crystallization of the zircon host. Employing these and several other isotopic systems for a variety of phases, Earth scientists are able to determine mineral or bulk-rock radiometric ages. Some workers refer to isotopic ages as absolute ages (quantitative) to distinguish them from fossil ages, which are all relative (qualitative).

Radiometric methods have allowed the assignment of times of formation to primitive stony meteorites and to pul-

verized light-colored rocks of the lunar highlands, which are as old as 4.55 and slightly in excess of 4.4 billion years old, respectively. These planetary materials, which are substantially older than the lunar maria (dark lava–floored lunar lowlands, thought by the ancients to represent seas), have not been appreciably reheated since condensation of the solar nebula – the time of origin of our solar system. Because they have not been heated to high temperatures, recrystallized, and homogenized since their formation, the ages of these primitive meteorites and lunar highland rocks reflect the time of original collapse of the interstellar gas cloud and origin of the Sun, as well as subsequent early accretion of the young Earth, the Moon, and the other planets. This subject is discussed further in Chapter 4.

Long before radioactivity was discovered, however, geologists had wrestled with the problem of the great antiquity of the Earth. A basic tenet of stratigraphy, the study of layered, sedimentary rocks and lavas formed at or near the sur-

Figure 3.1. The exponential decay of a radioactive element, showing several half-lives; each would be about approximately 1.5 billion years in the case of ^{40}K (*Source:* F. Press and R. Siever, *Earth,* W. H. Freeman 1994.)

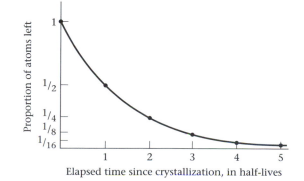

face of the Earth, states that, in undeformed sections, the oldest layer occurs at the base, with successively younger strata (layers, beds) overlying this unit, and the youngest stratum resting on top. Clearly, the oldest sediments in a section are laid down as a blanket resting on the basement of a depositional basin, with progressively younger strata deposited sequentially above. Similarly, the oldest lava bed occupies the basal portion of a stack of flows, with the most recent eruptive at the top. Thus, the relative ages of such surficially deposited units can be determined readily by recognizing which way is up in the stratigraphic section. This simple concept is known as the *law of superposition*.

Nearly two centuries ago, the geologic processes that result in accumulations of layered sediments in lakes and rivers and along the seashore were observed and the depositional rates roughly measured. It was found that sedimentary rock sequences apparently require hundreds of thousands – even millions – of years for the buildup of a few hundred meters of section. The enormous age of the Earth was thus inferred by postulating that current processes and rates of deposition were constant through geologic time. This is a statement of the *concept of uniformitarianism*, which means that the type of geologic activity taking place today also occurred in the distant past.

Support for the law of superposition was provided by the nature of fossil materials – evidence of prehistoric life – entombed in strata; it became clear early in the nineteenth century that the more ancient rocks contain strange, relatively primitive life forms, or none at all, compared with the progressively more familiar, organizationally more complex animal and plant remains contained in the overlying, younger beds. Indeed, these contrasting groups of fossils from different geologic ages are ubiquitous on Earth and have allowed geographically remote stratified rock series to be compared and relative ages of formation to be assigned. Darwin's theory of evolution provided a mechanism to explain the observed changes in life through the geologic ages and supported the concept of uniformitarianism. In the natural environment, the presently observable, measurable processes are the key to unraveling Earth history.

Prior to the introduction of the radioactive dating method, however, geologists used less accurate techniques, such as that based on sediment accumulation rates, with which to extrapolate and gauge the absolute ages of rocks. Moreover, larger fossils are virtually nonexistent in strata laid down before the Cambrian period, approximately 570 million years ago; consequently, only crude relative ages could be assigned to rocks older than the Phanerozoic Eon – the time of macroscopic life as we know it. These earlier, Precambrian times have been termed the Proterozoic Eon (time of primitive life), and the yet older Archeozoic Eon (Archean, time of ancient life). The unrepresented interval between the formation of the Earth and preservation of the oldest terrestrial rocks (4.5 to 3.9 billion years ago) is sometimes called Hadean time, in reference to the inferred hot, turbulent con-

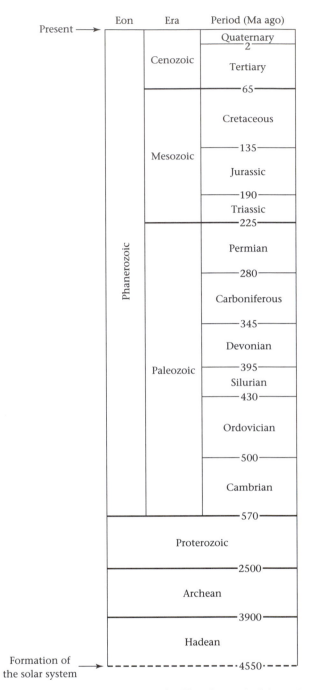

Figure 3.2. The geologic time scale. The Cenozoic, Mesozoic, and Paleozoic eras in aggregate constitute the Phanerozoic Eon, or time of macroscopic life. The Precambrian consists of the progressively more ancient Proterozoic, or time of primitive life, the Archeozoic (Archean), or time of ancient life, and the Hadean, lack of a rock record, or pregeologic time. Ma = time before present in millions of years. (*Source: Geologic time scale*, Geological Society of America, 1983.)

ditions of accretion of the planet. A simplified geologic time scale is provided for reference in Figure 3.2.

The ever-changing course of life through the ages is an exciting, engrossing topic in itself, but is treated only briefly

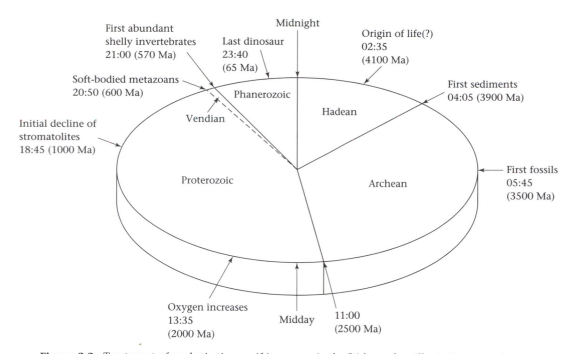

Figure 3.3. Treatment of geologic time as if it were a single, 24-hour day, illustrating some important events in Earth history. The earliest evidence of surficial processes (presence of liquid water, and sedimentary deposits) appears a few minutes after 4:00 A.M., and a recognizable shelly invertebrate fauna becomes evident only at approximately 9:00 P.M. The dinosaurs put in a brief appearance and died out 20 minutes before midnight. Early hominids arose in East Africa approximately 75 seconds before the strike of midnight, and if you blink, you'll miss civilization (the past 10,000 years). Ma = time before present in millions of years. (*Source:* G. Brown, C. Hawkesworth, and C. Wilson, eds., *Understanding the Earth*, Cambridge University Press, 1992.)

here. Because so few rocks have been preserved from the early stages of planetary development, clues to the history of life are progressively more fragmentary and obscure in the Proterozoic and Archean eons. Note that the Phanerozoic Eon, or time of macroscopic life, represents only about the last eighth of the interval that has lapsed since the formation of the Earth! Primitive bacteria were in existence at least 3.5 billion years ago, and their obscure origins surely must have been somewhat earlier; evidently, most of the Precambrian terrestrial history was characterized by infinitesimally slow evolution of primitive life forms prior to the spectacular flourishing of more advanced plants and animals during the Phanerozoic. A few landmarks in the development of life are illustrated in Figure 3.3, which attempts to convey the enormity of geologic time. As simple forms evolved toward more complicated organisms, the potential for diversity became greatly enhanced. Thus organic evolution appears to be an ever-accelerating process.

To sum up, most geologic processes operating on planet Earth are imperceptibly slow, and take place over millions or billions of years. As discussed in Chapters 4 to 6, these processes are driven principally by the transfer to the surface of deeply buried heat, and over the course of geologic time, thermal relaxation has resulted in a gradual deceleration of Earth dynamics. Our stately geologic clock is running down. As we now know, life changes, too, began extremely slowly, but unlike the inorganic world, the organic began to evolve more rapidly as time passed and as the complexity of biologic organization increased.

BIOLOGIC TIME

The evolution of terrestrial life corresponds rather closely to geologic time because traces of biologic activity extend well back into the Archean Eon. Bacteria that are anaerobic (i.e., that live without consuming oxygen) and photosynthetic have been described from 3.5-billion-year-old siliceous sedimentary deposits of western Australia, and must have existed even earlier. These filamentous single-celled organisms formed domical or mushroom-like colonies called stromatolites (Fig. 3.4). They flourished in tidal marine waters, and still do in protected lagoons such as Shark Bay in western Australia, and the Gulf of California. As fine clay washed over these ancient biogenic mounds and they were buried by later sediments, aqueous fluids percolated through the muds, hardening them into layered rock; in the process, the original organic structures were replaced by new mineral matter, thus preserving some such colonies as fossil stromatolitic reefs.

Figure 3.4. Australian stromatolites, past and present. (A) Archean stromatolite from northwestern Australia, with original structure replaced by chert (SiO_2). Laminated, concentric banding of growth layers can be observed. (B) Modern stromatolitic mounds along the west coast of Australia at Shark Bay. (*Source:* Photographs courtesy of Ken McNamara.)

Stromatolites were widespread in the early Earth. Minerals preserved in Archean strata, as well as computer models of planetary outgassing and paleoclimatology, suggest that early in its history, the planet was enveloped by a strongly reducing atmosphere characterized by the scarcity of free oxygen but rich in methane, hydrogen, and perhaps also ammonia.

As illustrated in Figure 3.5, cells without a nucleus – or prokaryotes – have dominated the parade of terrestrial life through the geologic ages. The rise of cells containing a separate nucleus – eukaryotes – occurred only after nearly 2 billion years of exclusively prokaryotic life (chiefly bacteria). The young Earth had a reducing atmosphere rich in methane, ammonia, and/or carbon monoxide/carbon dioxide, hence the most primitive, nonnucleated cells were necessarily anaerobes. Waste products of such organisms include oxygen, which led to the progressive oxidation of the near-surface, condensed portion of the planet, and ultimately to the gradual accumulation of free oxygen in the atmosphere.

This increasing abundance of molecular oxygen in turn promoted the emergence of other biochemical energy conversion systems, including aerobic bacteria and related forms. More energetic metabolism, allowing the sustenance of more complex life forms and enhanced diversity, thus became a possibility. Was free oxygen the fuel that ignited the evolution toward eukaryotes, then to multicelled metazoans that evolved into the plants and animals we see today? Possibly! In spite of the remarkable evolutionary blossoming that followed, it is important to remember that single-celled organisms constituted the exclusive forms of life for approximately 2.7 billion years.

Only toward the end of the Precambrian (approximately 600 to 800 million years ago) do we have any evidence for the presence of more complex, multicellular organisms and, subsequently, of primitive invertebrate life. The clues are present predominantly in the form of plant impressions, and fossil tracks and traces made by bottom-dwelling marine invertebrates. However, a rich collection of soft-bodied crea-

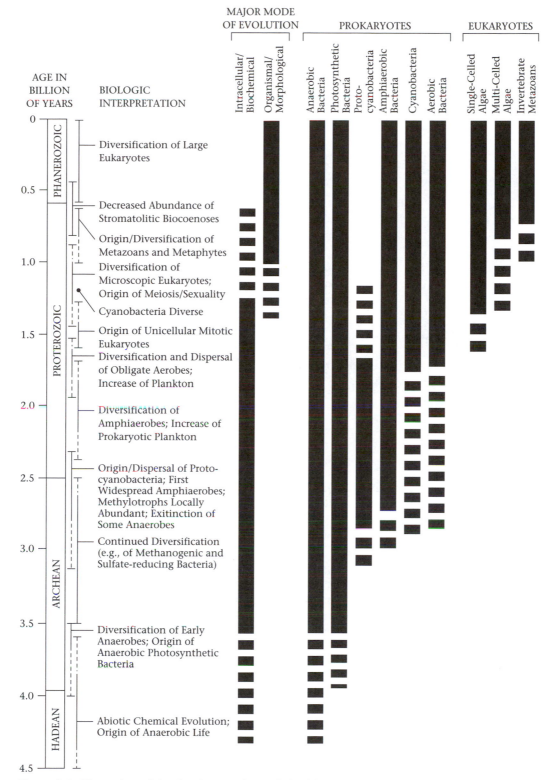

Figure 3.5. Illustration of the dominance of unicellular life (especially nonnucleated) throughout the geologic ages. (*Source*: J. W. Schopf, *Earth's earliest biosphere*, Princeton University Press, 1983.)

Figure 3.6. Evolution of the perissodactyls (horses). (A) The lineage of the horse family from Eocene to present. (*Source:* A. S. Romer, *Vertebrate paleontology,* W. H. Freeman, 1947.) (B) Three stages in the evolution of the skull and jaws, teeth, and feet in horses. (a, b, c) Skull and jaws, upper and lower molars, and hind foot of *Hyracotherium,* a primitive Eocene horse. (d) Skull and jaws of *Parahippus.* (e,f) Upper and lower molars and hind foot of *Merychippus,* a Miocene horse. (g, h, j) Skull and lower jaws, upper and lower molars, and hind foot of *Equus,* a Pleistocene and Recent horse. This figure shows the progressive increase in the length of the facial or preorbital portion of the skull, and the deepening of the skull and lower jaw to accommodate the increasingly higher cheek teeth; the progressive complexity of the molar crowns; and the reduction of the side toes. d and g, one-eighth natural size; a, c, f, j, approximately one-fourth natural size; b, approximately one-half natural size; e, h, approximately one-fourth natural size. (*Source:* A. S. Romer, *Vertebrate paleontology,* W. H. Freeman, 1947.) (C) Appearance of members of the horse family: a = *Hyracotherium;* b = *Orohippus;* c = *Mesohippus;* d = *Merychippus;* e = *Pliohippus;* and f = *Equus.* (*Source:* B. J. MacFadden, *Fossil horses,* Cambridge University Press, 1992.)

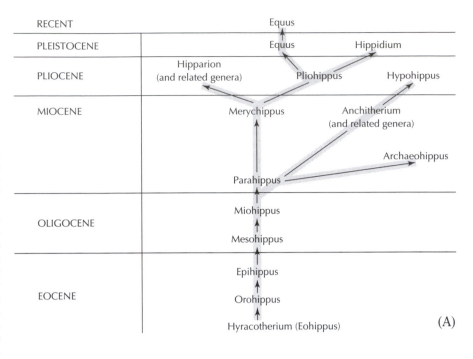

tures of perplexing lineage – the so-called Ediacara fauna – has been described from the youngest Proterozoic strata (termed "Vendian" in age) of eastern Australia. By the beginning of the Cambrian period, plants and animals displaying remarkable morphologic diversification and occupying a wide range of habitats burst onto the scene. To our knowledge, most life forms had developed shells or other preservable hard parts by this time. Such organisms are more likely to leave fossil evidence than are soft-bodied forms. Who knows how many floral and faunal types left no record at all? The Paleozoic Era, lasting 325 million years, witnessed a proliferation of invertebrates, fishes, and amphibians. The following 180-million-year-long Mesozoic Era is characterized as the age of dinosaurs. Reptiles actually arose in the late Paleozoic, and some have survived into the Cenozoic, or age of mammals. Similarly, primitive mammals existed throughout the Mesozoic Era; however, only after the demise of the dinosaurs – approximately 65 million years ago – did they inherit the Earth. Once the reptiles' reign of terror was largely concluded, mammals rapidly occupied the many ecologic niches previously controlled by their cold-blooded precursors, diversifying and evolving in the process – an activity that continues to this day. The general phenomenon by which organisms evolve in harmony with the changing environment is termed *adaptive radiation.*

As an example, the evolution of the horse, from a cocker spaniel–sized *Hyracotherium* (previously called *Eohippus*) that lived 50 million years ago to the extant *Equus* is illustrated in Figure 3.6. Imperceptible changes in the horse population have occurred over geologic time. Portions of an evolutionary continuum, are represented by individual species. A gradual increase in animal size took place, along with parallel structural changes in the dentition and especially in the limbs and digits. Specialization as a grazing animal whose best defense was fleetness of foot is evident. Especially significant is the gradual conversion of the five-toed *Hyracotherium* to the modern genus, which runs only on the middle toe (now a hoof), the four vestigial digits having become greatly reduced in size. Clearly, environmental stresses ensured that those individuals that survived long enough to perpetuate favorable genetic characteristics in the horse population had the attributes of size and speed, but not, necessarily, of wit. This subject is explored in some detail in Chapters 17 and 18.

We now turn from the measured sweep of biologic evolution through geologic time to the duration of an individual species. A species' life-span seems to be roughly proportional to the simplicity of organization, and is inversely related to the organism's complexity and degree of specialization. Adaptability and species longevity seem to be a function of simplicity. Some species of primitive bacteria have survived virtually unchanged for more than 3 billion years, whereas most complexly organized invertebrate and vertebrate species have persisted for roughly 5 to 10 million years. The dinosaurs ruled the Earth for approximately 180 million years; however, individual species existed for much shorter periods of time. The modern horse, *Equus,* arose from a more ancient form, *Pliohippus,* in the early Pleistocene, so modern horses have been galloping about for the past 2 million years, with no sign yet of an impending demise. Cro-Magnon man (*Homo sapiens*) has been in evidence since the third intergla-

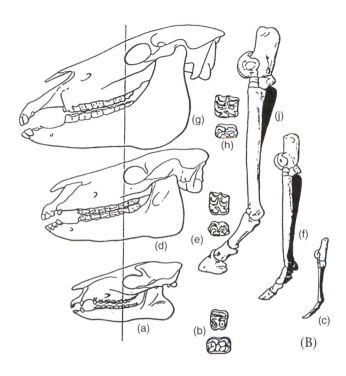

(B)

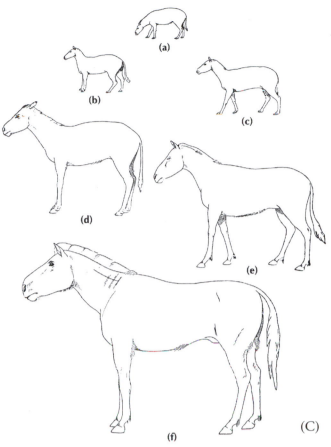

(C)

cial of the Pleistocene some 200,000 years ago. Although somewhat controversial, studies of certain types of DNA suggest that ancestors of the modern human race lived in Africa approximately 140,000 to 280,000 years ago.

If the average horse generation is approximately five years, approximately 400,000 generations of the modern *Equus* have trodden the Earth. Clearly, the life-span of the least biologic unit – an individual animal – is quite brief relative to the duration of the species. Assuming that a human generation is approximately 20 years, approximately 10,000 generations of anatomically modern man have existed since the late Pleistocene, only approximately 100 generations since the birth of Christ, and just 25 generations since the invention of the printing press and the beginning of the Renaissance. In the case of *Homo sapiens*, the enormously accelerated evolutionary trends in the biosphere now evident, including massive extinctions, are a reflection of conscious manipulation of the overall environment. Regarding biologic time, we may conclude that modern evolutionary trends take place over thousands to millions of years, many orders of magnitude longer than the life-span of biologic individuals, but far briefer than the vast sweep of geologic time. Early in planetary history, of course, the primitive single-celled organisms changed almost infinitely slowly, more in line with rates of the rather more stately solid-Earth processes.

POLITICAL TIME

Orders of magnitude shorter are the important time intervals associated with human civilization. Its entire duration may be thought of, generously, as approximately ten millennia

(back to approximately 8000 BC). Most major periods of technological development are far more abbreviated, and present growth in this area is accelerating at a remarkable rate.

Technological breakthroughs have allowed an incredible increase in the speed of communication and transportation. Magellan and Drake took two to three years to circumnavigate the globe, but we can travel to almost any place on Earth within twenty-four hours, unless we have to land at O'Hare Airport en route. The Viet Nam War was viewed on television in millions of American living rooms, we saw the Berlin Wall demolished in real time, and television commentators were analyzing the Gulf War on site even as hostilities commenced. Future technological advances in communication and transportation surely will occur; however, beyond increased diversity and broader availability, the impact of bodily travel on time probably will be modest. Telecommunication already is at the velocity of light, and that, representing a fundamental physical limit, will not change.

It took eight years to plan and successfully launch the Apollo Project, and astronauts landed on the Moon less than a decade after inception of the NASA lunar program. On the other hand, humanity has made very limited progress on environmental improvement since the publication of Rachel Carson's book, *The Silent Spring*, raised our awareness of widespread, insidious terrestrial pollution more than thirty years ago. Why such a contrast in effectiveness? The reasons are many. Sending humans to the Moon was a complicated engineering feat, but the technology was already at hand or on the drawing boards, and the goal was simple, clear, and spectacular. Broad-based support existed in the United States, reflecting the inherent excitement of the challenge, and because of both economic and national-defense incentives. The Apollo Project captured our collective imagination. In contrast, environmental protection and enhancement is a far more complicated worldwide problem due to the numerous interacting Earth systems that range in size from global to local – geosphere, hydrosphere, biosphere, and atmosphere – most of which are currently not fully understood. Although the technological aspects are formidable, they represent only a small part of the problem: Humanity's adverse impact on the natural surroundings is deeply and inextricably interwoven with political, economic, and social attitudes, as well as numerous, profoundly conflicting values and goals. Changes in socioeconomic behavior, resource utilization, and environmental management produce both winners and losers. Accordingly, progress has been slow even in the most favorable situations, and what has been accomplished is not easily and widely recognized. Moreover, solutions tend to be site- or phenomenon-specific, and are not necessarily readily transferable to other ecologic situations. The "ozone hole" of the Circum-Antarctic is a technologically tractable problem, but its abatement has been long in coming. Also, the solution to this particular problem sheds little light on how we might address other concerns such as the anthropogenic buildup of greenhouse gases, and the worldwide reduction in biodiversity.

It must also be emphasized that global environmental degradation is a transnational problem; its solution requires the long-term dedication of all the world's people. In the market-driven West and industrialized Pacific Rim countries, planning parameters for ongoing socioeconomic programs are equated with the time to the next election – say, two to three years. For the nations constituting the Eastern Bloc, which, until recently, were all managed economies, there is a profound lack of necessary financial resources, and feedback from the citizens was not customarily solicited. In the Third World, undeveloped economies and a burgeoning population hamstring efforts to increase the GNP and the standard of living. For all three types of sociopolitical system, the exploitation of environmental capital and nonrenewable resources to fuel the near-term economy promotes ever-increasing human impact and consequent ecologic degradation, without enhancing the sustainability of the environment on which the basic economic system depends. We are all borrowing from the future carrying capacity of the planet in order to address today's pressing need to earn a living from the Earth (see Chap. 22).

A moderately successful political effort to manage the production of chlorofluorocarbons (CFCs) and halon compounds, a class of manufactured greenhouse gases, resulted in the signing of the Montreal Protocol in 1989. These syn-

thetic, chlorine-bearing compounds were – and are – employed as inert propellants, refrigerants, cleaning agents, and aerosols. However, in 1974, the American chemists Molina and Rowland found that, although harmless in the lower atmosphere, these CFCs gradually diffuse into the stratosphere, where, in the presence of sunlight, they react with other gaseous species, liberate chlorine, and destroy ozone (O_3) in the process. Ozone is generated and consumed naturally at essentially the same rate, thereby maintaining a dynamic equilibrium. Its formation occurs primarily through the dissociation of molecular oxygen by absorption of solar ultraviolet radiation. Oxygen molecules can also combine with oxygen atoms to form ozone, if a suitable liquid or solid surface is present. Ozone is destroyed by several natural catalytic processes; approximately 70 percent of the natural destruction is due to the nitrogen cycle (see Chap. 19). Chlorine is believed to be the principal agent upsetting the natural balance of ozone production and destruction. As a consequence of the work of Molina and Rowland, concern has been raised that increasing concentrations of CFCs could add enough chlorine to the atmosphere to increase the net destruction rate and decrease the net amount of ozone. This subject is discussed in greater detail in Chapter 13.

An "ozone hole" (in actuality, abnormally low concentrations of ozone at high altitudes) has developed over Antarctica during the southern spring for the past few years, as shown in Figure 3.7. When present in small concentrations,

Figure 3.7. Vertical profiles of ozone using electrochemical ozonesondes from McMurdo Station in Antarctica, August–October 1992–1993. The numbers show the drop from Antarctic winter (August) to unusually low levels in Antarctic spring. By October, the total ozone over Antarctica had been reduced by more than 50 percent of its 1979 value. Local depletion was virtually 100 percent at altitudes of 15 to 20 kilometers. O_3 = molecule consisting of three oxygen atoms; DU = Dobson units, a measure of gas species abundance. mPa = millipascals. (*Source:* Data from NASA.)

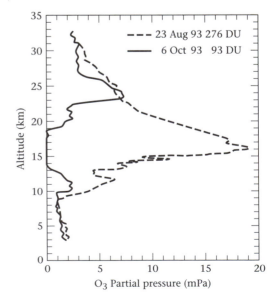

as it is in the stratosphere, ozone absorbs harmful ultraviolet radiation, thereby providing a shield for the Earth's biosphere. Large doses of such short-wavelength radiation cause skin cancer. The prospect of a widespread loss of the ozone layer galvanized an assembly of ninety-two signatory countries to regulate the production and usage of CFCs including halons. The resulting international accord is now in effect; it mandates a gradual decrease in usage of these synthetic gas species in the developed countries, with temporary exemptions for the developing nations, to be followed by subsequent programed reductions. This modest triumph of concern for the environment over unbridled exploitation was achieved because the threat to humanity was perceived as unambiguous, because exclusive chemical manufacturing patents were lapsing, and because relatively inexpensive substitutes were becoming available. Even so, fifteen years elapsed between the disquieting discovery by Molina and Rowland and the signing of the Montreal Protocol.

Of far greater consequence for the vulnerable web of life are the massive energy conversions of fossil and nuclear fuels, and the loss of the world's tropical and temperate forests. Climate change, loss of habitat, and diminished biodiversity are inevitable results of these trends. Inexpensive energy is the currency of civilization, and although many forms are readily utilizable, all have damaging side effects (e.g., acid rain, greenhouse gas buildup, radiation, thermal pollution). Because per capita consumption of energy is increasing and world population is growing rapidly, the consequences of ever greater energy utilization will further intensify these side effects in the future (Chap. 22). As of now, there are neither abundantly available alternative energy sources nor realistic prospects of diminishing demand. Fossil fuels represent a finite, nonrenewable resource on this planet, and although deposits remain to be discovered and utilized, our days of hydrocarbon usage are numbered (see Chap. 24). If a marked decrease in recoverable energy, and a severe, inescapable degradation of the environment are to be avoided, both highly desirable goals, we must ultimately reach equilibrium with the carrying capacity of the planet and its sustainable complement of natural resources.

How long will it take to achieve this goal? Centuries, judging by our experience in addressing the relatively straightforward problem of CFCs. However, are we gaining ground, rather than losing it at an ever increasing rate? Time will tell (political time, that is). Unfortunately, political time for action is very short indeed. As of now, degradation of the planet is continuing – even intensifying – not lessening. A hopeful sign is that worldwide awareness of the seriousness of widespread human-induced pollution, the systematic loss of ecologic habitat and biologic diversity, the real possibility of anthropogenic climate change, and the heightened need for the preservation of the environment have become first-order global issues. The fact that you are reading this text is testament to that increasing concern. What is required is thoughtful global action by knowledgeable, concerned elec-

torates, empowered and intellectually equipped to deal with these interconnected problems.

SEASONAL/DAILY TIME

Within yet another time frame, encompassing the period from one complete rotation of the Earth about its spin axis (24 hours), to the planet's orbit about the Sun (365¼ days), the amount of sunlight and therefore the temperature of the surface vary in a most familiar, understandable way. The oceans, atmosphere, and especially the biosphere respond in cyclic fashion to these periodic, predictable changes we know as day/night and seasonal variations. Can a systematic, long-term trend in climate change be discerned? Not as yet, but if one is under way, the coming conditions will have serious implications for the intricate, interactive life-support systems of the future biosphere.

INSTANTANEOUS GLOBAL CATASTROPHES

We have observed that the various changes on planet Earth proceed at vastly different rates, from the imperceptible, slow processes typifying geologic evolution, through the subtle but somewhat more energetic, interactive transformations of the dynamic web of life, to the rapid, sometimes disruptive changes characteristic of human activities. One receives the impression that gradual, relatively smoothly changing conditions characterize the geosphere, biosphere, and anthrosphere, and to a very great extent, this is so.

As described in Chapter 4, the evolution of the early solar system was initiated by the collapse of an interstellar gas cloud – our solar nebula. The Sun and orbiting protoplanetary bodies formed through the gravitational attraction of matter – ever more rapidly at first as discrete mass concentrations increased, and then more slowly as the diminishing amounts of interplanetary debris were swept up by the enlarging planets as well as the far, far more massive Sun. The accreting bodies acted much like mass "vacuum cleaners," attracting the matter within gravitative reach. The solar system debris, asteroids, or planetesimals, were (and are) chiefly solid meteorites, with the gassy cometary materials relatively more abundant near the outer limits of the solar system.

Because the Earth grew by accreting planetesimals, meteorite impacts occurred often during the primordial rapid-growth stage; however, such arrivals declined in both frequency and mass as time passed. For example, as shown in Figure 3.8, an ancient impact feature 900 kilometers in diameter occurs on the Moon (an even larger impact crater is present on Mars); a similar collision with the Earth would have released enough heat to kill any extant life-forms. Moreover, due to its greater mass, the early Earth must have been bombarded even more intensely than the Moon. Meteors still enter the Earth's atmosphere, and can be observed on clear nights (the middle of August is the best viewing time because of the annual arrival of an abundant swarm of planetesimals;

Figure 3.8. On the extreme western edge of the visible side of the Moon is the Mare Orientale impact basin; the crater is approximately 950 kilometers in diameter. Impacts of this scale would have been sufficient to sterilize the Earth's surface environments and cause the biochemical evolution of life to start over. (*Source:* Courtesy of NASA, *Lunar Orbiter IV* photograph.)

the display is known as the Perseid meteor shower). Most such objects burn up in the Earth's atmosphere, being too small to survive long enough to reach the ground. Larger asteroids still do impact the Earth from time to time, as the 1.2-kilometer-wide Meteor Crater in Arizona attests. That incoming projectile, approximately 40 meters in diameter, landed approximately 50,000 years ago. The scars of much larger impacts are abundantly visible on the Moon (Fig. 3.8) and the other inner planets (see Chap. 4); however, reworking caused by the surficial processes of erosion and deposition of sediments on Earth has obliterated most of the terrestrial evidence of past collisions with asteroids.

The Earth's biota experienced a global crisis 65 million years ago, manifested by the great dying out of the dinosaurs, as well as many other, unrelated plant and animal species. This time of mass extinction coincided precisely with the deposition of a ubiquitous layer of clay, as confirmed by radiometric dating. The clay contains unusually large con

Figure 3.9. (A) Index map of the Chicxulub Crater (circle), thought to have resulted from the impact of a massive comet or asteroid some 65 million years ago. Thick clay layers in Haiti and in Mexico contain fused pellets, possibly ejected by the massive impact. (B) Artist's conception of the instant of impact. K/T = Cretaceous–Tertiary boundary. See color plate section for color version of this figure. (*Source:* Courtesy of NASA.)

centrations of platinum-group elements, and locally, soot, glass droplets, and mineral grains display pressure-shock features. Platinum and iridium are very rare in the Earth's surficial rocks but occur in substantial trace amounts in common meteorites; the clay, therefore, contains components thought to represent the geochemical signature of the material of an incoming planetesimal, vaporized on impact and added to the ensuing dust cloud that circled the planet before settling out. The energy released on collision would have partially melted the rocks at ground zero, shocking some of the entrained mineral grains in the process. Furthermore, the heat generated by the impact probably would have ignited global forest fires, thus accounting for the soot. Megawave gravel deposits around the Caribbean contain variable amounts of these distinctive constituents, suggesting that the target area lies nearby. After a diligent search, the 65-million-year-old impact feature was recently discovered along the northern margin of the Yucatan Peninsula. Chicxilub Crater is 180 kilometers across, and probably was excavated by a meteorite approximately 10 kilometers in diameter. It is situated along the coast of the Gulf of Mexico and lies partly below sea

level; the on-land portion is overlain by more recent sedimentary strata, which obscure the true nature of the feature. Its location, is presented in Figure 3.9. An artist's conception of the Chicxilub collisional event, and a schematic cross section of such a structure are diagrammed in Figure 3.10. The resultant temporary but profound, worldwide climate change generated by the shock waves and longer-term influence of a sun-obscuring global dust cloud evidently were sufficiently severe that many types of life died out altogether. Could the dinosaurs have prevented this cataclysmic destiny? How about humans?

Among the time scales to be considered as we study Earth systems are those geologically instantaneous catastrophes that have resulted in the formation of the planets. Such events, numerous in the early solar system, have become progressively rarer, but they do occur. Thus, global change takes place over a range of temporal scales and attendant rates, spanning the duration of time since the condensation of the solar nebula involving infinitesimally slow processes, to mere seconds for some of the more spectacular, catastrophic events.

Figure 3.9. (B)

(A)

Geologic Cross Section of a Giant Impact Formation

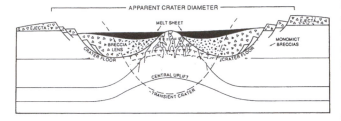

(B)

Deposition of Ejecta Material and Secondary Cratering

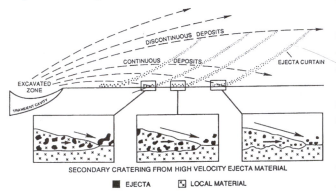

Figure 3.10. Diagrammatic cross-section of a terrestrial impact crater such as proposed for the Yucatan Peninsula based on NASA sources (an analogous structure, the lunar impact crater Humboldt, 200 kilometers in diameter, is approximately the same size as the proposed impact at Chicxulub). The structure of such a crater is illustrated in (A). In addition to shock features, fragmentation, and impact melting, the collision would expel an extensive spray of ejecta from the crater. Airborn debris from the spray, such as the glass droplets of melted rock found in Beloc, Haiti, eventually landed far from the point of impact, as shown in (B). See text for discussion. (*Source:* NASA pamphlet, 1993.)

QUESTIONS

1. Describe the methods that have been employed to pinpoint events in geologic time, and discuss the nature of the dating, including contrasting accuracies and degree of quantitativeness. What methods would you employ to date: (A) the lunar maria; (B) King Tut's sarcophagus; (C) a granitic rock from the Sierra Nevada?

2. Global change takes place at several contrasting rates. How are these rates related to the: (A) evolution of biologic species; (B) compositional change of the atmosphere; (C) adoption of environmental regulations?

3. Debate the influence of political time scales on implementation of environmental policy.

4. Why is the rate of biologic change accelerating, and how will it influence the global to local ecology?

5. What steps should the world population take if astronomers were to predict that in 631 days, a 15-kilometer-diameter asteroid in Earth-crossing orbit is likely to impact the interior of Alaska?

6. How would you go about testing the concept of uniformitarianism? Does it matter if this concept proves to be incorrect?

FURTHER READINGS

Alvarez, W. 1997. *T. rex and the crater of doom.* New York: Vintage Books.

Cloud, P. 1985. *Oasis in space.* New York: Norton.

Emiliani, C. 1992. *Planet Earth.* New York: Cambridge University Press.

Gould, S. J. 1989. *Wonderful life.* New York: Norton.

Hawking, S. W. 1988. *A brief history of time.* New York: Bantam Books.

Hazen, R. M., and Trefil, J. 1991. *Science matters.* New York: Doubleday.

National Research Council. 1990. *Confronting climate change.* Washington, DC: National Academy of Sciences.

Silver, C. S. (ed.,) 1990. *One Earth, one future.* Washington, DC: National Academy of Sciences.

NATURAL PROCESSES

THE GEOSPHERE

4 The Earth's Place in the Solar System

W. G. ERNST

OUR LITTLE CORNER OF THE COSMOS

Our planet formed along with the rest of the solar system during the collapse of a locally dense, mass-rich region of intergalactic gases. This event was a minor but typical perturbation that occurred approximately 4.5 to 4.6 billion years ago, long after the Big Bang, the time when the universe as we conceive of it began its rapid expansion and cooling. Just what happened? Consider an expanding cloud of interstellar gas, or nebula, rotating slowly and characterized by a nonuniform distribution of matter. Because of the enhanced gravitative attraction of a local mass concentration, more matter would be drawn into the enlarging clump. As the mass buildup of interstellar dust and gas molecules continued, the enhanced gravitative attraction of additional gas and particulates would cause further growth. Most of the material, dominantly hydrogen, would be concentrated in the central object, and the great pressure and attendant high temperature attending such a condensation would stoke the nuclear furnaces of what was becoming a newly born star. This nuclear fusion process, whereby hydrogen combines to form helium and liberates radiant energy, constitutes the principal energy source for our Sun and, in fact, for all stars.

The preservation of angular momentum requires that the massive central star rotate more rapidly than the primordial nebula; another consequence is that minor clumps of matter (i.e., planetesimals) not attracted by the central mass are sprayed into orbits defining a common plane at right angles to the rotation axis of the Sun. This zodiacal plane, or "plane of the ecliptic," as it is sometimes called, is a reflection of the "spin-up" of the solar nebula during its collapse.

The suitable environment for life that characterizes the Earth results from the felicitous coincidence of planetesimal accretion of matter to a suitable size at the appropriate distance from the local star (our Sun), such that abundant liquid H_2O was stabilized, resulting in the hydrosphere. The chemistry and energetics that can support biologic activity appear to have severe thermal limits. The condition involving a somewhat saline aqueous medium in equilibrium with an initial severely reduced, but increasingly oxygenated, atmosphere must have been just right from almost the beginning, or else humans would not be here to chronicle the history of terrestrial biologic evolution.

Clearly, a remarkably intricate interplay and continuous feedback among the hydrosphere, atmosphere, biosphere, and solid substrate must exist. Until we understand both qualitatively and quantitatively the interconnectedness of the many Earth systems, conscious human modifications of the planet will be rather analogous to a five-year-old trying to repair a Swiss timepiece. History shows that the environmental side effects of human activities have generally been unanticipated, commonly deleterious, and in many instances, absolutely disastrous. We must know in detail how the Earth's solid, liquid, and gaseous envelopes interact in order to save the planet for our posterity as well as for all other forms of life.

Look at the sky any night when the Moon is down, and you will marvel at the number of stars to be seen. Away from the glow of the cities, and at high elevations where the obscuring effects of atmospheric haze are minimal, the view of the heavens is still grander. Gaze through a good telescope, and the myriad visible stars are astounding, overwhelming in their profusion. In fact, as the viewing capabilities improve, we see ever farther across the universe, with no prospect, thus far, of an end. Is there a limit to the cosmos?

Cosmology is the quest to comprehend the origin of the universe. Its quantitative foundations are in high-energy physics and astrophysics as well as observational astronomy. According to theory, the so-called Big Bang occurred at a time when all matter was tightly compressed together. Over an instant (approximately 10^{-28} seconds), *matter* and *antimatter* annihilated each other at incredibly high temperatures, producing high-energy radiation. Matter exceeded antimatter by about one part per billion, and that surviving the annihilation now makes up the mass of the expanding uni-

W. G. Ernst (ed.), *Earth Systems: Processes and Issues*. Printed in the United States of America. Copyright © 2000 Cambridge University Press. All rights reserved.

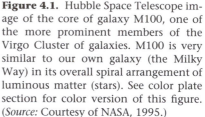

Figure 4.1. Hubble Space Telescope image of the core of galaxy M100, one of the more prominent members of the Virgo Cluster of galaxies. M100 is very similar to our own galaxy (the Milky Way) in its overall spiral arrangement of luminous matter (stars). See color plate section for color version of this figure. (*Source:* Courtesy of NASA, 1995.)

verse. By approximately 200,000 years after the Big Bang, the nuclear and extranuclear particles had cooled to 3000 K* and began to combine and reform to produce, first, ionized plasma (a charged stream of gaseous constituents) and then, with further cooling, atoms and molecules. Continued expansion and the slow, continued cooling has led to the low, 2 K background temperature of the modern interstellar medium. Stars represent gravitational attractors that self-aggregated due to heterogeneities in the original mass distribution; thus the repulsive forces causing the universe to fly apart are partially counteracted by gravitational collapse in regions of local mass concentration. This results in spiral-or disk-shaped clouds of stars known as *galaxies*, which themselves congregate into *clusters* and even *superclusters*. A spiral galaxy, similar to our own Milky Way, is illustrated in Figure 4.1. The intergalactic separation is approximately an order of magnitude (ten times) greater than the diameter of the galaxies; the intervening void contains traces of interstellar dust and widely dispersed gas molecules.

Stars are luminous because the energy generated in their interiors, through the combination of hydrogen molecules at very high temperatures and pressures to form helium, gives off a broad spectrum of radiation, including visible light. This process of hydrogen burning, known as *nuclear fusion*, may be written $2H_2 \rightarrow He + radiation$. Virtually all the visible matter in the universe is concentrated in stars and in tenuous nebular clouds. However, various types of evidence suggest that most of the mass of the universe actually consists of nonluminous *dark matter*. Recent interpretations of measurements employing the Hubbell Space Telescope suggest that dark material may exceed luminous substance by at least an order of magnitude, and perhaps more. If so, the phase of expansion the universe is now experiencing may, in the distant future, be succeeded by a contractional stage. This compression could be followed by another explosive expansion, invoking the image of a pulsating universe. The fact is, the nature and fate of the cosmos are largely unknown, and constitute one of the great mysteries and intellectual challenges of science.

When did the Big Bang occur? Measurement of a stellar phenomenon known as the *red shift* helps provide the answer. Imagine an object that emits radiation at a constant wavelength, and that is moving relative to an observer's detection system. As the object approaches the observer, the number of waves received per unit time increases (apparent frequency increases), whereas as the object recedes, the number of waves received per unit time decreases (apparent frequency decreases). For this same reason, using a sound-wave

* The absolute temperature scale is given in Kelvins (K), where the temperature is 273° greater numerically than the Celsius temperatures. Thus, 25°C = 298 K. Absolute zero K (= −273°C) is the condition under which all atomic translational, rotational, and vibrational motion ceases.

analogy, the throb of an approaching freight locomotive has a higher pitch than after it thunders past the observer. The phenomenon is known as the *Doppler effect,* and applies equally well to light radiation and to sound. Thus light waves are compressed (more waves, hence more vibrations per unit time) when emitted by an object approaching the viewer, and are extended (fewer vibrations per unit time – or, in terms of visible radiation, shifted toward the red portion of the spectrum) as the object recedes from the viewer. All stars have spectra displaced toward long (red) wavelengths, and the more distant ones possess greater red shifts – thus are moving away more rapidly – than nearby stars. Accordingly, we know that virtually every light source in the universe is receding from our vantage point in the solar system. Except for local collisions, this is true for the rest of the universe as well. In spite of the enormous velocities, because of the even more stupendous distances involved, the elapsed time since initiation of this centrifugal system places the time of the Big Bang as approximately 15 billion years ago. The size of the universe boggles the imagination.

Light travels at approximately 300,000 kilometers per second. Even so, the nearest star to our solar system, Proxima Centari, is 4.3 light-years away; that is, light from that source has been winging its way toward us for more than four years before we see it. Our Sun resides in one arm of the Milky Way galaxy, a fairly typical example of a spiral galaxy. The Milky Way is 50,000 light-years across. The closest major galaxy to us is Andromeda, approximately 600,000 light-years away. Farther out, in all directions, we see a somewhat inhomogeneous distribution of galaxies, globular clusters and superclusters of galaxies, the most distant of which are approximately 5 billion light-years away. With the advent of new terrestrial arrays of very large optical telescopes, and space-borne instruments, we expect to see approximately 10 billion light-years out; however, an important time/distance remains to be covered back to the time of the Big Bang. And beyond that?

ORIGIN OF THE SOLAR SYSTEM

Star formation results from regional gravitational collapse of an expanding solar nebula. Due to heterogeneities and mass concentrations in the primordial interstellar medium, self-attraction occurs as a local process and accelerates due to the increasing accretionary mass of the central attractor. The original gas cloud should, generally, have had a weak rotational component, and the preservation of angular momentum would require the mass condensation to result in the more rapid revolution – or spin-up – of the system, producing a rotating *zodiacal disk* (i.e., *the plane of the ecliptic*). Although most of the matter would infall to the massive, central attractor, creating enormous pressures, temperatures, and, consequently, hydrogen burning, clumpings of leftover gases and particulate matter would aggregate into planets and *planetesimals* orbiting in the plane of the ecliptic. These relationships are illustrated schematically in Figure 4.2. Be-

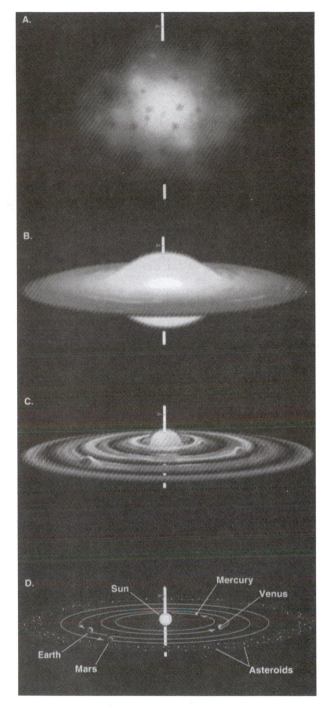

Figure 4.2. Inferred stages in the collapse of the solar nebula to form the Sun and planetesimal debris in the plane of the ecliptic (based on an illustration from Skinner and Porter, 1987). As the slowly rotating nebula shown in (A) began to condense, it spun up condensed matter in the zodiacal disk, as illustrated in (B) with progressively more evolved states shown in (C) and (D). (*Source:* B. J. Skinner and S. C. Porter, The dynamic Earth, John Wiley & Sons, © 1987.)

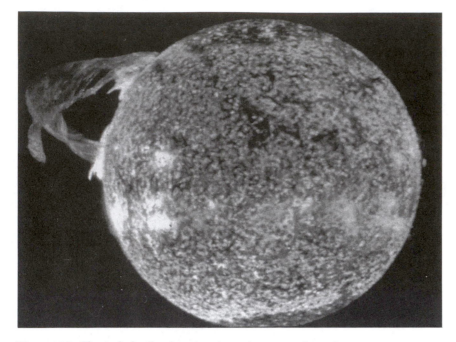

Figure 4.3. The turbulently churning, incandescent surface of the Sun seen in ultraviolet wavelengths; the image exhibits an enormous burst of ionized plasma, a solar flare. Such eruptions characterize sunspot activity and result in magnetic storms on Earth. Why do you suppose this is? (*Source:* Courtesy of NASA.)

yond the limits of gravitational attraction by the solar system, matter would remain dispersed in the *Oort cloud* as the gaseous interstellar medium and as icy cometary materials, primarily condensed gases. On a much smaller scale, formation of the planets occurred in the same fashion as the accretion of the Sun – by sweeping up the nearby matter as a consequence of gravitational attraction. Our local star, the Sun, being much larger, achieved far greater temperatures and pressures during growth; this, in turn, ignited the nuclear reactions that power it. Surface temperatures approach 6000 K, whereas the core of the Sun is at a temperature on the order of 12,000,000 K. A view of the Sun and a large solar flare is shown in Figure 4.3. The latter is a great stream of incandescent plasma rising from the surface of the Sun in a graceful arc along the lines of the solar magnetic field.

What are the dimensions of the solar system? Although the size of the universe is mind boggling in terms of light-years, our own insignificant little corner of the cosmos is nevertheless enormous relative to that of planet Earth. For instance, it takes light about a second to reach the Moon, and another to return to Earth. So when the Mission Control Center in Houston was communicating with *Apollo* astronauts in the vicinity of the Moon, radio conversations exhibited nearly a two-second delay in response. Looking more widely about the solar system, it takes sunlight 8.3 minutes to reach our planet, 40 minutes to reach Jupiter, 4 hours to arrive at Neptune, and up to 5.5 hours to reach Pluto. That is a long distance at the speed of light!

The planets in our solar system exhibit systematic and progressive variations in their chemical constitution and surface, as a function of their orbital distances from the Sun. Spatial relationships for the nine planets as well as that of a typical comet (a wandering "dirty snowball") possessing an orbit that crosses the inner solar system, and of a meteorite swarm (the *asteroid belt*) are shown in Figure 4.4. Some planetary satellites, such as the Earth's Moon, the four Galilean satellites of Jupiter, Saturn's Titan, Neptune's Triton, and the largest asteroids are larger than, or approach the sizes of, Pluto and Mercury, the smallest planets. The figure clearly shows that Pluto has an elliptical orbit that crosses that of the eighth planet, Neptune; consequently, while Pluto is generally farther from the Sun at certain times as it follows its orbital course, its trajectory crosses in side the path of Neptune, and it travels slightly closer to the Sun than does its much larger neighbor.

Table 4.1 presents additional solar system information in more quantitative fashion. The innermost planet, Mercury, is silicate- and, especially, metal-rich; the next three planets, Venus, Earth, and Mars, are fundamentally stony; the outer planets represent progressively more gassy giants, and finally, even farther out, proportionately more icy bodies. The Oort cloud, at the gravitational limit of our solar system, is the ill-defined, sparsely populated home of icy condensates, the *comets*.

The reason for these present-day contrasts in planetary compositions appears to be related mainly to distance from the Sun. Condensation of the solar nebula 4.5 to 4.6 billion years ago* might have led to initial compositional differences in the planets during their growth stages, reflecting chemical heterogeneities in the primordial gas cloud. On the other hand, even if homogeneous accretion of planetesimals (*meteorites*) occurred, the *solar wind* (high-energy atomic particles streaming away from the Sun) undoubtedly exerted a strong influence over the retention of the more volatile elements in the planets. Those bodies closer to the Sun now contain high proportions of refractory constituents of low volatility such as magnesium (Mg), calcium (Ca), iron (Fe), and silicon (Si), whereas, proceeding away from the Sun, the planetary masses become progressively enriched in low-temperature of condensation and low-melting-temperature elements such as hydrogen (H), carbon (C), oxygen (O), nitrogen (N), and sulfur (S), and compounds such as methane, ammonia, and carbon dioxide. Very probably, this thermal gradient encom-

---

* The ages of relatively pristine meteorites have been determined employing U/Pb radiometric dating, as described in Chapter 3, and their times of formation utilized to estimate the time of origin of the solar system.

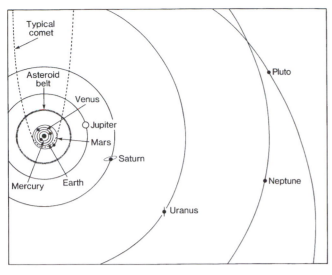

Figure 4.4. Orbits of the nine planets in the solar system, a typical comet in Sun-approaching orbit, and the asteroid belt, with radii properly scaled. Sizes of the planets and the Sun are illustrated diagrammatically. The ring systems of Saturn and Uranus are indicated schematically. (*Source:* W. K. Hartmann, *Moon and planets*, Wadsworth Publishing, 1983.)

passing the Sun and its environment has been largely responsible for the compositional contrasts of the planets. In effect, solar radiation "blew away" the more volatile gaseous constituents from the inner portions of the solar system, whereas the outer reaches are more nearly primordial in chemical composition.

Each planet is unique in its external appearance and evidently, also, in its internal constitution. The Earth is the third member of the four iron and stone, close-in planets. Similar to our earthly satellite, the Moon, all the inner planets are terrestrial-like bodies. However, they exhibit strikingly different surficial features, principally because of contrasting degrees of reworking of their outermost rinds.

TABLE 4.1. DISTANCES FROM THE SUN, DIAMETERS, AND TYPES OF PLANETS IN THE SOLAR SYSTEM

	Distance from Sun (in AU[a])	Diameter of Planet (in km)	Planetary Type
Mercury	0.4	4,878	Iron/stony
Venus	0.7	12,104	Stony
Earth	1.0	12,756	Stony
Moon	1.0	3,476	Stony
Mars	1.5	6,796	Stony
Asteroids	~2.8	small	Stony
Jupiter	5.2	142,796	Gassy
Saturn	9.5	120,660	Gassy
Uranus	19.2	50,800	Gassy/icy
Neptune	30.0	48,600	Icy
Pluto	39.4	2,400–3,800	Icy
(Oort cloud)	100–50,000	small	Icy

[a] One astronomical unit (AU) is the distance between the Earth and Sun, approximately 149,600,000 kilometers.

The interiors of all planets are hotter than their surfaces because of internal (primordial, radiogenic, self-compression, and kinetically generated) heat. The most efficient mechanism for removal of large quantities of buried heat is by material circulation (or convective overturn) within the planet, with cooling taking place at the surface. Larger planetary masses contain more heat, and inasmuch as they possess larger ratios of mass to surface, they rid themselves of thermal energy more slowly; hence, they remain hot and internally convective for a longer period of time. As the most massive of the terrestrial planets, therefore, Earth should be the most dynamic; this appears to be the case.

A BRIEF TOUR THROUGH THE SOLAR SYSTEM

The ancients studied the heavens and discovered that the positions of the stars are fixed relative to one another. Of course, the universe is expanding; however, because all objects are at virtually infinite distance and are moving directly away from each other, their relative positions with respect to an observer remain constant. Naturally, the early astronomers had no way of knowing about the red shift. They did puzzle over a small class of luminous objects they termed "wandering stars" because these bodies moved differentially through the firmament. We now know that these objects are other planets in our solar system. Why is Venus called the evening (or morning) star? It is because, having an orbit closer to the Sun than does our planet, observers on Earth can view Venus for only a few hours after sunset and before dawn. The innermost planet, Mercury, is so near to the Sun that it is detectable only in very favorable orbital positions just after sunset or immediately before sunrise.

Unlike the self-luminous stars, planets primarily reflect light rather than generate it. This is a consequence of the fact that they are not massive enough to generate the high internal pressure and temperature conditions requisite to initiate hydrogen fusion. Gassy giants such as Jupiter emit slightly more radiative energy than is received from the Sun; however, this is due to the energy released during ongoing self-compression, not to nuclear fusion. Jupiter is only 0.1 percent the mass of the Sun, which is probably a good thing for us. Were it substantially more massive, it might have begun hydrogen burning, and then our solar system would have been characterized as a binary star system. Such a configuration would have provided a somewhat different environment in the vicinity of the Earth; however, inasmuch as, at its closest approach to us, Jupiter is approximately four times as far away from the Earth as is the Sun, we would not have been incinerated (or more correctly, primordial life would not have been destroyed). Also, it must be noted that binary star systems are common in the universe.

An interesting but thus far unexplained relationship is that the distance to a planet from the Sun is approximately twice as far as that of the next inner one. This observation is known as *Bode's rule*. The sole exception is the gap between Mars and Jupiter; however, this region is the site of the previously mentioned asteroid belt (see Fig. 4.4). Such planetesimals, condensed rocky and iron-rich masses ranging from fine dust to 10 kilometers or more in diameter, may represent a fragmented planet. More likely is the hypothesis that planetary accretion in this zone of the ecliptic plane was inhibited due to gravitational disruption by the neighboring massive planet, Jupiter. Most meteorites currently circling about in our solar system are probably derived from the asteroid belt, flung into eccentric orbits by the gravitational pull of Jupiter.

Each planet in our solar system is made up of various proportions of matter similar to iron and stony meteorites, as well as of condensed gas species (predominantly hydrogen, methane, carbon dioxide, water, and ammonia). Where abundant, the metal is concentrated in the planetary core, the stony material constitutes the surrounding mantle and crust, and the volatile constituents are present as surficial condensed ice layers, liquid oceans, and a planet-enveloping atmosphere (e.g., see Fig. 5.1). This is a reflection of the fact that during accretionary heating, such bodies underwent a gravitative differentiation process, whereby the densest sub-

Figure 4.5. *Mariner 10* composite mosaic image of Mercury during the spacecraft's closest approach to the innermost planet. The plethora of impact craters can be observed. Why do you suppose there are so many? (*Source:* Courtesy of NASA.)

stances were concentrated toward the center of the planet, with concentric outer layers consisting of progressively less dense matter. In the inner planets, at least, this chemical stratification took place during early-stage partial melting occasioned by radioactivity- and meteorite impact–generated self-heating. Reflecting the higher densities of metals and metal alloys, and the lower fusion temperatures of iron and iron-nickel alloys compared with magnesium silicates and oxides, the molten iron would have migrated down the gravity gradient to form a dense planetary core, surmounted by the silicate "slag" of the mantle (see Fig. 5.4). As a continuing process, the deeper portions of the terrestrial planets have become progressively devolatilized due to the thermally driven escape of gaseous species trapped by the accretionary process. This thermal annealing, for instance, accounts for the more fluid envelopes surrounding the Earth, as well as the carbon dioxide–rich atmosphere of Venus. In cases where a liquid iron-nickel core is present, fluid motion within it is thought to be responsible for the generation of a planetary magnetic field (as discussed briefly in Chapter 5). The thermally driven chemical-density differentiation process described above, while generally applicable to the entire solar system, is exactly what scientists hypothesize to have transpired on Earth.

MERCURY. So close to the Sun that visual observation from Earth is difficult, Mercury is only 5.5 percent of the mass of our planet. Like Venus, Mercury has no orbiting natural satellites. It is slightly larger than the Moon, and, as shown in Figure 4.5, exhibits a similarly pockmarked surface, due chiefly to asteroid impacts. As on the Moon, dark lava flows floor the largest basins. Mercury has no atmosphere, and is characterized by a relatively low albedo. Its aggregate density is 5.4 grams per cubic centimeter; consequently, we surmise that it must have a large metallic core. (Masses of the individual planets can be deduced from their orbital paths around the Sun. Their specific sizes are determined by trigonometric measurements, conducted from Earth, and now, from space. Thus, the overall density of each planet is known, being simply the ratio of its mass to its volume.) Mercury possesses a weak dipolar magnetic field aligned with the planetary rotation axis.

VENUS. The second planet, illustrated in Figure 4.6, is Earth's virtual twin. Its dimensions and density, 5.3 grams per cubic centimeter, are comparable to terrestrial values, and it possesses 81.5 percent of the mass of the Earth. However, Venus has an atmosphere that, unlike ours, is very dense, ninety times that of the Earth, and predominantly consists of carbon dioxide. On Earth, water is ubiquitous as globe-encircling oceans. Carbon dioxide is not abundant in the terrestrial atmosphere; instead, it is stabilized by precipitation of calcium carbonate plus or minus calcium-magnesium carbonate from seawater, and largely held in the solid portion of the planet as the sedimentary rocks limestone and dolomite (see Chap. 5). Because Venus is closer to the Sun, quently it possesses a very low albedo; volatile species could readily escape, and did because of the very weak gravity field. Its density, 3.3 grams per cubic centimeter, is relatively low, and suggests that the Moon has only a very small core – a conclusion supported by the absence of a dipolar magnetic field. Craters formed by meteorite impacts are of considerable but variable abundance on the Moon. Views of this familiar satellite are presented in Figure 4.8. It is rather similar in surface features to Mercury, but is less dense. The dark *lunar maria* are underlain by lavas similar to those that issue from the Hawaiian volcanoes, and some of these flow fields are more sparsely pockmarked than are others. The relatively light-colored *lunar highlands* consist of pulverized, fragmented rocks that evidently have been repeatedly smashed by incoming asteroids early in lunar history. Radiometric data obtained for specimens retrieved during the *Apollo* Project have corroborated this hypothesis conclusively, inas-

Figure 4.6. *Mariner 10* photograph of Venus, seen from a distance of 720,000 kilometers. The image, shown for ultraviolet wavelengths (a small portion of the entire radiation spectrum), illustrates weather patterns developed in the dense Venusian atmosphere. (*Source:* Courtesy of NASA.)

its H_2O is completely volatilized; hence, CO_2 cannot be dissolved, transported, and precipitated to become sequestered in marine carbonate strata, but instead is concentrated in the dense atmosphere. This thick atmosphere provides the planet with a high reflectivity (high albedo). In contrast to the Earth, Venus has a very high surface temperature, on the order of 500°C. This phenomenon is due partly to its proximity to the Sun and primarily to its dense atmosphere, which traps incoming solar radiation (the *greenhouse effect*) – an extreme example of the phenomenon introduced in Chapter 1. Radar imagery reveals large, circular structures (impact craters) as well as curious pancake-shaped volcanic domes and intricately fractured terrains (Figs. 4.7A and B, respectively). Venus lacks the bimodal topography of ocean basins and landmasses that characterizes Earth, but it does have two prominent, continent-like areas. Overall, however, the topography is typically subdued. Judging from its aggregate density, Venus must have a metallic core. It lacks a magnetic field, possibly because of its slow rate of rotation (once every 243 Earth days). Its spin is termed *retrograde*, because it rotates more slowly than its revolution (orbit) about the Sun.

MOON. Within our solar system, humans are blessed with a rather large satellite, relative to the size of the central planet. Even so, the Moon contains only the equivalent of 1 percent of the mass of the Earth, it lacks an atmosphere, and conse-

Figure 4.7. Two radar imagery views and a synthetic composite of Venus, as determined by the *Magellan* spacecraft mission. (A) Fractured, pancake-like magma domes on Venus, approximately 25 kilometers in diameter; they appear to be similar to rhyolite domes on Earth. (B) Planar terrain on Venus, riven by two sets of intersecting fractures. (C) Topography of Venus; highlands stand slightly above lowlands, but relief is not as pronounced as on Earth. (*Source:* Courtesy of NASA.)

(A)

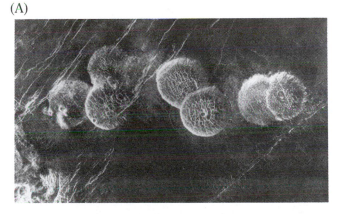

(B)

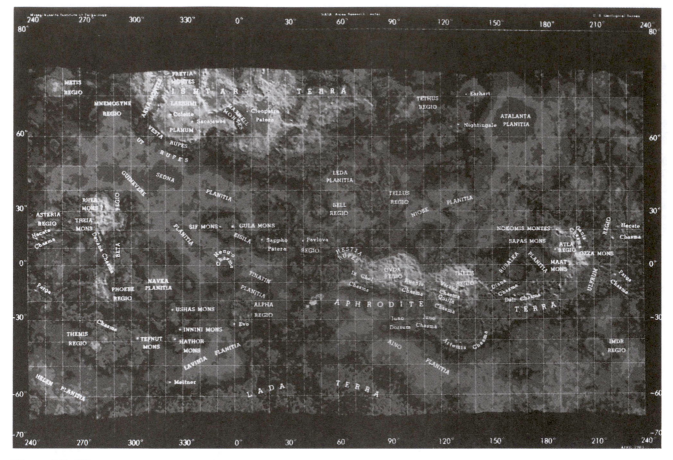

Figure 4.7. (C)

Figure 4.8. *Apollo 17* spacecraft photographs of the Moon. (A) The overall view. (B) (opposite) A close-up of the surface looking homeward (the crescent over the lunar horizon is Earth). (*Source: Courtesy of NASA.*)

much as some lunar highland fragmental rocks are as old as 4.4 billion years, whereas the maria lavas were erupted only 3.2 to 3.8 billion years ago.

EARTH. The third planet has an overall density of 5.5 grams per cubic centimeters, and a strong dipolar magnetic field, due to an iron-rich core. The Earth is distinct from all the other inner, rocky planets in possessing distinct continents and ocean basins. Perhaps most important, H_2O is present as a significant constituent of the atmosphere, as the dominant component in the liquid, globe-encircling oceans, and in the solid polar ice caps. This is shown dramatically in Figure 4.9. The Earth is quite appropriately called the blue planet. Its atmosphere is largely composed of nitrogen (78 percent) and oxygen (21 percent), with only minor amounts of carbon dioxide, and vanishingly small concentrations of noble gases and more reduced volatile species (methane, ammonia, and hydrogen). Would you anticipate that the Earth would have a high or low albedo? Earth certainly must have sustained a similar number of meteorite impacts as the Moon; however the surface of our planet preserves only a few of the geologically more recent impact sites; most meteorite craters presumably have been reworked and destroyed through erosion or mountain building. Volcanoes are of several distinct types

(A)

and are confined to roughly linear chains. Most mountainous regions also form distinct belts, especially those wrinkled zones exhibiting evidence of continental contraction (Chap. 6). Of course, what also differentiates the Earth from the rest of the solar system is biologic activity – the myriad and marvelously intricate forms of life.

MARS. Our outer solar system neighbor, the red planet, has two tiny satellites. Mars is often likened to Earth, but in fact it is much colder (as might be surmised by its brilliant white polar ice caps), is quite a bit smaller, and is less dense. Having only 10.7 percent of the mass of the Earth, its aggregate density is only 3.9 grams per cubic centimeter, indicating the presence of a relatively small metallic core; Mars lacks a dipolar magnetic field. The tenuous atmosphere is only approximately 1 percent as thick as that of Earth (Mars possesses a relatively low albedo), and consists chiefly of carbon dioxide. A good portion of this volatile component is present as frozen carbon dioxide ("dry ice"), along with H_2O ice, covering the martian polar regions. Like the Earth, its spin axis is titled slightly relative to the plane of the ecliptic, so Mars experiences seasons in which the ice caps reciprocally wax and wane. It also rotates at approximately the same rate as the Earth. The surface of Mars, which contains modest amounts of iron, is strongly oxidized, accounting for the distinctly red hue of the planet. Various features are shown in Figure 4.10.

The surface of Mars exhibits impact features and volcanic edifices. The southern hemisphere is heavily cratered; one such feature, Hellas, is 2000 kilometers in diameter, the largest known crater in the solar system. Huge, broad, basaltic shield volcanoes (so called, as discussed in Chap. 5, because their shape is reminiscent of a gently curved shield), similar

Figure 4.9. Familiar *Apollo 17* image of the Earth, with coverage extending from the Mediterranean Sea to the Antarctic ice cap. The blue planet, as it is sometimes called, contains H_2O in three physical states, solid, liquid, and gaseous (why is this?), dramatically illustrated in this photograph. See color plate section for color version of this figure. (*Source:* Courtesy of NASA.)

to Mauna Loa, Mauna Kea, and the lunar maria, are numerous, especially in the northern hemisphere. The largest, Olympus Mons, is 600 kilometers across at the base and rises 25 kilometers into the thin martian atmosphere (see Fig. 4.10C). The surface of Mars also displays canyons that have dendritic (tree-like) patterns. One such sinuous feature is Valles Marineris, which is 4000 kilometers long, 100 to 250 kilometers wide, and 7 kilometers deep. At some stage of the history of the red planet, this declivity appears to have resulted from erosion by running water. A less plausible alternative hypothesis of its origin is that this depression is a down-dropped block, bounded on either side by faults. Episodic dust storms and wind-blown sand have been encountered above and on the martian surface by the *Mariner 9* spacecraft and the *Viking* lander. Even though Mars possesses a very tenuous atmosphere, it does have energetic weather patterns; sustained wind velocities of up to 270 kilometers per hour have been measured.

ASTEROIDS. Judging from meteorites falling on and recovered from the surface of the Earth, planetesimals residing in this belt consist of two distinct types, stony and iron. Stony meteorites consist predominantly of magnesium (and fer-

(B)

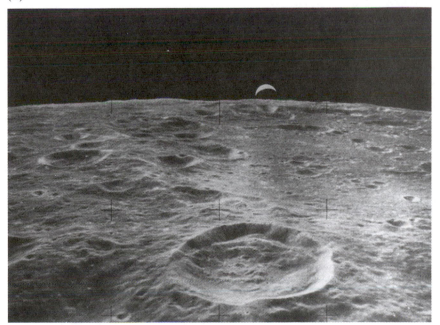

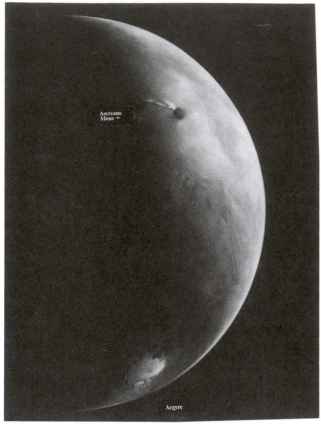

(A)

Figure 4.10. *Viking Orbiter 2* photographs of Mars. (A) At the top, with carbon dioxide ice on its western flank, is the giant Martian volcano, Ascreans Mons. The large, frost-encrusted, impact crater basin, Argyre, is located near the South Pole. (B) General view of cratered surface; the original shows the intense red color of Mars. See color plate section for color version of (A) and (B). (C) Computer-enhanced image of Olympus Mons, a giant shield volcano and the largest yet discovered in the solar system (see text). Obviously, volcanism was important during a stage in the development of this planet. (*Source:* Courtesy of NASA.)

(B)

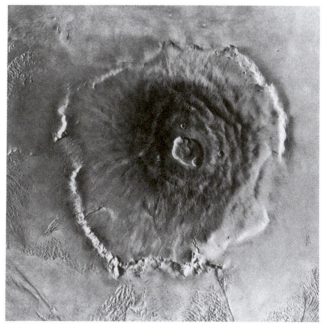

(C)

rous meteorites consist predominantly of magnesium (and ferrous iron) silicates, whereas metallic meteorites are made up of an alloy of native iron (90 percent) and nickel (10 percent). A few meteorites contain both silicate and metal, and an even rarer group of stony meteorites contains carbonaceous matter. The significance of meteorite compositions and their proportions is discussed in Chapter 5. The immense gravitational attraction of the fifth planet probably continually disrupts the self-accumulation of a planet-sized mass in this orbit, thus maintaining the asteroids as a swarm of small, condensed bodies orbiting about the Sun in a broad band.

JUPITER. Jupiter is the largest of the giant, gassy planets; its mass is more than twice that of all the other solar system planets, satellites, and asteroids combined. The overall density of this behemoth is 1.3 grams per cubic centimeter; it is an enormous ball of compressed gases and clouds of ice crystals, probably with a small rocky and/or metallic core. The dense atmosphere is composed principally of hydrogen (H_2), helium (He), ammonia (NH_3), methane (CH_4), and their condensates. Shown in Figure 4.11, it is famous for turbulent circulation patterns, including the enigmatic red spot. Slight compositional differences are reflected in the color contrasts, which help to identify convection cells. The red spot (Fig. 4.11B) is just such a feature and is one of numerous, immense Jovian hurricanes; this evolving cyclonic system has a long diameter of approximately 22,000 kilometers – twice the diameter of the Earth. Sequential photographs from space provide striking proof of the dynamic circulation of this stupendous weather system, divided into circumferential latitudinal, jet stream–like belts girdling the rotational axis of the planet. Jupiter, like most of the outer giants, spins rapidly, 360 degrees, in just under ten hours Earth time. Also, due to self-compression, Jupiter emits twice as much radiative energy as it receives from the Sun.

Jupiter is surrounded by a flock of sixteen satellites, of which the outer dozen are relatively minor bodies. The inner four, termed the Galilean satellites (so called because Galileo discovered them with his primitive telescope), are as large as, or larger than, our Moon. Under favorable night viewing conditions, you can see them with a good pair of binoculars. Closest to Jupiter, Io has a density of 3.5 grams per cubic centimeter, a surface covered by dark lavas and numerous sulfur-spewing volcanoes. Io lacks meteorite impact scars due to periodic repaving by volcanic activity. Farther out, Europa is characterized by a density of 3.2 grams per cubic centimeter, and displays a network of fractures in

Figure 4.11. *Voyager 1* spacecraft images of Jupiter showing (A) planetary circulation of the atmospheric system, dominantly ammonia; and (B) details of convecting cells surrounding and including the red spot, an enormous cyclonic system. (*Source:* Courtesy of NASA.)

what appears to be a frozen H_2O ocean. Meteorite impact craters are rare on this relatively smooth-surfaced satellite. The next outboard satellite, Ganymede, has a density of 2.0 grams per cubic centimeter, and possesses more numerous craters; in addition, it consists of faulted, grooved, down-dropped blocks partly overwhelmed by ice and frost. Callisto, outermost of the large Jovian moons, with a density of 1.8 grams per cubic centimeter, exhibits a surface consisting primarily of pockmarked H_2O ice. Clearly, the inner Galilean satellites have a higher proportion of stony materials in their planetary constitutions, whereas, farther from Jupiter, the satellites are progressively enriched in volatiles and condensed gases (ices).

SATURN. The sixth planet in the solar system is the second biggest, having 30 percent of the mass of Jupiter. Like its inner neighbor, but on a somewhat lesser scale, it emits small amounts of radiation (why?). Saturn has an aggregate density of 0.7 grams per cubic centimeter, hence consisting almost exclusively of gaseous constituents and icy equivalents. Its atmosphere is fully as turbulent as that of Jupiter. Its most famous feature, illustrated in Figure 4.12, is a series of equatorial, disk-shaped, nested rings. The radius of the disk is 65,000 to 75,000 kilometers, and the overall width of the rings, or bands, is approximately 10,000 kilometers; somewhat surprisingly, the thickness of the disk is less than 1 kilometer. The rings consist of particulate matter, including dust, rocky debris, ices, and oil droplets. One of Saturn's seventeen moons, Titan, is larger than Mercury. Titan possesses a hazy atmosphere rich in nitrogen, ethane, and methane, some of which appears to be condensed into an ocean of liquid nitrogen and hydrocarbon; it also displays polar ice caps.

URANUS. Uranus has a density of 1.2 grams per cubic centimeter and a thin system of narrow rings. Like Jupiter and Saturn, it is surrounded by a swarm of satellites, fifteen in number. Because of the abundance of methane and other hydrocarbons, its atmosphere is obscured by a photochemical haze similar to terrestrial smog. The spin axis of Uranus nearly coincides with the zodiacal plane. The change of seasons as well as day–night variations must be enormous on Uranus compared to Earth and most of the other planets in the solar system.

NEPTUNE. The eighth planet, nearly identical to Uranus, is slightly more dense at 1.7 grams per centimeter, and it has only eight moons. Both Uranus and Neptune consist chiefly of methane and condensed ices. Neptune exhibits two dark oval areas, one of which – the great dark spot – is comparable in size to the Earth. Like Jupiter's red spot, the great dark spot is thought to represent an enormous, counterclockwise rotating, cyclonic weather system. Neptune also appears to be surrounded by thin, incomplete rings or arcs of particulate matter, probably dust and/or ices.

PLUTO. We have relatively little information about this tiny planet, which is smaller than our Moon; our spacecraft have not yet explored it. Pluto has a markedly elliptical orbit and may actually be a large asteroid gravitatively captured by Neptune. Similar to Uranus, Pluto's spin axis nearly coincides with the ecliptic plane. Pluto has a density of 1.1 grams per centimeter and consists chiefly of solidified gases, probably predominantly frozen nitrogen. Its sole moon, Charon, is similar in composition and density to Pluto and represents approximately one-eighth the mass of that planet.

SUMMARY

Assuming that meteorite populations are (and were) relatively uniformly distributed throughout our portion of the solar system, the density of an impact crater that we observe on a rocky planet's surface provides a relative measure of the extent of surface reworking caused by the internal mecha-

Figure 4.12. *Voyager 1* spacecraft image of Saturn including the equatorial rings (planetesimals, dust, and ices). These rings are not continuous, two-dimensional disks; they consist of dispersed particulate matter. (*Source:* Courtesy of NASA.)

nisms of the planet. The more numerous the impact craters are, the slower the bodily circulation, volcanism, deformation, and consequent surficial reworking by mass flow within the planet must be. Intensely pockmarked planets are thus nearly dead, or at least are less active internally than those reworked, modified bodies that lack an abundance of such craters. Comparing the size of the Moon, Mercury, Mars, Venus, and the Earth with their abundance of impact features, it is apparent that most of our planet's surface has formed during the very recent geologic past, whereas internal engines of the Moon, Mercury, and Mars slowed or stalled and became almost totally inactive several billion years ago. Venus is least known because of its dense, obscuring carbon dioxide cloud cover; however, radar imagery has provided enough topographic control to tentatively conclude that surface activity, and therefore internal circulation, probably ceased there more than 500 million years ago. These relationships are illustrated diagrammatically in Figure 4.13.

Although the Earth possesses chemical characteristics similar to its sunward, torrid sister Venus, as well as its outer, frigid brother, Mars, it is obviously very different in surface aspect compared with these neighboring planetary siblings. This difference is understandable based on relative sizes and internal heat budgets, as discussed above. However, it is also a consequence of the position of our planet relative to the energy reaching the planetary surface – the radiant solar flux. Solid, liquid, and gaseous H_2O are stable at and near the Earth's surface. The proximity of Venus to the Sun is responsible for its higher surface temperatures and lack of liquid H_2O oceans; accordingly, as noted in the descriptions of the individual planets, the carbon dioxide compliment of Venus occurs as a gas and resides in the dense Venusian atmosphere. Mars, in contrast, is so distant from the Sun that, on its surface, H_2O frost is presently confined to seasonal polar caps (which actually consist mostly of "dry ice" – solid CO_2); the carbon dioxide on Mars is thus locked up in icy solid layers. Photographs taken by spacecraft of the inner solar system (e.g., compare Figs. 4.5, 4.6, 4.8, 4.9, and 4.10) clearly show that in comparison to the dense, hot, carbon dioxide–enshrouded Venus, and the arid, cold, nearly atmosphereless Mars, as well as deader-than-a-doornail Mercury and the Moon, the Earth's envelope is defined by a particularly hospitable and dynamic interplay among atmosphere, hydrosphere, and the variegated surface of the solid planet. The presence of liquid water has allowed for the storing of the terrestrial budget of carbon dioxide as carbonate strata (deposited from seawater) in the rocky crust, rather than in the atmosphere. In addition, the origin and sustinance of carbon-based life itself clearly require liquid H_2O. We have much to be grateful for, and many reasons to cherish the environmental conditions that characterize planet Earth.

From a terrestrial perspective, we may suspect that, contrary to popular belief, the mountains and rivers of our landscape are not the immutable, permanent features they seem to be. The forces of erosion, such as wind, running water, glaciation, and downslope mass movement, acting over the

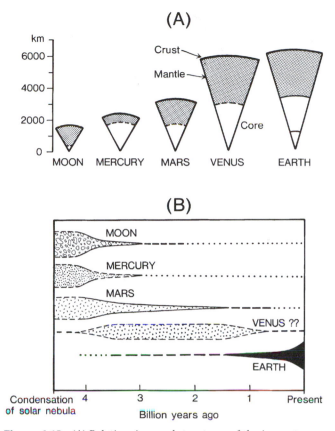

Figure 4.13. (A) Relative sizes and structures of the inner, terrestrial planets, and of the Moon (see also Table 4.1). (B) Inferred surface reworking of these bodies due to volcanism, erosion, deposition, and meteorite bombardment as a function of time. Evidently, surficial modification of most of the inner planets except the Earth (and possibly, to a lesser extent, Venus), virtually ceased 3 to 4 billion years ago. (*Source:* Adapted from J. W. Head and S. C. Solomon, *Tectonic evolution of the terrestrial planets, Science,* vol. 213, 1981.)

millennia, are reducing topographic promenances to more subdued outlines. Rivers, flowing in their ever-shifting courses, carry solid debris and other, dissolved materials supplied by the agents of erosion toward an ultimate rendezvous with the sea; there, the entrained particles spread laterally in the coastal waters and settle, and dissolved constituents eventually precipitate out. Over the billions of years of geologic time since the formation of the Earth, these processes must have planed off the continents many times over, and deposited vast quantities of sediment in the ocean basins.

Why, then is the Earth not a nearly featureless plain at or beneath sea level? A map of the surface elevations, Figure 4.14, provides graphic proof of what we already know: Our planet is typified by high-standing continents and depressed ocean basins. Apparently, thermally propelled constructional forces within the Earth, reflecting the motions of deeply buried solid rock, have swept the ocean basins nearly free of sediments, and have resulted in upheaval of vast segments of the Earth's crust: These hidden internal processes oppose the more obvious effects of surface degradation and continually

Figure 4.14. Relief map of the surface of the solid Earth. Our planet is divided into two major physiographic provinces – continents and ocean basins. The ocean basins are full "to the brim" with seawater (in fact, overly so); the continental shelves are submerged portions of the high-standing continents and islands that lie below sea level. See color plate section for color version of this figure. (*Source:* Courtesy of B. C. Heezen and M. Tharp, *World ocean floor*, U.S. Navy, 1977.)

reconstruct the topography. We still have mountain belts, just as they existed in the ancient geologic past.

To develop an appreciation for the processes at work today and, by inference, those that operated long ago (employing the principle of uniformitrianism), it is necessary to consider the internal structure of the Earth and the materials of which our planet is constructed – topics that are discussed in Chapter 5.

QUESTIONS

1. The planets in our solar system differ from one another in terms of composition, mass, and state of aggregation (solid, liquid, or gaseous). Explain why this is so.

2. How does the chemistry of a planet relate to its distance from the Sun? Describe four contrasting examples.

3. The ancients regarded the heavens as permanent and immutable, with a fixed spatial arrangement of the stars. Nevertheless, they recognized and puzzled over a small class of astronomic objects they called "wandering stars." How would you explain to them the true nature of these luminous bodies and the reasons for their motions?

4. In what way does the abundance of impact craters on a planetary surface provide us with constraints regarding the internal activity of the host body?

5. Imagine you are an extraterrestrial being searching for a new planetary home. Why pick the Earth? What conditions and substance(s) make the Earth more hospitable than all other planets in the solar system?

6. Lacking even a shred of positive evidence, some reputable scientists nevertheless are quite confident that life must exist elsewhere in the cosmos. Justify your view on this matter.

FURTHER READINGS

Beatty, J. K., O'Leary, B., and Chaikin, A. (eds.) 1990. *The new solar system.* New York. Cambridge University Press.

Hartmann, W. K. 1983. *Moons and planets.* Belmont, CA: Wadsworth Publishing.

Lightman, A. 1991. *Ancient light.* Cambridge, MA: Harvard University Press.

Skinner, B. J., and Porter, S. C. 1992. *The dynamic Earth.* New York: Wiley.

5 Earth Materials, and the Internal Constitution of the Planet

W. G. ERNST

SOLID PORTIONS OF THE BLUE PLANET

The origin and evolution of the Earth has resulted in a planetary body surrounded by a thin atmosphere enveloping a rocky crustal rind that, in turn, is partially covered by condensed volatile components (primarily water) in the world's oceans. Downward, below the familiar rocky exterior, the solid Earth passes by degrees into magnesium silicate – and oxide-rich shells of progressively increasing density; together, these concentric shells make up the mantle. Below this, the metallic core of the Earth consists of an iron-nickel alloy, a molten shell in its outer portions, and an iron-nickel solid sphere in the inner, central mass. The stuff of the planet – rocks and minerals – constitutes the building blocks from which the Earth and its various layers are constructed.

Except for sunlight, all the resources that support the incredibly rich web of life have been derived from the Earth itself; developing an appreciation for its physicochemical evolution and diversity is essential to earning a sustainable living from our planet. Natural hazards, too, are a reflection of the dynamically interactive Earth systems. Our attempt to live comfortably and harmoniously with nature, thus far only partly successful, depends on our recognition, understanding, and protection of the interconnected systems that operate on our terrestrial home. Failure to do so will diminish by degrees the future biologic carrying capacity of the planet, and all surviving life will be doomed to a progressively more meager existence.

The Earth's structure is a reflection of its original accretionary growth and subsequent physical and compositional evolution. Changes over time may be attributed predominantly to the heating, dynamic circulation, and escape of energy from the deep interior. What is the present configuration of the planet? The Earth consists of a series of concentric shells of different yet complementary physical and chemical properties. (Diagrams representing a sector from the surface to the center of the Earth are shown in Figure 5.1.) This chapter describes the Earth's overall structure first; then it details the kinds of geophysical and geochemical evidence that scientists use to deduce the nature of these shells. Finally, Earth materials – rocks and minerals – are discussed.

The globe is surrounded by a gaseous envelope consisting chiefly of nitrogen (78 percent) and oxygen (21 percent); the atmosphere is densest at sea level and becomes progressively rarefied upward. The lower zone, approximately 0 to 10 kilometers, is the *troposphere*, with the less dense *stratosphere* overlying it (see Chap. 13). No sharp break with outer space exists; however, the gas molecules are so widely dispersed above approximately 20 kilometers that we may regard this elevation as the effective limit of the Earth's atmosphere.

The ocean basins make up approximately two-thirds of the surface of the solid Earth; they are great depressions lying on average approximately 5 kilometers below sea level. Continents and islands, together constituting approximately one-third of the solid Earth's surface, rise above sea level only a few hundred meters on average. Because seawater more than fills the ocean basins, the hydrosphere laps onto the margins of the continents and islands, as, for example, at Hudson's Bay, the Grand Banks, and the East China Sea; these shallowly submerged regions are called *continental shelves*. Thus, the Earth's surface consists mainly of ocean water, covering slightly more than 70 percent of the exposed area; land, some of which is lake-or ice-covered, constitutes the remainder.

The contrasting tracts of continents and ocean basins seen in Figure 4.14 reflect an underlying structural and chemical difference clarified in Figure 5.1B. The outer rind of the solid Earth, or *crust*, is of two distinctly different types (see the end of this chapter, the section entitled "Rocky Earth Materials," for a description of crustal rock types). Continental crust, enriched in silica, alkalis, volatiles, and radioactive elements, is characterized by granitic rocks and is approximately 20 to 60 kilometers thick. Oceanic crust, enriched in calcium

(A)

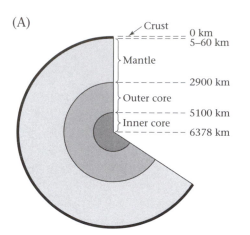

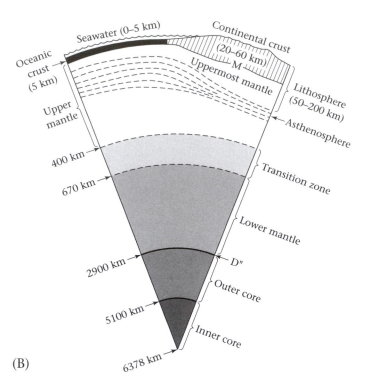

Figure 5.1. Schematic section through the Earth. (A) Overall structure, properly scaled. (B) Details of the crust and mantle are shown schematically, but not to scale. M = Mohorovicic Discontinuity, at the base of the crust; D″ = the narrow zone at the core–mantle discontinuity; km = kilometers. (*Source:* W. G. Ernst, *The dynamic planet*, Copyright © 1990, Columbia University Press. Reprinted with permission from publisher.)

(B)

(Ca), iron (Fe), and magnesium (Mg), contains only modest amounts of potassium (K) and sodium (Na), and lesser amounts of silicon (Si) compared to continental crust; it is almost exclusively basaltic in composition and is approximately 5 kilometers thick. The crust is a very thin skin, representing a little less than 1 percent by volume of the whole Earth, and only approximately 0.4 percent by mass.

The more or less horizontal boundary between the crust and the underlying *mantle*, both beneath the continents and ocean basins, is known as the Mohorovicic Discontinuity (or "Moho" or "M″"). This discontinuity is a zone of finite thickness in which lower crust and upper mantle materials are interlayered. It is no more than a few hundred meters thick at the base of the oceanic crust; however, the thickness of the zone is probably somewhat greater under the continents. Physical and chemical properties of the mantle below the Moho contrast markedly with those of the overlying crust.

The mantle in its entirety represents the largest proportion of the Earth, some 83 percent by volume and 68 percent by mass. Relative to the crust, it is rich in magnesium, is less rich in iron, and contains appreciably less silicon; the mantle is strongly depleted in volatile and radioactive elements. It consists of a number of distinct but intergradational layers. The uppermost layer is relatively rigid and, along with the overlying, comparably rigid crust, is termed the *lithosphere*, or rocky sphere. Beneath old portions of the continents, the lithosphere may be several hundred kilometers thick; however within ocean basins it may be as thin as 50 kilometers and beneath spreading centers (see Chap. 6), it is even thinner. The lithosphere, both mantle and crust, covers the entire planet and is divided into seven gigantic plates and numerous smaller ones (see Fig. 6.12), each of which moves as a coherent, integral unit relative to other plates. Which plate are you sitting on at this moment? Beneath the lithosphere lies the more ductile mantle layer known as the *asthenosphere* (glassy or weak sphere). Its upper boundary with the stiffer lithosphere is relatively sharp, on a scale of several kilometers; however, the asthenosphere gradually changes downward into more rigid mantle, at depths of 220 kilometers or more. The so-called upper mantle consists of lithosphere plus asthenosphere plus more rigid mantle material down to depths of approximately 300 to 400 kilometers. The transition zone of the upper mantle is situated approximately between the depths of 400 and 670 kilometers; in this region, physical properties of the upper mantle change fairly and continuously with increasing depth. The gradual emergence of nearly constant physical characteristics below the transition zone signals passage into the lower mantle. This innermost and most voluminous mantle shell appears to be relatively homogeneous to a depth of approximately 2900 kilometers at the boundary with the Earth's core.

The core–mantle boundary, and a thin, basal mantle zone, termed D″, represents a layer of profound change in the constitution of the materials that make up the Earth. The overlying mantle shells consist of magnesium (plus ferrous iron) and silicon oxide phases, whereas the outer core at depths between 2900 to 5100 kilometers consists of liquid iron-nickel alloy (primarily iron). From a depth of 5100 kilometers to the Earth's center at approximately 6378 kilometers, the inner core appears to be a solid iron-nickel alloy (chiefly metallic iron). The outer, molten core makes up approximately 15 percent of the Earth by volume, with the inner, crystalline core constituting the remaining 1 percent by volume of the planet. Because of its great density, the core represents more than 31 percent of the Earth's mass.

MAPPING THE STRUCTURE OF THE EARTH

Scientists have deduced the overall layering of our planet by employing numerous lines of evidence. In order to extract Earth resources, we sink mine shafts (3 kilometers) and drill bore holes (13 kilometers) into the upper part of the continental crust, but penetrate less deeply into the crust of the ocean basins. Accordingly, much of the information used to decipher Earth structures is model-dependent and circumstantial, and relies on remote sensing techniques. Despite these limitations, the evidence supporting the structure of the Earth as described above is quite robust.

SEISMIC DATA. Most of what is known regarding the layering of the deep Earth arises from study of the propagation of seismic waves. The transmission velocity of vibrational energy, such as results from an earthquake or a chemical explosion, is a function of the chemical and mineralogic composition of the material through which it travels, as well as the temperature, pressure, and physical state (i.e., liquid or solid). In general, transmission velocity increases with decreasing mean atomic number (see Table 3.1) and decreasing temperature, and especially with elevated pressure. The overall consequence is that, at greater depths in the Earth, hence higher pressures, the speed of propagation of seismic energy increases with depth within any layer of constant chemical-mineralogic constitution. The transmission velocity through the Earth decreases, however, when that part of the Earth becomes partially or completely molten. Seismic amplitude (the strength of the wave) diminishes with distance traveled; the intensity of shaking is especially reduced when the waves pass through partially or fully molten materials.

Several different types of seismic wave, as well as heat, are generated and transmitted by a release of kinetic energy within the Earth. Part of the vibrational energy is transmitted along the surface as relatively slow surface waves; the remainder of the mechanical energy passes through the Earth as body-transecting waves. As illustrated in Figure 5.2, two types of body waves may be distinguished: compressional (P) and shear (S) waves. P waves involve propagation of successive compressions and rarefactions through a medium, whereas slower S waves represent vibrations of material transverse to the direction of wave travel.

Seismographs are instruments that record the shaking of the Earth and its intensity as a function of time. Data supplied by two or more seismograph stations can provide the average velocity of a wave traveling through the material between them. Where a discontinuity – a change in physical and/or chemical parameters – is encountered, part of the seismic energy is reflected (bounced off), whereas the remainder is refracted (bent) into the newly penetrated layer. In either case, the energy is repartitioned into P and S vibrational modes. Reflection of seismic energy is good evidence that a discontinuity exists at depth, whereas the speed of

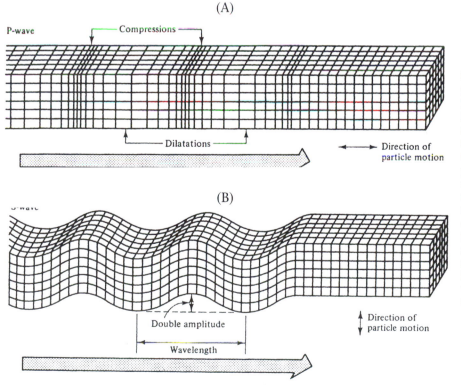

(A)

P-wave   Compressions

Dilatations   ← → Direction of particle motion

(B)

S-wave

Double amplitude

Wavelength

↑ Direction of
↓ particle motion

Figure 5.2. Distinction between (A) primary (P) compressional waves and (B) secondary (S) shear waves, two forms of vibrational energy. Transmission direction is to the right (stippled arrow). A unit volume of Earth material compresses and expands as the P wave passes through it, whereas the unit volume shears as the S wave is propagated through it. P and S are transmitted through solids; however, S waves are extinguished in a liquid medium. Both P and S are body waves; in contrast, surface waves (not illustrated) are propagated only along the surface of the Earth. (*Source:* G. C. Brown and R. C. L. Wilson, eds., *Understanding the Earth,* Cambridge University Press, 1992.)

transmission of the several types of wave provides constraints regarding the chemical and mineralogic constitution of the different layers.

The variations of P and S wave velocities with depth provide the most important information regarding the nature of deep crust, the structure of the mantle, and the core. Observations from seismograph stations around the world have produced the overall relationships presented in Figure 5.3 (the crust is too thin to be shown in this diagram). Although a general increase in speed of propagation with depth is evident, there are clear discontinuities and changes in the rate of increase, which mirror contrasts in mineralogy, bulk composition, stiffness, and state (solid, liquid, or mixed). At the base of the lithosphere, a slight decrease in the velocity of propagation of P waves (Vp) and a pronounced deceleration in the velocity of propagation of S waves (Vs) are evident. In the upper mantle region of low velocity, the attenuation (or decrease in amplitude during transmission) of seismic energy also increases markedly, especially for (S) waves. These observations are compatible with the hypothesis that this low-velocity region is characterized by small degrees of partial melting of the mantle. Evidently the Earth's *geothermal gradient* – the covariation of temperature and pressure as a function of depth within the planet – has encountered physical conditions appropriate for the onset of *incipient fusion* (melting); here, the grain boundaries of mantle minerals are wetted by a thin film of molten silicate. Such a phenomenon would also account for the observed weakness and plasticity of the asthenosphere relative to the overlying, cooler, more rigid lithospheric plates, inasmuch as the presence of a film of melt along grain boundaries would tend to lubricate and weaken the hotter asthenospheric material, and thus increase its ductility. Below a depth of approximately 220 kilometers, the upper mantle tends to become more rigid by degree, and the rate of attenuation of seismic energy gradually lessens. Why do you suppose this is?

Continuing downward, the region between approximately 400 and 670 kilometers is characterized by materials in which the speed of seismic waves increases rapidly with depth. This transition zone of the upper mantle is the region where magnesium (and iron) silicates are thought to transform to considerably denser oxide phases that have intrinsically higher Vp and Vs transmission values than the overlying, less dense mineral assemblages. A more homogeneous lower mantle occupies depths greater than approximately 670 kilometers; it is distinguished by smoothly increasing seismic velocities, suggesting simple compression of the high-pressure minerals produced in the overlying transition zone.

A profound discontinuity exists at a depth of approximately 2900 kilometers at the outer core–mantle boundary. Vp drops from approximately 14 to 8 kilometers per second, S waves are extinguished. The latter phenomenon indicates that the outer core is molten, for S waves cannot pass through a liquid. The decrease in P wave velocity is compatible with a relatively high mean atomic number; in addition,

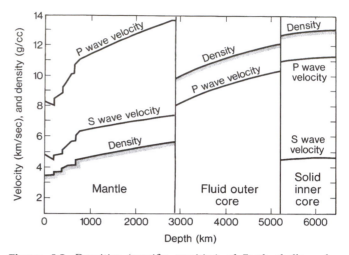

Figure 5.3. Densities (specific gravities) of Earth shells and transmission velocities of body waves through them as a function of depth within the Earth (National Research Council). As indicated, shear (S) waves do not pass through the liquid outer core of the Earth; however, S waves are transmitted through the solid inner core. km/sec = kilometers per second; g/cm³ = grams per cubic centimeter. (*Source:* W. G. Ernst, *The dynamic planet*, Columbia University Press. Copyright © 1990. Reprinted with permission from the publisher.)

the measured speed of (P) waves through liquid iron (plus minor nickel) at high pressures fits the observed Vp values quite well. An upward jump in P wave velocities marks the outer – inner core boundary at a depth of approximately 5100 kilometers. Moreover, S waves have been detected within the inner core. Together, these two observations indicate that the inner core of the Earth is solid, probably dominantly iron.

Not shown in Figures 5.1 and 5.3 are recent discoveries that the mantle is laterally heterogeneous as well as vertically stratified. Different propagation velocities of seismic waves in various regions indicate the existence of chemical-mineralogic or, more probably, thermal contrasts. Variation of wave velocity in different directions of the same mantle segment at depth indicates anisotropy of the material as well, implying differential flow at various levels within the deep Earth.

The seismic evidence is critical for an appreciation of the layered nature of the Earth. Clearly, our planet consists of a series of concentric shells. However, can we be sure that other Earth materials would not equally well satisfy the observed Vp and Vs behavior? We now turn to a consideration of other types of data in order to reduce and/or eliminate such ambiguities, and to refine the structural model of the Earth derived from seismic information.

DENSITY DATA. As determined from measurement of the gravitational constant, the surface gravitational acceleration, and volume, it is known that the overall density of the Earth is 5.5 grams per cubic centimeter. In addition, the planetary moment of inertia, a measure of mass distribution times the square of the effective radius, is less than that of a homogeneous sphere, suggesting that mass is concentrated toward the center. This relationship is in accord with the inferred densities of the various concentric shells of the planet. The volumetrically insignificant crust has a density of 2.6 to 3.0, whereas the major volume of the Earth – the mantle – possesses a density ranging from 3.2 measured on materials derived from just below the Moho, grading downward toward an estimated value of 4.6 for a mixture of silicon, magnesium, and iron oxides at the core–mantle boundary; therefore, the core itself must have an average density of approximately 10 to 11 in order for the entire Earth to average 5.5 grams per cubic centimeter. Extrapolated to core temperatures and pressures, such densities are appropriate for iron–minor nickel alloys, including small amounts of a dissolved low-density solute. Thus mass and inertial considerations corroborate the proposed layering of the Earth, with progressively denser units residing in the successively deeper levels of the planet (see Figs. 5.1 and 5.3).

METEORITIC DATA. Meteorites represent the most complete sampling of condensed extraterrestrial planetary materials in the solar system. Judging from measured orbital trajectories, the vast majority appear to have come from the asteroid belt between Mars and Jupiter. As previously described, meteorites are of two principal types, stony and iron. The latter are made up predominantly of a metallic iron-nickel alloy, and minor magnesium silicates occur in some. The former consist chiefly of magnesium-rich silicates, and many contain minor amounts of iron-nickel alloy and/or sulfides. Most stony meteorites contain numerous spheroidal silicate blobs termed *chondrules:* These droplets seem to have coalesced from the solar nebula to form accreting asteroids, hence such meteorites are appropriately known as *chondrites.* Stony meteorites of a rare class, the carbonaceous chondrites, contain hydrous minerals and carbonaceous matter in addition to the anhydrous silicates. Does the latter represent the former existence of carbon-based life in a surficial, crustal environment? No one knows conclusively; however, the association of organic-like material with H_2O-containing minerals is provocatively reminiscent of the terrestrial surface.

The different groups of meteorites are chemically and physically similar to the various concentric shells inferred to make up the Earth. The carbonaceous chondrites are analogous to the volumetrically insignificant crust. Chemically and mineralogically, the mantle is thought to be quite comparable to stony meteorites, and the core to iron meteorites. Also, their relative proportions are appropriate. Nearly 90 percent by volume of all meteorites retrieved after having been observed to fall are stony in composition – a value close to the volume proportions of the Earth's mantle. The slightly larger volume ratio of core material on our planet, 16 percent, relative to iron meteorite abundances, approximately 10 percent, suggests a more complete metal-silicate separation in the case of the more massive Earth compared to condensation, accumulation, and chemical differentiation of planetesimals in the asteroid belt. The similarities of meteoritic and terrestrial materials, and their comparable volume proportions, provide yet another compelling line of evidence regarding the internal constitution of the Earth.

HEAT-FLOW AND GEOCHEMICAL DATA. Presuming that the Earth began to accrete cold during gravitational collapse of the solar nebula, successive impacts by late-arriving planetesimals would have tended to heat the growing planet at an increasing rate. This is due to the fact that the energy released as shock waves and heat during collision is proportional to the masses of the colliding bodies. Nuclear fission also liberates thermal energy; consequently, radioactive isotopes would have contributed to this early-stage heating. Finally, as the planetary interior heated up at depth, the melting of particles of iron, coalescence of liquid iron droplets, and descent of molten metal to form an enlarging metallic core would release large amounts of heat – apparently enough to raise the temperature of the entire Earth an average of approximately 2000 K. All these phenomena would have acted in concert to produce high temperatures and a strong thermal gradient in the primordial Earth, with the interior much hotter than the exterior. Heat must have moved toward the cooling surface by a variety of mechanisms, chiefly radiative transfer, bodily flow of material (convection), and thermal conduction. Partial fusion and degassing would have resulted in the upward migration of both low-temperature melting constituents, fluids, and large-radius elements insoluble in dense minerals stable at great depths.

These processes would have enriched the crust of the Earth in fusible, large-radius atomic species, and in volatile constituents, such as those concentrated in exposed rocks, the oceans, and the atmosphere today. Most of the heat-producing, radioactive elements – those of potassium, uranium and thorium – appear to be localized near the surface of the solid Earth, judging from the large proportion of observed heat flow that is contributed by atomic fission in the crust. If similar quantities of these radioactive elements were sequestered in the mantle or core, amounts of heat vastly greater than measured would be emerging at the surface today; accordingly, we know that both mantle and core are nearly devoid of radioactive elements. Moreover, provided most of the heat-producing, unstable isotopes are concentrated in the crust, the computed terrestrial abundances of potassium, uranium, and thorium for the whole Earth are nearly identical to the proportions of these constituents in meteorites, the Sun, and throughout the solar system.

A schematic temperature-pressure diagram showing the modern thermal gradient of the Earth and the overall planetary differentiation into core and mantle is presented in Figure 5.4. Also illustrated are diagramatic curves for the melting of metallic iron (and minor nickel) and the incipient fusion of mantle material. The Earth's geothermal gradient evidently coincides with the partial melting curve for magnesium-rich silicates only in the upper 100 to 220+ kilo-

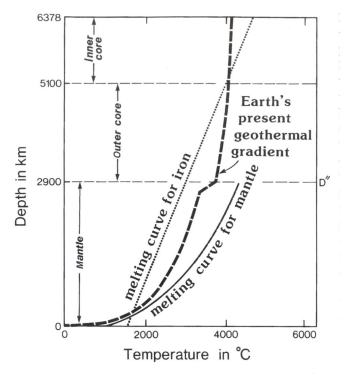

Figure 5.4. Diagrammatic temperature–depth relations in the present-day Earth. The melting curves for iron and for the onset of melting in the mantle are not smoothly varying; however, temperature–depth trends are approximately as illustrated. The illustrated abrupt temperature decrease upward through the D" Zone bounding the base of the lower mantle is conjectural. km = kilometers. (*Source:* W. G. Ernst, *The dynamic planet*, Columbia University Press. Copyright © 1990. Reprinted with permission from the publisher.)

meters of the mantle, the location of the ductile asthenosphere; at both greater and shallower depths, the mantle should be completely solid, and therefore more rigid. The inner–outer core boundary must be a common point on both the iron-nickel alloy melting curve and the Earth's geothermal gradient because, at the inner–outer core boundary, liquid and solid iron-nickel alloy are stable together. Near-surface temperatures are much lower than those in the deep Earth; hence rocks that make up the crust are far less refractory than the underlying mantle. Also, the presence of volatiles such as H_2O lowers the fusion temperatures in the outer rind of the Earth. Consequently, though largely solid, deeper portions of thick continental crust undergo partial fusion if buried too deeply – thus limiting the thickness of the crust and the depth to the Moho. In summary, the observed heat flow and geochemical concentrations of fusible, volatile elements near the surface of the Earth are quite compatible with its overall inferred structure and chemical constitution.

MAGNETIC DATA. The general nature of the global magnetic field is illustrated in Figure 5.5. It is dipolar, that is, it possesses two poles. At present, the magnetic and rotational poles do not coincide exactly. Moreover, the magnetic field

lines (directions of the magnetic force field) shift slowly with time. The origin of the Earth's magnetic field is not completely understood; the most widely accepted theory postulates that the outer, liquid core acts as a dynamo. Fluid flow within this conducting paramagnetic material, and differential motion between the outer, liquid, metallic core and the base of the solid, silicate mantle, are thought to induce the surrounding magnetic field. The latter, in turn, couples with the external solar wind (an ionic flux, or plasma, consisting of hydrogen and helium nucleii that stream outward from the Sun at 1 to 4 percent the speed of light). The Earth's internal dynamo appears to be governed by rotation of the planet, so that over long time periods the average magnetic pole positions probably coincide quite closely with the spin axis. In support of this hypothesis is the observation that rapidly rotating planets in our solar system that possess a dense, metallic core also appear to be enveloped by a magnetic field. Most are weaker than the terrestrial force field because the planetary iron-rich core is either small, or completely solidified, or both.

DYNAMIC NATURE OF THE EARTH

Study a modern relief map of the continents and ocean basins (Fig. 4.13), and you will be struck by the remarkable, congruent outlines of the continental borders framing the Atlantic Ocean. Equally astonishing is the existence along the north–south median of this ocean basin of a longitudi-

Figure 5.5. The magnetic field of the Earth, showing lines of equal intensity and orientation. The inclination of the field lines of magnetic force is proportional to latitude. We can approximate the field by imagining a simple bar magnet (shown in white) passing through the Earth's center and inclined 11.5° to the pole of rotation. (*Source:* W. G. Ernst, *The dynamic planet*, Columbia University Press. Copyright © 1990. Reprinted with permission from the publisher.)

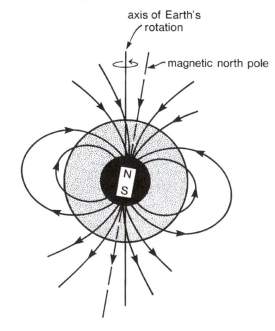

nal, undersea mountain range that faithfully reflects the outlines of the continental margins. Other, equally stupendous submarine mountain chains, or rises, characterize the Circum-Antarctic, Indian, and eastern Pacific oceans. They exceed the length and breadth, and rival the topographic ruggedness of on-land mountain belts. Major features within the ocean basins that are perhaps as impressive are the oceanic deeps, or trenches, that lie basinward from the island arcs, peninsulas, and continental margins of Indonesia, Japan, Kamchatka, the Aleutians, Central and South America, the Scotia Arc, and the Lesser Antilles. The most prominent mountain systems on the continental crust completely encircle the Pacific and link up with the eastern portion of the transverse Alpine-Mediterranean-Himalayan-Indonesian zone; just seaward from these mountain chains lie the oceanic trenches. What interplay of forces, so obviously lacking on our planetary neighbors, has resulted in this diversity and richness of topographic expression on Earth?

Geologists who studied the continental crust early in the nineteenth century recognized that strata of great antiquity are exposed on dry land. These old units are gradually being eroded and carried away piecemeal toward the sea. What drives this process? Energy from the Sun preferentially warms the tropical oceans, producing evaporation and moisture-laden clouds in the atmosphere, and currents in the hydrosphere. H_2O is transported via atmospheric circulation systems to the land, where cooling and precipitation result in the flow of water, ultimately back to the sea carrying entrained crustal debris (Chap. 9). Mechanical weathering, chemical solution, and erosion wear down the topography (Chap. 8). Yet the Earth is not a flat, monotonous plain; in fact, evidence indicates that our planet has always had mountains. Evidently, processes hidden from view are at work deep within the Earth, slowly, inexorably thrusting sections of the continental crust to higher elevations. This dynamic instability reflects mantle circulation fueled by the internal heat discussed earlier. The crustal manifestation of deep-seated convection is mountain building, an activity that deforms preexisting rocks, is commonly accompanied by volcanism, and generates lofty mountain ranges.

Uplifted segments of the Earth's crust are also a reflection of gravitative equilibrium, or mass equivalence, termed *isostasy*. Measurements of the Earth's gravitational field have shown that low-density, topographically high continental crust is underlain by a large root of comparably low-density materials; such roots extend downward into the uppermost mantle much like the subsurface portion of an iceberg floating in the sea. The elevated excess of continental mass is thereby gravitatively compensated through a displacement of denser mantle by light crustal material at depth, resulting in an equivalent mass deficiency. Consequently, the observed gravity field does not show a marked positive or negative geophysical *anomaly* (a departure from the expected field). Similarly, low-lying continental crust is relatively thin, and the distance to the Moho is less than that under high mountain ranges. Different portions of the continental crust

range from approximately 20 to more than 60 kilometers in thickness. The denser oceanic crust is even thinner than the thinnest continental crust (the Moho is approximately 5 kilometers below the sea floor); consequently, ocean crust is confined to deep basins.

Two different manifestations of gravitative compensation are thus recognized: (1) where crusts of the same aggregate densities are involved, the thicker crust rises to higher elevations than the thinner crust; (2) where crusts of similar thickness but contrasting densities are involved, the lower-density crust rises to higher elevations than the higher-density crust. Buoyancy is a reflection of Archimedes' principle, as illustrated schematically in Figure 5.6. Because the asthenosphere is weak, it easily readjusts to accommodate the masses above, and mass balance (isostatic equilibrium) is closely approached. Many vertical crustal movements are thus a direct consequence of isostatic compensation.

The nature of the Earth's crust is a reflection of both vertical and horizontal flow within the mantle, and the resultant differential movement of enormous lithospheric plates.

Figure 5.6. Archimedes' principle and isostatic balance of the Earth's crust: (A) blocks of relatively dense wood, low-density cork, and ice floating in a beaker of water; (B) blocks of continental crust (density, approximately 2.7) and oceanic crust (density, approximately 3.0) "floating" in a mantle substrate (density, approximately 3.2); (C) rugged continental topography and corresponding low-density roots. Oceanic crust rides lower than even thin continental crust. The mass of any column of unit cross section down to an arbitrarily deep level of compensation – for example, the top of the asthenosphere – is everywhere the same. All three examples show the principle of isostasy. s.g. = specific gravity = density = mass/volume. (*Source:* W. G. Ernst, *The dynamic planet*, Columbia University Press. Copyright © 1990. Reprinted with permission from the publisher.)

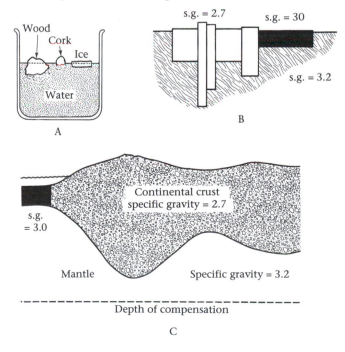

(This is discussed in more detail in Chapter 6.) The chemical heterogeneities and topographic relief expressed in the continents, ocean basins, mountains, and oceanic trenches are manifestations of the dynamic, internally circulating nature of our evolving planet, as well as modifications at the surface due to erosion, transportation, and deposition. That ever-changing landscape, in turn, is a manifestation of the planetary redistribution of buried terrestrial heat and solar energy.

METEORITE IMPACTS AND EARTH HISTORY

Thus far we have attributed the degradation of the Earth's surface solely to the action of terrestrial agents of erosion (mass slumpage, and transportation of Earth materials in media such as wind, glaciers, and, most important, running water). It is true that these erosion processes are dominant today; however, they were not always dominant in the past. As described in Chapter 4, the growth of the Earth from planetesimals during the collapse of the solar nebula resulted in rapid initial growth of our planet, followed by a gradual decrease in the number of impacting meteorites (e.g., planetesimals, asteroids) as time progressed. The same scenario probably holds throughout the solar system as gravitative sweep-up of the planetesimals by the enlarging planets took place.

Once the Earth had achieved a considerable size and mass, and a structure similar to the one in existence today, collision with a moderate-sized asteroid – for example 10 kilometers in diameter – would have drastically affected the terrestrial environment during and immediately after impact. The tremendous amount of energy released when a meteorite of this size hit the Earth would have caused excavation of a large crater, local partial melting of the upper mantle and crustal veneer, regional vaporization of seawater, and the injection of great quantities of dust into the atmosphere. This dense dust cloud would have enveloped the Earth, blocked out sunlight, and raised planetary reflectivity for decades, causing a pronounced global lowering of surface temperatures and the abrupt onset of a severe climatic crisis. The actuality of the occurrence of specific impact events and the magnitude of the effects are still being debated; however it seems likely that a major meteorite collision would profoundly and adversely influence the Earth's biosphere, as well as the physical environment. One such impact is hypothesized to have taken place 65 million years ago, at the Cretaceous (K)–Tertiary (T) time boundary, as described in Chapter 3. This Chicxilub cratering event is held responsible for worldwide mass extinctions among many species of plants and animals, including everybody's favorite group of organisms – the dinosaurs.

Planetary neighbors of the Earth that have less dynamic interiors, including the Moon, Mars, and Mercury, bear indisputable evidence of abundant meteoritic collisions (Fig. 4.2). Therefore, it is a virtual certainty that the Earth, too, was subjected to such bombardment, especially early in the history of our planet. However, evidence of such impacts is exists in only a few places. The course of biologic evolution almost certainly has been – and will continue to be – capriciously, and adversely, influenced by episodic collisions with Earth-approaching asteroids, as well as by other kinds of natural disasters (see Chap. 25).

MINERALOGIC EARTH MATERIALS

Much like police detectives, Earth scientists are obliged to piece together the origin and evolution of the planet from fragmentary remains contained in the substances of the planet itself. The long and involved history of the Earth is imperfectly preserved as clues in the rock record – and has been obliterated from the hydrosphere, atmosphere, and deep interior of the planet. The crust is, of course, the most readily accessible part of the lithosphere. Because rocks and soils are made up largely of minerals, we must develop an appreciation for these constituents in order to understand the diverse origins of the Earth's various units and their altered products at the surface. Of obvious practical value are terrestrial mineral resources (see Chaps. 23 and 24) – the largely inorganic materials, as well as fossil fuels – necessary for the continued support and well-being of civilization, and, indeed, a prerequisite for all life. In order to obtain a better understanding of the materials that constitute the solid Earth, it is necessary first to define them.

A *mineral* is a naturally occurring, inorganically produced solid having a systematic, three-dimensional atomic arrangement. It exhibits a restricted range of physical and chemical properties. Minerals are solids, in contrast to liquids or gases. They are compounds of invariant composition, such as quartz, SiO_2, or they have compositions that range between fixed values, such as the olivines, $(Mg,Fe)_2SiO_4$. The latter exhibits continuous variation, *solid solution*, from a pure magnesium silicate, Mg_2SiO_4, to a pure iron silicate, Fe_2SiO_4. Minerals are constructed of atoms, systematically located and rigorously repeated in three dimensions. The atomic configuration is termed the *crystal structure*; for most minerals, the arrangement of atoms and repeat distances are different in different directions. Substances having a regular, ordered atomic structure are said to be *crystalline*; hence, all minerals are crystalline.

A *crystal* is any atomically ordered solid bounded by planar growth surfaces, called crystal faces; the latter bear a definite geometric relationship to the atomic structure. Consider the beautiful terminations bounding "rock crystal" quartz, SiO_2 (Fig. 5.7). These crystal faces are a consequence of the manner of crystal growth and have specific, invariant angular relationships to the atomic periodicity. The symmetry exhibited by a crystal, therefore, reflects its three-dimensional atomic structure, or packing of atoms.

Bonding, the attraction of one atom to another, provides the coherence evident in minerals and other forms of matter. Atoms consist of a nucleus made up of one or more positively charged protons and (except for hydrogen and ^3He) an equal or larger number of uncharged neutrons. The number of pro-

Figure 5.7. Quartz, of the variety known as "rock crystal." (*Source:* W. G. Ernst, *Earth materials*, Prentice-Hall, 1969.)

tons defines the chemical element, and the number of protons plus neutrons equals the atomic mass. The nucleus is surrounded by a cloud of electrons, which are negatively charged particles approximately 1/1837 the mass of a proton; electrons are confined to specific orbital energy levels. Neutral atoms are characterized by equal numbers of protons and electrons. If electrons are removed from or added to the outermost electron shell, the atoms become charged ions (cations are positively charged, anions are negatively charged). *Ionic bonding* is the electrostatic attraction of unlike charged ions for one another. *Covalent bonding* results when electron clouds interpenetrate and the nucleus of each atom attracts the electron shells of the adjacent atom. *Metallic bonding* characterizes substances in which outer orbital electrons roam freely through a cation-dominated structure. *Van der Waals bonding* is the extremely weak attraction that results from nonuniform charge distribution of otherwise neutral atoms.

A *rock* is a naturally occurring multigranular aggregate of one or more minerals and/or amorphous solids such as volcanic glass. Units generally must be large enough to constitute a significant portion of the solid Earth. Gradations exist from a mountain range composed of, for example, granite, to granitic veins (fracture fillings) of infinitesimal thickness cutting through another rock. In both cases, the constituent grains, either of the same or of different minerals, owe their association to a common genetic process, the origin of the rock itself. The genesis and diversity of rocks are described in the following section, entitled "Rocky Earth Materials"; a few definitions are presented here.

Geologists recognize three main rock-forming processes, and therefore three principal classes of Earth-forming substances. In the first process, molten rock, or *magma*, solidifies

either to glass or to an aggregate of one or more minerals, or to some combination of glass and minerals; such rocks are termed *igneous*. *Ignis* is the Latin word for "fire"; igneous rocks are, in a sense, formed from fire deep inside the Earth. Lava flows and ash falls are good examples. *Sedimentary* rocks consist of mechanically accumulated fragments of preexisting Earth materials, as well as chemical or biochemical precipitations from a fluid medium. *Sedimentum* is the Latin for "a settling," and sedimentary rocks have had their constituents settle out from the transporting medium, either air or, more generally, water. Such processes take place at the Earth's surface. Examples of sedimentary deposits include shell beds and stream gravels; when these become cohesive rocks, they are called shelly limestone and conglomerate, respectively. *Metamorphic* rocks include all those rocks whose original minerals or textures, or both, have been altered markedly by recrystallization and/or deformation; metamorphism generally takes place at considerable depth within the Earth. The Greek word *meta* is translated as "successive," or "change"; a metamorphic rock represents a later configuration of minerals and/or textures different from those of the original material. Greenstone, marble, and slate are familiar examples of metamorphic rocks, derived from basalt, limestone, and mudstone, respectively.

Soils are the weathering products of rocks. Important enough to be treated separately (see Chap. 8), soils represent the chemical breakdown and mechanical disaggregation of preexisting Earth materials; most contain decomposing organic material as well. Virtually all land plants take their nutrients from soils; inasmuch as all terrestrial animals ultimately are dependent on the botanic cover for sustenance, soils are crucial to the survival of nonmarine life.

SILICATE MINERALS. More than 2000 mineral species have been described. Although the majority of silicates are rare, common groups do exist, such as oxides and carbonates. However, as mentioned earlier in this chapter, the Earth's crust and upper mantle consist chiefly of silicates. Rather than describing these extremely important minerals in detail, we will classify and discuss general characteristics of their structural diversity, then briefly mention some of their contrasting physical and chemical properties.

The crystal structures of the silicates to a large extent determine their properties. These minerals belong to a small number of structural types; hence, the range of physical characteristics of the silicates is rather limited. Silicates all possess rather intricate atomic arrangements, that represent the linking together (polymerization) of much simpler atomic building blocks. The most important cation in silicates is silicon. This small, highly charged positive ion, which is quadrivalent and has an ionic radius of approximately 0.26 angstroms, is surrounded by four much larger oxygens. Each oxygen carries a divalent negative charge and has an ionic radius of 1.35 angstroms. Because of their relative sizes, at crustal pressures only four anions can be packed around the central silicon. A net negative charge on the complex,

$(SiO_4)^{-4}$ necessitates bonding to other cations. The degree of tetrahedral connection provides the basis for structural classification of the silicates (see Figs. 5.8 through 5.11). For simplicity, we will consider bonding to be 100 percent ionic; however, atomic attractive forces are partly covalent in the silicates. Other bond types also characterize some of these and other mineral groups.

In some silicate minerals, such as olivines and garnets, individual silicon tetrahedra share oxygens only with cations other than silicon. Their chemical formulas therefore include $(SiO_4)^{-4}$ anion complexes. A typical olivine, forsterite, is written Mg_2SiO_4, and almandine garnet has the formula $Fe_3^{+2}Al_2(SiO_4)_3$. Each of these mineral groups shows extensive solid solution. Common olivines range in composition between forsterite and the ferrous iron end member fayalite, $Fe_2^{+2}SiO_4$. In the garnets, the ferrous iron of almandine may be replaced by calcium or manganese, and the aluminum by ferric iron. The silicon-oxygen arrangement in an independent tetrahedron is illustrated in Figure 5.8A.

In epidote and certain other less common silicates, two silicon tetrahedra share a common oxygen; for such paired tetrahedra, the silicon:oxygen ratio is not 2:8, but 2:7. The shared anion is known as a "bridging oxygen" because it cross-links two tetrahedrally coordinated silicon atoms. The valence orbital of the oxygen is completed by the donation of an electron from each of the two nearest-neighbor silicons.

Figure 5.8. Coordination models of (A) independent silicon tetrahedra such as occur in olivine and garnet; (B) paired tetrahedra as found in epidote; and (C) six-member tetrahedral rings characteristic of the mineral beryl. In this and the subsequent diagrams of Figures 5.9, 5.10, and 5.11, small black spheres represent silicon (and minor aluminum, in some cases); large open spheres represent oxygen. In addition to the coordination models, arrangements of silicons and oxygens are shown as individual tetrahedra. Because of strong bonds in all directions, minerals characterized by isolated, paired, and ringed silicon-oxygen tetrahedra are hard and refractory, and do not in general possess good cleavages. Å = angstroms. (*Source:* W. G. Ernst, *Earth materials*, Prentice-Hall, 1969.)

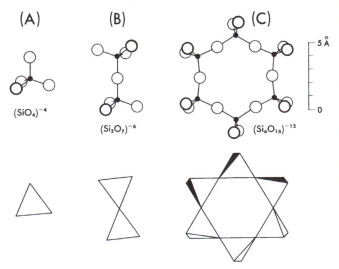

(A)   (B)   (C)

$(SiO_4)^{-4}$ $(Si_2O_7)^{-6}$ $(Si_6O_{18})^{-12}$

The other tetrahedrally disposed oxygens each receive one electron from their central silicons but must obtain the second electron from peripheral, coordinating cations. Because there are six such nonbridging oxygens, the anion complex has a net negative charge of six $(Si_2O_7)^{-6}$. Paired silicon tetrahedra are illustrated in Figure 5.8B.

Increased polymerization results in three-, four-, or six-member rings of silicon tetrahedra. Here, two oxygens in each tetrahedron are bridging oxygens, and the remaining two must each obtain an electron from other cations. Thus, the net negative charge on three-, four-, and six-member rings is six, eight, and twelve respectively. The arrangement in a six-member ring is shown in Figure 5.8C. For example, the $(Si_6O_{18})^{-2}$ group, indicating six silicon tetrahedra per ring, characterizes the mineral beryl, $Be_3Al_2Si_6O_{18}$; gem varieties of this mineral are known as emerald.

More extensive polymerization of silicon tetrahedra gives rise to chains, sheets, and frameworks. The two fundamental types of silicon-oxygen chains, single and double, are illustrated in Figure 5.9. In single-chain silicates, two of the oxygens in each tetrahedron are bridging oxygens, as Figure 5.9A shows. Only half of each bridging oxygen may be assigned to a specific tetrahedron, and the silicon:oxygen ratio reduces from 1:4 in the case of isolated silicon tetrahedra to 1:3 in single chains. Every tetrahedron contains two nonbridging oxygens, each of which receives an electron from a peripheral cation. As in ring silicates, the basic single-chain unit, then, is $(SiO_3)^{-2}$, which is typical of the pyroxene mineral group. Examples include enstatite, $MGSiO_3$ and diopside, $CaMg(SiO_3)_2$.

The most important double-chain silicates belong to the amphibole group. The structure consists of two single chains of silicon tetrahedra with alternate tetrahedra cross-linked to the adjacent chain, as depicted in Figure 5.9B. Within the chain repeat of four silicons, five bridging oxygens are shared between two tetrahedra – hence, there are only eleven instead of sixteen oxygens in the unit repeated, written $(Si_4O_{11})^{-6}$. The net negative charge is six, because there are six nonbridging oxygens per four silicons. Typical double-chain silicates include the amphiboles anthophyllite, $Mg_7(Si_4O_{11})_2(OH)_2$, and tremolite, $Ca_2Mg_5(Si_4O_{11})_2(OH)_2$.

Amphiboles and pyroxenes display essentially infinite silicon tetrahedral polymerization in the direction parallel to the chain length. In sheet silicates, the cross-linkage is carried further, and the tetrahedral units are extended in two dimensions to produce layers, as illustrated in Figure 5.10. Just as double chains may be described as six-member silicon tetrahedral rings polymerized in one direction, so sheet structures may be described as six-member tetrahedral rings cross-linked in a plane (compare Figs. 5.9B and 5.10). In sheet silicates, three of the four oxygens surrounding a central silicon are bridging oxygens. Therefore, per silicon, there is one nonbridging oxygen, and the equivalent of 1.5 oxygens whose charge requirements are completely satisfied by the tetrahedrally coordinated cations. The basic repeat unit includes four silicons and ten oxygens; hence, the formula of the tetrahedral unit is $(Si_4O_{10})^{-4}$. Talc and muscovite mica are sheet silicates, as indicated by structural formulas, $Mg_3(Si_4O_{10})(OH)_2$ and $KAl_2(Si_3AlO_{10})(OH)_2$, respectively. They are colorless, whereas the presence of ferrous iron accounts for the appearance of the black mica, biotite $K(Mg, Fe^{+2})_3(Si_3AlO_{10})(OH)_2$. Clay minerals and chlorite, important in sedimentary and metamorphic rocks, respectively, also constitute major groups of sheet silicates.

Although to this point we have tacitly assumed that all tetrahedrally coordinated cations are silicon, this is not true for many rock-forming silicates, as illustrated by the muscovite and biotite formulas presented above. In micas, one-fourth of the tetrahedral sites are occupied by aluminum and three-fourths contain silicon. Certain amphiboles and pyroxenes, too, contain aluminum in fourfold coordination. Tetrahedrally coordinated aluminum is absolutely essential in the feldspars.

Feldspars and quartz are the most abundant minerals in the Earth's continental crust. In these mineral groups, the tetrahedral units are cross-linked in three dimensions; consequently, every oxygen is shared between two silicons, or ranges up to a maximum of one aluminum and a minimum of one silicon per oxygen. This three-dimensional polymerization results in a tetrahedral framework analogous to that shown in Figure 5.11; minerals possessing this type of configuration are called framework silicates. For quartz, silicon is the only tetrahedrally coordinated cation. It is surrounded by four bridging oxygens, each counted as belonging one-half to each of two tetrahedra, so the structural formula reduces to quartz. In feldspar structures, however, between one-fourth and one-half of the tetrahedral sites contain aluminum; accordingly the framework is represented by portions of mineral formulas such as $(AlSi_3O_8)^{-1}$ and $(Al_2Si_2O_8)^{-2}$. The substitution of Al^{+3} for Si^{+4} generates a charge imbalance that must be compensated for by an equivalent number of monovalent cations or half as many divalent cations. Formulas of the plagioclase feldspars, are albite (Ab), $NaAlSi_3O_8$, and anorthite (An) $(CaAl_2Si_2O_8)$. They form a complete solid-solution series; here, three compositional parts of this plagioclase range are distinguished: sodic (less than 10 percent An); intermediate (10 to 50 percent An); and calcic (greater than 50 percent An). Yet a third feldspar end member potassium feldspar, orthoclase (Or, $KAlSi_3O_8$ exists). Here, potassium resides in the cation site that would be occupied by sodium in albite. The alkali feldspars represent solid solution between $NaAlSi_3O_8$ and $KAlSi_3O_8$.

With the exception of the forms of silica of nearly invariant composition, extensive substitutions in the different cat-

Figure 5.9. Coordination models of (A) single-and (B) double-chain polymerized silicons and oxygens; linked tetrahedra are also illustrated in (A). These linear structural units occur in (A) pyroxenes and (B) amphiboles, respectively. Two prismatic cleavages that avoid breaking the bridging oxygen bonds characterize chain silicates. Because of high bond strengths, these minerals are hard and have high melting points. Å = angstroms. (*Source:* W. G. Ernst, *Earth materials*, Prentice-Hall, 1969.)

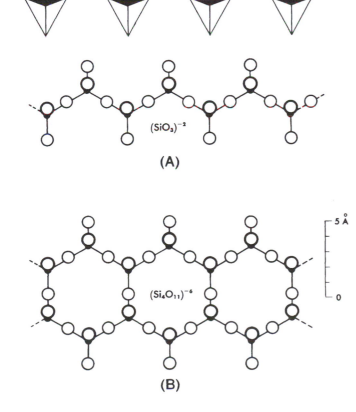

(A)

(B)

Figure 5.10. Coordination model of a tetrahedrally polymerized sheet. Perfect cleavage parallel to the silicon-oxygen tetrahedral layers totally avoids the bridging oxygens. These planar structural units characterize the micas, chlorites, and clay minerals. Weak bonding across the layers accounts for their softness and causes many sheet silicates to decompose readily at elevated temperatures. Å = angstroms. (*Source:* W. G. Ernst, *Earth materials*, Prentice-Hall, 1969.)

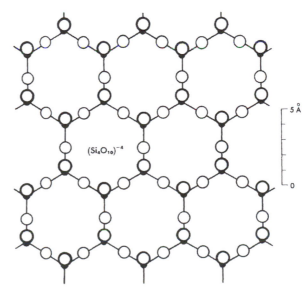

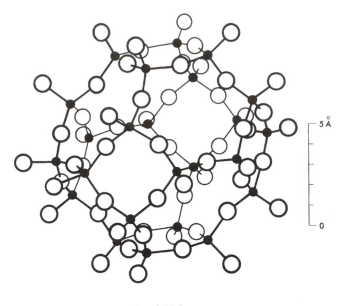

$(Si_{1-n}Al_nO_2)^{-n}$

Figure 5.11. Coordination model of a three-dimensional framework of polymerized tetrahedra. Framework silicates in general have imperfect or no cleavage because of strong bonds in all direction. They are relatively hard and refractory, except for species that contain hydroxyl, and/or water of hydration. The open atomic packing imparts low densities to most minerals of this structural type. The example illustrated is a complex hydrous framework silicate of the zeolite group; other common phases such as quartz and feldspars have similar, although more compact, structures. Å = angstroms. (*Source:* W. G. Ernst, *Earth materials*, Prentice-Hall, 1969.)

ion sites are responsible for the broad range of compositions found among most rock-forming silicates. Despite these chemical variations, crystal structures of a particular mineral group are rather uniform in their atomic arrangements because of high silicon-oxygen (and aluminum-oxygen) bond strengths. Accordingly, tetrahedral polymerization controls the physical properties of the individual silicate groups to a very large extent, regardless of chemical composition.

ROCKY EARTH MATERIALS

OCCURRENCE OF IGNEOUS ROCKS. These types of rock constitute approximately 75 percent of the continental crust, and well over 90 percent of the oceanic crust. Some have been derived directly from the mantle, whereas others represent partly remelted oceanic or continental crustal materials. The high temperatures required to fuse igneous rocks in the laboratory and that have been measured for erupting lavas (in excess of 900 to 1100°C, depending on the composition of the liquid and the attendant pressure), demonstrate that magmas must originate at considerable depths in the Earth, where such temperatures are characteristic.

Chemically speaking, igneous rocks exhibit rather limited compositional variations. Their principal constituent oxide is silica, SiO_2, which ranges from about 40 to 75 percent by weight in the common rock types. Where the silica content is low, magnesium and iron oxides commonly are major components; this is reflected by the abundance of dark, calcium-, magnesium- and iron-bearing minerals such as olivines, pyroxenes, amphiboles, and biotites, as well as anorthite-rich plagioclase. Silica-poor igneous rocks do not contain quartz, and are called *subsilicic*; ferromagnesian-rich ones are termed *mafic*. Generally speaking, most subsilicic rocks are mafic, and vice versa. The yet more magnesium- and iron-rich, silicon-, sodium-, and potassium-poor, olivine- and pyroxene-rich materials characteristics of the upper mantle are known as *ultramafic* rocks. Feldspar is never a major constituent, and where it occurs, it is dominantly calcic plagioclase. In igneous rocks in which the silica contents exceed approximately 60 to 65 percent by weight, quartz is typically present and is associated with potassium- and sodium-rich feldspars with or without muscovite, and only minor amounts of dark, ferromagnesian minerals. Such rocks are light-colored and are termed *silicic* and/or *felsic* (that is, feldspar-rich).

Occurrences of igneous rocks belong to two different categories: *extrusive* and *intrusive*. To the first group belong those that have reached the surface of the Earth in a largely molten condition. Lava flows, for example, are streams of magma issuing on the land surface or the sea bottom, and volcanic ash is magma that has been violently blown apart during extrusion by the explosive expansion of dissolved gases as external pressure is reduced due to removal of the overlying magma column. Volcanic activity is among the most catastrophic of internal geologic processes; it rivals earthquakes in terms of devastation and loss of life. The second group of igneous rocks consists of those that crystallized from magmas that solidified before reaching the surface of the Earth. Processes that formed them are thus hidden from view, and must be deduced from the study of uplifted and deeply eroded parts of the crust.

Two broadly contrasting modes of volcanic activity have been recognized: (1) the quiescent, or fissure, type; and (2) the more spectacular central-eruptive type. The great volumes of mafic lava that have poured out on the Earth's surface episodically through geologic time have issued principally from fractures, along which magma from considerable depth has gained access to the surface. Lavas of this type are dominantly dark, subsilicic basalts, that are characteristically quite fluid during extrusion. On continents, they accumulate to form broad plateaus, approaching a kilometer or more in thickness and tens of thousands of square kilometers in lateral extent; a familiar example is the Columbia River Plateau. Even more voluminous are the basalts (and gabbros, their intrusive equivalents) that originate at spreading ridges, and that floor the ocean basins. These accumulations average approximately 5 kilometers in thickness, as determined by seismic measurements.

More obvious but volumetrically less important are extrusive rocks of the central-eruptive type, which build distinctive volcanic features. Two major varieties are shield volcanoes and stratovolcanoes. To the former group belong

Figure 5.12. The great Hawaiian shield volcano, Mauna Loa, is a gently sloping topographic feature. Nevertheless, with a summit elevation 4170 meters above sea level, it dominates the skyline. In the foreground, a lava lake occupies a breached crater pit, Kilauea Iki, a satellite of Kilauea volcano. This igneous complex, located in a mid-plate regime, is a reflection of the passage of the Pacific lithosphere over a fixed hot spot in the deep upper mantle (see Chap. 6, and Figs. 4.13 and 25.4). (*Source:* Photograph by W. G. Ernst.)

massive, broad structures with a shield-like outline and gently sloping flanks, such as Mauna Loa, Mauna Kea, and Kilauea in Hawaii (Fig. 5.12). These constructs are produced by the episodic extrusion of highly fluid, subsilicic lava from a central fissure system; the low viscosity promotes the formation of a broad, gently sloped – and in some cases, enormous – eruptive structure. In contrast, steep-sided stratovolcanoes are built up through eruption of more viscous lavas having compositions intermediate between mafic and felsic melts; violent outbursts of ash are common to this type of volcanism. Famous photogenic cones include Mount Fuji (Fig. 5.13); Mayon, Vesuvius, and Rainier are also of this type.

Deep-seated igneous processes are not observable at the surface and must be inferred from old, eroded terranes. Based on shape, three principal types of intrusive igneous bodies can be distinguished: (1) downward-expanding masses; (2) tabular sheets; and (3) cylindrical pipes. Where the contacts between an intrusion and the intruded wall rocks are parallel to layering in the latter, the igneous body is termed *concordant*. In contrast, the marginal contacts of a *discordant* intrusion truncate layering in the host rocks.

The first type of intrusive igneous bodies is downward-expanding masses. The cross sections of many *plutons* (intrusives of irregular or unknown shape) show an increase in width at greater depth. Igneous masses are termed *batholiths* if their present outcrop areas exceed approximately 300 square kilometers, *stocks* if they are less extensive. The great majority of such bodies consist dominantly of intermediate and silicic rock types. Felsic batholiths are exposed in the cores of many eroded mountain systems around the world. A diagrammatic cross section of a long-since-solidified and partly eroded plutonic complex is illustrated in Figure 5.14. Roof pendants and inclusions are preexisting wall rocks partly or completely surrounded by the pluton. The upper crustal portions of many batholiths enlarge downward; however, they must terminate within the lower crust, because batholiths do not occur in the mantle. Hence, their overall cross sections – never well-exposed because of their enormous size – are probably teardrop- or mushroom-shaped.

The second type of intrusive body is tabular sheets. Tabular intrusions are of two distinct varieties, discordant *dikes* and concordant *sills*. Dikes are intruded along fissures cutting

Figure 5.13. The beautifully symmetric cone of Mount Fuji, seen from the southwest; elevation of the summit is 3778 meters above sea level. Similar to other northeastern Japanese volcanoes, Fuji is sited over the Pacific lithospheric plate. The globe's largesty lithospheric plate, it bends down and descends to the northwest beneath Honshu. (*Source:* Photograph by W. G. Ernst.)

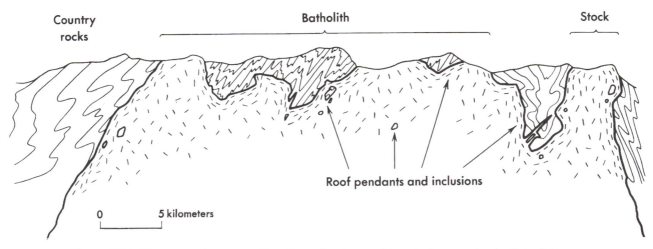

Figure 5.14. Diagrammatic cross section through a large subjacent, downward-enlarging pluton such as the Sierra Nevada batholith. Roof pendants are large masses of invaded wall rocks (country rocks) that hang down into the igneous body; inclusions are smaller fragments completely engulfed by the magma. km = kilometers. (*Source:* W. G. Ernst, *Earth materials*, Prentice-Hall, 1969.)

across the preexisting layering of the host rocks. Dike walls are nearly parallel, and their lengths far exceed their widths. Some dikes are very thin, having widths on the order of 1 centimeter; others are tens to hundreds of meters thick. Sills are intruded along fractures coincident with the preexisting layering of the *country rock* (the surrounding units into which the magma is intruded). As with dikes, examples of all principal magma types occur in sills; also similar to dikes, sills have lateral dimensions that far exceed their thickness.

The third type of intrusive body includes relatively small, discordant plutons only. The chief representatives of this group are cylindrical intrusions, plugs, or pipes, commonly less than 1 kilometer across. Many are probably the filled conduits of eroded volcanoes, or volcanic necks. As such, they may consist of any rock that occurs as lava. Volcanic necks commonly are associated with dikes that radiate from the central throat, as illustrated in Figure 5.15. A distinctly different type of pluton is represented by cross-cutting, roughly teardrop-shaped, buoyant masses that have worked their way upward in a largely solid or plastic condition. To this category belong a variety of ultramafic rocks representing fluidized mantle materials. Some of these bodies have risen from great depths, as indicated by the presence of trace amounts of diamond, a mineral that forms only under conditions of extreme pressure.

Intrusive rocks are coarser-grained than chemically equivalent lavas. Intergradations exist, but average grain sizes of the former generally exceed 4 millimeters, whereas the latter consist of interlocking crystals commonly smaller than a millimeter in length. The ultimate size of a specific mineral grain depends on the amount of material that reached its growing surfaces and was deposited during crystallization. This is a function of the concentration of the various components (e.g., silicon, magnesium, aluminum, oxygen, and sodium) in the molten solution, the ease with which fresh melt circulated to the environment of the crystal, the diffusion velocities of components through the magma, and the rate of cooling. High temperature, a low cooling rate, and the presence of volatile constituents that increase fluidity, circulation, and diffusion all promote large grain size.

Although lavas are extruded at temperatures approaching or exceeding 1000°C, depressurization during ascent causes dissolved volatiles to be given off abruptly and the melt cools rapidly; consequently most lavas are quite fine-grained. In contrast, deep-seated, intrusive magmas cool very slowly (on the order of a few tens of degrees Celsius per million years for large masses), and, through retention of volatiles, they

Figure 5.15. An eroded volcanic neck, exposed near Shiprock, Navajo Nation, New Mexico. Radial, nearly vertical dikes emanate from the volcanic throat. (*Source:* Courtesy of the American Museum of Natural History.)

remain fluid to lower temperatures. The cooling episode is much longer, accounting for the larger average grain size of plutonic rocks compared to volcanics. Especially among silicic intrusions, the last melt to freeze may be highly charged with whatever volatiles are retained. This pressurized but tenuous aqueous fluid crystallizes to an extremely coarse-grained quartz and alkali feldspar-rich aggregate known as pegmatite. Many gem minerals are quarried from such coarse-grained veins fillings.

CLASSIFICATION AND CHEMICAL DIFFERENTIATION OF IGNEOUS ROCKS. Igneous rocks are distinguished based on contrasting chemical compositions. Because the bulk-rock chemistry is reflected in the nature and proportions of the minerals present, we will utilize a simple classification based on the proportions of quartz, feldspars, and dark, ferromagnesian minerals, and on the composition of the plagioclase. The rock types are defined briefly in Table 5.1. Although percentages of minerals are listed, a nearly continuous spectrum of compositions is reflected in the igneous rock series, and the values presented have wide, somewhat overlapping ranges in proportions of constituents. The amount of dark minerals varies directly with the calcium content of the plagioclase, and inversely with the quartz content; accordingly the percentage of mafic minerals, which can be estimated from the overall appearance of a rock, is quite helpful in the identification of hand specimens.

The most refractory minerals (i.e., those stable at the highest magmatic temperatures) form first, and as the melt cools, progressively more fusible minerals crystallize, in some cases replacing or armoring earlier ones. This crystallization takes place *discontinuously*, as in the ferromagnesian minerals, in which olivine at high temperatures gives way first to production of pyroxene(s), then hornblende, next biotite, and finally, muscovite at relatively low igneous temperatures. Simultaneously, formation of another group of minerals proceeds *continuously*, as in the feldspars, where calcium-rich plagioclase is rimmed or replaced by intermediate, then sodic plagioclase, and finally sodium- and potassium-rich feldspar plus quartz. This *reaction series* was first elucidated by the Canadian geologist Norman L. Bowen. Relationships are portrayed schematically in Figure 5.16.

Bulk compositions reflect differences in mineral proportions among the various rock types. Compared to mafic intru-

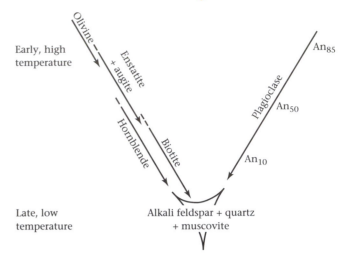

Figure 5.16. The reaction series of mineral crystallization from a cooling, chemically differentiating magma. The discontinuous sequence of minerals is illustrated on the left, the continuous sequence on the right. See text for explanation. An = anorthite. (*Source*: W. G. Ernst, *Earth materials*, Prentice-Hall, 1969.)

sives, felsic plutons possess greater amounts of silica and alkalis, and lesser amounts of lime, iron, and magnesia, reflecting the abundance of quartz, sodic, and potassic feldspar and the scarcity of ferromagnesian, calcium-rich minerals. Chemical analyses of lavas reveal compositional ranges corresponding to those of intrusive rocks. How do differences in bulk chemistries come about? This phenomenon is a consequence of the unequal distribution of elements between melt and minerals, followed by the partial or complete separation of liquid from the more refractory solid assemblage (Fig. 5.16). This can happen during heating and partial fusion of a preexisting rock. Once a melt is generated, it may leave behind the relatively dense, refractory crystalline residuum and, being buoyant, ascend into the upper crust, where it will lose heat. On cooling, solid accumulations of new, early-formed crystals tend to be initially enriched in magnesium, iron, and calcium, whereas the late, low-temperature magmas are charged with the more fusible components – alkalis, silica, and volatiles. The solid and the liquid can be separated at this stage as well. The process is known as *igneous differentiation* and explains the contrasts in compositions of melts and solids produced during either partial fusion and/or partial freezing.

TABLE 5.1. MINERALOGIC CHARACTERISTICS OF MAJOR IGNEOUS ROCK TYPES

| Rock Name | | Mafic Minerals | Approximate Percentages of Minerals | | | | Composition of Plagioclase |
Intrusive	Extrusive		Mafics	Quartz	K-feldspar	Plagioclase	
Granite	Rhyolite	Biotite ± hornblende	10	35	25	30	Sodic
Granodiorite	Dacite	Hornblende ± biotite ± augite	25	20	15	40	Intermediate
Diorite	Andesite	Hornblende ± enstatite ± augite	35	5	10	50	Intermediate
Gabbro[a]	Basalt	Augite ± enstatite ± olivine	45	0	5	50	Calcic
Peridotite	Komatiite	Olivine ± enstatite ± augite	80	0	0	20	Calcic

[a] Shallow intrusive dikes and sills are termed *diabase*.

OCCURRENCE OF SEDIMENTARY ROCKS. Estimates of crustal volumes for the major rock types include on the order of 5 to 10 percent sedimentary strata, which occur as a nearly ubiquitous, relatively thin surficial covering of both continents and ocean floors. The thicknesses of great sedimentary basin deposits rarely exceed 15 kilometers. Sedimentary units are of great economic importance, however, for in them is sequestered much of the world's readily extractable mineral wealth, including coal, petroleum, natural gas, nuclear fuels, aluminum, iron and manganese ores, and building materials such as stone, sand, gravel, marl, and limestone. Indeed, many metal deposits of, for example, copper, zinc, and lead, which appear to be causally related to the intrusion of magmas, are localized at the contacts between plutons and chemically dissimilar, metamorphosed sedimentary rocks.

Sediments are accumulations of materials that have been reworked by surficial processes from preexisting rocks of any origin. Geologists broadly distinguish between *clastic sediments,* or detritus, the products of mechanical accumulation of individual grains; and *chemical sediments*, materials that have precipitated from inorganic solutions. Bioclastic and biochemical deposits are analogous sediments produced largely through biologic interactions (e.g., shell beds and coral reefs). Of course, most clastic rocks contain some chemically precipitated material, and most chemical sediments carry clastic fragments. All sedimentary rocks result from the following processes: (1) weathering of source materials; (2) transportation, commonly in running water; (3) deposition by organic or inorganic means; and (4) *lithification*, that is, compaction and cementation whereby the original unconsolidated sediment is converted into a more cohesive, less porous rock.

The nature of a sediment depends in part on the source area from which it was derived, and on the attendant weathering. Weathering occurs by altering the surface of the rock; it involves mechanical and chemical breakdown of the original Earth materials, usually by interaction with the atmosphere or with water. When rocks are weathered, they commonly form a soil profile in which three intergradational horizons or layers can be recognized (see Chap. 8). The depth to which a soil profile develops is a function of the rates of chemical weathering and mechanical breakdown, versus the removal of the soil in solution and/or as particulate matter. High topographic relief favors rapid mechanical disaggregation and erosion, whereas low relief inhibits it. Chemical reactions are accelerated by high temperatures, and most require the presence of water; consequently, elevated temperatures and high rainfall characteristic of the tropics promote chemical weathering. Such climates promote luxuriant plant and animal life, and the rapid decay of organic matter. Cold, rigorous climates inhibit chemical activity because of low temperatures, lack of mobile aqueous solutions, and an impoverished flora and fauna; however, frost action and glaciation are effective erosive agents, through fracturing and through bodily movement during freezing and thawing of water and ice.

The nature of a sediment depends chiefly on the mode and length of transportation from the source area to the depositional environment. Mass movement is a response to the force of gravity, and includes the accumulation of *talus*, angular fragments of rock derived from disintegrating cliffs and steep slopes. Mud flows are another variety of mass movement but involve much finer-sized materials, including large amounts of clays. Avalanches and landslides have properties intermediate to these two extremes. Particles entrained in a transporting medium such as air, water, or ice are abraded as they rub against one another and are carried over the underlying material; accordingly, grain sizes and angularities are reduced as a function of the distance traveled. The dissolved products of chemical attack are transported from the parent rocks to a site of deposition exclusively in water.

The manner of deposition strongly influences the nature of the resultant sediment. Rapid accumulation of ill-sorted clastic debris of several size ranges generally results in a weakly laminated or massive deposit – for instance, coarse sands; on the other hand, slow buildup of fine-grained particulate matter, or of chemical precipitation alternating with the settling of fine clastic debris, tends to produce fine-bedded sediments such as laminated muds. The amount and rate of deposition from the transporting medium is a function of the flow velocity (the slower the flow, the greater the likelihood of the load being deposited) and the nature of the transported sediment (if particles are large they will be more easily deposited) – both, in turn, being related to topography, climate, vegetation, and the mineralogic and chemical constitution of the source, as discussed above.

Two principal types of sediment are distinguished, based on the site of deposition: continental (or terrestrial), and marine. Terrestrial sediments include the products of mass movements, stream gravels, lake beds, sand dunes, and glacial deposits. All such sediments, with the possible exception of some lake beds, are composed primarily of clastic units that grade laterally and vertically rather abruptly to other sedimentary rock types, reflecting rapidly fluctuating depositional conditions on land. Marine sediments, on the other hand, tend to be more widespread and continuous. (Why is this?) To this group belong shallow-water deposits of deltas, beaches, continental shelves, and slopes, as well as deep-sea sediments of the ocean basins. Both chemical precipitates and clastic materials are abundant in such sediments.

The process of *lithification*, whereby an unconsolidated sediment is converted into a coherent rock, involves several distinct but intergradational phenomena. Compaction of a sediment due to the weight of the overlying rocks decreases pore space and results in deformation of soft mineral grains, clast rotation, flattening, and squeezing out of water from the pore spaces, and also produces interlocking grain boundaries. Iron oxide, silica, and/or calcium carbonate minerals precipitate from pore solutions and cement the particles together. Reaction among the original, chemically dissimilar fragments produces new minerals that transect former grain boundaries and increase rock coherence; this recrystallization is known as *diagenesis*. With continued burial and consequent increase in temperature and pressure, diagenetic transformation gives way imperceptibly to metamorphic recrystallization. This latter process takes place generally at greater depths in the Earth than does diagenesis, although the two phenomena are completely intergradational.

CLASSIFICATION AND CHEMICAL DIFFERENTIATION OF SEDIMENTARY ROCKS.

Sediments are described on the basis of their grain sizes, chemical and mineralogic compositions, and modes of accumulation. Classification by grain size is most readily applied to clastic sediments because particle size is related to origin of the rock. Variations in the bulk compositions of clastic sediments are also related to grain size, reflecting contrasts in chemical and mechanical stabilities of different minerals. In chemically precipitated deposits, grain size is less useful because postdepositional recrystallization commonly causes a coarsening of the particles; grain sizes are not readily correlated, therefore, with composition for these sediments.

Among the clastic rocks, we recognize four main grainsize groups: (1) Conglomerates contain particles greater than 2 millimeters in diameter. (2) The most abundant grains of sandstones fall in the 1/16- to 2-millimeter size range. (3) Siltstones contain grains between 1/256 and 1/16 millimeter in diameter. (4) Claystones carry particles finer than 1/256 millimeter in diameter. Where planar fractures – or parting – along bedding surfaces is developed, claystones and fine-grained siltstones collectively are termed *shale*. If fractures parallel to bedding layering are not present in a deposit containing both clay and silt, the somewhat massive rock is termed *mudstone*. Where, as often happens, a sediment contains grains of several size ranges, the deposit is named on the basis of the most abundant particle size.

Limestones produced principally through the accumulation of calcareous fossil materials include coquina (shell debris), chalk (primarily exoskeletons of one-celled marine organisms), and reef limestones. Except for chalk, most fossil fragments display moderately coarse particle sizes. In contrast, the inorganically deposited (chemically precipitated) carbonates and siliceous sedimentary rocks such as lithographic limestone ($CaCO_3$), chert (SiO_2), and banded iron formation ($Fe_2O_3 + SiO_2$) are typically very fine-grained, and names are based on rock composition.

As is clear from the description presented earlier in this chapter, chemical variations among the igneous rocks are relatively modest, because unequal partitioning of elements between coexisting melt and crystals is not pronounced at elevated temperatures. Therefore, igneous differentiation produces a relatively narrow compositional spectrum of related magmatic rocks. In contrast to these primary rock types, sediments exhibit a strikingly broad range of bulk compositions, reflecting the fact that widely varying mechanical and chemical properties of the constituent minerals from a source rock, as well as differences in climate, erosion, and transportation history, markedly influence the composition of the final sedimentary deposit. The low-temperature process whereby source rocks are converted to secondary accumulations of contrasting compositions is termed *sedimentary differentiation*.

Weathering of preexisting rocks at or near the Earth's surface results in the solution of readily dissolved constituents and in the hydration and oxidation of others (Chap. 8). Materials transported as clastic grains are separated on the basis of mechanical properties; relatively soft minerals possessing good cleavages (fracturing along crystal-structural planes) are more rapidly and more finely abraded than hard mineral fragments lacking fractures and perfect cleavage. Therefore, grain size and mineralogic composition are related in a sediment. Sand grains consist predominantly of quartz, whereas clay-sized particles are primarily aluminous sheet silicates; as a result, the size sorting that occurs during transport results in a distinct chemical separation.

The process of deposition accentuates this chemical differentiation: coarser-grained feldspar-rich, and especially quartz-rich, gravels and sands are laid down in agitated, shallow water nearer to the source than the more finely divided silt and clay particles, which are carried successively farther from the parent rocks and ultimately settle out in quiet-water environments. Furthermore, near-shore reworking by wave and current action tends to break down all but the most stable grains, both chemically and mechanically; as a result, residual quartz is enriched in such deposits because of its superior hardness, lack of cleavage, and resistance to solution. Chemically and biochemically precipitated materials are diluted by the rapid accumulation of coarser-grained clastics in the near-shore regime; however, as the particle sizes of fragments diminish in quieter waters, generally at greater distances from the source, clastic sedimentation rates decline, and chemically precipitated materials become volumetrically more important. After deposition, the processes of dissolution and precipitation by aqueous fluid passing through the sedimentary layers may further change the rock bulk composition by dissolving unstable particles and precipitating an interstitial cement. *Multicycle* sediments are those that have been deposited, then eroded, transported, and redeposited – in some cases, several times. As a consequence, such mature sediments exhibit pronounced chemical differences from the source rocks. First-cycle, immature sediments, in contrast, have bulk compositions not greatly disparate from the parental terranes. Why should this be so?

As indicated in Figure 5.17, the average compositions of igneous source rocks of the continental crust and of typical sedimentary lithologies, sandstone, shale, and limestone, are markedly different. Sand-sized particles are enriched in quartz and feldspars, which accounts for the high silica contents of sandstones. In contrast, shales carry abundant clay minerals that are relatively low in silica but high in alumina, alkalis (especially potash), and H_2O. Limestones consist principally of calcite, $CaCO_3$ and dolomite, $CaMg(CO_3)_2$. However, many clastic rocks carry moderate amounts of carbonate, and most limestones contain minor quantities of quartz

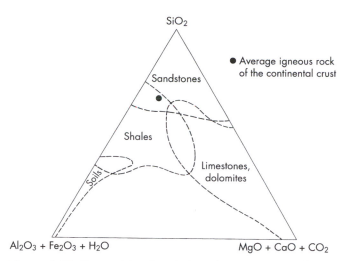

Figure 5.17. Compositional variations (as proportions of oxides) of some sedimentary rocks and residual soils. Sandstones are rich in quartz; limestones plus dolomites contain large proportions of carbonates; and shales plus soils carry abundant clay minerals and ferric oxides. (*Source:* W. G. Ernst, *Earth materials*, Prentice-Hall, 1969.)

and clay minerals. The average bulk composition of continental sedimentary rock, which is approximately 60 to 80 percent shale, 15 to 20 percent sandstone and conglomerate, and approximately 10 to 15 percent carbonate, corresponds relatively well to the average igneous rock of the continental crust, except for volatile constituents. These latter components, especially H_2O and CO_2, undoubtedly were initially present in the primary magmas; they were given off on crystallization and are thus concentrated in the atmosphere, in the seas, and in secondary rocks. The aggregate compositions of the sedimentary veneer and of the atmosphere plus hydrosphere seem to have evolved with time. The early Earth's atmosphere appears to have been strongly reducing (see Chap. 3) but gradually became more oxidizing due to biologic activity, which released oxygen through photosynthesis.

OCCURRENCE AND CLASSIFICATION OF METAMORPHIC ROCKS. Recrystallized and/or deformed preexisting rock units constitute approximately 15 percent of the continental crust. To a greater or lesser extent, the oceanic crust should also be considered as at least incipiently metamorphosed because it has been altered by reaction with heated seawater near spreading centers (see Chap. 6). Metamorphic rocks are products of pressure-temperature conditions intermediate between those of igneous and sedimentary environments. Thus, on the one hand, metamorphic processes merge with diagenesis, a sedimentary phenomenon; on the other hand, very intense crustal metamorphism leads to partial fusion of deeply buried continental and oceanic crust and in the generation of granitic and andesitic magmas.

Mechanical properties are of major significance under conditions of low-temperature metamorphism where chemical reactions take place slowly. In contrast, the conditions of medium- and high-temperature metamorphism tend to be characterized by more rapid recrystallization due to accelerated rates of chemical reactions. Two principal metamorphic processes may be distinguished: mechanical deformation and chemical recrystallization. The former process includes grinding, crushing, and ductile deformation of an initial rock, phenomena that reflect readjustment of the material, or *strain*, as a consequence of differential pressures, or *stress*. Recrystallization takes place because preexisting mineral assemblages are destabilized by changes in the temperature, pressure, or chemical milieu. Nearly all metamorphic rocks show the combined influence of both mechanical deformation and chemical reaction; they differ principally in the degree of development of these effects. Metamorphic rocks exhibit contrasts reflecting variations in bulk composition as well, due to differences in original rock chemistry or to alteration accompanying recrystallization.

Planar features are observed in specimens of intensely strained metamorphic rocks. Fine-grained metashales (slates) break along planar surfaces, termed *cleavage planes*, reflecting alignment of mineral grains. Where compositional layering is evident, such parting is called *foliation*. The orientation of flaky minerals is illustrated diagrammatically in Figure 5.18A. Highly aluminous rocks, such as slates and schists, carry large amounts of mica and chlorite and display well-developed foliation – planar layering – as a consequence of deformation. Prismatic and needle-shaped chain silicates, such as crystals

Figure 5.18. Sketches of rocks showing (A) platy mineral orientation or foliation, and (B) prismatic mineral orientation or lineation. (C) Photograph of small-scale folds in granitic gneiss. (*Source:* Photograph by W. G. Ernst.)

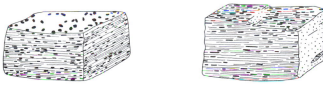

(A)   **(B)**

of amphibole, commonly are aligned to produce a rock lineation (Fig. 5.18B). If foliation is also developed, the lineation direction customarily lies within the plane of foliation. Silica-poor igneous rocks such as basalts have bulk compositions appropriate for the crystallization of large amounts of amphibole; hence, lineation is most marked in the metamorphic equivalents of these rock types. In the more intensely recrystallized, foliated rocks, an increase in grain size enables us to recognize individual mineral flakes and a faint tendency for compositional differences to develop in adjacent layers. Such rocks are termed *schists*, and the foliation is called *schistosity*. The coarsest-grained metamorphic rocks, gneisses, display distinct mineralogic and compositional banding. Their foliation is termed *gneissosity*. A contorted gneiss is illustrated in Figure 5.18C.

Mechanical deformation is called *cataclasis*, and metamorphic rocks in which reduction to small particle size is conspicuous are known as *cataclastic rocks*. Because recrystallization and chemical interaction are not pronounced, planar and linear features in cataclastic rocks must have formed principally through the milling down, shearing, and dislocation of preexisting materials. Surviving original grains show the effects of plastic strain and severe granulation. Mylonites are thoroughly sheared and milled-down lithologies with grain sizes in the range of approximately 0.01 to 0.1 millimeter, in general, foliations and lineations are well-developed in mylonitic rocks. *Fault gouge* is extremely fine clayey cataclastic material lining a zone of differential shear, whereas *fault breccia* consists of a mixture of coarse, angular blocky rock fragments as well as milled-down materials.

In contrast to cataclastic rocks, which are produced by dominantly mechanical deformation, *contact metamorphic rocks* form by a pronounced increase in temperature in the virtual absence of differential stress. Contact metamorphic rocks are localized as concentric zones surrounding hot igneous bodies emplaced at upper levels of the crust. In such environments, the difference in temperature between the wall rocks and an intrusive magma is substantial: shallow plutons are emplaced at temperatures approaching 800 to 1000°C, whereas the wall rocks a few kilometers below the surface are relatively cool, having initial temperatures prior to pluton emplacement on the order of 100 to 200°C. Therefore, a marked thermal gradient exists adjacent to the intrusive contact. Deeper in the crust, the temperature contrast is much less pronounced. H_2O-bearing silicic magma may be generated near the base of the continental crust at temperatures approaching 600–650°C; in the presence of a normal crustal geothermal gradient, deep-seated country rocks also are relatively warm, with temperatures exceeding 500°C at 30–40 km depth. Accordingly, deep levels of the continental crust lack strong horizontal and vertical thermal gradients, and are instead characterized by broad zones of similar recrystallization intensity.

Thus far we have contrasted the dominantly mechanical deformation of cataclastic rocks with the mainly chemical recrystallization of contact metamorphism. The most common varieties of metamorphic rocks, however, develop on a regional scale in response to both deformation and mineral reaction. Such widespread rocks are known collectively as *regional metamorphic rocks*, and are typified by oriented mineral fabrics. Foliations and lineations are widespread features reflecting differential motions (shearing) of the rock sections undergoing recrystallization. Similar to contact metamorphic rocks, regional metamorphics show progressive zonal sequences ranging from lower temperature, fine-grained, volatile-rich mineral assemblages to higher temperature, coarse-grained, volatile-poor mineral associations. In contrast to contact metamorphism, however, the zones developed under regional metamorphic conditions are quite broad, indicating gradual lateral and vertical changes in the physical conditions. The roots of mountain belts and old, stabilized continental nucleii are characterized by regional metamorphic rocks.

THE ROCK CYCLE. The erosion cycle discussed earlier in the chapter represents part of an even more extensive reworking of Earth materials, the *rock cycle*. The rock cycle relates the history of formation of various units to one another, to their immediate and ultimate sources, and to the continuing compositional differentiation and growth of the crust. Relations are presented schematically in Figure 5.19. As mentioned, this chemical differentiation involves crystal-melt fractionation (including partial fusion) of both uppermost mantle and lowermost crust, weathering, differential solution, mechanical and chemical transport, sedimentation, and metamorphic fluid-rock chemical exchange. Theories have been advanced to account for the primary differentiation of what was presumably an initially homogeneous planet into the present layered Earth – dense, metallic iron and nickel core, magnesium- and iron-rich silicate mantle, and low-density, alkalic, siliceous crust – surmounted by the excess volatile constituents concentrated in the hydrosphere and atmosphere Most scenarios propose an early stage of widespread heating and partial melting of the Earth and sinking of the heavy constituents, rising of the fluids and buoyant magmas, followed by a much longer recrystallization and cooling period.

This latter stage represents the present condition of the Earth. Nevertheless, the generation of silicate melts at considerable, but variable, depths, followed by magma ascent, continues today. After the mantle formed, nearly continuous incipient fusion of rising asthenospheric plumes and sheets beneath spreading centers and hot spots (see Chap. 6) can account for the outpourings of basaltic lavas that constitute the oceanic crust. Such materials have been produced throughout geologic time; however (as discussed in Chapter 6), most of the oceanic crust has been returned to the deep Earth during the subduction process attending continued mantle convection.

Beneath the continents, somewhat more evolved, andesitic magmas are being generated as a consequence of partial

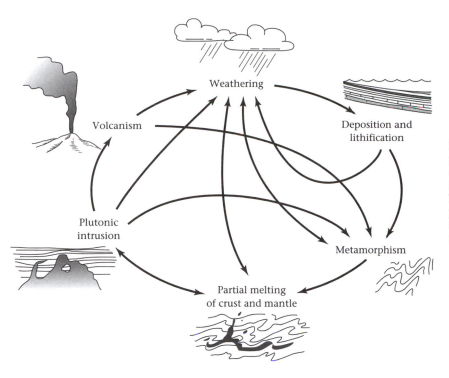

Figure 5.19. The rock cycle. Different paths and successions characterize the crustal histories of Earth materials from different geologic environments. The cycle has been repeated in variable sequence, depending on local geologic circumstances, since formation of the planet. (*Source:* W. G. Ernst, *Earth materials*, Prentice-Hall, 1969.)

melting along or above a descending lithospheric plate (Chap. 6). However they originate, primary intermediate and granitic melts more silicic and felsic than the parental mafic crust and ultramafic mantle material are being generated today. Furthermore, they, too, seem to have been produced episodically through the whole of recorded geologic time. In short, new material is being added to the continental crust semicontinuously.

Whether igneous rocks crystallize at considerable depths or on the surface, many eventually are uplifted, weathered, and eroded. The products of erosion are bodily transported and, after mixing with contrasting materials, deposited as sediments, then lithified during and after burial. In other environments characterized by the presence of aqueous fluids, preexisting igneous rocks may be metamorphosed or even partly remelted near the base of the continental crust. The products of these changes are metamorphic and secondary plutonic igneous rocks, respectively.

Both sedimentary and metamorphic rocks are derived ultimately from igneous parents – the primary sources of crustal material. Like igneous rocks, both sedimentary and metamorphic rocks may be subjected to the changes described above: (1) erosion and deposition as new sediments, (2) granulation, and recrystallization to produce metamorphic rocks, or (3) partial fusion (ultrametamorphism), giving rise to igneous rocks and residual, refractory metamorphics. This complex sequence of lithologic addition and reworking is repeated over and over in different orders as dictated by geologic circumstances. The process results in the increasing heterogeneity and compositional differentiation of the crust, as well as in a gradual increase in the total volume of continental mass due to the net addition of buoyant, low-melting material from the mantle. The entire cycle is intimately related to the deformational process – itself a consequence of buried heat and resultant mantle flow, a manifestation of the attempt by the planetary system to achieve gravitative equilibrium. Clearly, the rocks and minerals that constitute the dynamic Earth are constantly involved in the processes of breakdown and formation anew.

QUESTIONS

1. Describe the contrasting properties of the asthenosphere and lithosphere, and explain why they exist. List the observations that have been employed to constrain the physicochemical nature of these two Earth units.

2. Which represents a larger planetary mass, the continental crust or the oceanic crust? How did you arrive at your conclusion?

3. What do the chemical compositions and mass proportions of meteorites tell us about the internal constitution of the Earth?

4. What is the difference between a rock and a mineral? What is a crystal?

5. Why do you suppose it is that geologists refer to (some) igneous rocks as *primary*, whereas all metamorphic and all sedimentary rocks are termed *secondary*? What would secondary igneous rocks be, and how might you recognize their origin?

6. Is the depth to the Moho greater beneath the Himalayas or the Indian peninsula? Explain why this is so.

FURTHER READINGS

Bolt, B. A. 1993. *Earthquakes.* New York: W. H. Freeman.

Ernst, W. G. 1968. *Earth materials.* Englewood Cliffs, NJ: Prentice-Hall.

Ernst, W. G. 1990. *The dynamic planet.* New York: Columbia University Press.

Press, F., and Siever, R. 1994. *Understanding Earth.* New York: W. H. Freeman.

Tarbuck, E. J., and Lutgens, F. K. 1999. *Earth, an introduction to physical geology.* Englewood Cliffs, NJ: Prentice Hall.

6 Drifting Continents, Sea-Floor Spreading, and Plate Tectonics

W. G. ERNST

THE NOT-SO-SOLID EARTH

The uniqueness of the Earth's surface among the planets of our solar system reflects the complex interplay between external and internal forces. Differential vertical and horizontal motions – slow but sure – characterize the outermost skin of the planet. This remarkable mobility explains the growth and persistence of long-lived, high-standing continents and the relative youth of low-riding ocean basins, although the former are being planed down by erosion and the latter are being filled through sedimentary deposition. Such extremes of topography are unknown elsewhere in our solar system. The surface of the Earth is being constantly reworked, and a dynamic equilibrium has been established between the competing agents of crustal erosion and deposition (an external process) and of crustal construction (an internal process).

The dynamism of the planet, fueled by incoming solar radiation and the escape of buried heat, is responsible for oceanic/atmospheric circulation and, through mantle flow, the concentration of energy and mineral deposits in the crust, as well as all other substances necessary for life. Solar energy absorbance and transfer mechanisms are responsible for the terrestrial climate and its variations – as well as cyclonic storms and coastal flooding. Plate tectonic activity is the ultimate cause of many geologic hazards, including earthquakes and tsunamis (tidal waves), volcanic eruptions, and avalanches and landslides. More subtly, drifting continents interact with weather patterns, causing global climate change over time and influencing the course of biologic evolution. A wobbling planetary spin axis also contributes to climatic cycles. We need to understand as fully as possible the workings and feedback mechanisms of this not-so-solid Earth and its encircling organic and fluid envelopes in order to maximize the efficient utilization of its materials, to protect its resources for the sustenance of current and future life, and to shield ourselves from its natural destructiveness.

Over the past thirty years, Earth scientists have experienced a double-barreled revolution in the understanding of how our planet was born and how it has evolved to its present state. Scientific exploration of the solar system has illuminated planetary contrasts and has enhanced theories regarding how the inner, terrestrial bodies such as Earth, and the outer, gassy, icy giants such as Jupiter, formed. On Earth, the precepts of plate tectonics have lead to broad, realistic models for the interconnected origins of continents and ocean basins, hydrosphere and atmosphere, and mountain ranges and oceanic deeps. Previous chapters have dealt with the factual basis for our understanding of the evolution of the solar system (Chap. 4), and of the internal structure of the Earth (Chap. 5). This chapter describes advances in our understanding of the relatively thin but dynamic outermost rocky shell of the planet – the very substrate on which all terrestrial life depends.

Scientists have studied the on-land geology of planet Earth for nearly two centuries, and much is known concerning the diverse origins of the uppermost part of the continental crust, its structure, and constituent rocks and minerals. Until recently, however, the ocean basins were more poorly understood. Within the past three decades, the results of marine research have spectacularly elucidated the bathymetry, structure, and physicochemical nature of the oceanic crust; as a result, we have a far better appreciation of the manner in which various segments of the Earth's outer rind have been produced, have evolved with time, and have been altered or destroyed. A startling product of this work is the realization that portions of the mantle are slowly circulating, or convecting. Production of new oceanic crust along submarine ridge systems results from this plastic flow of the deep Earth (asthenosphere), as does addition to – and deformation of – the continental crust in the vicinity of seismically and volcanically active continental margins and island arcs. Oceanic ridges are sited over upwelling mantle columns known as plumes, whereas mantle currents are descending beneath active continental margins and island arcs. What propels this ductile asthenospheric flow is uncertain; however, mantle circulation on a majestic scale is clearly taking place in response to differences in density.

Both continental and oceanic crust form only the uppermost, surface layers of great, solid lithospheric plates; differential motions of these plates are a reflection of the convection of the underlying asthenosphere on which they rest. The continents of the eastern and western hemispheres are presently drifting apart from one another across the Atlantic, and have been doing so for more than 120 to 190 million years; locally, continental fragments came together in the past and others are presently on a collision course, especially around the Pacific Rim. The idea of *continental drift* was proposed seriously approximately eighty years ago by Alfred Wegener; however, the concept was almost universally rejected by scientists and was ignored for nearly half a century. Then, between 1963 and 1968, virtually all geologists and geophysicists quickly changed their minds about continental drift. How did such a profound reversal in thought take place so abruptly?

WEGENER'S CONTINENTAL DRIFT HYPOTHESIS

Alfred Wegener (1880–1930) was a German meteorologist and geographer who conducted much of his research in Greenland. He published and refined his hypothesis on the origin of continents and oceans repeatedly (1912, 1915, 1924, 1929). The basic idea was that the light, buoyant continents float on a denser but weaker substrate, the oceanic crust, much as icebergs float in the sea; similar to Arctic pack ice, the continents maintain structural coherence but episodically are fractured and drift apart and/or collide.

Wegener was impressed by the remarkable congruence of the eastern and western hemispheric coastlines across the Atlantic. A modern computer fit of these continents is illustrated in Figure 6.1. He reasoned that such a geometry could have resulted only from an irregular rupture and displacement – like pieces of a gigantic jigsaw puzzle – of portions of a once contiguous landmass. In hindsight, the geometric argument ought to have been sufficiently compelling; however, it was considered inconclusive at the time because the fit was imperfect. Unfortunately, Wegener chose the shorelines to represent the continental margins; he should have selected the outer limit of the continental shelves or midway down the continental slope, because these represent the actual edge of the continental crust, and they yield a much better fit.

Of course, the submerged extents of the continents were not accurately known in Wegener's time. The "pieces of the puzzle" can be restored closely but not exactly because continental shelves have variable widths. Moreover, the continents have been deformed subsequently by mountain-building processes; also, erosion and deposition over the millions of years following the initiation of drifting have

somewhat modified the edge geometries of the once contiguous continental fragments. Nevertheless, reexamine Figure 4.13, and you will be impressed with the remarkable correspondence of shape across the Atlantic. Note, too, that New Zealand and adjacent submerged plateaus fit readily into the east coast of Australia.

As shown in Figure 6.2, Wegener employed the geometric outline argument (and other evidence, discussed below) to reconstruct the entire continental assembly as it apparently existed 200 to 250 million years ago. This completely joined landmass he called *Pangea*. It consisted of two major supercontinental segments: (1) North America and Eurasia, or *Laurasia*; and (2) South America, Africa, India, Australia and Antarctica, or *Gondwana*. Some portions of this supercontinental assembly – such as Australia, Antarctica, and New Zealand, and Africa plus South America – seem to fit together readily, whereas ambiguity exists as to the exact point where other fragments of the continental crust join together.

The following lines of evidence, quite controversial at the time, led Wegener to his particular reconstruction. Fossil plants and animals found in 205 to 285-million-year-old strata exhibit striking similarities in southern South America, South Africa, peninsular India, and parts of both Antarctica and Australia plus New Zealand. Geographic relations are illustrated in Figure 6.3. The *Glossopteris* flora is confined to the more southerly (in present coordinates) parts of the Gondwana supercontinent, where it constitutes a distinctive fossil assemblage. Such fossil plants are totally lacking from equivalent ecologic environments represented by deposits of the same age in the Laurasian supercontinent, because they could not propagate, flourish, and be distributed through equatorial paleoclimatic zones.

Likewise, the distinctive *Therapsida* reptiles, which exhibit skeletal features transitional toward primitive mammals, are similarly confined to southern portions of Gondwana. Several distinctive fossil genera have been described: *Cynognathus* has been found in southern Brazil and in Zaire; *Mesosaurus* remains have been recovered from central Argentina and from South Africa; and *Lystrosaurus* fossils have been identified from eastern Africa, Madagascar, peninsular India, and the continental margin of Antarctica. Assuming that the continents and oceans were fixed in their current positions, early paleontologists evoked the hypothesis of land bridges – dry-land corridors, or chains of islands, now vanished beneath the sea – to account for the far-flung occurrences of these unique plants and animals. The necessities of explaining how the later sinking of such buoyant continental materials into the mantle transpired without leaving a trace, which flies in the face of the principle of isostasy (see Chap. 5), and why the biota failed to migrate northward into Eurasia and North America, were neatly avoided by invoking continental drift. While the controversy raged, however, paleontologists disputed the extent of faunal and floral similarity, and questioned the breakup of Pangea as a viable explanation for the wide geographic dispersal of these distinctive plants and animals.

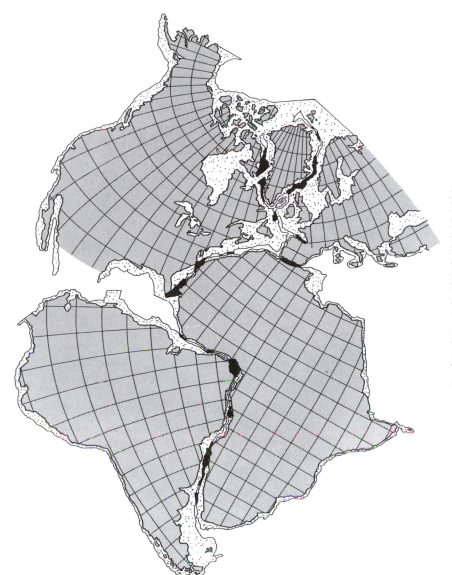

Figure 6.1. The congruence of continental landmasses across the Atlantic Ocean (employing shelf edges, not shorelines). The submerged edges of the continents are shown by a stippled pattern; gaps in the fit are denoted in white, the overlaps (primarily postrifting sedimentary deposits) in black. Present-day longitudes and latitudes are superimposed on the above-water portions of the continental crust for reference. (*Source:* Adapted from E. C. Bullard, J. E. Everett and A. G. Smith, The fit of the continents around the Atlantic, Philosophical Transactions of the Royal Society of London, 1965.)

Figure 6.2. Wegener's original continental assembly, Pangea, appropriate for the end of the Paleozoic Era, approximately 250 to 200 million years ago. Laurasia is the assembly of North America plus Eurasia; Gondwana makes up the remainder of the landmasses. (*Source*: Adapted from R. S. Dietz and J. C. Holden, Reconstruction of Pangea: breakup and division of continents, Permian to present, *Journal of Geophysical Research*, vol. 75, 1970.)

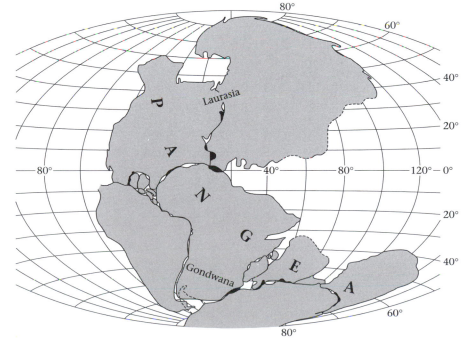

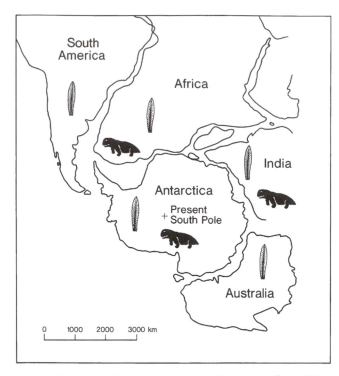

Figure 6.3. Fossil *Therapsida* fauna and *Glossopteris* flora, 200 to 250 million years old, unique to the Gondwana supercontinent (schematic). Illustrated mammal-like reptiles, now extinct, were thought to make good pets. They would have been very poor swimmers (look at those stumpy, little legs!); consequently, they could hardly have reached India by water, were it then in its present location. km = kilometers.

Bedded sedimentary rocks that are approximately 200 to 300 million years old contain evidence of climates at the times of deposition (paleoclimates) that seem strangely out of place with regard to their modern locations. For example, bouldery gravels and clays indicative of continental glaciation, such as those that characterize polar and circumpolar regions, occur in present-day temperate and tropical latitudes in South Africa, Patagonia, central Australia, and peninsular India. Even more surprising are the equally old occurrences of thick coal beds and sedimentary layers of salt, currently situated in high latitudes reported from Antarctica, from Spitzbergen, Germany, and from Alaska. Such accumulations are forming today principally at near-equatorial latitudes in excessively moist (rain forest) and arid (desert) climates, respectively. Although the drifting of continents across the globe readily accounts for the geographically linked changes in paleoclimatic zonation, many of the individual occurrences were disputed by the antidrift lobby as due to local, topographically induced climatic anomalies, whereas other deposits were said to be incorrectly described and interpreted.

Yet another supportive observation advanced by Wegener concerned the fact that certain geologic belts are truncated at the margins of Africa but are matched by similar geologic belts in South America, Australia, and India. Some of these abruptly terminated geologic provinces are illustrated sche-

matically in Figure 6.4. The trends of these belts are continuous when the scattered parts of Gondwana are restored to their predrift configurations. Nonbelievers countered that such correlations were highly interpretative, and that analogous trends across the North Atlantic were even more speculative. They faced a serious problem, however, in the abrupt termination of geologic features at some continental margins, and were obliged to erect the hypothesis of now-vanished borderlands, which, like the problematic land bridges, were obliged to disappear mysteriously beneath the seas, isostacy notwithstanding.

Compressional mountain belts contain rock units that show evidence of great crustal contraction and appear to have been folded and contorted much like the rumpling of a rug. Wegener argued that such structures form at the leading edges of continental masses that are "plowing through" the ocean basins. Examples include the Andes, Aleutians, Kamchatka Peninsula, Indonesia, and the Japanese Islands; the great Himalayas and the Tibetan Plateau were regarded as a product of the collision of the Indian Peninsula with the southern margin of Asia. The unconvinced responded that if the continents were strong, integral structures moving through the weaker oceanic crust, the latter, not the former, should have been deformed – and this is clearly not the case. Moreover, Wegener's model failed to explain the locations of great mountain chains within the interiors of continents, including the Rockies, the Appalachians, and the Urals.

The crucial argument, however, that delayed acceptance of the concept of continental drift involved the lack of an adequate mechanism to explain the phenomenon. Wegener believed that motions of the continents subsequent to the breakup of Pangea were largely toward the equator as a consequence of the rotation-generated oblateness (polar flattening) of the Earth; the postulated westward component was assumed to be a response to tidal forces. Laboratory deformation of rocks at rapid strain rates, however, demonstrated the apparent strength of both oceanic and continental crust;[*] for this reason, geophysicists argued that the outer portions of the Earth were much too strong to allow motions of the type envisioned by Wegener. Furthermore, the onset of rifting and drift allegedly began approximately 190 million years ago in the proto – North Atlantic, while separation of South America and Africa took place approximately 120 million years ago – both at times of near worldwide quiescence, not of widespread mountain building and crustal unrest. The geologic unlikelihood as well as the geophysical impossibility of Wegener's concept were thus emphasized by the antidrifters.

Yet now we know that continental drift is a reality. What had been missed? This insightful concept was rejected chiefly because scientists could not explain how Wegener-style drift

---

[*] More recent experiments have demonstrated that rocks behave weakly under conditions of long-term stress buildup, hence deform plastically at slow strain rates; however, this was not known at the time of the continental drift controversy.

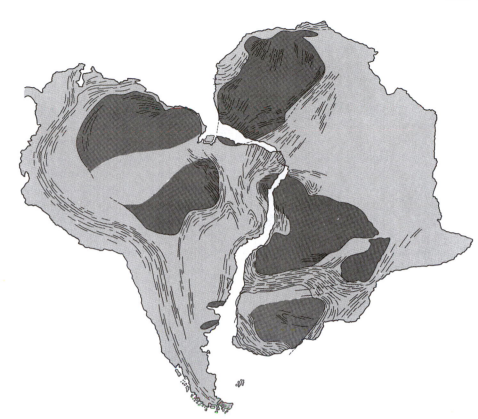

Figure 6.4. Geologic belts truncated at the present margins of the South Atlantic Ocean. Dark areas are early Precambrian shields (ancient continental nucleii); lineated patterns show the trends of late Precambrian and younger mountain belts. (*Source:* Adapted from P. M. Hurley, The confirmation of continental drift, *Scientific American*, 1968, vol. 218, pp. 52–64.)

could occur. Paradoxically, although the mechanisms of continental drift are still being debated today, we now agree that without a doubt, the continents are moving relative to one another.

SEA-FLOOR SPREADING AND PLATE TECTONICS

Thirty years after Wegener's death, new types of information became available, leading to a reconsideration of the continental drift hypothesis. At the close of World War II, classified bathymetric data were released; more important, scientists put out to sea with sophisticated new equipment and different kinds of data were gathered. These marine observations, employed in conjunction with the better-known on-land geology, have allowed a much fuller appreciation of processes at work on our planet. The new data are of several different types, including the amazing topography of the ocean floor, marine linear magnetic anomalies, belts of seismicity, and sediment patterns and ages of deposition in the ocean basins.

BATHYMETRY OF THE OCEAN BASINS. The sea floor lies, on the average, approximately 5 kilometers below the water surface. Although much of it is a flat, featureless plain, both oceanic deeps and submarine mountains transect the ocean bottom. *Oceanic deeps* (trenches) lie directly offshore from most island arcs and are especially abundant in the western Pacific. The most astonishing discovery has arisen from the charting of globe-encircling submarine *oceanic ridges* – or

rises. These are constructed predominantly of basalt and its deep-seated equivalent, gabbro. The Mid-Atlantic Ridge, illustrated in Figure 6.5, lies equidistant between Europe and Africa on the east, and the Americas on the west, and faithfully reproduces the Atlantic outlines of both continental groups. Knowledge of its existence surely would have supported Wegener's hypothesis, had it been discovered and accurately delineated during his lifetime. The East Pacific Rise – which seems to terminate within the Gulf of California but reappears for a short stretch off the coast of Oregon, Washington, and British Columbia – is in fact the northern extremity of another major ridge system that extends southwestward through the Pacific Ocean and westward between Antarctica and Australia, and then curves northwestward through the Indian Ocean to end in the Gulf of Aden–Red Sea rift complex. A splay trends southwestward around South Africa, separating it from Antarctica, and on eastward to the Mid-Atlantic Ridge (see Fig. 4.14). Yet another submarine mountain system, the northern extension of the Mid-Atlantic Ridge, transects the Arctic Ocean before terminating in eastern Siberia.

Other bathymetric features in the ocean basins include isolated volcanic islands, linear island chains, and sea mounts. The latter are islands that are now submerged to considerable depths; many are flat-topped. Except for a few subsea plateaus that may be fragmented blocks of relatively thin continental crust, all these features are compositionally similar to oceanic crust. Oceanic fracture zones, which manifest themselves as submarine escarpments, are particularly obvious and numerous as they tend to cut through and offset

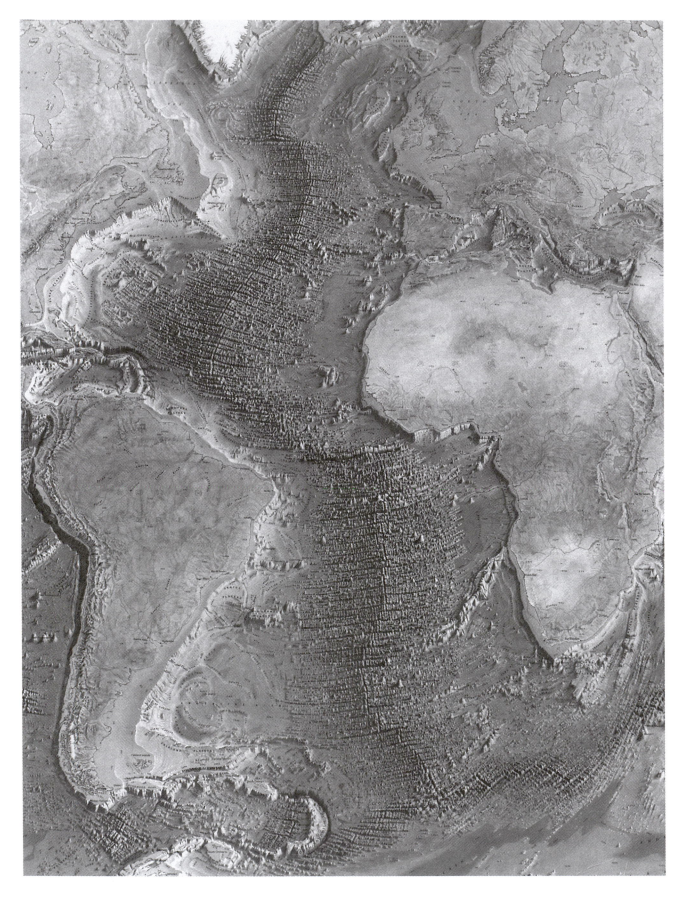

the ridges (see Fig. 6.5); however, such hydrographic features also transect island chains, and a few apparently terminate segments of island arc and trench complexes.

The ocean ridges and rises represent the near-surface expression of slowly, almost imperceptibly ascending mantle currents; upward velocities are on the order of a few centimeters per year. Whether this upwelling is due to part of a convection cell that returns asthenosphere to higher levels after it has been dragged to depths by the lithosphere sinking elsewhere, or is a consequence of deeply buried thermal anomalies that heat the asthenosphere, causing it to become buoyant and move upward, is not known – both processes are probably operative. Approaching the sea floor, the rising asthenosphere undergoes partial melting in response to the lowered pressure and produces modest amounts of basaltic liquid. The melt within the upwelling asthenosphere is less dense, and therefore more buoyant. Accordingly, it ascends toward the interface with seawater, where it cools and solidifies to form the oceanic crust, capping the stiffer and less buoyant, solid mantle. The potential void left behind is replaced by a rising complex of new basaltic melt and underlying, softened mantle material. Oceanic ridges are termed *spreading centers* because the cooling lithospheric slabs, or plates, that lie on top of the ductily flowing mantle currents also slowly move at right angles away from the axis; this lithospheric plate junction is known as a *divergent boundary*. A schematic view of its geometry is shown in Figure 6.6. This process is termed *sea-floor spreading*.

As it migrates away from the ridge axis and cools, the lithosphere gradually thickens with time at the expense of the upper part of the underlying asthenosphere. Heat is continuously lost from the upper mantle, so incipient melting is confined to greater and greater depths. The lithosphere–asthenosphere boundary (solid, rigid mantle above; incipiently molten, ductile mantle below), which is very close to the sea bottom beneath the oceanic rise, therefore descends to greater depths away from the spreading axis (Fig. 6.6). As the lithosphere cools and thickens with the passage of time, the overall density of the material increases; accordingly, due to isostatic compensation it sinks into the asthenosphere slightly, so that it lies farther below sea level than near the ridge axis. Unlike low-density continental masses floating on a denser substrate, the oceanic lithosphere has a higher density than the underlying asthenosphere. The situation involving cooler ocean crust–capped lithosphere on top of hot-

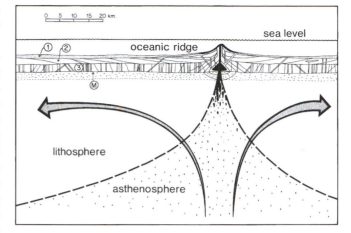

Figure 6.6. Diagrammatic cross section of a mid-ocean ridge or divergent plate boundary. Curvilinear mantle flow lines (arrows) lie in the plane of the paper. Basaltic magma is illustrated as black blobs. Layers 1, 2, and 3 are deep-sea sediments, basaltic flows, and gabbroic and diabasic intrusives, respectively (see Chap. 5). The base of the lithosphere, indicated by heavy dashed lines, is the region where the more rigid mantle above becomes more plastic downward (i.e., the asthenosphere), probably as a consequence of the presence in the asthenosphere of grain-boundary melt at higher temperatures. km = kilometers.

ter asthenosphere is therefore gravitatively unstable: The lithosphere will sink (resulting in overturn) where geometrically possible.

As lithosphere is being created, it must also be destroyed elsewhere, because the surface area of the Earth remains the same. A lithospheric plate moves away from the ridge axis until it eventually reaches a *convergent boundary* with another plate. One slab must return to the mantle in order to conserve volume; this process is called *subduction*. A diagram is presented as Figure 6.7. A bathymetric low, or trench, marks the point where bending of the downgoing slab is greatest. The oceanic crust–capped plate is denser than the underlying material; consequently, it usually descends back into the mantle. It is more difficult for a continental crust–capped lithospheric plate to sink, because it is less dense than the mantle below; however, it sometimes occurs as a consequence of descent of ocean crust–capped lithosphere dragging a segment of continental crust into and down the inclined subduction zone.

Although most of the volcanism in the ocean basins occurs along the spreading centers, isolated upwelling of the asthenosphere builds islands and island chains within oceanic–crust capped plates as well. This phenomenon is due to thermal plumes, or *hot spots*, thermal anomalies thought to be located at considerable depth in the asthenosphere. The volume of melt generated at hot spots is much less than that produced at the ocean ridges.

The "wake" of volcanic islands generated by the Hawaiian hot spot is shown in Figure 4.14. The hot spot is a fixed point in the mantle beneath the moving lithosphere. This thermal plume is being overridden by the west-northwestward-

Figure 6.5. (opposite). Bathymetry of the Atlantic Ocean, showing the Mid-Atlantic Ridge, the remarkable submarine mountain chain located between eastern and western hemispheric continents. This prominent bathymetric feature lies everywhere exactly half-way between the Americas and Europe and Africa. Why is this? This figure also shows that the congruence across the Atlantic refers to the fit determined by the edges of the shelves, not the shorelines of the continents. (*Source:* Courtesy of B. C. Heezen and M. Tharp, *World ocean floor*, U.S. Navy, 1977.)

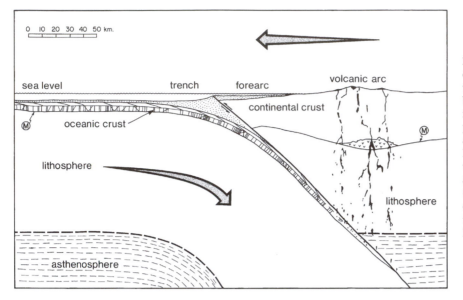

Figure 6.7. Diagrammatic cross section of an oceanic trench and island arc/continental margin, or convergent plate boundary. A large component of mantle flow (arrows) lies in the plane of the paper. Both basaltic and andesitic magmas are shown as black blobs. The base of the lithosphere, shown by dashed lines, is the region where the more rigid mantle above becomes more plastic downward (i.e., the asthenosphere), probably as a consequence of the presence in the asthenosphere of grain-boundary melt at higher temperatures. km = kilometers.

drifting Pacific lithospheric plate. The elbow or bend in the island chain was formed approximately 40 million years ago. In which direction do you suppose the Pacific lithosphere was moving immediately prior to this time?

When continental rifting occurs, the thickened crust tends to break apart along an irregular trend, as shown in Figure 6.8; in contrast, the process of mantle upwelling and lateral flow demands a straight spreading center and parallel flow lines. Therefore, the ridge is made up of short, straight segments of ridge offset by fracture zones. Think of each of the ridge segments as the junction of two back-to-back con-

Figure 6.8. Hypothetical initiation of the Mid-Atlantic Ridge above an upwelling asthenospheric current, the surface expression of which is assumed to have been an irregular, curvilinear feature. Parallel flow in the mantle and oceanic crust results in straight spreading-center segments offset by right-angle fracture zones, or transform faults (Mercator projection): (A) prior to spreading; (B) after a modest amount of divergent plate motion – for example, 1000 kilometers. Whether the plates are pulled apart by descent of a dense slab in the subduction zone, slide off the bathymetric ridge, or move passively due to flow in the underlying asthenosphere, the lithospheric motions and return flow in the ductile mantle are the same.

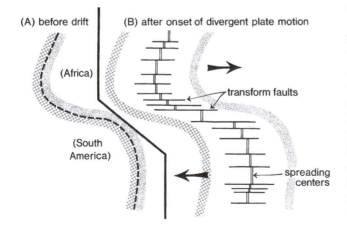

veyor belts, the strips of sea floor and underlying solid mantle as the moving belts. The oceanic fracture zones, or *conservative plate boundaries* (that is, plate boundaries involving neither creation nor destruction of lithosphere), are vertical planar discontinuities that offset ridge segments and are parallel to the flow of the lithosphere. These planar features are termed *transform faults*. Faults are fractures that have undergone differential movement of the opposite walls. In general, oceanic fracture zones may be described as arcs of *small circles* on a sphere, analogous to lines of latitude on a globe, whereas a ridge approximates an arc of a *great circle*, analogous to the equator, and to lines of longitude on a globe. Where transforms cut across a portion of a continent, we call such features strike-slip faults, because the movement is parallel to the linear trend (or strike) of the fracture.

For simplicity, we have treated plate motions as if they were linear translations on a flat surface. More accurately, the lithospheric plates are curvilinear segments (similar to pieces of the rind of an orange or, more properly to scale, segments of the skin of an apple) surmounting a spherical Earth. Plate motions therefore may be described as rotations about axes through the center of the globe. Of course, the poles of such rotations need not – and, in general, do not – coincide with the Earth's spin axis.

MARINE MAGNETIC ANOMALIES. Since the end of World War II, oceanographic surveying of the sea bottom has greatly improved our knowledge of submarine topography and oceanic crust dynamics. Concurrent with these investigations, the Earth's near-surface magnetic field was measured by shipboard geophysicists. The local magnetic field exhibits variations in intensity – magnetic anomalies – that are superimposed on, and deviate from, the regional magnetic field. As illustrated in Figure 6.9, such anomalies are arranged symmetrically about an oceanic ridge as alternating bands of greater and lesser magnetic intensity than the expected field (positive and negative magnetic anomalies, respectively).

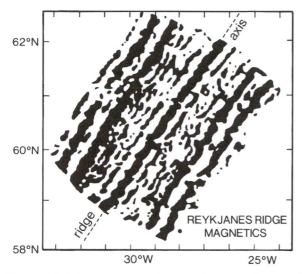

Figure 6.9. Measured magnetic anomaly pattern for a small portion of the Mid-Atlantic Ridge just southwest of Iceland. Positive anomalies are shown in black; negative anomalies, in white. The crude bilaterally symmetric striping continues both northeast and southwest, parallel to the extension of the ridge axis (see Fig. 6.5). (*Source:* Adapted from J. R. Heirtzler, X. LePichon, and J. G. Baron, Magnetic anomalies on the Reykjanes Range, *Deep Sea Research*, vol. 13. 1966.)

Clearly, these "zebra stripes" must be related to some process taking place in the vicinity of the spreading system. To understand this phenomenon, we must discuss the concept of *paleomagnetism.*

Approximately thirty years ago, it was recognized that the Earth's magnetic field changes its polarity episodically over a relatively short time (on a geologic time scale). This is called a *magnetic reversal* and means that the north-seeking end of a compass needle would point south during a period of reversed polarity. At present, we are living in a time of so-called normal polarity. The magnetic north pole is close to the Earth's rotational north pole. However, how do we know that the Earth's magnetic field has switched polarity at (irregular) intervals over geologic time?

Rocks laid down at or near the surface, such as lava flows, volcanic ash layers, marine sediments, and lake beds, contain minor concentrations of magnetic minerals. Magnetic domains within these minerals, or the individual grains themselves, become aligned with the Earth's magnetic field as it existed at the time of formation of these rocks (see Fig. 6.10). When measured in the laboratory, such specimens display weak *remnant magnetism*, which was imposed in response to, and coincident with, the Earth's magnetic field during deposition or crystallization of the rock. Both the magnetic north direction (azimuth) and inclination from the horizontal (dip) of the magnetic field lines at the time of origin can be recovered if the initial orientation during formation of the collected rock is known. The magnetic azimuth indicates normal or reversed polarity and the direction of the magnetic pole. The magnetic dip provides the magnetic latitude of the rock at the time of origin, in other words, the paleolatitude.

When the remnant magnetism of a sufficiently thick sequence of layered rocks is measured, it is found that groups of rock units possessing normal polarity alternate with layers characterized by reversed polarity. Moreover, where good fossil and/or radiometric data allow the precise assignment of ages to such a series, analogous stacks of layered rocks around the world can be demonstrated to provide exactly the same time versus magnetic polarity sequence. The geomagnetic time scale for relatively young rocks is illustrated in Figure 6.10. The Earth's magnetic field undergoes polarity changes on an irregular frequency but always maintains its dipole character.

In 1963, when the widespread magnetic anomaly patterns of the ocean basins began to be recognized, it was concluded that the anomaly stripes, shown in Figure 6.9, were due to the remnant magnetism of the oceanic crust superimposed on the regional field – positive in cases where the basaltic lavas were extruded during a time of normal polarity of the

Figure 6.10. The geomagnetic time scale for the past 80 million years, showing worldwide intervals of normal (black) and reversed (white) magnetic polarity, as deduced from remnant magnetic measurements of accurately dated layered rocks. Geologic periods and epochs, anomaly numbers, and age in millions of years (M.Y.), are indicated. (*Source:* Adapted from J. R. Heirtzler, G. O. Dickson, E. M. Herran, W. C. Pitman, III, and X. Le Pichon, Marine magnetic anomalies, geomagnetic field reversals and motions of the ocean floor and continents, *Journal of Geophysical Research*, vol. 73, 1968.)

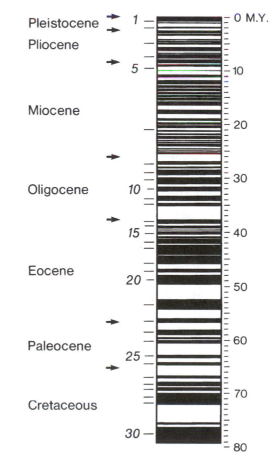

Earth's magnetic field, negative where they extruded during a time of reversed polarity. Inasmuch as the crust at the ridge axis is youngest, it follows that as one moves orthogonally away from the spreading center, progressively older basalts are encountered, producing the linear pattern of alternating positive and negative anomalies. A mid-oceanic ridge system, in effect, acts like a pair of back-to-back magnetic tape recorders, with the constituent rocks faithfully reproducing the fitful reversal of the Earth's magnetic field. The rate of horizontal transport of the oceanic crust away from the ridge is on the order of several centimeters per year, judging from application of the geomagnetic time scale to the observed magnetic anomaly pattern. When this relationship was realized, the concept of sea-floor spreading and the opening of the Atlantic were quickly accepted. Continental drift had been proved beyond reasonable doubt!

SEISMICITY AND LITHOSPHERIC PLATES. We have shown that measurements of the bathymetry and remnant magnetism of ocean basins such as the Atlantic are compatible with a model of rising currents in the plastic mantle beneath the divergent plate junctions of oceanic spreading systems, and lateral movement of the lithospheric plates at right angles to the ridge axis. The history of continued opening of the Atlantic stands in sharp contrast to the Pacific Basin, which is being constricted around its periphery and is

slowly shrinking. Convergent junctions occur around the margins of those plates where one lithospheric slab sinks beneath another. It is the study of seismic wave velocities and attenuation of shaking intensity that provided crucial evidence for this now universally accepted plate tectonic model. Seismic data have unambiguously demonstrated the existence of a stronger, more rigid lithosphere surmounting a weaker, more ductile asthenosphere; seismicity also provides information concerning the nature and relative motions of these lithospheric plates.

Earthquake activity is virtually confined to the lithosphere, which, on the rapid application of stress, behaves in brittle fashion compared to the underlying, softer asthenosphere. The *focus* of an earthquake is the site at depth where rupture of material releases seismic energy. Within-plate seismicity is much less intense than along slab margins, where frictional energy release reflects strong differential plate motions; for this reason, concentrated seismic activity tends to outline present-day lithospheric plate boundaries. Examination of a world seismic map such as Figure 6.11 provides us with an appreciation of the sizes and shapes of the plates. Locations on the surface of the Earth directly above the earthquakes, or *epicenters*, are shown in Figure 6.11, rather than the actual depth locations, or *hypocenters*. Seven major plates are outlined by this seismicity, as indicated by Figure 6.12. In addition, several smaller plates are evident. The near-

Figure 6.11. World seismic activity for the period 1961–7. Dots mark earthquake epicenters (surface locations directly above earthquake foci, or hypocenters). The ocean ridges are outlined by narrow zones of earthquake activity; in contrast, some convergent continental margins and island arcs are the sites of broad belts of intense seismicity. (*Source:* Adapted from M. Barrazangi and J. Dorman, World seismicity maps compiled from ESSA coast and geodetic survey epicenter data, 1961–1967, *Seismological Society of America Bulletin*, vol. 59, 1969.)

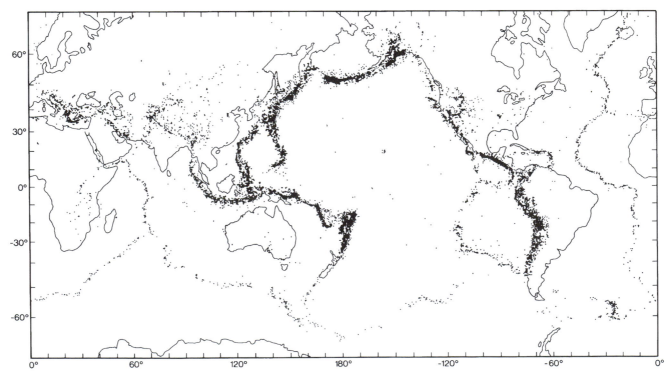

surface portions of some plates are dominantly oceanic crust (e.g., Pacific, Nazca, and Philippine plates), with others mostly capped by continental crust (e.g., Arabian and Eurasian plates); in general, however, the upper portions of the lithospheric slabs consist of both oceanic and continental crust. Here is where Wegener's hypothesis was incorrect. Of course, he had no way of knowing that continents and ocean basins were merely the uppermost parts of much larger, mobile lithospheric plates.

The boundaries of the plates are of three varieties, as previously described: divergent, convergent, and conservative. Oceanic spreading ridges, trench-subduction zones, and transform-fault fracture zones embody these three types, respectively. Where upwelling mantle currents approach the surface, high-temperature, solid but incipiently melted asthenosphere lies at shallow depths; thus, heat flow is high, and the rocks deform chiefly through ductile flow rather than by rupture. For this reason, seismicity associated with spreading centers and ridge-offsetting transform faults is confined to depths of less than approximately 50 kilometers. On the other hand, in subduction zones beneath oceanic trenches, which are regions of abnormally low heat flow, downgoing slabs sink to depths of 600 to 700 kilometers before finally losing their rigidity and integrity, judging from earthquake hypocenter locations. That is why the divergent, shallow plate junctions shown in Figure 6.5, 6.11, and 6.12 display narrow, well-defined bands of epicenters, whereas in contrast, epicenters from various portions of inclined subduction zones characteristic of convergent plate junctions are dispersed over a much broader region. Another indication

that lithosphere is descending into the deep upper mantle is gained from the observation that velocities of seismic waves in the neighborhood of the subduction zone are quite high, and attenuations of vibrational energy are unusually small for these mantle depths, suggesting the presence of a slab of relatively cold, highly conductive lithosphere.

The third type of plate boundary is the conservative or transform lithospheric plate junction. Here, two plates slide past one another in parallel without either diverging or converging. Although continental transforms are well known, the most abundant and clearest examples are found among the transform faults (i.e., submarine fracture zones) offsetting oceanic rise segments. A schematic example is illustrated in Figure 6.13. Differential motion between the plates is in the displacement sense opposite that anticipated from the apparent offset of the ridge crest, because oceanic spreading-center segments were formed more or less in their present relative positions (see Fig. 6.8) rather than having once been a continuous linear feature that was later displaced by faulting. Figure 6.13 also shows that beyond the ridge-ridge portion of the transform, no differential slip takes place between the plates. Inferred differential lithospheric motions are supported by the presence of shallow earthquake foci, but only along segments of transform faults that connect neighboring spreading centers.

SEDIMENTARY ROCKS IN THE OCEAN BASINS. Continents contain a thin, nearly continuous veneer of young rocks. However, they are made up of much older materials as well; the oldest sections of surviving continental crust appear to be approximately 3.8 billion years old. Ancient rocks are preserved in only fragmentary form; hence, the continents bear a rather indistinct, obscure, and incomplete record of early Earth history. In most cases the oldest rocks are confined to the stable continental interiors, which are referred to as *Precambrian shields*, or *cratons*.

Prior to the recognition of sea-floor spreading, geologists imagined that, in contrast to the imperfect preservation of rocks deposited on or thrust onto the land, the ocean basins must contain a well-preserved succession of sedimentary strata, providing a complete record of geologic time since the establishment of the hydrosphere. An implicit assumption was that the ocean basins represent permanent depressions in the Earth's crust, fixed on the planetary surface since earliest geologic time. We now know that the oceanic crust is formed along oceanic ridges, and that sea-floor spreading carries the crust, an overlying blanket of

Figure 6.12. Major lithospheric plates and their boundaries as deduced from worldwide seismicity. Convergent boundaries are shown with barbs on the upper plate. (*Source:* Adapted from J. F. Dewey, Plate tectonics, *Scientific American*, vol. 226, no. 5, 1972; and J. Francheteau, The oceanic crust, *Scientific American*, vol. 249, no. 13, 1983.)

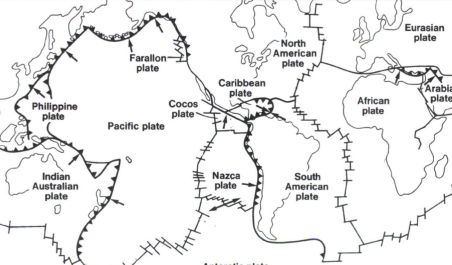

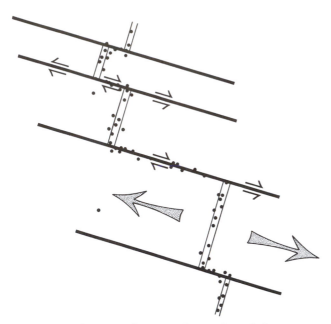

Figure 6.13. Diagram of an oceanic transform fault, or conservative lithospheric plate boundary. Earthquake epicenters are denoted by dots; motions of the plates relative to the spreading center are denoted by arrows. Transform faults are seismically quiet except for the portions linking adjacent ridge segments, where active differential slip is taking place. (Why?)

marine sediments, and immediately underlying mantle away to an eventual arrival at a trench (Fig. 6.7). Here the sediments, with or without oceanic crust, either are scraped from the downgoing slab and pushed up against the margin of the overlying plate or, more commonly, disappear back into the mantle by descent along the inclined subduction zone. This scenario reflects flow of the upper portions of mantle convection cells; such a process effectively removes the oceanic crust and overlying layers of sedimentary material deposited upon it approximately every 50 to 200 million years. Therefore, according to modern concepts, sediments laid down on oceanic crust should be exclusively young compared to the age of the Earth. Are they?

To answer this question, an oceanographic vessel capable of operating as a drilling platform on the high seas was commissioned in the late 1960s. Scientific results obtained during the following two decades abundantly verified the hypothesis of sea-floor spreading. It was demonstrated that only a thin layer of very young sediments occurs in the vicinity of a submarine ridge whereas, proceeding away in the presumed sea-floor spreading (and ocean-floor transport) direction, the sedimentary cover thickens gradually, and the age of the oldest, basal stratum resting directly on the submarine basalts increases commensurately. The most ancient sediments and underlying oceanic crust in the enlarging North Atlantic are approximately 190 million years old, reflecting the rifting of Laurasia, just as postulated by Wegener. The oldest sediments in the constricting Pacific Basin, situated directly in front of the Mariana Trench, are of roughly comparable age. Geo-

graphic relationships are shown in Figure 6.14. The sedimentary contents of the world's marine basins, as well as the underlying oceanic crust, are exclusively youthful, geologically speaking.

Sea-floor spreading rates can be determined by combining measured orthogonal distances from the ridge crests coupled with ages of the oceanic crust, deduced from the times of deposition of the immediately overlying sediments and paleomagnetic data. In general, stable lithospheric plates that are not subducting, such as the American and the Eurasian, are moving away from the mid-oceanic ridges at 1 to 3 centimeters per year; in contrast, plates characterized by descending portions, such as the Pacific, Nazca and Philippine, are moving 6 to 12 centimeters per year relative to hot spots and adjacent plates. This velocity relationship suggests that slab "pull" down the subduction zone is a more important factor influencing plate motions than is ridge-associated slip.

Thus, the ancient record of the Earth is written solely in the continents, because the continued sea-floor spreading of the ocean basins in response to ongoing mantle convection sweeps the oceanic basalt–gabbro complex and its thin sedimentary cover against convergent plate junctions on a cycle of approximately 100 million years. In the vicinity of the trench, the sedimentary debris and underlying oceanic crust are scraped off (at least in part) or underplated at depth, and both are deformed and stored as new portions of an accreting island arc or continental margin.

EARTHQUAKES AND VOLCANISM

If you compare a global map of historically active and geologically youthful volcanoes as shown in Figure 6.15 with a chart depicting worldwide seismicity such as Figure 6.11, you will be struck immediately by the nearly one-to-one correlation. Earthquakes and volcanoes crowd the margin of the Pacific Basin and form a less distinct, semicontinuous Alpine-Himalayan-Indonesian seismovolcanic zone as well. Other volcanic belts coincide with the oceanic rises; however, inasmuch as most eruptions are submarine, this activity has gone largely undetected. The residents of Japan, Chile, Hawaii, and southern Alaska have long been aware of the ubiquitous interrelationship: Earthquake activity and volcanic eruption are obviously associated geographically and temporally. But why?

Certain seismic events appear to be a result of frictional resistance accompanying the rise of partially molten, viscous material within a volcanic conduit. This relationship provides the explanation for some earthquakes preceding eruptions in mid-plate regimes such as Hawaii. Moreover, tensile rupture takes place where thin lithosphere is necking down, as along oceanic spreading centers such as Iceland. However, most earthquakes are a reflection of compressional stress buildup and episodic release in sections of solid rock sliding past one another along a fault zone, either in the upper mantle or in the crust, or both. Fault zones mark all lithospheric plate boundaries. Plate boundary faults, especially in

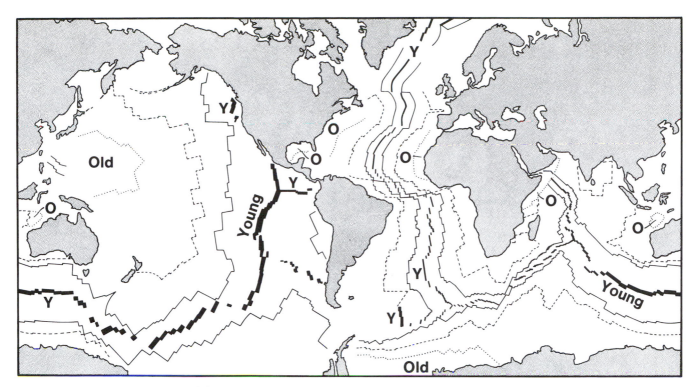

Figure 6.14. Age of the oldest sediments, resting directly on oceanic crust, in the present deep-sea basins. The rocks flooring the oceans are youngest adjacent to the spreading centers, and become progressively older in the portions of the basins far removed from the oceanic ridges. (*Source:* Adapted from S. W. Pitman, R. L. Larson, and E. M. Herron, *The age of the ocean basins*, Geological Society of America Map Series, 1974.)

Figure 6.15. Modern and recently active volcanoes of the world. Their locations relate spatially to the trenches (Fig. 6.12) and to earthquake epicenters (Fig. 6.11). (*Source:* Adapted from T. Simkin and L. Siebul, Explosive eruptions in space and time, *Explosive Volcanism: Inception, Evolution, and Hazards*, National Academy of Sciences, 1984.)

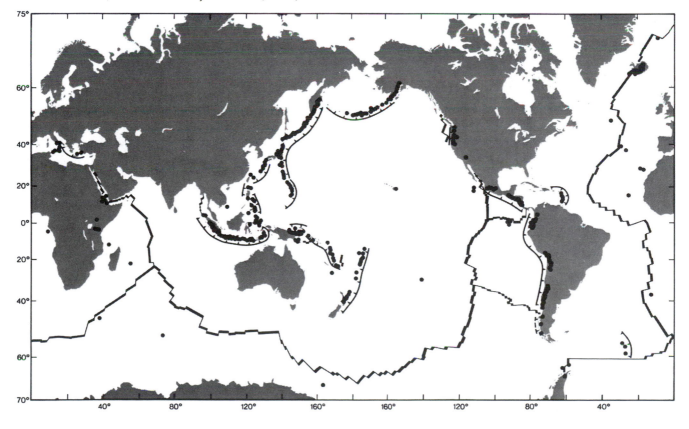

regimes characterized by lithospheric stretching and fracturing, provide a ready-made "plumbing system" whereby molten material at depth can gain access to upper levels of the crust and even reach the Earth's surface. The localization of volcanic rocks and their plutonic equivalents along linear seismic belts is therefore understandable in the light of plate tectonics.

Most earthquakes originate at depths of less than 50 kilometers. Appropriate lithospheric regimes include extensional or divergent boundaries, transform boundaries, and intraplate areas. Within the ocean basins, lavas associated with mid-plate and spreading-ridge tectonic settings are dominantly basaltic; these igneous rocks represent the partial fusion products of hot asthenosphere and rising mantle plumes. On the other hand, where continental masses have undergone stretching or transform motion, two chemically distinct lavas in close association are characteristic – basalt and rhyolite. Basalts are rich in magnesium, iron, and calcium, and come from the upper mantle. Rhyolites are silicic, alkali-rich volcanic rocks that form indirectly in stretched continental crust in an extensional setting where high-temperature asthenosphere or molten rock rises to shallow depths; many rhyolites are probably derived by partial melting of basal portions of the continental crust in these particularly hot environments.

Shallow- (less than 50 kilometers), intermediate- (50 to 200 kilometers), and deep- (200 to 700 kilometers) focus earthquakes typify convergent plate boundaries. Set back a horizontal distance of approximately 75 to 150 kilometers from the trench on the stable, nonsubducted slab, voluminous volcanics (of basaltic, andesitic, and dacitic compositions and their intrusive granitic equivalents), accumulate during the construction of island-arc and continental-margin magmatic provinces. Andesitic plus dacitic lavas and ash deposits are intermediate in composition between basalt and rhyolite and are essentially chemically equivalent to the average continental crust (granodiorite). Such magmas appear to be generated at depths of 100 kilometers or more along or just above subducting lithosphere.

It is apparent from this brief sketch that the compositions and areal dispositions of rocks, as well as their association with intense seismic activity, are closely correlated with plate tectonic regimes. Thus, the exclusively basaltic nature of the oceanic crust owes its origin to partial melting of ascending, depressurizing columns of asthenospheric material. In contrast, partial melting of subducting lithospheric slabs and/or the mantle above the subducting slab, beneath island arcs and continental margins, gives rise to the buoyant andesitic (e.g., Aleutian Arc, Andes) and granitic (e.g., Sierra Nevada) melts that characterize convergent plate tectonic settings. Extensional regimes within continental interiors (e.g., Columbia River Plateau) tap both primary basaltic magma originating in the upper mantle and, in some places (e.g., the Basin and Range Province of Nevada and western Utah), remelted, remobilized portions of deeply buried and heated continental crust (secondary rhyolites).

MOUNTAIN BUILDING

Some on-land mountain chains, such as the Andes, the Sierra Nevada, and all oceanic spreading centers, are constructed predominantly of lavas that have reached the surface, volcanic ash, and equally important molten equivalents that have solidified as intrusives at depth within the crust. In contrast to the oceanic ridges, however, continents are also typified by mountain belts that display evidence of great shortening and thickening of the crust (e.g., the Appalachian and Rocky mountains). These compressional mountains contain great tracts of preexisting layered rocks that have been contorted into a series of folds and fault-bounded blocks.

Wegener related the deformation of originally flat-lying strata to continental drift by postulating crumpling at the leading edge of the advancing continental crust. Today we realize that this large-scale contraction has been manifested in a way far more complex than that imagined by Wegener. Three different styles of mountain building may be recognized, each of which reflects a particular plate tectonic regime: convergent, divergent, and transform. A further complication involves the arrival and collision or departure and truncation of far-traveled crustal blocks (exotic terranes). All such mountain belts are correlated with the environs of present and ancient plate boundaries.

MOUNTAIN BELTS OF CONVERGENT PLATE BOUNDARIES. By far the most complicated, this type of mountain belt consists of three distinct structural portions: (1) a seaward trench complex and accretionary wedge; (2) a medial fore-arc basin; and (3) a landward volcanic/plutonic arc or continental margin. A cross section of such a complex, represented by northern and central California of approximately 80 to 120 million years ago, or by modern Chile, is presented as Figure 6.7. The now eroded and well-exposed Californian paleomargin is analogous to the currently active Chilean volcanic/plutonic continental margin.

The seaward portion consists of oceanic crust and a prism of deep-water sedimentary rocks, individual units of which have been carried beneath the overriding, nonsubducted lithospheric plate. Decoupling of this largely sedimentary debris on and beneath the landward side of the oceanic trench results in the accumulation of an *accretionary wedge*, or *subduction complex*. Low-angle faults in which the overlying blocks move upward relative to the underlying segments are termed *thrust faults*. A series of these subparallel thrust faults roughly parallels the boundary of the descending slab and causes imbrication and folding within the accretionary wedge. As this section thickens and buoyantly rises, faulting of opposite sense, or *normal faulting*, takes place.

Immediately landward from the trench offscrapings, and above the underplated region beneath the overlying wedge, lies a column of *fore-arc basin* sediments; these shallow-water sediments, like those of the seaward trench deposit, have been shed predominantly off the growing volcanogenic arc. Resting on the nonsubducted lithospheric plate, fore-arc strata in general are much less deformed than the contorted melange caught in the jaws of the subduction zone.

The landward *volcanic/plutonic arc* is yet more complicated because such belts typically contain remnants of the old pre-subduction continental margin, plus a host of igneous arc rocks. Magmas rise from the mantle above the downgoing slab and from the descending lithospheric plate itself, and transport heat into the crust, thereby causing partial melting of the most fusible, deeply buried portions of the continental crust. Some of the buoyant, molten material solidifies at depth as plutons; other portions gain access to the surface and vent as explosive, ash-laden clouds or as less violently extruded lavas. Continental crust is essentially andesite-dacite in bulk composition; the subduction process that generates these magmas (intrusives as well as extrusives) appears to be responsible for the formation and growth of continents. Thus andesitic and kindred melts represent the partial fusion products of subducted oceanic crust that buoyantly ascend either from the downgoing, heated slab, or from the nonsubducted mantle wedge overlying the sinking plate, or from both. Varying degrees of reaction with the overlying mantle and crust, and comingling of liquids may account for the observed compositional range of igneous rocks that make up the margins of the continents and island arcs. This constitutes the geologic regime where new continental crust is forming.

Fracturing, faulting, and deep-level ductile deformation are common within all three domains that constitute mountain belts constructed in the environs of a convergent plate boundary. Because substantial motions take place at an angle to the plane of cross section (illustrated in Figure 6.7), lateral shuffling of the various regimes also occurs. Moreover, sea-floor spreading may modify the accretionary margin by bringing exotic oceanic crust and fragments of unrelated continental crust to the convergent junction, thus providing additional complications to this compressional mountain system. Rearrangement due to transform movement results in the construction of a collage of terranes – individual fault-bounded blocks – some formed more or less in place, others truly far-traveled. Thus, continental growth may take place through the addition of unrelated terranes as well as the generation of new continental crust within the convergent plate margin regime.

MOUNTAIN BELTS OF DIVERGENT PLATE BOUNDA-RIES. A central rift zone, not pictured in Figure 6.6, characterizes oceanic ridges, reflecting the divergence of back-to-back lithospheric slabs. These spreading centers, of course, are the sites of production of almost all of the world's oceanic crust.

Where extensional zones occur within the much thicker continental crust and underlying mantle lithosphere, the attenuation typically is distributed over a much broader region. In these geologic realms, faulting involves jostling, lateral stretching, and downward relative motion of crustal sections, resulting in block-fault mountains and in tilting and thinning of the continental crust. Such faults are known as normal faults. Examples of on-land extensional zones include the Basin and Range Province of the American Southwest, and the East African, Red Sea, and Dead Sea rift systems. Rise of a column of asthenosphere beneath a continent, bifurcation near the surface, and horizontal flow are indicated in divergent plate tectonic settings. Thermal expansion due to close approach of asthenosphere to the surface causes broad uplift of the continental crust during early stages of mantle upwelling; ultimately, the rifted area may subside to form a new ocean basin, such as the Red Sea. Similarly, uplift followed by crustal stretching seems to have attended the breakup of Pangea at the beginning of the Mesozoic Era.

MOUNTAIN BELTS OF CONSERVATIVE PLATE BOUNDA-RIES. Perfectly transform-type plate junctions involve neither convergent nor divergent motion. However, transform plate boundaries that transect old, geologically heterogeneous continental crust tend to involve deflections around coherent blocks, and thus contrast markedly with the straight planar features of the oceanic fracture systems. The fault bends in complex continental crust are of two types: mass excess, and mass deficient. During movement along a curved fault, the former, or *constraining bend*, results in compressive deformation, localized mountain building, and thickening of the continental crust; the latter, or *releasing bend*, generates a topographic depression – pull-apart basin – during slip along the transform. Thus, continental transform boundaries (strike-slip faults) are characterized by straight, mass-conservative horizontal-slip segments, and by locally deformed zones showing adjacent upthrust blocks, compressional folds and high-angle thrust faults, and down-dropped blocks, sag basins, and normal faults. Where fractures and faults extend to great depth in the continental crust, and into the upper mantle, molten material may rise into the transform zone as intrusions and extrusions.

MOUNTAIN BELTS AND EXOTIC TERRANES. Differential motions of lithospheric plates result in the dispersal of some formerly contiguous crustal units, and in the juxtaposition of other, far-traveled geologic complexes of contrasting crustal origins and histories. The rifting of preexisting continents, rearrangement and suturing of microcontinental fragments, island arcs, and oceanic crust accompanying sea-floor spreading and drift results in the construction of new continental mountain belt collages. Portions of such assemblies consist of unrelated, fault-bounded terranes – some may be oceanic in origin, whereas others may be parts of preexisting continents.

The principles of plate tectonics, as illustrated in Figures 6.6, 6.7, and 6.8, require that plate junctions mark the dynamic boundaries of individual crust-capped lithospheric slabs, and that all plates move differentially with respect to each other. In this sense, the continents are themselves exotic, unrelated terranes (Figs. 6.1 through 6.4). On a smaller scale, the area west of the San Andreas fault is far-traveled relative to the rest of California (Fig. 6.12); so is India with respect to the rest of Eurasia. Some Earth scientists have interpreted much of the western cordillera of North America, and the Indonesian Archipelago, as constituting amalgamated assemblages of exotic terranes. Whether this is correct or not, we must regard the continents as consisting of both native and foreign crustal units, reflecting their unique individual histories of growth.

REGIONAL GEOLOGIC SETTING OF THE PACIFIC BASIN

Other than the exclusively near-surface earthquakes typifying oceanic spreading ridges, the seismicity of the Earth is spatially related in large part to continental-margin and island-arc features. Shallow-focus (0 to 50 kilometers) earthquakes occur in the vicinity of the oceanic trenches and range to more profound hypocentral depths (intermediate-focus, 50 to 200 kilometers; deep focus, 200 to 700 kilometers), moving progressively landward across the arc onto the margin and then well into the continental interior (see Fig. 6.11). Based on tectonic concepts involving sea-floor spreading, global tectonics, and continental drift, we know that the observed seismicity is a result of differential motions of adjacent lithospheric plates; typically, earthquake foci outline the edges of the plates themselves (Fig. 6.11).

Now we consider the on-land geology. Far from the stable, old continental interiors, the Pacific Rim is marked by predominantly Mesozoic and younger mountain belts. Evidence of pronounced crustal shortening across the edges of the ocean basin is provided by widespread contractional features,

Figure 6.16. The convergent suture complex of (A) Indonesia, where Pacific, Australian, Eurasian, and Philippine plates are impinging on one another, compared with (B) Mesozoic accreted terranes in western North America. For a map showing the world's major plates, refer to Figure 6.12. (*Source:* Adapted from E. A. Silver and R. B. Smith, Comparison of terrane accretion in modern Southwest Asia and the Mesozoic North American cordillera, *Geology*, vol. 11, 1983.)

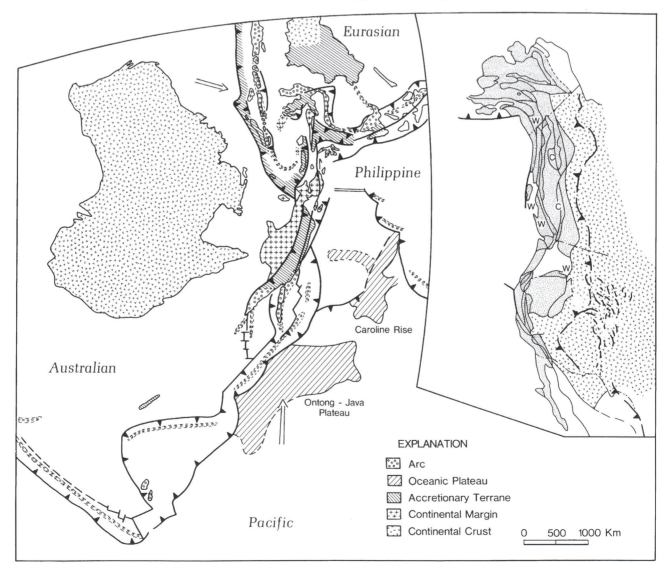

Eurasian

Philippine

Caroline Rise

Australian

Ontong - Java Plateau

Pacific

EXPLANATION

Arc
Oceanic Plateau
Accretionary Terrane
Continental Margin
Continental Crust

0   500   1000 Km

such as faults and folds. These mountain belts are replete with microcontinental slivers and yet more numerous scraps of oceanic crust; such phenomena attest to the drift of itinerant terranes and their ultimate stranding against an accreting continental- or island-arc margin. A portion of the Mesozoic plate tectonic assembly of North American accreted terranes is illustrated in Figure 6.16; also shown is a possible modern analog in the Indonesian area of the western Pacific. These belts are also the site of voluminous andesitic volcanism (see Fig. 6.15), and where deeply eroded roots of mountain chains are laid bare, huge granitic batholiths are revealed. Thus the geologic hazards of volcanism, seismicity, and land slumpage (see Chap. 25) can all be related to plate tectonic processes so intensely developed along the Pacific Rim.

The above are important manifestations of convergent lithospheric plate motions. The progressive constriction of the world's largest ocean basin is a consequence of the breakup of Gondwana and the opening of the Atlantic Ocean; expansion on a portion of the globe must be compensated for by contraction elsewhere. Sea-floor spreading occurs in the Pacific as well, however, and is reflected by the existence of the Southeast Indian Ocean Ridge–East Pacific Rise system; the velocity of divergence originating along this stupendous mid-ocean ridge is 6 to 10 centimeters per year, approximately three times as rapid as the spreading rate of the Mid-Atlantic Ridge. However, unlike the Atlantic, the Pacific is characterized by a nearly continuous peripheral ring of subduction zone – and where these are missing, transform plate junctions are present instead. The Circum-Pacific is the Earth's principal foundry in which new continental crust is forming and being recycled, as described for convergent plate boundaries in previous sections of this chapter. We therefore examine the geologic relationships of the U.S. West Coast as a typical example. By so doing, we lay the plate tectonic framework requisite for an understanding of the geologic hazards attending evolution of this young, vigorous mountain belt (see Chap. 25). Such phenomena are of wide application and can be readily extrapolated throughout the Circum-Pacific region.

PLATE MOTIONS ALONG THE PACIFIC RIM

The age of ocean-floor sediment lying directly above the basaltic crust of the Pacific Basin is shown in Figure 6.14. Chemically precipitated, fossiliferous materials and very fine-grained clays settle out slowly but continuously in the deep sea; the debris begins to be deposited on the oceanic crust immediately after the submarine basaltic lava cools and solidifies in the vicinity of a spreading center. Thus, the age of the basal, oldest portions of sedimentary layer 1 (see Fig. 6.6) marks the time of solidification and cooling of the underlying basaltic substrate.

The contrasting ages of sediments resting on oceanic basalts define submarine geographic bands, comparable to those defined by magnetic anomaly patterns. Both types of "zebra striping" parallel the spreading systems because each is a reflection of the time of formation of the oceanic crust. The basalts and overlying sediments are, of course, youngest at the presently active divergent plate junctions, and become progressively older in the directions of spreading orthogonal to the ridge axes. (Why?) The most ancient sea-floor sediments still preserved in the western Pacific Basin were deposited approximately 190 million years ago. The bilateral symmetry of spreading obviously has been preserved in the Atlantic as mirrored by the ages of the basal sediments (see Fig. 6.14), whereas in the Pacific, the original symmetry has been partially obliterated. How can this be explained? Because the surface area of the globe remains constant, equivalent amounts of convergent and divergent plate motions are required for the Earth overall. Consequently, the oceanic crust and its sedimentary cover must be destroyed sooner or later by the subduction process, and at the same global rate as they are being produced. In the Pacific, the conveyer belt–like motion of sea-floor spreading brings the crust and very fine-grained overlying strata to a trench at the basin margin within 100 to 200 million years as a maximum. Moreover, the rate of plate consumption in the Circum-Pacific trenches exceeds the production of new lithosphere at the spreading ridges; hence, the Pacific Basin is gradually shrinking. On the other hand, the Atlantic is essentially devoid of subduction zones, consequently, because of sea-floor spreading, this basin is enlarging.

The Southeast Indian Ocean Ridge lies equidistant between the continental masses of Antarctica and Australia–New Zealand; this spreading system came into existence approximately 80 million years ago when Australia–New Zealand split off from Antarctica and drifted northward. As the Southeast Indian Ocean Ridge passes farther eastward into the Pacific Basin, where its extension is known as the East Pacific Rise, it turns northeastward (see Fig. 4.14). The rise impinges on the North American continent in the neighborhood of the Gulf of California, reappearing again north of the Cape Mendocino fracture zone as the Gorda–Juan de Fuca Ridge. These are segments of the same divergent plate junction; however, in the vicinity of California, the spreading system has been overrun from the east by the North American continental crust–capped lithospheric plate.

We can now understand the loss of bilateral symmetry of the Pacific sea floor by reconstructing the Cenozoic movement history. Consider the East Pacific Rise as our spatial frame of reference. Today, as in the recent geologic past, eastern lithospheric limbs of the East Pacific Rise move eastward and southeastward, then turn downward and sink beneath the yet more easterly North and South American continental crust–capped lithospheric plates. The modern locus of subduction is marked by the Chile-Peru and Middle America trenches; plates moving into these oceanic deeps from the west include the Farallon, Cocos, Nazca, and Antarctic (Fig. 6.12). Propelled westward by sea-floor spreading in the Atlantic, the North and South American plates are slowly encroaching on the eastern limbs of the East Pacific Rise;

(A)

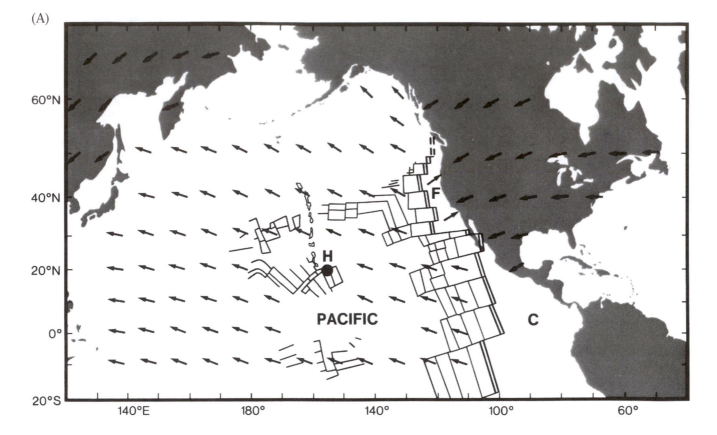

hence, convergent lithospheric plate boundaries bordering North and South America on the west gradually migrate westward relative to the divergent boundaries in the eastern Pacific Ocean.

Where the two have collided, as in California, trench and ridge are mutually destroyed, and a newly generated transform fault system accommodates the differential motion between North American and Pacific lithospheric plates. The initial encounter was along the central California coast approximately 29 million years ago. Since then, the transference of plate motions has propagated through western California as pairs of northwestward- and southeastward- migrating ridge-trench-transform triple junctions, with a continental transform connecting the two points of impingement where divergent and convergent lithospheric plate boundaries meet; this collision of ridge and trench results in their mutual annihilation. The continental transform – a conservative plate boundary – is actually a broad zone of subparallel strike-slip faults, the most prominent strand of which is the San Andreas fault. Three stages in this plate tectonic scenario, at 37, 20, and 0 million years ago, are depicted in Figure 6.17.

Most of California's ongoing seismicity (discussed in Chap. 25) has taken place by differential horizontal slip, as would be expected along the continental transform system described above. This displacement scenario is easily explained by the lithospheric plate interactions illustrated schematically in Figure 6.18 (see also Fig. 6.12). However, some recent earthquakes, such as the two San Fernando Valley

events (1972, 1994) and the Whittier Narrows earthquake (1987), involved thrust faulting; in all three cases, compression has been temporarily relieved in the so-called big bend segment of the San Andreas fault through crustal shortening and, in two out of the three cases, upward movement of the San Gabriel Mountains block. This phenomenon reflects activity associated with a constraining bend in the continental crust–transecting strike-slip fault zone.

THE DRIVING FORCES OF PLATES

Mountain building, earthquakes, volcanic activity, crustal accretion, sea-floor spreading, continental drift, and subduction are all surficial manifestations of flow within the deeper Earth. Upper mantle circulation, or convection, represents the mechanism that Wegener sought, but never discovered, in his attempt to explain continental drift.

Convection in a ductile medium results from the condition whereby a dense upper layer surmounts a less dense substrate: On perturbation, overturn occurs as a consequence of the gravitative instability. This density inversion and resultant bodily flow reflect the fact that as a solid (plus interstitial melt) cools, it contracts; accordingly, lithosphere, the outer, cooler rind of the Earth, is more dense than the hotter asthenosphere directly below it. With increasing depth in the Earth – for the sake of simplicity, we assume the same bulk composition and absence of mineralogic transformations – the progressively hotter mantle may become sufficiently less dense and more ductile so that it becomes buoyant and ca-

(B)

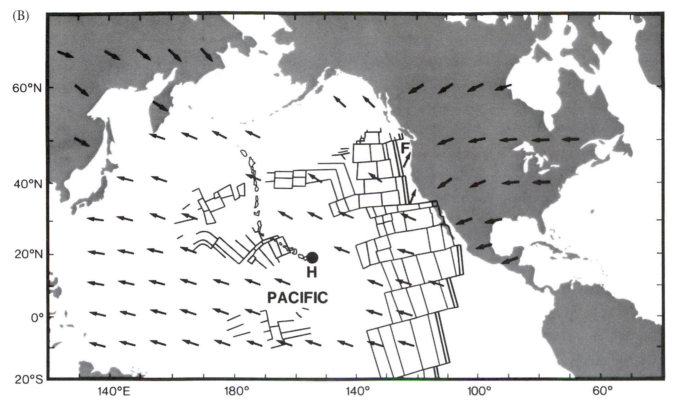

Figure 6.17. Dispositions and relative movement directions (vectors) of lithospheric plates in and adjacent to the Pacific Basin (A) 37, (B) 20, and (C) 0 million years (m.y.) ago. The present outline of North America is shown, although prior to 10 to 20 million years ago, Baja was nestled against the coast of mainland Mexico. Magnetic anomaly patterns are depicted, as is the Hawaiian hot spot (H) and the chain of islands formed by basaltic eruptions sited above this mantle plume. Prior to approximately 40 million years ago, the Pacific plate was moving more nearly northward, as indicated by the linear island chain, a hot spot track. Arrows indicate motion of the plates at the time designated. F = Farallon plate; C = Cocos plate; N = Nazca plate. (*Source:* Adapted from P. C. Engebretson, A. C. Cox, and R. G. Gordon, *Relative motion between oceanic and continental plates in the Pacific Basin*, Geological Society of America Special Paper No. 206, 1985.)

(C)

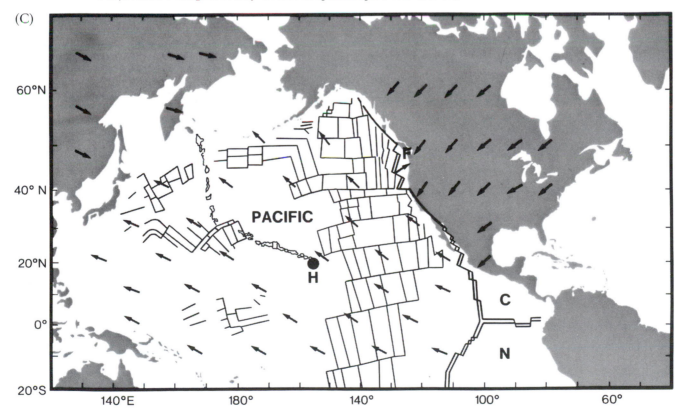

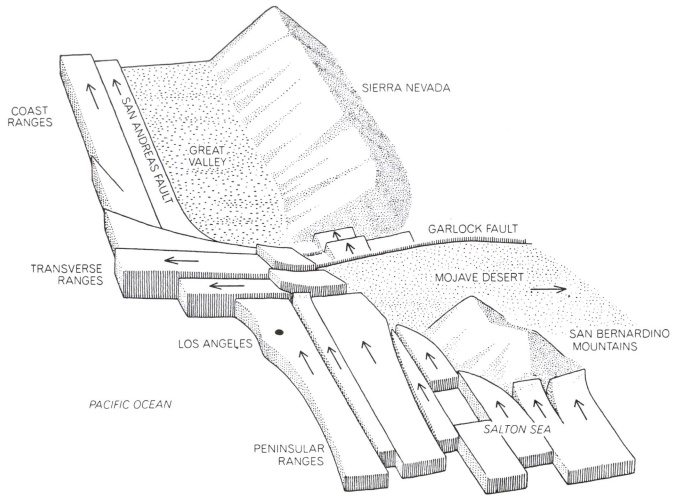

Figure 6.18. Diagram of differential movement of crustal blocks within southern California along the San Andreas continental transform system. The "big bend" area is defined by a more westerly trend of the San Andreas–San Gabriel fault system in the region between the Mojave Desert and the Transverse Ranges. The San Gabriel fault lies on the southwest side of the upthrust pair of crustal blocks (facing the Los Angeles Basin), the San Andreas on the northeast. (*Source:* D. L. Anderson, The San Andreast fault, *Scientific American*, vol. 225, no. 5, 1971. Copyright © Bunji Taqawa, 1971.)

pable of overturn. Below the transition zone, dense minerals are present in the lower mantle, which may also be richer in iron than the upper mantle; this configuration appears to be gravitatively stable. If this is so, convection may be confined to the upper mantle – however, this conclusion is still uncertain.

Two kinds of motive force appear to be responsible for upper mantle convection: (1) near-surface descent of dense, negatively buoyant lithospheric plates, occasioned by their slipping laterally off bathymetric elevations such as mid-oceanic ridges, as well as sinking in the vicinity of trenches; and (2) upward-welling, buoyant mantle flow due to deeply buried heat sources. The abundantly documented, ubiquitous occurrences of oceanic ridges, subduction zones, and mantle plumes (hot spots) demonstrate that both types of process are operating today, and undoubtedly also attended ancient plate motions. The relative magnitudes of the near-surface and deep driving forces that contribute to the upper mantle circulation cells are still a matter of debate. In any

case, the process allows thermal energy buried within the Earth to be bodily transported toward the surface and, ultimately, lost to outer space.

QUESTIONS

1. Why is the Earth's surface capped by continents and ocean basins, whereas the rest of the inner planets are devoid of such features?

2. How does the modern concept of sea-floor spreading and plate tectonics contrast with Wegener's hypothesis of continental drift? Given the knowledge base of the time (1920), can you defend the views of the nondrifters?

3. Describe a research plan, the goal of which is to document the time of opening of the South Atlantic Ocean.

4. Why are the ocean basins floored by relatively homogeneous basaltic crust, whereas the continents contain igneous rocks ranging from basalt to rhyolite (averaging granodiorite in composition)?

5. Discuss the fundamental cause of contractional mountain ranges. How does this compare with regions of continental (or oceanic) extension?

6. Explain the nature of transform faults. How does this type of break contrast with normal and thrust faults?

FURTHER READINGS

Allègre, C. 1988. *The behavior of the Earth.* Cambridge, MA: Harvard University Press.

Cox, A. (ed.) 1973. *Plate tectonics and geomagnetic reversals.* San Francisco, CA: W. H. Freeman.

Hallam, A. 1973. *A revolution in the earth sciences: from continental drift to plate tectonics.* New York: Oxford University Press.

Uyeda, S. 1978. *The new view of the Earth.* San Francisco, CA: W. H. Freeman.

Wegener, A. 1915. *Die Entstehung der Kontinente und Ozeane.* Braunschweig, Germany: Friedrich Vieweg & Sohn.

Wilson, J. T. (ed.) 1972. *Continents adrift: readings from* Scientific American. San Francisco, CA: W. H. Freeman.

7 Fluvial Landforms: The Surface of the Earth

KEITH LOAGUE

THE SURFACE OF THE LAND – A RIVER RUNS THROUGH IT

Fluvial landforms are the result of water moving at or near the Earth's surface. Thus, fluvial landforms can often provide an independent means to interpret past climates, because changes in global climate will eventually be propagated to the fluvial landscape. An understanding of hydrology is the linchpin to a process-based characterization of fluvial geomorphology. Land use decisions during the past 100 years have, in many instances, had an impact on the hydrologic response of a given region and subsequently accelerated changes in the surrounding landscape. For example, erosion caused by surface runoff has resulted in significant landscape denudation in both agricultural and urban areas throughout the world. The evolution of fluvial landscapes results in geologic hazards, such as landslides, that can have detrimental environmental and societal impacts. For example, there is currently considerable debate concerning the cumulative watershed effects (CWEs) that are associated with logging old-growth rain forests. It has been shown that when a watershed is clearcut, sediments in the streams draining the area are dramatically increased; this, in turn, is due to loading from erosion and mass wasting that have caused increased runoff and recharge. With more sediment in the stream, the dissolved oxygen levels decline, the temperature increases, and the habitat for fish is degraded. Land cover changes resulting from fire also facilitate accelerated landscape evolution by fluvial processes with the removal of vegetation. Detailed observations and heuristic simulations provide a framework for understanding fluvial landform processes. Modern geomorphology, the study of fluvial landforms, is an interdisciplinary field combining biology, climatology, geology, and hydrology. It is an exciting time to study fluvial landscapes. Land management decisions in the future will require that we understand the fluvial landscape, as well as the possible environmental impacts that may result from our interference with these natural systems.

Don't try to describe the ocean if you've never seen it.
Don't ever forget that you just may end up being wrong.

Song lyrics of Jimmy Buffett

One-third of the Earth's surface, standing an average of 840 meters above sea level, is composed of the continents and oceanic islands. Various geologic parent materials and climatic conditions combine to produce vastly different landscapes. Detailed observations, often of very long duration, are necessary to capture the dynamics of landscape evolution. In this chapter, the focus is an introductory characterization of the impact that water has on shaping landmasses above sea level. The occurrence, distribution, and movement of water, both at the surface and within the near surface, provides much of the energy needed to do the morphologic work that shapes the Earth's surface. The hydrologic cycle, depicted in Figure 7.1, is an integrating system of storage reservoirs among which water is transferred at different rates. The average residence times for water (i.e., liquid, solid, and vapor phases) in the different components of the hydrologic cycle are provided in Table 7.1.

This chapter is divided into three sections. The first section discusses the occurrence and characteristics of different fluvial landform types. The second section reviews the influences and processes that shape dynamic landscapes. The final section discusses landscape evolution, and includes a case history of the restoration of a misused insular system.

FLUVIAL LANDFORM TYPES

Moving water is the single most important agent in changing the configuration of the Earth's surface. Water detaches, transports, and deposits sediments. With gravity as the principal driving force, sediment is moved downhill by moving water. The result of erosion in steep upland catchments is the wearing down of the landscape and the creation of new landforms that have more subdued topography. In this sec-

W. G. Ernst (ed.), *Earth Systems: Processes and Issues*. Printed in the United States of America. Copyright © 2000 Cambridge University Press. All rights reserved.

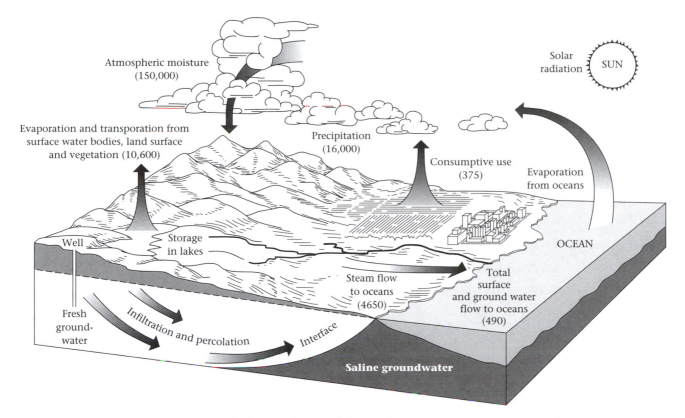

Figure 7.1. Hydrologic cycle, showing the gross daily water budget of the conterminous United States in millions of cubic meters of water per day. (*Source:* B. J. Skinner and S. C. Porter, *Physical geology*, Copyright © John Wiley & Sons, Inc., 1987. Reprinted by permission of John Wiley & Sons, Inc.)

tion, three fluvial landscapes are described: alluvial fans, deltas, and floodplains.

ALLUVIAL FANS. Alluvial fans are formed by the accumulation of sediments into cone-shaped deposits that enlarge downslope from the point where a stream leaves a steep mountain front or escarpment. As the slope of a typical mountain stream is reduced, when the valley plain is encountered, part of its sediment load is deposited because of the water's diminishing carrying capacity (which is proportional to the stream). As sediments are deposited, the stream

swings laterally back and forth across the fan from a fixed apex, such as a canyon mouth, seeking lower ground. Once on the fan, the stream typically splits into several channels. No two alluvial fans are exactly alike, varying in size and shape depending upon several factors, including: the area of the drainage basin; the lithology of the rocks in the source area; tectonic activity; the area available for the fan to spread into; and climate. An example of an alluvial fan is shown in Figure 7.2.

In general, fans consist of fluvial sand and gravel deposits interbedded with poorly sorted mudflows. The composition and sorting of a fan is a function of the upslope runoff-producing condition and, subsequently, the amount of water the stream discharges. Although alluvial fans are commonly associated with desert environments, they can also be found in humid regions. Typically, favorable infiltration characteristics and infrequent precipitation cause alluvial fans to be dry between storms, with only ephemeral stream flow and sediment deposition during heavy rainstorms.

DELTAS. A delta is an alluvial deposit built by the deposition and accumulation of sediment in shallow standing water. Deltas are formed where a stream enters an ocean or lake, and are composed of two components, the delta front and the delta plain. The delta front comprises the shoreline and the gently sloping submerged offshore zone. The delta plain forms an extensive lowland area lying on the terrestrial side

TABLE 7.1. RESIDENCE TIMES FOR WATER IN THE DIFFERENT COMPONENTS OF THE HYDROLOGIC CYCLE	
Components of the Hydrologic Cycle	Residence Time
Oceans and seas	~ 2000 years
Lakes and reservoirs	~ 10 years
Swamps	1–10 years
River channels	~ weeks
Soil water	2 weeks–1 year
Groundwater	2 weeks–10,000 years
Ice caps and glaciers	10–1000 years
Atmospheric water	10 days
Biospheric water	1 week

Source: Adapted from Freeze, R. A. and Cherry, J. A., *Groundwater*, Prentice Hall, 1979.

Figure 7.2. Alluvial fan in Copper Canyon, Death Valley, California; the fan begins at the upper center of the photograph, where the streams leaves the confined channel in the mountains for the broad lowlands. (*Source:* Photograph courtesy of U.S. Geological Survey.)

of the delta front, which is composed of active and abandoned channels. A delta can have a single channel or a diverging set of channels fanning across the plain. The area between channels within a delta can be occupied by floodplains, tidal flats, marshes, or lakes. The geomorphologic characteristics of an individual delta are a function of fluvial, wave, and/or tidal processes. An example of a delta is shown in Figure 7.3.

The velocity of a stream when it enters a body of standing water decreases rapidly, which in turn causes an abrupt change in the stream's energy and ability to transport sediment. A delta grows thicker and wider over time as long as there is a sediment supply. The weight of accumulating sediment in a delta can cause the underlying deposits to compress, eventually resulting in land subsidence.

FLOODPLAINS. A floodplain is a strip of relatively uniform, subdued relief on a low-gradient valley floor bordering a stream. Floodplains result from erosion and deposition caused by the dynamics of a flooding stream. Typically, an entire floodplain is inundated only during low-frequency floods (major events). Of course, portions of the floodplain can be flooded more often due to smaller and more frequent events. There are two types of sediment deposition on a fluvial floodplain: lateral-accretion deposits, and vertical-accretion deposits. Lateral-accretion deposits are laid down as a stream migrates across its floodplain. Vertical-accretion deposits accumulate beyond the confines of meandering and braided stream channels during flood events when bankfull discharge is exceeded. The most prominent depositional landform features in a floodplain are the levees (ridges), which lie parallel to the stream channel. The lateral accretion in a meandering channel is the result of the gradual development of point-bar deposits on the convex side of meander bends. In general, water moves more slowly on the inside bend of a channel. Therefore, the stream is unable to transport as much sediment, resulting in deposition. A meander cutoff is produced when one meander bend overruns another. Abandoned channels in a floodplain are caused by rapid changes in stream courses. When a meander loop is isolated from the main channel by sediment deposition at each end of the section, it becomes an oxbow lake.

Figure 7.3. View of the Nile River delta from the space shuttle. (*Source:* Photograph courtesy of Lunar and Planetary Institute.)

PROCESS AND INFLUENCE

The modern characterization of the surficial processes that shape fluvial landscapes can be categorized as either hydrologic or geomorphic. In this section, near-surface hydrologic and geomorphic processes are discussed relative to their influence on the evolution of fluvial landforms. The principal dialect of modern hydrology and geomorphology is mathematics. Therefore, the discussion here, although basic, is more quantitative than qualitative.

HYDROLOGY. Hydrology as a science embraces the full life history of water on the Earth and its interaction with the environment. The discipline of hydrology treats the occurrence, endless circulation between the ocean, atmosphere, and land (see Chap. 9), and properties (chemical and physical) of water.

Groundwater. Groundwater is the subsurface water that occurs beneath the water table in soil and geologic formations that are fully saturated. Groundwater movement, like heat flow, is controlled by a potential gradient, with flow moving from higher to lower fluid potential. The fluid potential in groundwater movement is hydraulic head. The hydraulic head (h) is the sum of the pressure head (ψ) and the elevation head (z). In the saturated subsurface, pressure heads are positive. The velocity in which groundwater moves is calculated by Darcy's law:

$$q = -K_s \frac{dh}{dl} \qquad (1)$$

where q is specific discharge (velocity), K_s is the saturated hydraulic conductivity, and dh/dl is the hydraulic gradient between different locations. The saturated hydraulic conductivity is a constant of proportionality; it is a function of the media and the fluid flowing through it. The fluid pressure, which, as discussed below, is important to slope stability analysis, is given by:

$$p = \rho\, g\, \psi \qquad (2)$$

where p is the gauge fluid pressure (i.e., with atmospheric pressure set equal to zero), ρ is the fluid (water) density, and g is the acceleration due to gravity. The partial differential equation that describes saturated subsurface fluid flow is based upon combining the continuity equation for fluid mass flow with Darcy's law. Analytic and numerical solutions to the various forms of the groundwater flow equation, although beyond the scope of this chapter, are now readily available. Groundwater flow models are very useful in developing new ideas and for asking "what if" questions related to applied problems.

Unsaturated (Vadose) Zone. The unsaturated zone occurs above the water table where the soil-pore spaces are only partially filled with water. The unsaturated zone is also referred to as the vadose zone. Soil-water movement in the unsaturated zone, like the saturated subsurface, is controlled by potential gradients, with flow moving from higher to lower potential. Pressure heads in the unsaturated subsurface are negative. Water in the unsaturated zone is held under tension; therefore, no natural outflow to the atmosphere can

Figure 7.4. Surface runoff on the ranch road crossing the 0.1-square-kilometer rangeland catchment known as R-5 (located near Chickasha, Oklahoma), resulting from a 65-millimeter rainfall event on November 17, 1984. (*Source:* Photograph from Keith M. Loague.)

occur except by evaporation or transpiration. Natural outflows, such as a spring, must come from the saturated zone. Darcy's law for unsaturated vertical flow is written as:

$$q = -K(\psi) \left(\frac{\partial \psi}{\partial z} + 1 \right) \tag{3}$$

where $K(\psi)$ is the nonlinear characteristic curve for hydraulic conductivity, for a particular unsaturated soil type, which is a function of the pressure head. The nonlinear partial differential equation describing unsaturated subsurface fluid flow is based upon a combination of the continuity equation for fluid mass flow and the unsaturated version of Darcy's law. For one-dimensional vertical flow, the unsaturated flow equation is known as the Richards equation. Most models of unsaturated flow are based upon numerical approximations.

Infiltration is defined as the entry of water into the soil profile at the surface. Many equations describe infiltration under various conditions. A common form for a one-dimensional vertical infiltration equation is:

$$i = 0.5\ S\ t^{-1/2} + 0.5\ K_s \tag{4}$$

where i is the infiltration rate, S is sorptivity, and t is time. The first term on the right-hand side of equation (4) represents the early-time tension-controlled component of the infiltration process; the second term is the gravity component. Infiltration is an important process in the development of most fluvial landforms.

Surface Water. Surface water occurs in channels, as overland flow, and in bodies of water such as lakes or reservoirs. In this chapter, the focus is on channels and overland flow because this is where surface water does most of its carving of the landscape. Estimates of stream discharge and flow velocities are critical to the sediment transport calculations in erosion analysis. The depth and velocity of surface water flow are functions of the land surface slope, roughness, and depression storage. The amount of precipitation available for surface runoff is the residual of recharge from infiltration and evapotranspiration. The large-and small-scale drainage patterns of surface water runoff are dependent upon the erosive energy of the moving water and the erodibility of the surface. An example of surface runoff on a rangeland road is shown in Figure 7.4. A hydrograph for a Horton-type rainfall-runoff event is shown in Figure 7.5. The transient flow of water over a surface or in a channel is described by the simultaneous solution of the partial differential equations of continuity and momentum. These hyperbolic equations are known collectively as the Saint Venant or shallow-water equations. Mathematically the shallow-water equations describe the propagation of a wave over a surface or in a channel and commonly are solved numerically.

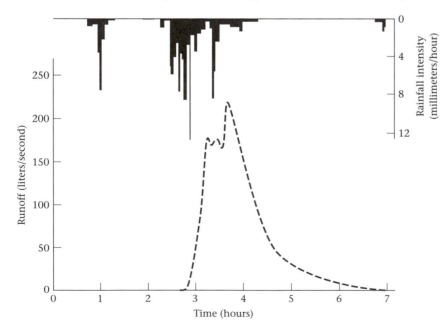

Figure 7.5. Observed hydrograph and hyetograph for the April 29, 1974, rainfall-runoff event on the R-5 catchment (the same catchment referred to in Fig. 7.4). (*Source:* Adapted from Keith M. Loague and R. Allan Freeze, *A comparison of rainfall-runoff modeling techniques on small upland catchments,* Water Resources Research, vol. 21, no. 2, paper number 4W1294.)

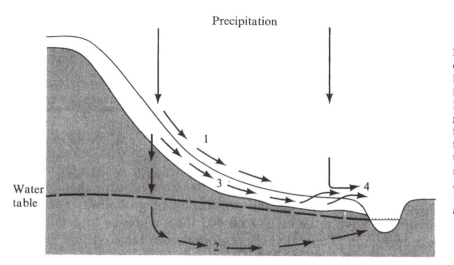

Precipitation

Water
table

Figure 7.6. Possible paths of water moving downhill. The unshaded zone indicates highly permeable topsoil; the shaded zone indicates less permeable subsoil or rock. Path 1 is Horton overland flow; path 2 is groundwater flow; path 3 is shallow subsurface stormflow; path 4 is Dunne overland flow (composed of direct precipitation on the saturated area plus infiltrated water that returns to the ground surface). (*Source:* Adapted from T. Dunne and L. B. Leopold, *Water in environmental planning*, Freeman and Company, 1978.)

Streamflow generation. The mechanisms of streamflow have been studied in some detail for at least 60 years. There is now considerable knowledge regarding rainfall-runoff processes and their controls. This understanding is the result of both careful observations and field experiments, and the heuristic simulations of hypothetical examples with rigorous mathematic models. It is now well known that there are three distinct forms by which water enters a channel segment in a catchment: (1) groundwater discharge, (2) subsurface stormflow (SSSF), and (3) overland flow. Groundwater discharge provides the sustaining baseflow component to a stream between storm periods. The flashy response of streamflow to individual rainfall events can be ascribed to either SSSF or overland flow. The conditions that lead to SSSF are quite restrictive. For example, SSSF can be associated with high-permeability preferential pathways such as mats of roots in shallow soils on steep hillslopes. At most locations, the primary source of rapid lateral inflow to a chan-

Figure 7.7. Schematic illustration of the occurrence of dominant near-surface streamflow generation mechanisms. (*Source:* Adapted from T. Dunne, Field studies of hillslope flow processes, in *Hillslope Hydrology* (ed. M. J. Kirby), Wiley, 1978, pp. 227–293; and from R. A. Freeze, A stochastic-conceptual analysis of rainfall-runoff processes on a hillslope, *Water Resources Research* [1980], 16: 391–408.)

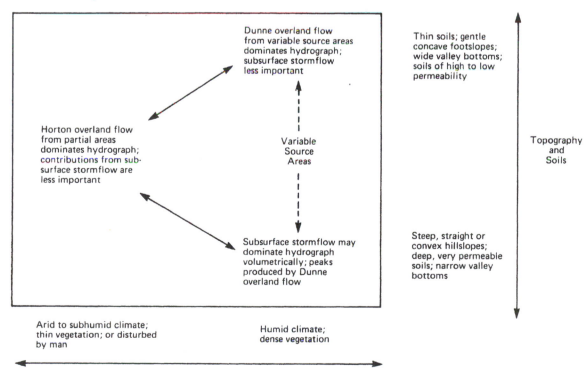

Figure 7.8. Headward erosion. (*Sources*: Photograph taken by G. K. Gilbert. Courtesy of U.S. Geological Survey.)

nel is overland flow. Overland flow can be generated on a hillslope only after surface ponding has occurred. It has been recognized for some time that surface saturation can occur by two quite distinct mechanisms: Horton overland flow and Dunne overland flow (each named after the researcher who best characterized these two separate processes). For the Horton mechanism, the rainfall intensity must exceed the saturated conductivity of the surface soil for a period of time sufficient for ponding to occur. For the Dunne mechanism, the duration of rainfall, at an intensity less than the saturated hydraulic conductivity, must exceed the period of time necessary for an initially shallow water table to rise to the surface. Horton overland flow is more common on upslope areas within a catchment, generated from areas where surface conductivities are the lowest. Dunne overland flow is more common in near-channel lowlands, generated from areas where the water table is the shallowest. Both the Horton and Dunne mechanisms can lead to variable source areas that expand and contract through wet and dry periods. The occurrence of the differ-

ent mechanisms of streamflow generation is schematically illustrated in Figures 7.6 and 7.7.

GEOMORPHOLOGY. Geomorphology is the study of the origin and evolution of topographic features that are a function of the physical and chemical processes operating at or near the Earth's surface. Landslides are a dramatic example of the hazards (see Chap. 25) that can be associated with landscape evolution. Humans have often facilitated disaster by building in areas susceptible to mass wasting. The photograph shown in Figure 7.8 is a classic example of headward erosion on a hillslope. Channel head initiation is currently one of the most intensely studied topics in geomorphology. Figure 7.9 schematically illustrates the climatic, vegetation, land use, and topographic conditions under which various geomorphic processes dominate hillslope erosion.

Weathering. Weathering is the physical and chemical alteration of rock or minerals at or near the Earth's surface. The products of weathering are discussed in some detail in Chapter 8. In physical weathering, the original rock disintegrates to

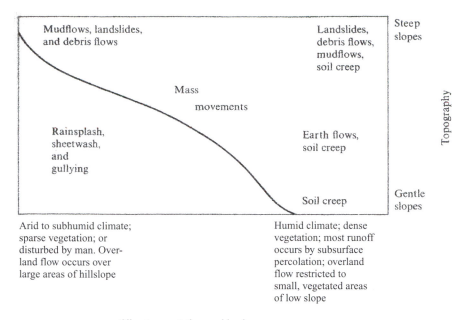

Mudflows, landslides,
and debris flows

Landslides,
debris flows,
mudflows,
soil creep

Steep
slopes

Mass

movements

Rainsplash,
sheetwash,
and
gullying

Earth flows,
soil creep

Topography

Soil creep

Gentle
slopes

Arid to subhumid climate;
sparse vegetation; or
disturbed by man. Over-
land flow occurs over
large areas of hillslope

Humid climate; dense
vegetation; most runoff
occurs by subsurface
percolation; overland
flow restricted to
small, vegetated areas
of low slope

Climate, vegetation, and land use

Figure 7.9. Conditions of climate, vegetation, land use, and topography under
which various geomorphic processes
dominate hillslope erosion. (*Source:*
Adapted from T. Dunne and L. B. Leopold, *Water in environmental planning.* ©
1978 by W. H. Freeman and Company.
Used with permission.)

smaller-sized material with no appreciable change in chemical or mineralogic composition. During chemical weathering, the chemical and/or mineralogic compositions of the original rocks and minerals changes. In nature, physical and chemical weathering occur together; therefore, it is difficult to separate the effects of one from the effects of the other.

The mechanism common to all processes of physical weathering is the establishment of sufficient stress within the rock so that it breaks. The processes that are most common to physical weathering are (1) unloading by erosion, (2) expansion in cracks or along grain boundaries by freezing water or crystallizing salts, (3) fire, (4) diurnal fluctuations in surface temperature, and (5) plant growth.

Chemical weathering occurs because rocks and minerals are seldom in equilibrium with near-surface waters, temperatures, or pressures. The products that form are more stable in the near-surface environment. Chemical weathering can be subdivided into congruent and incongruent dissolution. Congruent dissolution occurs when the mineral goes into solution and dissolves completely with no precipitation of other substances. Incongruent dissolution takes place when all or some of the ions released by weathering precipitate to form new compounds. Weathering of the near surface is a component of fluvial geomorphology, making sediment available for erosion and mass wasting.

■ **Erosion.** Erosion is an inclusive term for the detachment and removal of soil and rock by the action of running water, wind, waves, flowing ice, and mass movement. Climate and geology are the most important influences on erosion, with soil character and vegetation being dependent upon them and interrelated to each other. Virtually all significant erosion in fluvial landscape evolution is from the regolith (soil or weathered rock). Erosion is a function of erosivity and

erodibility. Erosivity is the potential to cause erosion (i.e., eroding power) and includes the impacts of raindrops, running water, and sliding or flowing earth masses. Erodibility is the vulnerability of the near surface to processes such as raindrop splash erosion, wash erosion, rill erosion, gully erosion, and pipe erosion. Erosion from unpaved roads, such as logging roads, is currently an important research topic because of the extremely high rates of sediment production and down-gradient impacts.

The basic erosion processes related to overland flow are: sediment detachment from surface runoff and raindrop splash; sediment transport from surface runoff and raindrop splash; and sediment deposition. The linkage between overland flow and erosion models is the flow depth and velocity for transporting sediment.

■ **Mass Wasting.** Mass wasting is the downslope movement of soil or rock debris in response to the pull of gravity. The gravitational stress acting on a mass is proportional to the weight acting upon it. Movement takes place when the component of gravity parallel to the slope exceeds the resisting strength of the material. Whether or not movement takes place is determined by the geologic and tectonic setting, the steepness and shape of the slope, by friction, by vegetation, and by the physical properties of the material. Geomorphologists classify mass movements by the nature of the material (rock or unconsolidated); the speed of movement (a few centimeters per year to many kilometers per hour); and the nature of the movement (falling/sliding or flowing). Where the dominant material is rock, the types of mass movement include rockslides, rockfalls, and rock avalanches. Where the dominant material is unconsolidated, the types of mass movement include creep, earthflows, debris flows, slumps, mudflows, debris avalanches, and debris slides.

Gravity is the principal driving force in mass wasting. The components of the gravitational force vector are the component normal to the slope and the component parallel to the slope, as shown in Figure 7.10. The magnitude of the component parallel to the slope, F_p, is related to the total gravitational force as:

$$F_p = F_g \sin \theta \qquad (5)$$

where F_g is the weight of the overburden, and θ is the slope angle. The gravitational force parallel to the slope varies from zero on a horizontal surface to a maximum on a vertical slope (i.e., as the slope angle increases, $\sin \theta$ increases). The magnitude of the component normal (perpendicular) to the slope, F_n is:

$$F_n = F_g \cos \theta \qquad (6)$$

The gravitational force is at a maximum on a horizontal surface and decreases progressively to zero on a vertical slope (i.e., as the slope angle increases, $\cos \theta$ decreases). The importance of the normal component is that as it increases, the frictional force (F_f) between an object and the surface it rests upon also increases. A convenient way of expressing stress (σ) on a mass of material on an inclined slope is to multiply its density by the vertical distance from the surface to the desired point of calculation.

The stress component parallel to the inclined plane, as illustrated in Figure 7.11, is:

$$\sigma_p = \sigma \sin \theta \qquad (7)$$

The stress component normal to the inclined plane is:

$$\sigma_n = \sigma \cos \theta \qquad (8)$$

The downslope and perpendicular components of the total stress are, respectively, the shear stress and the normal stress.

Resisting forces act in opposition to the gravitational forces (i.e., they oppose deformation or motion). The maximum resistance of material to stress is known as its strength, defined as the amount of stress required to cause failure of the material. The strength of materials varies considerably, depending on the physical properties of the material. The

Figure 7.10. Forces acting on a mass on an inclined surface. θ = slope angle; F = frictional force, F_g = force due to the weight of the overburden; F_n = force component normal to the slope; F_p = force component parallel to the slope.

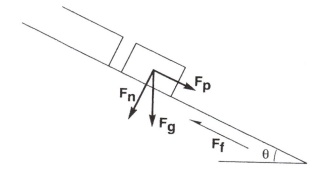

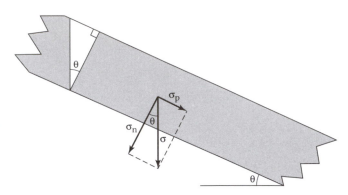

Figure 7.11. Stress components acting on an inclined plane. σ = total stress, σ_n = normal stress; σ_p = shear stress (downslope component); θ = slope angle.

resisting strength of a material may also vary for different kinds of stress (e.g., noncohesive unconsolidated material does not react in the same way as cohesive unconsolidated material). Stresses are transmitted through a soil or rock both by the soil skeleton and by the pore fluid. The soil skeleton can transmit normal stresses and shear stresses through the interparticle contacts; however, the pore fluid can exert only homogeneous pressure. The stresses transmitted by the soil skeleton through the interparticle contacts control the strength and deformation of the soil. Where the stresses applied to the soil profile are wholly supported by the pore-fluid pressure, they are not felt by the contacts between particles; hence, the soil behavior is not affected. This is the basis of the principle of effective stress. The effective stress (σ') acting on any plane (Fig. 7.12) is defined as:

$$\sigma' = \sigma - p \qquad (9)$$

The total stress is equal to the total force per unit area acting normal to the plane. The effective stress is approximately equal to the average intergranular force per unit area. The compressibility and shear behavior of a soil deposit are strongly influenced by its stress history. Soils derive their strength from the contacts between particles, which can transmit the normal and shear forces. In general, these interparticle contacts are primarily frictional, and so shear strength is directly governed by the effective stresses. The linear relationship for shear strength is:

$$s = c' + \Delta c + \sigma' \tan \phi' \qquad (10)$$

where c' is cohesion, Δc is cohesion from plant roots, and ϕ' is the angle of shearing resistance determined using effective stress. Cohesion binds soil together by electrostatic forces between clay particles or by chemical cementation. Plant roots also provide cohesion, which can be very important to slope stability. As the cohesion of a material increases, the total shear strength increases, according to equation (10). Cohesion is the residual shear strength when effective normal stress is zero. Sands and gravels are known as cohesionless soils because they have no shear strength when they are unconfined; their strength when confined arises from inter-

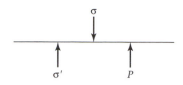

Figure 7.12. Graphic representation of the relationship between total stress (σ), effective stress (σ'), and fluid pressure (p) on a horizontal plane.

particle friction and the interlocking of particles. Clays are known as cohesive soils because they generally possess a significant shear strength when unconfined. Occasionally, some bonding occurs between clay particles; often, however, this strength is due to suction (i.e., the pressure head is less than atmospheric pressure). The tension results in a positive effective stress and some shear strength.

Two important differences between clays and sands distinguish their behavior in shear: permeability and shape. Clay is much less permeable than sand; this inhibits the movement of water. As a result, it may take years after a change of surface loading on a deposit of clay for excess pore pressures to dissipate and the effective stresses to reach equilibrium. The shape difference between clays and sands is straightforward. Clay particles are plate-like, and sand grains are spherical. When a shear plane forms in a clay and there has been substantial shear deformation across it, adjacent clay particles tend to align themselves parallel to the plane with significantly smaller shear strength. Obviously, evaluation of pore pressures is essential to an effective stress stability analysis. Because positive pore-fluid pressures decrease effective normal stress [see equation (9)], saturated soil on a slope has less internal frictional strength than a dry mass. Consequently, slope failures are commonly associated with heavy rains that saturate the soil (i.e., the fluid pressure increases as the shear strength decreases because the effective normal stress is reduced).

LANDSCAPE EVOLUTION

The evolution of landscapes through time is the central focus of modern geomorphology. In the late nineteenth century, pioneering explorers, such as John Wesley Powell and Grove Karl Gilbert, made major advances in characterizing geomorphic features and surface processes in the American West. Specifically, the roots of many of the basic, process-based, geomorphic principles used today can be traced to Gilbert's 1877 classic monograph on the Henry Mountains of southern Utah. Parallel to Charles Darwin's hypothesis for coral reef formation, William Morris Davis forged a comprehensive space–time concept of landform evolution known as the cycle of erosion. In its simplest form, the cycle of erosion consists of the incessant dissection of an initial surface high above base level, with the stream-carved landscapes evolving through a sequence of stages until the land is reduced to a

low plain. The stages in Davis' (1909) cycle of erosion are youth, maturity, and old age. The major limiting requirements of the cycle of erosion concept are stability in both sea level and the Earth's crust. Obviously, tectonic processes and sea level are dynamic; in fact, the fluctuations in sea level during the Pleistocene (i.e., the glacial epoch) were most likely between 100 and 150 meters.

In addition to Davis' cycle of erosion, several other conceptual models of landscape evolution have been posited. The modern era of geomorphology has, however, gone full circle, back to Gilbert, and has witnessed a growing emphasis on quantitative analysis of landform morphology. This new era is driven by detailed field characterization of near-surface processes of the type first synthesized by Robert Horton in his remarkable 1945 paper on the erosional development of streams and their drainage basins. Recently, geographic information system (GIS) and digital terrain and elevation model (DTM, DEM) technologies have been coupled to simple process models to characterize quantitatively regional-scale landscape evolution.

RESTORATION GEOMORPHOLOGY. The evolution of many landscapes has been accelerated in the past 100 years due to advances in technologic capacity coupled with human mismanagement. Large-scale open pit and strip mining, fire, road construction, residential development, and timber harvesting are only some of the many land management activities that have facilitated drastic regional-scale changes to the landscape. Currently a large environmentally driven effort is under way to fix, or ''heal,'' misused and devastated landscapes. This field is termed restoration geomorphology. The Kaho'olawe case study presented below is an example of a process-based study that falls into the niche of restoration geomorphology.

Kaho'olawe: A Case Study. The landscape of the Hawaiian Islands is incredibly dynamic and extremely active (see Fig. 7.13). Kaho'olawe, the eighth largest island in the Hawaiian chain (see the inset in Fig. 7.14), provides an excellent example of accelerated landscape evolution resulting from land mismanagement. The island is 18 kilometers long and 10 kilometers wide, with a maximum elevation of 454 meters; its climate is semiarid. Kaho'olawe is 1.5 million years old and 117 square kilometers in area. It was first occupied by humans, around 900 AD. Archaeobotanic evidence suggests that dryland forest gave way to grassland vegetation as a result of agricultural practices of native Hawaiians during the period prior to European contact. Abandonment of the traditional Hawaiian land stewardship system in favor of western land tenure in 1848 initiated rapid land use change on Kaho'olawe. The average annual rainfall for Kaho'olawe is estimated to be approximately 750 millimeters per year. Constant strong winds have resulted in the widespread redistribution of eroded soil in recent times. However, soil erosion had not been a significant problem on the island until traditional Hawaiian land use was replaced by grazing.

(A)

Figure 7.13. Upper Olokele Canyon is southwest Kauai, in the Hawaiian Islands, (A) prior to (B) following the landslide of October 1981. (*Source:* G. A. MacDonald, A. T. Abbott, and F. L. Peterson, *Volcanoes in the sea: the geology of Hawaii*, 2nd edition. Copyright © University of Hawaii Press, 1983.)

(B)

The U.S. Navy took possession of the island in 1941. Initially used for the training of military personnel during World War II, the island continued to be used as a bombing range and small-ordnance training site until 1990. The feral goat population on the island remained essentially unchecked during most of the Navy management period. The damage to the Kaho'olawe landscape, due to erosion of surfaces unprotected by vegetation, worsened as overgrazing continued. Bombing further degraded the vegetation cover and subsequently increased erosion. During the 1970s, Hawaiian protests regarding military use, and general mismanagement of Kaho'olawe led to the formation of an organization called Protect Kaho'olawe 'Ohana (PKO). Members of the PKO won a summary judgment in federal court against the Navy in 1977 and later negotiated a consent decree as an out-of-court settlement of a civil suit. In October 1990, by presidential order, bombing and munitions training on the island ceased.

■ **Physical Environment.** The geology of Kaho'olawe is depicted in Figure 7.14. Kaho'olawe is a shield-shaped extinct volcano composed chiefly of thin flows of unfractionated basalt poured in rapid succession from three rift zones and a vent at their intersection. Marine erosion has cut cliffs as high as 240 meters along the eastern and southern shores. The summit of the island has been eroded to bedrock. The slopes of Kaho'olawe are corrugated with gulches 15 to 60 meters deep, cut by ephemeral streams.

The island-wide land cover distribution for Kaho'olawe is shown in Figure 7.15. The land cover classifications shown in this figure were identified and mapped at a regional scale based upon a 1986 set of U.S. Navy–flown low-altitude color air photographs and ground-truth information. Measurements of near-surface hydraulic properties were made at 110 locations on the island and used with land cover information to explain the spatial distribution of near-surface hydraulic properties. "Land cover" is used here as a surrogate for "near-surface hydraulic property" distributions in space at unsampled locations. Figure 7.16 shows the land cover distribution for the Hakioawa catchment (the shaded catchment in Fig. 7.17).

■ **Study Objective.** The overall objective of this effort was to characterize the near-surface hydrologic response across the entire island of Kaho'olawe. Spatially and temporally variable water-balance estimates are crucial to the characterization of soil-water recharge and surface runoff for the entire island. Estimates of soil-water recharge and surface runoff are critical to planning revegetation activities and erosion control in order to accomplish the landscape restoration goals on Kaho'olawe. To quantify the surface water component of the Kaho'olawe water balance, a rainfall-runoff model was employed. The distributed simulation approach, coupled with land-cover characteristics, makes it possible to address the impact of spatially variable near-surface hydraulic properties on estimates of surface runoff.

■ **Rainfall-Runoff Model of Horton Overland Flow.** It is well understood that overland flow, caused by excess rainfall on exposed soil and subsoil surfaces, leads to sediment transport and reduced likelihood for reestablishing vegeta-

Figure 7.14. Geologic map of Kaho'olawe. (*Source:* Adapted from H. T. Sterns, *Geology and groundwater resources on the islands of Lauai and Kaho'olawe, Hawaii*, Bulletin #6, Hawaii Division of Hydrography, 1940.) AA' and BB' show cross-section lines, with cross sections illustrated at the bottom of the figure.

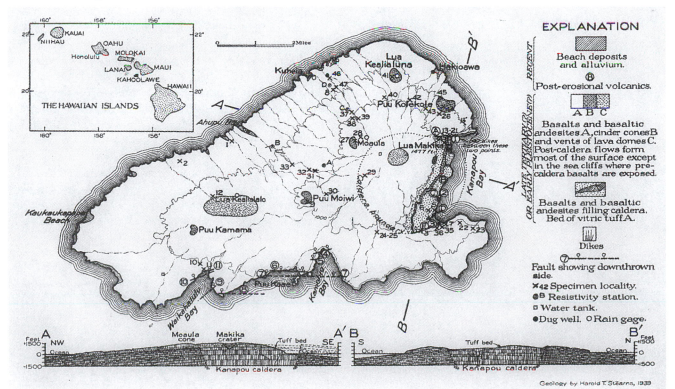

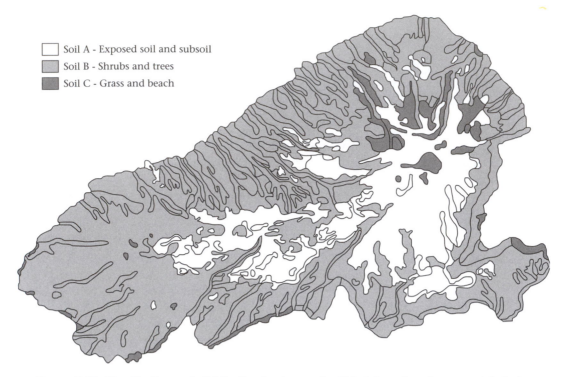

☐ Soil A - Exposed soil and subsoil

▨ Soil B - Shrubs and trees

▨ Soil C - Grass and beach

Figure 7.15. Classification and distribution land cover for Kaho'olawe, based upon aerial photographic interpretations. (Source information provided in Fig. 7.17.)

Figure 7.16. Classification and distribution of land cover for the Hakioawa catchment, after The Hakioawa catchment is the shaded catchment shown in Figure 7.17.

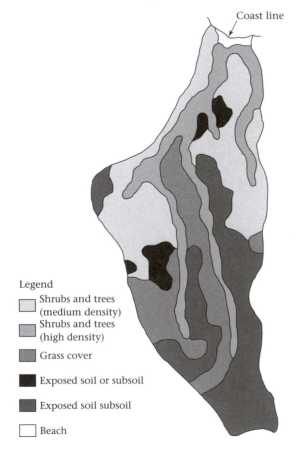

Coast line

Legend

☐ Shrubs and trees (medium density)

▨ Shrubs and trees (high density)

▨ Grass cover

■ Exposed soil or subsoil

▨ Exposed soil subsoil

☐ Beach

tion. The primary form of streamflow generation on Kaho'olawe, where erosion is significant, is overland flow. The Horton mechanism is the dominant overland flow process for many areas across Kaho'olawe because of low surface conductivities and a deep water table.

A quasi-physically based rainfall-runoff model (QPBRRM) was used in this study to investigate the magnitude of Horton runoff likely to occur. The operating algorithms for the QPBRRM are based on solutions to and/or simplifications of the full set of coupled partial differential equations that describe near-surface hydrologic response. The model has three major components: an infiltration algorithm that allows calculation of the rainfall excess; a routing algorithm that translates partial area rainfall excess, generated on the overland flow planes, into lateral inflow hydrographs at the stream channel; and a routing algorithm for tracking the streamflow hydrograph through the channel system. The model allows partial-source areas to expand and contract during a storm. Observed runoff events were not simulated in this study.

Implementation of the QPBRRM for a given Kaho'olawe catchment was accomplished using a set of overland flow planes that divide the areas of interest into segments. For this study the island of Kaho'olawe was divided into its seventy major catchments (Fig. 7.17). Based upon field measurements and aerial photographic interpretations, a scenario was developed to represent the near-surface hydraulic properties for Kaho'olawe; the three basic land cover designations were exposed soil or subsoil, shrubs and trees, and grass. The general and detailed characterizations of the overland flow planes used for QPBRRM simulations of Kaho'olawe and the Hak-

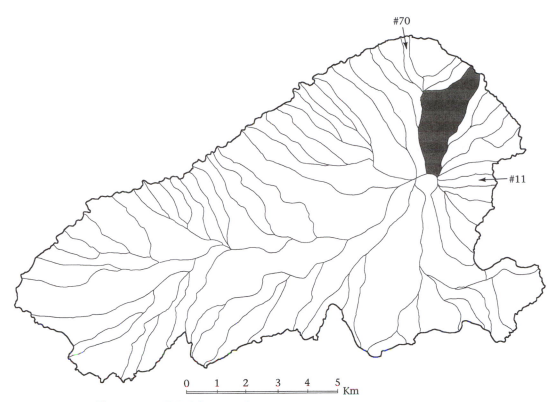

Figure 7.17. The seventy Kaho'olawe catchments; each drainage terminates at the ocean. The shaded catchment is the Hakioawa catchment. km = kilometers. Parcels #11 and #70 are discussed in the text. (*Source for Figs. 7.15–7.21:* K. Loague, D. Lloyd, T. W. Giambelluca, A. Nguyen, and B. Skata, Land misuse and hydrologic response: Kaho'olawe, Hawaii, *Pacific Science* [1996], 50: 1–35.)

Figure 7.18. Surface representations for quasi-physically based rainfall-runoff model (QPBRRM simulations). (A) Segments used to transform the Kaho'olawe catchments into overland flow planes.

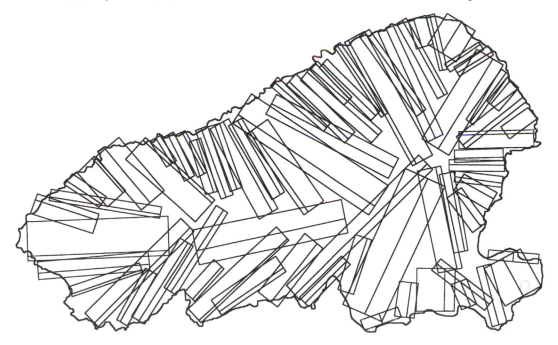

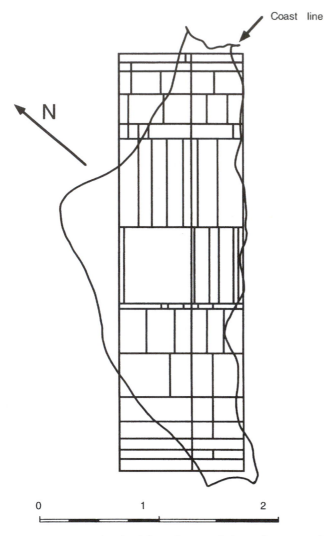

Figure 7.18.(B) Overland flow planes and channel segments for the Hakioawa catchment.

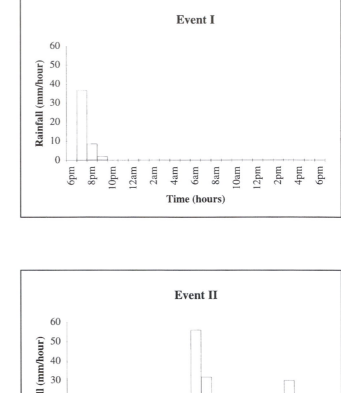

Figure 7.19. Rainfall events I and II used for the QPBRRM simulations of Horton overland flow on Kaho'olawe. mm/hour = millimeters per hour.

ioawa catchment are shown in Figure 7.18. The level of detail shown for Hakioawa in Figure 7.18B is the same for the other sixty-nine catchments. The two rainfall events used for this case study of Kaho'olawe are illustrated in Figure 7.19. The rainfall events in Figure 7.19 are typical for Kaho'olawe and serve to illustrate the type of Horton runoff response the island experiences as a result of large storms.

■ **Discussion.** The runoff response (runoff depth) for Kaho'olawe for rainfall events I and II is graphically summarized in Figure 7.20. On a coarse scale, the distributed nature of runoff response corresponds closely to the land cover distribution in Figure 7.15. As should be expected, the response (runoff depth) is much greater and more widespread for the larger of the two events (event II). The catchment with the largest response to rainfall event I is #70; catchment #11 has the largest response to rainfall event II. (See Figure 7.17 for

the locations of these land parcels.) The variability in runoff depth for the different events and the different catchments illustrates the differences (and changes) in erosion potential across the island.

Closer inspection of the Hakioawa catchment facilitates a more detailed, catchment-scale examination of the impact land cover variability has on runoff response to the two rainfall events. Figure 7.21 illustrates the distribution of excess rainfall, reinfiltration, and contributing overland flow partial source areas. The areas of excess rainfall, reinfiltration, and Horton overland flow in Figure 7.21 correspond closely to the distribution of land cover shown in Figure 7.16. Inspection of Figure 7.21 illustrates the potential location of sediment sources and deposition sites for the Hakioawa catchment for the two rainfall events; the specific catchments characterized as excess rainfall locations are taken here as a surrogate for sediment sources (erosion sites), whereas the reinfiltration catchments are taken here as a surrogate for

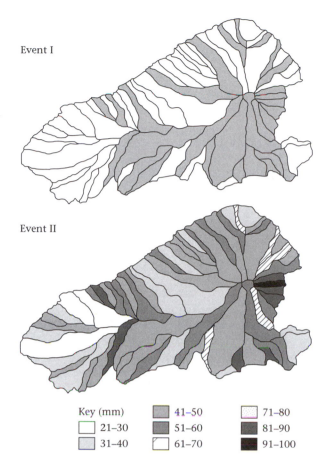

Key (mm)
☐ 21–30	▨ 41–50	▦ 71–80
☐ 31–40	▩ 51–60	▦ 81–90
	☐ 61–70	■ 91–100

Figure 7.20. QPBRRM simulated depths for the Kalo'olawe catchments: event I and event II. mm = millimeters.

deposition sites. It is clear from the response shown in Figure 7.21 that sediment source and deposition sites are a function of the size of the rainfall event. Aggregated over long time intervals, hydrologic response simulations of this type can be used to drive surface erosion–dominated landscape evolution models in order to reconstruct the past or to investigate the impact of future land-management decisions.

■ **Paradise Lost or Restoration Geomorphology?** The runoff estimates reported here provide a quantitative hydrologic response characterization for Kaho'olawe. Simulations of this type, combined with evapotranspiration estimates, could be used in a near-surface water balance to calculate, by difference, recharge, this number, in turn, could be used to estimate soil-water contents in space and time for Kaho'olawe. Information on potentially available soil-water content will be extremely useful for future revegetation efforts on Kaho'olawe.

A study such as this sets a foundation for simulating sediment transport on Kaho'olawe. The QPBRRM simulations of surface runoff by the Horton mechanism enable scientists to estimate the velocities and depths of overland flow for different land covers, which can be coupled with a process-based erosion model. The amount of sediment that can be moved from an upland source depends on the amount of sediment made available for transport by detachment processes and the sediment transport capacity of the moving water. Simulation of landscape evolution scenarios on a human time scale, under different management options, by coupling our QPBRRM simulations with an erosion model should be useful in the restoration and future management of Kaho'olawe.

■ **Epilogue.** Recently, Kaho'olawe was returned to the state of Hawaii. The U.S. government has set aside funds to begin the restoration process, which could take generations. The initial step in healing this island is to remove all the remaining ordnance.

SUMMARY

Landforms of fluvial origin represent a significant fraction of the terrestrial surface of the Earth. Three major fluvial landforms were discussed at the beginning of this chapter: alluvial fans, deltas, and floodplains. All the fluvial landforms at the Earth's surface are shaped by hydrologically driven geomorphic responses such as erosion and mass wasting. A cause-and-effect understanding of a dynamic fluvial landscape, which can be used to make predictions about future changes and perhaps guide management and restoration activities, is made possible only by coupling the disciplines of hydrology and geomorphology at the level of basic processes. Field experiments designed to induce linked hydrologic and geomorphic response, at plot and hillslope scales, and the physics-based simulation of these observed processes follow the measure and model approach pioneered by Gilbert and Horton, and facilitate extrapolations to larger scales for the characterization of evolving fluvial landscapes.

The geologic rates typically associated with landscape evolution can be greatly accelerated when the equilibrium conditions under which a landform is evolving are dramatically disrupted by changes that are either natural or the result of human activity. More than ever before, the landscape is being modified relative to the time scales associated with human activities. For example, the conversion of large land areas from forest to agricultural use has had a devastating impact on the tropical landscapes in Brazil and Southeast Asia, increasing erosion to alarming and perhaps irreversible rates. We know how to minimize or ameliorate the damage to the terrain – the question is, will we act before the harm is irrevocable?

QUESTIONS

1. What are the global geomorphologic implications of deforestation in the Amazon rain forest? How would this differ from deforestation of the Olympic Peninsula, or the Alaskan Range?

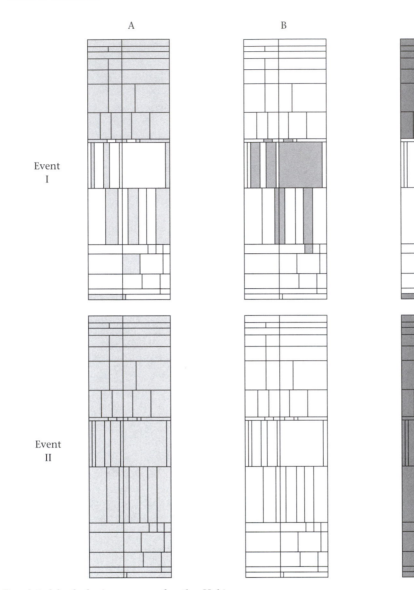

A   B   C

Event
I

Event
II

Figure 7.21. Simulated hydrologic response for the Hakioawa catchment for rainfall events I and II. (A) Overland flow plane segments experiencing excess rainfall (red); taken here as potential sediment source areas. (B) Overland flow plane segments experiencing reinfiltration (blue); taken here as potential sediment deposition sites. (C) Overland flow segments that are partial source areas (green). The bottom edge of each response representation is the coast portion of the catchment (in Fig. 7.17 and 7.19 (event II) the top edge is the coast).

2. What are the cumulative watershed effects (CWEs) that result from the local- and regional-scale clearcutting in the Pacific Northwest?

3. Why does the removal of vegetation lead to increased pore pressures in the subsurface and, in some cases, to slope instability?

4. What are the components of the hydrologic cycle that play principal roles in the evolution of fluvial landscapes?

5. What are the important hydrologic and geomorphic elements to consider in the restoration of a misused landscape?

FURTHER READING

Davis, W. M. 1909. *Geographical essays*, ed. D. W. Johnson. Boston, MA: Ginn.

Dunne, T. and L. B. Leopold. 1978. *Water in environmental planning*. San Francisco, CA: W. H. Freeman.

Freeze, R. A., and J. A. Cherry. 1979. *Groundwater*. Englewood Cliffs, NJ: Prentice Hall.

Gilbert, G. K. 1877. *Report on the geology of the Henry Mountains*. Washington, DC: U.S. Department of the Interior.

Horton, R. E. 1945. Erosional development of streams and their drainage basins: hydrophysical approach to quantitative morphology. *Bull. Geolog. Soc. Am.* 56: 275–370.

Selby, M. J. 1993. *Hillslope materials and processes*, 2nd edition. New York: Oxford.

8 Chemical Weathering and Soils:

Interface between the Geosphere, Hydrosphere, Atmosphere, and Biosphere

MARK J. JOHNSSON

GROWTH SUBSTRATE OF THE LAND

A thin film of soil covers much of the Earth's land surface. Within this film the geosphere, hydrosphere, atmosphere, and biosphere come together. Through chemical weathering, the constituents of the geosphere are altered and reorganized, and are combined with material from the other three spheres to produce the soils in which nearly all of the world's crops are grown. The rates of chemical weathering and of erosion determine the thickness and nature of this important resource. When humans affect these rates, they do so at great peril, for nearly all life on land ultimately depends on soils.

Chemical weathering is the process by which Earth materials are broken down and converted to new forms. Some of these new substances remain in virtually the same location as the original bedrock, and form soils. Others are dissolved and carried away by running water, to provide nutrients for aquatic ecosystems in rivers, lakes, and seas. The materials of soils are eventually loosened and removed, forming deposits of sediments that are gradually converted to sedimentary rocks. Soil and sediment transport conditions vary widely in both the intensity of the chemical weathering environment and in how long weathering proceeds. Given the diversity of microenvironments in which chemical weathering can occur, and the diversity of rock types that undergo weathering, a wide array of weathering products and soils can be formed.

The Earth system has experienced wide changes over geologic time. The chemistry of the atmosphere has undergone radical change, the evolution of different types of plants and the animals that feed on them have led to changing biodiversity, and the global climate system has continually varied. Many of these phenomena bring about changes in the nature of chemical weathering, which are in turn reflected in changes in the composition of weathering products. Chemical weathering is the single most important process in the transfer of material from the geosphere to the rest of the Earth system. Sedimentary rocks made up of materials that have undergone chemical weathering are our only direct record of these changes in the Earth system over geologic time. Thus, understanding the processes involved in chemical weathering and how chemical weathering affects the composition and distribution of sedimentary rocks is critical to our ability to chart changes in the Earth system over long time scales. Understanding how the Earth system has operated over geologic time also helps us predict how it will respond to changes in the future, especially those changes induced by humans. Given the role of soils in determining agricultural productivity, changes in this component of the Earth system will be of great concern as the human population continues to increase.

Many minerals found in rocks were formed under conditions of temperature and pressure very different from those at the surface of the Earth (Chap. 5). When uplift and erosion bring them to the surface, they are often chemically unstable under surficial conditions. The process by which minerals dissolve or break down to form new minerals that are stable under conditions at the Earth's surface is called *chemical weathering*. The majority of weathering occurs when bedrock is converted to soil; however, additional weathering can occur as soils are eroded, transported, and deposited as sediments. Chemical weathering is an efficient partitioning process, whereby rock constituents are apportioned between three principal classes

of materials: ions in solution, secondary weathering products (i.e., new minerals grown during weathering), and incompletely weathered debris (Fig. 8.1). Incompletely weathered debris may consist of material that was resistant to chemical weathering, or it may contain appreciable amounts of unstable material that escaped transformation due to weak chemical weathering conditions or to rapid erosion. Of the solid products of weathering, secondary weathering products typically are fine-grained (commonly of clay size), whereas incompletely weathered debris tends to be relatively coarse and is concentrated in the silt-and sand-sized fractions of sediments.

Chemical weathering is one aspect of *erosion*, the process

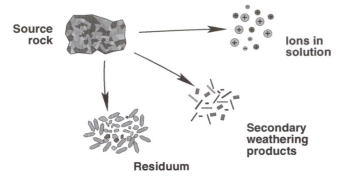

Figure 8.1. Chemical weathering results in the partitioning of rock components into three principal classes of weathering products: dissolved ions in solution, secondary weathering products, and incompletely weathered residuum.

by which the materials of the Earth's crust are dissolved or worn away and transported to sedimentary basins. Erosion almost always involves chemical weathering; however, under conditions of very rapid uplift, or in areas of very dry climate, chemical weathering is slow. Under these conditions, chemically unstable materials of the Earth's crust can be eroded and transported with relatively little change in their composition. Processes that can break down rocks without altering their composition include wedging by ice or salt crystals in fractures or between mineral grains, pounding by surf, temperature variations, and abrasion (grinding) in river beds or beneath glaciers. These processes are referred to collectively as *physical weathering*. Chemical weathering is distinct from physical weathering in that actual chemical alteration of minerals occurs, in addition to physical disaggregation. In reality, however, they typically occur together, and both contribute to the mass removed from the land surface by erosion. In this chapter we are concerned primarily with chemical weathering because chemical processes are most effective in transporting materials from the geosphere to the hydrosphere and biosphere.

Chemical and physical weathering are important parts of the rock cycle (see Chap. 5). All rocks exposed at the Earth's surface – whether igneous, metamorphic, or sedimentary – are broken down by weathering. Disaggregated and altered components are transported to low-lying areas by such processes as landslides, rivers, glaciers, and wind. Thus erosion removes weathered materials from upland areas to sedimentary basins where they are deposited. As sediments gradually accumulate, sometimes to great thicknesses, they are compacted by the weight of the sediments above and cemented by minerals precipitated in the spaces between grains to form sedimentary rocks. With sufficient burial, sedimentary rocks may be altered by heat and pressure to become metamorphic rocks, or even melted to form igneous rocks. When tectonic conditions change (e.g., through continental collision – see Chap. 6), these deeply buried rocks may be uplifted to the surface to begin the cycle of weathering and erosion anew.

BASIC WEATHERING PATHWAYS

The common silicate minerals form a weathering stability series (Fig. 8.2) analogous to Bowen's reaction series, the sequence in which minerals crystallize from a silicate melt (see Chap. 5). Minerals that crystallize at higher temperatures (olivine, amphiboles, pyroxenes, calcic plagioclase) are markedly less stable under chemical weathering conditions than are minerals that crystallize at lower temperatures (sodium plagioclase, potassium feldspar, micas, and quartz). In other words, the first minerals that crystallize from a melt are the first to decompose at the surface. The further a silicate mineral is from its pressure and temperature environment of stability, the more susceptible it is to chemical weathering. Nonsilicate minerals (e.g., carbonates, oxides, sulfides) are commonly even less stable under chemical weathering conditions than are most silicate minerals. Chemical weathering preferentially destroys the most unstable minerals, causing a relative increase in the proportion of the more stable minerals in soils and sediments. Because quartz is a particularly common rock-forming mineral and is also quite resistant to chemical weathering, the most pronounced effect of chemical weathering is commonly an overall enrichment of sediments in quartz.

Another way to view chemical weathering is to consider the relative mobility of various elements under chemical weathering conditions. Some elements are more mobile than others, largely because of the relative strengths of the metal–oxygen bond found in most minerals. Thus, of the most abundant elements in the Earth's crust, titanium, silicon, aluminum, and iron are relatively immobile, and magnesium, manganese, sodium, potassium, and calcium are increasingly more mobile under chemical weathering conditions (Table 8.1). Nonmetals are generally quite mobile during weathering, and sulfur and carbon are quickly removed from rocks exposed at the Earth's surface. Minerals that are least stable under chemical weathering conditions

Figure 8.2. The stability of minerals in soils is proportional to the order in which they crystallize from a silicate melt; a weathering stability series resembles Bowen's reaction series (see Chap. 5). (*Source:* Adapted from S. S. Goldich, A study in rockweathering, *Journal of Geology*, vol. 46, 1938, pp. 17–58.)

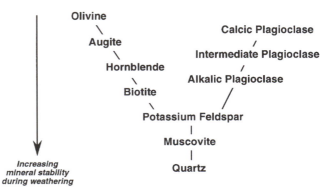

TABLE 8.1. RELATIVE STRENGTHS OF CATION-OXYGEN BONDS IN COMMON ROCK-FORMING MINERALS	
Bond	Relative Strength
Si—O	2.4
Ti—O	1.8
Al—O	1.65
Fe^{+3}—O	1.4
Mg—O	0.9
Fe^{+2}—O	0.85
Mn—O	0.8
Ca—O	0.7
Na—O	0.35
K—O	0.25

Source: G. D. Nicholls, Environmental studies in sedimentary geochemistry, *Science Progress*, vol. 51, 1963, pp. 12–31.

of a variety of mafic minerals such as amphibole or mica. From the discussion above, one might expect that continued weathering of a granitic source rock would first result in a selective depletion of the mafic minerals, then of plagioclase, followed by potassium feldspar, and finally quartz. Simultaneously, secondary weathering products might form, including clay minerals (e.g., kaolinite and montmorillonite) and iron oxides. Indeed, examination of weathering sequences, such as are preserved in soils, shows just such a trend.

As discuss in the section "Structure of Soil," a simple weathering profile produces a layered sequence, with the most weathered material at the top and the least weathered at the bottom of the profile. Figure 8.3 shows an ancient weathering profile developed on granodiorite in Colorado that displays a systematic decrease in the abundance of unaltered mafic minerals and plagioclase upward in the profile, and a systematic upward increase in clay minerals and iron oxides. These trends are typical of the weathering of felsic rocks; plagioclase and minerals rich in calcium, magnesium, and iron are preferentially decomposed relative to quartz and potassium feldspar, while clay minerals form as a result of that decomposition.

Granitic rocks are rather impermeable; that is, water does not flow through them easily. Chemical weathering depends to a large extent on the presence of water; consequently, weathering of impermeable granitic rocks commonly progresses inward from external exposed surfaces and internal fractures, and the rock mass erodes into rounded forms (Fig. 8.4). Such "spheroidal weathering" is common in other homogeneous rocks that contain regularly spaced fractures as well.

are composed of the more mobile elements. The more mobile elements are those that are most important to supporting life, and the ways in which these elements cycle between the geosphere, biosphere, atmosphere, and hydrosphere are important in controlling the distribution of life (see Chap. 19).

The clearest studies of chemical weathering involve the weathering of a homogeneous igneous source rock. In these cases, primary minerals that obviously have not been subjected to multiple cycles of weathering can be correlated unambiguously with their weathering products. Although the exact reactions and products vary with the temperature and chemical conditions under which weathering occurs, we can identify general pathways that are commonly followed by the weathering of igneous rocks of felsic (silicon-rich) and mafic (silicon-poor) composition.

WEATHERING OF GRANITE. Granitic rocks are felsic and consist primarily of quartz, potassium feldspar, and plagioclase feldspar, although they also may contain one or more

WEATHERING OF BASALT. Weathering of mafic igneous rocks such as basalt differs from the weathering of felsic rocks such as granite in three important ways. First, the mineral components of basalt typically are far less stable under Earth

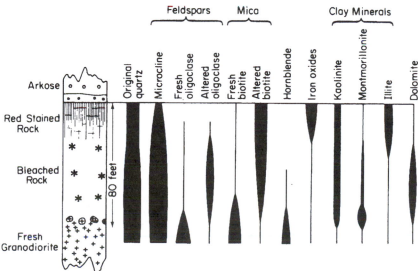

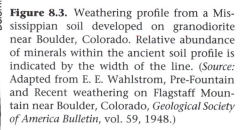

Figure 8.3. Weathering profile from a Mississippian soil developed on granodiorite near Boulder, Colorado. Relative abundance of minerals within the ancient soil profile is indicated by the width of the line. (*Source:* Adapted from E. E. Wahlstrom, Pre-Fountain and Recent weathering on Flagstaff Mountain near Boulder, Colorado, *Geological Society of America Bulletin*, vol. 59, 1948.)

Figure 8.4. Spheroidal weathering of relatively impermeable, homogeneous rocks such as granite results from weathering concentrated along fracture planes, such as in this granite from the Sandia Mountains of New Mexico. (*Source:* Photograph by Mark Johnsson.)

surface conditions than are those of granitic rocks. Second, basaltic rocks are generally layered due to differences in composition and texture of different lava flows. Third, extrusive rocks such as basalt are much finer grained than intrusive rocks, so that the individual mineral grains have a large ratio of surface area to volume. For these reasons, basaltic rocks tend to weather more quickly than granites.

Basaltic rocks, which are common in volcanic arcs and are the principal substrate of the ocean basins, crystallize at higher temperatures than do granites; they are composed principally of minerals that are less stable under chemical weathering conditions than those in granites. Indeed, quartz, the most resistant common mineral, is absent from basalts. Further, they are extrusive igneous rocks that cool quickly; they contain abundant glass (noncrystalline material), which is extremely unstable in the weathering environment. Because of their susceptibility to weathering, basalts form well-developed soils with little unweathered residual material. In intense weathering environments, these soils tend to be enriched in the relatively immobile elements aluminum and iron. In tropical environments, where chemical weathering is especially intense, soils developed on basalt are tinged a vivid red color due to the iron minerals they contain.

The sequence of weathering of basalts generally results in the progressive loss of glass, mafic minerals such as olivine, and plagioclase (Fig. 8.5). Secondary minerals formed during weathering include the clay minerals montmorillonite, halloysite, and kaolinite. These minerals form progressively during weathering: Basalt first weathers to form montmorillonite, then montmorillonite is converted to halloysite, and finally halloysite alters to kaolinite if weathering continues. Under intense weathering conditions, clays such as halloysite and kaolinite, which are rich in the relatively immobile elements aluminum and silicon, are more stable than montmorillonite, which is richer in the more mobile elements magnesium, calcium, and sodium.

MECHANISMS OF WEATHERING

We have discussed some of the general trends observed during chemical weathering; however, we have not yet examined how these trends are produced. Chemical weathering is the result of complex interactions between the geosphere, biosphere, atmosphere, and hydrosphere, taking place at the

Figure 8.5. Mineral transformations with the progressive weathering of basalt. (*Source:* Adapted from D. Carroll and J. C. Hathaway, Mineralogy of selected soils from Guam, U.S. Geological Survey Professional Paper 403-F, 1963, 51 pp.)

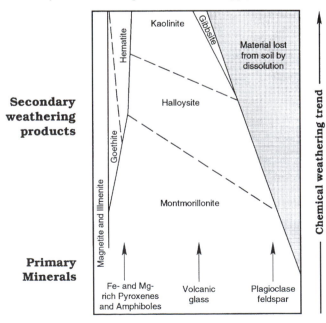

only point where all these "spheres" come in contact – at the surface of the Earth. In order to appreciate the complexity of these interactions, it is necessary to understand the general mechanisms of weathering from a chemical point of view.

WEATHERING PRODUCTS. The solid products of weathering take on many forms; the most important are the *clay minerals*, members of a diverse group of sheet silicates (similar to micas) that typically have grain sizes of less than 2 micrometers (1 micrometer equals one one-thousandth of a millimeter). The basic building blocks of the clay minerals are sheets of tetrahedra, consisting of silicon and aluminum bound to oxygen atoms, and sheets of octahedra, consisting of magnesium and iron bound to oxygen and hydrogen groups (Fig. 8.6). Two tetrahedral sheets linked by an octahedral sheet give rise to *three-layer clays*; linked tetrahedral and octahedral sheets are known as *two-layer clays*. Three-layer clays are generally richer in more mobile elements, such as magnesium, calcium, sodium, and potassium, than are two-layer clays. Three-layer clays that are important in chemical weathering reactions are montmorillonite, $(Ca, Na, Mg)Al_{1.7}(Mg, Fe)_{0.3}Si_4O_{10}(OH)_2 \cdot H_2O$, and illite, $K_{0.8}Al_{1.6}Mg_{0.4}Si_{3.6}Al_{0.4}O_{10}(OH)_2$. Two-layer clays, generally found in more intense chemically weathering environments, include kaolinite, $Al_4Si_4O_{10}(OH)_8$, and its hydrated form halloysite, $Al_4Si_4O_{10}(OH)_8 \cdot 2H_2O$.

Other important weathering products are members of the oxide and hydroxide groups of minerals. Such weathering products typically are of clay size (less than 2 micrometers) but are not sheet silicates. Most important are oxides and hydroxides of iron and aluminum. Particularly abundant are the minerals goethite (FeOOH), hematite (Fe_2O_3), diaspore (AlOOH), and gibbsite ($Al(OH)_3$).

WEATHERING REACTIONS. The means by which the hydrosphere and the geosphere interact during chemical weathering can be described by chemical reactions. In most cases, however, these reactions are gross oversimplifications of the complex processes operating during weathering. Although we may understand the quantitative relationship between products and reactants, the exact means by which one is converted to the other is currently the subject of much research.

Water is a powerful reactant in chemical weathering. This is partly because the water molecule can break down, or dissociate, into its constituent ions, which are then free to attack mineral surfaces:

$$H_2O \rightarrow \quad OH^- + \quad\quad H^+ \qquad\qquad (1)$$
$$\text{water} \quad \text{hydroxyl ion} \quad \text{hydrogen ion}$$

The hydrogen ion, H^+, is especially effective in attacking mineral surfaces. The quantitative measure of hydrogen ion concentration is pH: Low pH values (less than 7) indicate high hydrogen ion concentration, or acid conditions; high pH values (greater than 7) indicate low hydrogen ion concentration, or alkaline (basic) conditions.

Water can also combine with carbon dioxide in the air or soil to form carbonic acid:

$$CO_2 + \quad\quad H_2O \rightarrow \quad H_2CO_3 \qquad\qquad (2)$$
$$\text{carbon dioxide} \quad \text{water} \quad\quad \text{carbonic acid}$$

Some of the carbonic acid thus formed breaks down to release hydrogen ion as well as the bicarbonate ion, HCO_3^-:

$$H_2CO_3 \quad \rightarrow \quad H^+ + \quad\quad HCO_3^- \qquad\qquad (3)$$
$$\text{carbonic acid} \quad \text{hydrogen ion} \quad \text{bicarbonate ion}$$

Figure 8.6. Basic crystal structures of the clay minerals. (A) two-layer clays; (B) three-layer clays. See also Figure 5.10. (*Source:* H. Hamley, *Clay sedimentology*, Copyright © Springer-Verlag, 1989. Used with permission. Adapted from R. E. Grim, *Clay mineralogy*, McGraw-Hill, 1968.)

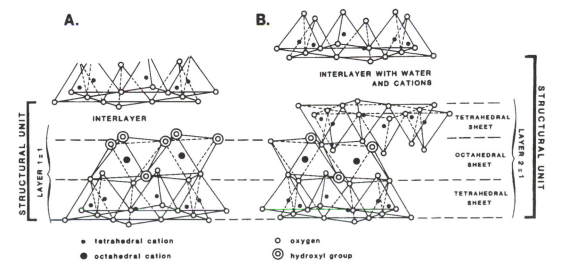

thus increasing the hydrogen ion concentration, and lowering the pH to create acidic conditions. Because reactions (2) and (3) occur whenever rain falls through the atmosphere, rainwater is naturally acidic, with a pH of approximately 5.6. Other natural acids make soil waters slightly more acidic, to a pH of approximately 5.2. This acidity is partly responsible for rainwater's ability to dissolve minerals. So-called acid rain is rainwater that has even lower pH, due to reactions similar to (2) and (3) that involve anthropogenic pollutants.

The pH of water in soils is often naturally much lower than 5.2, because it has been in contact with additional carbon dioxide produced from the decay of organic matter. The extent and type of vegetation are thus important in determining the reactivity of soil waters. Furthermore, since both the organic matter itself and the organisms responsible for decay are the products of evolution, the nature of chemical weathering in soils may have changed along with the evolving biosphere over geologic time.

EXTENT OF CHEMICAL WEATHERING. The partitioning of the constituents found in minerals among the three classes of chemical weathering products – ions in solution, secondary weathering products, and residual unweathered material – is controlled by the nature of the weathering reactions themselves and by the extent to which the weathering reactions are allowed to proceed. The extent of chemical weathering is a function of both weathering intensity and weathering duration.

Weathering Intensity. Because chemical weathering reactions must take place in the presence of water, the intensity of chemical weathering is controlled to a large extent by the quantity of water in the weathering environment. Climate obviously is of primary importance; therefore, areas with high precipitation such as tropical regions have the most intense chemical weathering conditions on Earth. However, chemical weathering is less intense in temperate regions with abundant rainfall, indicating that factors other than high precipitation are important in promoting chemical weathering in tropical climates. Elevated temperatures generally increase reaction rates. In addition, higher temperatures enhance evapotranspiration, which may act to increase precipitation further, creating a positive feedback loop as long as there is abundant water in the climate system. Thus, the combination of high rainfall and high temperatures ensures that tropical climates promote intense chemical weathering.

As important as the volume of water moving through the weathering environment is the composition of that water; acidic waters promote weathering. Carbon dioxide, which reacts with water to form carbonic acid (H_2CO_3), is a powerful reactant in the weathering of carbonate and silicate minerals. In addition to atmospheric contributions, carbon dioxide enters soil waters when they percolate through organic debris near the ground surface. The extent and nature of that organic layer is important in determining the reactivity of soil waters. In addition to contributing carbon dioxide to soil

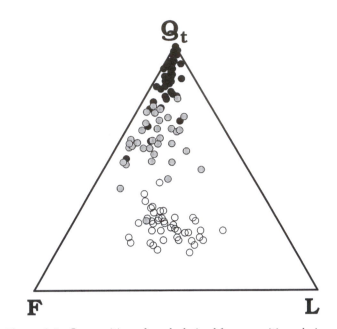

Figure 8.7. Composition of sands derived from granitic rocks in contrasting climatic regimes. Sand compositions are here expressed in terms of the relative proportions of quartz grains (Q_t), feldspar grains (F), and grains consisting of more than one mineral, or lithic fragments (L). These three components adequately describe the composition of most sandstones, and are plotted on a ternary diagram. Ternary diagrams are useful for portraying compositional information in terms of three components, one more component than can be conveniently expressed on most other types of diagrams. The proximity of a point to an apex of the triangle indicates the relative contribution of that component (e.g., pure quartz would plot at the Q_t apex, a 50–50 mixture of quartz and feldspar would plot midway between the Q_t and F apices, and an equal mixture of quartz, feldspar, and lithic fragments would plot at the center of the triangle). Sands produced in semiarid climates (indicated by open circles) are richer in feldspar and lithic fragments than are those from humid temperate climates (indicated by filled gray circles). Sands from tropical climates, indicated by filled black circles, are extremely quartz-rich. (*Source:* Adapted from A. Basu, Petrology of Holocene fluvial sand derived from plutonic source rocks: implications to paleoclimatic interpretation, *Journal of Sedimentary Petrology*, vol. 46, 1976; and M. J. Johnsson, R. F. Stallard, and N. Lundberg, Controls on the composition of fluvial sands froma tropical weathering environment: sands of the Orinoco River drainage basin, Venezuela and Colombia, *Geological Society of America Bulletin*, vol. 103, 1991.)

and ground waters, organic matter can increase the acidity of reacting waters through the introduction of organic acids, themselves an important source of hydrogen ions. Finally, anthropogenic air pollutants such as SO_x and NO_x gases also dissolve in water to form powerful acids, and elevated mineral weathering rates are found in areas experiencing human-induced acidic precipitation.

High precipitation and temperatures, coupled with generally high organic acid concentrations, lead to intense chemical weathering in tropical and humid temperate environments. The intensity of chemical weathering in soils is preserved in the nature of the sediments derived from these soils; sediments from intense chemical weathering environments are generally enriched in chemically resistant minerals, such as quartz. As an example, we compare the compositions of sands from small streams draining granitic rocks in the Venezuelan Guayana Highlands, the southern Appalachian Mountains, and the northern Rocky Mountains. Although these three regions are very similar in terms of bedrock and topographic relief, they vary tremendously in terms of climate and vegetation. The Guayana Highlands represent a humid tropical climate, the southern Appalachians represent a humid temperate climate, and the northern Rockies are characterized by a semiarid temperate climate. Sands produced in the humid tropical climate are enormously enriched in quartz compared to those derived from similar rocks in semiarid settings (Fig. 8.7). The humid temperate climate produces sands of intermediate composition.

Weathering Duration. The rate of chemical weathering during soil formation is determined largely by the characteristics of the erosional environment. This environment can be viewed in terms of *transport-limited* and *weathering-limited* erosion regimes. Material loosened by weathering moves downslope by transport processes such as landslides and slow creep. If the transport processes removing weathered material from an area are potentially more rapid than the weathering processes generating the material, then erosion is limited by the rate at which chemical weathering proceeds (Fig. 8.8A), and only a thin soil is developed. On the other hand, in cases where the maximum weathering rate exceeds the ability of transport processes to remove material, erosion is limited by the rate at which transport processes remove material, and much thicker soils generally develop (Fig. 8.8B).

A principal difference between weathering-limited and transport-limited conditions is that the solid weathering products have a much longer time to react with the soil and groundwaters under transport-limited conditions. When exposed to the weathering environment for long periods of time, soils are extensively leached and weathering products are correspondingly poor in mobile elements such as calcium, sodium, magnesium, and potassium. Detritus is greatly enriched in resistant minerals, especially quartz. Relatively immobile elements, such as iron and aluminum, may be retained in soils, and in lateritic crusts. Under weathering-limited conditions, where potential sediment transport rates exceed the rate at which weathering can generate loose material, detritus spends much less time in the soil profile. Under these conditions, weathering products are richer in mobile elements, and immobile elements do not accumulate disproportionately in the soil. Residual detritus consists of a varied suite of unstable as well as stable minerals; the proportions of these minerals reflect the composition of the parent bedrock.

An example of a place undergoing both transport-limited and weathering-limited erosion is the Guayana shield of Venezuela. Although large tracts are underlain solely by granitic rocks, two distinct provinces of the Guayana shield can be differentiated: a southern region of very low relief, and a

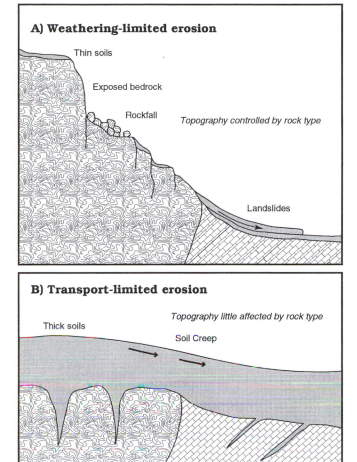

Figure 8.8. Schematic cross-sectional diagrams exhibiting important characteristics of regions experiencing (A) weathering-limited erosion, and (B) transport-limited erosion. (*Source:* Adapted from M. J. Johnsson and W. A. Nierenber, eds., Chemical weathering controls on sand composition, *Encyclopedia of Earth System Science*, Academic Press, 1992.)

northern region where steep scarps separate flat erosion surfaces. Transport-limited erosion on the low-relief southern portion of the Guayana shield leads to protracted weathering of detritus in the thick soils of the region, and sands draining this region consist of little other than quartz (Fig. 8.9). In contrast, weathering-limited erosion on the steep slopes of the elevated Guayana shield leads to sands much richer in feldspar.

As soils are eroded, their components are separated and transported, usually by river processes. Rivers temporarily store sediment through deposition in alluvial deposits on floodplains. During the time sediment resides in such deposits, it continues to be exposed to chemical weathering, and further compositional modification may occur. Rivers normally rework and erode their floodplains during channel migration, thereby reincorporating older material that has been stored in alluvial sequences. If sediments have been intensely weathered during storage, then the result will be a dilution of the sediment a river carries by older material that has been

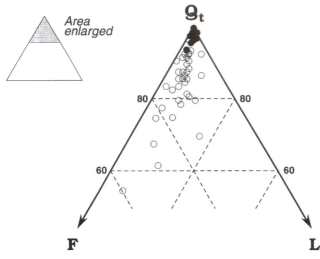

Figure 8.9. Compositions of sands derived from granitic rocks of the Guayana shield, Venezuela, expressed in terms of total quartz, feldspar, and lithic fragments. See Figure 8.7 for an explanation of ternary diagrams and abbreviations. The Guayana shield is divided into two regions of contrasting relief and erosion regime: a northern region of high relief undergoing both transport-and weathering-limited erosion; and a flat southern region experiencing only transport-limited erosion. Sands from rivers draining areas undergoing only transport-limited erosion (indicated by filled circles) are markedly more rich in quartz than are those from regions marked by both transport-and weathering-limited erosion (indicated by open circles). Only the upper (quartz-rich) portion of the ternary diagram is shown. (*Source:* Adapted from M. J. Johnsson, R. F. Stallard, and N. Lundberg, Controls on the composition of fluvial sands from a tropical weathering environment: sands of the Orinoco River drainage basin, *Geological Society of America Bulletin*, vol. 103, 1991.)

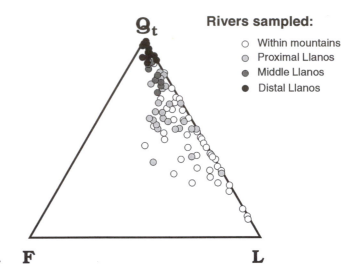

Figure 8.10. Compositions of sands from rivers reworking the broad alluvial plains of the Venezuelan Llanos, expressed in terms of quartz, feldspar, and lithic fragments. See Figure 8.7 for an explanation of ternary diagrams and abbreviations. Sands collected within the mountains where the streams originate are much poorer in quartz than those that have crossed the alluvial plains and reworked the material stored on them. (*Source:* Adapted from M. J. Johnsson, R. F. Stallard, and N. Lundberg, Controls on the composition of fluvial sands from a tropical weathering environment: sands of the Orinoco River drainage basin, *Geological Society of America Bulletin*, vol. 103, 1991.)

altered by chemical weathering. The relative proportion of the older material increases in the downstream direction, resulting in a progressive downstream increase in the proportion of weathered material carried by the river. This is reflected in an increase in the proportion of resistant minerals, as is seen in the sands carried by rivers traversing broad alluvial plains on the Venezuelan Llanos (Fig. 8.10).

Ultimately, sediment comes to rest in a setting where it is buried and isolated from the weathering environment. The rapidity with which this takes place may be important in further shaping sediment composition. Relatively low sedimentation rates tend to leave sediment exposed to continued weathering, producing a further loss of unstable phases. Sediments deposited in basins with high sedimentation rates typically preserve a greater proportion of unstable constituents.

Slope is an important variable controlling the duration of weathering during soil formation, transport, and deposition. During each of these processes, steep slopes reduce the time that sediment is exposed to weathering. Steep slopes correlate with rapid downslope transport rates, favoring weathering-limited erosion conditions. Steep canyons afford little opportunity for alluvial storage, decreasing transport times

and limiting the exposure of sediments to chemical weathering during transport. Furthermore, steep slopes tend to lead to high sedimentation rates in adjacent sedimentary basins, limiting the exposure of sediments to weathering in the depositional environment. The duration of chemical weathering therefore is least in tectonically active areas where rapid uplift maintains steep slopes.

SOILS

Rocks are exposed at the Earth's surface only discontinuously and sporadically. In most regions, bedrock is covered by a veneer of detritus and weathering products, or by alluvium produced by the redistribution of that material by rivers and streams. It is this veneer that represents the interface between the geosphere and the biosphere, atmosphere, and hydrosphere, and it is here that chemical and physical weathering do most of their work to convert bedrock to sediments. This veneer is familiar to us as soil; however, defining what we mean by "soil" is difficult. To the engineer, soil is all unconsolidated material at the Earth's surface. To the geologist, soil is the physical substance covering bedrock. To the agriculturist, soil is any material at the Earth's surface in which crops can be grown. A soil scientist may take a more abstract view of soil, regarding it as the differentiation of this material into various horizons through a complex suite of processes, regardless of whether the parent material is bedrock, alluvium, or an older soil.

A soil may be defined as the product of a weathering "front" passing through bedrock. Detritus and secondary products created by weathering accumulate during this process. Material is continually removed from the top of the soil column by erosion, while bedrock at the base of the column is continually being converted to soil. If the rate of removal and the rate of weathering are in equilibrium, a soil profile of constant thickness will persist through time. The weathering front producing the soil profile moves relatively downward through bedrock as the land surface is lowered by erosion. This means that material at the base of the profile has experienced less weathering than material at the top of the profile; the intensity and duration of weathering varies systematically from the bottom to the top of the soil profile. Thus, by investigating mineralogic changes in an *in situ* soil column, we can evaluate the progression of reactions occurring as bedrock is altered. This can tell us much about how material is cycled through the Earth system.

STRUCTURE OF SOIL. Prominent features of most soils are layers, or *horizons*, which subdivide a soil profile (Fig. 8.11). These layers are not the result of sedimentation, and the law of superposition (Chap. 3) does not apply to soils. They are instead the result of physical and chemical processes that

Figure 8.11. Example of a typical simple soil profile, showing an upper horizon rich in organic material (A_o or O horizon), an underlying leached layer (A horizon), a zone of accumulation (B horizon), and a lower zone of altered bedrock (C horizon). Subscripts on the horizon notation refer to finer subdivision of the principal soil horizons. (*Source:* Adapted from M. J. Johnsson, *Soil processes*, 1996.)

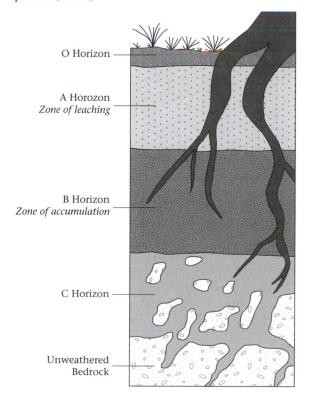

O Horizon

A Horozon
Zone of leaching

B Horizon
Zone of accumulation

C Horizon

Unweathered Bedrock

occur at or near the Earth's surface, giving soils their distinctive structures.

Most soils contain a thin layer at their top that contains abundant organic matter. This layer, called the O horizon, is commonly dark gray or black, contains abundant partially decomposed plant material and living roots, and has distinct structures reflecting its high organic content. It may vary in thickness from a few centimeters to as much as 50 centimeters, depending on the climate and the fecundity of vegetation in the area. The O horizon is especially important because it is a storehouse for most of the nutrients available to natural vegetation or crops. The soil underlying the O horizon is dominated by weathered minerals and secondary minerals formed by chemical weathering. The uppermost layer of this material, the A horizon, is a zone of leaching. Downward-percolating waters remove soluble ions, carrying them to lower portions of the soil profile. In addition, clays and organic matter may be physically transported from the A horizon to lower in the profile by vertical movement of water, a process known as *translocation*. The A horizon typically is 10 to 150 centimeters in thickness. The zone of accumulation beneath this zone of leaching is known as the B horizon. Here, the products of leaching and translocation may accumulate, to form either mineral precipitates within the soil, or clay films lining voids. The B horizon may be from 10 centimeters to several meters in thickness. Below the B horizon lies a zone little affected by these vertical water movements, and which commonly preserves the structure of the bedrock. This so-called C horizon ranges from a few tens of centimeters to over ten meters in thickness, and is gradational with the unaltered bedrock below.

Soils are zones of primarily vertical water movement, leaching materials from the upper horizons and depositing them in the B horizon below. It is important to note, however, that not all material leached in the A horizon is deposited in the B horizon; lateral flow of soil waters removes much material from the soil profile in the form of soluble ions. Soil processes result in the development of complex structures, such as the distinctive blocks – known as peds – into which soil crumbles. The shape, nature, and distribution of such soil structures is a reflection of the balance between accumulation and leaching of both mineral and organic matter.

Although we are dealing primarily with mineral weathering, we must also consider the transformations of organic matter within the soil column and how these transformations affect the soil weathering environment. Organic matter is continually degraded through oxidation (conversion to carbon dioxide) within most soil profiles. The rate of this degradation can be expressed as the change in the amount of carbon per unit time, denoted dC/dt. This rate can be determined by:

$$dC/dt = A - rC$$

where A is the annual addition of organic matter, C is the amount of organic matter already in the soil, t is time, and r

is the decomposition constant, or the fraction of organic carbon decomposing each year. The value r is of interest because this is a direct measure of carbon dioxide production within the soil, which is important in determining the intensity of weathering in the soil environment. If a soil is in equilibrium, $dC/dt = 0$ and $A = rC_e$, where C_e is the equilibrium concentration of carbon. Thus r can be calculated ($r = A/C_e$) if C_e and A can be measured. Where r is high (e.g., 0.3 or more), oxidation of organic matter is rapid, carbon dioxide production is great, and the soil provides an intense weathering environment. Such conditions are typical of tropical regions where, although organic production is high, rapid degradation leads to acidic soils impoverished in organic matter. Where r is low, oxidation or organic matter is slower, less carbon dioxide is produced, and the soil weathering environment is less intense.

SOIL PROCESSES AND COMMON SOIL TYPES. Many variations exist on the idealized soil profile described above. These variations lead to fundamental differences among soils and provide a basis for their classification. Many different processes can contribute to the structure of a soil; which processes are most important at a particular location depends on the local environmental conditions. Some of the more common and important processes include: leaching and acidification, translocation of clays, podzolization, desilicification, and reduction of iron.

As described above, when sufficient precipitation exists, the more soluble ions are removed from the upper horizons. Because precipitation is naturally acidic, and becomes more so through contact with carbon dioxide produced by respiration and decomposition of organic matter in the soil column, leaching is accompanied by acidification. The extent of acidification not only is important in determining the intensity of mineral weathering within the soil, but also in influencing the decomposition of organic matter. When the pH of water in soil drops much below 5, the structure of the soil ecosystem is changed such that the decomposition of organic matter is slowed. Thus, a negative feedback is set up between soil water pH and organic matter decomposition that keeps soil waters from becoming too acidic. However, when sources of acidity other than carbon dioxide are introduced, such as anthropogenic acid rain, this feedback system does not counteract the added acidity introduced by these sources. In this case, acidity builds up, disturbing the soil ecosystem and reducing the overall productivity of the soil.

Translocation of clays through the soil column leads to well-developed A and B horizons, marked respectively by the removal and the accumulation of clays. The translocated clays take the form of ordered clay skins on the walls of pores in the A horizon, and they give the soil a blocky structure. Such clay-rich B horizons, known as *argillic horizons*, are common throughout the world and are characteristic of the groups of soils known as *alfisols* and *ultisols*.

Podosols are soils that contain an intensely leached A horizon. They form in acid soils where soluble cations are aggressively removed from the uppermost soil layers. In addition to carbonic acid, organic acids combine with such insoluble ions as aluminum and iron, helping to remove them from the A horizon. These organic/inorganic combinations are deposited in the B horizon, which is sharply and dramatically delineated in such soils. Podzolization is most pronounced in areas with high organic acid production, such as beneath pine and oak forests, typically in cool to temperate humid environments, such as on the Russian steppes.

A different form of leaching is known as *desilicification*. When organic acids are not present to combine with aluminum and iron, these elements are the least soluble of the common constituents of bedrock and soils. Most other ions – including silica, which is intermediate in solubility between iron and aluminum on the one hand, and alkali ions such as potassium, sodium, and calcium on the other – are leached and removed from the A horizons, leaving accumulations of iron and aluminum oxyhydroxides (goethite, hematite, gibbsite, and related minerals). Such soils, termed *oxisols*, are common in the tropics, where their accumulations of iron impart red or yellow hues to the tropical landscapes. Desilicification takes time, and well-developed oxisols may take tens of millions of years to develop. Accordingly, they are commonly restricted to the oldest land surfaces. Extreme desilicification results in the removal of nearly all components other than iron or aluminum oxides and hydroxides, and soils may become hard, brick-like deposits known as *laterite* (or *bauxite*, if particularly rich in aluminum), important economic sources of these elements.

When a soil contains little oxygen, the opposite situation may develop. Under such *reducing* conditions, iron is more soluble and may be transported to lower horizons. Furthermore, under reducing conditions, iron does not form the warm-hued iron oxyhydroxides, but instead impart a gray, blue, or green cast to the soil. Such horizons are called *gleyed horizons* and are common where circulation of water through the soil is poor and where large amounts of decomposing organic matter are present.

Additional processes are important in particular environments. For example, in arid regions with abundant calcium carbonate in the bedrock, capillary action draws soil waters *upward* in the soil column, where evaporation at the surface can cause the concentration and precipitation of minerals. One of the most common such deposits is *caliche*, a nodular form of calcium carbonate that is common in arid regions around the world.

SOIL-FORMING FACTORS. The process of soil formation is controlled by five important variables, known as *state factors*: parent material, climate, organisms, relief, and time. These state factors form a convenient framework in which to evaluate soils and soil-forming processes. They are already familiar to us, because each of these variables has proved to be of great importance in chemical weathering. "Parent material" refers not only to various types of bedrock on which a soil might be developed, but also to such reworked material as alluvium or older soils. Parent material exerts a strong control on soil formation in young soils and in areas where soil formation processes (including chemical weathering) are slow. Climate is one of the most important factors in determining the intensity of chemical weathering. In addition, particular climates might favor the formation of distinctive soil accumulations such a bauxite, laterite, or caliche. Organisms (including vegetation, animals, microorganisms, and humans) are important in contributing organic matter, including organic acids, to soils and thereby promoting reducing, acidic conditions conducive to chemical weathering. In addition, burrowing organisms open the soil to increased water infiltration, enhancing dissolution rates. Humans have a profound capacity to alter soils, either by changing the composition of the vegetation over a soil, often promoting erosion and soil removal, or by compaction through construction or introduction of grazing animals. Relief is important in soil formation in a number of ways. First, soils are transported down slopes by processes such as soil creep and landslides; steep areas are marked by young, thin soils on hilltops and upper slopes, and by accumulations of transported soils, called *colluvium*, on lower slopes and valleys. In addition, lateral water movement through soils promotes weathering and removal of constituents from upland soils; however, stagnant water conditions in lowland soils retard weathering. Finally, we have seen how slope affects the fifth variable, time, by determining whether erosion is transport-limited or weathering-limited. Time is an important variable in determining the structure of a soil; well-developed horizons will not form until dynamic equilibrium conditions have been reached, and some soil types, such as oxisols, require long periods of time to develop.

These state factors provide a useful framework for thinking about soils – however, they are only a departure point. Most soils have far more complicated histories than implied by this discussion. Many of the state factors vary over time, often in an irregular or oscillatory manner. When several state factors change simultaneously, analysis of their influence on soil structure becomes extremely complex. In addition, the discussion above has dealt with soils that have reached an equilibrium with environmental conditions. Not only is equilibrium impossible when the state factors are constantly varying, but catastrophic events may interrupt soil formation altogether. Any one location on Earth is likely to be periodically subject to sudden erosion as the result of, for example, landslides or floods (see Chap. 25). Alternatively, sudden deposition events may bury a soil. When this happens, a new soil begins to develop in the overlying material and, with time, this new weathering front is superimposed on the older one. Such situations are common, and result in the complex set of soil patterns that occur throughout the world.

HUMANS AND SOILS. It is clear from studies of soil development that soils take hundreds to thousands of years to form. Human development occurs on a far faster time scale, and many human activities result in soil removal, leaching of important nutrient elements, or buildup of toxic contaminants. For example, removal of natural vegetation to allow the growth of crops usually results in increased soil removal through runoff or wind. The dust bowls of the 1930s in western North America were the result of stripping the land of natural vegetation for agricultural purposes, and for grazing. This is a common occurrence worldwide; as a result, large areas have lost their fertile soils. In most cases, these soils will not be renewed for generations because anthropogenic soil erosion rates vastly exceed the rate at which soil is regenerated. Even when soil erosion is not a problem, overuse can result in the removal or leaching of nutrient elements (especially potassium, nitrogen, and phosphorous) on which plant life depends. In most agricultural areas, the rate of removal of these components far exceeds the rate of their regeneration through weathering or nitrogen fixation, and they must be introduced in the form of chemical fertilizers (see Chaps. 28 and 29). Application of such fertilizers brings its own problems, including the perturbation of natural ecosystems through runoff containing fertilizer or pesticide residue. Increased evaporation within soils that have been stripped of natural vegetation can lead to the buildup of salts and the concentration of poisonous trace elements such as selenium. Compaction of soils is a common result of overgrazing. Rainwater does not penetrate compacted soils well, slowing or stopping soil formation. Stripped of vegetation as a result of overgrazing, such compacted soils are easily eroded. These trends are alarming because soil is one of our most important natural resources, forming the basis for nearly all of our agriculture. We must learn to recognize soil as a natural resource – one that is essentially nonrenewable on human time scales.

CHEMICAL WEATHERING AND PALEOCLIMATE

An important goal of Earth systems science is to learn how the surface environment has changed over the course of Earth history. By understanding such changes, we are in a better position to predict environmental changes resulting from human perturbation of the Earth system. Perhaps the most important aspect of the Earth's environment is climate. A number of clues to past climate exist, including the distribution of fossil organisms, the distribution of rock types diagnostic of particular climates (e.g., coal and evaporite deposits), and rare samples of ancient air trapped in ice and amber. Evidence from chemical weathering provides far more ubiquitous clues to past climates; unfortunately, however, this evidence is commonly difficult to interpret.

The most direct record of past chemical weathering regimes is preserved soils, called *paleosols* (Fig. 8.12). Normally, soils are removed by erosion before deposition of sediments begins in an area. Not uncommonly, however, soils are buried before they can be eroded, and may be preserved in the geologic record. Although they may be altered by chemical

Figure 8.12. A preserved ancient soil, or paleosol. This paleosol is developed on granite of early Proterozoic age, and is overlain by a sandstone of Cambrian age. The paleosol records the conditions under which the granite was weathered, at a time shortly before the deposition of the overlying sandstone. (*Source:* Photograph by Mark Johnsson.)

and physical processes within the Earth's crust before they are returned to the surface by uplift, they may preserve trends in the distribution of minerals and elements from the original soil. These trends can tell us much about the chemical weathering environment and, by inference, about past climate.

The compositions of sedimentary rocks themselves provide information regarding the chemical weathering environment, provided that we can interpret the rock record correctly. Chemical weathering can vastly alter the composition of sediments from that of the original bedrock. In an ancient sequence, if we know the original bedrock's composition, chemical differences between bedrock and sedimentary rocks derived from them can tell us much about past climates. Understanding how to read environmental information from the compositions of sedimentary rocks is especially exciting because such strata have been preserved through much of geologic time and are common throughout the world. Accordingly, they have the potential to provide us with a far more complete picture of past environments than do scattered data from paleosols, fossils, and the like. However, the controls on the composition of sedimentary rocks are extremely complex, and we are only now beginning to understand them.

CHEMICAL WEATHERING AND ATMOSPHERIC CARBON DIOXIDE

The processes of chemical weathering, erosion, and sedimentation are important aspects of the rock cycle. In addition, chemical weathering is important to the cycling of reactive chemical components such as carbon. It is through weathering and erosion that material is removed from the land, introduced into the ocean, buried, lithified, and cycled through

the system once again. Along the way, important partitioning processes operate, introducing feedback mechanisms between sedimentary rocks, the hydrosphere, the biosphere, and the atmosphere.

The global geochemical cycling of carbon (Fig. 8.13; see also Chap. 19) is of particular interest because carbon dioxide is not only a primary determinant of the intensity of chemical weathering but also an important greenhouse gas (i.e., it allows short-wavelength solar radiation – light – to enter the atmosphere but traps most of this energy when it is reradiated at longer wavelengths in the form of heat). Therefore, the level of atmospheric carbon dioxide may be a primary control on long-term variation in global climate. Since the industrial revolution, humans have significantly increased the level of atmospheric carbon dioxide through the burning of fossil fuels (see Chap. 1). Understanding natural controls on the global geochemical carbon cycle is crucial to the evaluation of the potential effects of humanity's perturbation of that cycle.

Sedimentary rocks represent the Earth's largest reservoir of carbon, containing more than 1500 times the carbon found in the atmosphere, ocean, and biosphere combined. This sedimentary carbon reservoir is divided between carbon in carbonate rocks (limestones and dolomites) and carbon in the form of organic matter preserved in sedimentary rocks (of which coal and oil are small subsets). Carbon dioxide is absorbed from the atmosphere by plants through photosynthesis, and fixed in the soil in the form of decaying organic matter. There it reacts with water to form carbonic acid (H_2CO_3). The amount of carbonic acid in soil waters strongly influences the intensity of chemical weathering in the soil environment. Both carbonate and silicate minerals react during weathering with carbonic acid to release bicarbonate (HCO_3^- and soluble metallic cations. These dissolved weath-

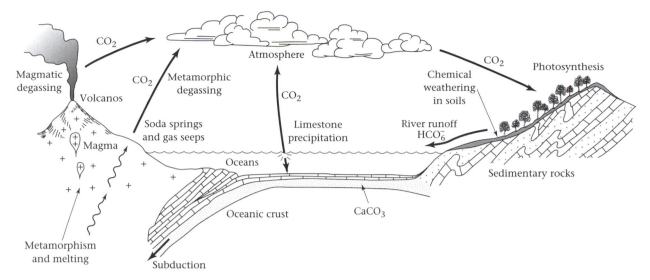

Figure 8.13. The global carbon cycle, as it operates on geologic time scales. (Compare with discussion in Chapter 19.)

ering products are transported by ground and river waters to the ocean, where marine life (principally foraminifera and coccolithophores) utilizes the calcium and bicarbonate ions to precipitate exoskeletons (shells) of calcium carbonate:

$$Ca^{+2} + \quad 2HCO_3^- \quad \rightarrow \qquad\qquad (4)$$

calcium ion   bicarbonate ion

$$CaCO_3 + \quad CO_2 \quad + \quad H_2O$$

calcite   carbon dioxide   water

These shells ultimately sink and accumulate to form new carbonate rocks. The biologic precipitation of calcium carbonate releases carbon dioxide into the ocean (see reaction (4), and because the ocean and atmosphere maintain a balance through gas exchange, this carbon dioxide quickly finds its way back into the atmosphere. Because of the amounts of carbon dioxide consumed to weather silicate versus carbonate rocks, the result of reaction (4) is that weathering of carbonate rocks results in no net loss of atmospheric carbon dioxide, whereas in silicate weathering, only half the carbon dioxide consumed in chemical weathering is returned to the atmosphere. Equilibrium within the global carbon cycle is maintained when rocks containing carbon, including those formed through reaction (1), are deeply buried and subject to metamorphism or melting. Under these conditions, carbon dioxide is released through degassing reactions such as

$$CaCO_3 + SiO_2 \quad \rightarrow \quad CaSiO_3 \quad + \quad CO_2 \qquad (5)$$

calcite   quartz   wollastonite   carbon dioxide

Carbon dioxide released by igneous and metamorphic processes enters the atmosphere principally through volcanic eruptions and at gas seeps. Volcanic degassing of carbon dioxide is, in turn, controlled by the global plate-tectonic network: variations in worldwide sea-floor spreading rates largely control temporal variations in carbon dioxide degassing by controlling the intensity of volcanism and recrystal-

lization associated with the creation and destruction of oceanic crust at mid-ocean ridges and subduction zones.

Chemical weathering plays an important role in the global cycling of carbon, in that it largely determines the rate of carbon dioxide consumption from the atmosphere. Atmospheric carbon dioxide is replenished through volcanism and metamorphic degassing. Thus, relative variations in chemical weathering and global tectonism govern the level of dissolved carbon dioxide in the atmosphere and ocean. Because the reservoir of carbon in the atmosphere is small relative to the fluxes described above, even minor variations in these fluxes can have a profound effect on the concentration of atmospheric carbon dioxide. Owing to the importance of carbon dioxide as a greenhouse gas, such variations in atmospheric carbon dioxide may be a dominant control on temperatures at the Earth's surface over geologic time.

A feedback loop involving the cycling of carbon exists between chemical weathering and global climate. Chemical weathering, itself strongly influenced by climate, plays an important role in determining global climate. The composition of sediments provides a detailed record of chemical weathering through geologic time. As we become proficient at separating the effects of the many parameters shaping sediment composition, the sedimentary record may provide us with a way of directly observing long-term changes in the intensity of chemical weathering and may allow for a better understanding of the interactions among atmosphere, hydrosphere, and biosphere over geologic time.

QUESTIONS

1. A soil can form in any Earth material, from bedrock to deposited alluvium. Imagine a soil developed in alluvium on a floodplain. Every year, an additional layer of alluvium is deposited; consequently, the uppermost parts of the pile of alluvium have been exposed to weathering for only a short time, but the bottom parts have been exposed for longer periods. Given that leaching of ions is most intense in the upper part of the alluvium, what effect would the different ages of the material in the alluvial pile have on the level of alteration due to chemical weathering? How would varying the long-term average sedimentation rate and/or the intensity of chemical weathering affect your answer?

2. If atmospheric carbon dioxide is indirectly controlled by the global extent of chemical weathering on land through the global carbon cycle, what effect would a rising sea level have on atmospheric carbon dioxide? How might this be reflected in the rates of production of carbonate rocks? Can you define a feedback loop operating in this system? Is it positive or negative?

3. We might expect land use changes to affect chemical weathering, soil production, and erosion because of the influence of vegetation on these processes. What might be the effect on each of these processes of the sudden replacement of a pine forest by cropland? If this change persisted for a significant length of geologic time, how might these effects be preserved in a basin receiving sediments derived from these soils?

4. For most of Earth's history, there was no life on land. How do you think chemical weathering may have been different prior to the advent of land plants approximately 430 million years ago?

5. Prior to approximately 2500 million years ago, the Earth's atmosphere contained no free oxygen. How might chemical weathering have been different under such an atmosphere?

6. Humans have, over the past 150 years or more, greatly increased the levels of carbon dioxide, nitrous oxide gases, and sulfur oxide gases in the atmosphere. Presumably, this should be reflected in overall greater intensity of chemical weathering.

How might this be manifested in soil development, water quality, and crop yields? How could you test for the overall effect?

7. Typical natural erosion rates in lowland areas are on the order of 0.5 millimeter per year. When chemical weathering rates and erosion rates are balanced, a soil of constant thickness is developed and persists through time. If erosion rates suddenly increase without a concomitant increase in the rate of chemical weathering, soil erosion occurs. This is the situation over much of the Earth's croplands, where human activities have accelerated erosion rates. If a soil 1.5 meters thick is in equilibrium with chemical weathering on a forested patch of ground experiencing an erosion rate of 0.5 millimeter per year, how long would it take for that soil to be completely removed if erosion rates were doubled due to the removal of the forest to make way for cropland, assuming that the rate of chemical weathering remains unchanged? In reality, chemical weathering rates could decrease in such a situation, because organic acids and carbon dioxide from organic matter would be reduced when the forest was removed. How long would it take to remove the soil completely if, in addition to doubling of the erosion rate, there was a halving of the rate of chemical weathering?

FURTHER READING

Berner, R. A., and Lasaga, A. C. 1989. Modeling the geochemical carbon cycle. *Scientific American* 260: 74–81.

Drever, J. I. 1985. *The chemistry of weathering.* Dordrecht, The Netherlands: Reidel.

Fyfe, W. S. 1989. Soil and global change. *Episodes* 12: 249–254.

Jenny, H. 1980. *The soil resource.* New York: Springer-Verlag.

Wild, A. 1993. *Soils and the environment.* Cambridge: Cambridge University Press.

THE HYDROSPHERE

9 The Hydrologic Cycle

PATRICIA A. MAURICE

EUTROPHICATION AND DE-EUTROPHICATION OF LAKE ERIE

Water covers two-thirds of the Earth's surface and is fundamental to life on Earth. However, water, both below and on the surface of the Earth, is as vulnerable as it is valuable. Humanity's impact on this natural resource grows ever more apparent, from the overuse of limited water supplies in arid areas to the pollution of vast stretches of the ocean. The Great Lakes region of the United States and Canada represents the world's largest freshwater (non-ice) resource. Lake Erie, which is volumetrically the smallest of the Great Lakes, has been severely affected by human waste and pollution. During the fifty years from 1910 to 1960, the human population of the Lake Erie basin swelled from 3.8 million to 11.2 million. Domestic wastes were discharged to the lake and its tributaries without first undergoing filtration or treatment. Industrial wastes also burgeoned, particularly during and after World War II. Farming activities and the use of fertilizers and pesticides grew to help supply the growing population. Together, these changes led to widespread biologic activity. Between 1948 and 1962, the concentrations of soluble phosphate in the lake increased from 7.5 to 36 micrograms per liter – an approximate 500 percent increase in a mere decade and half! In many ecosystems, phosphate is a "limiting nutrient," which means that it is essential for life but is present in such short supply that it limits the amount of growth in a community. This huge spike in the nutrient supply fueled a rapid increase in planktonic productivity. As large numbers of plankton eventually died and fell to the bottom of the lake, decomposition of their bodies consumed oxygen and led to anoxic conditions. Lake-water oxygen levels in the bottom water decreased, leading to a decline in populations of sport fish, including blue pike, cisco, and whitefish. Sheepshead and carp replaced these species as water quality continued to plummet. The situation became so severe that in the 1960s and 1970s, Lake Erie was known as the Dead Sea of North America.

Fortunately, the public outcry over the fate of this beautiful and important natural resource finally resulted in establishment of agencies and guidelines to help clean it up and restore it to its natural beauty and splendor. The Clean Water Act of 1972 mandated that municipalities and industries refrain from dumping untreated or poorly treated wastes into public waters. Construction of sewage treatment plants eventually resulted in greatly reduced phosphate loads. As a direct result of pollution controls on effluents over the past twenty-five years, phosphate concentrations have declined substantially. The oxygen depletion trend appears to be reversing, as well. Both of these trends are good news, and they are reflected in a decrease in several nuisance species of plankton. Nonetheless, Lake Erie is far from being safe and clean. During storm events, sewage is sometimes still released directly into the lake. No one knows the long-term fate of the pesticides, heavy metals, and industrial chemicals that accumulated over the course of the twentieth century. As in many environmental dilemmas, our Earth appears to be both resilient and vulnerable. The outcome for Lake Erie's habitability is still in doubt.

The term *hydrologic cycle* (from the Greek *hydro-*, meaning "water") refers to the complex system whereby water circulates among its various reservoirs at and near the surface of the Earth. These reservoirs include the oceans, the atmosphere, underground water, surface water, glaciers, and the polar ice caps. The hydrologic cycle pervades our terrestrial existence, playing a key role in many natural phenomena. It is directly coupled to the Earth's energy cycle, because solar radiation combines with gravity to drive the global circulation of water. This circulation, in turn, plays an important role in the heat balance of the Earth's surface. The hydrologic cycle is also closely linked to the geosphere and its rock cycle (see Chap. 5). Water erodes geologic materials, and the breakdown of these materials releases many chemical constituents that in turn define the chemical nature of the water. Water can also build geologic formations, through both chemical and mechanical depositional processes. Water is essential to all life forms in the

W. G. Ernst (ed.), *Earth Systems: Processes and Issues*. Printed in the United States of America. Copyright © 2000 Cambridge University Press. All rights reserved.

planetary biosphere. Even the slightest differences in the volumes and chemical compositions of natural waters can have a tremendous impact on biologic communities. As such an essential natural resource, water is of major concern in local, regional, and international law (see Chap. 30), and has been at the root of many international conflicts.

Thus far in the text, we have discussed the solid Earth – the geosphere – including its place in the solar system, as well as its composition, structure, and dynamic nature. From the vantage point of our neighbor planets, however, the most striking feature of Earth is the water that covers more than two-thirds of the planet's surface, and that surrounds the planet in vapor form. Chemical and density stratification during and immediately following accretionary heating of the Earth's primordial planetary body resulted in its present layered configuration. Internal heat and chemical reactions caused water, which was originally bound as oxygen and hydroxide in minerals, to diffuse from the Earth's interior toward its surface. This *degassing* of both water and other volatile species resulted in the accumulation and eventual condensation of the fluid envelopes of the Earth – the globe-encircling oceans and the atmosphere. Of course, water was also delivered to our planet by infalling comets and other H_2O-bearing planetesimals. In this chapter, we examine the hydrologic cycle with a systems approach, focusing on how it is coupled to the Earth's energy cycle and how the chemical characteristics of water vary through the hydrologic cycle, giving several examples of water-related environmental and legal issues.

FLUIDS IN MOTION

MOVEMENTS WITHIN THE HYDROLOGIC CYCLE. Water at the surface of the Earth exists in a state of dynamic equilibrium, circulating among the oceans, the atmosphere, and terrestrial environments in a complex system known as the *hydrologic cycle*. The principal components, or reservoirs, of the hydrologic cycle (Fig. 9.1) include the Earth's oceans, atmosphere, polar ice caps, ice fields, surface water (lakes, rivers, and wetlands), and underground water (including both soil water and groundwater). *Flow paths* between these reservoirs include *precipitation, evapotranspiration, recharge, discharge,* and *runoff.* Figure 9.2 shows the volume of water in each of the reservoirs, as well as the percentage of the total water supply that each represents. As evident from this diagram, most of the world's water resides in the global oceans. It is important to remember that although water is conserved within this closed system, this fact does not suggest a static picture. The system is in a state of dynamic equilibrium wherein local net gains or losses must be offset by equivalent net losses or gains elsewhere. Evaporation exceeds precipitation over the oceans, for instance; however, the converse is true for the terrestrial environments. Thus, evaporation and precipitation result in a net transfer of water from the oceanic realm to the terrestrial realm. Under steady-state conditions, and over the long term, this transfer is compensated

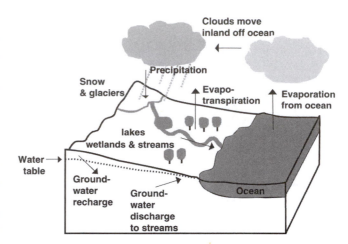

Figure 9.1. Schematic illustration of the hydrologic cycle.

for by the runoff of water into lakes, into rivers, and ultimately back to the ocean. By "steady state," we mean that over the long term, the amount of water in a reservoir does not change, even though there is a continuous inflow and outflow. The living world (flora and fauna) takes up some of the water on this journey; in the steady state, however, this uptake is matched by an equivalent release of water through death, decay, respiration and evapotranspiration. Dynamic equilibrium is thus maintained. The hydrologic cycle is thus responsible for the presence of life on land.

The rates of movement of water within each of the various components of the hydrologic cycle vary greatly. Atmospheric water can travel as much as hundreds of kilometers per day, while in the same time span, stream water travels only a few tens of kilometers. Water trapped in solid form in ice caps and glaciers crawls along at a rate measured in centimeters, or perhaps meters, per day. Underground water moves even more slowly, at a rate of meters or less per year. To understand the larger picture of movement of water within and between basins, one must grasp the concept of

Figure 9.2. The global hydrologic cycle, reservoirs of water (10^{15} kilograms and fluxes 10^{15} kilograms per year). (*Source:* Data from National Research Council, 1986.)

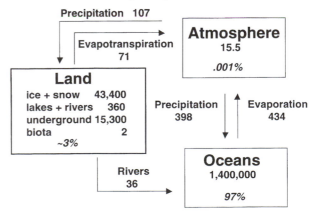

mean residence time (*MRT*), defined as the amount of time an average water molecule spends within any one reservoir (e.g., the atmosphere) before moving, via some flow path, to another reservoir (e.g., as rainfall to a stream). Because MRTs are affected directly by the highly variable rate of travel of water, they vary greatly. The following examples provide a perspective on both the magnitude of the numbers and their variability. For atmospheric water, MRT is approximately ten days. For the oceans, MRT varies from a few days or weeks for shallow water to hundreds or even thousands of years for deep water; taken as a whole, the MRT for oceanic water is more than 3000 years. The average MRT of groundwater is 10,000 years; however, as you might expect, MRT for individual aquifers varies greatly as a function of depth and geologic properties: typical MRTs for shallow aquifers are 1 to 5 years, whereas MRTs for deep basins can vary from approximately 10,000 to as much as 1,000,000 years.

Thus far, we have stressed the dynamic element of this cycle. You may be wondering why water moves at all. The following section addresses the question of what drives the global circulation of water known as the hydrologic cycle.

ENERGY TO DRIVE THE HYDROLOGIC CYCLE. The hydrologic cycle is driven by the flow of energy to, within, and around our planet. The interior of the Earth is hot (see Chap. 5), reflecting the kinetic (impact) energy imparted by accreting planetesimals, the infall of the iron-nickel core, and the decay of long- and short-lived radioactive isotopes. This energy gradually reaches the planetary surface through convection within the Earth and thermal conduction through the lithosphere. The proportion of this internal heat is very small, on the order of 0.1 percent, when compared to the energy arriving in the vicinity of the Earth's surface from the Sun (greater than 99 percent). Despite this huge amount of incoming solar energy, not all of the Sun's available radiation reaches the Earth's surface. Approximately 35 percent of the incoming flux is reflected back into space by cloud cover, atmospheric dust, and haze. Some is absorbed and re-emitted by the so-called greenhouse gas species (including water vapor, H_2O; carbon dioxide, CO_2; ozone, O_3; methane, CH_4; and nitrous oxide, N_2O). The radiation that actually falls on the planetary surface is in part reflected back into the atmosphere, and in part absorbed by the Earth's surface materials as heat. The percentage of incident sunlight bounced back – the reflectivity of a substance – is called the material's *albedo*. Albedo is influenced not only by the surface type but also by the angle of incidence, which is the angle at which the Sun's rays strike a surface. In order to better understand this concept, consider these typical examples of albedo values: liquid water, 2 to 5 percent (Sun near zenith) and 50 to 80 percent (Sun near horizon); forests, 5 to 10 percent; bare ground or rock, 10 to 25 percent; crops and grasslands, 20 to 25 percent; and polar caps and ice fields, 50 to 100 percent. The different albedo values for solid versus liquid water can lead to an interesting positive feedback loop. Considering that the polar caps and ice fields have a much higher albedo than liquid water, if global warming causes the ice caps to melt, then high albedo ice will be converted to lower albedo liquid water. A greater amount of incident sunlight will be absorbed at the poles, which will work to increase temperatures and further melt the ice caps. This positive feedback loop may continue to operate until other factors such as changes in oceanic circulation patterns operate to break the cycle.

To understand the flow of heat energy in the atmosphere, we must recognize that the Earth's lower latitudes (near the equator) receive far more solar energy than they re-emit, whereas the polar regions emit more energy than they receive. This situation creates an energy imbalance that nature seeks to correct, driving circulation patterns of wind and water in its quest to right the disequilibrium. Because the Earth's surface at tropical latitudes is almost normal (perpendicular) to the incoming solar energy, the amount of energy intercepted per unit area of the Earth's surface is maximized in equatorial regions. In contrast, proceeding toward the Earth's spin axis, the planetary surface is progressively inclined to (at an angle to) the arriving radiant flux. At the poles, the angle of incidence reaches zero (grazing incidence). Thus, the sheer quantity of potential light available to a given square kilometer diminishes greatly toward the poles (Fig. 9.3A and B). The low Sun angle in polar regions means that its radiant energy must travel longer paths through the atmosphere than it does in near-equatorial regions. Reflection and absorbance/re-emission in the atmosphere are increased in polar regions, which contributes to their chilly climate. Variations in the Earth's surface albedo are another important factor in the absorption of solar heat energy. Tropical zones are characterized by large oceanic expanses where the low albedo of liquid water (2 to 5 percent) means that large amounts of energy can be absorbed. In contrast, polar regions are characterized by ice caps and pack ice, the high albedo of which (50 to 100 percent) ensures very high reflectivity and a return of heat to the atmosphere.

As a consequence of this differential energy absorbance, strong latitudinal thermal gradients tend to be generated near the planetary surface. This would be the case *even if the Earth did not experience seasonal variations in solar flux*. These temperature gradients drive both the coupling and the longitudinal circulation of the hydrosphere and atmosphere. If the Earth did not rotate, a simple, convective circulation pattern between the equator and the poles would result, with warm air rising at the equator, moving northward and southward to the poles, cooling and descending, and ultimately making its way back to the equator (as if the air were traveling along a giant conveyor belt). Such a simple circulation pattern (Fig. 9.4) was proposed in the mid-1700s by George Hadley. So-called Hadley cells do exist; however, they do not extend all the way from the equator to the poles, but are broken up into several segments of latitude, each 30° wide. The simple pattern proposed by Hadley is complicated by the *Coriolis effect*, which, as a result of the Earth's rotation, deflects flows of both water and air toward the right in the Northern Hemisphere and toward the left in the Southern

(A)

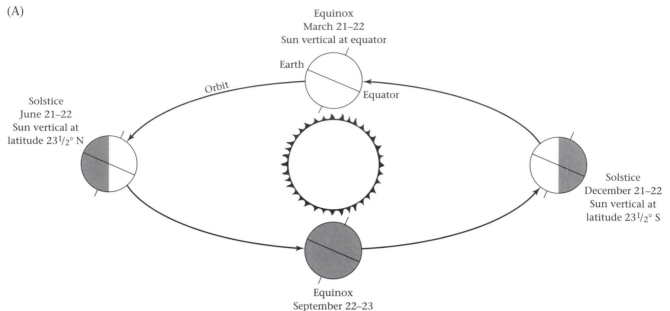

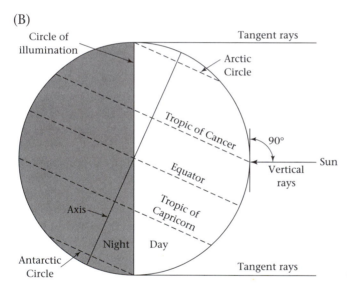

Figure 9.3. (A) Earth–sun relationships. (B) Characteristics of the summer solstice (Northern Hemisphere).

Hemisphere. The Coriolis effect breaks up the simple flow of air between the equator and the poles into belts. On an idealized, rotating, water-covered globe, the circulation pattern would be as shown in Figure 9.5; however, the actual system is far more complex as a result of interactions with Earth's scattered landmasses. Generalized patterns of global atmospheric and oceanic surface circulation are presented in Figures 9.6 and 9.7, respectively. An understanding of the concept of ocean–air mass energy transfer is important for an understanding of the hydrologic cycle (climate dynamics are discussed more fully in Chapter 14).

THE PARTS OF THE WHOLE: COMPONENTS OF THE HYDROLOGIC CYCLE

This section discusses the individual components of the hydrologic cycle in terms of their contribution to the cycle as a whole. Three-fourths of the Earth's surface is covered with water; however, only a small portion of it is fresh. Of the fresh water, 77.5 percent is locked in ice fields and glaciers, inaccessible to the biosphere unless global warming causes it to melt (a subject discussed in Chaps. 10 and 15). In this chapter, we briefly discuss the importance of the oceans and the atmosphere to the hydrologic cycle (we leave the bulk of the discussion of these complex reservoirs to Chapters 10,

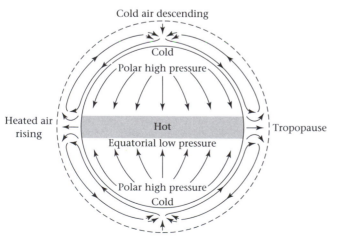

Figure 9.4. The general circulation of the atmosphere: the atmospheric circulation pattern that would develop on a nonrotating planet. (*Source:* A. Goudie, *The nature of the environment,* 2nd ed. Copyright © Basil Blackwell Ltd., 1989. Used with permission.)

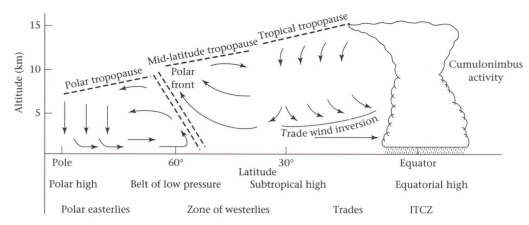

Figure 9.5. The general circulation of the atmosphere: the actual atmospheric circulation shown schematically in a vertical section from polar regions to the equator. ITCZ = intertropical convergence zone; km = kilometers. (*Source:* A. Goudie, *The nature of the environment*, 2nd ed. Copyright © Basil Blackwell Ltd., 1989. Used with permission.)

11, and 14). Emphasis here is on the fresh water found in lakes, soils, streams, and groundwater aquifers, because this is the water so essential to the biosphere and the quality of life for all living things.

OCEAN WATER. Oceans cover more than 70 percent of the Earth's surface. Not surprisingly, they are the dominant reservoirs of the hydrologic cycle, constituting 97 percent of the total volume of water in the hydrosphere. Ocean water contains many dissolved constituents, making it nonpotable to humanity and many terrestrial species. It does, however, support an abundance of oceanic life. Average salt content of the oceans is approximately 35 parts per thousand. The major dissolved components of seawater are shown in Table 9.1. chloride (Cl^-) and sodium (Na^+) are by far the most abundant ions; sulfate (SO_4^{2-}), magnesium (Mg^{2+}), calcium (Ca^{2+}), potassium (K^+), and bicarbonate (HCO_3^-) are also present as major components. Surface ocean waters are somewhat basic (alkaline), with an average pH of approximately 8.1. Typically, concentrations of the major components of ocean water do not vary greatly from one location to another; this uniformity of composition contrasts with the pronounced variability of surface water and groundwater chemical compositions. However, concentrations of minor constituents, including trace and heavy metals and nutrients, vary with depth and location, a fact that contributes to dramatic differences in biologic productivity from one part of the ocean to another (see Chap. 12).

THE ATMOSPHERE AND PRECIPITATION. The atmosphere contains a much smaller proportion of Earth's water than do the oceans, a mere 0.001 percent by volume of all the water in the hydrosphere. It is nonetheless a crucial component. Through evaporation and various forms of precipitation (rain, snow, sleet, fog), the atmosphere is the most significant conduit for the transfer of water between the oceans and the continents, a role implied by its low MRT for

water. Atmospheric water is crucial to life on Earth. Considering its importance, it is startling to realize the vulnerability of this system, the relatively tiny size of which means that even small anthropogenic effects can greatly affect the equilibrium of the system.

The average annual global precipitation rate hovers near 105 centimeters per year, or approximately 0.3 centimeter per day. As any traveller knows, however, volumes of precipitation are not uniform over the surface of the Earth. Some areas, such as the Na Pali Coast of Hawaii (Chap. 2), receive many meters of rainfall per year, whereas rainfall over other regions, such as the arid Sahel region of northern Africa, is negligible. Differences in precipitation are due to global circulation as well as to more local or regional factors. A disproportionately large amount of solar energy warms the equatorial regions of our planet as compared to the poles, which greatly influences precipitation. The great influx of heat in equatorial regions causes hot, moist air to rise from the abundant oceans of the region. As this air rises, it expands and cools, causing the moisture to condense and precipitation to occur. Thus, fully 30 percent of global precipitation occurs near the equator. Air continues to flow poleward in each hemisphere (see Figs. 2.1, 10.8, and 14.1), cooling and losing additional moisture. The now relatively cool and dry equatorial air sinks at approximately 30° latitude, contracting and warming again. This combination of low humidity and high temperature creates and supports a band of subtropical deserts between 15° and 30° latitude in both the Northern and Southern hemispheres, including the Sahara and the Kalahari deserts. At yet higher latitudes, air flow tends to be more complex and turbulent as a result of the combined influences of the Coriolis effect and the next poleward set of circulation cells. Superimposed on these global circulation patterns are significant smaller-scale effects such as the *rain shadow effect* (see Chap. 2).

Despite the fact that water that evaporates from the oceans leaves behind most of its salt components, the result

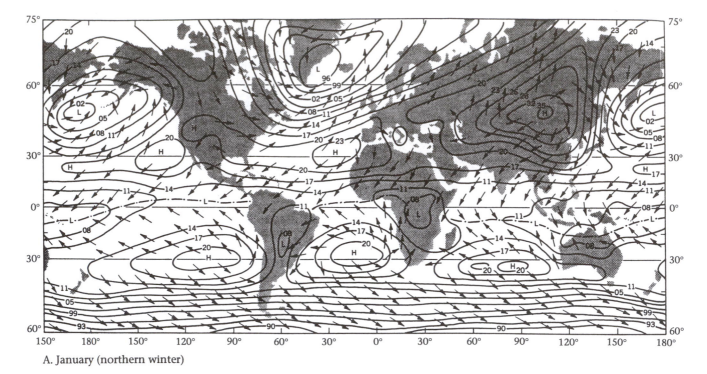

A. January (northern winter)

B. July (northern summer)

Figure 9.6. Average position and extent of the principal surface ocean currents. (A) January (northern winter), (B) July (northern summer). H and L indicate high- and low-pressure areas, respectively; numbers refer to pressures in excess of 1000 millibars. (*Source:* Adapted from Arthur N. Strahler, *The Earth sciences*, Harper Collins, Copyright © 1971, by Arthur N. Strahler.)

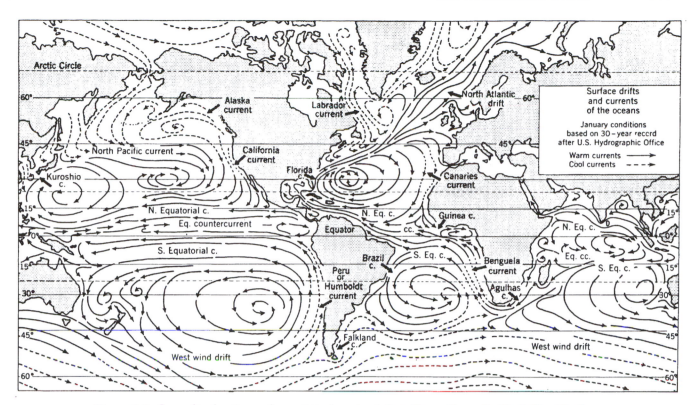

Figure 9.7. Generalized patterns of oceanic surface circulation. (*Source:* Cesare Emiliani, *Planet Earth,* Cambridge University Press, 1992.)

TABLE 9.1. AVERAGE COMPOSITION OF SEAWATER, IN MILLIGRAMS PER LITER

Major Element	Ion	Concentration in PPM
Calcium	Ca^{2+}	410
Magnesium	Mg^{2+}	1,350
Sodium	Na^+	10,500
Potassium	K^+	390
Chloride	Cl^-	19,000
Sulfate	SO_4^{2-}	2,700
Bicarbonate	HCO_3^-	142
Bromide	Br	67
Strontium	Sr	8
Silica	SiO_2	6.4
Boron	B	4.5
Fluoride	F	1.3

Source: Edward D. Goldberg, The fluxes of marine chemistry: Proceedings of the Royal Society of Edinburgh, Section B, vol. 72, 1972, pp. 357–364.

$$Cl^- = Na^+ > Mg^{2+} > K^+ > Ca^{2+} > SO_4^{2-} > NO_3^- = NH_4^+$$
Primarily marine ⟵⟶ Primarily continental

The chloride ion (Cl^-) in rain generally derives from sea-salt aerosols, although, increasingly, some chloride may be a product of industrial pollution. Sodium (Na^+) and magnesium (Mg^{2+}) also have a predominantly marine origin; however, some sodium and magnesium may be due to forest fires, and small quantities are derived from soil dust. Potassium (K^+) and calcium (Ca^{2+}) are almost entirely continental in origin. Sulfates (SO_4^{2-}) and nitrates (NO_3^-) are primarily continental and are derived largely from the burning of fossil fuels. Ammonium (NH_4^+) has an almost entirely continental origin, from fertilizers, biologic processes, and burning of coal.

The various particulate and gaseous components of the atmosphere may have an impact on terrestrial environments either as dry, deposited substances sorbed onto the surfaces of vegetation or human structures, or as components of precipitation. Sea salt from the oceans enters the atmosphere as small bubbles on the tops of breaking waves, projecting small salty particles into the atmosphere. The MRT in the atmosphere for such salty particles is several days. During this time, some of the salts make their way onto the continents as dry deposition or precipitation. Such inputs, known as *cyclic sea salt*, can have important effects on the compositions of surface waters and groundwaters, particularly in coastal regions. But there are anthropogenic effects as well (see box entitled *Acid Deposition*).

ing water vapor is never pure H_2O, for it picks up a variety of chemical constituents while being transported through the atmosphere. The major sources of chemical components of the atmosphere include natural aerosols in the form of sea salts, natural aerosols in the form of terrestrially derived particulates, natural gaseous constituents, and anthropogenic (pollutant) particles and gases. Scientists have described a hierarchy of ions in precipitation based on whether the source is predominantly marine or continental:

ACID DEPOSITION

The burning of fossil fuels has altered the chemical composition of precipitation, most notably by increasing the amount of sulfuric and nitric acids in the atmosphere. Both these acids are found naturally in the atmosphere and in precipitation. Sulfuric acids are derived primarily from oxidation of the dimethyl sulfide and hydrogen sulfide produced by oceanic biologic processes, and the nitric acid precursors ammonium (NH_4^+) and nitrate (NO_3^-) appear to be released naturally from terrestrial vegetation. Together, sulfuric acid, nitric acid, and carboxylic acid (from carbon dioxide) impart acidity to rainfall. This natural acidity is greatly increased by anthropogenic influences. Burning of sulfur-containing fossil fuels, most notably high-sulfur coal, releases sulfur dioxide (SO_2) into the atmosphere. This sulfur dioxide is slowly oxidized and hydrated to form sulfuric acid (H_2SO_4). Nitrous oxide gases, collectively termed NO_x, are derived from the burning of fossil fuels as well. These anthropogenic acids eventually have an impact on terrestrial environments and human structures either through *dry deposition*, wherein the solutes directly sorb to or are taken up by surfaces, or through acid precipitation (rain, sleet, and snow). Together, these wet and dry forms are known as *acid deposition.*

Acid deposition can greatly enhance the rates of dissolution and increase municipal costs. Priceless monuments have been severely eroded worldwide. Acid deposition can also lead to changes in the chemistry of soil, lake, and stream waters. When acid deposition affects an area rich in carbonate rocks, the acidity can be buffered successfully by the rapid dissolution of soluble carbonate minerals. However, when acid deposition affects soils developed on granitic and some metamorphic rocks, as in the Adirondack Mountains of New York and the White Mountains of New Hampshire, the buffering capacity is not adequate, and the acid constituents remain mobile. Figure 9.8 presents a map showing the pH of precipitation and the portions of the United States that containing rock types, climate, soils, and vegetation such that lakes and streams are most vulnerable to acidification.

A classic long-term study conducted in Hubbard Brook, New Hampshire, has documented the effects of increased acid deposition on such sensitive terrestrial ecosystems. One of the most harmful effects appears to be the increased mobilization of aluminum (Al) from aluminosilicate minerals. Some forms of aluminum are toxic to plant and aquatic life; increased aluminum levels may be largely responsible for observed harmful effects on biologic communities. The most dramatic observed effects are fish kills associated with acidic pulses. Many studies of these fish kills have revealed that the sensitivity of fish to acidic episodes in lakes and streams appears to vary with both the species and the life stage of the fish. Typically, young fish (feeding fry) appear to be the most sensitive, although large mortalities also have been associated with hatching eggs; for some species, even yearling and adult populations have shown sudden mortalities.

The Clean Air Act of 1963, as amended in 1970, has lead to a decrease in sulfur dioxide emissions in the United States; however emissions of nitrous oxide gases have increased in many parts of the United States over the past two decades. Although "acid rain" is commonly recognized as a potentially devastating contributor of non–point source pollution, continued acidification remains a pressing problem.

Acid deposition problems have been even more severe in Europe than in the United States. In 1979, in recognition of these problems, thirty-three countries in Europe and North America signed a treaty, the Convention on Long-Range Transboundary Air Pollution, under the auspices of the United Nation's Economic Commission for Europe (ECE). This treaty committed signers to conduct joint research on and monitoring of acid rain, and to set up initial negotiations to stabilize or reduce sulfur dioxide and nitrous oxide discharges. The international agreements have greatly decreased the sulfur dioxide emissions in Europe; for example, from 1980 to 1990, emissions were reduced in Austria by 75 percent in Sweden by 60 percent, in France by 60 percent, and in Italy by 37 percent. In fact, the reduction of sulfur dioxide emissions in Europe has been one of the most encouraging examples of international environmental agreements to date. Unfortunately, several nations refused to sign the treaty, and in a few, such as the former Yugoslavia, sulfur dioxide emissions increased by up to 20 percent in the same period. Clearly, there is more to be done in the battle against acid deposition.

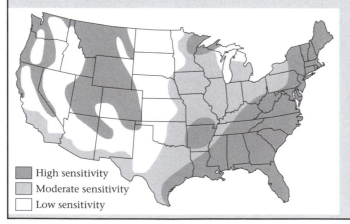

High sensitivity
Moderate sensitivity
Low sensitivity

Figure 9.8. Sensitivity of different regions of the United States to acid deposition. (*Source:* Courtesy of U.S. Environmental Protection Agency.)

UNDERGROUND WATER. All water beneath the land surface is referred to collectively as *underground water*, or subsurface water. Underground water occurs in two distinct zones (as shown in Fig. 9.9): the *unsaturated zone*, which occurs directly beneath the land surface and in which pores between grains are only partially filled with water; and the *saturated zone*, which generally lies beneath the unsaturated zone and in which all of the pores and fractures are completely filled with water. The top of the unsaturated zone is the *soil zone*, an area where plant roots, decomposing plant matter, and microbial processes work chemically and physically to transform rock to soil (Chap. 8). Water in the saturated zone, commonly termed *groundwater*, is an important source of fresh water for many portions of the world. The *water table* separates the unsaturated zone (above) from the saturated zone (beneath). Directly above the saturated zone is the *capillary fringe*, where water is pulled by capillary forces upward against the force of gravity from the saturated zone.

A rock unit containing a volume of groundwater useful to humans is termed an *aquifer*. The ability of a rock unit to serve as an aquifer depends on both the porosity and permeability of the geologic materials. *Porosity* is defined as the fraction of the bulk volume of rock or sediment that is composed of voids (which may be filled with air and/or water) and may be represented in the form of "percent void." *Permeability* is a measure of the interconnectedness of the pores. A rock unit can have high porosity but low permeability, if, for example, the individual grains are cemented together. Thus blocked, they would prevent passage of water from one pore space to another.

For an *unconfined aquifer*, the *water table* is defined as the level in the saturated zone at which the hydraulic pressure (the weight of an overlying column of water per unit cross section) is equal to the atmospheric pressure (pressure induced by the weight of the overlying atmosphere per unit cross section). The water table generally follows the overall pattern of topography of the land surface. If we drilled a well into an unconfined aquifer, the natural level of water in the well would represent the water table. This phenomenon is commonly referred to as a *water table aquifer*. In some places, however, impermeable rock units known as *confining beds* may occur above or between aquifers. Aquifers bound by confining beds are referred to as *confined aquifers*. If a well is drilled into a confined aquifer, the water level in the well may stand above the surface of the upper confining boundary, sometimes even above land's surface, in which case it is referred to as an *artesian aquifer*.

Like all components of the hydrologic cycle, groundwater is in flux and always moving, albeit generally at a very slow rate. For an unconfined aquifer, the direction of groundwater movement is indicated by the slope of the water table, which can be determined by drilling several wells and measuring the difference in relative pressure between them. Such a measurement is known as *head*, and can be thought of in simple terms as the potential energy of the mass of groundwater at a given point relative to a specified base level (e.g., sea level).

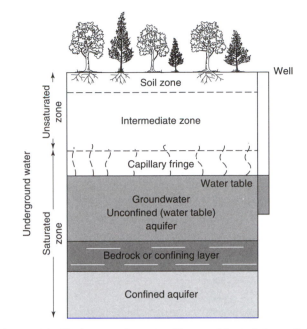

Figure 9.9. Underground water. (*Source:* Adapted from R. C. Heath, *Basic groundwater hydrology*, U.S. Geological Survey Water Supply Paper 2220, 1983.)

Because the water table is defined as the location where pressure equals zero (referenced to atmospheric pressure), for an unconfined, water table aquifer, the total head at any location is equal to the elevation above sea level of the water table at that point.

A fundamental law of hydrology is that water always moves from locations of greater total head to locations of lesser total head (see Fig. 9.10). This rule is summarized in *Darcy's law*, named for the French hydraulic engineer Henri Darcy. Darcy's law simply states:

$$Q = K \times A \times I$$

where Q (flux) is the quantity of water moving through a given area over a given time; K is the hydraulic conductivity, a parameter that depends on the flow properties of the geologic materials and of the water; A is the cross-sectional area through which the water flows; and I is the *hydraulic gradient*, a measure of the change in head between two wells located at a given distance (dh/dl). In effect, Darcy's law is a confirmation of our everyday observation that water always flows downhill, with the qualification that for confined aquifers, the pressure on groundwater caused by overlying aquifer materials must be considered.

In the groundwater system shown in Figure 9.10, if we were to begin pumping water from one of the wells, the water level in the well would begin to decline. The head in the well would fall below the head in the surrounding aquifer, causing water to begin to flow from the aquifer into the well. With continued pumping, the level of water in the well would continue to decline, and the rate of flow from the aquifer into the well would continue to increase. This move-

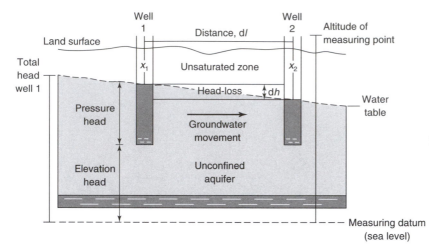

Figure 9.10. Groundwater heads and flow. (*Source:* Adapted from R. C. Heath, *Basic groundwater hydrology*, U.S. Geological Survey Water Supply Paper 2220, 1983.)

SALTWATER INTRUSION

Many coastal regions, especially where human development has placed a burden on groundwater resources, experience a serious underground encroachment of seawater. This problem is known as *saltwater intrusion*. Freshwater falling on the land surface does not mix with saline groundwater at depth because freshwater is less dense than saltwater and so floats on top of it. Hence, along coastlines, a fresh groundwater lens tends to float atop deeper, denser salt water (see Fig. 9.11). When freshwater heads are lowered by withdrawals of the aquifer water from wells, the contact between fresh and salt water migrates upward in the region of the well, saltwater intrusion, also termed *saltwater encroachment*, occurs. Saltwater intrusion is a serious problem throughout much of the North Atlantic coastline. Some coastal cities in California also experience this problem. However, saltwater intrusion is not restricted to coastal regions; it may also occur in mid-continent areas that are underlain by rock units containing ancient salty water. Unfortunately, once a reservoir is contaminated by salt water, cleanup requires several cycles of replacement by fresh water.

Figure 9.11. Saltwater intrusion by excessive pumping of groundwater near the coast. (*Source:* Adapted from R. C. Heath, *Basic groundwater hydrology*, U.S. Geological Survey Water Supply Paper 2220, 1983.)

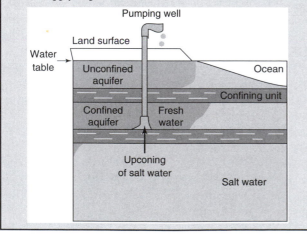

ment of water out of the aquifer and into the well results in a *cone of depression*, a conical depression in the water table around the well. In regions where extensive pumping occurs, the resultant cone of depression can cover an enormous area. Differences in the responses of unconfined versus confined aquifers to pumping, and of aquifers in different types of geologic settings, constitute a major topic of study in groundwater hydrology.

In many parts of the world, withdrawal of groundwater from wells for agricultural, industrial, and municipal use exceeds the natural recharge capability of the aquifer. This excess withdrawal is known as *groundwater mining* (see Chap. 30); in addition to depleting freshwater resources for future generations, it can lead to compaction of aquifer materials, decreased porosity and permeability, and subsidence of the land surface. The aquifer rocks, no longer saturated or supported by pore waters, are compacted by the weight of overlying rocks. In low-elevation coastal regions throughout the world, such land surface subsidence often leads to flooding of the land surface by saltwaters, and to increased erosion. Examples include the Houston–Galveston Bay area of Texas in the United States and Venice, Italy, where the Adriatic Sea is encroaching on irreplaceable Renaissance monuments plazas, and palaces. Another is subsurface invasion by marine waters (see box entitled *Saltwater intrusion*).

SURFACE WATER. All the water at the Earth's surface moves in response to two forces: heat and gravity. Evaporation can transport water from the gravitational low of the ocean basins to elevated land surfaces. Precipitation falling upon terrestrial ecosystems may follow several different flow paths before accumulating in one or more rivers, streams, and lakes. Together, these bodies are referred to as *surface water*, and are distinct from *groundwater*. Although surface water makes up only approximately 0.3 percent of the Earth's complement of water, it is crucial to the sustenance of life. Streams and rivers eventually carry water back to the oceans and complete the cycle begun by the imbalance of evaporation and precipitation over the oceans and the landmasses.

Rivers. Rivers and streams are important resources, sup-

Figure 9.12. World's largest drainage, basins, showing the location of deltas at their mouths. (*Source:* Adapted from B. J. Skinner and S. C. Porter, *The Dynamic Earth*, John Wiley, 1987, figure 10.22, p. 234.)

plying fresh water to many of the world's great cities and providing natural highways for transportation of foodstuffs and other products essential to humans. Rivers are a source of electricity in the form of hydroelectric-generated power, and are a critical source of cooling water for nuclear power plants. Rivers, of course, come in many sizes, and even the flow rate of a single river can vary dramatically with seasonal and other effects. A few rivers stand out on a global scale because of their immense size and power, carrying billions of tons of sediment and enormous volumes of water to the oceans each year (see Fig. 9.12). The Amazon River in South America has the greatest annual discharge, over 6000 cubic kilometers of water per year; the Mississippi River has the greatest discharge of any river in North America, with slightly less than 600 cubic kilometers of water per year.

The source of all river water is precipitation that falls upon terrestrial environments and travels by a variety of different pathways to streams and rivers. These different hydrologic *flow paths* are illustrated in Figure 9.13. Direct surface runoff, or *overland flow*, occurs when the soils become saturated and can no longer take up additional water. The excess rainfall runs off the soil surface and flows overland to the stream channel. Overland flow may also occur over bare rock surfaces, for example, across impermeable granitic outcrops exposed by glacial scouring. *Subsurface flow* is another important contributor to streamflow and includes contributions from *interflow*, which is shallow flow through only the soil zones, and *groundwater discharge*, which is water that emerges from the saturated zone. Finally, some water in a stream may accumulate from direct precipitation onto the surface of the stream. The relative importance of the different sources of stream water varies, of course, with precipitation patterns (precipitation intensity, duration, magnitude) and with the hydrogeologic characteristics of an area (shape and size of the watershed basin, thickness of soils, and type of aquifer materials). Rivers are extremely important in that they can stretch great distances from their source, bringing water into areas

Figure 9.13. Hydrologic flow paths leading to a river.

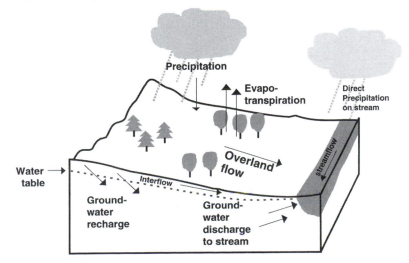

that are otherwise moisture-poor, as does the Nile River.

Quantification of the roles of different flow paths to streamflow is important in order for humans to make the best use of this potentially rich resource. Such quantification requires compilation of a stream *hydrograph*, a continuous record of streamflow as a function of time. To record a hydrograph, hydrologists define a transect across a stream bed, measure the cross-sectional area of this transect, record the height of water in the stream, and then use a current meter to measure the flow rate at several locations within the transect. Changes in flow over time are then studied. Between storm events, the hydrograph is relatively flat; this portion of the hydrograph is known as the *base flow*. Discharge from an aquifer makes up a large component of the base flow. Much research has been devoted to analyzing hydrographs, particularly the controls on hydrograph shape, as a means of predicting the flood potentials of different types of basins (e.g., basins with low versus high relief, basins underlain by "hard" rocks versus sediments, small versus large basins). As shown in Figure 9.14, storms cause a dramatic change in the hydrograph. Typically, a storm hydrograph contains a rising limb, a crest, and a recession limb, related to the increase in water flow caused by the storm, and the eventual decrease as storm waters ebb. During and immediately following storm events, the contributions of overland flow and interflow may be particularly important, as may some shallow groundwater flow. The resulting increased streamflow caused by the storm often results in a typical crest, as shown in Figure 9.14. The exact shape of a storm hydrograph depends on many factors, including: the size and shape of the basin; the basin's slope, geology, soils, and vegetation; the type of precipitation (snow, sleet, or rain); the intensity and duration of the storm; and the amount of time that has passed since the last storm. Storm hydrographs in small, headwater streams tend to respond rapidly to increased precipitation and runoff, showing a sharp peak. In contrast, storm hydrographs further downstream tend to respond more slowly, with a time lag between the onset of the storm and the storm crest, and a broader, shallower crest.

The changing shape of the hydrograph caused by a storm

Figure 9.14. Components of a typical hydrograph.

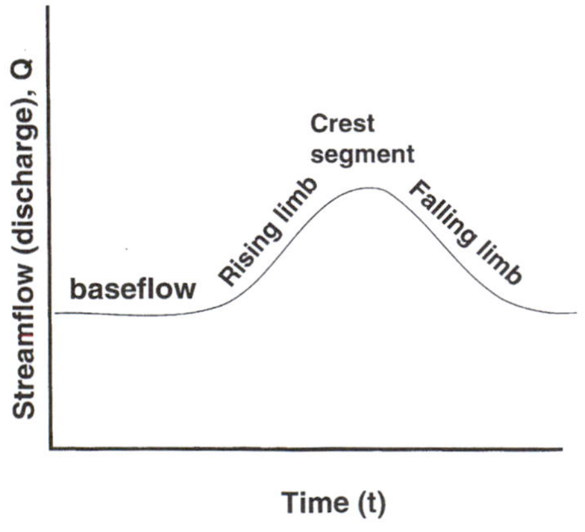

may be accompanied by changing stream chemistry. Contributions from overland flow and interflow show a predominance of organic acids and inorganic components such as iron (Fe) and aluminum (Al) derived from the litter layer and the soil zones. Base flow has chemical characteristics derived from the slow reaction of aquifer waters with aquifer materials. In mountainous areas receiving acidic deposition from pollutants in the atmosphere in the United States, such as the Adirondack Mountains, there is a characteristic change in the stream water hydrograph and in stream chemistry that occurs in the spring, as the snowpack melts. Over the winter, pollutants captured in snowflakes accumulate in the snowpack. For some as yet not fully understood reasons, the pollutants typically are concentrated in the first meltwater coming off the snowpack, such that 50 to 80 percent of the acidity in the snowpack is released during the first 30 percent of the melt. The first pulse of meltwater reaching streams can therefore be extremely acidic, often leading to fish kills in the early spring.

Lakes. Lakes constitute a relatively small portion of the hydrologic system, containing only approximately 0.01 percent of the Earth's water. Yet, they are extremely important sources of fresh water, and they will tend to become increasingly important in the United States as groundwater "mining" continues to decrease available groundwater resources. For example, the Great Lakes are the Earth's largest reservoir of fresh water, and they can be expected to play an increasingly important role in supplying our nations' freshwater needs. The study of lakes, *limnology* (derived from the Greek *limne*, meaning "pool" or "marsh"), has merited a great deal of research because of the importance of lakes in maintaining biologic communities, and as both a drinking-water and recreational resource. Lakes occur wherever water pools in substantial amounts, whether through natural means, such as within glacial depressions, or through anthropogenic means, such as the damming of rivers. Most lakes contain relatively impermeable basements; therefore, input is limited to precipitation, streamflow, and direct runoff or interflow. Some lakes, however, gain some fraction of their water from groundwater. In addition, lakes with permeable basements may drain partially into underlying aquifers and hence serve as groundwater recharge basins. In the San Francisco Bay area, many recharge ponds have been constructed to promote such accelerated recharge. These ponds serve a dual purpose in providing water supplies to combat fires – a particularly important function when one considers the increased fire danger due to earthquakes in the Bay area. Lake hydrology may be complex; depending on the underlying geology, lakes may contain different zones of recharge and discharge, or have varying recharge/discharge properties depending on fluctuations in precipitation volumes and hence in the level of the water table. Other important hydrologic variables include inflow and outflow of river water to and from lakes, and evaporation from the lake surface. Both natural and constructed lakes serve as sediment traps. As river water discharges into a lake, the velocity of the water decreases, and suspended sediments settle out. Thus, although dams can be constructed to establish reservoirs, the lifetimes of such reservoirs are limited by the accumulation of sediments that ultimately fill the lake basin. The same phenomenon leads to the eventual silting in and filling up of natural lakes and ponds.

The chemistry of lakes depends on several factors, including their hydrology (e.g., relative importance of surface water and groundwater inputs, and amount of evaporation), the surrounding geology (e.g., carbonate versus granitic terrain), temperature-driven circulation patterns, and various anthropogenic influences (e.g., impact of acid rain, agricultural and industrial pollutants). Temperature is a particularly important variable in lake chemistry, because of the role that temperature plays in determining the density of water. In temperate zones, thermally driven mixing typically occurs within lakes. During the summer (Fig. 9.15), the surface of a lake is heated. The thickness of this heated zone depends on the velocity of winds that stir the water at the surface. The well-stirred uppermost portion of a lake is referred to as the *epilimnion*. Below the epilimnion is a zone of rapid temperature decrease known as the *thermocline*, and below this region is the cold *hypolimnion*. In the autumn, as air temperature drops, water in the eplimnion is cooled; as the water temperature drops, it eventually becomes so heavy that the lake *overturns*. When the temperature at the surface drops below 4°C – the temperature at which water is most dense – the water density decreases until a temperature of 0°C is attained, at which point it forms ice. Hence, before the surface of the lake actually freezes, the lake develops a stable thermal stratification wherein overturn ceases. As we might guess, there is a similar spring overturn. Some temperate lakes are so shallow, however, that stratification and overturn never occur, whereas in the tropics, where large temperature changes do

not occur, lake overturn – if it occurs – is driven by more complex variables.

Lakes are often classified by their capacities to support aquatic life, known as their *potential productivities*. Productivity, in turn, is related to the available supply of essential nutrients. Photosynthesis is the starting point of the ecologic cycle in lakes. In most aquatic ecosystems, phosphorus is the essential nutrient in limiting supply; nitrogen is the next most likely limiting nutrient. *Eutrophic* lakes are characterized by a good nutrient supply, especially of phosphorus and nitrogen, and hence a high rate of biologic productivity. In contrast, *oligotrophic* lakes have a restricted nutrient supply and hence a lower rate of productivity. The term *eutrophication* is applied when the anthropogenic input of nutrients is excessive and increases productivity to the extent that the lake may achieve very high biologic activity. Eutrophication can lead to a large increase in phytoplankton, known as an *algal bloom*. Such algal blooms can, in turn, lead to accumulation of toxic substances, depletion of lake oxygen, and release of noxious gases such as hydrogen sulfide. The depletion of lake oxygen occurs as organisms die and their bodies settle downward. In the lower portions of the lake, the organic matter from the organisms is broken down or degraded according to a reaction that is essentially the opposite of the photosynthesis reaction. Oxygen depletion kills off benthic (bottom-dwelling) organisms. The entire food web of a lake can be altered by eutrophication, leading to loss of fisheries and to further degradation of drinking-water quality.

Wetlands. Wetlands are often overlooked as components of the hydrologic cycle; however, they are in fact ecologically, hydrologically, and geochemically important. Wetlands tend to have negative connotations, particularly during the industrialization phase of a country's growth when they are looked upon as a nuisance (e.g., as a breeding ground for mosquitoes and other pests). Until the past few decades, so-called wetlands management programs in the United States concentrated on destroying wetlands to limit diseases such as malaria and yellow fever, and to provide increased land for agriculture and urban growth and development. Unfortunately, the U.S. Fish and Wildlife Service has estimates that more than half of the approximately 104 million square kilometers of wetlands in the contiguous forty-eight states have been lost since the arrival of the first European settlers. We may never be able to appreciate the magnitude of this loss because we are only now beginning to recognize the number of extremely important functions, crucial to the health of ecologic systems, which wetlands provide.

It is difficult to define wetlands, since they are so diverse that no simple definition can fully encompass their variability. In general, they are recognized as land transitional between terrestrial and aquatic systems where the water table is at or near the surface or the land is covered by shallow water. Wetlands can be further categorized according to their unique complement of soils and vegetation particularly adapted to growth in wet conditions. The most important distinguishing characteristic of wetlands, however, is the

Figure 9.15. A typical thermal trough a temperate freshwater lake in summer. m = meters, (*Source:* Adapted from E. K. and R. Berner, *The global water cycle*, Prentice-Hall, 1987.)

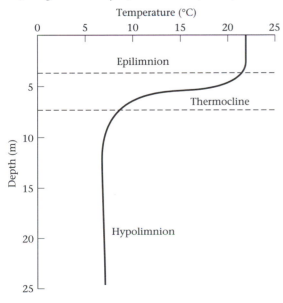

presence – and at least temporary storage – of water. Wetlands are dynamic ecosystems, the boundaries of which fluctuate seasonally with changes in rainfall and evapotranspiration. Because wetlands are sponge-like, able to store differing amounts of water, they are extremely important natural resources for flood control. For purposes of calculating reservoirs of the hydrologic cycle, wetlands usually are lumped together with lakes and streams; however, they have recently been recognized for their essential ecosystem services and are beginning to attract the intense study they deserve.

Wetlands come in many shapes and sizes. *Coastal* or *marine wetlands* can be continuously or periodically submerged by ocean waters. In estuarine areas, wetlands are subject to varying interactions between fresh water and salt water, depending on differences in tidal influences and stream water runoff. The term *palustrine wetlands* is applied to all nontidal wetlands, excluding those wetlands that are adjacent to or part of deep river channels or the edges of large lakes. *Riparian wetlands* are associated with streams having relatively deep channels, or are located around lakes.

Nontidal wetlands are classified according to their hydrologic properties and vegetation. Here are some examples:

■ **Swamp.** Wooded wetlands where standing or gently flowing surface water persists for long periods of time. Vegetation is dominated by trees and shrubs.

■ **Marsh.** Frequently or continually inundated wetland characterized by emergent vegetation adapted to saturated soil conditions (hydrophytes). By definition, marshes do not accumulate large amounts of peat.

■ **Bog.** A peat-accumulating wetland that has no significant inflows or outflows of surface or groundwater and that supports acidophilic ("acid-loving") mosses such as *Sphagnum* moss. Bogs typically are underlain by impermeable clay lenses that cut off the bog waters from association with groundwaters. Precipitation is the major source of both water and nutrients; consequently, bog waters tend to be strongly acidic and deficient in mineral nutrients.

■ **Fen.** A peat-accumulating wetland that receives some drainage, but is maintained primarily by groundwater. Fens tend to be less acidic than bogs. Sedges, grasses, and reeds are dominant forms of vegetation; *Sphagnum* mosses and trees are rare or absent.

As one might guess from these definitions, wetlands provide excellent examples of the interrelationships between hydrology, water chemistry, and biology in natural systems. The development of diverse vegetation types in wetlands is related to the flow, level, and chemistry of groundwater and the interactions between surface waters and groundwaters. Water chemistry can be related to a particular water source, with regard to nutrient concentrations, oxygen content, and pH, which provide a "fingerprint" for the source. These three factors are crucial to the wetlands' ability to support life. Bogs that are perched and removed from contact with groundwater are fed primarily by rainwater, whereas fens that receive some drainage from soils and aquifer materials are to some

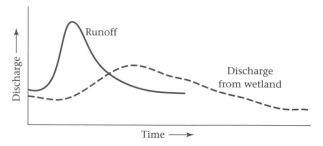

Figure 9.16. Diagrammatic illustration of the mediation of storm flow by a wetland.

degree dependent on surface flow. Characteristic plant communities in wetlands are related to the nature of the flow of the water. This is not only a chemical and nutrient effect, but also a temperature effect. In temperate zones, aquifer waters are colder in the summer and warmer in the winter than are surface waters, and different plant communities respond differently to these temperature influences.

One of the most important functions served by palustrine and riparian wetlands is modification of runoff peaks and the resultant decreased likelihood of floods. Figure 9.16 illustrates how wetlands can serve to "smear out" storm hydrographs, leading to gradual release of water rather than rapid release and flooding. Wetlands also may serve as sinks for excessive nutrients, such as nitrogen. Wetlands have been accurately dubbed "the kidneys of the landscape" because they tend to cleanse waters of noxious metals and excessive nutrients, retaining these pollutants within their borders and allowing cleaner water to pass into underlying aquifers and nearby streams. In fact, wetlands are being used in a number of states for treatment or posttreatment of waste water.

In addition to these benefits, wetlands may recharge aquifers, protect shorelines, trap sediments, and provide irreplaceable habitats for waterfowl, other animals, and unique forms of vegetation. Some wetland-specific plants provide new medically important drugs. Finally, wetlands are important recreational resources; fishing, boating, bird-watching, and other recreational activities abound.

In California, a new twist to the struggle between environmental, governmental, and industrial-agricultural groups is centered around artificial wetlands in the form of rice fields in the Sacramento River Valley. Growing rice consumes huge quantities of water. During drought episodes in the late 1980s and early 1990s, rice farmers, accused of growing "monsoon crops in the desert," were the object of substantial pressure from environmental groups to stop growing their crop in order to alleviate stresses on available water resources. Some rice farmers argued that their artificial wetlands were, in fact, providing habitats for waterfowl (ducks, geese, etc.) that had been displaced by the drainage of natural wetlands in the area. Eventually, the farmers' argument gained credence. A coalition among farmers, environmental groups such as the Nature Conservancy, and governmental agencies has been formed to help maximize the environmental and agricultural benefits of rice growing in the Sacramento River Valley. Environmentally friendly practices, such as relying on waterfowl to rework rice debris left over after harvesting into the wetland basins (with natural fertilizing) rather than burning it to produce thick billowing clouds of smoke, are being worked out and implemented. This is a pilot program for the interaction between different coalitions; its wisdom and success remain to be seen.

HOW DO WE STUDY COMPONENTS OF THE HYDROLOGIC CYCLE?

WATER BUDGETS. One of the best means of understanding hydrologic, geochemical, and biologic processes in forested ecosystems, and of understanding the effects of different processes on the chemical characteristics of natural waters, is known as the *small watershed approach*. This approach was developed by a group of biologists, chemists, geologists, and hydrologists working at the Hubbard Brook Experimental Forest in the White Mountains of New Hampshire (USA). In this approach, illustrated schematically in Figure 9.17, a small watershed was selected with what was thought to be an impermeable granitic basement, such that groundwater inputs and outputs to and from the watershed were considered to be negligible. The existence of an impermeable basement meant that differences between chemical inputs in precipitation and outputs in surface water could be attributed to processes that occurred solely within the confines of the watershed. The watershed-scale cycling of chemical constituents was determined by monitoring, over the course of many years, the volume and chemistry of precipitation into the basin and the volume and chemistry of stream water leaving the basin, and computing the difference. Although it was eventually shown that even solid, granitic rocks can have important groundwater contributions by *fractured flow* through cracks and fractures in the rocks, the small watershed approach has proved to be very helpful in deciphering hydrologic and chemical changes within watersheds.

Calculating a Water Budget for a Forested Ecosystem. Calculation of a yearly water budget is one of the primary steps in any watershed study of chemical cycling through forested ecosystems. Water budgets provide quantification of the movement of water into and out of a basin, and hence of the cycling of dissolved chemical constituents. Yearly water budgets for a forested ecosystem may be represented by the following equation:

$$P = SW + ET \pm GW \pm dST$$

where P is the precipitation input to the basin, SW is the surface water discharge (outflow), ET is the evapotranspiration, GW is the volume of groundwater recharge ($-$) or discharge ($+$), and dST is the change in water storage in the basin, including changes in surface water, soil water, and

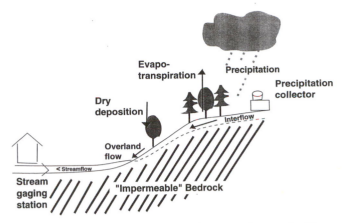

Figure 9.17. Schematic illustration of the "small watershed" approach to hydrogeochemical cycling through a forested ecosystem.

groundwater storage. *Evapotranspiration* is a flux component of the hydrologic cycle. It consists of the combined evaporation from open bodies of water, evaporation from the surfaces of soils, and transpiration by plants.

Over the long term, when many years of data are averaged, the net change in storage (dST) in a basin should be negligible (unless humans are mining the water). If the basin's borders are chosen carefully, groundwater recharge and discharge can be neglected as well, leading to a much simplified equation:

$$P = SW + ET$$

In general, measurement of P and SW is relatively straightforward. Typically, precipitation influx to a basin is measured by an event-recording rain gauge, and surface water outflow is measured at a stream gauging station. Estimating evapotranspiration directly, however, often proves to be a difficult task. In the classic watershed study at the Hubbard Brook Experimental Forest, scientists estimated evapotranspiration as equaling the difference between precipitation and streamflow, assuming that the watersheds were underlain by impermeable bedrock that minimized the groundwater component of the budget, and that the change in storage was relatively unimportant over the long term. These assumptions are not always borne out in field locations, so several methods of calculation of evapotranspiration have been developed. Nevertheless, evapotranspiration generally remains a significant source of error in water budget calculations.

Calculating Chemical Input/Output Budgets. In basins in which groundwater inflow and outflow are negligible, if we know the volume and chemistry of stream water leaving the basin, we have some idea of whether processes occurring within the basin cause a net gain or a net loss of different chemical components. This type of approach has been particularly important in studies of acid deposition. Key questions include whether ions such as sulfate and nitrate added to a basin through acid deposition cause acidification of stream and lake waters, and whether this acidification is accompanied by the mobilization of harmful materials, such as aluminum from soils. By monitoring sulfate, nitrate, and aluminum levels in precipitation and surface water, along with the water volumes, we can determine whether there is a net uptake or release of these components within a basin. Of course, the ideal of an impermeable basement, wherein groundwater is not a factor, is seldom attained in the real world; consequently, the contribution of groundwater to hydrogeochemical cycling is an important topic of ongoing research.

Based on our discussions of hydrology and geochemistry, you should be able to anticipate what will happen to water geochemically and hydrologically as it enters a basin as precipitation and moves through the soils, aquifers, streams, and lakes toward the ocean. The initial precipitation will be relatively dilute water, with the chemistry varying according to the storm track (for example, cyclic salt inputs, pollution inputs). As this precipitation passes through the canopy, it will acquire dry deposited materials (e.g., sulfates) and will leach natural organic acids and other constituents from the leaves. The changes that occur will be related to the vegetation type; for example, water passing through coniferous stands tends to accumulate more sulfate and aluminum than does water passing through deciduous vegetation. Upon entering the soil zone, water undergoes a variety of simultaneous transformations. Evapotranspiration acts to increase the concentrations of chemical constituents; simultaneously, plants selectively tend to remove certain nutrients, especially phosphorus and nitrogen. Water chemistry is influenced by interactions with soil minerals, decaying organic debris, and soil microbes, and by changing oxidizing/reducing conditions. Organic acids leached from the litter layer play important roles in soil-mineral weathering.

As water continues along a flow path to an underlying aquifer, it reacts with aquifer rocks and minerals, changing chemically according to the aquifer lithology. From our discussion of stream hydrology, and the diverse flow paths by which water may enter a stream, we can expect that the aqueous chemistry will vary with the hydrograph. The ultimate composition of the water leaving the basin is influenced not only by the precipitation chemistry but also by the local climate, hydrology, geology, soil, and vegetation. Because of the complexity of natural systems, detailed local studies are required to determine all the complex processes that together control the cycling of chemical constituents through any given forested ecosystem.

SUMMARY

In this chapter, we have discussed the principal components of the hydrologic cycle. As an example of an Earth systems approach, we have shown how the water cycle interacts with the Earth's energy budget and how the chemistry of water varies with hydrogeologic setting. Because water flow and chemistry are closely related, emphasis has been placed on how the chemistry of water varies through the hydrologic cycle and with different hydrogeologic conditions. Over the course of this chapter, we have endeavored to make the reader aware of complexities involved in dealing with water-related problems, and why hydrologists, geochemists, and other Earth scientists cannot always provide easy answers to problems of water use. Because water represents a fundamental limit to growth, and because human activities have such profound influences on global water quality, a deeper understanding of the hydrologic cycle is critical for the future well-being of the planetary biosphere.

QUESTIONS

1. Why might the preservation of wetlands along the coasts of Lake Erie be important for the health and vitality of the lake?

2. Some state and local laws now require construction of holding ponds, also known as trapping basins, next to construction sites, to capture increased sediments released by construction and to prevent these sediments from entering local rivers. What types of problems might be associated with such artificial ponds? Why might long-term studies be needed to assess their impact?

3. Why do we need to study both the chemistry and the amount of water entering a basin to determine whether the basin as a whole is accumulating or releasing constituents such as sulfates and nitrates? Why would such studies need to be conducted over the course of many years, and all seasons?

4. In determining whether there is a net accumulation or release of chemical constituents within a forested ecosystem, why is it important to sample stream chemistry over all portions of the hydrograph, including both base flow and storm flow?

5. In the United States, recent recognition of the importance of wetlands is leading to legislation for their protection. Many states now prohibit the draining of wetlands. Some, such as Florida, require the reestablishment of artificial wetlands to "replace" any natural wetlands that are drained. Although the construction of artificial wetlands is receiving much attention, the wisdom of allowing further destruction of natural wetlands needs to be considered. Natural wetlands are unique, complex hydrologic and ecologic systems that take thousands of years to develop. It seems unrealistic that artificially constructed wetlands could replace all the functions of their natural counterparts. Although the concept of artificial wetlands is enticing, the often unpredictable behavior of natural systems has the potential to thwart the best-laid plans of engineers. What of the indigenous or endemic species of flora and fauna? Will episodes of flooding increase near the site of the original wetland? What other questions or potential problems can you think of?

FURTHER READING

Chahine, M. T. 1992. The hydrological cycle and its influence on climate. *Nature* 359: 373–379.

Hem, J. D. 1985. Study and interpretation of the chemical characteristics of natural water. U.S. Geological Survey Water-Supply Paper 2254, 263 pp.

Freeze, R. A., and Cherry, J. A. 1979. *Groundwater*. New York: Prentice-Hall.

Berner, E. K., and Berner, R. A. 1987. *The global water cycle: geochemistry and environment*. New York: Prentice-Hall.

Drever, J. I. 1988. *The geochemistry of natural waters*, 2nd ed. New York: Prentice-Hall.

Likens, G. E., Bormann, F. H., Pierce, R. S., Eaton, J. S., and Johnson, N. M. 1977. *Biogeochemistry of a forested ecosystem*. New York: Springer-Verlag.

10 Atmosphere–Ocean Coupling and Surface Circulation of the Ocean

JAMES C. INGLE, JR.

WINDS AND WAVES

Ancient Greek philosophers viewed the ocean beyond the Mediterranean Sea as a great river, Oceanus Fluvius, which they considered to be directly related to the Earth (Ge) and the sky (Uranus). Today, we recognize that surface circulation of the global ocean is largely the product of so-called zonal winds, the Earth's rotation, differences in the density of seawater, and the spatial constraints imposed by continents. The direct link between the motion of the atmosphere and the surface ocean is apparent to anyone watching wind-whipped storm waves crash on a beach. Even when local winds are calm, drifting objects offer direct evidence that the surface of the sea is in constant motion. Viewed on a global scale, the Earth's intimately coupled gaseous atmosphere and liquid ocean constitute a single dynamic circulatory system undergoing constant convective motion driven by radiant heat from the Sun. The importance of this system to the well-being of our planet cannot be overemphasized. The constant physical and chemical exchanges between the ocean and the atmosphere govern the tempo and mode of both marine and terrestrial environments on all scales, and thus control the fundamental character of the surface of the Earth. The goal of this chapter is to provide an understanding of the dynamic interplay between the atmosphere and the ocean, with an emphasis on the basic processes controlling the surface circulation of the ocean. The following chapter reviews circulation of the deep ocean and considers ocean circulation in its entirety.

Although the ocean covers 71 percent of the Earth's surface and constitutes the dominant and defining environment of our planet, we are only now becoming familiar with how it circulates and what lies beneath its surface. The very dimensions of the ocean have until very recently stood as a primary obstacle to understanding the circulation of this immense body of water, despite its obvious importance to human affairs ranging from sustenance and commerce to climate prediction. The earliest written records make it clear that the seagoing Phoenicians applied their knowledge of wind and current patterns in the Mediterranean Sea and elsewhere for both war and trade, information no doubt hard won at sea and patiently accumulated to their advantage over the centuries. Prior to the launching of Earth-observing satellites, any attempt to gain an overview of the surface circulation of the ocean involved the same strategy presumably employed by the Phoenicians some 3000 years ago – laborious compilation of observations made aboard individual ships at sea in order to obtain the larger picture. In fact, knowledge gained in this manner by American sailors allowed Benjamin Franklin to publish the first map of a major ocean current, the Gulf Stream, in 1769, aiding merchant shipping from North America to England in the process.

By 1837–40, the German geographer Heinrich Berghaus had assembled the first truly comprehensive maps depicting major surface currents of the world oceans aided by the observations of other scientists including Alexander von Humboldt. At approximately this same time, the importance of commercial whaling led the government of the United States to support an "exploring expedition." This global voyage was led by Commander Charles Wilkes, who published a five-volume report in 1845. This study included a remarkably accurate map of global surface circulation, which, for various reasons, received little attention. Subsequently, one of Wilkes's younger cohorts, Lieutenant Matthew F. Maury, set about synthesizing information on winds, currents, and ocean temperatures recorded in a huge collection of ships' logbooks that had been stored, but little used, by the U.S. Navy. Maury's efforts yielded a fresh set of maps and track charts depicting global patterns of winds and surface currents. These maps and charts were widely distributed and had an immediate positive impact on maritime trade and selection of sailing routes (Fig. 10.1). The usefulness of Maury's

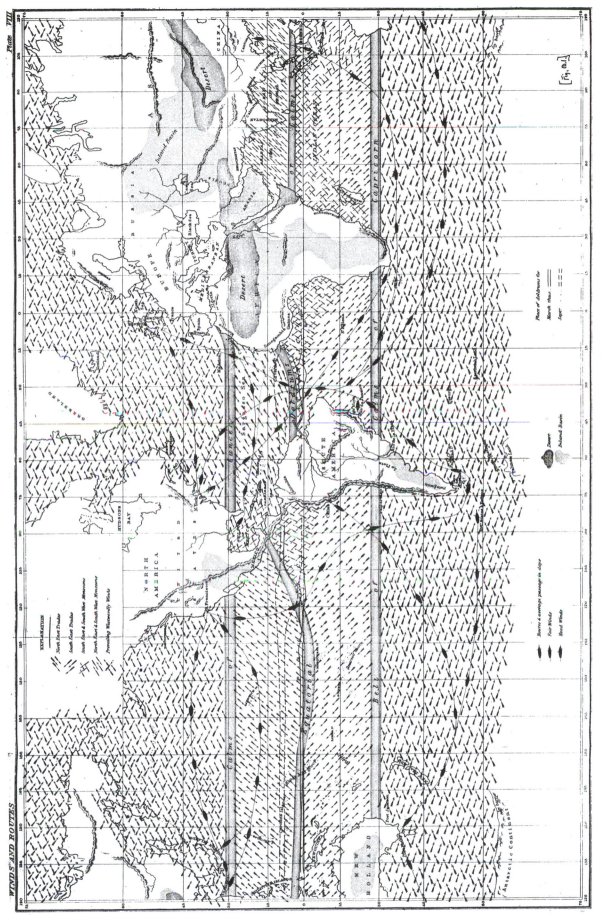

Figure 10.1. Early map of surface winds over the ocean and suggested best sailing routes, compiled by Lieutenant M. F. Maury of the U.S. Navy, and published in 1855. Although Maury's attempts to explain wind and ocean circulation were controversial, his maps and charts proved immensely useful to merchant shipping and stimulated the first attempt to standardize oceanographic observations on an international level. (*Source:* M. F. Maury, *The physical geography of the sea*, Harper Brothers, 1855.)

Figure 10.2. Satellite view of the Earth centered over Africa and emphasizing the dominance of the ocean environment. Thick swirls of clouds in the lower half of this photograph mark storm systems and winds driving the West Wind Drift (also known as the Circum-Antarctic Current) eastward around the ice-covered continent of Antarctica (clearly visible at the bottom of the globe). The arc of spotty clouds ranging across the Atlantic Ocean, central Africa, and the Indian Ocean just above the midline of the photograph represent the equatorial zone of rising warm, moist air and low-pressure storm cells associated with the intertropical convergence zone (ITCZ). Areas of clear sky over southwest Africa and the desert areas of North Africa, the Red Sea, and the Persian Gulf constitute mid-latitude zones of descending dry air, which increases in temperature through conduction and high pressure (adiabatic compression) as it approaches Earth's surface. Clear air over Antarctica is due to extremely cold, dense, dry air descending over the polar region. (*Source:* Photograph courtesy of NASA.)

charts led to the first attempt to standardize oceanographic observations aboard all sailing vessels. Among his other accomplishments, Maury also produced the first bathymetric (depth) chart of a portion of the Atlantic Ocean based on primitive sounding by U.S. Navy vessels under his direction. In 1855, Maury summarized his work in a volume entitled the *Physical Geography of the Sea*, which was reprinted several times to meet demand and translated into several languages. However, many of his attempts to explain ocean and atmospheric circulation were naive and speculative, and therefore were much criticized by the scientific community, despite the value of his charts for improved sailing.

Although progress continued in amassing data on surface circulation, little was known about the deep sea, and another century passed before the first modern maps depicting the floor of the ocean appeared. In the intervening years, the modern science of oceanography took shape, spurred by the results of the HMS *Challenger* expedition, which circled

the globe from 1872 to 1876. Among the many scientific firsts accomplished during the *Challenger* expedition were measurements of temperature, salinity, and density at 362 deep-sea stations, yielding some of the first insights into the layered structure and circulation, of the deep sea. This circumglobal cruise was the impetus for a number of later expeditions, and the collection of many additional measurements.

However, it was the advent of surface and submarine warfare during World Wars I and II that accelerated research into all aspects of physical oceanography and ocean circulation. This latter research peaked during the Cold War years of 1950–90, as ever more sophisticated acoustic technology was employed to determine the detailed density structure of the ocean as a means to detect enemy vessels and operate effectively in the submarine environment (as dramatically fictionalized in Tom Clancey's 1984 novel *The Hunt for Red October*).

Despite all the advances in sampling and measuring properties of the surface and deep ocean from the 1950s to the present, it is the synoptic observations of the Earth from satellites that have allowed a quantum leap in our understanding of coupled atmosphere–ocean circulation (Fig. 10.2). Satellite observations became routinely available in 1960 with the launching of *TIROS*, which photographed the Earth's weather patterns and forever changed our view of the ocean and the world. Monitoring of electromagnetic radiation from various Earth environments was initiated by the *LANDSAT* satellites beginning in 1972, followed by the *NIMBUS* satellites from 1978 through 1984. *NIMBUS* instruments provided quantitative measurements of ocean productivity, water vapor in the atmosphere, ice coverage, and a host of other parameters reflecting aspects of ocean circulation and climate. *SEASAT*, the first satellite devoted exclusively to oceanographic observations, was also launched in 1978 but operated for only three months. Despite the premature failure of *SEASAT*, its radar provided the first synoptic observations of ocean wave patterns and surface winds along with measurements of sea surface temperature and elevation.

In 1992, the French–U.S. *TOPEX/Poseidon* satellite was launched into orbit for the purpose of observing and sensing the ocean – the first such satellite since ill-fated *SEASAT*. The high-altitude orbit of *TOPEX/Poseidon* allows it to observe over 95 percent of the ice-free ocean every ten days. Instruments aboard the satellite are capable of measuring a range of ocean parameters with unprecedented accuracy, including wind speed and direction, surface currents, and minute variations in the height of the sea surface reflecting both aspects of wind-driven ocean circulation and the effect of the Earth's rotation and gravity. Significantly, the first analyses of *TOPEX/Poseidon* data are only recently appearing in the scientific literature; it is fair to say that 1992 marked a watershed moment in our ability to observe and understand ocean circulation, comparable to the scientific threshold crossed when the *Challenger* expedition set sail over a century ago.

THE AIR AND SEA IN MOTION

The most important role of the atmosphere–ocean system is the redistribution of excess solar heat that the Earth receives in the equatorial and mid-latitude regions (Fig. 10.3). The atmosphere carries heat in the form of latent heat of evaporation (e.g., water vapor) as well as sensible heat (the sort measured with a thermometer), whereas the ocean carries heat only in the sensible form. Most of the Sun's radiant heat energy arriving at the sea surface goes into breaking the weak hydrogen bonds between individual water molecules and evaporating seawater, not into raising the temperature of the water. This process is a function of the polar character of the water molecule and reflects the fact that the heat capacity of water (e.g., heat absorbed divided by temperature rise) is the highest of all solids and liquids, with the exception of liquid ammonia. Thus, the great volume and heat capacity of seawater allows the ocean to store enormous amounts of solar heat and release it slowly back to the atmosphere by conduction at the air–sea interface. This process, along with the continuous evaporation and condensation of seawater by the atmosphere, serve to keep the surface temperature of the Earth within a range allowing life as we know it to thrive. Clearly, the ocean plays a critical role in maintaining a life-friendly climate on our planet.

Circulation of the atmosphere and ocean represents the never-ending quest nature of these two systems collectively to establish thermal equilibrium between the poles and the equator – a goal they will never reach thanks to unequal distribution of the Sun's heat over the curved surface of the Earth, the Earth's constantly changing climate, and the slow but unceasing tectonic rearrangement of continents and ocean basins. Air and water heated in the tropics and subtropics are transported poleward, while water and air cooled at high latitudes move equatorward. Complex and little-understood feedback loops characterize these processes and assure that any change in the behavior of the atmosphere or ocean will have profound consequences for the circulation of both systems. The hypersensitive interrelationships between the atmosphere and the ocean are dramatically illustrated by the global weather extremes associated with El Niño/Southern Oscillation (ENSO) conditions, which appear approximately every seven years in the Pacific region and elsewhere.

At the initiation of an El Niño event, the large, atmospheric high-pressure zone normally present over the South Pacific weakens while the large, low-pressure system operating over the Indian Ocean becomes stronger. In turn, equatorial trade winds weaken, and the thick mound or wedge of warm water normally maintained by these winds in the western Pacific is allowed to flow eastward toward the Pacific coasts of the Americas. The arrival of the warm surface water in the eastern Pacific disrupts normal upwelling of cold nutrient-rich waters, causing the temporary collapse of fisheries, exceptional storms, high rainfall and flooding, and even a rise in local sea level. The consequences of a sustained El Niño event include associated changes in the positions of the atmospheric jet streams and disruption of normal weather patterns on a global scale, translating to billions of dollars in storm, drought, and agricultural damage, as well as lost lives and subsequent years of recovery. There is clearly a need to understand the details of ocean–atmosphere interactions, and these processes represent frontier areas of ongoing ocean research.

Aspects of atmosphere–ocean coupling are intuitively straightforward; the wind blows and the sea's surface is moved, resulting in *wind-driven surface circulation.* However, the wind effectively stirs only a relatively thin layer of surface water, commonly no more than 100 meters in thickness. Contrary to common thinking, the bulk of the water in the oceanic bowl is not stirred by the wind but is gravity-induced and circulates as a function of variations in the temperature, salinity, and density of individual water masses. Therefore,

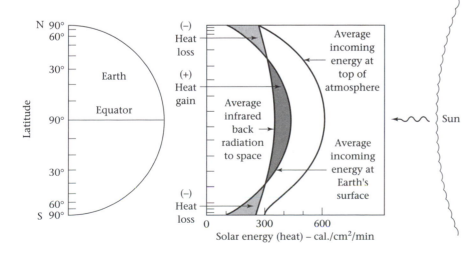

Figure 10.3. Net gain and loss of solar energy (heat) at the top of the Earth's atmosphere and at the surface of the Earth. Solar energy received at the surface (insolation) is unevenly distributed across latitude as a function of the curvature of the Earth and atmospheric effects. See text for discussion.

circulation of the deep ocean is referred to as *thermohaline circulation* ("thermo" means "temperature," and "haline" means "related to salinity or salt content"), a subject considered in Chapter 11. Although only a relatively thin layer of the surface ocean is directly moved by the wind, global patterns of surface circulation and air–sea interaction control much of the ocean's physical, chemical, and biologic character at all depths.

Convective and advective processes are responsible for heat transport and motion of discrete air and water masses within the atmosphere and the ocean (Fig. 10.2). *Advection* refers to changes in the property of an air mass or water mass by virtue of bodily motion. The term is often restricted to the horizontal motion of air or ocean water but can also apply to vertical motion. Water or air in motion is advectively transporting heat or other properties of the fluid regardless of how the motion was initiated, whether through mechanical means (e.g., one moving air mass pushing another) or by gravity-driven convection. A simple analogy is the initiation of motion by stirring water in a pot with a spoon. In contrast, convective motion is self-initiated whenever a fluid or air mass experiences changes in density (e.g., change in mass per unit volume, measured in grams per cubic centimeter) as a function of variations in temperature, composition, or pressure and is a response to gravity acting upon these changes. As a given mass of air or water is heated, molecular activity increases and it expands, taking up more space (e.g., specific volume or volume per unit mass), with a consequent decrease in its density. The change to lower density causes the air or water mass to rise within the density-stratified column of the atmosphere or ocean to a level commensurate with its new density. Conversely, a parcel of cold, dense air or water will sink in the density-stratified column to a level where it is surrounded by air or water of similar density and underlain by fluid of greater density.

The high angle of the Sun's rays or solar beam with Earth and the round shape of the Earth ensure that most of the radiant solar heat or energy is received between 30° N and S of the equator (Fig. 10.3). In contrast, the polar regions experience a constant heat deficit due to the low angle of the Sun's rays at high latitudes and heat loss through back radiation to space. The unequal distribution of heat results in a significant difference in temperature, or thermal gradient, between the equator and the polar regions regardless of whether the Earth is in a glacial or nonglacial climatic mode. The greater the difference in temperature between the poles and the equator, the steeper the thermal gradient. The ever changing pole-to-equator thermal gradient, together with subtle but critical changes in the density layered structure of the atmosphere and the ocean, maintain constant convective motion in both systems. The steeper the pole-to-equator thermal gradient, the faster the rate of atmosphere–ocean convection and circulation, and vice versa. (See the box entitled *The Dynamic Energy Balance of the Earth.*)

The overall pattern of atmosphere–ocean convective circulation thus mimics the familiar convective motion seen in

a pot of heated water on a stove. Water heated at the bottom of the pot expands as molecules become more active, decreases in density, and rises to the surface where it cools, increases in density, and sinks to the bottom of the pot, where the process begins again. With sufficient heat and time, the result of this process is a rolling boil – *vigorous circulation will have been induced through changes in the temperature and density of the water without any mechanical help* from a spoon. Just as the rate of boil in a pot of water can be modulated by adjusting the stove-top flame, any changes in the pole-to-equator thermal gradient over time result in increasing or decreasing rates of convective circulation within the atmosphere and ocean. The steepest pole-to-equator thermal gradients occur when the Earth is in a glacial mode, and evidence suggests that these periods are indeed characterized by increased rates of atmospheric and ocean circulation. Conversely, periods in Earth history when the polar regions remain ice-free and relatively warm are marked by shallow pole-to-equator thermal gradients and relatively slow ocean–atmosphere circulation.

The convective and advective processes operating in the atmosphere cause air masses to rise and sink, generating variations in atmospheric pressure at the surface of the Earth. In turn, winds are generated as air rushes from areas of high pressure to areas of low pressure. Winds developed in the lower atmosphere transfer their energy and momentum to the surface layer of the ocean through friction at the air–sea interface. Not surprisingly, global patterns of surface circulation in the ocean reflect the direction and strength of winds (i.e., wind stress) in the lower atmosphere. This is the case no matter what the configuration of continental masses might

---

THE DYNAMIC ENERGY BALANCE OF THE EARTH

The highest amounts of solar energy per unit area – measured in calories per square centimeter per minute (cal/cm²/min) – are received in the equatorial and mid-latitudes in the form of short-wave, visible radiation. Alternatively, the amount of heat radiated back to space from the Earth's surface is approximately the same at all latitudes and is largely in the long-wave, infrared portion of the spectrum. The difference between incoming and outgoing radiation across latitude results in a net deficit of heat in the polar regions and a net surplus of heat in the middle and lower latitudes. This pattern accounts for the pole-to-equator thermal gradient of the planet, which in turn drives the convective circulation of the atmosphere and ocean. Only approximately 50 percent of the solar energy received at the outer edge of the Earth's atmosphere reaches the surface of the Earth. The amount of solar heat entering the surface layer of the ocean every twenty-four hours is balanced by heat lost through back radiation to space, evaporation of seawater (loss of latent heat), and conduction of heat from the ocean to the overlying atmosphere, which, together with the high heat capacity of water and mixing of the surface ocean, maintains relatively stable surface ocean temperatures over time.

be over geologic time. However, tectonically induced changes in the number, shapes, and locations of continents and ocean basins over geologic time play an equally large role in modulating ocean circulation, in turn emphasizing the intimate link between the behavior of internal (endogenic) and external (exogenic) Earth systems over periods of millions of years.

THE LAYERED STRUCTURE OF THE ATMOSPHERE AND OCEAN

Many people enjoying a summer swim in a lake or the ocean have experienced the strange sensation of having their upper body in comfortably warm surface water while their legs dangle in cold water below – a clear example of the thermal and density stratification of a water column. The warmer, less dense water floats on top of denser, cold water with a sharp temperature and density boundary dividing the two layers. Convective circulation of the atmosphere and ocean is dependent upon variations in the density of the various air and water masses. Both systems exhibit a layered structure, with the heaviest or densest air and water residing in the bottom of the atmosphere and the bottom of the ocean, respectively. Because of density stratification, horizontal motion dominates both systems. Areas of vertical motion are limited to zones where advective or convective processes introduce instabilities in the air or water column through rapid changes in temperature or other parameters, causing the air or water to rise or sink. Although zones of vertical motion in the atmosphere and ocean are limited in areal extent, they are critically important.

The Earth's gravity maintains the highest density of atmospheric gases immediately adjacent to the Earth's surface, accompanied by high atmospheric pressures (this phenomenon is manifest in the annoying ear pressure change felt during an airplane landing or a rapid descent down a mountain road). Atmospheric gases thin and pressure and density decrease with increasing altitude above the Earth's surface. The atmosphere can be divided into four major layers based on the systematic changes in its temperature, beginning with the near-surface troposphere and followed by the stratosphere, mesosphere, and thermosphere. The lowest layer, the troposphere, contains 75 percent by mass of all the gases in the atmosphere, along with most of the water vapor and the majority of clouds, dust, and so forth. The lower atmosphere is heated both by release of latent heat when water vapor condenses into clouds and rain, and by conduction from the surface of the Earth (in the same manner that your hand is heated when you place it on a hot surface). Latitudinal variations in the surface temperature of the Earth as well as seasonal variations in temperature at any given location maintain constant thermal instability in the troposphere. Convective motion in the troposphere spawns the Earth's weather, and this layer contains the major surface wind systems of primary importance to oceanographers studying ocean circulation.

Density stratification in the ocean is the product of variations in temperature, salinity, and pressure with depth. Density (measured in grams per cubic centimeter) increases with decreasing temperature, increasing salinity, and increasing pressure; density decreases with increasing temperature, decreasing salinity, and decreasing pressure. A plot illustrating the range of variation in temperature and salinity demonstrates that the modern ocean has an average temperature of only approximately 3.5°C, reflecting the high-latitude origin of most of the water filling the ocean basins (Fig. 10.4). Temperatures in the open ocean range from a low of −2°C to a high of approximately 32°C, exceeding the much narrower range of salinity. Salinity represents the amount of salt in a given volume of water and is routinely measured in terms of parts per thousand (‰). The ocean exhibits an average salinity of approximately 35 parts per thousand. Pressure due to the weight of the overlying water column becomes a significant factor affecting the calculation of density only at depths greater than 1000 meters and is generally discounted at shallower depths. Thus, *temperature and salinity constitute the key parameters governing the density of seawater*, and both vary horizontally and vertically in the ocean (Fig. 10.5). Even a minute change in either or both of these parameters in a given parcel of water translates to a significant change in its density. The relatively narrow range of temperatures and salinities in the ocean means that scientists must precisely calculate very small differences in density, a calculation typically carried out to five places and then converted to a density factor termed *sigma-t* for convenience. For example, a calculated density of 1.02532 grams per centimeter would be converted to a sigma-t value of 25.32 for ease of plotting and manipulation.

Clearly, any process controlling the temperature or salinity of seawater has the capacity to change the density of the

Figure 10.4. Range of temperature and salinity in the global ocean, as illustrated by contours enclosing values for 99 and 75 percent of all the water in the ocean. The range of salinity is relatively narrow compared with that of temperature. The very cold average temperature of the ocean reflects the high-latitude origin of most of the water in the deep ocean. (*Source:* Adapted from M. G. Gross and E. Gross, *Oceanography*, 7th edition. Englewood Cliffs, NJ: Prentice-Hall, 1996.)

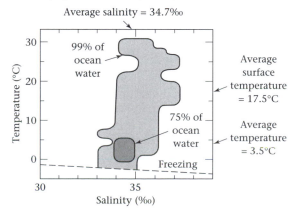

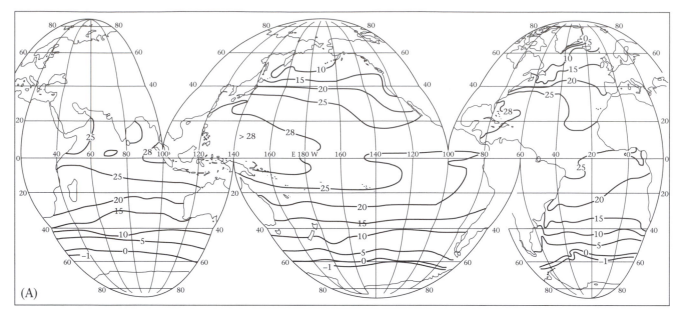

Sea Surface Temperature (°C)

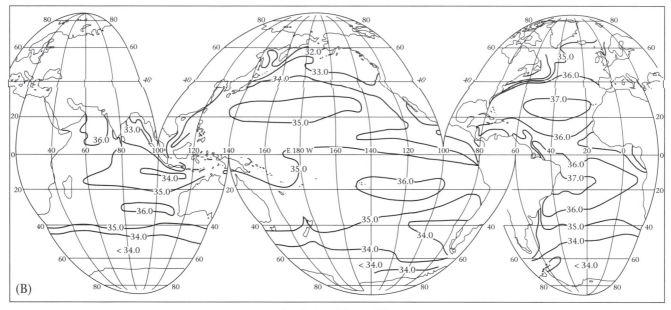

Sea Surface Salinity (%)

Figure 10.5. Generalized patterns of ocean surface (A) temperatures and (B)salinities in August (Northern Hemisphere summer). Lines of equal temperature (isotherms) and equal salinity (isohalines) tend to parallel latitude in the open ocean. Alterna- tively, isotherms are torqued north and south along continental margins, reflecting displacement by surface currents in these regions. (*Source:* Goode Base Maps, courtesy University of Chi- cago.)

water and in turn, its position in the oceanic water column. Significantly, most variations in temperature and salinity in the ocean as a whole initially occur at the sea surface through heating, cooling, evaporation, precipitation, and freezing (Figs. 10.5 and 10.6). The distribution of surface temperature reflects the pole-to-equator thermal gradient and patterns of wind-driven surface circulation. The temperature of a parcel of surface water can vary as a function of its latitudinal posi- tion, mixing with water of different temperature, or resi- dence time in a given locality. Warm, light water character-

izes the equatorial areas, cold, dense water is formed in polar regions as surface water arriving from lower latitudes is cooled. Variations in salinity also play a major role in creat- ing differences in density of individual water masses. In the highest latitudes, seawater is frozen; sea ice begins to form at a temperature of approximately −1°C. Because salt is ex- cluded during the formation of ice crystals, the salinity of the remaining unfrozen water increases, in turn lowering its freezing temperature and increasing its density, due to this increase in density, the unfrozen water sinks. The salinity of

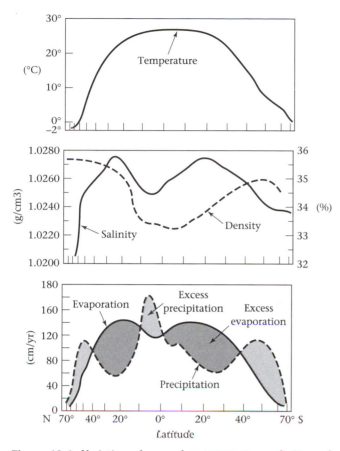

Figure 10.6. Variation of sea surface temperature, salinity, and density with latitude and average annual patterns of precipitation (rain) and evaporation at the sea surface. A correlation exists between areas of excess precipitation beneath the polar frontal zones (approximately 50 to 60° N and S) and the equatorial region with areas of depressed surface salinities. Similarly, there is a clear relationship between areas of excess evaporation associated with mid-latitude high-pressure zones (approximately 30° N and S) and relatively higher surface salinities. The slight offset of the equatorial zone of excess precipitation to the north of the geographic equator reflects the northward displacement of the so-called meteorologic equator (e.g., the east-west line of hypothetical thermal equilibrium between hemispheres located approximately 5° north of the geographic equator) and the inter-tropical convergence zone (see Fig. 10.8). g/cm³ = grams per cubic centimeter; cm/yr = centimeters per year; ‰ = parts per thousand.

surface water is also increased in the mid-latitudes as a function of evaporation. Conversely, high precipitation in equatorial and subpolar areas decreases the salinity of surface waters in these regions. The mixing of water masses of different temperature and salinity can also alter the character of the newly formed water mass (as discussed in Chap. 11). Finally, it is important to note that salinity can be the dominant factor controlling the density of seawater in some shallow-marine settings, such as over continental shelves or within semienclosed coastal lagoons and estuaries, and in larger enclosed bodies of water, such as the Mediterranean Sea and the Persian Gulf.

Variations in temperature, salinity, and density with depth in a hypothetical mid-latitude column of ocean water clearly define the *three basic layers or zones* common to much of the world ocean: the surface, intermediate, and deep layers (Fig. 10.7). In some areas, a fourth layer is present in the form of Antarctic Bottom Water, representing the very cold and relatively saline, high-density water derived from the freezing of seawater around the margins of Antarctica. The *surface* or *mixed layer* is the product of stirring and turbulent mixing by the wind. Although the surface layer can vary between 50 and 500 meters in thickness, frictional decrease in velocity with depth on average limits wind-induced motion to only the upper 100 meters of the ocean and in the process defines the limits of the mixed layer. The range of temperature and salinity in the surface layer are relatively constant at any given location, reflecting latitudinal location, mixing with the atmosphere, wetness of the overlying atmosphere (i.e., humid versus dry air), and seasonal variations. For example, a shallow seasonal thermocline or temperature gradient commonly develops within the surface layer as summer warming heats near-surface water. Although the surface layer contains only approximately 2 percent of all the water in the ocean, it is arguably the most important part of the ocean in terms of the physical, chemical, and biologic processes that control the activity and character of the ocean as a whole.

In contrast to seasonal variations in temperature within the surface layer, the *permanent thermocline* ("thermo" means "temperature," and "cline" means "gradient") begins at the base of the surface layer and extends on average to a depth of approximately 1000 meters (Fig. 10.7). The top of the thermocline is commonly associated with the 15°C *isotherm* (i.e., the line of equal temperature) in mid-latitude locations, with the base of this zone generally marked by the 4 or 5°C isotherm. A well-developed permanent thermocline is present in most of the ocean but is absent in polar regions where surface temperatures remain very cold throughout the year. The permanent thermocline is commonly accompanied by an equally dramatic *halocline* representing a significant increase in the salinity of water across these same depths. The rapid changes in temperature and salinity associated with the thermocline and halocline combine to produce an accompanying gradient in the density of the intermediate water termed the *pycnocline*. The pycnocline layer or zone contains approximately 18 percent of the water in the ocean and serves to separate the relatively dynamic surface layer from the very cold, dense, and relatively stable water residing in the deep ocean.

The deepest of the three primary zones constituting the oceanic water column is appropriately termed the *deep layer* or *deep zone*, which is characterized by very cold, high-density water resulting from its origin in higher-latitude regions (Fig. 10.7). Water in the deep layer has an average temperature of less than 4°C. The deep layer includes 80 percent of the water in the ocean and consequently plays a major role in global heat distribution (as discussed in Chap. 11). That water in contact with the deep-sea floor is termed

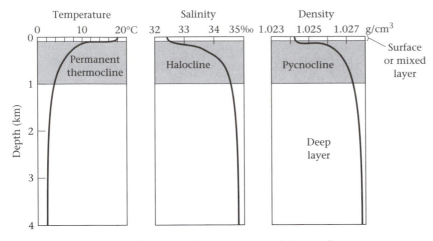

Figure 10.7. Generalized patterns of temperature, salinity, and density through a mid-latitude water column, emphasizing the basic three-layer density-stratified character of the ocean. The thermocline, halocline, and pycnocline zones mark steep gradients in these parameters within the intermediate layer of the ocean. The depth to the base of thermocline, halocline, and pycnocline can vary between 700 and 1500 meters, but is shown here as an idealized 1000 meters. The surface or mixed layer extends to only approximately 100 meters due to the frictional decrease in wind-induced motion with depth. g/cm^3 = grams per cubic centimeter; km = kilograms; ‰ = parts per thousand.

bottom water. Discrete water masses traveling along the bottom of the deep sea include the Antarctic Bottom Water, the coldest and densest water in the world ocean, with an average temperature of $-0.4°C$.

ZONAL WINDS AND CIRCULATION OF THE LOWER ATMOSPHERE

Convection of the atmosphere is driven by the unequal distribution of the Sun's heat over the surface of the Earth. Given a nonrotating Earth, air heated in the equatorial region would expand and rise, creating a low-pressure zone, and would flow toward the poles. As the air approached the polar regions, it would cool, contract, increase in density, and sink, creating high pressure and flow equatorward along the Earth's surface, completing a simple convective loop. Under these conditions, surface winds would blow from the polar regions of high pressure to the equatorial region of low pressure. Moreover, the flow of surface winds would be oriented due north-south, at right angle to the lines of equal atmospheric pressure or isobars.

In the real world, the Earth is rotating from west to east, masses of air moving freely over the Earth's surface must be viewed in a rotating frame of reference that in turn imparts an apparent torque to their motion. One can think of the solid Earth as rotating out from beneath a moving air mass and in the process creating an apparent clockwise deflection of its motion (to the right) in the Northern Hemisphere and a counterclockwise deflection (to the left) in the Southern Hemisphere. The right-and left-handed hemispheric deflections are due to the *Coriolis effect* or "force," named

after the French mathematician Gaspard G. de Coriolis, who first quantitatively explained the effect of a rotating frame of reference in 1835. For example, if a person riding on the inside ring of a rotating carousel throws a ball to another rider traveling on the outside edge of the carousel, the ball appears to travel in a curved path from the perspective of the ball thrower. In fact, the ball travels in a straight line between the two riders, as observed by a person standing next to the carousel and viewing the action from a fixed frame of reference. The curved path of the ball observed by the thrower results from the ball catcher moving at the same time the ball is in flight – an apparent deflection of the ball's path imparted by the rotating frame of reference of the thrower. No "force" (or acceleration) is involved in creating this apparent deflection of motion.

Any object with mass moving horizontally and freely over the surface of the rotating Earth is subject to the apparent deflection of its motion due to the Coriolis effect, although no force has been applied. The apparent deflection of motion occurs when the speed and direction of the object are viewed or measured in reference to the underlying surface of the rotating Earth. Hence, it is more correct to speak of the Coriolis effect rather than the Coriolis force. However, the Coriolis effect or "force" is relatively weak and typically does not influence the motion of small masses over short distances where other forces are dominant. For example, in a bathtub, the water does not always swirl to the right as it exits down a drain in the Northern Hemisphere. Another example is the fact that you do not have to compensate for the Coriolis effect as you walk down the street or toss a Frisbee to a friend. In contrast, as huge masses of air in the atmosphere or water in the ocean travel great distances over the rotating solid Earth, the Coriolis effect is very significant relative to other forces acting on these masses.

The differential velocity of the eastward spinning Earth increases with increasing latitude; therefore, the amount of deflection imparted by the Coriolis effect or "force" and experienced by a moving air mass (or water mass in the ocean) is dependent upon its velocity and latitude. The Coriolis effect is zero at the equator and increases with increasing latitude toward the poles, while the magnitude of deflection increases with the increasing velocity of the object in motion. Hence, both the circulation of the atmosphere and the surface ocean are profoundly influenced by the Coriolis effect. The phenomenon is clearly seen in the tendency of the surface winds to turn to the right of their motion in the Northern Hemisphere and to the left of their motion in the Southern Hemisphere (Figs. 10.2 and 10.8). These deflections, together with convection and variation in the velocity of air masses as they travel poleward, result in six cells or "tubes" of rotating air that encircle the Earth and define

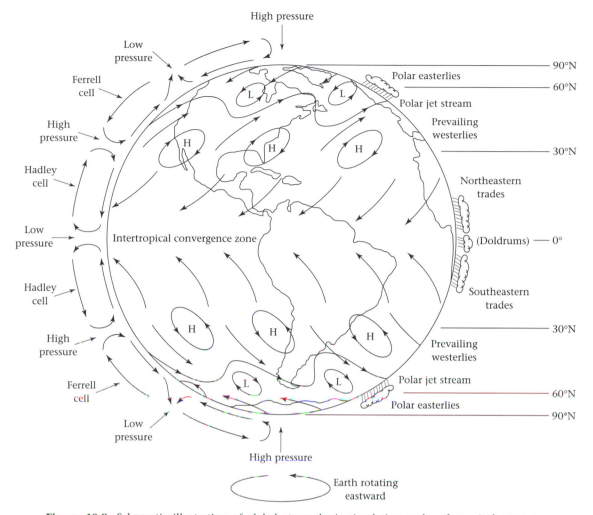

Figure 10.8. Schematic illustration of global atmospheric circulation and surface wind patterns (arrows on the Earth's surface). Three large convecting cells of air (shown in cross section on the left-hand side of the globe) define circulation of the lower atmosphere in each hemisphere. The surface components of each atmospheric cell form the zonal wind belts that drive surface circulation of the ocean. The limbs of the atmospheric cells include: (1) zones of rising moist air, low pressure (L), and high rainfall in the equatorial zone and along the polar fronts (at 50 to 60° N and S), and (2) zones of descending dry air, high pressure (H), and low rainfall over the polar regions and in the mid-latitudes (at approximately 30° N and S). Eastward rotation of the Earth and the Coriolis effect cause surface winds to veer to the right of their motion in the Northern Hemisphere and to the left of their motion in the Southern Hemisphere. The polar jet streams are *not* surface winds but rather flow eastward at high tropospheric altitude along the polar fronts in wave-like patterns and influence the position of individual high- and low-pressure systems north and south of these fronts.

global atmospheric circulation. The number of convecting cells is in part a function of how fast the Earth is spinning, and (as noted earlier) only two large convection cells would be in operation on a hypothetical nonrotating Earth.

A brief description of the major atmospheric cells will assist in explaining the formation and pattern of surface winds responsible for wind-driven surface circulation of the ocean. Solar radiation is at a maximum in the equatorial region, resulting in high sea surface temperatures and high rates of evaporation of seawater. As the warm, moist air expands and rises, it ultimately reaches an elevation where it cools to the point that water vapor condenses to form clouds and rain, releasing latent heat in the process. The result is a

persistent band of clouds and low pressure along the equatorial region. Small, low-pressure cells form continually in this region, resulting in heavy rainfall throughout the year and hot and humid weather with irregular breezes (Fig. 10.8). The high rainfall associated with this zone offsets the high rates of evaporation and depresses the salinity of equatorial surface waters (Fig. 10.6). In short, heat-driven vertical motion dominates the equatorial atmosphere, resulting in a zone of weak and variable surface winds termed the *doldrums*, a phenomenon that presented a major obstacle to ancient sailing vessels and still hinders modern sailing vessels attempting to cross the equator. At the same time the cooler and drier air aloft in the upper troposphere moves north and south away from the equator and continues to cool as it travels poleward, gaining density as its temperature drops.

At approximately 30° N and S of the equator, the now cold and relatively dense air derived from equatorial areas sinks toward the surface of the Earth, creating a zone or belt of high pressure (Figs. 10.2 and 10.8). As the air descends, it is heated by conduction (i.e., by transfer of heat from the relatively warm surface of the Earth) and by compression (i.e., increasing adiabatic pressure). As the temperature of the air rises, its capacity to hold moisture increases, with the result that the mid-latitude high-pressure belts are characterized by cloudless skies and low rainfall. The high rates of evaporation associated with the 30° high-pressure zones increase the salinity of underlying surface water and create deserts on land. Upon reaching the Earth's surface, some of this air flows toward the low-pressure belt of the equatorial region. The Coriolis effect deflects these winds to the right of their motion in the Northern Hemisphere and to the left in the Southern Hemisphere, forming the northeast and southeast *trade winds*, which are separated by the *intertropical convergence zone* (ITCZ) and the doldrums. The trade winds blow continually westward except during the unusual conditions associated with El Niño events, when they slow, stop, or even reverse their direction.

Commonly during the late summer, isolated low-pressure disturbances within the tropical trade wind belts between 5° and 20° latitude grow into increasingly large and violent storms termed *hurricanes, typhoons,* or *cyclones* (except in the equatorial South Atlantic Ocean). These storms rapidly transport large amounts of latent heat into higher latitudes accompanied by winds in excess of 118 kilometers per hour and heavy rains, often with tragic consequences where they meet land. A hurricane derives its energy from the latent heat released as water vapor, rising off the tropical ocean, condenses into clouds and rain around the low-pressure center of rapidly rising warm air marking the eye of a storm. Thus, ocean surface temperatures play a key role in the formation, travel, and ultimate death of these storms. Evidence indicates that sea surface temperatures between 26 and 29°C are necessary to initiate the rapid vertical convection characteristic of a hurricane and that this process cannot be sustained when a storm arrives over water of less than 20°C.

At the same time the trade winds are blowing westward and equatorward, some of the air descending at 30° N and S flows poleward and is deflected eastward by the Coriolis effect, forming the *prevailing westerlies** in both hemispheres (Fig. 10.8). The warm, dry air of the westerlies aggressively evaporates seawater and increases the humidity of these air masses as they sweep poleward. Meanwhile, the very cold and dense air formed at higher altitudes over the north and south poles sinks in these regions, forming high-pressure zones marked by cold, dry air that flows westward and equatorward, constituting the *polar easterlies*. The warm prevailing

westerlies and cold polar easterlies collide at approximately 50 to 60° N and S, forming the wave-like *polar front* in the Northern Hemisphere and the *Antarctic front* in the Southern Hemisphere.

The polar frontal zones include associated polar *jet streams* of high-velocity winds in the upper troposphere that travel eastward around the world in ever changing sinusoidal patterns. These convergences or collisions result in the advection of the relatively warm and lower-density air of the prevailing westerlies up and over the cold, dense air of the polar easterlies. The rising, warm, moist air is rapidly cooled and water vapor condenses, forming prominent zones of clouds and high rainfall, and continually producing low-pressure storm systems. The high rainfall associated with the polar frontal zones dilutes underlying surface water, imparting characteristic lower salinities to surface currents formed in these regions (Fig. 10.6). The low-pressure storm systems formed in these zones move from west to east along the polar fronts, guided by the associated jet streams. As is the case in the mid-latitudes, the drier air at higher elevations flows equatorward and poleward away from 60° N and S latitude, closing the convective cells on either side of the polar front.

Thus, six global atmospheric cells are responsible for the basic pattern of surface winds no matter what the Earth's climatic state, and regardless of the geologically transient locations of the continents and oceans (Fig. 10.8). Although horizontal motion prevails in these convecting cells, it is the relatively narrow zones of vertical motion marking the limbs of the cells that are of special importance. Rising warm, moist air, condensation, and high rainfall are associated with the belts or zones of low pressure at 0 to 20° and 50 to 60° N and S, whereas zones of descending air and high pressure mark zones of little precipitation at 30 to 40° and 80 to 90° N and S. As the surface winds blow from areas of high to low pressure, they set in motion the large-scale surface circulation of the ocean and transfer a portion of their energy to the ocean.

WIND-INDUCED MOTION OF THE SEA SURFACE

Small capillary waves created by surface tension constantly roughen the surface of the ocean and allow the wind to grip the sea surface, transfer energy and momentum via frictional drag, form waves, and sustain wind-driven surface circulation and major surface currents of the global ocean. However, wind waves per se simply represent the transfer of energy along the air–sea interface via orbital motion and involve very little transport of water. This section is concerned with the long-term momentum imparted to the surface waters of the ocean by the combined effects of wind, the rotation of the Earth, and gravity; these effects are responsible for initiating and sustaining the major surface currents of the ocean and their transport of truly enormous volumes of water on a global scale. Once set in motion by the wind, momentum carries the surface ocean forward in the direction of the dominant wind pattern even after local winds have slackened or

* Winds are labeled according to the direction *from which they blow*; hence, westerlies arrive from the west and travel eastward.

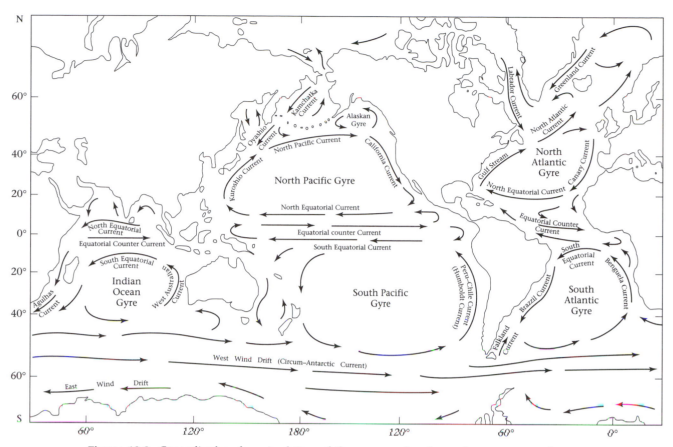

Figure 10.9. Generalized surface circulation of the ocean and major surface currents and gyres. Compare surface currents shown on this map with general patterns of surface winds shown on Figure 10.8.

died. However, frictional effects cause a rapid decrease in current velocity with depth, and they restrict wind-induced motion on average to the uppermost 100 meters of the water column – the so-called surface layer. As with moving air masses, surface water masses in the ocean are subject to the Coriolis effect and drift to the right of their motion in the Northern Hemisphere and to the left of their motion in the Southern Hemisphere.

GYRES AND BOUNDARY CURRENTS

A quick comparison of maps depicting generalized patterns of surface winds and major surface currents clearly illustrates that zonal winds (e.g., latitudinal wind belts) are a primary factor controlling surface circulation of the ocean (Fig. 10.8 and 10.9). Other factors also affect surface circulation, including differences in the temperature, salinity, and density of individual water masses, variations in the elevation of the sea surface from place to place, and the Coriolis effect. Surface circulation is also affected by the size and shape of individual ocean basins and the positions and configuration of gateways between oceans, such as Drake's Passage between South America and Antarctica, which allows free communication between the Atlantic and Pacific oceans.

On an ocean-covered Earth without continents, surface winds would produce a series of six east- and west-flowing surface currents, each of which would continually circle the world beneath the six zonal wind belts. In reality, the only region of the ocean displaying this laboratory-like configuration lies below 40° S and constitutes the Southern Ocean surrounding Antarctica, where no continents are present to block or divert wind or surface water motion. The result is the dreaded "Roaring Forties," where winds have an infinite fetch and the West Wind Drift (also known as the Circum-Antarctic Current) continuously transports water around the globe driven by the unimpeded Southern Hemisphere westerlies. Elsewhere, continents act like walls blocking wind-driven east-west motion and force the surface ocean to move north and south along continental margins.

The combined result of zonal winds and flow constraints imposed by continents is the formation of large surface *gyres* in each ocean basin, representing essentially closed current loops or rings. Wind motion and the Coriolis effect produce clockwise subtropical gyres in the Northern Hemisphere and counterclockwise subtropical gyres in the Southern Hemisphere (Fig. 10.9). Each gyre includes four major surface currents: two east-west currents driven by the zonal wind belts (e.g., the trades, westerlies, and polar easterlies) forming the northern and southern limbs of a given gyre, and two north-south boundary currents that flow parallel or subparallel to the adjacent continental margins. Surface currents in the Indian Ocean are more complex due to changes in seasonal wind patterns associated with the monsoonal climate in its northern reaches, and due to the fact that this ocean is located largely in the Southern Hemisphere and bounded by the West Wind Drift. Nevertheless, north and south equatorial currents and a subtropical gyre also characterize the Indian Ocean. The major exception to this general pattern of closed surface gyres is the West Wind Drift, or Circum-Antarctic Current, which represents unconstrained surface flow around Antarctica.

Polar regions represent special cases because of their unusual geography and freezing temperatures. A slow but sustained west-flowing gyre prevails in the Arctic Ocean under the influence of the polar easterlies, as demonstrated by studies of drifting ice. Polar easterlies in the Southern Hemisphere also drive a west-flowing current around the margin of Antarctica, the East Wind Drift in contrast to the dominant and much larger eastward-flowing West Wind Drift (Figs. 10.2 and 10.9).

The trade winds, prevailing westerlies, and polar easterlies in both hemispheres are all responsible for sustaining major east- and west-flowing surface currents. The *north and south equatorial currents* in the Pacific, Atlantic, and Indian oceans are clearly the product of the trade winds in these regions. These latter wind systems drive equatorial water westward until it reaches a continental margin, where it is deflected north or south and eventually encounters the prevailing westerlies. Significantly, some of this water returns eastward in the form of *equatorial countercurrents* that flow in the narrow zone between the prevailing trade wind belts and beneath the doldrums, driven by west-to-east pressure gradients rather than by wind (Fig. 10.9). In addition, some equatorial water moves eastward within submerged undercurrents such as the Pacific Equatorial Undercurrent located just beneath the North Equatorial Current. Although there is some "leakage" of surface waters across the equatorial regions of the major ocean basins, most significantly in the Atlantic Ocean, the Coriolis-controlled clockwise and counterclockwise circulation of northern and southern hemispheric gyres generally separates Northern and Southern Hemisphere surface circulation regimes. This is *not* the case in the deeper ocean (as discussed in Chap. 11).

East-west surface currents are characterized by relatively slow and steady velocities of between 3 to 6 kilometers per day. However, the dimensions of the ocean's surface currents dictate that they transport enormous volumes of water regardless of whether they are moving fast or slow. Indeed, the volumes moved by these currents are so large that a special flow unit named the *sverdrup* (after the famous oceanographer Harold U. Sverdrup) is applied to their measurement. A sverdrup (Sv) represents a flow of 1 million cubic meters of seawater per second. Although relatively slow, east-west currents such as the North Pacific Current and the North Atlantic Current flow at rates of 10 to 16 Sv.

Eastern and *western boundary currents* represent the north-south limbs of the principal surface gyres and display subtle to exaggerated western intensification due to the Earth's eastward rotation, the conservation of angular momentum, the effect of west-blowing trade winds, and the consequent pileup of water against continental "walls" forming the western margins of the ocean basins. These currents are responsible for transporting enormous volumes of warm tropical water poleward and bringing cool water equatorward from higher latitudes, as emphasized by the position of the 20°C isotherm on opposite sides of the Pacific and Atlantic oceans (Fig. 10.5). One need only compare the hot and muggy summers experienced by the inhabitants of Tokyo (35°41' N) with the cool and foggy summers of San Francisco (37°47' N) to grasp the impact of boundary current asymmetry on the climate of adjacent continental margins. Because of the westward intensification of flow within individual current gyres, western boundary currents are characterized by high velocities (2 to 5 kilometers per hour) and relatively narrow and deep profiles, as typified by the Gulf Stream in the western North Atlantic Ocean and the Kuroshio Current in the western North Pacific Ocean. In contrast, eastern boundary currents such as the California Current in the Pacific and the Canary Current in the Atlantic, are commonly slow moving (0.1 to 2 kilometers per hour) and have wide, shallow profiles. Despite the differences in speeds and cross-sectional geometries of western and eastern boundary currents, they transport approximately the same amount of water, thus maintaining continuity of flow within a given gyre.

The temptation is to view the eastern and western boundary currents as gigantic rivers; however, these flows are not confined to rigid and fixed channels. Although they are bounded on one side by a solid continental margin, their seaward and subsurface boundaries are simply other water masses. Hence, the tracks of these currents change in shape and position with variations in speed and volume of flow. For example, the Gulf Stream experiences exaggerated meandering as it jets past Cape Hatteras, in turn creating isolated rings or patches of warm Gulf Stream water that separate from the main current and take on a life of their own for up to several months or a year before mixing with surrounding water. Similarly, studies of the Kuroshio and California currents demonstrate that they also meander and form large eddies in response to seasonal and long-term changes in winds and climate. Satellite monitoring has revealed that these mesoscale features of surface currents are far more dynamic than previously understood and have practical importance for both local weather forecasting and fisheries predictions.

Although surface wind stress is the primary driving force maintaining surface circulation, other factors enter into this process. No one would argue with the concept that water runs downhill, and that the interaction of wind, gravity, variations in density of seawater, and the Coriolis effect combine to enhance horizontal surface circulation through the creation of "hills" and "valleys" on the ocean's surface. In addition, surface winds in some areas force the density-stratified ocean to do something it fiercely resists – move vertically. There is more to wind-driven surface circulation than wind alone.

EKMAN SPIRAL AND EKMAN TRANSPORT

We have already noted that as the wind blows across the sea surface, there is a decrease in velocity of the water with depth due to frictional effects. As the wind drags the thin veneer of water at the air–sea interface, momentum is lost in transferring energy and motion to the next layer below and so on down through the water column until a point of essentially zero motion is reached, on average a depth of approximately 100 meters at mid-latitude locations. At the same time that a given layer is moving horizontally, it is also under the influence of the Coriolis effect and therefore is deflected slightly to the right of its motion (in the Northern Hemisphere), leading to a systematic change in the direction of flow with depth. The combined result of these two processes is the *Ekman spiral*, named after the Swedish scientist V. W. Ekman, who first quantitatively described this important interaction in 1905 and initiated modern concepts of wind-driven ocean circulation (Fig. 10.10).

Ekman demonstrated that under the influence of a steadily blowing wind and given a homogeneous column of water (e.g., an ocean of uniform density and viscosity), surface wa-

ter moves at an angle of 45° to the right of the wind's motion in the Northern Hemisphere and 45° to the left of the wind in the Southern Hemisphere. As momentum is lost through friction with depth and as each layer is deflected farther to the right or left, an end point is reached at which an extremely weak current is moving in the opposite direction to that of the surface motion. This latter depth averages approximately 100 meters and marks the base of the surface or mixed layer. Summing the individual vectors of each layer yields a net direction of motion for the column of 90° to the right or left of the prevailing wind. Thus, under ideal conditions, *the net horizontal motion of the entire wind-driven surface layer (approximately 0 to 100 meters) is perpendicular to the direction of the wind* (Fig. 10.10). This motion is commonly referred to as *Ekman transport* or Ekman drift. Although actual measurements of wind and surface current vectors deviate from these ideal or theoretically constant angles, the Ekman spiral relationship offers a powerful predictive tool for dealing with the dynamics of the surface ocean and has special significance for understanding gyre circulation and upwelling. The largest deviations from the ideal Ekman spiral relationship occur in shallow areas of the ocean, such as over continental shelves where frictional dissipation along the sea bed occurs. On the other hand, angular relationships predicted by the Ekman spiral effect commonly prevail in open ocean areas, allowing oceanographers confidently to forecast and hindcast motion of the surface layer based on wind direction and speed.

GEOSTROPHIC CURRENTS AND DYNAMIC TOPOGRAPHY

As the trade winds and westerlies blow across the North Pacific Ocean, they not only set in motion the North Equatorial Current and the North Pacific Current, but also result in the net drift of the surface layer at 90° to the right of their motion, representing a clear example of Ekman transport. This process slowly moves warmer, less dense surface water toward the center of the North Pacific subtropical surface gyre, creating an area of convergence. Keeping in mind that specific volume (measured in cubic centimeters per gram) is the inverse of density (measured in grams per cubic centimeter), the less dense water takes up more space or volume than relatively higher-density water (Fig. 10.11). Hence, as Ekman transport forces warm, less dense water into the center of the gyre, it stands at a slightly higher elevation than the surrounding sea surface, forming a large, low hill or mound. One can easily envision the relationship between den-

Figure 10.10. Idealized view of the wind-driven Ekman spiral and Ekman transport within the surface layer of the ocean (in the Northern Hemisphere). The lengths of the solid arrows depict the frictional decrease in velocity with depth; the directions of the arrows illustrate the results of the Coriolis effect on the motion of each succeeding layer down through the column. The Coriolis effect accounts for a 45° angle (to the right of motion in the Northern Hemisphere) between the direction of the wind and the direction of the wind-driven surface current, whereas the net drift of the entire surface layer is at 90° to the right of the wind. m = meters. (*Source:* Adapted from P. R. Pinet, *Oceanography*, West Publishing Company, 1992.)

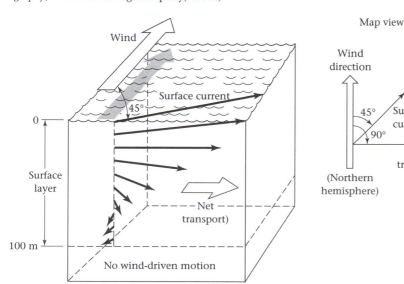

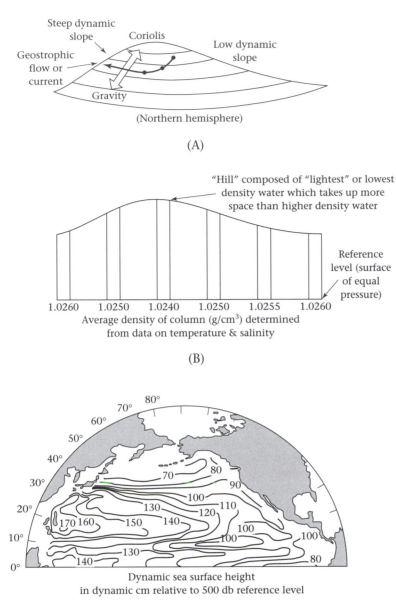

(A)

(B)

(C)

Dynamic sea surface height
in dynamic cm relative to 500 db reference level

Figure 10.11. Aspects of dynamic topography of the (A) sea surface, geostrophic flow, and geostrophic surface circulation (in the Northern Hemisphere). (A) The idealized motion of a particle of water on a large dynamic "hill" in the center of an oceanic gyre as it attains a position of perfect balance between (1) wind-driven uphill motion imparted by Ekman transport (Coriolis effect) to the right of west-blowing trade winds, and (2) downhill motion induced by gravity (pressure) and the slope of the "hill." When a position of balance is achieved, the particle (and others of similar density) travels to the right and around the dynamic "hill" representing a geostrophic flow or current. Dynamic topography of the sea surface is created when wind-driven Ekman transport pushes warmer and less dense water toward the centers of an oceanic gyre. (B) As shown in this hypothetical cross section, the higher specific volume of the less dense water causes these water masses to stand at a slightly higher elevation above a given reference level than surrounding waters of lower density. g/cm³ = grams per cubic centimeter. (C) An example of the dynamic topography of the North Pacific Ocean in terms of height differences measured in dynamic centimeters above a reference level of 500 decibars (db) (the decibar represents the standard measure of pressure in the ocean and is defined as 100,000 dynes per square centimeter). The eastward rotation of the Earth has caused the dynamic "hill" of warm surface water to shift westward, in turn dictating that the steepest dynamic slopes and highest velocities of geostrophic flow occur along the western Pacific margin.

sity, specific volume, and sea surface elevation by selecting an arbitrary reference depth (i.e., a level or depth of equal pressure) and comparing water columns of different character above the given reference level (Fig. 10.11). A column of warm, lower-density water will obviously take up more space (volume) and hence stand higher above the reference level than will an adjacent column of cooler, denser water.

Satellite-based measurements have confirmed that the mounds of water marking the centroids of gyre circulation in fact stand as much as 2 meters higher than the level of the ocean forming the margins of a given gyre. Although the slopes of these giant "hills" are very gentle, they result in pressure gradients with water moving downhill toward areas of lower pressure. In effect, the high-standing mound of water responds to the horizontal pressure gradient and attempts to "flatten out" the sea surface. Individual water particles are

acted upon by gravity pulling them down the "hill" at the same time that the wind-induced Coriolis "force" is pushing them uphill toward the center of the gyre as a result of Ekman transport (Fig. 10.11). As gravity, density, and the horizontal pressure gradient cause a particle to move downhill, the Coriolis effect again acts to deflect motion to the right (in the Northern Hemisphere). The particle thus moves down and around the hill until reaching an elevation where the effects of gravity (and density) and the uphill Coriolis "force" are in precise balance. When balance is achieved at a given elevation or position on the slope, the particle then travels continually around the hill with other particles of similar density, forming an integral part of an ensuing surface current.

Currents and gyre motion generated in this manner are termed *geostrophic currents* and *geostrophic circulation*, respectively. The word "geostrophic" literally means "Earth turned" and refers to the fact that the motion of the water is largely controlled by the Earth's rotation and the Coriolis effect, in balance with the effects of density and gravity. The variations in the topography or elevation of the sea surface brought about by these interactions, and which in turn govern geostrophic currents, are logically termed *dynamic topography*. Using data on variations in temperature, salinity, and density, and an arbitrary reference level (commonly a depth representing an equal pressure of 500, 1000, or 1500 decibars), oceanographers routinely construct contour maps of the dynamic topography of the sea surface (Fig. 10.11). The resulting patterns of dynamic topography can then be used with great confidence to predict current flow from the orientation and shape of contour lines and the appropriate Coriolis effect (left or right), and the estimated velocity of a given current can be derived by knowing the slope of the sea surface – the steeper the slope, the faster the current. Areas of closely spaced dynamic contours depict steep slopes and hence high-velocity currents, whereas areas of widely spaced contours correspond to low slopes and hence slow-current motion.

Because of the Earth's rotation from west to east, the centers of major surface gyres and the associated elevated mounds of less dense water are shifted toward the western sides of individual ocean basins. Because of this shift of mass and momentum within the gyres, the steepest dynamic topographies are found along the western margins of the surface gyres leading to the so-called western intensification of currents and typified by the Gulf Stream and Kuroshio currents (Figs. 10.9 and 10.11). In contrast, the eastern margins of the gyres exhibit low dynamic slopes and relatively sluggish current flow.

Surface winds not only maintain the ocean's dynamic topography and the resulting geostrophic circulation but also govern the location and magnitude of one the most important physical processes of the surface ocean – upwelling.

UPWELLING AND DOWNWELLING

Upwelling represents the vertical movement of subsurface water to the surface of the ocean, commonly from depths within the upper thermocline or pycnocline layers. This process is critical to the recycling of key nutrients in the ocean (e.g., phosphorous, nitrogen, and silicon) and their transport to the surface of the sea, where they can be utilized by phytoplankton through photosynthesis. Upwelling areas are characterized by exceptionally high rates of primary biologic productivity along with secondary productivity by the grazers through top carnivores – in fact, the entire food chain – drawn to these areas. Predictably, upwelling areas support rich and important fishing industries. A 1969 study estimated that upwelling zones account for 50 percent of all marine fish production in the world despite the fact that they constitute less than 1 percent of the surface area of the global ocean. Although more recent studies have lowered this esti-

mate to 25 percent, these remarkable numbers emphasize the importance of upwelling to the biologic health of the ocean.

Coastal upwelling can be triggered when local winds blow in an offshore direction for a long enough period to push surface water away from the coast, allowing deeper water to move upward to replace it. In some areas, advective collision of two surface water masses causes upwelling, such as occurs where the Oyashio Current meets the Kuroshio Current off northern Japan (Fig. 10.9). Vertical motion and upwelling can also occur where a surface current flows over a shallow submerged bank or seamount or where a surface current flows past a large coastal prominence. The most significant upwelling processes in both coastal zones and the open ocean involve wind-driven Ekman transport (Fig. 10.10).

Where winds blow parallel with a coastline, Ekman transport can induce either upwelling or *downwelling* (Fig. 10.12). As water in the surface layer is moved horizontally (at 90° to the direction of the wind) it is replaced by water from below, commonly from depths of 100 to 300 meters, within the upper part of the thermocline or intermediate layer. If Ekman transport is in the offshore direction, upwelling results. Alternatively, if transport of the surface layer is toward the coast, downwelling occurs due to the wall-like effect of the continental margin. Ekman coastal upwelling and associated zones of high biologic productivity are common along the western coasts of continents (i.e., the eastern sides of ocean basins) where sustained seasonal winds blow north and south. Clear examples of these settings include the Pacific coasts of North and South America and the Atlantic coast of South Africa and adjacent Namibia. All three regions experience seasonal winds that cause vigorous upwelling of cold, nutrient-rich water during the spring and early summer seasons, in the turn producing fog and cool weather during these periods. Mark Twain's much paraphrased statement that "the coldest winter he ever experienced was a summer in San Francisco" neatly sums up this interplay between upwelled cold water and the overlying atmosphere.

Upwelling also takes place away from coastal regions in the open ocean where the directions of wind and current motion together with the Coriolis effect cause Ekman drift in opposing directions, allowing water to well up from below. This type of wind-induced vertical motion is termed *divergent upwelling*, a process that characterizes the equatorial region between the northern and southern trade wind belts (Fig. 10.12). Ekman transport forces surface waters to the north (to the right of the west-blowing trade winds) in the Northern Hemisphere and to the south (to the left of west-blowing trade winds) in the Southern Hemisphere. The result is a north-south divergence of surface layer motion away from the equator and the upwelling of subsurface water in the intervening area.

Divergent upwelling also takes place in the Circum-Antarctic region in the area between the east-flowing West Wind Drift and the west-flowing East Wind Drift (Polar Current). In this case, northward Ekman transport associated with the east-flowing West Wind Drift is in opposition to the

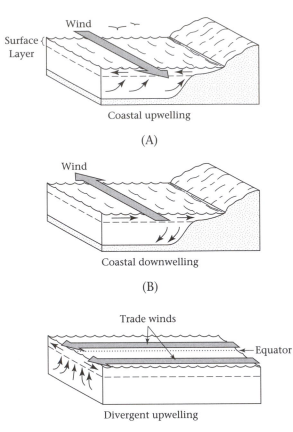

Figure 10.12. Schematic illustrations of (A) coastal upwelling, (B) coastal downwelling, and (C) open ocean divergent upwelling (in the Northern Hemisphere). All three types of vertical motion occur as a function of wind-driven Ekman transport (see Fig. 10.10). As winds blow equatorward (south) along the western side of a continent in the Northern Hemisphere, Ekman transport forces surface water seaward away from the coast (at 90° to the right of the wind direction). This latter water is in turn replaced by nutrient-rich intermediate water upwelled from below the surface layer, triggering high primary and secondary productivity. Conversely, the direction of Ekman transport is reversed if winds blow poleward in the same area, with the result that surface water undergoes downwelling (sinking) as it is forced against the coast. Equatorial divergent upwelling results where west-blowing northern trade winds induce northward Ekman transport at the same time that parallel southern trade winds (in the Southern Hemisphere) are inducing southerly Ekman transport. Thus, subsurface water is upwelled in the zone between the two opposing surface flows.

southward Ekman drift associated with the west-flowing East Wind Drift, resulting in the so-called Antarctic Divergence, a prominent zone of high primary productivity. Upwelling in this region is also assisted by the offshore flow of winds from the Antarctic continent. In addition, upwelling of intermediate water takes place in this area to compensate for the sinking of the extremely cold and dense surface water produced by the freezing of seawater around the Antarctic margin, a density-driven process.

QUESTIONS

1. Humans have been using the ocean for exploration, trade, and harvesting of marine resources for thousands of years (not to speak of the ocean's past and future role in global politics and war). What advantages to human well-being can you ascribe to an increased knowledge of ocean circulation?

2. List as many fundamental differences as you can think of between eastern and western boundary currents, and their effects on the adjacent continental margins, including their climates and cultures.

3. Describe the surficial conditions of the Earth, assuming the ocean (and the rest of the hydrosphere) had been removed some 100 million years ago.

4. What changes in global ocean surface circulation would you predict might take place if the Isthmus of Panama were removed, allowing Pacific and Atlantic-Caribbean water and circulation to be connected along the equator? This is not a moot question; the Isthmus of Panama was in fact not present prior to 3 million years ago.

5. What do you think the surface circulation of the ocean would look like if the Earth were rotating westward rather than eastward?

FURTHER READING

Bearman, G. (ed.) 1989. *Ocean circulation.* Oxford: The Open University and Pergamon Press.

Garrison, T. 1999. *Oceanography, an invitation to marine science,* 3rd edition. Belmont, CA: Wadsworth Publishing Company.

McLeisch, W. H. 1989. The blue god. *Smithsonian* 19(2): 44–58.

Pickard, G. L., and Emery, W. J. 1990. *Descriptive physical oceanography,* 5th (SI) enlarged edition. Oxford. Pergamon Press.

Wunsch, C. 1992. Observing ocean circulation from space. *Oceanus* 35(2) 9–17.

11 Deep-Sea and Global Ocean Circulation

JAMES C. INGLE, JR.

MOTIONS IN THE DEEP

The prevailing scientific view of the deep sea in the mid-nineteenth century was that it was a cold, stagnant, unmoving body of water devoid of life. It took more than a century of deep-sea exploration to erase these erroneous notions, save one. It is indeed a very cold place thanks to the fact that most of the deep water is derived from polar regions. Contrary to what some scientists had anticipated, the 1872–6 *Challenger* expedition found plenty of life in the deep sea, along with evidence that the deep ocean circulates as a function of gravity acting upon differences in the density of seawater imparted primarily by variations in temperature and salinity. We now refer to this process as *thermohaline circulation* ("thermo" means "temperature," and "haline" means "salt"), which, in the modern ocean, is driven principally by the sinking of cold, dense water in the polar and near-polar regions. Calculations based on very limited temperature and salinity data from the Atlantic Ocean in the 1930s predicted the presence of swift, density-driven flows in the deep sea. However, the first reliable measurements demonstrating the presence of deep, high-velocity currents were not made until the 1960s, attesting to the slow progress in gaining hard scientific information on how the deep sea operates.

Part of the problem lies in the shear size and volume of the water forming the deep ocean, and our inability to collect synoptic measurements of its character and circulation on a global scale in a manner akin to satellite surveillance of the surface ocean. We are overcoming these limitations by employing increasingly complex computer programs that can model the circulation of the deep sea, and that can evaluate the deep ocean's response to global, regional, and local changes in temperature and salinity. Both direct measurements and computer modeling demonstrate that the deep sea plays a fundamental and perhaps dominant role in atmosphere–ocean circulation over the long term. This makes intuitive sense when one considers that wind-driven circulation is restricted to the shallowest 2 percent of the ocean (by volume), whereas density-driven thermohaline circulation prevails in the remaining 98 percent of the ocean below these depths. For example, we know that the alternating glacial and interglacial climates of the past 2 million years involved major changes in rates of deep-sea flow. However, we are only now approaching a clear understanding of how this volumetrically dominant portion of the global ocean behaves during these contrasting climatic states.

Quantitative knowledge of how the oceans work is a prerequisite for understanding the cause of global climate and climate change. Can the oceans adsorb the increasing levels of carbon dioxide that humans are injecting into the atmosphere and, in the process, slow or halt serious global warming? Where and how are the deep sea and atmosphere most tightly coupled? Can detailed measurements and computer models of heat transport by the deep and shallow ocean assist in predicting long-term changes in weather and climate? Answers to these and many other important questions hinge squarely on our understanding of deep-sea circulation. This chapter highlights the basic processes controlling deep-sea circulation, and it summarizes the larger picture of global ocean circulation.

Although recent great strides have been made in understanding the surface circulation of the ocean, the dynamics of the deep sea remain little known. Moreover, the deep sea is a hostile environment to explore. The high pressures, low temperatures, and saline water encountered in the deep sea severely test every instrument devised by marine scientists and engineers to probe this realm. These factors, along with others such as absolute darkness, have prevented humans from routinely exploring this environment in manned vessels. Only six manned deep submersibles are currently operational, and none can descend to the greatest depths in the ocean. In fact, only two humans have ever visited the deepest part of the ocean (the Challenger Deep, the lowest part of the Mariana Trench) and that only for a matter of minutes in 1960, aboard the bathyscaphe *Trieste*. The character and circulation of the deep sea

W. G. Ernst (ed.), *Earth Systems: Processes and Issues*. Printed in the United States of America. Copyright © 2000 Cambridge University Press. All rights reserved.

are most practically measured using small instrument packages deployed from oceanographic vessels, remotely operated deep-diving vehicles (ROVs), and stationary buoys (floating platforms), or are remotely sensed using acoustic (sound) techniques.

Given the harsh conditions of the deep sea, the design criteria for undersea instruments capable of directly measuring this environment are similar to those applied to building instruments destined to explore outer space and other planets. While satellites circle the globe and serve up daily views and measurements of most of the ocean surface, individual ships at sea still lower instruments over the side attached to cables hundreds to thousands of meters in length, one station at a time, to garner even the most basic measurements of temperature and salinity at depth. Not surprisingly, there are many areas of the deep sea that have yet to be probed with modern oceanographic instruments. In 1972–7, a major effort to increase knowledge of the deep sea was carried out. The Geochemical Ocean Sections Study (GEOSECS) used newly designed equipment, including conductivity-temperature-depth devices (CTDs), with the capacity to record changes in these and other parameters to depths of 5000 meters. The data resulting from these studies, along with many subsequent measurements, have allowed construction of detailed cross-sectional views of ocean temperatures, salinity, dissolved oxygen, nutrients, and so forth from top to bottom through entire ocean basins. Moored (stationary) instrument packages designed to measure temperature, salinity, and so forth are also deployed on the ocean's surface and at depth, with data collected remotely via a satellite link or retrieved by ship. In addition, weighted subsurface current

drogues (i.e., floats) and current meters are routinely used to monitor velocities and directions of deep-water flow. Currently, arrays of sound-emitting and-receiving devices are being used to analyze subtle variations in the velocity of sound in ocean water. These variations reflect differences in the density of the water and, in turn, differences in water temperature, salinity, and so forth. This technique is termed *acoustic tomography*. Perhaps the most dramatic ongoing application of sound involves the measurement of the temperature of the entire ocean through use of a low-frequency sound emitter in deep water and oceanwide listening stations, a technique termed *acoustic thermography*.

Regional and global-scale patterns and rates of deep-sea circulation, the age of deep waters, and the time needed to replace the water in entire ocean basins are commonly analyzed using chemical tracers. Radioactive tritium, originally introduced into rivers and the ocean as a by-product of nuclear bomb testing in the 1950s and 1960s, is a particularly useful tracer of deep-water flow. The changing distributional patterns of chemical tracers such as dissolved oxygen and carbon dioxide, as well as nutrients including nitrate, phosphate, and silica, are also used to monitor the movement of intermediate and deep water.

Accelerating interest in the deep sea has recently spurred two international research programs aimed at understanding deep-sea circulation (the World Ocean Circulation Experiment) and the ocean's role in the global carbon cycle (the Joint Global Ocean Flux Study). Data emerging from these programs are being incorporated into computer models of ocean circulation with a special focus on ocean's role in modulating and forcing global climate change.

THE NATURE OF DEEP-SEA CIRCULATION

All the data available from the deep sea collectively confirm the density-stratified character of the water filling the oceans. The basic three-layer structure of the world ocean consists of a surface layer, intermediate layer (marked by thermocline and pycnocline), and deep layer (Figs. 10.7 and 11.1). The entire ocean undergoes continual convective circulation, with wind-driven surface currents carrying warm water poleward from the tropics where it is cooled at high latitudes. As discussed in Chapter 10, some of this water subsequently returns equatorward within surface currents. At the same time, enormous volumes of cold, dense water, produced through cooling and freezing in polar latitudes, sink rapidly in these regions and spread out in the deep sea. Much of the cold, dense water formed in the Arctic Ocean is trapped within the bowl-like Arctic basin, except where it literally cascades over the deep-sea ridges joining Greenland, Iceland, and Scotland at depths of 500 to 700 meters and spreads out into the adjoin-

Figure 11.1. A schematic cross section through an idealized ocean basin illustrating the basic three-layer, density-stratified structure of the ocean including the surface layer, the intermediate layer (marked by the pycnocline and thermocline), and the deep layer. The intermediate layer or pycnocline zone protects the deep layer from changes in temperature and salinity in low- and mid-latitude locations where density stratification is best developed. In contrast, the thermocline is absent in the polar regions where cold, dense water is formed at the surface through cooling and freezing, and subsequently sinks and spreads out from these areas forming the deep layer. km = kilometers. (*Source:* Adapted from M. G. and E. Gross, *Oceanography*, 7th edition, Prentice-Hall, 1996.)

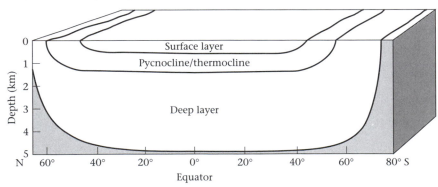

ing Atlantic Ocean. In contrast, cold, dense water formed as the overlying sea ice freezes in the Antarctic region, sinks rapidly, and flows freely northward into the adjacent deep sea and beyond. Mixing of cold and warmer surface water masses in subpolar areas also leads to the formation and sinking of distinctive intermediate and deeper water masses, adding complexity to the layered structure of the ocean. Together, freezing, cooling, and mixing of surface waters in the polar and subpolar latitudes contributes the bulk of the water filling the modern ocean.

Although the atmosphere, surface, and deep ocean are linked at the air–sea interface (see Fig. 10.8), the processes and patterns of circulation of the intermediate and deep layers of the ocean differ fundamentally from the system of surface gyres driven by east-west zonal winds. For example, the rate of thermohaline deep-sea circulation is much slower than that of wind-driven surface circulation, with all that this implies regarding the lengthy residence time of water in the deep sea, as it moves from ocean basin to ocean basin. Once a given parcel or mass of dense water sinks into the deep sea, it becomes isolated from the processes modifying its character at the air–sea interface. Therefore, the temperature and salinity of deep-water masses vary little and repre-

sent *conservative* properties that can be utilized as a means to identify and trace the motion of the water over great distances. In contrast, *nonconservative* properties of deep water, such as levels of dissolved oxygen, carbon dioxide, and nutrients, can vary significantly over time as a function of how long a given parcel of water has been isolated from the air–sea interface, and the rate at which biologic and chemical processes alter its character as it travels from one region to another.

In contrast to surface gyres, which are generally restricted to the Northern or Southern Hemisphere, density-driven flow in the deep sea involves massive transport of water across the equator, creating ocean-scale convective loops joining surface and deep-sea circulatory systems in both hemispheres (Fig. 11.2). Also in contrast to the wind-driven east-west flow of the surface ocean, density-driven flow in the deep sea is generally oriented north-south, constrained by the shape and orientation of the major ocean basins except in the Circum-Antarctic region, where both surface and deep currents flow around this isolated continent. An early theoretic study of deep-sea circulation by Henry Stommel emphasized the effect of the Earth's rotation (i.e., the Coriolis effect) on this process and predicted the presence of deep geostrophic boundary currents along the western sides of the Pacific, Atlantic, and Indian ocean basins, flows later confirmed by observation. Stommel also noted that most of the deep and bottom water in the global ocean is formed in only two areas, the Weddell Sea adjacent to Antarctica and the North Atlantic Ocean. Similarly, source areas for intermediate water are limited to high-latitude surface convergences where warm, salty water meets cold, low-salinity water (Fig. 11.2) In short, the rates and patterns of deep-sea circulation over the long term are primarily controlled by (1) the never ending and climatically influenced production of waters of varying temperature, salinity, and density at the sea surface (Fig. 10.8) from which they sink, and (2) the slowly evolving shape and latitudinal positions of the ocean basins governed by plate tectonic processes.

The density-driven nature of deep-water circulation assures that variations in the topography of the deep-sea floor play a large role in controlling patterns of flow, especially the movement of dense bottom water. In fact, local tectonic changes in the shape of the deep-sea floor can have profound global consequences for the flow of deep water from one area to another, emphasizing the intimate link between ocean circulation and crustal dynamics when viewed on geologic time scales. As with surface circulation, the locations of oceanic "gateways" such as Drake's Passage, between the Antarctic Peninsula and

Figure 11.2. A schematic illustration of deep-sea circulation at depths below 2000 meters, as originally proposed by H. Stommel. The wide arrows mark relatively high-velocity and high-volume western boundary currents (sometimes called contour currents), the thin arrows depict slower flows. The circled Xs mark the location of areas where cold, dense water is created through cooling and freezing, and sinks to form the deep and bottom waters of the world ocean. This water is formed in only two areas: the northern North Atlantic Ocean, and along the margin of Antarctica. Small dots mark areas where warm, saline surface water converges with cold, low-salinity surface water, resulting in the creation and sinking of intermediate water through mixing. (*Source:* Adopted from D. E. Ingmanson and W. J. Wallace, *Oceanography,* 5th edition, Wadsworth Publishing Company, 1995.)

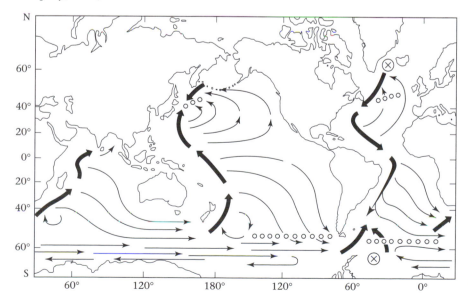

the southern tip of South America, allow communication from one ocean basin to another. Clearly, changes in the depth of the sea floor within these passages are critical in terms of either blocking or allowing deep-water flow to occur.

For example, evidence indicates that as the Isthmus of Panama was tectonically uplifted beginning 10 million ago, there was an initial cutoff of deep flow, then intermediate flow, and finally shallow-water flow across this slowly rising barrier, with the connection between Pacific and Caribbean-Atlantic waters completely severed approximately 3 million years ago. The consequences of this event included the strengthening of the Gulf Stream as water formerly flowing between the Atlantic and Pacific oceans was blocked and redirected northward, leading to changes fostering the initiation of Northern Hemisphere glaciation and ensuing Pleistocene (i.e., Ice Age) climate. Farther back in time, approximately 53 million years ago, sea-floor spreading along that portion of the mid-ocean ridge and rise separating Antarctica and Australia gradually permitted increasing flow through this passage, with deep water moving freely through the area approximately 15 million years later. This latter event represented the initial step in the oceanographic isolation of Antarctica, which was later completed with the opening of Drake's Passage, accompanied by a massive buildup of ice on Antarctica and a major change in global climate.

Because the deep sea forms part of the convectively driven ocean–atmosphere circulatory system, it is also directly affected by any change in the global pole-to-equator thermal gradient over time. Rates of deep-sea circulation increase during periods of steep thermal gradients (i.e., periods of glacial climate) as a function of increased production and sinking of cold, dense water in the polar regions. In contrast, deep circulation presumably slows during periods of relatively warm climate, ice-free polar areas, and depressed pole-to-equator thermal gradients. More extreme states of deep-sea circulation are also possible. There is accumulating evidence that a peak period of sustained warm, nonglacial climate approximately 58 to 57 million years ago was marked by increased evaporative production of high-salinity surface water in mid-latitude regions, which, together with the absence of polar sources of deep water, resulted in a drastic reorganization of deep-sea circulation. The mid-latitude water apparently attained densities high enough that it sank and *flowed poleward*, carrying significant heat to the polar regions and creating a situation in which the deep ocean literally "ran backward" relative to the modern equatorward flow of deep water driven by the sinking of cold water in high-latitude regions.

No matter what climate mode the Earth experiences, it is a certainty that the basic physical processes governing surface and deep-sea circulation in the modern ocean also governed the flow of ancient oceans, and that we have much to learn by studying and understanding how the modern ocean operates, especially that part we know least about – the deep sea.

WATER MASSES

A key concept in analyzing and understanding the circulation of the global ocean is the formation of distinctive parcels of water, or *water masses*. Once formed, individual water masses have the capacity to mix with others, forming new water masses of different temperature, salinity, and density, ultimately governing the character of the ocean as a whole. Although the majority of the water in the ocean lies at depths below the surface or mixed layer, the interactions and processes at the air–sea interface are responsible for imparting the distinctive temperature, salinity, and density marking each water mass. These distinctions allow (1) specific source areas to be identified, and (2) the movement of individual water masses to be traced as they descend into the deep sea and spread out into the ocean basins.

A simple definition of a water mass is a large volume of ocean water that can be identified by its distinctive range of temperature and salinity as a function of having a common source area or origin. Although water masses can also be

Figure 11.3. A typical temperature-salinity (T-S) scatterplot illustrating the presence of four distinctive water masses forming the water column within a 10° square of the Pacific Ocean directly south of the Hawaiian Islands (10 to 20° N and 150 to 160° W). The pattern of dots is schematic and simply indicates the general distribution of individual pairs of analyses; in fact, some 9429 T-S pairs were originally plotted to form this diagram. The approximate depths of each water mass are given in parentheses. The T-S pattern of North Pacific Intermediate Water includes Pacific Equatorial Water, with which it is mixing at depth. The solid line running through the pattern of dots represents the position of the mean T-S curve for this location. m = meters; ‰ = parts per thousand. (*Source:* Adapted from G. L. Pickard and W. J. Emery, *Descriptive physical oceanography*, 5th edition, Pergamon Press, 1990.)

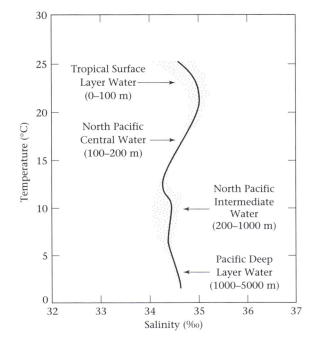

identified and traced using nonconservative parameters, such as nutrient content and dissolved oxygen, they are characterized and identified primarily by the simple plotting of temperature and salinity (Fig. 11.3). Multiple analyses of temperature and salinity at various depths through a given water column routinely yield definitive patterns on temperature-salinity (T-S) diagrams. These plots constitute a valuable oceanographic tool to "fingerprint" individual water masses and to analyze the mixing of water masses. Major water masses identified in this manner (Table 11.1) are commonly given names based on their source area (e.g., Mediterranean Water, Antarctic Bottom Water) or area of principle residence in the ocean (Antarctic Circumpolar Water, Pacific Equatorial Water, etc.).

Six basic processes control the temperature and salinity of seawater and lead to the formation of major water masses of varying density, including (1) freezing, (2) cooling, (3) warming, (4) evaporation, (5) the addition of fresh water via rain (condensation and precipitation) and the melting of ice, and (6) the mixing of two or more water masses at the surface or at depth. Processes such as fresh water entering the coastal ocean from a river can also affect local variations in temperature and salinity; however, this discussion deals exclusively with open ocean and regional effects.

As might be expected, the warming of surface water takes place primarily in equatorial through low mid-latitude regions (0 to 30° latitude) where the highest amounts of solar insolation or heat are received (Fig. 10.8). Alternatively, surface waters are continually cooled in high-latitude regions where they give up heat to the atmosphere through conduction and evaporation (i.e., latent heat) as cold winds sweep over their surface. Cooling reaches extremes in the polar and near-polar regions and results in the extensive freezing of surface water, especially during the winter months. The freezing process occurs in stages from the initial formation of individual crystals ("grease ice") to slush, pancake ice, sheets, and finally pack ice, which covers much of the Arctic Ocean and surrounds Antarctica. Significantly, when seawater freezes, the resulting ice crystals incorporate no more than 15 to 30 percent of the available salt, causing the salinity of the remaining unfrozen water to increase. This process creates water of exceptionally high density (e.g., water that is both very cold and very salty), exemplified by the production of Antarctic Bottom Water during winter freezing in the Weddell Sea adjacent to Antarctica (Fig. 11.2).

Although high rates of evaporation mark the equatorial region, the high rates of rainfall from tropical low-pressure systems lower the salinity of surface waters in these regions (Fig. 10.6). This is not the case in the mid-latitudes (approximately 30° N and S) where high-pressure systems prevail and rates of evaporation far exceed rates of precipitation, leading to increased surface salinites in these regions. In contrast, surface waters lying beneath the tracks of low-pressure storm cells associated with the polar frontal zones (approximately 60 to 70° N and S) receive high amounts of rainfall that significantly dilute surface salinities in these regions (Fig. 10.8).

Marginal seas can also represent source areas of distinctive water masses, as clearly illustrated by the Mediterranean Sea (Fig. 11.4). In this case, North Atlantic surface water of normal salinity enters the Mediterranean Sea through the Straits of Gibraltar. As the water circulates eastward, high rates of mid-latitude evaporation increase its salinity and density to the point that the water sinks in the eastern Mediterranean Sea and flows westward at depth. Figure 11.4 shows that salinity becomes the *dominant* factor controlling density and that the water sinks despite its relatively warm temperature (Table 11.1). The high-salinity Mediterranean Water ultimately spills over the Gibraltar sill and spreads out as a distinctive and easily identified deep-water mass in the North Atlantic Ocean. (The term *sill* refers to a topographic feature of the sea floor (commonly a submarine ridge) that effectively separates one ocean basin from another, or a part of one basin from another part.) Thus, both surface processes in the Atlantic Ocean proper and evaporation of the Mediterranean Sea conspire to make the Atlantic Ocean a relatively warm, salty ocean in contrast to the Pacific Ocean, which, by comparison, is relatively cool and fresh.

Finally, the temperature, salinity, and density of seawater can also be altered through the mixing of two or more water masses of differing temperature and salinity. The mixing process can occur in areas where two surface currents converge, or at depth where two or more water masses mix via sinking or collision along a given density or pressure gradient. When lines of equal density (isopycnals) are plotted on a standard T-S diagram, it becomes immediately apparent that water masses of contrasting tempera

TABLE 11.1. CHARACTERISTICS OF SELECTED MAJOR WATER MASSES IN THE WORLD OCEAN

Water Mass	Temperature (°C)	Salinity (‰)	Density (g/cm³)
Antarctic Bottom Water	−0.4	34.6	1.02786
Antarctic Circumpolar Water	0–2	34.7	1.02775–1.02789
Antarctic Intermediate Water	3–7	34.2–34.4	1.02682–1.02743
North Atlantic Central Water	8–19	35.1–36.7	1.02630–1.02737
North Atlantic Deep Water	2.5–3.1	34.9	1.02781–1.02788
South Atlantic Central Water	6–18	34.5–36.0	1.02606–1.02719
North Pacific Central Water	10–18	34.2–34.9	1.02521–1.02634
North Pacific Intermediate Water	4–10	34.0–34.5	1.02619–1.02741
Mediterranean Sea Water	8–17	36.5	1.02592–1.02690

°C = degrees Celsius; ‰ = parts per thousand; g/cm³ = grams per square centimeter.

Source: Data from D. E. Ingmanson and W. J. Wallace, *Oceanography*, 5th edition, Wadsworth Publishing, 1995.

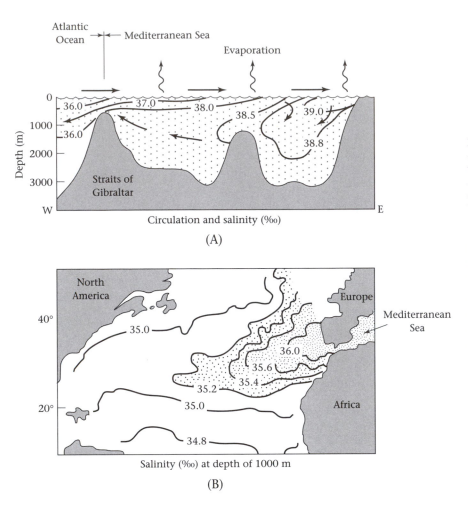

(A)

(B)

Figure 11.4. (A) A cross-sectional view of the creation and circulation of dense, warm saline water in the Mediterranean Sea through evaporation and its exit at depth over the Gibraltar sill into the North Atlantic Ocean, together with (B) a plan view of isohalines (contour lines of equal salinity) at 1000 meters in the North Atlantic Ocean illustrating the spread of Mediterranean Water westward at this depth. m = meters; ‰ = parts per thousand. (*Source:* Adapted from G. L. Pickard and W. J. Emery, *Descriptive physical oceanography*, 5th edition, Pergamon Press, 1990; and P. R. Pinet, *Oceanography*, West Publishing Company, 1992.)

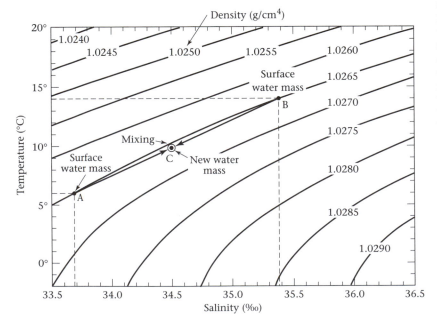

Figure 11.5. Temperature-salinity-density nomograph for seawater, with an example of how two water masses of dissimilar temperature and salinity can in fact have exactly the same density and hence travel at the same depth along the same pressure or density gradient. The straight arrows constitute a mixing line on the nomograph and illustrate what happens when two hypothetical surface water masses (A and B) of different character but the same density converge (i.e., collide) and mix, producing a new mass (C) of slightly higher density than either of the two parent water masses – a process termed *caballing*. Because new water mass C has a higher density than water masses A and B, it will sink. Water mass A is cold and has low salinity, whereas water mass B is relatively warm and possesses high salinity, similar to water mass relationships inherent to actual convergence zones such as the North Atlantic Convergence (see Figs. 10.9, 11.2, and 11.6), where caballing produces the North Atlantic Intermediate Water. g/cm³ = grams per cubic centimeter; ‰ = parts per thousand.

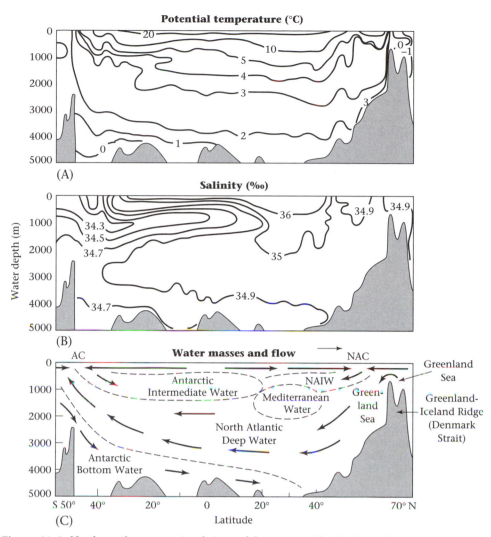

Figure 11.6. North-south cross-sectional views of the western Atlantic Ocean illustrating the general distribution of (A) temperature, (B) salinity, (C) major water masses, and the directions of flow at the surface and at depth. The blocking effect of the Greenland-Iceland and Iceland-Scotland ridges, which confine most of the extremely cold water formed in the Greenland and Norwegian seas (and the Arctic Basin proper), can be observed. Spillage of this water over the ridges or basin sills into the North Atlantic occurs at depths of only 500 to 700 meters. Compare the patterns of isotherms (contours of equal temperature) and isohalines (contours of equal salinity) with the generalized boundaries of labeled water masses. *Potential temperatures* depicted in the top figure represent values that have been corrected to remove the slight increase in the *in situ* temperature brought about by adiabatic compression of the water column at depth (e.g., the increase in temperature induced by compression or pressure of the overlying water). In effect, *in situ* water temperatures are normalized to what they potentially would be if the temperature of each water sample were measured at sea level. Use of potential temperatures allows temperatures at all depths to be readily compared and allows temperature to be treated as a truly conservative (stable) property. AC = position of the Antarctic Convergence where caballing produces the Antarctic Intermediate Water; NAC = location of the North Atlantic Convergence where North Atlantic Intermediate Water (NAIW) is formed through the same process; m = meters; ‰ = parts per thousand. (*Source:* Adapted from G. L. Pickard and N. J. Emery, *Descriptive physical oceanography*, 5th edition, Pergamon Press, 1990. Based on GEOSECS data.)

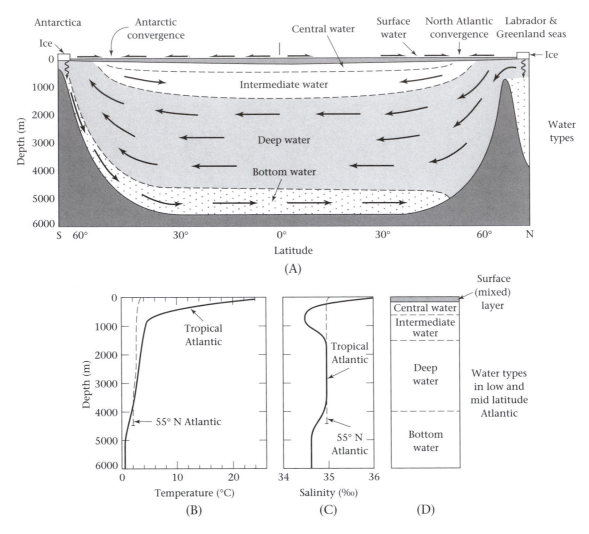

Figure 11.7. (A) Idealized north-south cross section of the Atlantic Ocean showing the general distribution and flow of basic water types, together with figures illustrating variations in (B) temperature and (C) salinity with depth at low- and high-latitude sites, and (D) the approximate positions of water types present in the mid-and lower-latitude regions of this ocean. There is relatively little variation of temperature and salinity with depth at high latitudes, with stratification best developed in low-and mid-latitude areas. A *water type* represents a homogeneous body of water displaying a sustained specific temperature and salinity. m = meters; ‰ = parts per thousand.

ture and salinity can in fact have the same density (Fig. 11.5). Thus, water masses of dissimilar temperature and salinity, but the same or similar density, can and often do travel at the same levels or depths in the density-stratified ocean.

The relationship illustrated on Figure 11.5 demonstrates how, when water masses of markedly different temperature and salinity but the same *in situ* density meet and mix, a new water mass is created that has a higher density than either of the parent water masses. Predictably, the newly created water sinks as it is continuously formed through mixing – a process termed *caballing*. Caballing is particularly important to the formation of intermediate water masses in the global ocean and commonly occurs in regions where relatively warm, saline surface water meets cool, dilute surface water (Fig. 11.2). For example, the warm, high-salinity water of the north-flowing Gulf Stream in the western North Atlantic Ocean is produced through equatorial heating, low- and mid-latitude evaporation, and the addition of Mediterranean Water that makes its way into the western North Atlantic at depth (Fig. 11.4). Simultaneously, and far to the north, cool and rela-

tively low-salinity water of the south-flowing Labrador Current is produced through polar cooling, melting of ice, and high rainfall as this current passes beneath the polar front. These two surface currents of contrasting character ultimately converge and mix in the northwestern corner of the North Atlantic Ocean (Fig. 10.9), inducing caballing along the so-called North Atlantic Convergence zone. The newly formed water resulting from the mixing process is colder than Gulf Stream water and more saline than Labrador Current water; therefore, it has a slightly higher density than the water in either of these two surface currents. Consequently, the new water sinks and spreads southward at depths commensurate with its density, forming the North Atlantic Intermediate Water (Figs. 11.6, 11.7, and 11.8). This same process is in operation along the Antarctic Convergence, leading to the formation of Antarctic Intermediate Water and, in the North Pacific where the Kuroshio and Oyashio currents meet, mix, and form, the North Pacific Intermediate Water (Fig. 11.2).

All six basic processes leading to the formation of distinctive water masses are well displayed in the Atlantic Ocean, and provide an excellent perspective on how surface, intermediate, and deep flow operate to maintain convective thermohaline circulation of the ocean on a global scale.

Figure 11.8. Schematic illustration of surface and subsurface circulation in the Antarctic region and primary water masses and currents. The Subantarctic Surface Water is warmer and hence has a higher dynamic height (exaggerated in this view) than the colder and denser Antarctic Surface Water (see Fig. 10.11). Caballing occurs along the Antarctic Convergence, leading to the formation of the Antarctic Intermediate Water. Dense Antarctic Bottom Water is created primarily as a product of the formation of sea ice in the Weddell Sea area during the winter season. The south-flowing North Atlantic Deep Water joins the Circumpolar Deep Water flowing continuously eastward around Antarctica. Shallower portions of the North Atlantic Deep Water upwell in this region to replace the dense, cold bottom water lost through sinking along the Antarctic margin. High pressure over Antarctica and the resulting west-blowing winds (i.e., the polar easterlies) sustain the East Wind Drift, while the prevailing westerlies maintain the eastward flow of the West Wind Drift, creating the Antarctic Divergence and associated high rates of upwelling along the boundary between these two opposing currents. (*Source:* Adapted from H. U. Sverdrup, M. N. Johnson, and R. H. Fleming, *The oceans*, Prentice-Hall, 1942, and J. P. Kennett, *Marine geology*, Prentice-Hall, 1982.)

THE ATLANTIC OCEAN: AN EXAMPLE

Figures 11.6 and 11.7 present north-south cross-sectional views of the Atlantic Ocean illustrating generalized patterns of major water masses, water types, flow directions, and the horizontal and vertical distribution of temperature and salinity. Even a casual comparison of these illustrations demonstrates the density-stratified structure of this ocean, and the fact that intermediate and deep water is derived primarily from the polar regions along with a significant contribution from the evaporative Mediterranean Sea.

Surface waters warmed in the equatorial and low latitudes of the Atlantic Ocean are transported poleward by major western boundary currents off North America (the Gulf Stream) and South America (the Brazil Current). The warm, saline water of the Gulf Stream encounters the cold, dilute surface water of the Labrador Current in the northeastern Atlantic Ocean, creating the North Atlantic Convergence. The resulting caballing produces North Atlantic Intermediate Water (NAIW), which spreads out and flows south at depths between 200 and 1500 meters, eventually mixing with Mediterranean Water traveling westward within the same depths. Simultaneously in the Circum-Antarctic region, warmer Subantarctic Surface Water moving eastward and poleward meets cold, dilute surface waters moving north from the Antarctic margin along the Antarctic Convergence. As in the North Atlantic, this process results in caballing and the continual formation of Antarctic Intermediate Water (AAIW), which sinks and moves north across the equator and slowly mixes with Mediterranean Water and North Atlantic Intermediate Water (Figs. 11.6 and 11.8). In contrast, the production of Antarctic Intermediate Water is less vigorous in the Pacific realm, and it flows northward only as far as the equator (Fig. 11.2). Intermediate water is also formed by caballing in the North Pacific Ocean along the convergence between the warm, salty Kuroshio Current and the cold, dilute Oyashio Current. This water sinks and spreads out within this region, forming the North Pacific Intermediate Water (NPIW) (Fig. 11.2).

The formation of bottom water in the Northern Hemisphere is restricted to the North Atlantic Ocean where cold, dense water formed in the Arctic basin flows into the Atlantic basin, forming the North Atlantic Deep Water (NADW). Alternatively, low surface salinities in the Bering Sea hamper the formation of cold, dense bottom water that might otherwise form there and flow into the adjacent North Pacific Ocean; flow in this region is further restricted by the blocking action of the Aleutian Arc and Ridge. Because the northern reaches of the Indian Ocean are in tropical latitudes, no dense bottom water is formed in this region.

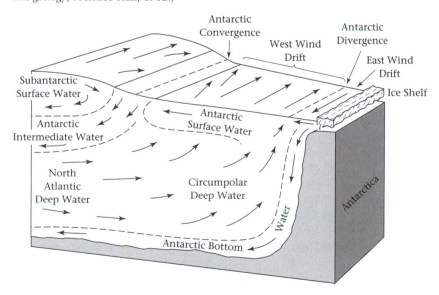

Early ideas about the origin of the North Atlantic Deep Water attributed its formation to the caballing process now known to be responsible for the formation of North Atlantic Intermediate Water. More recently, it has been recognized that the North Atlantic Deep Water is formed primarily through the rapid and severe cooling of relatively saline surface water arriving in the Greenland and Norwegian seas via the distal strands of the Gulf Stream. The cold, dry polar winds that sweep over these waters as they flow into this region cause them to give up heat to the overlying atmosphere through evaporation and conduction. The surface water thus experiences a rapid drop in temperature from 10 to 2°C, increasing its density to the point that it sinks. Additional cold, dense water is formed in this region through the seasonal formation of sea ice. These processes assure that the Greenland and Norwegian seas are continually filled with extremely cold, dense bottom water (i.e., Arctic Bottom Water) that is held in place by the deep ridges (sills) that form submarine dams between Greenland and Iceland, and between Iceland and Scotland (Fig. 11.6). Because these basins are full to the brim, excess cold, dense water flows over the ridges and spills into the adjacent North Atlantic Ocean. These latter flows reach velocities exceeding 1 meter per second and transport an average of 5 million cubic meters of water per second into the North Atlantic Ocean, forming the North Atlantic Deep Water (Figs. 11.6 and 11.7). Recent research suggests that these undersea "waterfalls" are not continuous but rather experience pulsating flows, likely reflecting climatically driven variations in the rate of production of the dense surface water in this region. Indeed, this same process is thought to be operating in the Labrador Sea, adding to the volume of bottom water entering the North Atlantic Ocean. Thus, formation of the bulk of the North Atlantic Deep Water is dependent upon the relatively high salinity of North Atlantic Surface Water arriving in the subarctic region, together with rapid, high-latitude cooling and evaporation of this water through open ocean surface processes. As the North Atlantic Deep Water travels southward at depth, other water is added to this flow including Mediterranean Water, thus slightly changing the character of the water over time.

The freezing of seawater and the formation of sea ice is primarily responsible for the production of Antarctic Bottom Water (AABW) in the southernmost part of the Atlantic Ocean or Southern Ocean.* This water subsequently flows northward into the North Atlantic Ocean as well as into the deeper areas of the Indian and Pacific oceans, where it forms a component of bottom water formed from several source areas. It has been estimated that approximately 59 percent of the water in the global ocean has its origin in the Antarctic or Southern Ocean region, with the Circumpolar Deep Current primarily responsible for promoting the exchange of water between the Pacific, Atlantic, and Indian oceans (Fig. 11.8).

When seawater freezes, salt is excluded, increasing the salinity of the remaining unfrozen water. Winter freezing of sea ice in the Weddell Sea and the Ross Sea embayments of Antarctica is generally thought to be responsible for the creation of most Antarctic Bottom Water, with the Weddell Sea considered to be the dominant source of this water. Antarctic Bottom Water has an average temperature of −0.4°C and a salinity of 34.66 parts per thousand, and constitutes the densest water in the global ocean (Figs. 11.6, 11.7, and 11.8). As the sea ice forms in these areas, the resulting unfrozen water of increased salinity sinks onto the underlying shelves and ultimately flows down the continental slope into the deep sea where it spreads out along the sea bottom, invading the deepest areas of all three major ocean basins (Figs. 11.2 and 11.8). In fact, Antarctic Bottom Water represents the most widespread water mass in the global ocean, and its distinctive temperature, salinity, and dissolved oxygen content allow it to be easily identified and traced, even in high-latitude abyssal areas of the North Atlantic Ocean (approximately 45 to 50° N). Finally, it is important to note that the sinking of dense Antarctic Bottom Water is compensated by the slow upwelling of North Atlantic Deep Water and Circumpolar Water in the Southern Ocean.

The fact that Antarctic Bottom Water is restricted to the deepest areas of the ocean floor dictates that topography plays a significant role in constraining its flow. For instance, the northward flow of Antarctic Bottom Water in the eastern South Atlantic Ocean is blocked by a large, deep sea ridge (i.e., the Walvis Ridge), whereas in other areas, high-velocity bottom currents are created where this water is funneled through narrow, deep-sea channels within fracture zones or ridge segments.

THE GLOBAL OCEANIC CONVEYER BELT

The prevailing patterns of surface, intermediate, deep, and bottom circulation within the Atlantic Ocean clearly demonstrate the nature of the linked convective flows responsible for global ocean circulation and the resulting transport of water, heat, and salt from low to high latitudes, from pole to pole at depth, and between major ocean basins. Atlantic circulation, together with convective thermohaline flow in the Pacific and Indian oceans, leads to the continuous ventilation of the global ocean with rates of deep-sea circulation dictating the residence time of water at depth before it resurfaces and begins another trip around a given circulatory loop. Estimated average residence times of deep water (e.g., water below the thermocline layer) is 460 years in the Atlantic Ocean, 830 years in the Indian Ocean, and 1030 years in the Pacific Ocean, largely reflecting differences in the volume of the Atlantic Ocean (338×10^6 cubic kilometers) versus that of the Pacific Ocean (1354×10^6 cubic kilometers) and the fact that the Indian Ocean is restricted to the Southern Hemisphere. However, there are basic differences be-

---

* The so-called Southern Ocean encompasses the southernmost portions of the Atlantic, Pacific, and Indian oceans, where the West Wind Drift and Circumpolar Deep Current flow continuously around the Antarctic continent (Fig. 11.8).

tween the deep circulation of the Atlantic, Pacific, and Indian oceans.

The warm, salty surface water created in the Atlantic Ocean loses heat to the atmosphere and gains density as it flows into Arctic latitudes, where it is transformed into cold, saline North Atlantic Deep Water. It then sinks into the deep sea and flows south across the equator into the Southern Hemisphere. As the North Atlantic Deep Water enters the South Atlantic Ocean, it rides up and over the northward-flowing and denser Antarctic Bottom Water, mixes with the Circumpolar Deep Water, and ultimately forms the so-called Oceanic Common Water, representing a deep-water blend of Antarctic Bottom Water, North Atlantic Bottom Water, and Antarctic Intermediate Water (Fig. 11.8), and constituting the largest water mass (by volume) in the world ocean. A portion of this water flows north into the Indian Ocean; however, most of it continues to travel eastward and eventually flows into the southwestern Pacific Ocean in the area between New Zealand and Antarctica, and finally into the Central and North Pacific Ocean, where it is referred to as the Pacific Deep Water (Fig. 11.2). Antarctic Bottom Water proper remains as a readily identified water mass in the deepest parts of the Pacific (i.e., areas deeper than 4000 meters) as it makes its way northward through abyssal channels aligned with the volcanic chains characterizing much of the Pacific Ocean floor.

Because the Pacific Ocean has neither a major northern source of deep water nor a major source of warm, salty Mediterranean-like water, it exhibits relatively sluggish intermediate and deep circulation. The slow circulation of the deep Pacific stands in marked contrast to the dynamic and complex thermohaline circulation of the Atlantic Ocean, driven by the continual sinking of dense water formed in the Arctic Basin, the Mediterranean Sea, and the Antarctic region. Pacific Deep Water represents an enormous and remarkably homogeneous water mass, exhibiting an average temperature of 1.5 to 2°C and a salinity of 34.68 parts per thousand; hence it is much colder and fresher than corresponding North Atlantic Deep Water, which has an average temperature of 2.5 to 3.1°C and a salinity of 34.9 parts per thousand.

Viewing circulation of the deep ocean on a global scale, it is apparent that the flow initiated and sustained by the continual sinking of cold, dense surface waters in the North Atlantic Ocean and Antarctic margin ultimately ends with the pooling of a large portion of this water in the deep Pacific Ocean (Fig. 11.2). This pattern is clearly reflected by the distribution of dissolved oxygen in the Atlantic and Pacific oceans (Fig. 11.9). The rapid sinking of North Atlantic Deep Water carries relatively high amounts of dissolved oxygen from the surface ocean into the deep sea. Over time, dissolved oxygen is depleted through respiration of deep-sea organisms and the breakdown of organic material in the water column as the water "ages" and flows from one ocean basin to another. The result of this process is the concentration of the oldest and most oxygen-depleted deep water in the North Pacific Ocean (Fig. 11.9). Because of the inverse relationship between dissolved oxygen and the capacity of ocean water to retain nutrients (e.g., nitrate, phosphate), the deep waters of the North Pacific are nutrient-rich and those of the North Atlantic are relatively nutrient-poor, with all that this pattern portends for the overall biologic productivity of these two oceans.

On a grand scale, ocean circulation can be viewed as a giant conveyor belt whose purpose is to redistribute the excess solar heat received in its equatorial regions via wind-driven surface currents and density-driven thermohaline circulation of the deep sea, in a continual attempt to achieve thermal equilibrium – an impossible goal the ocean shares with the similarly convecting atmosphere (see, e.g., Fig. 10.8). The conveyor belt analogy has been championed most effectively by Wallace Broecker of the Lamont-Doherty Earth Observatory, who has put forward a very simplified but useful model of heat and salt transport by the ocean, illustrated in Figure 11.10. The arrows depicting the hypothetical oceanic conveyor belt *do not* represent actual current stream

Figure 11.9. The distribution of dissolved oxygen (measured in micrometers per kilogram) μ/kg) in the Atlantic and Pacific oceans and the general pattern of deep flow between these two oceans (arrows). Relatively high levels of dissolved oxygen characterize the newly formed North Atlantic Deep Water as it is created at the surface and rapidly sinks in the northernmost Atlantic Ocean. Once isolated from the air–sea interface, any oxygen utilized at depths below the mixed or surface layer is not replaced. Thus, dissolved oxygen can be used as a measure of the relative "age" of the water. In addition, as dissolved oxygen decreases, the capacity of the water to retain key nutrients (e.g., phosphate and nitrate) increases. As deep water slowly flows from the Atlantic Ocean to the Pacific Ocean, dissolved oxygen is depleted, with the result that the oldest and most oxygen-depleted water ultimately accumulates in the North Pacific where it slowly upwells, assuring a rich supply of nutrients to this region. m = meters. (*Source:* Adapted from A. L. Gordon, *The role of thermohaline circulation in global climate change 1990–1991 report*, Lamont-Doherty Geological Observatory of Columbia University, Columbia University Press, New York, 1991.)

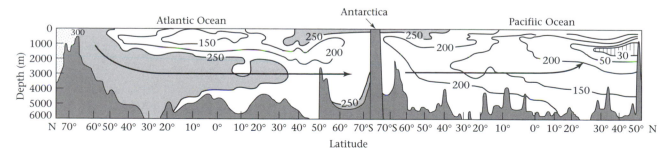

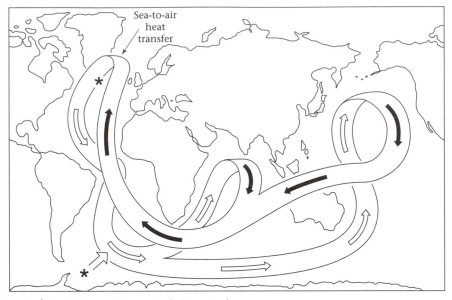

Cold deep water ⟹   Warm surface water ⟶
*Areas where cold, dense surface water sinks.

Figure 11.10. Highly simplified pattern of heat transfer via the "oceanic conveyer belt." The patterns of flow illustrated in this figure *do not represent actual surface or deep currents;* they depict the generalized path of warm surface water as it is transported from the Pacific and Indian oceans to the North Atlantic where it is cooled, increasing its density to the point that it sinks. The resulting cold, saline water then travels south at depth to join cold, dense water derived from the Antarctic margin and flows eastward. Although some of this water flows into the Indian Ocean, most of it continues to move eastward around Antarctica, with a large portion flowing into the Pacific Ocean. This latter water eventually makes its way into the North Pacific Ocean and slowly rises to the surface, where it begins its return trip to the North Atlantic. * = Areas where cold, dense surface water sinks. (*Source:* Adapted from Michael S. McCartney, Toward a model of Atlantic Ocean circulation, *Oceans,* vol. 37, no. 1, 1994.)

lines; they illustrate generalized surface and deep transport in a manner emphasizing the most likely flow pathways. It has been estimated that it takes approximately 1000 years for water to complete the circuit from initial descent into the deep sea in the North Atlantic Ocean to its return to the surface of the ocean.

It is important to emphasize that the generalized picture of oceanic heat transfer depicted in Figure 11.10 involves not only the transport of sensible heat by the ocean, but also the exchange of latent heat and water between the ocean and the atmosphere via evaporation and condensation (e.g., rain), as well as the exchange of heat through conduction at the air–sea interface. For example, the relatively large differences in the salinity of Atlantic Central Water (approximately 36.3 parts per thousand) and Pacific Central Water (approximately 34.9 parts per thousand) reflect the fact that: (1) the Atlantic Ocean is characterized by high rates of evaporation, which increase its salinity; and (2) the resulting wa-

ter vapor and associated latent heat is continuously exported from the Atlantic to the Pacific Ocean via the atmosphere. Moreover, the asymmetry of the South Equatorial Current in the Atlantic Ocean (Fig. 10.8) allows South Atlantic Ocean heat to be transported directly into the Northern Hemisphere. This latter process increases the temperature of the North Atlantic Ocean, which, in turn, transfers heat to the overlying atmosphere, enhancing its capacity to evaporate seawater and increasing the water vapor and latent heat being transported out of the region by the lower atmosphere. In contrast, large-scale cross-hemisphere transport of heat by equatorial surface currents does not take place in the Pacific (Fig. 10.8). As a result, the mean surface temperature of the North Pacific Ocean is relatively low, which in turn cools the overlying air, reducing its capacity to evaporate seawater and increase surface salinity, further strengthening the differences in salinity between the Pacific and Atlantic oceans. An additional consequence of lower evaporation rates in the Pacific Ocean is that the resulting lower surface salinities prevent this water from achieving sufficiently high density to sink and form a deep-water mass, even though the water is cooled significantly in the higher-latitude area of the North Pacific Ocean.

Heat is also continuously added to the Atlantic Ocean through the transport of warm surface water from the Pacific and Indian oceans by way of the Indonesian seaways and the Agulhas Current at the southern tip of Africa (Figs. 10.8 and 11.10). This process, together with the cross-hemisphere transport of warm equatorial water, cause the average temperature of the Atlantic Ocean to be much higher than that of the Pacific, as reflected by the average rates of evaporation in these two oceans. The rate of evaporation in the Atlantic Ocean is approximately 103 centimeters per year, whereas it averages only approximately 55 centimeters per year for the Pacific Ocean. These differences also reflect the net transfer of fresh water from the Atlantic Ocean to the Pacific Ocean as evaporation in the Atlantic yields water vapor that is later delivered to the Pacific by transport through the lower atmosphere from which it condenses and falls as rain. The result of these various processes is that the Atlantic Ocean is warm and salty, whereas the Pacific Ocean is relatively cool and fresh. This is the case despite the fact that the influx of fresh water from rain, rivers, and melting ice to the relatively small Atlantic Ocean is much greater than the input of fresh water from these same sources to the much larger Pacific Ocean. Thus, the evaporation/precipitation cycle between the ocean and atmosphere is critical to ocean–atmosphere heat transport and represents a process that is hypersensitive to both short-term variations in ocean temperature as well as longer-term changes in climate.

Looking back in time, one can envision other large-scale perturbations of salinity that likely affected the oceanic "conveyer belt," such as the rapid melting of the large continental ice sheets in the Northern Hemisphere approximately 12,000 years ago, as the last episode of glacial climate came to an end. In fact, there is evidence to suggest that the production of North Atlantic Deep Water was much reduced during periods when large volumes of glacial meltwater entered the North Atlantic Ocean, resulting in decreased surface salinities and slowing the formation of dense, cold saline water in this region. In turn, a slowdown in the formation and sinking of North Atlantic Deep Water would have slowed the flow of warm Gulf Stream water into the North Atlantic, curtailing the accompanying transfer of heat to the atmosphere in this region. Thus, a change in the salinity or temperature of water along any given portion of the oceanic conveyer belt can be expected to trigger upstream and downstream responses and to have global consequences for coupled ocean–atmosphere circulation.

Although the majority of heat transported by the ocean is assumed to represent heat originally received from the Sun, it has been proposed that heat generated by the internal Earth and emitted via vents and volcanic processes along the mid-ocean ridge and rise system has the potential to affect the character and circulation of oceanic bottom water. Could large volumes of buoyant warm water simultaneously emanating from widely distributed deep-sea vents modulate deep-sea circulation during periods of increased mantle convection and heat flow from the Earth's interior? Clearly, we still have much to learn about the deep sea and its role in global climate change.

QUESTIONS

1. What changes in deep-sea circulation do you think would occur if tectonic processes formed a large gateway or passage, allowing free circulation between the Arctic Basin and the North Pacific Ocean?

2. Can you think of a scenario in which deep-ocean circulation might come to a complete halt?

3. What do you think the long- and short-term biologic consequences might be of the general patterns of flow in the intermediate and deep waters of the ocean?

4. High rates of evaporation in the Mediterranean Sea create a distinctive warm, saline water mass that can be identified easily as it flows into the adjacent Atlantic Ocean. What other marginal seas do you think might constitute sources of distinctive water?

5. Why are temperature-salinity (T-S) diagrams so useful for identifying and tracing the movement and mixing of water masses?

FURTHER READING

Broecker, W. S., and Denton, G. H. 1990. What drives glacial cycles? *Scientific American* 262:48–58.

Hollister, C. D., and Nowell, A. 1984. The dynamic abyss. *Scientific American* 250:42–53.

Stommel, H. 1987. *A view of the sea.* Princeton, NJ: Princeton University Press.

Sverdrup, H. U., Johnson, M. W., and Fleming, R. H. 1942. *The oceans.* Englewood Cliffs, NJ: Prentice-Hall.

Whitehead, J. A. 1989. Giant ocean cataracts: *Scientific American* 260:50–57. pp. 50–57.

12 Chemical Oceanography

PETER G. BREWER

CHEMICAL CONSTITUTION OF THE SEAS

The ocean waters represent by far the largest and most complex chemical solution on the surface of the planet. Seen from space our Earth *is* primarily blue water, and the oceanic realm provides by far the largest habitat on Earth for living creatures. Life itself arose in the chemical soup of the sea, and the ocean appears to have been nature's greatest incubator for experimentation with many modes of life, from the smallest cells to the largest living animals; whether warm or cold, dark or light, productive or barren, niches provide the opportunity for evolutionary adjustment and unique geochemical signatures.

Benign to life, and corrosive to metals, the chemistry of seawater varies from the familiar common salt to vanishingly small concentrations of trace elements that provide wonderful clues to the forces that shape our Earth. Oceanographers have learned to trace the passage of time from the radiochemical clocks, or tracers, in the sea, and to unravel the chemical signatures of living organisms. They have discerned the patterns of circulation, witnessed oceanic change from the invasion of humanity's industrial activities, revealed the enormous buffer against chemical perturbation that stabilizes our planet, and are learning to place the changes of today in the context of the lessons of Earth history.

Oceanographers often deal with paradoxes. The total volume of seawater is truly enormous, at approximately 1.37×10^{21} liters; yet concentrations can be amazingly dilute, and many chemical species at the parts-per-trillion level can readily be detected and used for scientific discovery. The physical transport of the ocean moves this chemical solution in massive amounts; the Gulf Stream off the coast of Florida has a flow 1000 times greater than that of all the world's rivers combined. However, the global circulation of deep ocean water is ponderously slow, taking approximately 550 years on average to restore a parcel of abyssal water from its cold, dark sojourn to the sunlit upper ocean. The ocean plays a key role in the heat and water budget of our world. Transport of warm water by the ocean from low latitudes poleward by currents results in massive radiation of heat to the atmosphere: In the North Atlantic, the ocean radiates approximately 1×10^{15} watts per year to the air, moderating the climate of northern Europe. The cycles of evaporation and precipitation of the fresh water we need to support human life are dominated by the ocean; compared to this, the transport by rivers is small. Figure 12.1 shows the north-south transport of fresh water by sea and air from the evaporation and precipitation cycle.

There are no great forests in the ocean, apart from a few coastal kelp strands; instead, the dominant green plants are microscopic phytoplankton. These are not long-lived like vegetation on land; they have a doubling time of about a day and are consumed quickly by small, grazing plankton. Yet the photosynthetic activity of these plants exerts an extraordinary effect on ocean chemistry, depleting the upper ocean in nutrients and carbon dioxide, and comparably enriching the deep ocean as they are consumed, sink, and decay. Although these organisms are individually minute, their ensemble effect is so large it can be seen from space. Figure 12.2 shows a color image of the distribution of phytoplankton in the ocean. Produced by scientists at the NASA Goddard Space Flight Center from data taken by the Coastal Zone Color Scanner on the *NIMBUS* 7 satellite, the image is an ensemble of data from different seasons. The fertile green stripes along the equator and over the shallow continental shelves, the blue of the barren ocean deserts in the subtropical gyres, and the seasonal green of the northern-latitude subpolar gyres testify to the contrasts of the distribution of life in the surface ocean. The amount of carbon dioxide fixed annually by green plants through photosynthesis is about the same in the ocean and on land.

The calcium carbonate skeletons of many such plankton rain down to the ocean floor, recording Earth's geochemical history. Over the eons, they have provided a vast alkaline buffer against the chemical change from rising levels of at-

W. G. Ernst (ed.), *Earth Systems: Processes and Issues.* Printed in the United States of America. Copyright © 2000 Cambridge University Press. All rights reserved.

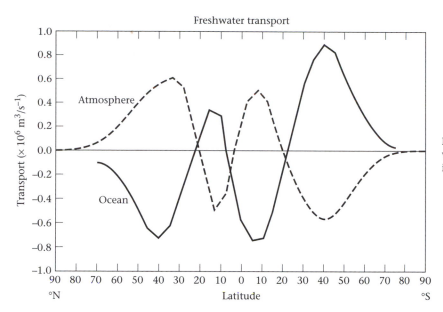

Figure 12.1. The meridional transport of water by the global ocean and atmosphere, in sverdrups (10^6 cubic meters per second).

mospheric carbon dioxide. The ocean is, and will remain, the largest sink for rising levels of atmospheric carbon dioxide, and the interaction of air and sea provides one of the greatest stabilizing factors for the support of life on this planet.

THE CHEMICAL COMPOSITION OF SEAWATER

Measuring and understanding the ocean is made easier by the simple fact that "sea salt" is the same everywhere. This

Figure 12.2. Distribution of phytoplankton in the ocean. Produced by scientists at the NASA Goddard Space Flight Center from data taken by the Coastal Zone Color Scanner on the *NIM-BUS 7* satellite, the image is an ensemble of data from different seasons. See color plate section for the color version of this illustration. (*Source:* Courtesy of NASA.)

concept of constancy of composition of the major ions that make up the salinity of seawater was discovered more than a century ago, and it holds remarkably true. For example, although the Persian Gulf is far saltier than the Arctic Ocean, the ionic ratios of the salts are exactly the same. This means that we can measure the salinity, and calculate the density (see Chaps. 10 and 11), of seawater with great accuracy from a simple bulk measurement. This constancy provides life in the ocean with a marvellously stable chemical matrix over the vast distances established by currents and migration patterns (Fig. 12.3).

Ocean water contains approximately 35 grams per kilogram (commonly noted as ‰, or parts per thousand) of "salt": about a 3.5% solution, or approximately 0.5 molal in ionic strength. The range varies from approximately 40 grams per kilogram in semienclosed tropical seas, such as the Red Sea, to 20 grams per kilogram in less saline areas such as the Baltic Sea. In practice it is very difficult to determine the salinity accurately by the obvious method of evaporating the water and weighing the remaining salt. All modern methods rely upon very careful measurement of electrical conductivity, and ocean salinity is thus easily determined to a precision of approximately one part per hundred thousand. A section of the distribution of temperature and salinity in the western Atlantic Ocean derived from the Geochemical Ocean Sections Study (GEOSECS) expeditions of the 1970s is shown in Figure 11.6. It reveals the strong layering with depth of the various ocean water masses.

Table 12.1 represents the basic composition of seawater for a sample of approximately 35 grams per kilogram of salinity. Only eleven major ions, those present in sufficient quantity to affect the salinity measurably (i.e., greater than 1 milligram per kilogram of seawater) are listed. From the data in Table 12.1, we can conclude that a solution of sodium chloride, magnesium sulfate, and calcium carbonate would come close to yielding the bulk major ion composition. Those

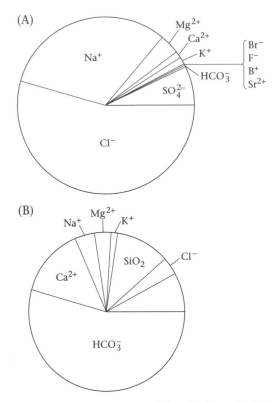

Figure 12.3. Average proportion of dissolved ions in (A) seawater and (B) river water.

TABLE 12.1 COMPOSITION OF 1 KILOGRAM OF SEAWATER	
Species	**Grams**
Na⁺	10.7822

Species	Grams
Na^+	10.7822
Mg^{2+}	1.2837
Ca^{2+}	0.4121
K^+	0.3991
Sr^+	0.0079
Cl^-	19.3529
SO_4^{2-}	2.7124
HCO_3^-	0.1135
Br^-	0.0672
CO_3^{2-}	0.0116
$B(OH)_4^-$	0.0013
F^-	0.0010
$B(OH)_3$	0.0203
Total	35.1709

Notes:

(1) The contributions of the borate and carbonate systems are approximately given at surface water pH.

(2) For many years the salinity of sea water was measured by titration of the chloride ion with a silver nitrate solution. This formed the international standard of measurement, and it is often reported as the "chlorinity." The ratio is salinity = 1.80655 × chlorinity. Chlorinity = 19.374 grams per kilogram.

chemical properties that vary strictly and only with salinity are dubbed "conservative" properties of seawater; in contrast, the multitude of trace species such as dissolved gases, trace metals, organic compounds, and radionuclides are "nonconservative" and can vary widely. We use the salinity of seawater for tracing the flow and mixing of oceanic water masses, and the trace species for diagnosing the variety of chemical and biologic processes that occur.

Seawater is a mildly alkaline solution: The pH of surface seawater today is close to 8.1, and the range of pH within the oceans is approximately 7.6 to 8.3. The total amount of carbon dioxide dissolved in seawater varies with salinity and biogeochemical processes; it is 2×10^{-3} molar. However, the oceanic carbonate system is not simply dissolved carbon dioxide gas; it is made up of the sum of the species (CO_{2gas} + H_2CO_3 + HCO_3^- + CO_3^{2-}). This system has a buffer capacity (the ability to resist acidic change), together with the contribution of borate ion, of approximately 3 millimoles per kilograms. It is this buffer capacity, given by the alkalinity of seawater, that is critical in the role of the ocean in moderating the rise of today's atmospheric carbon dioxide levels. We define the alkalinity as the amount of strong acid required to titrate 1 liter of seawater to the equivalence point of the carbonic acid system. It is measured in that way by a classic acid titration. The alkalinity is related to the sum of electrical charges, and two acid protons are required to neutralize one carbonate ion; therefore, the carbonate alkalinity is given by the sum:

$$\text{Carbonate alkalinity} = 2(CO_3^{2-}) + (HCO_3^-)$$

The alkalinity is a mass ionic property. Therefore, changing the total amount of neutral carbon dioxide gas through gas exchange with the atmosphere or through photosynthesis will not affect it; addition or dissolution of calcium carbonate, $CaCO_3$, from the shells of marine organisms, or from a strong acid, will. Thus the alkalinity and the total dissolved carbon dioxide content of seawater can vary independently of each other, even though they are closely related and the oceanic pattern of their distribution is of fundamental importance in ocean chemistry.

ACIDS AND BASES

Although Table 12.1 provides a rudimentary list of the composition of seawater, it begs the question of how this composition arose and why it is so stable. While the shape of the continental landmasses, and therefore the configuration of the ocean boundaries and the pattern of circulation, have changed dramatically throughout Earth history, a vast salty aqueous chemical solution has existed on the surface of this planet for most of geologic time. The chemistry of this solution was broadly established early in the 4.6-billion-year Earth history. The very early atmosphere was reducing and devoid of oxygen. However, with the evolution of photosynthesis, probably within the first 1000 million years of geologic time, the level of atmospheric carbon dioxide and oxygen rose, and a chemistry grossly similar to the one we have today was gradually established. Therefore, the ocean has existed long enough for weathering processes to have leached and dissolved huge amounts of material from the continents. How is a constant composition maintained over the vast stretch of geologic time? Either this material has remained dissolved, slowly accumulating in concentration in the ocean until an equilibrium saturation for a particular element is reached and it precipitates, or it has been continuously removed to the sediments as fast as it is supplied, keeping the oceanic concentration low and in a steady state. Rarely is there a solubility equilibrium control, and it is the rate of the removal process that is dominant; chemical elements are continuously stripped from the ocean.

The Swedish chemist Lars Gunnar Sillen was the first person to attempt to calculate the chemistry of the ocean from first principles through an equilibrium model. He considered the chemistry of a liter of seawater, in chemical equilibrium with its share of atmospheric gases, and its reaction with the rocks and minerals to which it was exposed. Sillen pointed out in 1961 that we may consider the origin of the ocean as the result of a gigantic acid–base titration. Acids that have diffused from the Earth's interior – HCl, H_2SO_4, H_2CO_3 – have combined with bases that have been set free by the weathering of primary rock. In this acid–base titration, involving the weathering of igneous rocks, the world oceans are approximately 0.5 percent from the equivalence point. This balance of nature is achieved extraordinarily well; most beginning chemistry students find it hard to carry out a laboratory titration with this precision!

Sillen was the first scientist to introduce rigorous chemical equilibrium theory to the field, and the chemical model he examined is an excellent starting point. He recognized that the pH of seawater could, in the very long term, be controlled by equilibria with silicate minerals in addition to the capacity of the carbonate system. The degassing that is described occurred for the most part very early in Earth's history. Sillen uncovered several problems, such as balancing the contribution of several components of weathered material today (he did not live to see the fascinating recent discoveries of active venting of potent, hot, reducing, mineral-rich hydrothermal fluids at oceanic ridge crests, and their role in balancing the chemistry of the ocean (see box entitled *Hydrothermal Circulation*). He also pointed out the dramatic role of even traces of dissolved oxygen in maintaining the oxidation balance of the sea.

Principles of acid–base chemistry work very well in ocean science because there are few kinetic barriers in these reactions. For example we know that the reactions involving the chemistry of carbon dioxide in seawater, such as the sequence

$$CO_2 + H_2O \leftrightarrow H_2CO_3 \leftrightarrow H^+ + HCO_3^- \leftrightarrow 2H^+ + CO_3^{2-} \quad (1)$$

can be exactly described by the rules of physical chemistry. Important experimental and theoretic conclusions are based on these principles. The chemistry of blood, and the chem-istry of sea water have much in common, and we can use closely similar equations for each.

OXIDATION AND REDUCTION

In marked contrast, calculations of "redox" equilibria – that is, those reactions in which the chemical elements are oxidized or reduced – often give results that are very different from those observed in nature. The deviations tell us much about important processes to look for. For example, in the oxidizing atmosphere of the Earth today, an equilibrium calculation would show that nitrogen gas should react with oxygen on a vast scale to form nitric acid that would reside in the ocean. The fact that this reaction does not dominate, and that we have an abundance of nitrogen in the atmosphere, implies large energy barriers and the presence of important biologic nitrogen-forming pathways that return nitrate to the atmosphere as nitrogen gas. The presence of oxygen in our atmosphere forces the presence of oxidized states, such as the rusting of iron and the decomposition of organic matter, to the oxidized form of carbon, carbon dioxide. If this is so, then how does redox disequilibrium occur?

Most redox disequilibria in the sea are driven by the photosynthesis of phytoplankton; in this strongly reducing reaction, the cell uses sunlight to split the water molecule into hydrogen and oxygen, and to fix carbon dioxide into organic matter. Other chemical species in the complex soup required to support life pass through the cell and are also transformed. The slow return path of many chemical elements from their reduced state after transit through a living cell provides an important factor in ocean chemistry; for example, the trace element iodine exists in seawater as both the stable oxidized form iodate (IO_3^-) and as the reduced form iodide (I^-) as testimony to its biologic role. The very few photochemical reactions that take place outside the living cell are driven directly by the action of sunlight on the molecules in the surface ocean. Examples include the "bleaching" or photochemical attack on dissolved organic matter to produce trace quantities of carbon monoxide (CO) in surface waters, and the photoreduction of nitrate; however, these processes are of minor importance.

GEOCHEMICAL TIME SCALES AND THE OCEAN

The concept of the residence time of a chemical species in the ocean introduces ideas of time dependence beyond the reach of equilibrium models. This is critical knowledge, and it allows us to begin to answer important questions. If we disturb the chemistry of the ocean, how long does it take to recover? If we introduce a pulse of foreign material, then how will this pulse evolve and decay with time? What is the fate of river-borne material?

The mean residence time (MRT) of an element (measured in years) is defined as the mean oceanic concentration (measured in grams) divided by the rate of supply by rivers (mea-

HYDROTHERMAL CIRCULATION

Circulation of seawater also takes place through the sediments and volcanic structures on the ocean floor. With the discovery in the 1960s of hot saline fluids venting in the deep rifts of the Red Sea, and the discovery in the 1970s of submarine hydrothermal vents in the deep Pacific Ocean (Fig. 12.4), scientists embarked upon a new chapter of ocean discovery. The tectonic structure of ocean basins is such that new hot oceanic crust is brought up from deep within the Earth at the Mid-Ocean Ridge, which stretches for 70,000 kilometers throughout the major ocean basins. Cold seawater – percolating within the sediments and fractured ocean crust and coming into proximity with this heat source – is chemically altered by the rocks and convected upward, setting up a large-scale hydrothermal circulation. This is seen as intense plumes of mineral-rich water venting at ridge crests. The heat loss helps balance the thermal budget of the sea floor. At these pressures, the boiling point of seawater is close to 350°C, and plumes of water close to this temperature are vented.

The chemical signature is unique, involving the reduction of seawater sulfate to toxic sulfide, removal of magnesium, and enrichment in metals and silica. The hot metal-rich and sulfide-rich fluids that vent mix instantly with cold, oxygenated seawater, and the plumes precipitate a dense cloud of metal-rich sulfide and oxide minerals. Large chimneys of these minerals form around the venting fluids; the hot waters may rise several hundred meters above the sea floor and give an ocean temperature anomaly observable over great distances.

Although the fluids are toxic to normal life, these sites are not barren wastelands. They are home to dense concentrations of extraordinary animals – many of them new to science – that derive their energy not from consumption of the products of *photosynthesis* but from the chemical energy in these systems by *chemosynthesis*. Bacteria are able to catalyze, and capture the energy of, the reaction between vented sulfide and seawater dissolved oxygen. These bacteria, either free in the water or symbiotically living within vent animals, provide the fixed carbon necessary for supporting the vent community.

From analysis of the thermal and chemical budgets of these systems, and from extrapolation of the findings at key sites, scientists estimate that the entire volume of the world's oceans passes through the ocean ridge hydrothermal system every 6 to 8 million years (Fig. 12.5).

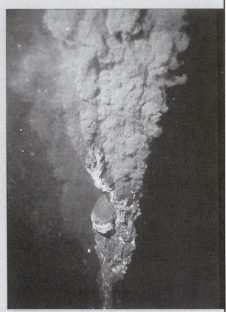

Figure 12.4. Black smoker. (*Source:* Courtesy of Woods Hole Oceanographic Institution; photograph by Robert D. Ballard 1979.)

Figure 12.5. Flux of chemicals from black smokers.

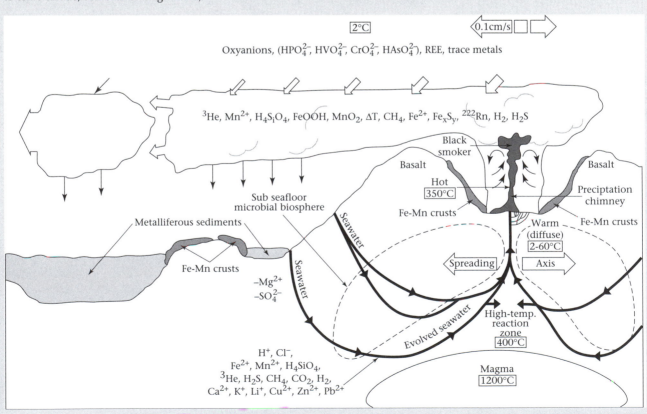

sured in grams per year). MRT may vary from millions of years for the highly soluble species that make up the major ion saline background solution, to approximately 100,000 years for a soluble but highly bioactive element such as phosphorous that is intensively cycled by marine organisms, to a few hundred years or less for elements such as the rare earths or thorium. MRTs for many chemical species are provided in Table 12.2. The information derived from the strongly varying distributions of elements with short MRTs is of great diagnostic value to ocean geochemists, because the sources and sinks revealed by their patterns in the ocean provide invaluable clues to fundamental processes.

What is the oceanic MRT of water itself? We saw earlier that enormous evaporation and precipitation fluxes to and from the sea drive the water cycle of the planet. The rate of supply of water by rivers is approximately 36,000 cubic kilometers per year, and the volume of the ocean basins is approximately 1350 million cubic kilometers. Thus the MRT of water, or the time that would be taken to "fill" the ocean basins, is 37,500 years.

MRTs for chemical species cannot be determined exactly, for river processes are too variable in space and time, and the reliability of data is often open to question. There are also other pathways for chemical input, by the atmosphere or

TABLE 12.2. CONCENTRATIONS OF THE RELATIVELY ABUNDANT ELEMENTS IN AVERAGE RIVER WATER AND IN AVERAGE OCEAN WATER, PLUS CORRESPONDING MEAN OCEANIC RESIDENCE TIMES

Atomic No.	Element	Conc. Mean River (10^6 moles per kilogram)	Conc. Mean Sea (10^6 moles per kilogram)	Mean Residence Time (Years)
1	H (as H_2O)	—	5.4×10^7	
2	He	—	1.8×10^{-3}	
3	Li	1.7	2.5×10^1	5.7×10^5
5	B	1.7	4.2×10^2	9.6×10^6
6	C (inorganic)	—	2.3×10^3	—
	(organic)	—	4	—
7	N (dissolved N^2)	—	5.8×10^2	—
	(NO^3)		3.0×10^1	
8	O (as H_2O)	—	5.4×10^7	—
	(dissolved O_2)		2.2×10^2	
9	F	5.3	6.8×10^1	5.0×10^5
10	Ne		7.5×10^{-3}	
11	Na	2.2×10^2	4.7×10^5	8.3×10^7
12	Mg	1.6×10^2	5.3×10^4	1.3×10^7
13	Al	1.9	(3×10^{-2})	6.2×10^2
14	Si	1.9×10	1.0×10	2.0×10^4
15	P	1.3	2.3	6.9×10^4
16	S	—	2.8×10^4	
17	Cl		5.5×10^5	
18	Ar		1.5×10^1	
19	K	3.4×10^1	10.2×10^3	1.2×10^7
20	Ca	3.6×10^2	10.3×10^3	$I., \times 10^6$
21	Sc	1	(1.5×10^{-5})	
22	Ti	2.1×10^{-1}	$(<2.0 \times 10^{-2})$	3.7×10^3
23	V	2.0×10^{-2}	2.3×10^{-2}	4.5×10^4
24	Cr	1.9×10^{-2}	4×10^{-3}	8.2×10^3
25	Mn	1.5×10^{-1}	5×10^{-3}	1.3×10^3
26	Fe	7.2×10^{-1}	(1×10^{-3})	5.4×10^1
27	Co	3.4×10^{-3}	(3×10^{-5})	3.4×10^2
28	Ni	3.8×10^{-2}	8×10^{-3}	8.2×10^3
29	Cu	1.6×10^{-1}	4×10^{-3}	9.7×10^2
30	Zn	4.6×10^{-1}	6×10^{-3}	5.1×10^2
31	Ga	1.3×10^{-3}	(3×10^{-4})	9.0×10^3
32	Ge		7×10^{-3}	
33	As	2.3×10^{-2}	2.3×10^{-2}	3.9×10^4
34	Se	2.5×10^{-3}	1.7×10^{-1}	2.6×10^4
35	Br	2.5×10^{-1}	8.4×10^2	1.3×10^8
37	Rb	1.8×10^{-2}	1.4	3.0×10^6
38	Sr	6.9×10^{-1}	8.7×10^1	5.1×10^6
42	Mo	5.2×10^{-3}	1.1×10^{-1}	8.2×10^5

through the industrial activities of humans. Chemical elements can be contained within complex natural organic molecules that are carried by rivers and are not easily assessed. Many chemical species are quickly removed at the saltwater front within estuaries and do not have a chance to take part in the broad oceanic cycle.

Rivers transport the chemical products of weathering to the sea. What processes cause their removal and keep the ocean clean? The dominant removal pattern balancing the river input is from biologic uptake by living organisms, and from adsorption onto the surfaces of the particles that sink to the ocean floor. This rain of particulate matter scavenges the ocean water column, sweeping it clean of reactive compounds as it falls to the ocean floor. The pulse of radioactive material introduced into the atmosphere and the ocean by the nuclear weapons tests of the 1960s can be observed working its way down through the ocean water column by mixing and by this scavenging process (see box entitled *Ocean Chemical Tracers*). Thus the signals of tritium (water) and reactive species such as plutonium become very strongly separated. Ocean scientists have deployed sediment traps, collecting devices in the shape of funnels or cones, in the ocean water column to intercept this rain of particles and decode the signal.

Some elements have very long MRTs and are therefore transported over vast distances, whereas other elements have short MRTs and are fixed in the sediments close to their point of introduction to the sea. The reason appears to lie in their "particle reactivity" or surface chemistry. Each settling particle presents a reaction surface for absorption, and the dominant molecular surface species on the solid is simply bound water. Consequently, those elements that react most strongly with hydroxyl ($-OH$) functional groups are chemically attracted to the particles and are stripped from the ocean waters as the particles sink. The fate of pollutants introduced into the ocean is monitored by studies of their particle reactivity, MRT, and transport by ocean currents.

To estimate reaction rates, we use the series of natural radionuclides and their chain of daughters. Several nuclides have half-lives so long that they remain in the Earth almost 5 billion years after the formation of the solar system. These nuclides include uranium-238 (^{238}U), uranium-235, (^{235}U), and thorium-232 (^{232}Th). With no perturbing process, the decay series of short-lived daughters from these radioactive parents would have reached "secular equilibrium" long ago; for every atom of the parent that decayed, an exact match in decay of the daughter chain would also occur regardless of the half-life. This is the inevitable natural balance, and we could therefore predict the chain of nuclides in nature with great accuracy. If a difference in chemical reactivity between parent and daughter occurs as nature transforms one element into another, the nuclides are transferred to different oceanic reservoirs, and the radioactive clock tells us how fast this transfer occurs. For instance, the parent uranium-238 is quite soluble in seawater and is mixed throughout the ocean, whereas its short-lived daughter thorium-234, born by alpha-particle decay of the parent, reacts with particle surfaces so strongly that it is rapidly transferred to particles and sinks to the ocean floor. The uranium-thorium disequilibrium observed in the ocean is a dramatic example of the powerful effects of particulate scavenging. Using this disequilibrium and the radiochemical clock, we can diagnose the rate of formation and decay of living particles in the ocean.

CHEMISTRY AND LIFE IN THE SEA

The sea teems with life. From the microscopic plankton arises a food chain of amazing diversity that continues all the way to the great whales, the largest animals on Earth. Only the sunlit upper 100 meters or so of the ocean can support photosynthesis (primary productivity), because little light penetrates beyond such depths. The total amount of plant material synthesized annually in the ocean is about the same as that on land, approximately 10^{16} grams of carbon per years. This process is intimately linked to the seasonal cycles of physics and chemistry that drive it. The productivity of the ocean is highly variable. (Figure 12.2 is a composite satellite image of the global ocean, revealing the distribution of chlorophyll and related pigments.)

The acquisition of such satellite ocean color information is an enormous scientific and technical challenge. The sea is far less reflective of light than is the land; it has a much lower albedo. The light "seen" by an orbiting satellite is dominated by the scattering of light back from the atmosphere, and the spectral signature of sunlight scattered back by the oceans to space is a very weak signal superimposed on this. The recovery of the oceanic signature is a technical triumph.

What does the satellite signal show? The northern and southern gyres above approximately 45° latitude are highly productive in the spring and summer, as are the coastal seas; the vast subtropical gyres are blue and lacking in chlorophyll; and a highly productive narrow stripe is observed along the equator, resulting from the upwelling of deep, nutrient-rich water there caused by the trade wind divergence. The shallow continental shelves are highly productive and contain some of the world's great fisheries. What causes such a distribution of life?

In high latitudes, winter cooling and wind-forcing deeply mix the upper layers (200 to 400 meters in depth), bringing the nutrient species phosphate, nitrate, and silicate to the surface; however, these factors also mix phytoplankton cells deeply, and out of the reach of sunlight so that they cannot readily grow. With the onset of spring, seasonal warming stabilizes the upper ocean, sunlight increases, and a thin, stably stratified "mixed layer" of water occurs, providing the physical environment that supports the plankton within the sunlit zone. This leads to rapid phytoplankton growth in a bloom, seen as a chlorophyll signature in the satellite image. This rapid bloom depletes the nutrients, and growth slows as summer progresses. The return of winter completes the cycle. The formation of the ocean mixed layer is of critical importance in the seasonal cycle of ocean physics. It results from

OCEAN CHEMICAL TRACERS

Ocean geochemists have developed marvelous skills to measure the vast array of chemical elements and their compounds in seawater. But how do we deduce, from a simple chemical analysis, critical time-dependent information, reaction rates, the balance of natural forces, the pace and pattern of deep ocean mixing, and so on? The ocean is so large, and the sum of chemical, physical, and biologic forces simultaneously at work so complex that at first this seems an impossible task. The tools that perform these measurement tasks are known as "tracers," and they come in many forms.

To estimate ocean mixing, tracers that are chemically inert, but whose input to the ocean has changed over time, are critical. Examples include the injection of tritium (^{3}H, the radioactive isotope of hydrogen, which has a half-life, $t_{1/2}$, of 12.32 years) as a pulse from nuclear bomb tests, which is carried as the water molecule (see Fig. 12.6, the distribution of tritium in the North Atlantic Ocean). For longer time scales radiocarbon, ^{14}C ($t_{1/2}$ = 5600 years), is used, but with some difficulty. Radiocarbon is produced naturally in our world by cosmic rays smashing into nitrogen atoms as they enter the upper atmosphere. The radiocarbon atoms are rained out on the surface of the ocean. This natural signal was disturbed by the same nuclear bomb tests that produced tritium, and the pulse of bomb radiocarbon is working its way into the ocean deeps, tracing the circulation patterns as it does so. We can visualize these tracers as chemical dyes that "color" the waters they touch, and thus allow us to see the evolution of their signal over the decades as the vast sweep of ocean waters around our planet proceeds. The bomb signatures will not be reproduced, for the United States has wisely recognized the environmental damaged caused by atmospheric testing of nuclear weapons.

The production of other tracers, particularly the chlorofluorocarbons used as refrigerant gases until recently (Chap. 13), provides another label. The rising atmospheric levels of these compounds are carefully documented; through gas transfer at the sea surface, seawater becomes tagged with the gaseous atmospheric signature last seen before becoming entrained in the deep ocean flows.

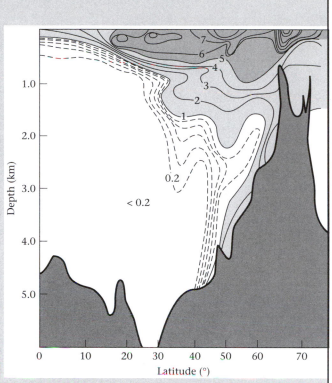

Figure 12.6. Tritium distribution in the western Atlantic Ocean. km = kilometers.

By measuring the chemical tracer content of deep waters, we can obtain a remarkably good picture of their age and origin. Here either the radiochemical clock, or the measured march of our industrial society, are put to good use.

---

the balance of input of heat from solar radiation, which stabilizes the warm upper layer, and the input of turbulent energy by wind, which mixes the heat downward. The mixed layer depth even varies with day and night.

The chemical signature of phytoplankton growth is conventionally given by an equation of the form:

$$106CO_2 + 16NO_3^- + HPO_4^{2-} + 122H_2O + 18H^+ \rightarrow \qquad (2)$$
$$(CH_2O)106(NH_3)16(H_3PO_4) + 138O_2$$

which represents photosynthetic production of organic matter from carbon dioxide and the nutrient elements, nitrogen and phosphorous. The form of this equation, first provided by the oceanographer Alfred Redfield in 1940, is called the Redfield equation; and the proportions of the chemical elements involved are called the Redfield ratios. Although various improvements to this basic form have been made over the years, either by adding more chemical species or by changing the numerical coefficients, the basic equation remains valid on a global scale. This constancy of the Redfield

ratios indicates the existence of a rain of fast-sinking particles to the ocean floor, not a slow drizzle of fine material that would take years to sink. The rapid grazing of all particles by organisms explains why the sand and dust blown into the Atlantic Ocean from the Sahara Desert is laid down as a narrow stripe on the ocean floor, not as a diffuse band.

Carbon dioxide is never in limited supply in seawater because the presence of large quantities of bicarbonate ion provides a vast reservoir of oceanic carbon dioxide for plants to draw on. However, nitrate and phosphate are present in trace quantities, and present very real limitations for plant growth. It is commonly believed that the broad chlorophyll signature revealed in Figure 12.2 is driven by the ready availability of nitrate at high latitudes through replenishment by seasonal mixing, and that the vast, warm, blue subtropical gyres are oceanic deserts resulting from the inhibition of the seasons and the lack of nutrients. The narrow green stripe at the equator is caused by the unique physical forces acting there that produce upwelling of deep water (Chap. 10).

Very recently, this view has been challenged by the work of the late John Martin and colleagues, who realized that other trace species can also exert control over oceanic fertility. In many areas, excess phosphate and nitrate, but little chlorophyll, are present in surface waters. The question of why the plants do not grow well there has been a long-standing geochemical puzzle. The trace element iron is essential to plant growth; however, it is so insoluble in seawater that only nanomolar quantities are present. In elegant laboratory experiments, Martin showed that plant growth is likely to be stimulated in many oceanic regions presumed to be "deserts" by the addition of minute amounts of dissolved iron. He hypothesized that the input of iron-containing dust from wind, or other sources, could have a major impact on plant growth. His colleagues proved this by direct experiment shortly after his death. It is possible to turn blue water dramatically green in many areas by stimulating plant growth with trace additions of iron.

The biogeochemical gases carbon dioxide and oxygen are important parameters of seawater. If we consider seawater as a simple solution, then the solubility of gases, and atmospheric equilibrium values, may be determined in the laboratory very accurately as a condition of temperature, salinity, and pressure (e.g., seawater at 24°C would contain 5.0 milliliters of oxygen per liter at atmospheric equilibrium). We can use these data as a starting point for examining the implications of the Redfield equation. In photosynthesis, oxygen is produced, and the sea surface very typically has oxygen values slightly above those predicted by atmospheric equilibrium, resulting in a transfer of oxygen from sea to air. In some cases, where the excess oxygen is trapped by a stable layer and prevented from escaping to the atmosphere, the excess may be several percent over equilibrium. Calculations on the oceanic oxygen budget provide a powerful constraint on estimates of photosynthetic activity. In equation (2), oxygen is produced and carbon dioxide is removed. Here the chemistry is far more complex.

The lifetime of ocean phyoplankton is short, with a doubling time every one or two days, and the rate of grazing by higher organisms is such that the material is rapidly devoured. So efficient is this process that even inorganic particles are rapidly scavenged and sent sinking into the abyss. Approximately 90 percent of the organic material devoured is utilized for energy and is efficiently returned the reverse of equation (2) as dissolved carbon dioxide, nitrate, phosphate, and so forth to the oceanic water column in the upper few hundred meters. Only approximately 10 percent remains to sink to the ocean floor.

Sediment traps deployed by oceanographers intercept this material and decode its chemical signature. This sediment is therefore quantitatively well characterized. As we might expect, the rain of material is linked to the pattern revealed by the satellite image of Figure 12.2. Beneath zones of high productivity, a high settling rate of particles occurs, and the flux observed decreases exponentially with depth as sinking material is intercepted and utilized by marine animals and bacteria throughout the oceanic water column.

The most marked result of this respiratory activity is the strong depletion of oxygen and the equivalent production of carbon dioxide, nitrate, and phosphate that occurs below the euphotic zone. The mid-waters of the ocean are characterized by a strong oxygen minimum. In some isolated basins around the world (such as the Black Sea) this depletion is so strong that bacteria turn to other chemical means to oxidize the organic matter they need to sustain life, for example, by seeking to use the next most readily available oxygen atoms, those bound to nitrogen as nitrate, and those bound to sulfur. The reduction of sulfate (SO_4^{2-}) ion to hydrogen sulfide gas then occurs, with radical chemical consequences, and a highly reducing system is created. This process is common in marine sediments. The rotten egg smell of many organic-rich muds is quite familiar; however, the large-scale circulation of the oceanic water column is poised today just above this chemical threshold. In contrast to the Atlantic Ocean, which is relatively well oxygenated, large areas of the mid-water depths in the Indian and North Pacific oceans have very small traces of oxygen and are close to becoming reducing. The distribution of dissolved oxygen tells us much about the pattern and time scale of ocean circulation. Because the ocean is nowhere sterile, respiration proceeds at all places and at all times, although the consumption of oxygen is very slow in the abyss.

THE DEEP CIRCULATION

How are the links maintained between the cycle of photosynthesis at the surface and decay and respiration at depth? The link is provided by ocean circulation a continuous loop that is in constant motion similar to a conveyor belt. The ocean is in constant motion on all scales, from the molecular to waves hundreds of kilometers long that take years to cross an ocean basin. A profile of deep-ocean water sampled at the equator reveals that most of the oceanic water column is cold and of polar origin as a result of the thermohaline circulation (Chap. 11).

The forces of wind, the rotation of the Earth, and the exchange of heat and salt drive the ocean circulation. Heated air in the tropics initiates convection in the atmosphere; as these giant convection cells move poleward, the force of Earth rotation imparts angular momentum. The result is the easterly pattern of the trade winds at approximately 20° latitude and the strong balancing westerly patterns at approximately 40 to 50° latitude that occur in both hemispheres (Chap. 10). The ocean surface waters are blown by the winds, but not in a straight line along the direction of flow. Instead, the force of Earth rotation, the Coriolis force, acts to deflect the water to the right of the wind direction in the Northern Hemisphere, and to the left in the Southern Hemisphere. This Northern Hemisphere motion to the right is transferred sequentially with depth, and the net result is that, on average, the upper layers of the water column will move at right angles to the direction of the wind. Wind blowing along the length of a coastline will cause water to move at right angles to the shore and can cause upwelling of colder, deeper water.

The wind-driven circulation dominates the ocean flows; however, the pattern of deep circulation results from the thermohaline forcing. Intense winter cooling and ice formation in regions such as the Norwegian, Greenland, and Labrador seas results in the creation of cold, salty, dense water masses that sink to great depths. The North Atlantic Deep Water thus formed by modification of the density of surface waters is compelled by planetary physics to flow south as a deep current in the western Atlantic, below the Gulf Stream, along the deep boundary of the South American continent, and to upwell in the Antarctic Ocean where it forms part of the great Southern Ocean westerly circulation. Winter cooling there also forms dense Antarctic waters, which sink and may be traced in the deep circulation of all three major ocean basins. The size of the Pacific Ocean makes this reservoir dominant, and the MRT of the deep-ocean circulation is dominated by abyssal Pacific flows. As this deep circulation progresses, the density of the bottom waters is reduced by gradual mixing with waters above; as a result of mixing, the abyssal Pacific waters gradually lose their identity and become entrained in the shallower circulations of Intermediate Waters. This mass eventually follows a return path from the Pacific Ocean via surface flow of water through the Indonesian Archipelago into the Indian Ocean, around South Africa and into the surface waters of the South Atlantic for return flow northward to complete the tour. Figure 12.7 portrays this circulation in schematic form (see also Fig. 11.10).

What is the time scale for this planetary tour? The ocean is too large, the circulation is too slow, and there are too many dynamic complexities for simple direct measurement; the problem of measuring ocean circulation is one of the most difficult in geophysics. The tools used by geochemists are important in deciphering this problem; they include the natural radioisotope carbon-14 (^{14}C), the synthetic tracers such as radiocarbon and tritium introduced by nuclear weapons tests, and chlorofluorocarbons from industrial activity (see the box entitled "Ocean Chemical Tracers"). We cannot "date" the ocean waters as one would a piece of wood that has been buried on land and remained as a single unit, because the constant mixing of complex water types of different regional origins and "ages" prevents this. Nonetheless, great progress has been made by combining geochemical data and physical models. From these tracers, we estimate that the time taken for North Atlantic Deep Water to transit from its formation point to its exposure in the Antarctic

Figure 12.7. Ocean circulation of cold, salty, deep current and warm, less salty, shallow current. (*Source:* On the ocean circulation, Volume II, WHOI Technical Report WHOI-96-08, Woods Hole Oceanographic Institution; photograph by Bill Schmitz, 1996.)

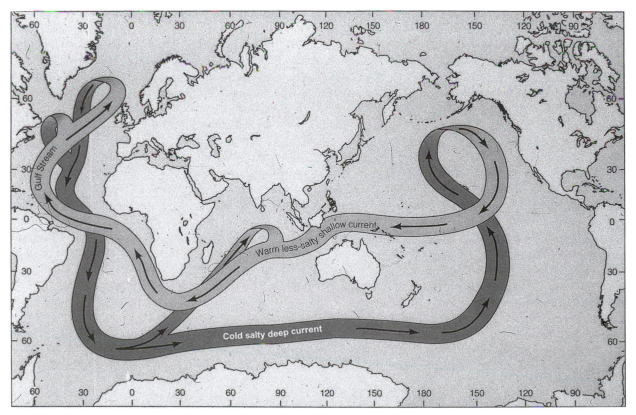

westerly circulation is approximately 250 years, and that the mean age for a molecule of deep-ocean water, before returning to the sunlight, is approximately 550 years. This signal is dominated by the vast Pacific Ocean, and portions of the North Pacific deep waters may have been isolated from the surface for almost 1000 years.

Today we are learning a great deal about ocean circulation from the deployment of satellites that measure winds and the height of the sea surface, subsurface drifters that are balanced to float in the mid-waters and regularly reveal their position with a pulse of sound, and large-scale physical and geochemical surveys combined with numerical models.

THE DEEP-OCEAN GEOCHEMICAL CYCLE. By combining our knowledge of chemistry and the physics of ocean circulation, we can piece together a fairly complete picture of the processes at work in the deep-ocean geochemical cycle. A very simple example would be to consider a parcel of seawater in the surface of the North Atlantic, rich in nutrients from winter mixing. In spring, an intense bloom of phytoplankton will occur, lowering the partial pressure of carbon dioxide gas, producing oxygen, and depleting the nutrients to very low levels. The grazing animals will consume this material, which sinks rapidly as waste-product particles. The water at depth, beginning its abyssal tour, and initialized with the wintertime signal at the point of its origin, will be exposed to a constant rain of organic matter that depletes its oxygen reserves and builds up its nutrient and carbon dioxide contents. As the circulation progresses, this geochemical aging is played out on a vast scale. Although the rate of deep-ocean oxygen consumption is slow and decreases exponentially with depth, it is always positive, because the sea is nowhere lifeless. The parallel buildup of carbon dioxide from deep respiration slowly makes the deep water more acidic, consuming carbonate ion and forming bicarbonate, as in equation (1).

The organic matter of equation (2) is by no means the only biologic product of the upper ocean. In addition to organic tissue, calcium carbonate ($CaCO_3$) and opaline silica (SiO_2) are produced as the skeletons, or tests, of marine organisms. This has enormous consequences, because approximately half of all ocean sediments are made up of the opal and calcite remains of these skeletons; yet their distribution is quite different and reflects their origins, accumulation rate, and sensitivity to dissolution. From the species of organisms recorded in the sediments, and from their isotopic signatures, we can read the record of past climates.

Seawater is everywhere undersaturated with respect to opal, and the deep ocean is strongly enriched in dissolved silicate, $Si(OH)_4$, as a consequence. Therefore, organisms, such as diatoms and radiolarians, that create a siliceous skeleton, do so against the chemical gradient. When these skeletons sink, they immediately start to dissolve. Once this material reaches the ocean floor, it is a race – dissolution rate versus burial rate by accumulating sediment – to see if the opal record can be preserved. Equilibrium saturation with the surrounding pore water prohibits further dissolution once the material is in the sediment. We find opal accumulation in the Antarctic, and under the Equatorial Pacific upwelling region where these conditions are met.

In contrast, surface seawater is supersaturated with respect to the biogenic carbonate minerals calcite (made by foraminifera and coccolithophores) and aragonite (made by pteropods). These minerals do not spontaneously precipitate inorganically because magnesium ions blocks the reaction; however, once formed by organisms, they are stable. Calcium carbonate is unusual in that it is more soluble in cold water; as the calcareous skeletons, or tests, sink, they are exposed to water that is progressively colder and depleted in carbonate ion due to the release of carbon dioxide from the abyssal respiration. The calcium ion content of seawater is very nearly fixed, so that the control lies almost totally with the carbonate system. At some point the decrease in the ion product, $[Ca^{2+}] \times [CO_3^{2-}]$, is such that the solubility limit is reached, and the tests begin to dissolve. Although the mineral phase aragonite is more soluble than calcite and dissolves at shallower depths, the principles are the same. The range of variability of the total carbon dioxide content of seawater is only approximately 20 percent; however, the small fraction of the total dissolved carbon dioxide that is carbonate ion ($CO_3^=$) can vary by approximately a factor of 6. Uncovering the chemical signature of the carbon dioxide system in seawater is critical to solving this puzzle.

The result is that the deep distribution of calcite in sediments reflects topography. Carbonate sediments are found draping ridge crests, and on continental shelves, and are absent from the floor of the deep basins. Moreover, when we piece this together with the deep circulation, it is clear that carbonate sediments are more readily preserved in the North Atlantic, where younger waters are relatively rich in carbonate ion, than in the older North Pacific, where the relentless process of carbon dioxide addition from respiration has depleted the carbonate ion content.

THE IMPACT OF HUMANS. The ocean is so large and its chemistry is so robust, we may be tempted to think that it is beyond the capacity of humans to change it. This is not so. We can recognize human activities in many ways: through the presence of trace quantities of exotic radionuclides such as plutonium-239, cesium-137, strontium-90, and tritium from nuclear programs, through the presence of chlorofluorocarbons, and dominantly through the buildup of carbon dioxide from the burning of fossil fuels.

The ocean is a vast reservoir of carbon. If we sum all components, approximately 98.5 percent of the ocean–atmosphere carbon inventory is in the ocean. Over the last century or so we have, through our industrial activities, raised the atmospheric carbon dioxide levels from approximately 280 parts per million to approximately 360 parts per million today. Left long enough, approximately 85 percent of this excess carbon dioxide will eventually be transferred to the ocean through gas exchange and reaction with carbonate

TABLE 12.3. CALCULATION OF THE CO_2 SYSTEM PROPERTIES OF SURFACE SEAWATER AT CONSTANT ALKALINITY AND TEMPERATURE FOR CHANGING PCO_2 (IN UNITS OF MILLIMOLES PER KILOGRAM, MMOL/KG)

	pCO_2 = 280 parts per million (Preindustrial value)	pCO_2 = 355 parts per million (1996 value)
TCO_2	2000.	2048
pH	8.188	8.103
$CO_3^=$	214	183
HCO_3^-	1776	1851
CO_{2gas}	10	13

Note: Alkalinity = 2300 mmol/kg; salinity = 35.000; temperature = 15°C. Over the past century or so we have consumed approximately 31 mmol/kg of carbonate ion in surface sea water by reaction with invading industrial CO_2, or approximately a 15 percent decline, and we have lowered the pH by approximately 0.05 pH units.

ion in the surface ocean. The reason that the transfer will not be closer to 100 percent lies in the fact that a change in the carbonate ion concentration occurs with the addition of carbon dioxide, and, by Le Chatelier's principle, the system will resist change. The effect was first described by Roger Revelle and Hans Suess in 1957, and is known as the Revelle (R) factor. It relates the fractional change in partial pressure of carbon dioxide gas (the atmospheric signal) to the equivalent fractional change in the total dissolved carbon dioxide content of seawater. The numerical value varies with temperature; the oceanic average today is approximately 10. This means that a large perturbation in the chemistry of the atmosphere (pCO_2) will generate only a small change in the total carbon dioxide content of the ocean.

The rate of turnover of the oceans is slow, the vast reservoir of carbonate ion is in the deep ocean, and the chemistry of the carbonate system contains many pitfalls. What is occurring today? From our knowledge of the oceanic circulation, from the natural cycle of carbon by plants, and from the basic chemistry of the oceanic carbonate system we can place the rise in atmospheric carbon dioxide in context. The increase in atmospheric carbon dioxide levels of approximately 75 parts per million has forced the surface ocean toward equilibrium with this value, raising the dissolved carbon dioxide gas content and consuming carbonate ion. A specimen calculation is given in Table 12.3.

What will happen next? The slow times inherent in gas exchange and the ocean carbonate system have lead to an imbalance, so that the ocean lags behind the atmosphere by approximately eight years. The slow rate of mixing of the large-scale thermohaline circulation precludes the involvement of the vast deep-ocean carbonate reservoir on other

than century time scales, and the rapid annual overturn of the surface layers down to a few hundred meters is the dominant short-term control. The carbon dioxide inventory in the upper 300 to 400 meters of the ocean is rapidly building up, and the excess carbon dioxide content is now approximately 48 millimoles per kilogram – an impressive signal. Over long time scales the calcium carbonate contained in the ocean sediments will react with the newly acidic water and dissolve, restoring the carbonate ion capacity. This chemical invasion of the deeps will take place first in the North Atlantic Ocean.

QUESTIONS

1. Discuss the factors influencing the chemistry of seawater.

2. What evidence indicates the compositional variation of ocean waters over geologic time?

3. Trace the geochemical pathways of calcium (Ca), sodium (Na), and phosphorus (P).

4. How does the composition of seawater influence the marine biota?

FURTHER READING

Broecker, W. S., and Peng, T. H. 1982. Tracers in the sea. Palisades, NY: Eldigio Press.

Millero, F. J., and Sohn, M. L. 1992. Chemical oceanography. Boca Raton FL: CRC Press.

Turekian, K. K. 1976. Oceans, 2nd edition. Englewood Cliffs, NJ: Prentice-Hall.

THE ATMOSPHERE

13 Atmospheric Composition, Mixing, and Ozone Destruction

ROBERT CHATFIELD

DYNAMICS AND CHEMISTRY OF THE AIR

Again and again the news reports the extent to which our Earth's air has been degraded by a variety of threats, for example, pollution, acid rain, ozone destruction, and greenhouse warming. Three basic physical observations have strong links to our air's seeming fragility among the many systems involved and affected by global change: (1) Air is clear, (2) air is thin, and (3) air is compressible. Several qualitative details and quantitative expressions of these obvious physical facts help explain the distinctive layering of our Earth's atmosphere, which allow some regions to become partially isolated and subject to unique threats. Two more subtle observations are that: (4) Air has been degraded by a variety of threats, for example, pollution, acid rain, ozone oxidation; and (5) air is significantly out of chemical balance with itself and the Earth's surface. These latter reflect the destruction of protective gas species, and the incremental addition of greenhouse molecules. The transparency, thinness, and compressibility of the terrestrial atmosphere are strongly linked to the air's seeming fragility among the many systems involved. Over a short period of geologic time, its components react with themselves and with the Earth's surface elements; the current atmospheric composition seems very unstable. The reason for the seeming imbalance is that air is continually being reconstituted as part of a living global system.

The deep links of these physical and chemical basics with the global dangers that face the human population are a major part of this chapter. This chapter and Chapter 14 describe the Earth's atmosphere, composition, temperature, and motions, as well as threats to its livability. We end with a discussion of atmospheric ozone and an introduction to the ways that the natural balance of ozone concentrations are threatened by chemical compounds released with changing industrial and agricultural activities.

This chapter first examines the composition and properties of clean air in our unique, living planet. Next, we explore how air mixes vertically, and we identify the origin of some local urban pollution problems, where degradation is often visible. We then discuss the global pollution problem of ozone destruction.

ATMOSPHERIC COMPOSITION

MAJOR COMPONENTS OF DRY AIR. Any discussion of global changes in the atmosphere must begin with its remarkable composition. Chapters 19, 20, and 21 explore how the chemical composition of our present atmosphere reflects the immense effect life has had on the whole planet. A major concern is the fact that the exceptional clarity of the atmosphere's major constituents allows climate changes to be caused by relatively small perturbations. The perturbations are due to changes in human-generated pollution and biologic change. The Earth's atmosphere is composed of 78.1 percent molecular nitrogen, N_2, and 20.9 percent molecular oxygen, O_2 – compounds that are much less abundant on other planets (Fig. 13.1). These percentages are extremely constant, as is the nearly 1 percent of the air that is argon (Ar). These proportions are the percentages of the gaseous species *molecules* to the total of all atmospheric molecules. This means, for example, that there are 41.8 times more oxygen *atoms* in a cubic centimeter of air than there are argon atoms.

Earth's high oxygen level is a global consequence of plant photosynthesis; oxygen is consumed by the respiration of animals, plants, and microbes. A near balance is maintained. A small portion of oxygen is consumed by burning. Episodic large-scale grass and forest fires in the drier seasons have been part of Earth's history until suppression measures became common, primarily within the past century. The 20.8 percent of molecules of oxygen is very high, and makes the atmosphere quite oxidizing. Indeed, at some level between 30 and 40 percent the ease of starting fires among the Earth's plant materials increases dramatically, although the practical upper limit for a world such as the one we know is not yet clear. Recent evidence points to high levels of oxygen, perhaps 32 percent during the Cretaceous Period, when dino-

W. G. Ernst (ed.), *Earth Systems: Processes and Issues.* Printed in the United States of America. Copyright © 2000 Cambridge University Press. All rights reserved.

(A)

(B)

saurs roamed the Earth. The very high supply of oxygen to tissues might help explain the high-energy metabolism necessary for rapid movement of the bulky predatory meat eaters.

Typically we exclude water molecules, H_2O, in the accounting mentioned because water varies considerably. That is, it ranges up to 3 to 4 percent over the warm tropical rain forests to 0.01 percent over the colder poles, and down to 0.0006 percent, or 6 parts per million in the ozone layer beginning 15 kilometers overhead. (Including the water in the other proportions quoted above would unnecessarily disturb the apparent constancy by a few percent.) Water vapor is constantly involved in the cycle of evaporation, condensation, and precipitation. Besides water, a host of minor or trace atmospheric species are described in Table 13.1. What

is remarkable about this long list is not only that many species are present in several parts in 10^6 (parts per million), several parts in 10^9 (parts per billion), or several parts in 10^{12} (parts per trillion), but also that these species are listed as important greenhouse gases (*g*). In other words, very tiny amounts perceptibly affect the Earth's radiation budget.

WATER VARIABILITY AND CLOUD SEEDING. The H_2O content of the atmosphere is a sensitive function of temperature, as Figure 13.2 shows. At each temperature, an air parcel can just hold so much water vapor. Below that point, water molecules are energetically more stable together as a liquid. Imagine taking free air and progressively cooling it. If air is cooled even a few tenths of a degree below a certain temperature, the water molecules will condense together into liquid

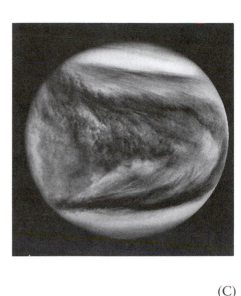

(C)

Figure 13.1. The atmospheres of (A) Earth, (B) Mars, and (C) Venus differ immensely and determine the surface conditions and landscape. Only the atmosphere of Earth is greatly out of chemical equilibrium and is conducive to life. This disequilibrium is a signal of life that is probably detectable to any potential observers nearby in our galaxy. (*Source:* All photographs, with the exception of (A), courtesy of NASA. (A) courtesy of Catherine Flack.) Atmospheric pressures (proportions) are: Earth, 1 bar (N_2 = 77%; O_2 = 21%; Ar = 1%; H_2O ~ 1%); Mars, 0.007 bars (CO_2 = 95%; N_2 = 2.7%; Ar = 1.6%); Venus, 90 bars (CO_2 = 96%; N_2 = 3.5%). See also the color plate section.

droplets rather than remaining spread throughout the gas parcel. This lower temperature at which water forms is also a property of the air, besides the ambient temperature, and is called the dew point temperature of the air. It is a nighttime cooling process similar to this that forms dew at the Earth's cold surface, hence the term *dew point.*

The rapid increase – in fact an exponential increase – of air's holding capacity for water vapor is shown in Figure 13.2. The dotted line below 0°C shows the trace of the dew point for droplets that cannot nucleate to form ice. The vertical arrows show the capacity to precipitate water into ice particles by natural or artificial ice nuclei. In this discussion, we have been thinking about H_2O coming out of the atmosphere to form liquid. The same physical law also describes the evaporation of water vapor into the air from the sea surface. The rapid increase of H_2O in the atmosphere that corresponds to a warming ocean temperature is an extremely important aspect of the theory of global average temperatures and global warming. We return to the exponentially rapid dependence of the H_2O vapor equilibrium on temperature in Chapter 14, where we consider these issues.

At an even colder temperature, the most stable form is ice, a solid. In the air, the process of condensation to liquid in clouds always takes place on tiny particles, appropriately

called *cloud condensation nuclei.* It is apparently difficult for molecules to form into well-ordered ice crystals; consequently, it is common for air parcels to cool 10 to 40°C below the ice formation or "frost point," with clouds still composed of liquid water. Water at the Earth's surface has no such inhibition: Some feature or portion of the solid surface apparently helps set the pattern for the ice-crystal-ordering process in the liquid at the frost point.

For decades it has seemed obvious that humans could dramatically alter the atmosphere by introducing tiny particles – artificial *ice nuclei* – to initiate the ice-forming process. Natural ice nuclei are much less prevalent than ordinary cloud condensation nuclei for water droplets. Wintertime cloud seeding with silver iodide sprayed from airplanes has been tried repeatedly, in hopes of producing more snow and, consequently, more springtime water runoff. Although the statistics of cloud-seeding efficacy have never been extremely convincing, seeding programs continue. Cloud-seeders make a telling point that is heard often in environmental debates. Although there is great uncertainty, we should spend the money to do something; the potential economic impact, even weighted by the likelihood of the effect, still justifies the expense of taking action! In western areas of North America, this argument, with its intent to bring about environmental change, has often succeeded.

AN ATMOSPHERE FOR LIFE. The Earth's atmosphere is somewhat anomalous among its sister planets in that it is composed of gases that readily react with each other and with the elements of the surface over geologic time. Additionally, the major constituents are simple, symmetric two-atom molecules such as oxygen and nitrogen or, even simpler, atoms such as argon. The energy-absorbing and chemical reaction capacities of gases are complex but typi-

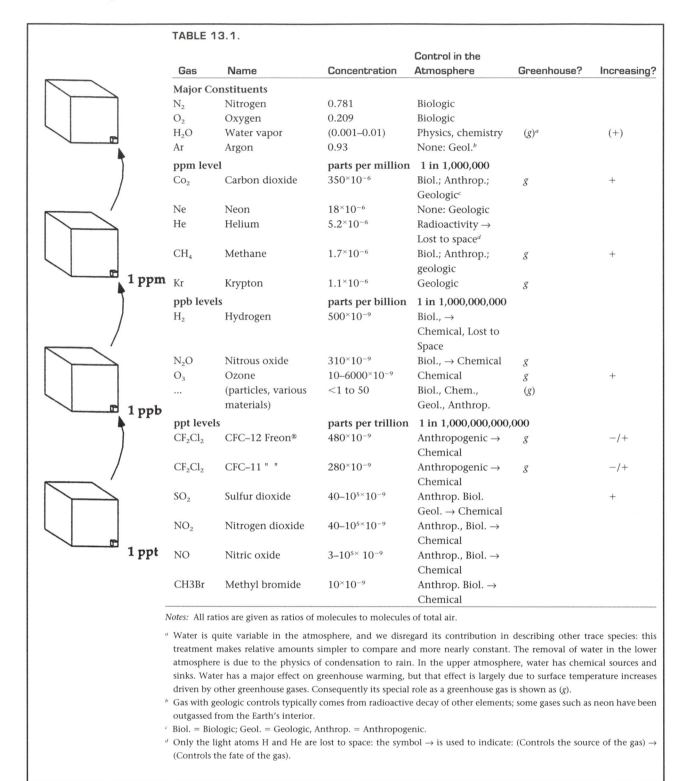

TABLE 13.1.

Gas	Name	Concentration	Control in the Atmosphere	Greenhouse?	Increasing?
Major Constituents					
N_2	Nitrogen	0.781	Biologic		
O_2	Oxygen	0.209	Biologic		
H_2O	Water vapor	(0.001–0.01)	Physics, chemistry	$(g)^a$	(+)
Ar	Argon	0.93	None: Geol.[b]		
ppm level		**parts per million**	**1 in 1,000,000**		
Co_2	Carbon dioxide	350×10^{-6}	Biol.; Anthrop.; Geologic[c]	g	+
Ne	Neon	18×10^{-6}	None: Geologic		
He	Helium	5.2×10^{-6}	Radioactivity $\rightarrow$ Lost to space[d]		
CH_4	Methane	1.7×10^{-6}	Biol.; Anthrop.; geologic	g	+
Kr	Krypton	1.1×10^{-6}	Geologic	g	
ppb levels		**parts per billion**	**1 in 1,000,000,000**		
H_2	Hydrogen	500×10^{-9}	Biol., $\rightarrow$ Chemical, Lost to Space		
N_2O	Nitrous oxide	310×10^{-9}	Biol., $\rightarrow$ Chemical	g	
O_3	Ozone	$10–6000 \times 10^{-9}$	Chemical	g	+
...	(particles, various materials)	<1 to 50	Biol., Chem., Geol., Anthrop.	(g)	
ppt levels		**parts per trillion**	**1 in 1,000,000,000,000**		
CF_2Cl_2	CFC–12 Freon®	480×10^{-9}	Anthropogenic $\rightarrow$ Chemical	g	–/+
CF_2Cl_2	CFC–11 " "	280×10^{-9}	Anthropogenic $\rightarrow$ Chemical	g	–/+
SO_2	Sulfur dioxide	$40–10^5 \times 10^{-9}$	Anthrop. Biol. Geol. $\rightarrow$ Chemical		+
NO_2	Nitrogen dioxide	$40–10^5 \times 10^{-9}$	Anthrop., Biol. $\rightarrow$ Chemical		
NO	Nitric oxide	$3–10^5 \times 10^{-9}$	Anthrop., Biol. $\rightarrow$ Chemical		
CH3Br	Methyl bromide	10×10^{-9}	Anthrop. Biol. $\rightarrow$ Chemical		

Notes: All ratios are given as ratios of molecules to molecules of total air.

[a] Water is quite variable in the atmosphere, and we disregard its contribution in describing other trace species: this treatment makes relative amounts simpler to compare and more nearly constant. The removal of water in the lower atmosphere is due to the physics of condensation to rain. In the upper atmosphere, water has chemical sources and sinks. Water has a major effect on greenhouse warming, but that effect is largely due to surface temperature increases driven by other greenhouse gases. Consequently its special role as a greenhouse gas is shown as (g).

[b] Gas with geologic controls typically comes from radioactive decay of other elements; some gases such as neon have been outgassed from the Earth's interior.

[c] Biol. = Biologic; Geol. = Geologic, Anthrop. = Anthropogenic.

[d] Only the light atoms H and He are lost to space: the symbol $\rightarrow$ is used to indicate: (Controls the source of the gas) $\rightarrow$ (Controls the fate of the gas).

cally increase with the number of constituent atoms and the bonds between atoms of a different nature (specifically, the polarity of the atomic bonds). These one-atom and two-atom gases cannot absorb the Sun's light, the radiation that heats the Earth system: Typically, gas-phase molecules can absorb the Sun's energetic radiation only by becoming chemically (electronically) excited, consequently possessing chemical energy. Even the minor species listed in Table 13.1 absorb little of the Sun's energy; some absorb efficiently per molecule but are simply too dilute. A major exception is ozone, O_3, which is described later in the section on "Vertical Mixing and Layering."

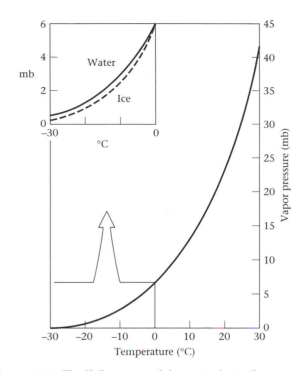

Figure 13.2. The H_2O content of the air is physically measured by its vapor pressure, that portion of the pressure exerted by all air molecules that is due to only the water molecules. (See the box entitled "Pressure, Density, and Temperature Relationships" to see how vapor pressures can measure the amount of gas-phase H_2O.) However, the presence of water in the atmosphere is limited; above a certain partial pressure, water molecules tend to stay condensed as liquid or ice. The curve shows the maximum or *saturation vapor pressure* of water for any given temperature. The region below the curve represents possible amounts of gas-phase H_2O in the atmosphere. A few light lines trace the relative humidity for differing temperatures. When the temperature drops for an isolated parcel of air, its *relative humidity* goes up. Below freezing at 0°C, there are two curves, representing saturation with respect to ice crystals and saturation with respect to supercooled liquid water droplets. These droplets can be induced to freeze by natural or artificial ice nuclei, as the text indicates. (*Source:* Leslie F. Musk, *Weather systems*, Cambridge University Press, 1988.)

There are many absorbers of various colors of light at the Earth's surface. For example, the green areas of our planet as seen from space indicate the pervasive activity of chlorophyll. Besides chlorophyll, the rich variety of differing chemical bonds here is largely responsible for this effect. The clarity of the atmosphere is excellent for transmitting the radiation energy for life to the planetary surface, and life is correspondingly known to provide the chemical energy that replenishes this kind of atmosphere.

The Earth's atmosphere is clearly very conductive of light energy, in that it allows radiation through to the surface. Consequently, a primary aspect of advanced life, the capture of light and its temporary storage as chemical energy, can be carried out easily using the blue and red portions of the spectrum, and reflecting the green. Part of this photosynthetic capture of chemical energy is the production of both food (carbon-hydrogen compounds) and its necessary counterpart (oxygen). This process makes the whole atmosphere – Earth system "alive" in the sense of being very self-reactive and temporary – unless there is life present. No wonder the search for a "living" atmosphere on remote planets, perhaps in other solar systems, is primarily the search for signs of free, reactive oxygen. Finding oxygen in light affected by distant planets is most easily accomplished by looking not for ordinary oxygen, O_2, but for ozone, O_3, for reasons we now discuss.

MINOR CONSTITUENTS, INCLUDING GREENHOUSE GASES. The Earth gains energy primarily in the visible wavelengths from the Sun, and energy exits the Earth system in the less energetic infrared or thermal wavelengths of light. (More details of this process are provided in Chap. 14, where we examine the temperature of the Earth system and the greenhouse effect.) The major atmospheric gases such as oxygen, nitrogen, and argon are extraordinarily clear: They transmit essentially all the sunlight in and allow most of the thermal radiation to escape. Not all gases are so clear: In Table 13.1, which provides a list of the minor gas species, many gases are marked with a *g*, indicating that they interact significantly with the Earth's heat radiation and are "greenhouse gases."

Complex molecules such as methane, nitrous oxide, ozone, and the fluorocarbons are significant greenhouse gases, even at levels of 1 part per million and 1 part per billion. The next most prevalent complex molecule, carbon dioxide, CO_2 is commonly considered Earth's main greenhouse gas. The salient fact for greenhouse activity is that gas molecules can interact with electromagnetic radiation because they are composed of different atoms, each atom holding onto electrons more or less tightly. Electromagnetic radiation interacts strongly with these small charge separations (technically, electric dipole moments). The shear number of atoms, bonds, and geometric structure possibilities in these more complex molecules also helps. For this reason, the many-atom chlorofluorocarbon compound $CFCl_3$, present at less than 1 part in 10^9, is a consequential greenhouse gas as well as a compound implicated in the destruction of the ozone layer.

All these species tend to mix together, over weeks in the lower atmosphere below 10 kilometers (e.g., in storms) and over decades into the stratosphere. When they are not mixed, and show local maxima and minima, that is an indicator of a source or sink of material that is significant enough to overcome the mixing processes. Some examples of chemical sources are new molecules originating from pollution, volcanoes, lightning, natural emissions from plants, and chemical reactions within the atmosphere.

The components of the atmosphere that interact powerfully with radiation are small, suspended bits of liquids and solids (paticulate matter, otherwise called *aerosols*). Cloud droplets and ice crystals strongly reflect visible light and can both absorb and emit infrared radiation. Only rarely does more than 1 part per 1000 of the atmosphere occur in the form of water droplets; however, those levels are enough to alter substantially the radiative balance. In fact, the lower atmosphere appears to our eyes quite "thick" and, we surmise, polluted, at much lesser amounts. This occurs when more than a few tens of parts per billion of aerosol particles are present. For example, Los Angeles, California, when as few as approximately 50 parts per billion (5×10^{-8}) of the atoms in the air occur as particles (i.e., aerosol droplets or solids), the air appears very smoggy: Under such conditions, the "visual range" rating is less than approximately 8 kilometers. (At this 5-mile range, objects become indistinguishable from the background haze.)

THE GAS LAW AND THE COMPRESSIBILITY OF THE AIR

The behavior of the atmosphere as a gas is well described by the ideal gas equation (see the box entitled *Pressure, Density, and Temperature Relationships*). The ideal gas equation summarizes neatly and quantitatively some of the important basic facts about air, or any gas that is not too compressed: (1) As pressure increases, so does density, and (2) higher temperatures tend to cause the density to decrease rapidly. This ideal gas behavior makes the atmosphere very different from the oceans, although there are some strong similarities. Being fluids, both the atmosphere and ocean are powerful in transporting heat around the climate system. Additionally,

ocean–atmosphere heat exchange and fluid motion forces interact intimately, but each has different, characteristic speeds. This multiply linked interaction of fast (atmosphere) and slow (ocean) components of a system is common and a very applicable description of what is needed for the evolution of complex, indeed "chaotic," interactions in physics. These interactions make the prediction of weather and climate a very challenging physics problem.

The variation of atmospheric density causes the atmosphere to have very different layers. Some of the layers contain motions and chemistry that seem quite exotic – for example, smog layers, the ozone layer, or the layers producing the aurora (in the extremely tenuous ionosphere). Figure 13.3 illustrates several of the key properties of the atmosphere. The plot shows the dramatic fall-off of pressure and density upward; a logarithmic scale is necessary to plot the vertical variation of pressure and density with altitude. If the atmosphere was more similar to the ocean and remained at a constant surface density, such as 0.0012 grams per square centimeter as computed in the box below, the top of the atmosphere would be reached at 8.5 kilometers, just above Mount Everest! In fact, the graph is extended to approximately 100 kilometers, and the "edge of the atmosphere" is typically set somewhere between 100 and 400 kilometers. One practical way of defining that edge and the "beginning of space" is where air drag on satellites becomes so small that their orbits are not greatly affected in a week or a month. The space shuttle orbits at an altitude of approximately 300

Figure 13.3. Plots showing the variation of Earth's density and pressure as a function of altitude. (A) The same line serves to indicate density, which rises to approximately 1.17 kilograms per square meter (kg/m³) at the surface, and also pressure, which reaches approximately 1 atmosphere (or 1.01 × 10⁵ Nt m², equivalent to the weight of 10.1 kilograms per square meter). The compressibility of gases makes this variation extreme. (B) By contrast, the temperature of the atmosphere (which is shown in degrees Celsius and also degrees absolute, or Kelvin), varies considerably, depending primarily on where light energy is absorbed by the air or the planetary surface. km = kilometers.

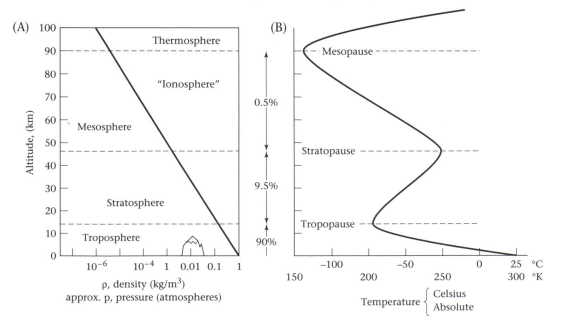

PRESSURE, DENSITY, AND TEMPERATURE RELATIONSHIPS: THE IDEAL GAS EQUATION

For purposes of studying the atmosphere, the ideal gas law can be read as a recipe for finding the density of a gas, n, expressed as the number of molecules per cubic centimeter:

$$n = p\,\frac{N_{Av}}{RT}$$

or

Number density = pressure

$$\frac{N_{Av}}{R \cdot temperature}$$

We may use this when we have measurements of the temperature and pressure. The temperature, T, needs to be on the absolute, Kelvin (K) scale, $T_{Celsius+273^\circ}$. Our application of the gas law is unique, because pressure is the effect of the rest of the atmosphere, around, and above, with no "container" except the Earth's surface. Pressure, p, is measured in hecto-pascals (hPa), which used to be called "millibars" (mb) on weather maps and are equivalent to physicists' pressure units of 10^{+2} newtons per square meter, or pascals. One thousand hecto-pascals or 1 pascal or newton per square meter is approximately what chemists call "1 atmosphere," or 1013 hecto-pascals. Either 1000 or 1013 hecto-pascals is near the middle of the range of pressure found on surface weather maps, which range from 990 to 1030 hecto-pascals. In the equation above, R is the ideal gas constant ($R = 8.31 \times 10^{+4}$ hPa cm^{+3} / K / mol -1) and Avagadro's number is N_{Av}, which allows us to calculate our number of molecules. (Recalling that the number density is only the number of molecules per unit volume, you can recover the more common expression, $PV = NRT$, where N is the number of moles of gas being studied.) Using this formula, we can become more familiar with conditions that are typical at the Earth's surface pressure of 1000 hecto-pascals and 293 K (20°C, 68°F). Substituting, we find that there are approximately $2.47 \times 10^{+19}$ molecules per cubic centimeter of air.

Check the ideal gas equation again and note how it quantitatively summarizes two important basic facts about air: (1) As pressure increases, so does density; and (b) higher temperatures tend to cause the density to decrease rapidly. Also, notice that because there is a simple relation between pressure and density, components of the atmosphere can be measured as partial densities or partial pressures. For example, the U.S. national standard for summertime ozone pollution is commonly expressed as 120×10^{-9} of an atmosphere (a pressure, or 120 parts per billion), although it is legally described as 240 micrograms per cubic meter (a density), figures that sum to essentially the same amount of ozone, at least at normal summertime temperatures.

kilometers, that is, within what can be described as the rarified uppermost thermosphere. As Figure 13.3 shows, atmospheric density and pressure tail off smoothly and rapidly from their surface values down to values less than a millionth their surface magnitudes at 100 kilometers. The box on pressure density describes this close relationship in more detail.

PRESSURE: THE WEIGHT OF THE AIR. What is the density of air in grams per cubic centimeter? We may use an average molecular weight for air of 29 grams per mole, appropriately interpolated between the 28 grams per mole for nitrogen and the 32 grams for oxygen. Hence, the density of air, r, is approximately 0.0012 gram per cubic centimeter, which is approximately one-thousandth the density of water. Compared to the Earth's hydrosphere, the atmosphere seems thin indeed! That low density, 0.0012 gram per cubic centimeter, occurs even though the 1000-hecto-pascal pressure is approximately the same pressure exerted by the weight of 10 meters of water above you. For example, if you swim down to a depth of 10 meters, you have just doubled the pressure on your body, and you can certainly feel it. Still, at the Earth's surface, the atmosphere cannot be compressed by all this pressure to a density of a little more than approximately 1 kilogram per cubic meter; consequently in the atmosphere there is approximately ten times more distance between molecules than in liquid water. One effect mentioned here will be very important in Chapter 14. According to the principles of quantum mechanics, simple gases have a very limited capacity to absorb and emit energy, whereas liquids and solids can often "borrow" from or "pass off" energy to other molecules nearby. The result is that simple gases tend to be "clear" and transmit the Earth's radiation energy in the infrared, whereas the terrestrial surface absorbs and emits heat radiation easily.

This atmospheric pressure at the Earth's surface is the physical expression of gravity, the effect of gravity on the $5.27 \times 10^{+21}$ grams of the Earth's atmosphere overlying its $5.1 \times 10^{+18}$ square centimeters of surface. These are huge numbers that are hard to comprehend; however, they are useful when it is necessary to verify a proposed new effect that is purported to cause or negate global atmospheric change.

THE LAPSE RATE OF TEMPERATURE WITH ALTITUDE. Figures 13.3 and 13.4 both show plots of temperature with altitude. Temperature exhibits considerably greater variations than the density or pressure variations illustrated in Figure 13.3. These temperature variations are used to define different regions such as the troposphere, stratosphere, and so forth. Figure 13.4 also indicates the names of these layers, which are closely tied to the energy balance of the atmosphere. The layers and the temperature fluctuations originate primarily because the atmosphere responds to the input of the Sun's radiant energy in several distinct ways. We begin at the bottom. Most of the Sun's light energy reaches the surface; nearly half of it is ordinary visible light, coming

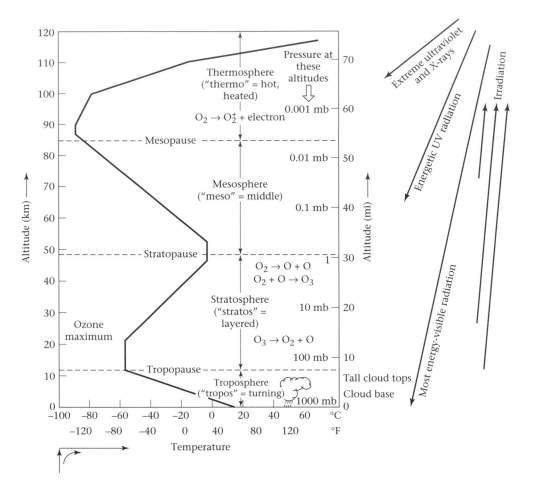

Figure 13.4. An expanded display of the variation of atmospheric temperature with altitude, showing how the temperature maxima correspond to absorption of the Sun's radiative energy, and also how the temperature patterns define the major atmospheric layers or "spheres": the troposphere, stratosphere, mesosphere, and thermosphere. Approximately the lowest kilometer of the troposphere is not named a separate sphere; however, this region, the "planetary boundary layer," exhibits strong effects from the daily cycle of solar cooling and heating. The troposphere above this region generally varies only a degree or so due to the daily cycle of heating; however, it does vary greatly with the effect of passing weather systems. km = kilometers, ml = millibars, mi = miles. (*Source:* Adapted from D. C. Ahrens, *Meteorology today; an introduction to weather, climate, and environment*, West Publishing Company, 5th edition, 1994.)

through a very transparent atmosphere (at least, on a clear day). The effect of this energy is to heat the surface, and that is what causes the daily temperature maximum at the surface. During the day, in sunlit areas all over the world, small air parcels near the surface become heated. Figure 13.5 shows a sample air parcel. As this heating process occurs, according to the ideal gas equation, T increases slightly and the density drops inversely; the low-density or buoyant air thus tends to rise. The air moves immediately away from that warming surface, and little or no energy can enter or escape the air parcel. A glider pilot or a passenger descending to land in an aircraft can feel these near-surface buoyant parcels or "thermals"; they are tens of meters across. As it withdraws from the surface, a parcel must cool; its pressure, density, and temperature must all change. We need a description beyond the density formula of the ideal gas law. What happens can be described by the concept of the conservation of energy:

Heat energy + gravitational energy
$$mc_pT + mgz = \text{constant}$$
$$\downarrow \rightarrow \uparrow$$
$$\begin{pmatrix}\text{mass of}\\\text{parcel}\end{pmatrix}\begin{pmatrix}\text{heat}\\\text{capacity}\end{pmatrix}(\text{temperature}) +$$
$$\begin{pmatrix}\text{mass of}\\\text{parcel}\end{pmatrix}\begin{pmatrix}\text{acceleration}\\\text{of gravity}\end{pmatrix}\begin{pmatrix}\text{height}\\\text{above ground}\end{pmatrix}=\text{constant}$$

The heat energy of the parcel $mc_p T$ is measured in joules of energy. This energy must decrease so that the gravitational potential energy of the parcel can rise (c_p describes the increasing heat energy of a mass of air, m, as temperature is raised, approximately 1003 joules per kilogram Kelvin), and

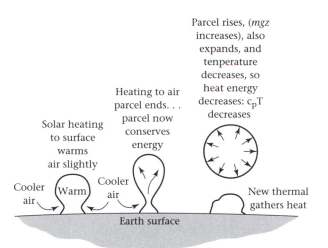

Figure 13.5. A close-up look at a thermal forming in the planetary boundary layer. The surface is heated by the Sun's rays; when the air parcel of the thermal detaches, it no longer can absorb energy. Energy is conserved: The buoyant parcel starts with more thermal energy, mc_pT, than its surroundings, then begins to gather some energy of motion and expand; when it has expanded and risen sufficiently, the motion stops, and all its excess heat energy has been converted to the gravitational energy at its higher altitude, mgz. This tradeoff of heat and gravitational energy is typical of many vertical motions in the atmosphere, and helps define the adiabatic lapse rate. c_pT = heat capacity.

g is the familiar gravitational constant that gives the gravitational force on that air-parcel mass, 9.8 joules per kilogram $\times$ meter, which is equivalent 9.8 meters per second per second). This can happen only if the temperature falls. We can use this formula for conservation of energy to derive an expected drop-off of temperature with altitude. Apply the conservation of energy from immediately prior to the parcel lifting off from the ground (call it position 1) to where it comes to rest at higher altitude (position 2):

$$mc_pT_1 + mgz_1 = mc_pT_2 + mg_2$$

$$T_2 - T_1 = -\frac{g}{c_p}(Z_2 - Z_1)$$

The factor g/c_p describes the 9.8 K drop-off in temperature per kilometer in this equation. Eventually, of course, the parcel can mix with the surrounding air; this basic process helps to establish the character of that surrounding air. The general temperature distribution is the result of many of these mixing processes. That is why the temperature of an air mass tends to decrease away from the surface. This temperature drop-off, or lapse, relationship is called the *adiabatic lapse rate*. The important assumption that the parcel neither receives nor loses a consequential amount of heat energy for the computation is called an *adiabatic assumption*. (*Adiabatic* was coined from Greek forms meaning "not moving through." This is a process in which no energy is moving in or out; i.e., other outside effects are not changing the temperature).

The quantitative relationship is not valid at night or during winter near the surface, because at these times the surface is cooler than the air above, and small air-parcel thermals that mix the air are not initiated. The temperature structure of the atmosphere in these situations can become very complex; however, temperature generally rises away from the surface.

Two more points concerning this conservation argument extend the applicability of this principle described for the adiabatic lapse rate. First, we could have included the energy of motion for a moving parcel of air with very little effect: Typically, the amount of energy associated with motion (so-called kinetic energy) is much less than the amount involved with heat or gravitational energy. Kinetic energy contributions are less than one-hundredth the other terms, and even the difference of terms is affected by less than 10 percent while the parcel is moving. Second, the same kind of argument can be made anywhere in the atmosphere; consequently, portions of the atmosphere freely exchanging in the vertical, without other effects, tend to attain this steep fall off of 9.8 K per kilometer, or approximately 1°C every 100 meters.

Figure 13.6 shows the rapid drop-off of temperature with height in the lower portion of the atmosphere. It also shows the important effect of clouds on the temperature of the atmosphere. Very similar arguments may be advanced concerning the tradeoff of temperature (heat energy) with altitude (gravitational energy), even when a parcel continues rising to the point that water must begin condensing, that is, when the temperature of the parcel must fall below the "dew point." It is a common observation that clouds provide an important effect at the surface by shielding the Earth from the Sun's heat during the day. However, there is another effect within the body of the atmosphere that is more subtle. This phenomenon involves the heat that accompanies evaporation and condensation. A familiar experience is the cooling effect on the surroundings when water evaporates. This cooling absorbs "latent heat" during evaporation. The opposite effect must occur when water condenses. One familiar but painful experience of this effect is the extreme heating when steam burns our skin: Not only is the gaseous water vapor very hot, but as it condenses it releases all the "latent heat" again. This heat-release effect is shown in Figure 13.6. When air parcels ascend into the cloud base, condensation begins, and accompanying it there is some warming of the parcel. The heating is now only "latent"; that is, at a given elevation, this heating results from the condensation of water. The figure shows not a very high temperature, but a change in the cooling of the atmosphere in and around cloud with further altitude. The effect of clouds is so common around the world that the appropriate "lapse rate" of temperature with height describing most of our cloudy lower atmosphere is not 9.8 K per kilometer but only approximately 6 K per kilometer.

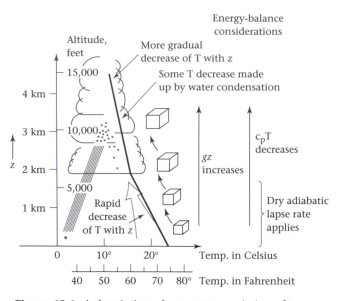

Figure 13.6. A description of a common variation of temperature during the afternoon in the lower few kilometers of the troposphere. The fast stirring of parcels by thermals in the boundary layer below the base of the cloud results in a temperature structure that approximates the adiabatic lapse rate. When some of the thermals push up through cloud base, they begin to condense cloud water, some of which falls as rain. The condensation of water tends to warm the air parcels or, more accurately, to limit their further cooling as they rise. This is an example of the release of latent heat that accompanies condensation. The effects on the rising parcels determine the temperature within the cloud, and ultimately set the basic c_pT = heat capacity, temperature structure of the whole troposphere. Effects such as the release of latent heat and also some absorption of solar radiation throughout the atmosphere tend to warm the middle troposphere to temperatures higher than the theory of the adiabatic lapse rate would predict. Consequently, the troposphere tends to show an average lapse rate of approximately 6°C per kilometer, not the 9.8°C per kilometer, which describes the strict tradeoff of heat energy and gravitational energy in the absence of these other heating effects.

VERTICAL MIXING AND LAYERING

The transport and mixing process goes by the meteorologic term *convection*. The large amount of convective motions give the lower major region of the atmosphere the name *troposphere*. *Tropos* is a Greek word, meaning "turning" or "changing"; these overturning and mixing motions suggested the name for this relatively turbulent lower region. Farther up in the troposphere, the buoyant mixing process is aided by clouds, and the whole region extends up to 8 to 16 kilometers. This is the realm of essentially all the dramatic changes we call weather events. If we take into account other effects such as radiation within the atmosphere and the condensation of water in clouds (mentioned in the box entitled *Pressure, Density, and Temperature Relationships*), we would expect a somewhat less steep drop-off of temperature; in fact, the observed values approach 6 K per kilometer. This is called the average *lapse rate*, named for the lapse or decline of tem-

perature with height. The region of the atmosphere where the overturning stops is extremely abrupt; this, the top of the troposphere, is called the *tropopause*. Above it lies the stratosphere, or *ozone layer*.

During the sunny parts of most days, the lowest few hundred meters to 1 kilometer of the atmosphere becomes mixed up by thermals, which heat, rise, and homogenize to bring the air to the adiabatic lapse rate, 9.8 K per kilometer. The process works rapidly, mixing within one hour, and establishing what is called the "mixed layer" of the troposphere. When the weather is windier, then wind-driven mixing generated by trees, buildings, waves – indeed all the Earth's surfaces – becomes as important or more so in generating the turbulence and mixing. It is often easy to see the depth of the mixed layer, for it can be filled with moist haze or smog and it has an abrupt top. Repeated observations demonstrate that the mixed layer rises through the day; sustained heating raises it to a kilometer height in the afternoon. It typically rises to another easily visible feature, the flat level defining the base of clouds.

In other situations, however, warmer air parcels are transported in layers over cooler ones, or the surface cools, as at night and in winter. Then, parcels cannot become buoyant and freely move upward. If the lapse rate is much below the adiabatic lapse rate near the surface, the atmosphere becomes very stable. Even the wind cannot mix down and cause stirring, stabilizing the motions further. This stability can become extreme, as over the polar ice caps and even continental areas downwind. The severe restriction of mixing makes possible the extreme temperatures of Siberia, but usually only in shallow layers near the surface. Even in the presence of general global warming, we may expect some very cold temperatures to be reported. These temperatures will reflect the statistics of weather extremes: Local variability in temperature may reflect the extreme calm conditions in the dark winter, not the global physics of greenhouse warming.

Because the natural tendency is for mixing to cause temperatures to decrease gradually with height, the two other atmospheric zones of increased temperature in Figure 13.4 must mark regions where the atmosphere itself is able to absorb incident solar radiation. This is just the case. Above 100 kilometers, extremely energetic ultraviolet (UV) radiation is able to ionize molecules, for example:

$$O_2 \rightarrow O_2{}^+ + e^-$$

However, many of the electrons can recombine electrically with the ions, releasing a large amount of energy – in fact, more than enough energy to split the oxygen molecule apart:

$$O_2{}^+ + e^- \rightarrow O + O + \text{heat energy}$$

When the oxygen atoms recombine, they release even more energy. Temperatures rise dramatically, from 400 to 1000 K; consequently, the region is called the *thermosphere*. All this energetic ionizing radiation is absorbed, protecting all of the atmosphere, and people and crops, beneath. There are also other ways that the thermosphere is heated, as organized eddy motions from below are turned into heat.

Far below, at an altitude of approximately 50 kilometers, longer-wavelength but still very energetic radiation can dissociate molecular oxygen to atoms, breaking bonds without separating charges; this process requires only half the energy described above. These energetic ultraviolet (UV) wavelengths produce ozone in the stratosphere. This is our first glimpse of the processes that generate the ozone layer (see Fig. 13.7).

(1) O_2 + UV photon $\rightarrow$ O + O

(2) O + O_2 + neighboring molecule $\rightarrow$ O_3 + molecule (more energy)

(2) O + O_2 + neighboring molecule $\rightarrow$ O_3 + molecule (more energy)

$3\,O_2 \rightarrow 2\,O_3$ (net reaction: result of above three equilibria)

Reaction (2) shows the production of ozone, O_3, and is written twice to illustrate the fate of each oxygen atom in reaction (1). It also introduces us to the fact that atmospheric chemical reactions do not necessarily occur simply because the reactants can release energy. When atomic and molecular oxygen, O and O_2, collide, they normally typically fall apart again because the ozone molecule cannot accommodate all the energy without breaking a bond. In this case, only a neighboring nitrogen or oxygen molecule is required to remove most of the excess energy by moving faster, which in turn increases the energy content of the stratospheric air – and it heats. Throughout this region, the ozone that is produced can also be destroyed by absorbing radiation, or photolyzing.

(3) O_3 + UV photon $\rightarrow$ O_2 + O

Eventually the oxygen atoms can rejoin to form molecular oxygen, O_2, again requiring a helper molecule,

(4) O + O_3 $\rightarrow$ $2O_2$ + molecule (more energy)

We can write these also as a result:

O_3 + UV photon $\rightarrow$ O_2 + O
O + O_3 $\rightarrow$ $2O_2$

$2\,O_3 \rightarrow 3\,O_2$ (net reaction: result of above three equilibria)

These four different reactions (numbered 1, 2, 3, and 4) take us around the full cycle,

$3\,O_2 \rightarrow 2\,O_3 \rightarrow 3\,O_2$

The part of the stratosphere that photochemically determines ozone cycles oxygen into and out of ozone once a week, in some place once a day or faster. The cycle maintains a certain amount of ozone in the atmosphere, and also has the net effect of absorbing ultraviolet energy and converting it to heat. The ozone formed can mix: ozone diffusing upward is immediately broken up by the more intense ultraviolet wavelengths available. However, ozone carried with downward-moving air is increasingly protected by the

Ozone production in the stratosphere

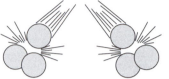

High-energy ultraviolet radiation strikes an oxygen molecule . . .   . . . and causes it to split into two free oxygen atoms.

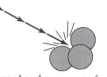

The free oxygen atoms collide with molecules of oxygen . . .   . . . to form ozone molecules.

The Chapman process of ozone destruction

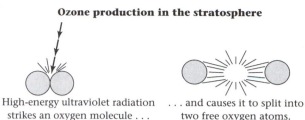

Ozone absorbs a range of ultraviolet radiation . . .   . . . Splitting the molecule into one free oxygen atom and one molecule of ordinary oxygen.

The free oxygen atom then can collide with an ozone molecule . . .   . . . to form two molecules of oxygen.

Figure 13.7. The basic chemistry of the stratosphere unaffected by human perturbation. The panels of the top half show the effect of short-wavelength, especially energetic ultraviolet light–splitting oxygen, resulting in the formation of two ozone molecules. The panels of the bottom half show the return process, with longer-wavelength, less energetic ultraviolet light–splitting ozone. The combination of a free oxygen atom with ozone is somewhat inefficient, in that many collisions result in no reaction. This recombination of oxygen and ozone can be catalyzed (effectively, speed up) by other stratospheric radical species, even in the unperturbed stratosphere. (*Source:* Courtesy of NASA, R. C. Goddard.)

absorbing oxygen and ozone above; this transport explains the presence of ozone in the lower stratosphere. Most of the ozone molecules of the stratosphere are in this protected, sequestered region of the stratosphere, their concentration determined by the chemically very active, light-absorbing region above. Typically, the most important ozone-determining regions of the atmosphere are between 30 and 45 kilometers above the Earth's surface.

This same process of absorbing molecules that shield molecules of all species below continues all the way from the thermosphere down to the troposphere.

Other processes can change the dynamic balance that is formed. One well-known example is the chlorine reaction

Figure 13.8 (right). A plot of mixing times versus altitude. The curve marked "diffusion" shows effects of the natural separation of heavier and lighter molecules by molecular-scale processes, the curve marked "mixing" describes effects where organized air motions stir and homogenize the atmosphere. Shorter times correspond to faster mixing. The point at which diffusion can separate components is at approximately 100 kilometers. Consider these examples: In the region from 0 to 10 kilometers, thermals, clouds, and winter storms mix the atmosphere "thoroughly" over periods averaging approximately 1 to 3 × 10⁶ seconds, or 10 to 30 days. In the lower stratosphere, the times reach thousands of days, up to ten years. Mixing times become faster in the upper stratosphere and mesosphere. As a result of all the limitations to mixing, the time requires to move ozone-destroying contaminants to the active part of the stratosphere where they are released, destroying stratospheric ozone, is approximately forty years. This fact of the slow but inexorable response of the atmosphere to emissions has presented a significant difficulty in reaching national and international accords. km = kilometers; sec = seconds. (*Source:* Adapted from James C. G. Walker, *Evolution of the atmosphere ICB*, Macmillan, 1977.)

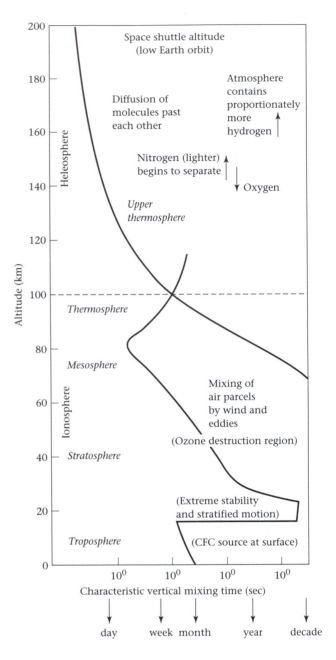

(the chlorine originates from the ultraviolet photolysis of chlorofluorocarbons).

$$Cl + O_3 \rightarrow ClO + O_2 \dots \text{etc.}$$

These reactions can affect both the very active upper stratosphere, and also the lower stratosphere. This is why it is not possible simply to "Put some more ozone up there if it's missing," as has been repeatedly suggested. If we were going to fix the ozone layer by adding ozone, we would have to fix it every day, and in doing so, compete with the energetic ultraviolet input of the Sun showering down on the entire sunlit stratosphere, which undoes the fix. Remember that the reaction O_2 + UV photon $\rightarrow$ O + O begins the natural chain of events that creates ozone. The energy needed to amend this production rate throughout the stratosphere requires something like four times the world electricity-generating capacity!

The heating of the upper regions of the stratosphere causes the same type of vertical stirring and exchange as that occurring near the Earth's surface; the stirring tends to follow the same adiabatic lapse rate, but with mixing. The region above, between the stratosphere and thermosphere heating regions – 55 to 85 kilometers – is called the *mesosphere*. *Meso* is the Greek word "intermediate." Although ionizing radiation is important for heating the thermosphere, so many ions are formed in the upper mesosphere and lower thermosphere that the entire region is termed the *ionosphere*. This overlapping "sphere" is named not for its thermal structure but for the presence of the ions. Much of the early research on the upper atmosphere focused on the ionosphere because the electric charges interact strongly with radio waves, and disturbances associated with the aurora recurrently interrupted normal radio communications. Episodes of frequent aurora – the result of more ion input from the Sun – tend to recur every eleven years, as part of the sunspot cycle. One early explanation for the Antarctic ozone hole focused on that

cycle. The theory was that the ion chemistry disturbed during sunspot maxima could have some perturbing effect lower in the stratosphere at 40 kilometers and below. Although we now understand that the ionosphere can contribute nitric oxide, NO, to the stratosphere, we now are fairly sure that the air motions and chemistry fail to explain either the ozone hole or the general decrease of stratospheric ozone.

Possibly the greatest threat to the ionosphere was the plan to explode nuclear weapons during a nuclear war in order to disrupt communications severely: in fact, the electromagnetic pulses (EMPs) produced could destroy most modern printed-circuit radio and control equipment. A peacetime threat would accompany attempts to beam down power from space. Giant solar-energy converters have been proposed that would beam back energy in powerful and concentrated microwave beams, with massive local disruption of the ions of the mesosphere, not to mention birds flying into a microwave beam down in the troposphere – cooked in flight! Most of the heating of the stratosphere occurs at its top. The effect is that parcels of air that move upward, cooling with the adiabatic lapse rate, are much more dense than the surrounding heated air; they either return to their origin or simply do not move. This means that air motions remain at the same level in very flattened layers. Hence the name *stratosphere*; the Greek word *strata* means "layers." In fact, the stratosphere does get mixed vertically, as a result of continual churning motions passed upward from large-scale weather systems moving around below in the troposphere.

The thermosphere has the same heating profile as the stratosphere: It is hot on top due to heating of the extremely ionizing radiation and other effects. However, other effects do come into play. The large number of ions can produce "ion drag" on the remaining neutral particles. Additionally, the temperatures become so high and the spaces between the molecule so great that molecular diffusion becomes an important factor. Figure 13.8, which shows the speed of transport processes combines many of the concepts presented in this section. Clearly, the transport is rapid where temperature decreases with height – the "unstable" regions – and is slower where temperature increases. On the other hand, the effect of processes at the molecular level – molecules driven by thermal energy – continue to increase, according to a general formula:

$$T_{mixing} = \frac{constant}{n}, \left(\frac{molecular}{mixing\ time} \right) = constant \times \left(\frac{volume}{molecules} \right)$$

Why do the components of the atmosphere remain mixed, when each gas has its own density, nitrogen having 28/32 the density of oxygen, and H_2O having only 18/32 the density of oxygen? In fact, how can carbon dioxide pollute the lower atmosphere, or the fluorocarbons the upper atmosphere, given that each of those gases is, in isolation, so much more dense than normal air? Indignant letters have been written to the *New York Times* pointing out the apparent absurdity of chlorofluorocarbon effects on the stratosphere, given the great mass of compounds such as $CFCl_3$ (approximately 136 grams per mole). It is indeed true that the thermal or molecular diffusion effects tend to cause the atmosphere to unmix, with the lighter molecules traveling to the top. The answer shown in Figure 13.8 is that there are sufficient motions of the types we have described, and that not much turbulent motion is required to keep the air mixed. The atmosphere becomes significantly unmixed only above 100 kilometers, where the air is much less dense. Consequently, the distances between molecular collisions become so large and the instances so infrequent that each molecule travels a long distance before colliding; there, each species begins to seek its own gravitational level.

URBAN AND REGIONAL POLLUTION. Table 13.2 is an overview of the most common problems associated with conventional air pollution. These pollutants are "conventional" in that they were recognized in the 1950s and 1960s, if not earlier, and affect urban and regional areas (multistate or province); with the exception of ozone and perhaps some fine-grained particles, these do not have intercontinental effects. Some pollutants cause problems primarily in the winter, and are restricted to particular valleys and streets, while others cause problems in the summer and may spread to affect many states, provinces, or even neighboring countries. Carbon monoxide – which originates almost solely from automobiles in industrial countries and biomass burning in less developed tropical countries – causes problems, and is the most local and weather-dependent, idiosyncratic of pollutants. Similar to the much rarer toxic industrial gases such as chlorine, carbon monoxide is dangerous because it tends to pool in very calm conditions and drain and disperse slowly over a day or so.

DESTRUCTION OF THE OZONE LAYER

PERTURBATIONS TO THE BASIC CYCLE. In the survey of atmospheric layers, we learned how stratospheric ozone is constantly being produced and destroyed by a set of reactions described in Figure 13.7.

Production

O_2 + UV photon → O + O

O + O_2 + neighboring molecule → O_3
  + molecule (more energy)

Destruction

O_3 + UV photon → O_2 + O

O + O_3 → 2 O_2

This cycle of reactions that create and destroy ozone was hypothesized by Sidney Chapman in the 1930s. There reactions seemed to describe adequately the balance of the ozone and ordinary oxygen in the stratophere.

$$3\ O_2 \rightleftarrows 2\ O_3$$

The rates of the individual reactions became more easily measured in physical chemical laboratories over the following decades, and understanding of the attenuation of the photolysis radiation improved as similar laboratories measured the absorption properties of oxygen and ozone at various wavelengths. The sciences of laboratory and atmospheric photochemistry developed together. Numerical calculations of the steady-state (or balancing) ozone concentration appeared to be reasonably accurate, given the uncertainties. By the 1970s, however, it was clear that the atmosphere had somewhat less ozone than the laboratory measurements of these reactions would indicate.

The laboratory science of photochemistry supplied an idea that proved useful: For compounds such as ozone that have excess energy to give away, catalytic reactions involving trace catalysts can shift the balance dramatically. For example, nitrogen compounds other than N_2 were known to exist

TABLE 13.2. U.S. NATIONAL AIR QUALITY STANDARDS THAT ARE FREQUENTLY EXCEEDED

Pollutant	Sources	US National Standard (short term)	Danger	Provenance, Special Circumstances	Comments
Carbon Monoxide	Automobile tailpipe emissions, burning of tropical agricultural wastes	9 parts per million in 8 hours	Asphyxiation, binds with human hemoglobin	Winter, cities and valleys	Stable conditions, often confined to a few hundred meters in streets, valleys
Ozone	Formed in air from emissions from automobiles, industries	120 parts per billion in 1 hour	Cell damage to human lungs and plant tissues	Summertime, cities, regions, and intercontinental	Formed in air, increases with temperature, from hydrocarbon and nitrogen oxide emissions
Particles	Formed from industrial, automotive emissions, dust	150 micrograms per cubic meter in 24 hours	Damage to lung tissue, visibility degradation	Cities; sulfate and nitrate also contribute to acid deposition; carbon-containing particles effects unclear; causes some greenhouse cooling during daylight!	Most apparent, arouses most comment; different size and composition particles have different effects
Acid deposition: nitric and sulfuric acid also in rain	Industrial (esp., coal-burning), automotive	Complex, effects determine emission	Destroys lake fish populations, kills trees	Regional and intercontinental: E. North America, Northern Europe, Venezuela	Much deposition is rain/snow, some is "dry"

in the stratosphere in trace quantities, originating from ionospheric reactions, the breakup of the pervasive compound nitrous oxide (N_2O), and other sources. Laboratory reactions and calculations confirmed that reactions such as

$$NO + O_3 \rightarrow NO_2 + O_2$$
$$NO_2 + O \rightarrow NO + O_2$$

can destroy ozone without consuming the catalysts themselves. The interplay between laboratory and atmosphere converged to provide a satisfactory and quantitative understanding of stratospheric ozone. More broadly, the physical chemistry of the process was quickly recognized as a form of a general kind of destruction pathway,

$$X + O_3 \rightarrow XO + O_2$$
$$XO + O \rightarrow X + O_2$$
Net: $O + O_3 \rightarrow 2 O_2, X \rightarrow X$

in which X stands for the cluster of two atoms, NO. The cluster, when not attached, typically is a fairly reactive, transient compound, with unbonded electrons that are available to attach to another molecule such as ozone. In chemical terminology, species such as NO are "free radicals" – "radical" meaning a "root" of some larger molecule, and "free" meaning that they are loose, with their bonding electrons unbound, that is, very reactive and likely to bind. In fact,

both X and XO are free radicals, because the XO compound also has an unpaired electron at the other end of the O.

The net reaction is identical to the ozone-destroying "Chapman reaction" described above. The role of the catalyst, X = NO, is to help distribute the excess energy of reaction; the energy further oscillates the larger molecules (creating more energetic rotations and vibrations). This role is similar to the role of the excess molecules in the reactions

$$O + O_2 + \text{neighboring molecule}$$
$$\rightarrow O_3 + \text{molecule (more energy)}$$

or other reactions in which chemical energy is released.

Another candidate for X is hydrogen. When fuel combusts, it creates a large amount of H_2O as well as carbon dioxide. A very early environmental worry was that the water from supersonic jet exhaust might affect the stratospheric ozone balance because of the additional X = H provided by the water. The low temperatures of the uppermost troposphere exclude almost all of the normal water supply to the stratosphere, freezing it out.

Lower ozone levels are also the result of transport. A portion of the ozone flows out of the stratosphere into the troposphere, where two important processes destroy ozone. Reactions within the troposphere allow ozone to be destroyed. The troposphere is composed of much higher water concentrations, which makes the following reactions very important:

O$_3$ + UV photon → O* (chemically excited oxygen)

O* + H$_2$O → OH + OH (two hydroxyl radicals)

The sequence of these reactions does two things. First, it destroys ozone (it is not energetically possible for ozone to reform from hydroxyl radicals). Second, it provides the troposphere with extremely reactive OH radicals, known as the troposphere's "cleanser" compound. Furthermore any remaining stratospheric ozone could react with the Earth's surfaces. Here, ozone levels were once thought to be very low, at least outside polluted cities and industrial regions. However, recently it has been discovered that these pollution processes, the smog ozone processes, are widespread, and the troposphere creates most of its own ozone.

THE SCIENCE OF THE FIRST "ENVIRONMENTAL NOBEL PRIZE." In the 1970s and 1980s, meteorologist Paul Crutzen and chemist Harold Johnston, were each studying the role of nitrogen oxides and discovered nearly simultaneously the probable existence of a relatively large new source of NO free radicals in the atmosphere: supersonic aircraft, such as the Concorde passenger aircraft of the 1970s. The convenience and perceived economy of the supersonic aircraft were obtainable only if they flew higher into the stratosphere. However, the radical X, nitric oxide, seemed to work much more effectively in the stratosphere than hydrogen. This realization launched the first major "ozone war" between environmentalists and industrialists with physical scientists raising the unwelcome specter of a general decline in the ozone layer. The need to define this environmental threat quantitatively engendered a new science of atmospheric chemistry. Scientists with widely different backgrounds came together to communicate: some created sophisticated laboratory instruments that could measure reaction rates and physical properties; others invented an undreamed-of type of analytic chemistry to measure reactive species at levels of parts per billion and parts per trillion; others described the mixing of different reactants by weather; and still others honed their numerical skills in order to calculate the effects.

One of the new instruments, applied by James Lovelock (who also invented the concept of Gaia – the world, as a living organism), measured trace compounds at the 10^{-12} (parts per trillion) level. As he sailed the seas on a small yacht, he measured two odd molecules, CFCl$_3$ and CF$_2$Cl$_2$. He found evidence that the two compounds increased continuously, apparently spreading out primarily from the Northern Hemisphere. The source of these molecules was well known because they were used in the refrigerators, air conditioners, and "aerosol" spray cans that are found predominantly the industrialized countries of the Northern Hemisphere. A limited number of manufacturers made these products, called chlorofluorocarbons (named for the chlorine, fluorine, and carbon they contain; the trademarked term that DuPont used, "Freon," was more commonly used.

Another term, "CFCs" is not trademarked). Using crude calculations, Lovelock discovered that nearly every fluorocarbon molecule emitted remained in the atmosphere.

Lovelock shared this discovery with Sherwood Roland, who was using his instruments to measure a variety of compounds. What might be the atmospheric fate of Lovelock's ultra-trace compounds? Roland turned this academic question over to Mario Molina, who had recently joined him as a postdoctoral researcher at the University of California at Irvine. Molina returned an astounding two-part answer. The first part predicted that the molecules would be very stable, because C–Cl and C–F bonds in CFCl$_3$ and CF$_2$Cl$_2$ are strong and resistant to chemical attack, and because the molecules are insoluble in rainwater. They would then mix, as do other atmospheric compounds. The fate of the CFCs that he outlined is shown in Figure 13.9. They would continue, having not reacted, to the region of the atmosphere where the ultraviolet light was sufficiently strong to cause the following reaction:

CFCl$_3$ + UV photon → CO$_2$ + HF + 3Cl

This occurs at approximately 40 kilometers in the stratosphere, the photochemically active region. CF$_2$Cl$_2$ would have a similar fate. These predictions have proved to be correct. The second part of Molina's conclusion was of greater importance: The destruction of the CFCs placed chlorine at-

Figure 13.9. Catalytic destruction of ozone due to the chlorofluorocarbon CFC1$_3$. Energetic short-wavelength radiation prevalent at 40 kilometers and above in the stratosphere splits the strong–C-C1 bond. That chlorine, and others released from this and other chlorine reservoir species, becomes available to catalyze ozone destruction. The effect of the reactions shown in the last four panels is the same as that of the O + O$_3$ reaction shown in Figure 13.8. (*Source*: Courtesy of NASA.)

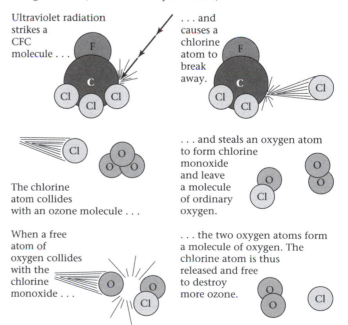

Ultraviolet radiation strikes a CFC molecule . . .

. . . and causes a chlorine atom to break away.

The chlorine atom collides with an ozone molecule . . .

. . . and steals an oxygen atom to form chlorine monoxide and leave a molecule of ordinary oxygen.

When a free atom of oxygen collides with the chlorine monoxide . . .

. . . the two oxygen atoms form a molecule of oxygen. The chlorine atom is thus released and free to destroy more ozone.

oms precisely where they could have considerable catalytic activity in destroying ozone. That is, he said imagine that chlorine behaves as a free radical (X = Cl), and consider

$$Cl + O_3 \rightarrow ClO + O_2$$
$$ClO + O \rightarrow Cl + O_2$$

Net: $O_3 \rightarrow 2\,O_2$

The nonreactivity of the CFCs below 40 kilometers was in fact a hindrance: they were in fact *source gases* of the "ozone-killer" *radical* Cl.

The publication of this discovery in the journal *Science* in 1974 had two major effects. First, it launched a second, decade-long "ozone war" that resulted in the convincing of environmental and industrial scientists around the world that there was a need to convert away from CFCs to slightly less economical alternatives. Aerosol cans were converted first to hydrocarbon or other propellants; then, refrigerants were replaced with similar compounds (hydrochlorofluoro-carbons, or HCFCs, such as CF_3CHCl_2) that could react away before reaching the stratosphere.

The second result of the publication of Molina's discovery occurred in 1995, when the Nobel Prize in chemistry was awarded to Molina and Rowland, and also to Crutzen, who had quantified the NO, free radical catalytic effect of nitric oxide, and presented principles and techniques for the solution of both stratospheric and tropospheric chemistry problems. This was the first Nobel Prize awarded for work in environmental science. It cheered a large group of meteorologists, geophysicists, and planetary physicists: A Nobel Prize had been awarded for extremely complex work integrating many specialties in the physical sciences, rather than for work in a narrowly defined specialty.

POLAR OZONE CHEMISTRY: RADICALS AND "RESERVOIR SPECIES." Numerical calculations of the stratospheric ozone content made it clear that the ozone-destruction chemistry was proceeding in the lower stratosphere (20 kilometers) as well as in the 40 kilometer region. This presented a problem, because the reaction

$$ClO + O \rightarrow Cl + O_2$$

worked most effectively at 40 kilometers, where the more energetic "hard UV" produced monatomic oxygen; the ultraviolet radiation broke up both ozone and oxygen molecules to make atomic oxygen, as shown in the basic Chapman ozone chemistry.

Mario Molina and his wife and co-worker Luisa Molina found an alternative that worked well at lower atmospheric levels:

$$2\,[Cl + O_3 \rightarrow ClO + O_2]$$
$$ClO + ClO + \text{any molecule} \rightarrow Cl_2O_2 + \text{same molecule}$$
$$Cl_2O_2 + \text{soft UV photon} \rightarrow Cl + Cl + O_2$$

Net: $O + O_3 \rightarrow 2\,O_2$

This reaction sequence did not follow the standard X = Cl pattern, but emphasized several consecutive steps. Another complexity was introduced: the "free radical" chlorine oxide, ClO, reacts with itself to make a somewhat less active compound, Cl_2O_2 (actually, Cl-O-O-Cl). Light is required to decompose this species, and eventually this allows ozone destruction to continue. Other complexities exist in the lower stratosphere. Free radicals such as ClO and NO_2 may react not only with themselves but also with each other:

$$ClO + NO_2 + \text{any molecule} \rightarrow ClONO_2 + \text{same molecule}$$

Two ozone-killer radicals can neutralize each other. However, similar to the manner in which the compound Cl_2O_2 is the combination of two ClOs, described above, the neutralization is not permanent. Similarly,

$$ClONO_2 + \text{UV photon} \rightarrow ClO + NO_2$$

This compound, "chlorine nitrate," is an example of a reservoir compound: It holds the killer radicals for a while and them releases them, sometimes in another part of the stratosphere. The manner in which molecules fall apart appears somewhat arbitrary: This decomposition creates ClO, whereas Cl_2O_2 creates Cl. Photochemistry is indeed subtle; even experts must rigorously check their assumptions about processes and their rates in the laboratory.

The mystery of the notorious Antarctic "ozone hole" (Fig. 13.10) was unraveled when it was realized chlorine nitrate can break down even in the Antarctic winter, when there is no ultraviolet light available in the stratosphere over the South Pole. In simple terms, the air is so cold that a variety

Figure 13.10. Percentage decline in the ozone layer, by latitude. Differences from latitude to latitude reflect primarily the transport and chemistry differences in the stratosphere. For example, approximately an 11 percent decline is found near the ozone hole region, showing the broader effect of the chemistry in that region. Destruction in the lower stratosphere over Antarctica affects the entire Southern Hemisphere. (*Source:* (Data from United Nations Environmental Programme.)

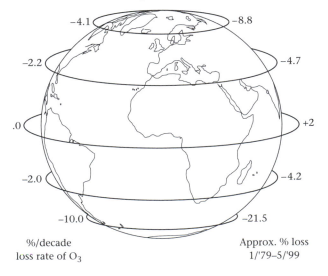

%/decade
loss rate of O_3

Approx. % loss
1/'79–5/'99

of forms of tiny crystals form from water, nitric acid, and sulfuric acid. These particles provide many liquid and solid crystal surfaces in the stratosphere, a region that typically has low levels of condensation. Vastly different chemistry can occur in the liquid and gas phases; one result is that chlorine nitrate can break down, freeing more active chlorine compounds. It is these active compounds that initiate an ozone-destruction process using Cl_2O_2 photolysis when the faint spring sunlight returns to the Antarctic region. The lower tropospheric process, acting with plentiful chlorine radicals, rapidly destroys most of the Antarctic ozone layer.

In summary, thus far we have discovered how apparently innocuous *source gases* (e.g., $CFCl_3$ and other CFCs) can carry free radicals through the atmosphere, and how they can release active catalysts (e.g., Cl) that change the balance of ozone and ordinary oxygen. The free radicals themselves, however, may be entrained temporarily in *reservoir species*.

Natural source gases for both nitric oxide and chlorine in the stratosphere have been assessed. The gas nitrous oxide, which we consider in Chapter 14 as a greenhouse gas, is important for the nitrogen budget of the atmosphere; it originates in soil ecosystems, where the emissions are determined by the net balance of soil-based production and destruction processes. Natural chlorine can be carried aloft by the ocean emission methyl chloride (CH_3Cl): it may originate primarily from seaweed. Other sources of chlorine have been suggested: for example, sea salt (NaCl) and volcanic hydrochloric acid gas (HCl). These compounds are known to be almost completely removed from the atmosphere by solution into rainfall. Because very little water enters the stratosphere, species dissolved in cloud water cannot enter either. Checks on this process have been carried out by NASA ER-2 stratospheric sampling aircraft. Measurement flights made over volcanic plumes and thunderstorms in the stratosphere have

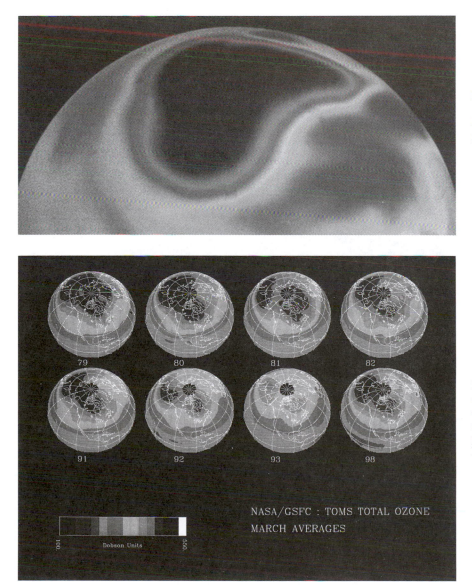

Figure 13.11A. The notorious Antarctic "ozone hole" centered over Antarctica. *Source:* Courtesy of NASA.) See color plate section.

Figure 13.11B. A series of maps of total ozone concentrations for eight months of March for the Northern Hemisphere springtime ozone. TOMS = Ozone Mapping Spectrometer. (*Source:* Courtesy of NASA.)

verified that these phenomena do not contribute substantial amounts of chlorine to the upper levels of the stratosphere.

THE PRESENT SITUATION. The decline in stratospheric ozone continues more or less in accordance with the details of global computer models that can simulate all the details of the emissions trends, transport from the surface to and from the upper stratosphere, and the gradual removal of chlorine from the stratosphere as the international protocols limiting source gases are implemented. Figure 13.11 shows the extent of the decline in the global ozone layer from 1985 through 1995. Declines averaged 3 to 4 percent in the total stratospheric ozone over northern industrialized countries, and 5 and 10 percent at southern latitudes, which lie closer to the ozone hole. Theory and ideal clear-sky measurements suggest that changes in the ultimate concern, ultraviolet radiation, should be double the changes in ozone, yielding 6 to 8 percent increases over industrialized countries. Other factors also affect local exposure to ultraviolet radiation, for example possible increases in cloudiness and tropospheric aerosol pollution, both of which can reflect radiation upward and thereby limit ultraviolet radiation exposure.

Scientists are paying particular attention to the region over the northern industrial countries. Theory describes a general decline; however, the details that determine ozone overhead in the springtime are more subtle. Figure 13.11B shows a series of maps of total ozone for several months of March. There may be somewhat more decline during this period than the general theory predicts. As previously described, ozone-hole chemistry may begin to have more obvious, detrimental effects. As the pictures suggest, there could be a large-scale decline in the Northern Hemisphere springtime ozone content of the atmosphere at high latitudes.

QUESTIONS

1. Give one example each of a compound that contribute: (a) many percents of the air's molecules; (b) 1 to 100 parts per thousand or more; (c) several parts per million; (d) a few parts per billion; (e) parts per trillion.

2. If the temperature of an air parcel increases by 1 percent, and the pressure remains the same (because it is reacting to the weight of the air above it, which remains the same), in what direction does the density go? If some complicated atmospheric effect causes both the temperature and the pressure increase by 1 percent at the same time, what happens to the density?

3. If the temperature is 25°C at sea level, what does the adiabatic lapse rate suggest the air temperature would be on the peek of a nearby 1000-meter mountain? (Ignore any local heating of the surface of the mountain, and adjacent air, by the Sun's rays.)

FURTHER READING

Graedel, T. E., and Crutzen, P. J. 1993. *Atmospheric change, an Earth system perspective.* New York: W. H. Freeman.

Turco, R. P. 1997. *Earth under siege.* Oxford: Oxford University Press.

World Meteorological Organization. 1995. Scientific assessment of ozone depletion: 1994. Global Ozone Research and Monitoring Project, Report Number 37, Geneva, Switzerland: World Meteorological Organization.

14 Atmospheric Motions and the Greenhouse Effect

ROBERT CHATFIELD

ATMOSPHERIC FLOW AND EVOLUTION

Greenhouse warming! It seems that with every summer heat spell, newspapers warn of its peril; with every snowy cold spell or record low, columnists and "instant experts" debunk it. Scientists who study the Earth–atmosphere system, meanwhile, describe it as an inexorable, more obvious phenomenon, brought about, at least in part, by human activities. They admit that there is just enough weather and other variability to make the exact extent of greenhouse warming that has already been realized difficult to gauge. There are even some reasonable scientific suggestions that the variability of regional weather – for example, more hurricanes or heavier snowstorms – could be tied to the effects of increasing greenhouse heating.

It is difficult to believe that these are the same atmospheric scientists who also teach that the Earth's weather must be at some point unpredictable, even as they struggle to improve the accuracy and length of weather forecasts. They admit the basic soundness of the teachings of Edward Lorenz of the Massachusetts Institute of Technology, a founder of what physicists now call "chaos" theory. He claimed that if mete-orologists neglected in their computations even the stirring of a flying seagull in the South Pacific, weather forecasts in North America would be fatally affected by error within a span of only a few weeks.

In this chapter, we examine the basic language of some of these varied suggestions and sketch out the basis of the argument. First, we look at some of the origins of weather and its variability. We describe the general circulation of the atmosphere and the origin of mid-latitude storms. Despite these uncertainties, we explain why the greenhouse effect itself is incontrovertible: There are simple physical arguments that, without the effect, the Earth would be unliveably cold, much like Mars. Further, we make a first basic estimate of the temperature of the Earth–atmosphere system based on the Earth's total heat balance, and we learn how the greenhouse effect makes the planetary surface comfortably warmer than the apparent Earth system temperature that is observed from space. Finally, we glimpse the "greenhouse-warming weather" and make reasonable speculations about the future intensity of hurricanes and droughts.

"Blown" weather forecasts are so familiar to us that we are very aware of the difficulty of making an accurate weather prediction; consequently, people have natural doubts about any scientist's ability to describe climate. Let us begin by distinguishing among the terms. What are we expecting in a weather forecast? *Weather* is commonly defined as the instantaneous state of the atmosphere all over the Earth; most basically, it consists of the temperature, pressure, speed and direction of the wind, and the humidity at each spot on the Earth. Additionally, the amount of sunshine, the position and thickness of clouds, and the amount and kind of precipitation are normal weather variables. It is increasingly common for descriptions of the weather to include the extent of haze (anthropogenic aerosol cover), smog, and exposure to the Sun's ultraviolet radiation. Common experience suggests that the values of these weather variables follow familiar patterns, usually with slow changes, or *weather persistence*, for hours and days at a time. Of course, occasional sudden changes, such as the onset of cold fronts with their frosts, as well as violent storms, are the most important weather developments to predict correctly. The most fundamental test of the accuracy of a weather forecast technique is that it must do better than prediction of "more of the same." For forecasts made a few days in advance, it is best to abandon the forecast of weather persistence and rely on the tabulated thirty-year climatology for the daily high and low of temperature, with simply a random guess, yes or no, concerning the *probability of precipitation*. Making the equivalent odds on a toss of the dice is sufficient to verify that the random guess matches the historically recorded probability of precipitation. The idea of weather also includes physical explanations for temperatures, winds, and so forth. For example, rain may be attributed to a winter storm, a hurricane, or a foggy drizzle. Each of these tends to follow a familiar cycle, as weather formed into a common structure passes by overhead.

Climate is the average long-term weather for an area and season. Most basically, it is the average temperature – the average daily maximum temperature and so forth. Climatology, the recording of climate, also includes the variation of these temperatures: for example, the standard of the expected range of daily maximum; or the historical record maximum. Thus climatology describes the "odds of the dice" mentioned above. Descriptions of climate strive to move beyond the most variable features of weather; a "well-behaved" climate statistic is the worldwide average temperature. As with weather, the idea of climate also includes attempted physical explanations; for example, an explanation for the decade-to-decade changes in maximum temperatures. There was an sequence of worldwide average maximum temperatures in the decade 1985–95, with clear cooling during the years 1992 and 1993. By far the most popular and accepted explanation for this phenomenon is that the Earth retained increasingly more radiation due to the increase of global greenhouse gas concentrations, which in turn increased the average temperature. Also, aerosols from the extraordinary explosion of Pinatubo volcano in 1991 veiled the Earth from sunlight for approximately two years, reflecting energy out to space. These are examples of energy-balance explanations for the global climate, examples that we examine in the section. "Atmospheric Heat Balance and the Global Greenhouse."

PROCESSES DRIVING THE CIRCULATION OF THE AIR

Here we examine some of the scientific components of the processes that control our Earth's typical weather. These components help explain why recurrent kinds of winds and other weather occur in different regions, and how the regions relate to each other. This chapter examines the basics of the atmosphere's *general circulation*. We build a description of weather based on the following concepts.

AIR MASSES. The tendency of clouds to favor certain areas in which air repeatedly becomes warm and moist illustrates the origin of *air masses*. Air masses are large volumes of air with fairly uniform properties; Most have been heated or cooled, moistened or dried by prolonged residence over a land or ocean area. Air masses are named to describe geographically their temperature and humidity. For example, maritime tropical air is warm and moist; polar continental air is cool and drier; Arctic continental air originates even closer to the poles and is frigidly cold and drier. Winds tend to move air with these properties away from the source regions, and the air then becomes progressively modified (e.g., becoming hotter and wetter over several days as it moves). Thus, maritime tropical, polar continental, and even

Arctic continental air masses regularly affect the mid-latitudes. (The northern mid-latitudes are defined as the region between 30 to 60° N, and are called the temperate latitudes, although some "temperate" regions such as Siberia may alternate between unpleasant extremes.) The air masses are most clearly distinguishable near the Earth's surfaces that influence them.

REGIONS OF ASCENT. In Chapter 13 we considered how the areas of the Earth's surface that are locally warmest produce hot-air thermals that rise 1000 meters or more and stir the daytime boundary layer. Extra heating and moistening of the air in some regions allows large numbers of deep, convective clouds to form. Warm water areas of the ocean, tropical rain forests, and the eastern regions of the temperate latitudes are common sites for this cloud formation. General upward motion occurs in these regions. Storms move around, bringing rain and upward motion to somewhat drier areas. The upward motion leaves less air mass locally – that is, lower atmospheric pressure – in the lower atmosphere. Consequently, air must begin to rush in horizontally in response to pressure. These horizontal forces are simply the expression of the tendency to return to the normal atmospheric pressure of surrounding regions. Chapter 13 explains that the calm, or rest state of the atmosphere, never quite achieved, is one in which atmospheric pressure has variations only in the vertical. That is, the pressure simply holds up the ever thinner air above.

DIRECT CIRCULATIONS. The cycle of upward motion and pressure-driven filling would be simple if we could neglect the fact that the Earth rotates. See Figure 14.1 for an example in which the Sun's heating is strongest at the equator, and the planet does not rotate. In a *direct* circulation like the one illustrated, heated air rises, cooled air sinks, and north-south winds move mass simply to compensate. This is the most basic example of a cell – or a repeating, nearly closed path – of atmospheric circulation. The eighteenth-century English scientists Halley (for whom Halley's comet is named) and Hadley described these circulations; the Hadley atmospheric circulation (with some added effects described below) is an example of the basic principle.

THE CORIOLIS EFFECT. The Coriolis effect is relevant because the Earth does indeed rotate. According to this effect, Newton's laws about objects continuing in a straight line do not apply, because Newton's law applies to observations made in a nonrotating, nonaccelerating system. Confusions we experience in an accelerating or rotating reference frame are most obvious in the familiar "forces" we feel when a car accelerates or swerves around a corner. The backward or sideways motions we perceive appear to come from forces, for example, "centrifugal forces." Actually, our bodies are following Newton's laws with an appropriate frame, not the swerving, accelerating car's. Most physicists refuse to refer to centrifugal and Coriolis effects as "forces." However, these

(A)

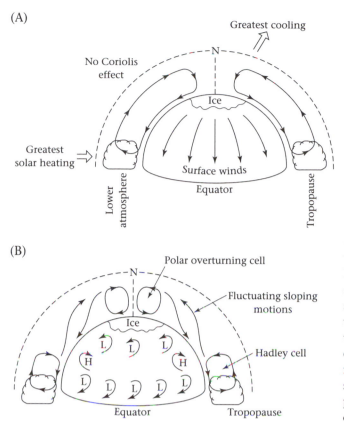

(B)

Figure 14.1. Air motions as they would occur on an imagined idealized Earth that does not spin on its axis, but is heated evenly in the tropics. In a direct circulation, as shown here, there is rising motion in the warm areas and sinking motion in the cooler areas. Poleward motions in the upper atmosphere and equatorward motions in the lower atmosphere act to enclose the circulation. Motions on the real Earth tend to begin in a similar manner near the Equator, where the effect of the Earth's rotation is small; similar cells, called Hadley cells, show this same direct circulation.

effects do *seem* to cause objects to change their motion, at least in our reference frame. On the rotating Earth, we can expect some "normal" behavior for motions that occur within seconds or minutes, but somehow a different behavior for objects that continue moving while the Earth turns substantially – that is, for some appreciable fraction of the Earth's rotation period, a day. In the northern hemisphere, moving objects always seem to curve to the right. In the nineteenth century, Frenchman Gaspard Coriolis was the first to relate a paradox of naval long-range gunnery, wherein shells landed to the right of their aimed target as a result of the rotation of the Earth. The shells were only minutes in the air, yet fine precision was needed to hit a distant target. This small rightward deviation occurred regardless of the direction in which the cannons were aimed, although the deviation did seem to depend on latitude. The effect disappears toward the equator. To our eyes on the rotating Earth, the effect looks like a "Coriolis force," which is (1) always perpendicu-

lar to an object's direction and (2) proportional to its speed. The latter condition means that nonmoving objects do not feel any such force.

To better understand the Coriolis effect, consider not only air parcels and artillery shells, but also the motion of satellites as humans attempt to make them follow a constant level in low Earth orbit just above virtually all of the Earth's atmosphere, without any extra steering motions. A satellite serves as an example of an object moving under Newton's laws, but without some of the distractions that arise from motions within the atmosphere. Both air parcels and satellites are very much under the control of the Earth's gravity: An air parcel is typically at 1.001 times the Earth's radius from the center of the Earth, whereas a satellite such as the space shuttle is still only at 1.05 times the Earth's radius. Both are very close to and under the strong control of Earth's gravitation. In Figure 14.2, gravity and centrifugal acceleration maintain a satellite at approximately the same distance above the Earth's surface. (From the physicist's perspective, the satellite continues to fall toward the center of the Earth while the surface of the Earth continues to fall away and the direction of pull changes.) Familiar coordinates of latitude and longitude that mark the satellite's progress are coordinates that refer to the spinning Earth. When the satellite moves northward, it appears to have a tendency to a rightward drift as seen from our rotating Earth. At the northernmost point of the orbit, it seems to be pulled southward; when it moves into the Southern Hemisphere, the tug seems to be pulling it eastward. From a fixed point not located on the spinning Earth, the satellite is making a simple circular orbit. The physical laws of gravitation dictate that in a lower orbit, the satellite moves faster in order to stay in a circle; in a higher orbit, the satellite moves more slowly. Regardless of these laws, the Coriolis force seems to be proportional to the velocity, and the orbits take the same form! Mathematical calculations are necessary to show that the apparent rightward drift has the same form and magnitude in all cases, depending only on the object's velocity in our spinning-Earth reference system. In the case we are considering, the velocity effect is easy to see. More apparent "Coriolis" force is needed for a faster object to maintain the ground track of the satellite – which it must, from the off-Earth viewpoint!

The same laws of motion apply to satellites, airplanes, artillery shells, and everything embedded in the Earth's atmosphere – indeed, even to the moving parcels of the atmosphere itself. Air parcels within the atmosphere apparently are surrounded by pressures from adjacent air, to "keep them moving with the rest of the Earth." Such forces can be considered in the physics of airflow, but they do not alter the parcel's need to follow Newton's laws of motion in a nonrotating reference frame. The force on any moving air parcel is (1) perpendicular to the direction of motion, and (2) proportional to its speed, rightward in the Northern Hemisphere, leftward in the Southern Hemisphere, maximizing at the poles, and diminishing to no effect along the equator.

Motion as observed from space is nearly circular, reflecting the balance of gravity and centrifugal effects, at a constant height above the surface.

Motion as plotted on a map

❶ Satellite (or air parcel) moves slightly faster than Earth at Equator, path appears straight

❷ Satellite moves ever faster than Earth at this northern latitude ... on map appears to turn to right.

❸ Eastward moving satellite moves straight, but Earth's lines of latitude curve away. Again, it appears to turn to right

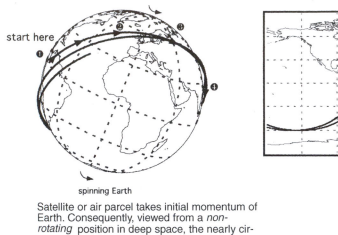

start here

spinning Earth

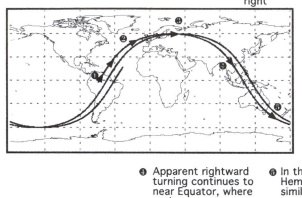

❹ Apparent rightward turning continues to near Equator, where path appears to straighten

❺ In the Southern Hemisphere, similar apparent turning, to the left

Satellite or air parcel takes initial momentum of Earth. Consequently, viewed from a *nonrotating* position in deep space, the nearly circular orbit seems to spin around with approximately the rotation rate of Earth.

Figure 14.2. The Coriolis effect is observed when the reference point is the rotating coordinate frame defined by the spinning Earth. The left-hand figure shows a satellite in low Earth orbit, as seen from space. The satellite makes a simple circular orbit under the force of gravity alone. It was launched from the Earth and carries along some of the Earth's angular momentum plus whatever was added by the initial rocket power. The right-hand figure shows this motion from the perspective of someone moving with the reference frame of the rotating planet. As the satellite moves northward, some effect causes it to turn continuously to the right, passing regions on the Earth. The satellite is conserving angular momentum. The apparent rightward forcing then brings the satellite to a due west-east trajectory at the northernmost portion of its orbit. The satellite continues in a straight line; however, the parallels of Earth's latitude circles curve away from the straight-line motion of the satellite. Place a pencil along a latitude circle (tangent to it) to observe this effect clearly. Again there is apparent rightward motion in the flight track. Exactly the same laws of physics apply to a cannon projectile or an airplane flying within the atmosphere. The same laws of motion apply to air parcels, even as they are affected by pressure forces from surrounding parcels. In addition to the various reasons for these apparent forces, the basic physical laws of the rotating Earth's coordinate system dictate the following: There is an apparent "force," always to the right (in the Northern Hemisphere) and always proportional to the speed of the parcel. (*Source:* Adapted from J. T. Houghton, L. G. Meira Filho, B. A. Callander, N. Harris, A. Kattenberg, 2d K. Maskell, eds., *Climate change 1995: the science of climate change*, Cambridge University Press, 1996.)

DIRECT CELLS WITH THE CORIOLIS EFFECT. Pressure-gradient forces can first accelerate a parcel from high to low pressure; however, within a number of hours, due to the Earth's rotation, the parcel is moving almost exactly perpendicular to the pressure force! See Figure 14.3. The forces become balanced. In equatorial regions, the Coriolis effect is minute, and the forces can accelerate parcels directly downwind.

The removal of air from some regions and the mass concentration in others tends to create regions of high and low pressure. Consider what happens when air begins to flow in response to the pressure effects, in order to resolve pressure differences (see Fig. 14.3). The balance of pressure-gradient force and the Coriolis effect causes air to move counterclockwise around a low-pressure system, and clockwise around a high-pressure system, in the Northern Hemisphere. In either hemisphere, a low forms a *cyclone* (same sense of rotation as the Earth) and a high-pressure system an *anticyclone*. This motion, paradoxically almost perpendicular to the force applied, is called *geostrophic flow*. *Geo* is from the Greek word

for "Earth," and *strophic* relates to turning. (In English, a "strophe" is also a "turn," or section, for example, a turn of phrase in poetry.)

CELLS, EDDIES, AND JETS

THE HADLEY CELL. The average circulation of the real atmosphere is strongly shaped by the Coriolis effect. Figure 14.4 shows two direct cells. The "Hadley cell" near the equator is the most obvious. (George Hadley was an English scientist, originally a lawyer, who appreciated the basics of the Coriolis effect long before Coriolis made the ideas explicit.) This cell acts much like the simple direct cell shown in Figure 14.1; however, it is restricted to tropical latitudes, where the Coriolis effect is weak and therefore motions do not tend to become contorted. The upward motion is not a general movement of all the air, as shown in Figure 14.1; rather, it is concentrated within intense, deep thunderstorms at latitudes with the most solar heating. The air pulled into these storms creates a region of air convergence: the intertropical conver-

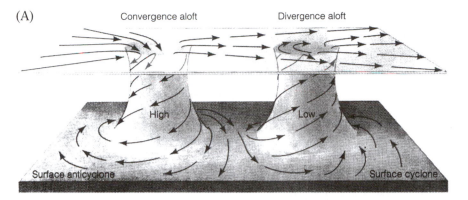

(A)

Convergence aloft

Divergence aloft

High

Low

Surface anticyclone

Surface cyclone

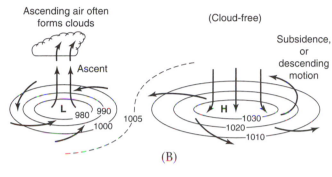

Ascending air often
forms clouds

Ascent

(Cloud-free)

Subsidence,
or
descending
motion

(B)

Figure 14.3. Motions of air into a low-pressure region in the Northern Hemisphere are shown at the right in (A) and at the left in (B). Air begins to move in from higher to lower air pressure, accelerated by the pressure-gradient force. However, the Coriolis effect acts continuously, causing the air to move to the right of the accelerating force; within hours, the motion becomes a slow spiral inward. The low-pressure region is called a *cyclone* because the air is moving cyclically, and in the same direction as the Earth's rotation. Similarly, in high-pressure regions, air moves outward from high to low pressure; however, the Coriolis effect moves it to the right to form nearly circular spiraling motions. These are called *anticyclones*: The motion is cyclic or nearly circular but is opposite to the direction of the Earth's spin. (*Source:* (A) Brian J. Skinner and Stephen C. Porter, *The blue planet*, Wiley, 1995, p. 342.)

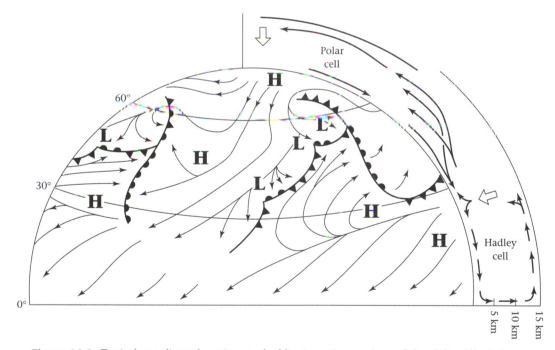

Figure 14.4. Typical coordinated motions and eddies in various regions of the globe. All winds are shown at the surface. Around the sides, cross sections show the vertical motions and great cells, such as the Hadley cells in the tropics. In the tropics, motions are driven by strong heating, and are *thermally direct*. At the poles, the strong chilling effects of the ice packs drive strong overturning motions. More complicated motions are typical in the mid-latitudes. From the Earth system perspective, these complicated motions are called *eddies*, or variations on the flow; from a more local perspective, they are *organized cyclonic storms*. These include winter storms and hurricanes or typhoons. Between these localized rainy regions are larger regions of fair weather, the great anticyclones, or high-pressure systems. (*Source:* Adapted from L. Musk, *Weather systems*, Cambridge University Press, 1988.)

gence zone (ITCZ). Over the continents, the ITCZ "follows the Sun" and the subsolar point tends to be hottest. This is not the case over the oceans, which retain heat for months and thus stay warm where the total heating throughout the year is most concentrated (i.e., the ITCZ within approximately 10° of the equator).

Surface air pulled into the ITCZ from the north is pushed to the right by the Coriolis effect and forms the trade winds. The trade winds are easterlies (winds are named from the direction from which they come, whereas ocean currents are named for the direction in which they are going). The trade winds are so called because their steadiness made for dependable trade routes in the days of exploration by sailing ships. Conversely, the Coriolis effect causes north-moving air at the top of the Hadley cell to produce strong west winds aloft, forming a westerly jet, a region or stream of fast-moving air.

Similarly, polar easterlies tend to form near the ice caps of the Earth: The high albedo and cool surfaces produce south-moving winds toward the warmer regions away from the caps; the Coriolis effect then deflects them rightward, to the east. There are variations in this flow near the poles, as Figure 14.4 shows. The idea of atmospheric cells as stable, repeating flows of air is accurate only in the tropics, the Hadley cell regions. In contrast to the three-cell patterns shown in many textbooks, mid-latitude and even polar "cells" are inadequate simplifications of very variable motions. Our common experience with the vagaries of weather and wind directions in these regions bears this out.

WEATHER EDDIES IN THE DISTURBED WESTER-LIES. Between the two regions illustrated in Figure 14.4 lies a region of intense mixing. The overturning motion becomes slanted: Warm, moist air typically rises and slowly moves north over thousands of kilometers; again, the northward motion and the rightward curving of the Coriolis force tend to create a jet, the polar jet. Winds may surpass 50 meters per second (100 miles per hour). The polar jet air tends to have the momentum of its origin, nearer the equator. Toward the pole, the high-momentum air becomes visible as high winds: The motion of the air is observed relative to this slower-moving mid-latitude portion of the Earth's surface. Cool, dry air sinks southward over thousands of kilometers (60° N to 30° N), and its original west-to-east motion seems to slow or even reverse.

The jets contain weather systems, and their winds tend to guide the weather systems originating from the surface and to be affected by them in turn. Thus, the jets and the paths of storms both become convoluted. Similar to water motions in a flowing stream, these wind motions are called *eddies*. The eddies affect the main flow, and vice versa. Surface winds turn often with the passing weather. The intricate, volatile behavior of the westerly polar jet and its attendant eddies led meteorologist Edward Lorenz to become an early expounder of chaos theory (as we now call it). Chaos theories focus on particular subject areas of physics that are characterized by

difficulty in prediction using physical equations. Lorenz applied chaos theory to real weather, explaining that minute differences in the present moment – for example, the flap of a seagull's wing in the South Pacific or the flutter of a butterfly in a tropical rainforest – can eventually change the course of large events – for example, the track of a storm – within a period of two weeks or so. He also applied the theory to weather forecasts, suggesting that minute differences in descriptions of the current state of weather can amplify into completely different weather forecasts ten days in the future. Lorenz also contributed insight into constancies amid chaos: He emphasized global energy balances as stable and reproducible. This description of the constant aspects is the study of the general circulation of the atmosphere; the attempt to make the best, practicable, detailed description of motions and phenomena for the future is the study of weather forecasting. Despite frequent and well-publicized local failures, there has been progress toward truly helpful forecasts ranging from five to ten days.

Theories regarding chaotic phenomena such as the weather can be judged by the patterns they do predict. By viewing the eddies over time, we notice some repeated features of the general circulation: High-pressure systems tend to lie over the cold eastern oceans, in other words, just off the west coasts of continents. Examples are the Pacific High and the Bermuda High. Low pressure tends toward other regions: examples include the southwestern oceans and the northeastern oceans (north of the highs). Among the latter are the Icelandic Low and the Aleutian Low, illustrated in Chapter 10. These common positions of highs and lows are determined by the complex interaction between ocean and atmosphere, including heat exchange between the two, and wind forcing of the ocean.

Changes in these positions can influence local climate significantly for entire states and provinces. Consequently, possible changes occurring as part of the global warming phenomenon are of great scientific and economic interest. These possible movements in the typical positions of the major highs and lows can result from a predicted geographic variation in greenhouse warming, as well as from the complexity of the ensuing response of air and ocean. These effects are the least understood physical effects of the expected global climate change.

CYCLONIC STORMS. Influenced by these very large patterns – highs and lows that often span half an ocean – smaller low-pressure areas move in the patterns commonly observed on television satellite animation loops. These moving lows take the form of cyclonic storms that tend to follow a "storm track," from southwest to northeast (in the Northern Hemisphere). They lack a regional name (such as the "Icelandic Low"). They are the most notable eddy feature of the mid-latitudes on any satellite picture or weather map. They are called winter storms or extratropical cyclonic storms. ("Extratropical" calls attention to contrasts between these temperate zone weather phenomena and hurricanes, which are tropical cyclonic storms characterized by intense rainfall, but a different origin; "winter storm" as a weather description contrasts with summertime features such as thunderstorms or hurricanes, but they are not restricted to winter.) Figure 14.5 shows the stages in the formation of cyclonic storms as the simple overturning motion of warm and cold air masses is made complex by the Coriolis effect, taking simple north-south motion and "spinning up" a cyclonic storm (winter storm).

TROPICAL CIRCULATIONS

The tropics have their own circulations. The Coriolis effect is weak; consequently, the basic motions respond quickly to new high- and low-pressure areas generated, for example, by concentrations of thunderstorms. However, there are also complex, partially understood forcings between the tropical ocean and the tropical wind systems. El Niño, or the El Niño–Southern Oscillation (ENSO), is a major example. El Niño – "the child" – is so named because fishermen along the coast of Peru noticed changes in the wind, and even rain in their extremely dry deserts, around Christmastime. This change of weather was not a Christmas present, however; it was accompanied by the cessation of ocean circulations that supported abundant fish populations in the Peruvian upwelling regions. After a period of time, typically a year, the rains revert to other regions of the Pacific Ocean. The name "ENSO" is used because this phenomenon affects the entire equatorial and South Pacific region, and often larger regions.

El Niño oscillations affect the mid-latitudes also, and this may change the route of storm tracks. The presence of an El Niño tends to produce disturbed weather in North America and southern Africa; however, the presence of El Niño alone does not necessarily signal a drought. One theory is that the El Niño shift of rains sends a "signal" (pattern of pressure changes in the atmosphere) but that other features of Northern Hemisphere weather can guide the signal toward or away from, for example, drought-sensitive regions of the west coast of North America. The current understanding is that the mere presence of an El Niño is insufficient to produce rain on the West Coast, although some hard-to-quantify "unusual weather" is commonly observed. Researchers continue to seek usefully predictive long-distance connections between weather and slowly changing, easily measured phenomena such as sea-surface temperatures. Some such connections may enhance our understanding of droughts and floods; however extreme, persistent phenomena may be due to other aspects of the "chaos" of weather phenomena that Lorenz described.

Recent years have brought an unprecedented frequency of El Niños, and a recent occurrence has lasted an unheard-of three years. The eruption of Pinatubo volcano probably has some subtle connection with this most recent episode, through the modification of the Earth's energy budget. This

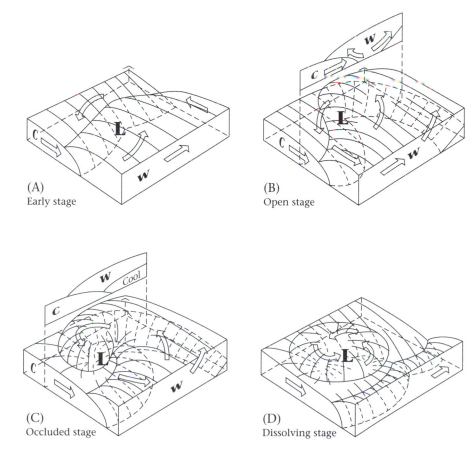

(A) Early stage

(B) Open stage

(C) Occluded stage

(D) Dissolving stage

Figure 14.5. A simplified view of the development of a typical mid-latitude weather system, known as an *extratropical cyclonic storm*, also called a *winter storm*, although it can occur during any season. (A) The storm begins anywhere along a front, or boundary between cooler (polar-influenced) and warmer (more tropically influenced) air. These boundaries tend to be distinct. (B) Warm air flows in from the south and rises, producing an area of low pressure. (C) Meanwhile, cold air sinks as it flows around the storm from the north; eventually some air will curve around the rear of the storm, along what is called the *cold front*. Cold air occupies progressively more of the surface, whereas warm air collects aloft. (D) As a result of the Coriolis effect, warm and cold air tend to take spiraling paths around the low-pressure center.

volcanic event is the largest eruption whose effects could be studied in detail using sophisticated satellite data. The energy budget effects and the secondary El Niño effects are complexly combined in the world's recent weather variations. Several questions are now preoccupying scientists: Is this change in the El Niño pattern related to global warming? Have the aerosols produced by El Niño and the Pinatubo eruption somehow temporarily slowed the buildup of greenhouse gases?

Tropical cyclones cause the most violent and unpredictable weather in the tropics. They are referred to by various names in different parts of the world: "hurricanes" in the New World, "cyclones" in the Bay of Bengal, and "typhoons" in East Asia and the Pacific. "Tropical cyclones" is the term that covers all of them. Tropical cyclones always begin in the warmest waters of the tropics, just north or south of the equator in the ITCZ. They begin as a typical loose organization of tall rain clouds, or cumulonimbus clouds. Usually these same clouds are called "thunderstorms" over the land, where they produce much more lightning. Atlantic hurricanes often start just south or west of the bulge of Western Africa. Particularly strong pumping of air from the lower atmosphere produces a low-pressure center there. Air rushes are the result of the pressure-gradient force; even at these low latitudes there is a slight cyclonic turning, causing a slight clockwise motion in the Northern Hemisphere. The winds eventually become so strong that centrifugal forces become significant, and the air whirls into a spiral (similar to water rushing down a bathtub drain). Unlike the winter storm phenomenon, the circulation becomes very tight and concentrated (several hundred kilometers across) and very nearly circular. Eventually the local effects of component thunderstorms interact with the larger circular structure and the hurricane takes shape. Remarkably, an "eye" forms, with very low pressure, calm winds, and often a cloudless sky – precisely in the middle of the storm. Tropical storms become classified as hurricanes when winds exceed 37 meters per second; at these speeds, winds whip moisture out of the ocean into the air, renewing the storm's energy source.

There are good reasons to expect that the frequency and perhaps intensity of hurricanes will increase if there is global greenhouse warming. It is observed that tropical cyclones almost never form unless sea surface temperatures are above 27°C (80.6°F), and are favored by even higher temperatures. This seems reasonable, because warmer seas evaporate water vapor to the atmosphere at an ever-increasing rate, and water vapor powers the intense oceanic cumulonimbus clouds that initiate and maintain the hurricane process.

ATMOSPHERIC HEAT BALANCE AND THE GLOBAL GREENHOUSE

This section broadens the focus on the way that energy produces today's greenhouse effect, and how this effect is increasing with time. To begin, we consider the Earth and the Sun as they interact with each other in space. Clearly, every heavenly object is emitting radiation – heat radiation. That heat radiation varies in two important ways. Most basically, the total rate of emission depends strongly on temperature, according to an extremely simple relationship described by the Stefan–Boltzmann law:

$$F = s_B T^4; \quad \begin{pmatrix} Total\ energy \\ emitted \\ per\ \mathrm{m}^2 \end{pmatrix} = \begin{pmatrix} Stefan- \\ Boltzmann \\ constant \end{pmatrix} (Temperature)^4$$

The flux of energy, measured in joules of energy per second per square meter, tells us how to calculate how energy is directed toward some specific area, for example, the Earth's surface. What is remarkable about the equation is the steepness of the rise with temperature, a fourth power. This means that as some effect raises the temperature of a patch of the Earth's surface, for example, by 1 percent that patch begins to radiate away energy 4 percent faster. You might expect that this effect would prevent much global temperature change, and it is indeed a powerful natural damper. This excess radiation with increased temperature works in a singular way: instead of simply emitting more photons at the same wavelengths, the radiation primarily shifts to activate more energetic varieties of radiation, from long-wavelength infrared (IR) to short-wave IR, to red to yellow to blue, and so forth. The increased temperature stimulates progressively more energetic types of atomic motions, and these can then excite more energetic electromagnetic radiation. Figure 14.6 shows the effect, and also the rapidly increasing T^4 effect of the Stefan–Boltzmann law. The emitted intensities at each

Figure 14.6. The intensity of emitted radiation is shown as it varies with wavelength. As the temperature of a solid body increases, more ways to emit radiation become available – an effect that derives from quantum mechanics. As a consequence: (1) the wavelength at which maximum emission of light occurs becomes shorter (and so, more energetic); and (2) the intensity of radiation emitted at all energies increases. Wien's law, as indicated in the figure, describes how the maximizing amount of radiation moves to shorter and shorter wavelengths with increasing energy. Both scales use powers of ten: The effect is very strong; in fact, it is proportional to the fourth power of temperature. μm = micrometers. (*Source:* Adapted from Richard Turco, *Earth under siege: from air pollution to global change*, Oxford University Press, 1997.)

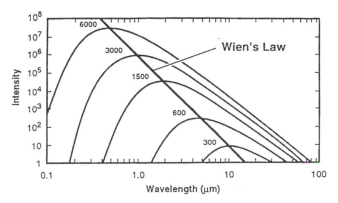

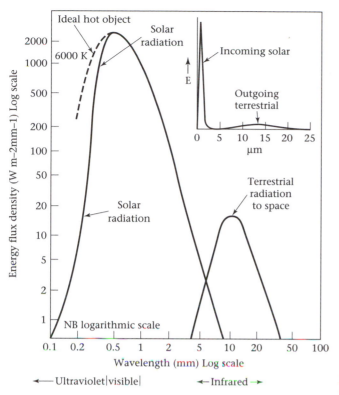

Figure 14.7. Thermal radiation from the Sun and Earth is compared. The main figure has log scales. The insert shows a linear scale of same the information. μm = micrometer, W m^{-2} = energy (E). (*Source:* Adapted from L. Musk, *Weather systems*, Cambridge University Press, 1988.)

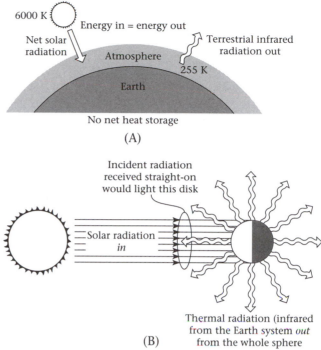

Figure 14.8. (A) An energy-balance argument describes how radiation that travels from the Sun to the Earth system and the Earth's own heat radiation must balance. That balance equation determines the temperature of the Earth system. (B) The geometry of solar and terrestrial radiation. The Earth System radiates in all directions, whereas the Sun's distant rays illuminate only half the Earth, with polar and dawn/dusk regions receiving a wide-ranging beam. The correct budgeting of radiation is calculated by considering full solar illumination on a disk of the Earth's radius oriented perpendicular to the Sun. (*Source:* Adapted from Richard Turco, *Earth under siege: from air pollution to global change*, Oxford University Press, 1977.)

wavelength increase so rapidly that special graphics scales are needed to capture all the variation.

The Earth system and the Sun have very complicated structures and do not have a single emitting surface; however, each can be assigned a temperature based on its energy emission to space (shown in Fig. 14.7). The appropriate temperature for the Sun is approximately 6000 K, and the temperature of the Earth system is approximately 255 K. The emitted flux F is related to the area under the intensity curves shown in Figure 14.6 (along with some detailed concern for geometry): The ratio of the temperature of the Sun to that of the Earth is 23.5; however, the ratio T^4 is over 300,000. Measurements by several NASA and European satellites confirm these temperatures and energy emissions. These satellites also show many regional variations, due to weather, but only tiny variations in the total energy emitted over time.

The basic climate fact we will use is that the Earth system must export as much energy as it receives; otherwise, the energy content and temperature would have to increase. Satellite observations show that the effective surface temperature of our planet is virtually constant. Considering the entire Earth system up to the top of the atmosphere, we can understand the 255 K figure, approximately −18°C (0°F). This does seem to be a very low overall temperature for our planet, and it requires some explanation.

The energy balance of the Earth depends on the Sun's radiative flux as the energy is spread out to the Earth's position on its orbit. That places the Earth in a beam of sunlight (shown in Fig. 14.8) whose area is πr_{Earth}^2, with a flux that is 1360 joules per square meter per second, or 1360 watts per square meter. Figure 14.8 also shows two other basic details of the Earth's heat balance. Only the portion of the Earth's surface facing the Sun receives solar radiation, which diffuses near the poles. The area of the Earth is calculated as follows: $A_{Earth} = 4\pi r_{Earth}^2$. (If you have a mind for details, you may correctly object that the sunlit half of the Earth is always $2\pi r_{Earth}^2$; however, then you must account for the diffusion of light over the surface near the poles, and also at dawn and dusk. Common experience confirms the idea that these are times and places of very weak solar heating. The easiest way to account for the fact that less solar energy is received is to look at the area of a circle oriented perpendicular to the Sun just at the Earth's orbital distance. The area of that circle, $A_{full\ Sun} = \pi r_{Earth}^2$, is fully exposed to the solar flux; the expression $F_{Sun} A_{Earth}$ correctly integrates the Sun's energetic effect.) On the other hand, the entire Earth's surface, $4\pi r_{Earth}^2$, and atmosphere $4\pi r_{Earth}^2$ radiate energy out to space. Considering the geometric details (i.e., substituting $A_{Earth} = 4A_{full\ Sun}$), we arrive at

$$F_{Sun} A_{Earth} = 4s_B T^4 A_{Earth} \cdot \text{(incomplete equation)}$$

However, as is clear to anyone who has seen the Earth's disk as photographed from space, the Earth reflects a tremendous amount of radiation. The clouds appear to be very white, and reflect 80 percent of the incident radiation or more. So do the polar ice caps and areas covered with fresh snow. Cloud-free regions of oceans are impressively dark, reflecting only approximately 10 percent of radiation; dense forests act similarly. Grasslands are intermediate, reflecting 10 to 30 percent of percent, and sandy deserts reflect up to 45 percent. Each of these reflection-to-space percentages is called and *albedo*, or "whiteness," and is given the symbol a. By adding all the geographic effects appropriately, we obtain an average number. The appropriate global average albedo is approximately $a_{Earth} = 40\%$. Table 14.1 shows a sampling of these albedos. Subtracting out the reflected energy, we can write a very simple equation for the conservation of energy that underlies the greenhouse effect:

$$F_{Sun}A_{Earth} - a_{Earth} F_{Sun}A_{Earth} = 4s_B T^4 A_{Earth}$$

Eliminating the A_{Earth} factors and summing terms, we have

$$(1 - a_{Earth})F_{Sun} = 4s_B T^4$$

$$\left(\begin{array}{c} Solar \\ energy\ flux \end{array}\right) \left(\begin{array}{c} Correction\ for \\ reflected\ light \end{array}\right) = \left(\begin{array}{c} Geometric \\ factor \end{array}\right) \left(\begin{array}{c} S.\text{-}B. \\ radiation \\ factor \end{array}\right)$$

$$\left(\begin{array}{c} Earth\ system \\ temperature \end{array}\right)^{\left(\begin{array}{c} Large \\ power! \end{array}\right)}$$

which leads to a formula for calculating the temperature of the Earth system:

$$T = \left[\frac{F_{Sun}(1 - a_{Earth})}{4\ s_B}\right]^{1/4}$$

TABLE 14.1. AVERAGE ALBEDO OF VARIOUS SURFACES AS THEY REFLECT THE SUN'S RADIATION

Surface	Albedo
Water	0.08
Fresh snow	0.8–0.9
Melting snow	0.4–0.6
Sand	0.3
Grass	0.2
Deciduous forests	0.15
Evergreen conifer forests	0.10
Tropical rainforest	0.10
Thunderstorm clouds	0.07–0.15
Low, broken stratocumulus	0.6
High, thin, cirrus	<0.4–0.5
Total, Planet Earth	0.31

These formulas suggest that the Earth is extremely resistant to changes in temperature: If the Earth's emission of heat radiation is extremely sensitive to the temperature of the Earth system, the converse is that a large change in the input from the Sun, or the Earth's albedo, or any such energy-related factor is necessary to change the temperature. A 4 percent change in these factors is required to make a 1 percent change in the temperature. Thus, there is a very powerful brake on global climate change. In other words, using the general terminology applied to interacting systems, this is a strong "negative feedback." Any temperature rise activates a process that strongly promotes cooling.

What causes this cold temperature of $-18°C$? Figure 14.9 shows the answer. The Earth's atmosphere is able to emit radiation in some infrared regions, which means that it can transmit radiation in all directions – some to ground, and some to space. The radiation that can interact with molecules can and does reabsorb and re-emit a fair amount; consequently, energy is passed back and forth between atmospheric layers. Most of the tradeoff occurs in the troposphere, which is cooler than the Earth's surface, as shown in Figures 13.3, 13.4, and 13.6. The net effect of the radiation that can escape from the warm surface, as well all the contributions from the cooler air layers, is a mix yielding an equivalent temperature of $-18°C$ (0°F).

Conversely considerable radiation originates in the atmospheric layers and travels down to the Earth. The planetary surface is considerably warmer than $-18°C$. Our energy analysis demonstrates that without the presence of the atmosphere, that would be the average temperature of the Earth. (Actually, at $-18°C$, a great deal more frozen water would form on the surface as ice, and the planetary albedo would be very high. The resultant increased albedo would be unstable, the world would freeze, and a prolonged ice age would ensue.)

The overwarming is due to the greenhouse effect. Radiation back from the atmosphere, at some wavelengths, warms the Earth's surface far above its simple equilibrium temperature of $-18°C$. The atmosphere acts like a blanket. This warming can be called the "atmosphere effect." Real greenhouses act primarily to retain energy in another way: the glass prevents heat from escaping by buoyant air motions; that is, it stops the thermals.

Figure 14.10 shows the actual phenomenon in more detail. Minor atmospheric constituents with complex molecular structures have energy bands (assemblages of emitting wavelengths) that allow them to interact with infrared radiation – to both absorb and emit at that frequency. (Absorption and emission are always inextricably linked – if a molecule can absorb at a wavelength, then its atoms and electrons can also rearrange to emit at that particular wavelength.) The important roles of water, carbon dioxide, and ozone are shown. Clearly, the Earth system temperature of 255 K is simply an intermediate figure between 300 K and 225 K in the figure. The region with wavelengths of approximately 8 to 10 micrometers shows some hatching for possible other

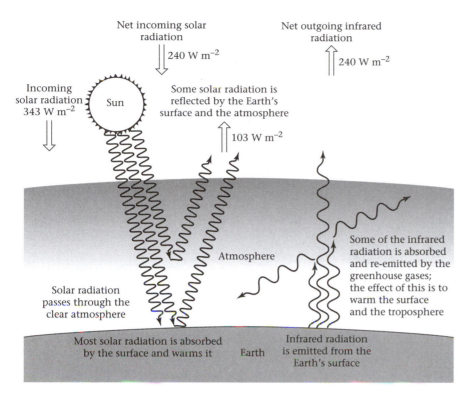

Net incoming solar radiation

240 W m^{-2}

Net outgoing infrared radiation

240 W m^{-2}

Incoming solar radiation 343 W m^{-2}

Sun

Some solar radiation is reflected by the Earth's surface and the atmosphere

103 W m^{-2}

Atmosphere

Some of the infrared radiation is absorbed and re-emitted by the greenhouse gases; the effect of this is to warm the surface and the troposphere

Solar radiation passes through the clear atmosphere

Most solar radiation is absorbed by the surface and warms it

Earth

Infrared radiation is emitted from the Earth's surface

Figure 14.9. The basic energy exchange processes involved in the Earth's radiation budget and greenhouse warming. The radiation of energy back from the central atmosphere to the Earth's atmosphere helps keep the system warm, even though the central atmosphere is emitting at only 255 K. In the energy balance of the Earth system as observed from space, the temperature also appears to be an average of 255 K, corresponding to an emission of 240 watts of energy per square meter (W m^{-2}) (*Source:* Adapted from J. T. Houghton, L. G. Meira Filho, B. A. Callander, N. Harris, A. Kattenberg, and K. Maskell, eds. *Climate change*, Cambridge University Press, 1996.)

absorbers/emitters, that is, other greenhouse gases. Figure 14.10 shows that some of the possible emitters are methane (CH$_4$), nitrous oxide (N$_2$O), the chlorofluorocarbons (CFCs), and ozone. Carbon dioxide builds up along the side. In fact, the fluctuations in absorption are much too subtle to depict on this graph; even the width of the boundary line is too coarse to show all the possible structures.

Graphic illustrations of the Earth's radiation budget are routinely available to us in the various satellite photograph used in weather discussions, as shown in Figure 14.11. A less commonly used satellite picture, the whole-Earth visible image, shows clearly the variation of albedo over the entire globe, and the light available to warm the Earth system. Daily and seasonal variations of the light and the important effect of snow and, especially, clouds on the albedo are clear. The

most commonly used image shows not the variable light of the Sun but instead the heat radiation emitted from the Earth. In fact, to aid in the interpretation of the continuously "lighted" visible image, black and white as sensed by the satellite are inverted by computer processing: Hot ground surfaces are portrayed as dark, whereas very high, very cold cloud tops are portrayed as very bright. Low clouds, with temperatures near the surface, are barely visible.

Figure 14.11 also shows a less common image – the infrared radiation emitted by water in one of the window regions shown in Figure 14.10. The wavelength is chosen to empha-

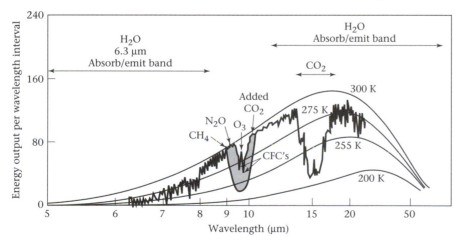

Figure 14.10. How greenhouse gases are "filling in the window" through which Earth's thermal radiation escapes the atmosphere. Relative absorption features contributed by a variety of trace gases are shown. Absorption of radiation at these wavelengths (and re-emission downward) is increasing as human activities increase these greenhouse gas concentrations. Carbon dioxide is a very strong radiation absorber on the left and right sides of the window; however, increases in the carbon dioxide concentration have small effects, increasing absorption at the lower-absorbing sides of the main absorbing regions. μm = micrometers. (*Source:* Adapted from J. T. Houghton, L. G. Meira Filho, B. A. Callander, N. Harris, A. Kattenberg, and K. Maskell, eds. *Climate change*, Cambridge University Press, 1996.)

Figure 14.11. Three satellite views of the Earth, using radiation emitted or reflected from the Earth system in the (A) visible, (B) "window" infrared, and (C) water vapor infrared bands. The black areas and white areas in the latter two views have been reversed, to appear comparable, that is, similar to the visible. The "window" infrared, so named because the atmosphere is nearly transparent, allows infrared radiation in and out freely. In fact, the surface of the Earth is brightest, because its surfaces are the warmest. The water vapor view shows a "blanketing" of the Earth; it shows graphically how a greenhouse gas can fill up the window, emitting radiation at a cool (mid-tropospheric) temperature to space while also returning Earth-surface radiation downward. As global warming continues, these satellite views of water vapor will become increasingly blanketed with white. (*Source:* Images from the Atmospheric Chemistry and Dynamics Branch, NASA Ames Research Center.)

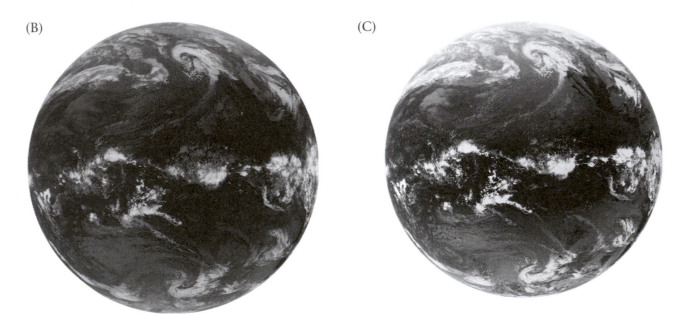

size water vapor in the middle of the troposphere (liquid water and ice surfaces in clouds also contribute where clouds are present). The mechanism of the global greenhouse effect can be clearly seen here: As water vapor increases with greenhouse warming, the blanketing effect in this channel will become more pronounced. The Earth's surface will be less visible in the photograph, thermal radiation from the middle troposphere will be seen at the sensor, and also more thermal radiation will return to the ground. A slightly different picture could be made at each wavelength; the sum effect of energy from all wavelengths and all regions of the planet defines the global Earth system temperature described above. Images made by using radiation from those infrared energy bands where carbon dioxide is currently absorbing, but not completely absorbing, will also become more filled in and

featureless as the greenhouse gas blanketing effect becomes more pronounced. The water vapor images also help to illustrate the fact that greenhouse warming will not be uniform everywhere – radiation will continue to escape from some areas of the Earth more effectively than from others.

It should now be apparent why the stratosphere is cooling as part of the greenhouse effect, and why the upper stratosphere is essentially disappearing into the mesosphere. This region is above the blanketing effect of the atmosphere; as excess carbon dioxide accumulates at 50 kilometers, it is in fact emitting more effectively to space. The upper stratospheric cooling is proceeding at approximately 8°C per decade. Paradoxically, one of our most direct confirmations of global change and general greenhouse warming is the predicted stratospheric greenhouse cooling! That process is easiest to observe because, despite the effect of ozone depletion, this region is also the most isolated from the greater complications of oceans, clouds, and land surface change.

A central uncertainty in predicting future global temperature warming is the role of clouds. They reflect not only light to space but also a large amount of heat energy back to the Earth. Their strong effect is noticeable at night: Cloudy nights remain comparatively warm, while clear nights tend to cool rapidly. The greenhouse role of water vapor is also noticeable: On clear nights, humid areas cool much less than the drier deserts, which start at the same early evening temperature – often, the deserts are even hotter at the outset. The deserts lack the water vapor molecules overhead necessary to radiate infrared back to the surface. These regions that often appear distinct in the water vapor images such as Figure 14.11 constitute a place where radiation escapes to space.

A similar concern for local climates has been recognized for years: anthropogenic change in the albedo of large areas of the Earth. One hypothesis concerned the change of the albedo in sub-Saharan Africa, the Sahel, where it was feared that degradation of the soil by more intensive agriculture and pasturage, occasioned by rising populations, would make the region brighter, as dark vegetation turned to lighter bare soil. Similar changes can be envisioned where darker forests are cleared, possibly leaving degraded grassy land – also a change for the lighter. The study of effects of global climate change that might be occasioned by more intensive clearing and agriculture have become even more complex. For example, recent studies have emphasized the importance of the dust lifted into the air as soils are cleared and bared: The dust so lifted and the smoke associated with land clearing and crop preparation seem to be the major visible effects of humankind on the aerosol particles in the atmosphere. These turn the clear global atmosphere more turbid, and thus affect the albedo. These effects are much more clearly visible from satellites than are the effects of sulfate pollution from industrialized countries. Another result is less obvious: a primary effect of clearing large areas of rain forest on wind systems and temperatures in the tropics is caused by friction. The uneven tops of tree canopies slow winds more effectively than pastureland or bare soil. Studies of these effects necessarily involve many scientific disciplines and accurate projections of human activities.

How do changes in atmospheric gases cause substantial effects? The $s_B T^4$ effect apparently acts to "brake" most changes in the radiation budget – a negative feedback. However, a strong accelerator exists, also. This is the way water vapor in the atmosphere responds to temperature increase. The law for the evaporation and recondensation of water is even more dramatic than T^4:

$$n_{H_2O} = \frac{n_{Reference}}{T/T_{Reference}} \, 10^{\left(-\frac{L_{vap}}{2.303 \cdot RT}\right)}$$

This expression employs R from the ideal gas equation; L_{vap} represents the latent heat of vaporization, and $T_{Reference}$ is the (reference temperature) of 273 K (equivalent to 0°C). This complex expression is simply based on a comparison of the energy that is available as thermal energy, $R\,T$, and the energy that is required to separate a water molecule from the liquid, L_{vap}. (The technical term for this common type of equation that describes energy partitions between different forms is a *Boltzmann law*. Such laws involve a comparison of energies and an exponential dependence. The reference density of water vapor in the equation, $n_{Reference}$, is 9.93×10^{27} molecules of water per cubic centimeter.) We are interested in the behavior with T, and this is more complex than any we have seen. Consider the exponent of 10, namely,

$$\left(-\frac{L_{vap}}{2.303 \cdot RT}\right)$$

The fact that the inverse of T is involved (T is located in the denominator of the fraction) causes this exponent to decrease as T increases. The minus sign in front of the fraction suggests the same behavior. The combination of the minus sign and the fact that T is located in the denominator means that the entire exponent expression increases T. This combination of effects implies that the number of water molecules increases with temperature. The fact that the entire expression is used to provide a power of 10, an exponential relationship, means that the increase is rapid.

The meaning of the algebraic expression is quite significant: while a 1 percent increase in T produces a 4 percent increase in the radiation emitted to space (the seemingly powerful "brake" mentioned above), a 1 percent increase in T produces a 20 percent increase in the water vapor concentration! The next 1 percent increase in T produces an even greater increase in water vapor concentration. That excess water vapor provides the Earth with much more of the greenhouse blanketing power. The implications are that there is a powerful accelerator on global temperature change. The oceans provide a huge source of water for the atmosphere, and as the ocean–atmosphere system increasingly warms, water is ever more rapidly evaporated to the atmosphere.

As the understanding of the evolution of similar planets – Venus, Earth, and Mars – developed, this exponential increase suggested the possibility that Earth could experience a "runaway greenhouse effect" and all its water could evaporate into a massive hot atmosphere. The implication was that Earth could end up with 400°C surface temperatures such as those on Venus!

What about clouds? Remember how bright they are and how they directly affect the Earth's albedo and the direct reflection of energy to space. If the temperature is hotter at the surface, and more water evaporates, will there be more clouds? Temperature must be cooler somewhere farther up in the atmosphere, which would create still more clouds. Or if the Earth cools, and the air has more regions that can cool and condense water, will there be more clouds? In fact, both kinds of effects can occur. The powerful influence of clouds on climate is a subject of intense study, both theoretical and experimental. To make matters more complicated, recall that more clouds not only reflect light to space but tend to trap

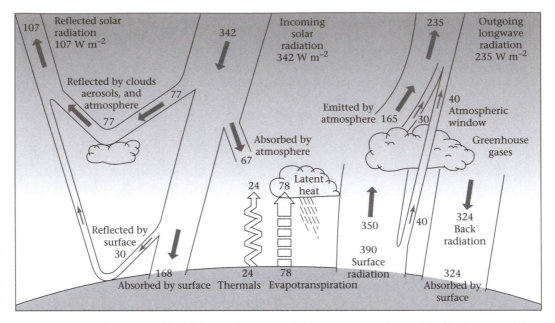

Figure 14.12. Proportional breakdown of energy paths flowing into and out of the Earth system. W m^{-2} = watts per square meter. (*Source:* Adapted from J. T. Houghton, L. G. Meira Filho, B. A. Callander, N. Harris, A. Kattenberg, and K. Maskell, eds. *Climate change 1995: the science of climate change,* Cambridge University Press, 1996.)

infrared radiation near the Earth, as the greenhouse gases do. So both the blanketing (insulation) and the reflecting of solar radiation by clouds require much more scientific study. However, some basic facts are known. For example, clouds are thought to provide a powerful limit, preventing a "runaway greenhouse effect."

The balance of these large effects is clearly governed by an immense number of details; thus far we have only briefly glimpsed the titanic countervailing forces of the Earth–atmosphere system. We can begin to see how detailed calculations of global climate change may be carried out (Fig. 14.12). Our arguments have been based on simple energy balances and have not considered the immense variation over the Earth's regions. Later chapters describe the detection of global warming and the possibilities for other associated global changes.

QUESTIONS

1. Describe mountain breezes, in which air is heated up by morning and midday sunlight on mountain slopes and mountain peaks, as an example of a direct circulation. What are the directions and nature of the winds near the surface, near the peaks, and aloft? In the absence of other weather effects, where does the warmed air above the peak eventually go?

2. Refering to this chapter and Chapter 13 (see also Chaps. 10 and 11) list several ways in which the atmosphere and ocean differ. Both fluids, the atmosphere and seawater carry heat (energy) with their motions. Which absorbs visible light most easily? Which emits infrared energy most easily? Where are they heated? How does this effect their stability and motions? Describe pressure effects on their density.

3. In the High Plains just east of and leading up to the long line of peaks and highlands called the Front Range of the Rocky Mountains, mountain-generated winds occur; however, by afternoon these winds blow predominantly from south to north along the range. Explain.

4. Name the main positive and negative feedback effects on global atmospheric temperature changes described in this chapter. Can you conjecture additional feedbacks based on albedo or water vapor effects? What are your ideas about the response of the Earth's surface and vegetation to global temperature warming?

5. Describe the motions of air in a cyclone, especially in the trailing cold area. Besides moving from north to south, this air is being forced downward. Does this help you explain the sudden clearing of clouds when such cold air masses move over an area? *Hint:* Even though the air feels cool at the surface, what is happening to the temperatures of the air parcels in the air mass as a result of the downward motions?

6. Most aerosol particles are clear and do not absorb light – they only scatter it. The effect of adding such aerosol to the atmosphere is to scatter more of the Sun's light away from the Earth, increasing the Earth's albedo, a_{Earth}. It has been suggested that aerosol in the Earth's atmosphere has been increasing in the past thirty years due to industries that manufacture the same sulfur compounds that produce acid rain (sulfuric acid and sulfates). What would be the qualitative effect on the Earth system's temperature of increasing the albedo by 1 percent? What would be the quantitative effect of increasing the albedo from 0.40 to 0.41 on an Earth system temperature of 255 K?

7. Using only the adiabatic lapse rate value provided in

Chapter 13, estimate the height in kilometers of the −18°C region in the atmosphere above the surface (−18°C is approximately the average overall Earth–atmosphere system temperature). You may assume additionally that a typical Earth surface temperature is 15°C.

FURTHER READING

Barry, R. C., and Chorley, R. J. 1992. *Atmosphere, weather, and climate*. London: Routledge Press.

Houghton, J. T., Meira Filho, L. G., Callander, B. A., Harris, N., Kattenberg, A., and Maskell, K. (eds.) 1996. *Climate change 1995: the science of climate change*. Contribution of Working Group I to the Second Assessment Report of the Intergovernmental Panel on Climate Change. Cambridge University Press. 564 pp.

Trenberth, K. E., ed., 1995. *Climate system modeling*. Cambridge: Cambridge University Press.

Turco, R. P. 1997. *Earth under siege*. Oxford: Oxford University Press.

Wallace, J. M., and P. V. Hobbs. 1977. *Atmospheric science an introductory survey*, San Diego, CA: Academic Press.

15 Can We Forecast Climate Future without Knowing Climate Past?

STEPHEN H. SCHNEIDER

SEA-LEVEL AND WARMTH IN THE CRETACEOUS

In the past twenty years, a number of theories have developed that attempt to link geology, climate, and sea-level rises. One theory was developed by Pennsylvania State University paleoclimatologist Eric Barron, who worked with a three-dimensional computer model of the atmosphere. He melted all the ice on Earth (in the mathematical equations of the model, of course) and assumed that the Sun had the same energy output then as it does today. He then calculated what the temperatures would have been 100 million years ago, appropriate for the high-sea-level paleogeography (shown in Fig. 15.2) during the reign of Tyrannosaurus rex.

First, Barron compared the present land area with that of 100 million years ago. Back then, there was only about 20 percent land area globally. Because land is more reflective than water, the Earth would have absorbed more sunlight then. Without bright white snow and ice at the poles, there is a so-called positive feedback that amplifies the warming. Taking all these factors into consideration, Barron calculated a temperature that was approximately 5°C warmer 100 million years ago than it is today. However, the paleoreconstruction of climate from fossil evidence suggested that it was actually more like 10°C warmer in those long-ago years. Obviously, something was missing – or else the model was insufficiently sensitive to the paleogeographic conditions assumed. Several explanations for this discrepancy have been proposed. As he explored the options, Barron found that if he increased the carbon dioxide in the computer model by four times, there was a much closer fit to the paleotemperature data. However, that is only circumstantial evidence – not proof that an enhanced natural greenhouse was the missing factor.

In fact, the entire business of testing climatic theories is a matter of collecting and comparing various pieces of evidence. One piece of evidence proves little by itself. However, when you perform five or eight different "validation experiments," and the evidence begins to build, confidence in your ideas is much higher than if the evidence from only one or two experiments supports your hypothesis. For the Cretaceous Period, several factors are consistent: It was approximately 10°C warmer, 80 percent rather than 70 percent of the Earth's surface was covered by oceans, sea levels were approximately 300 meters higher, warmth-loving plants and animals were widespread, and plant productivity was high (consistent with all the fossil fuels that were laid down). Carbon dioxide enhances plant productivity, less ocean basin volume coincides with active volcanism at mid-ocean ridges, and the latter causes more carbon dioxide emissions. Does all this "prove" that the carbon dioxide–driven computer models are correct? Not necessarily, of course, because all these factors could be coincidental or related to not-yet-understood additional factors. But is it a likely scenario? The story hangs together well. The proof is circumstantial but, to this observer, quite persuasive that many basic concepts we now have are reasonable.

In the field of Earth systems science, we look at how human activities change the composition of the atmosphere, the shape of the Earth's surface, and the amount of heat that is added to the system. We then try to forecast how those activities might change natural conditions over time. However, to make any forecast, one must have a forecasting tool. The "crystal ball" that scientists use is a computer model of the Earth system. Computer models are not yet well validated for conditions far from the present – for example, when we double carbon dioxide (CO_2) production over the next half-century. As a result, we must constantly search for ways to test these models. The best physical laboratory we have for such testing purposes is not a laboratory built with glass and steel and wire, but the very Earth itself.

FLUCTUATIONS IN CLIMATE PAST

The Earth's climate is historically quite variable (Chap. 3). We have evidence of past ice ages, as well as ice-free epochs lasting tens of millions of years. Scientists have even found 1 or 2 billion years of Earth history with little or no oxygen in

W. G. Ernst (ed.), *Earth Systems: Processes and Issues*. Printed in the United States of America. Copyright © 2000 Cambridge University Press. All rights reserved.

the atmosphere. There are other differences, as well: Continents were located in different places, the amount of energy from the Sun was different, the composition of the atmosphere was different. In other words, there have been natural "experiments" in the past that have been part of extremely large-scale changes – even bigger changes, in many cases, than humans could effect by altering the chemical composition of the atmosphere in the coming decades. On the other hand, the natural rates of these historic changes have usually, but not always, been extremely slow. The anthropogenic atmospheric change evidently now in progress will happen in a comparative eyeblink.

The questions we must address are: Can we quantitatively reconstruct what happened in the past, and how can we use that knowledge to calibrate the theories that we need in order to forecast credibly the climate of the future? Answers to these questions will enable us to evaluate better a number of public policy issues such as whether we should be switching from coal to gas, or even to solar energy. Should we have vigorous conservation or nuclear energy development? Should coal miners be retrained for other jobs? Do we need mandatory population programs or reforestation efforts? Or, can we safely wait for market forces to adapt smoothly to global changes? All these are fundamental public policy issues that raise trillion-dollar development questions and could affect the fates of many species (see Chap. 26).

To get a glimpse of the far-reaching impact of these issues, step back in time in America's heartland. The Sand Hills of Nebraska are predominantly grass-covered agricultural lands today; however, those hills were sandy between 3000 and 8000 years ago because that part of the American Plains was very dry then. The humid corn belt in Iowa and Illinois that we know today was much dryer back then as well – it was what paleoclimatologists call a "prairie peninsula," a tongue of aridity that penetrated a few hundred kilometers eastward relative to its present location.

Approximately 20,000 years ago, prior to the Holocene (the geologic name given to the present interglacial time interval that began approximately 10,000 years ago and continues today), there was no corn belt possible anywhere in the American Midwest because it was too cold. Spruce trees, today typically found hundreds of kilometers north in Canada's Boreal forests, were dominant in the present corn belt. As the ice gradually receded north and ultimately disappeared, the climate warmed and the natural vegetation patterns underwent disorganization, movement, and transformation, settling down several thousand years ago to the patterns we recognize today, with grassland in the Western Plains and hardwood forests covering the Eastern Plains and Northeast states. However, between approximately 3000 and 8000 years ago, when summers were perhaps a few degrees Celsius warmer than today, the extensive aridity of the prairie peninsula reigned in the Mississippi Valley. If a few degrees of warming were to recur today – this time as a result of anthropogenic greenhouse gases – would the Sand Hills of Nebraska become sandy again? Such a dramatic change

would not be good for agriculture as currently practiced in the American Midwest or for the overall economy in that part of the hemisphere. This is why climatologists want to know what caused the original natural change – to see if there is some common response to warming per se, or if the cause of the warming matters. If the cause does matter, and we know what it is, can we then "hindcast" the aridity of the prairie peninsula using the same tools we use to forecast a twenty-first century enhanced greenhouse effect? If so, our confidence in the accuracy of the forecast is significantly increased.

It is likely that changes in the Earth's orbit around the Sun arising from the gravitational tugs of Jupiter and other planets redistributed the amount of solar heating between winter and summer some 8000 years ago relative to now – with approximately 5 percent more summer sunlight and 5 percent less in the winter. This could account for a few degrees Celsius summer warming. Warming caused by increases in greenhouse gases (these gases create heat in both winter and summer) is not a good metaphor for what occurred during the warm summers of the prairie peninsula era. However, knowledge of that period does provide lessons for the twenty-first century. If we can apply the same climate models we use to project future anthropogenic changes to the study of past natural changes, and if they respond well, that validation process helps us to develop credibility in the model. Once we have corroborated the model, we can use it more confidently to forecast the climate. We return to this 8000-year-ago test case later, in the section dealing with climatic cycles and feedback (glacial and interglacial ages).

While geologists are digging up rock records, paleobotanists are in the field looking for other clues to climate past. They extract cores of soil and lake sediment and bring them back to the laboratory. There, scientists count the kinds of pollen grains remaining in each level of the core and determine their age by carbon dating materials at that level or by other methods (e.g., counting levels backward if they are annual strata). They figure out what trees, grass, and herb pollens existed and in what relative abundances, date their ages, and then infer from these relative abundances how the climate may have changed (e.g., different species prefer hot or wet climates). The association between species range limits and large-scale environmental factors, such as temperature and rainfall, constitute part of the subdiscipline called *biogeography* (see Chap. 21). Biogeographers can map on a large scale (hundreds of kilometers) what types of vegetation groupings are likely to be present in a place simply by knowing temperature and precipitation there. For example, tundra is predicted if summer temperatures are less than 10°C, tropical rain forests if temperatures and rainfall are high, deserts if the climate is arid, and so forth. Of course, local factors such as soils, competition, and herbivory keep such biogeographic "forecasts" very general (i.e., they can't accurately predict the relative abundance or ranges of individual species); however, they are useful as broad approximations.

Researchers also look at marine or glacial deposits where

chemical or isotopic compositions of rocks and fossils, or gas bubbles in ice, can serve as proxy indicators of temperature, sea level, and atmospheric composition. By taking samples from many locations, paleoclimatologists can look for signs of coherent patterns of change – requirement for confident quantitative paleoclimatic reconstructions. Researchers have been able to deduce that there was a prairie peninsula, that it occurred during the middle of the Holocene, and that there were many other changes taking place in the world at the same time – although annual global average temperatures have been, within a degree or two, quite stable for the past 10,000 years. Evidence of this global change includes fossil soils in the current deserts of Africa and India, which indicate that the monsoon precipitation zones in India and Africa were much wetter between 5000 and 9000 years ago than they are now or during the Ice Age. Although relatively little change occurred in the humid tropics 6000 years ago, the now-arid tropics underwent major changes. River runoff and the lake levels in Africa also were dramatically higher 5000 to 9000 years ago.

THE LINK TO CLIMATE FUTURE

How can paleoclimatic data help to validate modern climate forecasting tools? Computer models have been remarkably successful at explaining these enhanced mid-Holocene monsoons as a consequence of the increased summer heating associated with the altered planetary alignments that occurred then. Although these solar system patterns will not recur in a greenhouse-warmed future world during the twenty-first century, it is important for the credibility of our global change models that they can reproduce the major features of past climates once we know what caused those climates to differ from today's conditions. Later in this chapter, we will see how well these models performed (see the box entitled *Paleoclimatic Reconstructions and Model Simulations*).

Can we learn anything quantitative by taking a step much further back in time? We know from events today that gases emerge from the Earth's crust, particularly from volcanic fissures. It is very likely that these gases changed the Earth's atmospheric composition over the eons as the rates of volcanic activity varied. Geologists can estimate these rates by using a number of tools, including studies of ancient volcanic rocks and other deposits on land and in the seas. A change in volcanic outgassing rates, in turn, would have an impact on the greenhouse effect, which determines how much infrared radiative energy is trapped near the surface layers. Because these processes have been ongoing for billions of years, it is important to try to understand how the rates of

---

PALEOCLIMATIC RECONSTRUCTIONS AND MODEL SIMULATIONS

When researchers build climatic models using many factors, they can indeed reproduce that observed "sawtooth" pattern of Figure 15.3 in the computer output. However, this remains circumstantial evidence because we do not know for sure that the mix of these mechanisms operated in nature the way we have incorporated them into our model. Although we are able to simulate how ice ages come and go, we do not know conclusively how these events occur. Yet, there are enough consistencies between paleoclimatic reconstructions and model simulations to lend considerable credibility to the basic ideas.

What we do know from the remaining polar ice itself is that when temperatures were colder, there was less methane and less carbon dioxide in the atmosphere, and when they were warmer, there was a higher atmospheric concentration of these two greenhouse gases (see Fig. 1.1). These are natural changes, but in the expected direction: more greenhouse gases in interglacials, less in glacials – another set of positive feedbacks. In the early 1990s, climatologists Martin

Hoffert and Curt Covey studied the Cretaceous Period and looked at the differences between the last ice age and the current interglacial for clues to the sensitivity of climate to greenhouse gases. They also studied the previous interglacial and its ice age. They know what the carbon dioxide concentrations were in those periods because CO_2 values have been measured in ice cores. They assumed what those concentrations were in the Cretaceous Period, based on enhanced volcanism and sea-floor spreading estimates. They then calculated what the temperature difference was between the present time and various paleoclimatic periods with and without the paleo–greenhouse gas changes. They used these data to calibrate a most important relationship: how a doubling of atmospheric carbon dioxide changes the global temperature – known as the *climate sensitivity*. That figure remains controversial (it typically ranges from 1 to 5°C). Some scientists have calculated a 2 to 3°C warming for an equivalent doubling of carbon dioxide based on paleoclimatic situations—right in the

middle of the range from most climatic models: 1–5°C.

Is this calibration definitive proof that the sensitivity is 2.5°C? No, because other factors need to be considered. This ancient association between carbon dioxide and temperature might be a coincidence, not a physical linkage. However, it is encouraging that both these paleoclimatic methods provided approximately the same answer – the sensitivity of the climate to a doubling of carbon dioxide is approximately 2 to 3°C warming – a fairly coherent (but not conclusive) result. Regardless of the specific answer, this process illustrates a primary method used to test computer models. Because life influences and is influenced by carbon dioxide, which in turn is strongly associated with ancient climatic changes, this particular model reinforces the metaphor that life and climate have coevolved, each indelibly different because of their natural interactions over 4 billion years of Earth's history.

volcanic activity have changed over time. Such fluctuations might be affecting the atmospheric concentrations of some climatically or biologically important gases, such as carbon dioxide. Life – in particular, the algae and bacteria that have existed on Earth for nearly 4 billion years (Chap. 3) – also changes the chemical composition of the atmosphere, by using carbon dioxide and producing oxygen. This process of enriching atmospheric oxygen alters atmospheric composition and climate, which, in turn, affects life in a potentially endless feedback chain.

I developed a metaphor in my book, *The Coevolution of Climate and Life*, that may be helpful in visualizing how some of these processes interact. Coevolution is the biologic phenomenon whereby the evolutionary paths of separate species are altered by the presence of each other (e.g., butterflies eat plants that may evolve chemical defenses to which butterflies may develop resistance). The metaphor suggests that the physical, chemical, and biologic components of Earth all mutually interacted over geologic time in ways that altered their individual evolutionary paths.

Imagine compressing the 4.6 billion years of Earth's traceable existence into the proverbial six days of creation (see Fig. 15.1). Midnight on "Sunday" morning might be the first moment in Earth time, marking the beginning of the Precambrian Era, some 4.5 billion years ago. After a few hundred million years of cataclysmic planet forming processes, average surface temperatures, as inferred from sparse geologic evidence, wandered in the 10 to 25°C range until approximately 2 billion years ago (currently the global average surface temperature is approximately 15°C). This inferred temperature range is very large because the further back we peer into geologic history, the less geologic evidence there is, the more isolated it is, the more it can be influenced by local events, and the less we can conclude anything global. Consequently, we know very little quantitatively about what occurred much before 2 billion years ago. We do have some records of life (e.g., bacteria and algae) in rocks, and we believe that geologic processes such as outgassing from volcanoes went on; however, it is very difficult to determine global temperatures with any accuracy.

Figure 15.1. The "six days" of the coevolution of climate and life are shown here schematically. Although primitive life can be traced back to late Sunday, it is not until early Friday that rapid evolution of a large diversity of species begins in the Phanerozoic Era. All of art, science, religion, and our cultural heritage are crammed into the last second of the sixth day since the Earth's formation. (*Source:* S. H. Schneider and R. Londer, *The coevolution of climate and life*, Sierra Club Books, 1984.)

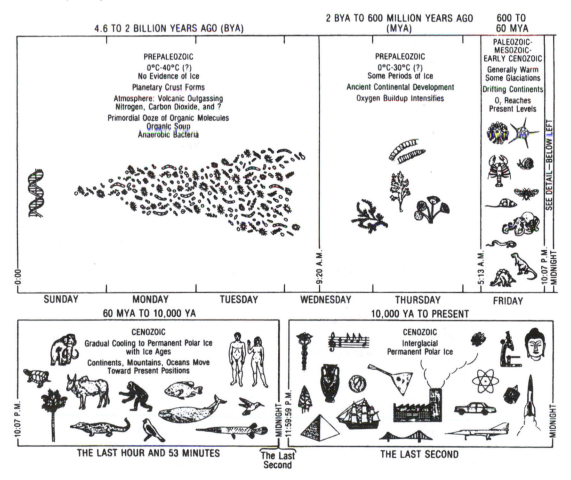

Two billion to 600 million years ago, which was still the Precambrian Era, we have more geologic and fossil evidence from which to infer what temperatures were. That evidence suggests temperatures ranging from very warm periods to very cold periods – again, with lots of question marks. This time frame, most geologists suspect, is when ancient continental development took place. Oxygen buildup intensified, and from one-celled life in the seas, biota evolved toward multicellular creatures by the end of the Precambrian.

The Paleozoic and Mesozoic eras span from 600 million to 65 million years ago. Biologic evolutionary development accelerated in an oxygen-rich atmosphere. However, these also were times of mass extinction. Nearly half of the very simple plankton in the oceans and most of the very large complicated creatures – most notably dinosaurs – disappeared at the end of the Mesozoic, approximately 65 million years ago. The debate is ongoing as to whether these extinctions resulted from Earth-bound biologic processes, an asteroid collision, or a period of particularly intense volcanic activity (Chap. 3). Many arguments focus on whether and why there were several such intense, short-period extinctions. Some scientists, studying the rate at which species seem to be driven toward extinction today as a result of habitat fragmention and other human changes, have argued that over the span of a few hundred years, human global change pressures may push global extinction rates to a level comparable to that of previous major natural extinction events ostensibly caused by extraterrestrial catastrophes.

Finally, the current geologic interval encompasses the 65 million years of the Cenozoic Era from the extinction crisis at the end of the Mezosoic until now – which represents approximately the final hour and fifty-three minutes of our metaphoric six-day time compression of geologic history. The Cenozoic is the period during which most of the mammals and many of the plants and animals that we know today evolved. It is also the time when the climate changed most dramatically, from a relatively warm, essentially ice-free Earth, through tens of millions of years of gradual cooling, to a world with permanent polar ice. In the last few million years, periodic ice age/interglacial cycles developed in which permanent ice was joined by major expansions of glaciers, especially at mid-latitudes and high altitudes. The arrival of widespread permanent polar ice was, at the earliest, perhaps 25 million years ago, and at the latest, approximately 10 million years ago. This approximation is difficult to make more precise, in part because the Antarctic continent is covered with ice, which tends to scrape away a large fraction of geologic evidence.

The final "second" of time on our six-day scale covers 10,000 years ago to the present, which is the Holocene Epoch. This 10-millennium period is the time in which modern ecosystems and human civilization developed, all enjoying a relatively stable (+/−1°C) interglacial climate era.

We focus here on a period about which we know a surprising amount, 100 million years ago in the middle of the Cretaceous Period. This is the time of Tyrannosaurus rex and the age of dinosaurs. It is a very well studied time, not only because scientists enjoy digging up dinosaurs and benefactors are willing to finance such paleontologic work, but also because governments and the oil industry have invested a great deal of money in research on this period. Fossil fuels, the hydrocarbon remains of ancient life that power modern industrial society (see Chap. 24), were laid down in extensive deposits in part during the Mesozoic Era. To find them beneath present-day landforms, we need to understand something about conditions then. For example, it is useful to know whether the sea level was rising or falling, as well as something about climatic conditions – whether it was arid or not. In shallow, Persian Gulf–like seas, phytoplankton and other biota would have formed in a productive zone near the surface and then expired and fallen to rest on the relatively oxygen-depleted sea floor. There, the hydrogen- and carbon-containing biotic remains were not as likely to be eaten by other creatures, or easily dissolved; as a result, they left behind substantial organic sediments, which after burial and compression later turned into beds rich in fossil fuels.

Knowledge of the paleogeography of that era is important in order to find fossil fuel deposits today, which in turn has been helpful to climatologists who have piggybacked onto that well-funded exploration work. This knowledge has allowed us to draw maps of what the continents might have looked like in the distant past (Fig. 15.2).

We know, for example, that the oceans were more extensive 100 million years ago (covering 80 percent, not 70 percent, of the Earth), and that sea level was approximately 300 meters higher than it is today. How could that be? There are four possibilities. First, there might have been more water then; however, most experts do not believe that to be likely. Second, perhaps there was no permanent ice then – a quite likely assumption. However, if all the ice on land were melted today and the meltwater were allowed to run into the seas, this would result in a sea-level rise of only approximately 70 meters. Therefore, the absence of ice 100 million years ago is only a small part of the reason for the 300-meter sea-level rise. A third possibility is that the continental masses were all somehow vertically depressed by hundreds of meters – a very unlikely event! The fourth possibility is that the volume of the ocean basins was less. How could such a thing happen? A likely explanation is that the mid-ocean ridges were larger. The fraction of the area of the Earth that was covered by land then was smaller – approximately 20 percent, compared to 30 percent today. This explanation is consistent with the 300-meter-higher sea levels, the absence of permanent ice, and less voluminous ocean basins with high and widespread mid-ocean ridges.

Why might this fact be important, other than to explain a mechanism for an ancient rise in sea level? If the mid-ocean ridges were particularly high (Chap. 6), that tells us that the subsurface volcanic activity that created them must have been very great. Carbon dioxide emerges along with volcanic materials, a fact that suggests that more carbon dioxide was injected into the oceans, from which it would have diffused

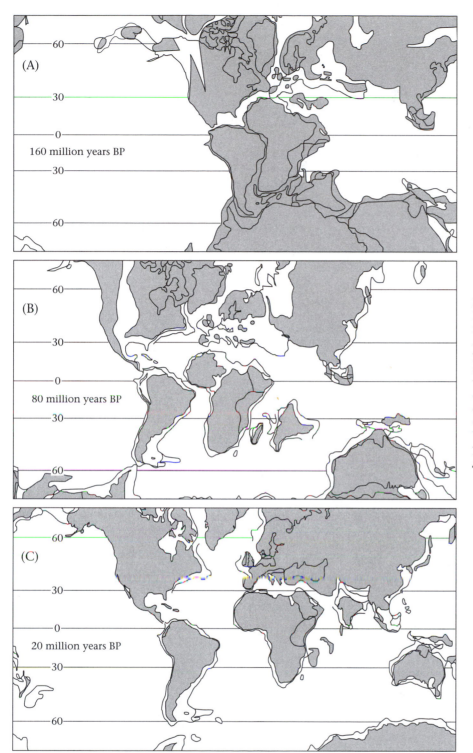

(A)

160 million years BP

(B)

80 million years BP

(C)

20 million years BP

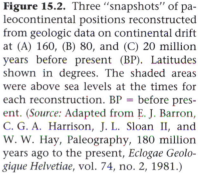

Figure 15.2. Three "snapshots" of paleocontinental positions reconstructed from geologic data on continental drift at (A) 160, (B) 80, and (C) 20 million years before present (BP). Latitudes shown in degrees. The shaded areas were above sea levels at the times for each reconstruction. BP = before present. (*Source:* Adapted from E. J. Barron, C. G. A. Harrison, J. L. Sloan II, and W. W. Hay, Paleography, 180 million years ago to the present, *Eclogae Geologique Helvetiae*, vol. 74, no. 2, 1981.)

upward into the air – suggestive of more plant production from carbon dioxide–enhanced photosynthesis and a warmer climate. The latter are both consistent with the rich fossil fuel and fossil dinosaur remains of the warm Mesozoic Era.

Can the Cretaceous warm period serve as a mirror for a greenhouse-warmed twenty-first century? My answer to that question is "not directly," because the "experiment" that we are currently performing on the Earth – the rapid increase in emissions that are the global change concern discussed in Chapter 1 – has no known precedent in the geologic past. Carbon dioxide and methane may have been doubled by nature in a century's time in the distant past; however, even if that had occurred, we cannot yet uncover clear evidence of it from either rocks or the paleobotanic record. In short, we have not discovered how to recognize this phenomenon if it did occur. The best we can do is to look at the paleoclimatic conditions that we know about, try to determine the causes, and test our modern tools on those past situations.

What we have found in our models depicting conditions of 100 million years ago is that although they do consistently explain the warmth of the Cretaceous Period as potentially related to carbon dioxide, they simulate too-warm temperatures in the ancient tropics when carbon dioxide is quadrupled. This discrepancy contributes to a large controversy: Some experts suggest that during the Cretaceous, the tropics did not warm up nearly as much as the high latitudes did. Could that be because there was some special feedback effect in the clouds? Or were there super-hurricanes driven by the warm waters that were transferring heat out of the tropics? Thus far, no convincing explanation exists. Should we then scrap the entire global climate model (GCM) approach? Of course not – such debates form the backbone of scientific research, pushing participants to find better explanations and models. This is how researchers work. We look at these geologic time frames and try to reconstruct them. The computer model that Barron and other paleoclimatologists employed to study the past (see the box entitled "Paleoclimatic Reconstructions and Model Simulations") is very similar to the model that other scientists use for forecasting the effects of carbon dioxide buildup in the near future. The primary differences are paleogeography and ice cover. The paleoclimatic studies are encouraging but not definitive – a very typical finding in the emerging practice of Earth systems science.

CLIMATIC CYCLES AND FEEDBACK: GLACIAL–INTERGLACIAL AGES

If we look at more recent times, for example, 700,000 years ago to the present, a series of climatic cycles is evident (Fig. 15.3). Approximately every 100,000 years or so, an interglacial interval occurs that lasts approximately 10,000 to 20,000 years, followed by a transition into a very deep ice age tens of thousands of years later. During most of the time between the interglacials and deepest glacials, it has been colder than now. Interglacials tend to give way to maximum glacial more slowly: a 70,000-year period of fluctuating ice buildup, then a 10,000-year, intense ice age maximum, and finally a very quick release (i.e., 10,000 years of "termination" followed by a full interglacial). Paleoclimatologists call this a "sawtooth" pattern. There is much debate about what could have caused the slow buildup of the ice in contrast to its far more rapid decay. The following is a simple example of one possible sequence of events.

Figure 15.4 maps the extent of ice since the last ice age. Twelve thousand years ago, the northern half of England was covered by ice; today none of this ice is left. The bulk of permanent ice in the interglacial age covers Antarctica and Greenland. Most ice did not disappear until 8000 to 10,000 years ago, and in Europe the nonpermanent ice sheet did not disappear until approximately 6000 years ago. What are the possible explanations for the buildup and decay of that much

ice? First, a heavy snow is required – so heavy that some can last through the summer. This can create a self-sustaining positive feedback. That is, the snow surviving the summer will reflect away more sunlight than snow-free surfaces, thereby furthering the chilling – and amplifying the entire process. However, it takes a very long long time for snow to build up into sheets that are several kilometers high. Because it is crucial for snow to survive the summer sunshine, scientists have struggled to understand how this might be possible. The orbital theory becomes relevant here, because when the tilt and precession of the Earth's axis changes, reducing summer sunshine in the Northern Hemisphere, a favorable time occurs for ice sheets to grow.

When the surviving fields of snow begin to build up, they compress into ice. The resulting dome-shaped glacier then

Figure 15.3. Variation in the oxygen isotope ratios in the shells of fossil forms living near the ocean floor, taken from a deep-sea core in the Pacific Ocean. When all other factors are constant, less negative values of this oxygen isotope ratio index (δO^{18}) indicate colder climates corresponding to increased ice volumes. However, a decrease in bottom water temperature would also cause a less negative δO^{18}. Each major change of direction in the oxygen isotope ratio curve is called a *stage*, as indicated on the figure. Inasmuch as similar stages are found from deep-sea cores taken all over the world, many paleoclimatologists believe that these major shifts in oxygen isotope ratio index indicate a record of global climatic change over the past million years or so. The only certain globally synchronous dates, however, are the present and the time when the Earth's magnetic polarity reversed approximately 730,000 years ago. ka = 1000 years. (*Source:* Adapted from N. J. Shackleton and N. D. Opdyke, Oxygen isotopes and paleomagnetic stratigraphy of equatorial Pacific core: oxygen istotope temperatures and ice volumes on a 105 and 106 year scale, *Quaternary Research*, vol. 3, no 1, 1973, pp. 39–55.

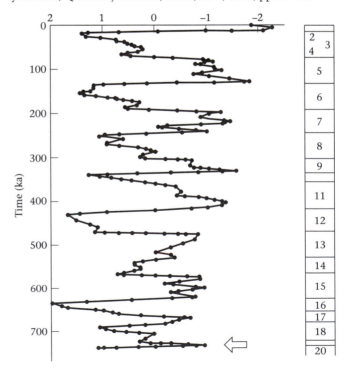

RETREAT OF ICE (thousands of years ago)

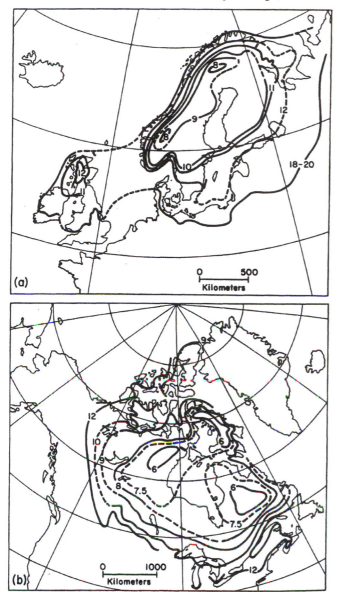

Figure 15.4. The Holocene: As the last Ice Age ended and the (A) Fenno-Scandinavian and (B) Laurentide ice sheets melted, climatic conditions generally moved toward their interglacial states. By examining the earliest dates of remains suggesting that the ice sheets had gone, scientists have reconstructed maps such as those shown here. However, it would be misleading to think that the ice sheets simply melted in place following the progression of time lines seen here. Large fractions of the Pleistocene ice sheets, particularly the North American one, could well have melted in the oceans after having been discharged into the sea more than 13,000 years ago as icebergs. Thus, these maps could show simply when the final remnants of snow and ice cover melted on land. (*Source:* Adapted from J. I. S. Zonneveld, Some notes on the last deglaciation in northern Europe compared with Canadian conditions, *Arctic and Alpine Research*, vol. 5, 1973; and R. A. Bryson, W. M. Wendland, J. D. Ives, and J. T. Andrews, *Radiocarbon isochrones on the disintegration of the Laurentide Ice Sheet, Arctic and Alpine Research*, vol. 1, 1969.

begins to flow outward, and the glacier becomes progressively larger, building itself into a continental-scale ice sheet, as shown in Figure 15.4. Because the Earth's crust below is not rigid, but viscoelastic, it buckles and sinks under the weight of the ice. Eventually, this force lowers the top of the ice sheet. By this time, the massive ice sheets have already moved equatorward – to England, New York, and Wisconsin. The cold air blowing off the ice freezes the high-latitude oceans, reducing evaporation and thus depriving the ice sheet of fresh snow supplies. The positive feedbacks that tended first to build the ice sheet to glacial maximum size can then proceed in reverse.

In order for the snow to accumulate, it must go through the atmosphere, which means oceans need to be warm enough to evaporate the moisture needed for snowstorms. This is why ice ages are believed to begin when high-latitude oceans are warm. However, in order to disappear, the ice must be starved of potential new ice-forming precipitation because the seas are cold and do not give up evaporative moisture to the atmosphere as easily. The ice sheets have expanded their margins to the south, and the bedrock has begun to sink. The latter activity lowers the top of the ice sheets, exposing them to warmer air. As the orbital configuration switches to allow more summer sunshine, the melting accelerates. As the ice sheets melt, and because the bedrock was depressed, the remaining blocks of ice may sit in puddles called periglacial lakes. Climate modeler David Pollard has calculated that periglacial lakes could be very effective at chopping off edges of ice sheets and making icebergs. This positive feedback is one way in which an ice sheet might disappear quickly. Another way might be that as the climate rewarms, melting snow around the glacier uncovers bare land, creating a dust source. When dust blows on top of ice or snow, instead of having the "clean" glacier's albedo properties and reflecting 80 percent of the sunlight that hits it, the darkened, dirty ice grains reflect only approximately 40 percent of the solar energy. Furthermore, as the climate warms and the ice recedes, larger plants begin to grow where snow-covered tundra existed previously. Tundra vegetation is replaced by taiga (trees and bushes) as warming begins. The dark trees reduce the reflectivity of the snow-covered land from approximately 80 percent to approximately 40 percent. Have you ever compared the reflectivity of a snow-covered field to a snow-covered forest? The wooded area is much darker. So growth of vegetation enhances the melting of the ice and snow. This newly formed vegetation absorbs more sunlight than tundra in another ice-age-ending, positive feedback.

At the end of this scenario, the ice age born in positive feedback kills itself with a different set of positive feedbacks. After favorable solar conditions return, some 10,000 years later, ice builds itself up by positive feedback mechanisms and eventually makes the world very cold and arid. As a result, most vegetation near the ice disappears, further enhancing the reflectivity, causing more cooling. After the ice sheet becomes very large, it extends southward and sinks.

The warm oceans chill, evaporation is limited, and several factors work together to eliminate the ice. Periglacial lakes, orbital changes, and dust are also part of the story. The word "story" is chosen deliberately, for the tale of positive feedback is not a certain explanation of ice age interglacial cycles – it is only one plausible sequence.

We now refocus on the recent past, and attempts to model it. Figure 15.5 illustrates the work of John Kutzbach and colleagues in the Cooperative Holocene Mapping Project (COHMAP) in 1988. This figure includes a calculation of the percentage of Northern Hemisphere solar radiation, a measure of the extent to which the solar energy input to the Earth from the Sun changed from the last ice age to now. Compare the Earth today with the Earth 18,000 years ago. Recall our earlier discussion of the prairie peninsula 3000 to 8000 years ago and the climatic optimum 6000 to 9000 years ago, when summer temperatures were perhaps 2°C warmer than those of the past milennium. Figure 15.5 shows that 8 percent more sunlight entered the Northern Hemisphere 9000 years ago in the summer, and that a proportionate amount less entered in the winter.

These and other data help to explain why mid-continents in Africa and Asia were a bit hotter in the summer some 5000 to 8000 years ago. The reason appears to be that greater amounts of sunlight reached the Earth in summer because of changes in its orbit. The Earth's spin axis is inclined to the ecliptic plane, and the orbit is an ellipse. Today the Earth is closer to the Sun in January; however, 9000 years ago the Earth was closer to the Sun in July. This time period of change is certain,

based on calculations from celestial mechanics. Because we know how the Earth's orbit has changed, we can calculate how the amount of energy from the Sun reaching each latitude zone each month has changed. We know that the Earth's orbital changes do not alter the total annual amount of energy reaching the entire Earth by more than a few tenths of a percent. However, the orbital effect changes by as much as 10 percent the latitudinal and seasonal distributions of solar heating – what we call *solar-orbital forcing* – of the climate system.

As all these examples indicate, scientists know a great deal about the Earth's past climate and probable causes of these changes. Paleoclimatologists have known for a long time, by studying cores taken from the sands of Africa, that Lake Chad, which is a very small lake today, once covered a larger portion of the sub-Sahara. Also, we know from cores taken from the mouths of rivers in India that it was much wetter there, too, approximately 5000 to 10,000 years ago. With recent advances in computer modeling, we can now begin to simulate and explain these changes in past climates. We can take what we know about ice cap changes, particles in the atmosphere, carbon dioxide, and energy from the Sun, and feed all these "forcings" (Fig. 15.5) into models. We then produce simulations of the climate 18,000 years ago, 12,000 years ago, or 9000 years ago. We can dig in sand and soils to reconstruct lake levels in Africa, showing that they were much higher 9000 to 6000 years ago, that they were lower in the ice age, and that they are at their lowest level now (Fig.

MODEL DESCRIPTION

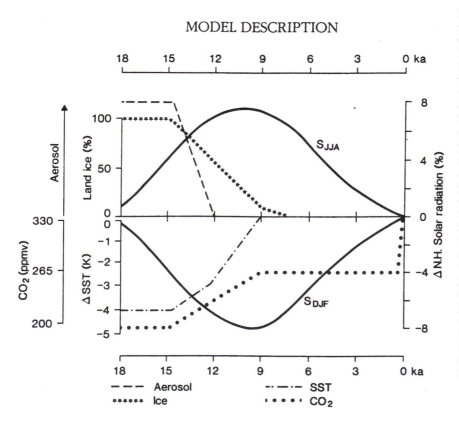

Figure 15.5. Boundary conditions for the COHMAP simulation with the community climate model for the past 18,000 years. External forcing is shown for Northern Hemisphere (N.H.) solar radiation in June through August (S_{JJA}) and December through February (S_{DJF}) as the percentage difference from the present. Surface boundary conditions include land-ice volume (Ice) as a percentage of the 18-000-year value (with the "maximum model" ice sheets of Denton and Hughes (1981) and global mean annual sea-surface temperature (SST, K) as departure from present. The horizontal scale indicates the times of the seven sets of simulation experiments. An additional experiment for the 18,000-year time included a lower glacial atmospheric carbon dioxide (CO_2) concentration of 200 ppmv (dots); the other deglacial experiments used a modern value of carbon dioxide concentration (330 ppmv). Experiments including glacial-age aerosols have not yet been completed. K = Kelvins; ka = 1000 years, ppmv = parts per million. (*Source:* Reprinted by permission from *Nature*, vol. 317, pp. 130–134, copyright 1985, Macmillan Magazines Ltd.)

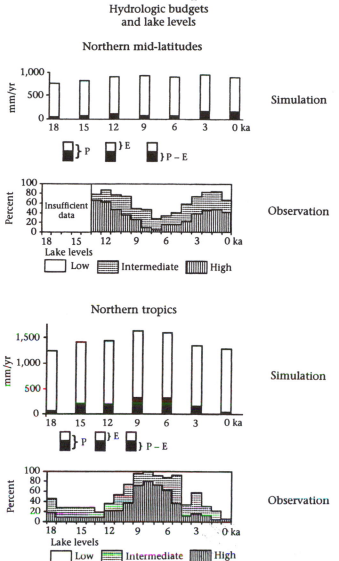

Figure 15.6. Orbital forcing of fluctuations in the level of tropical lakes from 18 to 0 kyr BP. Observed status of lake levels (low, intermediate, high) and model-simulated moisture budgets (precipitation (P), evaporation (E), and precipitation minus evaporation (P-E) for each 1000-year (observations) or 3000-year (model) interval from 18 to the present. The temporal variation in percentage of lakes with low, intermediate, or high levels is taken from the Oxford lake-level data bank. In both northern mid-latitudes (top) and the northern tropics (bottom), the shape of the high-lake-level curve corresponds roughly to the shape of the P-E curve. That is, lake levels and P-E were low at approximately 9000 to 6000 years ago in northern mid-latitudes and high approximately 9000 to 6000 years ago in the northern tropics. ka = 1000 years; mm/yr = millimeters per year. (*Source:* Modified from J. E. Kutzbach and F. A. Street-Perrott, 1986, reprinted by permission from *Nature*, vol. 317, pp. 130–134, copyright 1985 Macmillan Magazines Ltd.; and from the Office for Interdisciplinary Earth Studies, University Corporation for Atmospheric Research, Boulder, CO.)

15.6). We can also study fossil pollen in lake sediments in Iowa and observe how spruce forests moved north.

Using a climate model's prediction of climate change, plus another model that simulates forest changes in relation to temperature or rainfall changes, we can couple climate science with ecosystem science. This involves using fossils for reconstructing the temperature and rainfall patterns over time and forcing factors for calculating them in the model – or both. However, although the comparisons are encouragingly similar with regard to broad patterns, individual details are not in very good agreement. For example, we have learned from paleontologic studies that at the continental scale, enhanced monsoon rainfall can be accurately predicted or that ice age aridity can be well-simulated. However, at any one given location at any one point in time, our models do not provide credible information concerning specific details.

The fundamental question in the face of all this uncertainty becomes: Does the correlation between model and nature concerning monsoon changes or temperature and carbon dioxide and methane variations over 160,000 years, or in the Cretaceous Period, validate a quantitative cause and effect relationship between temperature rise and greenhouse gas buildup in the past 100 years? The answer is, simply, "Not yet." Other explanations remain possible, involving factors such as atmospheric dust, which could have blocked some of the greenhouse gas effect. However, the correlations are sufficiently consistent to support the argument that such associations are quite plausible, and that projected magnitudes and global rates of change are reasonable estimates with a substantial chance of being valid. I estimate (subjectively) that there is an 80 to 90 percent probability that the twentieth-century warming trend is driven in large measure by global change disturbances.

CONCLUSION

I must assert one obvious truism: Humans are forcing the Earth's environmental systems to change at a rate that is more advanced than their knowledge of the consequences. Under such conditions of uncertainty, surprises are virtually certain to occur. Once again, the paleoclimatic record may contain some clues. We have argued that *sustained, globally averaged* rates of surface temperature changes from the past ice age to the present were approximately a 1°C increase per 1000 years, a factor of 10 or more slower than the expected increase of several degrees Celsius per 100 years projected for the twenty-first century as a result of human activities. However, note the italicized words "*sustained, globally averaged*." Approximately 13,000 years ago, after warm-weather fauna had returned to northern Europe and the North Atlantic, there was a dramatic return to ice age–like conditions in less than 100 years. This miniglacial lasted for 500 years before the relatively stable Holocene Epoch was established. What happened? We do not know. Moreover, this was at least a regional change (i.e., the entire North Atlantic, including northeastern Canada and most of Europe). However, this change included dramatic ecologic retrenchment to ice age – like plants and animals. Globally, what occurred is less clear. No major climate change is evident in Antarctic ice cores. However, studies of fossil remains in the North Atlantic suggest that the warm Gulf Stream current was directed many degrees of latitude southward and that the overall structure of deep-ocean circulation reverted to near ice age form over the course of only a few decades.

Could such a rapid change be induced today by pushing the current climate system by human disturbances, including the buildup of greenhouse gases and sulfur oxides? This potential "surprise" is speculation, of course, yet its plausibility has concerned many scientists. Today, scientists have calculated that very rapid and sustained increases in greenhouse gases could trigger such a reorganization of the North Atlantic "conveyor belt" circulation, which would have dramatic and long-lasting consequences for Europe. Could the anthropogenic global warming of 2°C expected in the decades ahead trigger such surprise climatic instabilities as may have occurred 13,000 years ago? Again, this is speculation. The prospect of climatic surprises in general is chilling, and lends considerable urgency to the need to accelerate and deepen our understanding, to decelerate the rates at which we are forcing nature to change, or, better (in the author's value system), to do both. The extent to which these courses of action – or the present lack of concerted action – might cost humanity is addressed in Chapters 26 and 27.

Our confidence in projecting at least the magnitude and rates of global change in the decades ahead stems from the fact that we have an immense backdrop against which to calibrate the tools we use to look at the future: the geological, paleoclimatic, and paleoecologic past. Using the same tools to analyze known events of bygone eras and to forecast climatic changes provides an important component of valida-tion, though not perfect confirmation. As noted at the beginning of this chapter, we cannot forecast the future accurately without understanding and modeling a great deal of the Earth's past.

QUESTIONS

1. If you were in charge of allocating a United Nations fund for "environmental research," what percentage of your budget do you think you would set aside for research on global climate models and the paleoclimate? For what other environmental research concerns would you allocate funds? Which three do you consider most urgent? Explain briefly.

2. Why are rates of climatic change important?

3. How can ecologists and climatologists work together?

FURTHER READING

Crowley, T. J., and North, G. R. 1991. *Paleoclimatology*. New York NY: Oxford University Press.

Hoffert, M. I., and Covey, C. 1992. Deriving global climate sensitivity from paleoclimate reconstructions. *Nature* 360:573.

Rahmstorff, S., 1999. Shifting seas in the greenhouse? *Nature* 399:523–524.

Root, T. L., and Schneider, S. H. 1995. Ecology and climate: research strategies and implications. *Science* 269:334–341.

Kasting, J. F. 1993. *Earth's early atmosphere. Science* 259:920–926.

Schneider, S. H., and Londer, R. 1984. *The coevolution of climate and life*. San Francisco, CA: Sierra Club Books.

Wright, H. E., Jr., Kutzbach, J. E., Webb, T., III, Ruddiman, W. F., Street-Perrott, F. A., and Bartlein, P. J. 1993. *Global climate change since the last glacial maximum*. Minneapolis, MN: University of Minnesota Press.

16 Can We Predict Climate Change Accurately?

STEPHEN H. SCHNEIDER

VARIABLE WEATHER VERSUS CLIMATE CHANGE

Over the past forty years, the weather forecasts have improved greatly. In bygone days, many forecasters attempted to portray themselves as humorous characters, probably because early weather predictions were laughably imprecise, and the public needed assurance that these predictions were not important, right or wrong. However, rapid strides in the oceanographic and meteorologic sciences, coupled with weather stations located around the world, and, most important, a flotilla of weather satellites, have vastly increased our synoptic analysis of the present status of global weather patterns, and have provided hourly updates on developing trends. Local weather forecasting has become a mature science, and clear, usually accurate weather predictions on a twenty-four hour basis are now a reality.

In contrast, the process of understanding and thereby predicting climate change – the variation of short-term weather patterns spatially aggregated, and summed over decades or longer – is a field still in its growth stage. As we saw in Chapter 15, climate consists of an exceedingly complex interplay among numerous internal and external forcing factors, some providing positive feedback, others negative. Many factors such as cloud cover, are of first-order importance but as yet are poorly understood quantitatively. The origins of natural fluctuations in climate have not yet been fully identified and thoroughly characterized, which is also the case for the modern input of anthropogenic sources of change such as increased levels of greenhouse gas emissions. Consequently, the issue of climate change is a contentious one, even among scientists. The world needs this information; for example, much of the burgeoning human population lives along or near the seacoast. If warming is in progress, so is sea-level rise. Agricultural belts or forests could be moving too, with species extinctions and/or economic dislocations a distinct prospect. We urgently need to understand physical and biologic processes at work, to quantify their resultant influence on the global climate, and to learn how we might ameliorate the adverse effects of warming. If some negative effects are to be reduced, the time for action is now, and such action will be far more cost effective if it is based on accurate projections of our climatic future.

In Chapter 15 we argued that looking at evidence of climate change in the past is a useful technique for validating the tools we use to make forecasts about climate change in the future. However, in order to predict climate reliably, we first need to specify the processes controlling climate. We also need to understand and be able to model the factors that might induce changes in climate, factors we call *forcings*. An example of climate forcing, mentioned in Chapter 15, is the cyclic variation in Earth-intercepted solar flux. The ellipticity of the Earth's orbit and the inclination of its axis of rotation control the amount of sunlight reaching the Earth at each place and time. This solar heating forces the seasonal cycle, and Figure 16.1 shows how the basic atmospheric circulation is driven by solar forcing. As sunlight comes in, some of it is immediately reflected – primarily by clouds, deserts, and ice caps. The Earth's reflectivity, called its *albedo*, determines the amount of incoming solar energy that can be absorbed. Because of the Earth's spherical shape, 50 percent of the surface area is between the 30° parallel North and the 30° parallel South. However, much more than 50 percent of the sunlight reaching the surface of the globe enters in this 60° wide band because of the Earth's orbital geometry. As a result, the tropics and poles are heated differentially.

If the only thing controlling the climate were solar radiation, the absence of sunshine at the poles in winter would force the poles to be even colder than they are now. There must be other processes at work. An obvious one is the heat that is constantly being transferred around the planet by fluids in motion, notably the atmosphere and the oceans (see Chaps. 10, 11, 13, and 14). All of us have had the experience on a hot day of seeing ripples in the heated air near the surface of the ground when the Sun beats down – this is the

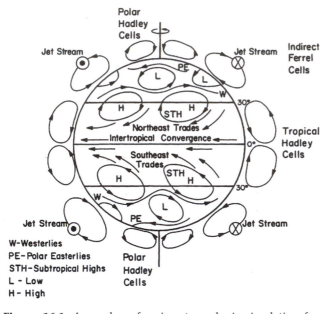

Figure 16.1. A number of major atmospheric circulation features are shown here in schematic form. The dots at the left-hand side of the figure labeled "Jet stream" indicate airflow upward, seemingly out of the page, whereas the cross through the circles representing the jet streams at the right-hand side of the figure represents flow into the page. In reality, the jet meanders across the mid-to high latitudes in both hemispheres, with predominantly west to east winds. The highs and lows shown are primarily lower atmospheric features, extending up only a few kilometers into the atmosphere, whereas the Hadley cells, Ferrel cells, and jet streams can extend vertically as much as 10 to 20 kilometers (6.2 to 12.4 miles) through much of the stratosphere. H = high; L = low; PE = polar easterlies; STH = subtropical high; W = westerlies. (*Source:* Adapted from S. H. Schneider and R. Londer, *The coevolution of climate and life*, Sierra Club Books, 1984.)

visual effect of rippling, buoyant plumes of lower-density, warmed air rising into cooler air above. Warm air also does the same thing on a much larger scale: It rises in the tropics and moves upward and outward into colder areas and sinks thousands of kilometers poleward from the equator. As a current of heated air moves, there is also a return flow moving alongside or underneath it. This circulation is known as a Hadley cell. However, there is an important added complication: The Earth is rotating.

If you were riding along with a parcel of air moving first upward and then toward the poles, you would carry with you the speed of the spinning equator and you would take this momentum with you and transfer it to the mid-latitudes, whichever way you went – north or south. You would appear to deflect to the right in the Northern Hemisphere, and to the left in the Southern Hemisphere. A scientist named Coriolis described this motion in a mathematical equation centuries ago – it is called the *Coriolis effect.* Some people also call it the "merry-go-round effect." For example, if you and a friend were sitting opposite each other on a rotating merry-go-round and you tried to throw a ball to your friend, the ball would go straight relative to the ground, but not to you

or your friend, because you are riding on the rotating platform. By the time the ball reached your friend on the other side, he or she would have moved. From your point of view or that of your friend, the ball appears to have been deflected, as if there were some magical force. However, there is no magic involved. The ball was simply traveling straight when thrown at the same time that you and your friend were rotating. Similarly, the Earth rotates too (see Fig. 16.1), and as a result westerly (i.e., blowing from the west) winds occur in mid-latitudes in both hemispheres because of the rising hot air in the tropics that deflects owing to Coriolis forces.

Many weather maps show the jet stream blowing at high altitudes predominantly from west to east. The jet stream is typically drawn as a wavy line because of a second factor that involves the fluid mechanical behavior of the gases we call the atmosphere. When certain velocities are reached in the atmosphere, velocities that are largely determined by how vigorously the Hadley cell is operating, the river of air can become unstable. This cell is most vigorous when the temperature difference between equator and pole is large – that is, in winter. Winds are produced by the temperature difference between places, this difference creates pressure and density differences, which create rising air, wind currents, and so forth. The summertime jet stream is relatively weak, whereas in the winter it is much stronger because the poles cool down in winter but the tropics stay relatively warm all year long. Thus, temperature contrast is greatest in winter, the Hadley cell is stronger, more air and heat are transported poleward, the circulation is more vigorous, and the jet stream is more variable.

When the large-scale circumpolar winds reach a certain speed on a rotating body, they can become unstable. When the jet streams are unstable, eventually they break into high- and low-pressure eddies, also known as *synoptic weather systems.* The motions of the atmosphere obey physical laws: conservation of mass, momentum, and energy. These laws form a coupled set of equations whose solutions can simulate mathematically the workings of the traveling weather systems. These simulations can explain why the mid-latitudes typically experience weather patterns that change every few days, and why the tropics can have weather that is often constant for months at a time. These simulations can also explain why even the mid-latitudes can sometimes have weather that is relatively constant for months at a time in the summer: because in the summer, jet streams are weaker, circle the poles farther from the equator, and are more stable. The jet stream is extremely important because it tends to be the dividing line between cold air to the poleward side and warm air to the equatorward side.

What occurs in abnormal years? For example, what occurred in the winter of 1976, when the western United States experienced a drought that was dramatic and intense, and Alaska was frequently warmer than Virginia? During that winter, the jet stream was uncharacteristically far north in the West, far south in the East, and not variable enough – it was "blocked," as meteorologists say – in both places. There is debate as to whether definable forcings outside the atmosphere such as unusual patterns of ocean temperatures result in such abnormal wind patterns or whether unpredictable, random internal atmospheric fluid motions are paramount. Regardless, the point of this discussion is that the position of the jet stream is dramatically important for local and regional climatic conditions.

Figure 16.2 shows that there are important departures from the *zonal* – or average conditions in each latitude zone – climatic patterns. Regional *centers of action* (as meteorologists used to call them fifty years ago) exist such as the Aleutian or Icelandic lows, which have counterclockwise circulation with a low-pressure center, or the clockwise high-pressure centers found in Bermuda and the Azores. At a low-pressure center, the surface air rushes in and (in the Northern Hemisphere) deflects to the right from the Coriolis effect, creating counterclockwise circulation. These lows appear as spirals, not absolute circles, because there is friction between the winds and the surface, slowing the circulation and creating the spiraling flow patterns.

A center of action such as the Icelandic Low shown in Figure 16.2 actually represents a statistical average of many traveling lows and highs, not an instantaneous snapshot of the typical weather map. It also shows the intertropical conversion zone (ITCZ). As shown in the illustration of Hadley cells (Fig. 16.1), the heated air rises in the tropics. Because of the law of the conservation of mass, what goes up must come down, and what goes out must come back. When the air comes back, it is blowing back toward the equator at the surface. The Coriolis force dictates that the air is going to deflect to the right in the north, and to the left south of the equator. The result is the northeastern *trade winds* (or *trades*) in the Northern Hemisphere and the southeastern trade winds in the Southern Hemisphere. When the trades collide at the "meteorologic equator," the air at the surface, which is loaded with moisture gathered from the long journey over water, is forced upward and expands as the pressure decreases with height. This expansion causes cooling of this moisture-laden air, resulting in massive thunderstorms. This is why the ITCZ is the principal thunderstorm belt of the world. If you have been in Florida in the summer, you know what occurs almost every afternoon there, approximately 20 miles inland. Moisture-laden air comes in off the ocean and hits a warmed land, causing the air to rise. Because the air does not need to travel upward very far before it cools sufficiently to reach the maximum saturation point of this humid air, the result is frequent afternoon thunderstorms.

This local example of a daily sea breeze is analogous to what occurs on a continental, seasonal scale. You have no doubt heard of the *monsoons* of Africa, South America, and Asia. The Portuguese word "monsoon" means "wind." The surface temperatures of oceans change by only a few degrees

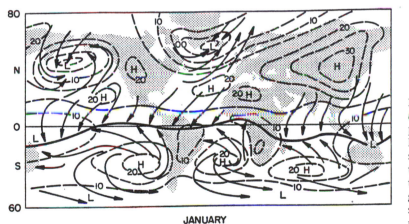

JANUARY

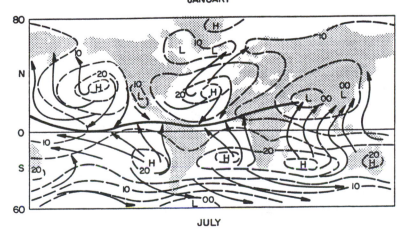

JULY

Figure 16.2. The geographic distribution of land and sea causes significant distortion of atmospheric circulation systems away from the purely latitudinal or zonal pattern so often shown on schematic figures. The solid dark line shows how the intertropical convergence zone (ITCZ) is distorted out of a zonal pattern because of the geographic effects of unequal heating of the land and sea. Typical patterns are shown for the extreme months of January and July. Surface features such as the Bermuda High and the Icelandic Low can also be seen, as well as prevailing surface trade winds that converge to form the ITCZ. The numbers associated with high and low pressure above the 1000-millibar level are typically found at that geographic location for the January or July season. H = high; L = low. (*Source:* S. H. Schneider and R. Londer, *The coevolution of climate and life*, Sierra Club Books, 1984.)

from winter to summer because of large thermal inertia (a deep column of water has a very high heat content). However, the temperatures of landmasses can change seasonally by tens of degrees (the surface layers of rocks and soils have low heat contents). Thus, the centers of the Asian or African continents heat up during the summer relative to the surrounding seas. In the winter the land is relatively colder and the air aloft becomes dense and sinks, hits the surface, and spreads outward. In the summer the land is relatively warmer and the buoyant air over it rises. When it rises, something must fill the space left behind. Air flows in; however, because it typically blows in from over an ocean and is laden with moisture, the result is summer monsoon rainfall.

Another common pattern of ocean temperatures results in cold, upwelling water off the west coasts of continents. The winds contact the oceans and create ocean currents. Along the west coast of North America, winds typically come from the northwest. These winds seem to push the ocean against the shore; however this is not what actually occurs. The Coriolis force affects the ocean, deflecting the flow of ocean water to the right in the Northern Hemisphere. That is, wind coming from the northwest in the Northern Hemisphere causes surface waters to deflect to the right, in effect pushing water away from the west coast (see Chap. 10). As the surface ocean water is deflected toward the southwest, that is, away from the coast, conservation of mass requires that water from below, which is much colder, fills in. That cold, coastal upwelling is linked to high biologic productivity because upwelling brings up nutrients from below. Upwelling is also conducive to the formation of low stratus clouds and fog. The warm air from the Pacific Ocean travels across and hits the cold, upwelling currents, causing atmospheric moisture to condense into low, essentially nonprecipitating clouds (e.g., San Francisco's famous fogs). Cold, upwelling waters also occur off the coast of Peru; as a result, typically, little rain falls. Air rises in large-scale patterns in the western tropical Pacific, not in the eastern tropical Pacific, where it sinks. This is the expected pattern: It rains in Indonesia and north Australia, and it is very dry off the coast of Peru – an east-west pattern known as the *Walker circulation*.

In addition to monsoon rains and cold upwelling currents off the coasts of western North and South America, climatologists study the effects of a phenomenon called El Niño, which means "the child." The name derives from the "Christ child" because it is a recurrent phenomenon that is most noticeable in winter, near Christmas. One explanation holds that there is a large, deep wave in the ocean that sloshes back and forth across the equatorial Pacific just under the surface. This is called an *internal wave*, and every few years or so it contributes to making the waters uncharacteristically warm off the coast of Peru, while they cool at the eastern end of the tropical Pacific. As the waters become warmer off Peru's coast, the heated air rises, reversing the typical dynamic of sinking air over cool, upwelling waters. Reversing the Walker circulation leads to torrential rainfall in Peru, drought in Australia, and fires in New Guinea, which normally experi-

ences a humid rain forest climate. Moreover, El Niño has global repercussions. The fluctuation between normal and El Niño Walker circulation patterns is called the Southern Oscillation signal. Computerized atmosphere, ocean and coupled sea and air models are now beginning to simulate these natural factors successfully.

Climate can be simulated on a range of different scales – on a 100,000-year scale (with ice ages and interglacials alternating) or on a scale of a few years, (simulating, for example, the Southern Oscillation). Both scales are needed in order to understand and accurately forecast global changes. The global change question relevant to Earth systems science primarily concerns human-induced climate change, whereas El Niño/Southern Oscillation (ENSO) is a natural climatic fluctuation. Some scientists even argue that the twentieth-century warming trend, especially the record warmth of the 1980s, is simply a fluctuation. It might be helpful to look at the kinds of characteristic changes that can be identified in various time series (see Fig. 16.3).

One kind of variation is periodic change in which the time series oscillates up and down around some mean value (see box entitled *Time Series Changes*). Figure 16.3 also shows an "impulsive" change between two long-term means. The eruption of a volcano – throwing dust into the stratosphere, that blocks sunlight and causes a rapid cooling below – is an impulsive change. The effect typically remains intense for a year or so, and then temperatures will "sawtooth" back up again in a warming trend over a period of a few years. Downward trends also exist. In contrast, during the past 100 years, the global surface temperature has been experiencing an upward trend. Superimposed on this are accompanying "bounces" in temperature, both from year to year and over

Figure 16.3. Various types (A – D) of climatic variations typically encountered in nature. (*Source:* Adapted from F. K. Hare, Proceedings from World Climate Conference, Geneva, 12–23 February 1979, World Meteorological Association, Publication No. 537.)

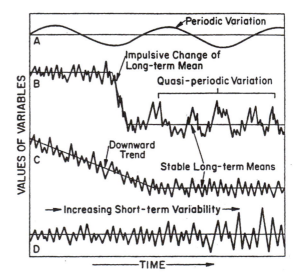

TIME SERIES CHANGES

A very interesting case, illustrated in Figure 16.3, is the curve in which variability increases with time although the long-term mean is constant. This variability can have a huge impact on living things. A corn plant, for example, can be killed when it cools below 0°C, even for a few hours. Crossing below the freezing threshold is, for that plant, a catastrophic event, regardless of whether it is a random fluctuation or a forced impulsive change. Similarly, a bird or an insect whose physiology does not tolerate temperatures above 30°C could perceive a trend toward increasing variability as a major event, even though a climatologist looking only at long-term "means" might argue that no climatic change is evident.

These examples suggest that when we talk about "change" we must be careful to define the time scale of the change and, if possible, the potential cause. Confusion concerning the definition of what constitutes a change versus what is a fluctuation has triggered an endless debate in some scientific circles. Statistics alone may not be able to resolve the debate. It is always desirable to have explanations for a variation; credible explanations can help to distinguish systematic "change" from random "fluctuation."

---

decades. Researchers argue about whether these bounces are natural, random fluctuations, or whether they are caused by definable, even if small, forcings outside the climate system such as volcanic dust veils.

EXTERNAL AND INTERNAL DRIVERS

In discussing the reasons for climate change, we distinguish two basic categories: external and internal causes. "External" means occurring outside of the system and not influenced much by the system – although external processes do not have to be physically external to the Earth (as the Sun is). If our focus is on change in the atmosphere on a one-week time scale (i.e., the weather), then the oceans, land surfaces, biota, and human activities that produce carbon dioxide are all classified as external – they are not greatly influenced by the atmosphere on that short a scale. However, if we focus on 100,000-year ice age interglacial cycles, the oceans and ice sheets are classified as part of the internal climatic system and vary as an integral part of the Earth's climatic systems. On this longer scale we must also include as part of our internal climate system the "solid" Earth, which in fact is not solid, but viscous and elastic (see Chap. 6). Which components are external and which are internal to our climatic system are not absolute; their classification depends on the time and space scales as well as the phenomena to be considered.

External climate drivers do not respond to feedback mechanisms and may be extraterrestrial, or native to our planet, whereas internal climate drivers are strongly influenced by feedback mechanisms and are exclusively terrestrial. Fluctu-

ations in solar emissions, perhaps related to a cycle of varying sunspots, are certainly external to the climate system. Influences of other planets' gravitational tugs on variations in the Earth's orbit are also external. Nothing that could happen to the Earth's climate could change either one of these forcings perceptibly. How about changes in levels of volcanic dust or carbon dioxide? On short time scales, these changes are largely external because the state of the climate system presumably does not have much influence on these levels in the short term. On the other hand, it might in the long term. Even atmospheric dust associated with volcanoes might be affected by the climate state in the long term because when the weight of ice sheets distorts the Earth's crust, the propensity for volcanic activity might change. If the climate is changed such that a once-vegetated zone becomes dry, dust can more easily be raised. Thus, on a time scale encompassing hundreds of years, dust generation is internal. Carbon dioxide and methane levels rise and fall with ice age cycles – clearly an internal change on a 10,000-year time scale. However, on 20- to 50-year scales, these greenhouse gases become largely an external cause of climate change (e.g., when they change as a result of people burning fossil fuel) because changes in climate have little effect on anthropogenic emissions.

How about changes in the character of the land surface? If these changes are due to human activities, land use change is largely an external factor. If they are due to climatic change – induced vegetation cover change, land surface change becomes an internal cause, because of positive and/or negative feedback.

Anomalous ocean surface temperature patterns such as the El Niño demonstrate the internal dynamics of the coupled atmosphere–ocean system: An ocean wave changes ocean temperatures, which in turn modify atmospheric circulation. However, the atmosphere rubs against the ocean, which causes the ocean to modify its temperature pattern, which forces the atmosphere to adjust, which changes the winds, which changes the way they contact the ocean, and so forth. Thus, energy and water interact in one large, internal dynamic system – similar to a bunch of blobs of Jell-O connected by elastic bands or springs, all interacting with one another while also being pushed from the outside (by solar or volcanic or anthropogenic forcing, in the case of climate systems). Salinity, which affects changes in sea ice and the density of seawater (which helps control where ocean waters sink), may also need to be explicitly included in the internal climatic system that we must model, particularly if long-term climatic changes including deep-ocean circulations are being studied (see Chap. 11).

These are only a few examples that illustrate the complex process of deciding which factors to include as part of the internal climate system, and which factors are external to that system. The importance of this debate between internal versus external cause was identified in the 1960s by Edward Lorenz, a pioneer in the discovery of chaos theory. Lorenz noted that a complex, nonlinear system can exhibit several types of behavior. "Nonlinear" means that the response of a system to some forcing is not a simple multiple of the strength of that forcing – that is, if a nonlinear system is pushed with one unit of force, it responds with one unit of response; if it is pushed with two units of force, it might respond with six (or one-half) units of response.

One mode of behavior is called *deterministic*. This means that the system responds to forcing in a one-to-one way (even if it is nonlinear). That is, a push causes a determinable response, and a double push causes another (but not twice as much, unless the system is "linear") determinable response. For example, a volcanic dust veil that reflects 1 percent of the solar energy back to space will cause a unique cooling that is determinable in principle (although in practice, various unknown factors may make the precise response uncertain).

Another type of system behavior is *stochastic*, meaning that the system behaves according to some statistical rules. For example, a pair of dice are not deterministic because no relationship can reliably predict the outcome of any individual roll; however, a *statistical distribution* that gives the probability of any combination of faces for unloaded dice can be determined, at least in principle.

Lorenz added a new kind of system behavior that was later dubbed *chaos theory* by mathematicians. He suggested that a nonlinear system can exhibit preferred modes of behavior – such as an ice age or interglacial climatic state – that are neither deterministic nor stochastic. We might perceive these modes as resulting from a forced change, but they are not necessarily forced. Rather, they could simply be two preferred tendencies of an extremely complicated, nonlinear system. Thus, an ice age or interglacial may simply be internal states that are the result of the internal dynamics of a complex climate system. Moreover, chaotic behavior does not produce exact replicas of various states of a system, although there is a tendency for the system to cluster around particular states, that Lorenz labeled *strange attractors* – the ice age and interglacials, in this metaphor.

In summary, a debate is raging in the scientific community over whether the climatic record results from external or internal causal factors, and whether the system is deterministic (responds continuously to forcing), stochastic (obeys statistical laws, such as unloaded dice), or chaotic (may respond in a nonrandom, and nondeterministic way). It is also likely that global climate possesses all three characteristics for different variables and at different scales.

External factors are the most commonly advanced explanations among scientists exploring global change, perhaps because they hold the promise of predictability. For example, there is a sunlight detector at the Mauna Loa Observatory, which is located more than 3000 meters above the Pacific Ocean on the island of Hawaii. The detector typically indicates that approximately 93.5 percent of the radiant energy from the Sun reaches the surface at that pristine location. Why not 100 percent? Primarily because very small molecules in the atmosphere – small in size relative to the wavelength of light – cause a phenomenon known as *Rayleigh scattering*. This process is preferential to blue light; it scatters approximately 6 percent of the direct solar beam. This scattering of sunlight by air molecules is the reason a clear sky is blue.

In 1963 there was a noticeable drop – approximately 2 percent – in the amount of solar energy reaching Mauna Loa's sunlight detector. This was the result of the eruption of Mount Agung on Bali, which threw sulfur dioxide into the stratosphere, where it photochemically converted to sulfuric acid aerosol particles, spread worldwide, and then slowly fell out over approximately a five-year period. This dust veil scattered away from the lower atmosphere approximately 1 percent of the solar energy. The Earth should have become cooler, and indeed it did by a few tenths of a degree Celsius. Volcanic sunsets are spectacularly hued because after the Sun sets, the high-altitude dust particles catch the rays of the recently setting Sun and the sky re-brightens. Although the Sun appears to have disappeared below the horizon, it is still shining through the upper atmosphere and illuminating all the excess dust that has been ejected from the volcano. The tell-tale sign is re-brightening and a pale purple color on clear nights or dawns. The orange that is sometimes visible lower toward the horizon can be the purple scattered light reflecting through long layers of the atmosphere, thereby losing its purple coloration.

In 1983, El Chichón, a volcanic mountain in Mexico, erupted throwing off a sizable chunk of the mountaintop. Although the ash from El Chichón injured people in the local vicinity and blocked sunlight for a radius of approximately 100 kilometers, the ash itself did not cause significant climatic change because it fell out of the atmosphere within weeks. It was the sulfur dioxide in the volcanic blast that was the significant climatic factor. The force behind El Chichón was equivalent to a multimegaton nuclear explosion – shot off like a cannon vertically into the stratosphere – of millions of tons of sulfur dioxide, which remained there subsequently for a year or two and became sulfuric acid. Most of Los Angeles' infamous smog is composed of sulfuric acid – even though its source is a continuous "human volcano," not an explosive natural one. And how did the eruption of El Chichón register at the Mauna Loa Observatory? Agung had looked large there; however, El Chichón was much larger in terms of the dust that was thrown into the stratosphere in the Northern Hemisphere.

Sulfur levels from the eruption of Mount Pinatubo in the Philippines in 1991 were even higher, and remained even longer. Probably as a result of the Pinatubo eruption, globally averaged surface temperatures in 1992 and 1993 were approximately 0.25°C colder than those in previous years. In fact, 1992 was the first year in the previous dozen years that had not registered in the record-high grouping of the 1981–91 decade. On the other hand, had 1992 been warmer than the rest of the twentieth century, those of us using computer models to calculate global warming effects would have been concerned. When volcanic dust reduces solar heat in the lower atmosphere by several watts per square meter, the sur-

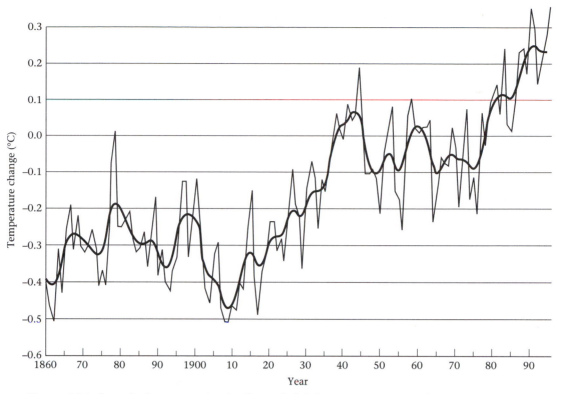

Figure 16.4. Record of past warming in observed global mean temperature changes, 1860–1995 (Intergovernmental Panel on Climate Change, 1996). The thick line is the five-year running mean; the thin line represents the value. (*Source:* Adapted from *Houghton et al., Climate change 1995: the science of climate change*, Cambridge University Press, 1996.) Although "off the chart," and not illustrated, 1998 was 0.2°C warmer than the previous record warm year, 1997. Thus the warming trend has continued to the present.

face should cool, our theory says, just as it should warm when a few watts per square meter of heat is added via excess greenhouse gases. In the former case, one of nature's own experiments allowed us to test our predictions of short-term climate change. In 1995, record warming returned as the Pinatubo dust cloud settled out of the atmosphere.

Figure 16.4 records global surface air temperatures. The dips coincide with the Krakatoa eruption of 1893 and that of Mount Agung in 1963. The eruption of El Chichón is reflected in another dip in 1983, and that of Mount Pinatubo in 1992 and 1993. Comparable year-to-year fluctuations do occur in noneruption years, commonly termed climate *noise* – presumably, much of it internally generated. Internal noise is superimposed on temperature changes caused by external forcing factors such as volcanoes. Many other kinds of external factors exist, such as land use. Numerous studies are currently being conducted to explore how climate would be affected if humans were to deforest an area such as the Amazon at a rapid rate. Experts are also trying to discover the extent and nature of the climate change that has already resulted from global deforestation (see Chaps. 28 and 29).

Figure 16.5 shows the results of an experiment in North Carolina comparing forested and deforested slopes. Throughout the seasons, the forested area evaporates more water than the cleared area. Trees have roots that tap into deep soil

moisture, and the stomates in the leaves open up to take in carbon dioxide (for photosynthesis) and release moisture in a process called *evapotranspiration*. Deforestation changes evapotranspiration rates, which are an important component in feeding moisture into the atmosphere. Human activities that change the land surfaces and the biota can also change the very nature of the weather upon which the biota depend. This feedback loop is called *biogeophysical feedback* and constitutes yet another set of processes that we need to account for in Earth systems science.

Another major process involved in the Earth system is the biogeochemical cycling of sulfur. Sulfur is produced not only by volcanoes but also by phytoplankton in the oceans (see Chap. 19). Sulfur produced by certain phytoplankton can end up as particles in the atmosphere that form *cloud condensation nucleii* upon which water droplets condense. They, in turn, can affect the amount of sunlight that reaches the oceans, which may influence surface temperatures or phytoplankton viability. Thus, chemical transformations can proceed simultaneously with atmospheric transport or biologic production. The interactions between the transport, the transformation, and the biologic sources and sinks of sulfur help to determine how sulfur cycles among the various reservoirs. In addition, a major component (more than half) of the sulfur cycle is attributable to human activities such as

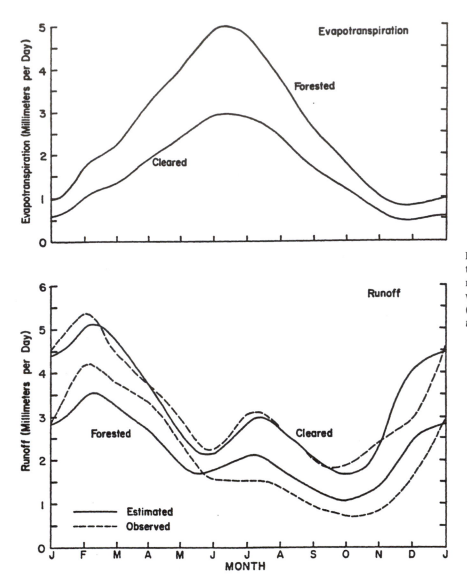

Figure 16.5. Estimated evapotranspiration (top) and estimated and observed runoff (bottom) from cleared and forested watersheds near Coweeta, North Carolina. (*Source:* William D. Sellers, *Physical climatology*, University of Chicago Press, 1965.)

coal or oil burning, particularly in the Northern Hemisphere. This process constitutes yet another example of what "global change" implies: Human activities are changing stocks and flows of important biogeochemical or nutrient cycles such as sulfur, carbon, water, nitrogen, and phosphorous. At the same time, they may be changing the climate, which is the principal transport mechanism for nutrients and determinant of habitats for biota.

The opacity – or haziness – of the atmosphere is another factor associated with global change. Particles suspended in the air (noted earlier, in connection with volcanic explosions) can affect the radiation balance of the atmosphere directly by scattering or absorbing radiation, as well as indirectly by being cloud condensation nucleii. Many particles are produced by human activities – for example, sulfur dioxide emissions from coal burning, and dust raised in agriculture or from desertification processes in which formerly vegetated lands are degraded to bare soils. Such particles typically cool the climate.

The principal cycle of interest to Earth systems scientists is the carbon cycle, because it involves the greenhouse effect, photosynthesis, respiration, and decay – all natural processes (Chaps. 19 and 20). The burning of fossil fuels and deforestation are global changes directly affecting the carbon cycle. The stocks and flows of carbon, which are mediated by the weather, have been influenced significantly by human activity. The evidence is overwhelming that the 10 percent increase in the carbon dioxide level that has been observed at Mauna Loa Observatory and other remote locations from pole to pole since 1957 is due to human activities (Chaps. 13 and 24).

Today, a mix of internal and external factors, as well as natural and anthropogenic processes, affect Earth systems. If we want to forecast the climate, even prior to writing the first equation in order to simulate how the system works, we must specify what may be forcing the climate to change and then try to trace the effects of that forcing in the simulation model's results.

The question is, "Which of all these internal and external factors is the most important influence on climate?" The answer depends critically on the time scale of variation involved. Clearly, changes to very slow cycles such as carbon stored in soils are not going to affect next year's weather because the latter is associated with rapid changes such as the seasonal cycle. A volcanic eruption, however, is a very important external forcing factor for a one-year average global temperature (typically cooling a hemisphere up to 0.50°C). However, at the time scale of a century (comparable to the time it will take humans to double atmospheric carbon dioxide concentrations), volcanic dust veils mimic the appearance of short-term noise. Consequently, the answer to the question is an unsatisfying, "It depends on which factor is most important relative to the scale of forcings and the characteristic response times of various subcomponents of the Earth system." Moreover, Earth system responses to various human disturbances could be any combination of several factors.

The task for scientists is to separate out, quantitatively, the cause-and-effect linkages from the many factors of comparable magnitude and opposite signs pushing on the interconnected Earth system. It is a controversial effort because so many subsystems and so many forcings are operating simultaneously. How do scientists do it? First, we examine the observations of changes in temperatures, ozone levels, and so forth. This allows us to identify correlations among variables. Correlation is not necessarily cause and effect; however, it can lead to a hypothesis of cause and effect that can be tested. The testing is commonly based on simulations with

WEATHER AND CLIMATE FORECASTING MODELS

The most comprehensive weather simulation models solve for the three-dimensional details of temperature, winds, humidity, and rainfall globally. A weather map generated by such a computer model typically appears to be quite realistic; however, it is never faithful in every detail. To create a computer model weather map, we need to solve six partial differential equations describing the fluid motions in the atmosphere and include well-known physical laws, such as conservation of mass, conservation of energy, and the ideal gas law. This would seem to be a problem-free process: After all, we know that these equations work in the laboratory, and we know that they describe fluid motions and energy and mass relationships. So why, then, are the models imperfect simulations? For weather forecasting, one question mark is the chaotic nature of the atmosphere, which limits predictability of specific weather events to a few weeks at most.

Climate is the average of weather; however, climate models also cannot be perfect simulators – why? The answer is that nobody knows how to solve such coupled, nonlinear partial differential equations precisely. There is no known technique. They are not simple algebraic equations for which the exact solution can be obtained by performing some manipulations. We can only approximate the solutions by taking these equations, which are continuous, and breaking them down into discreet chunks called *grid boxes*. A typical grid size is about half the size of the state of Colorado horizontally and hundreds of meters of atmosphere deep in the vertical direction.

Previously, we discussed clouds, noting that they are very important, that they reflect sunlight away from the Earth while trapping infrared heat. However, because no one has ever seen a single cloud the size of Colorado, we have a problem of scale: How can we treat processes that occur in nature at a smaller scale than we can resolve by our approximation technique of using large grid boxes? For example, we cannot treat cloud processes explicitly because individual clouds are the size of a dot in this grid box. Yet we can put forward a few reasonable propositions: If it is a humid day, for example, it is more likely to be cloudy. If the air is rising, it is also more likely to be cloudy.

These climate models can predict not only the average humidity in the grid box but also whether the air is rising or sinking on average. Then we can write what we call a parametric representation, or *parameterization*, to correlate large-scale variables that are resolved by the grid box with unresolved small-scale processes. For example, average cloudiness can be made equal to a parameter (which is a number), times the grid box average relative humidity, plus a parameter (another number), times the grid box average vertical velocity. We arrive at a prediction of grid box–averaged cloudiness through this parameterization. However, although the models are not ignoring cloudiness, the scale of individual clouds is not resolved explicitly. Instead, modelers try to approximate the average effect of small-scale processes they cannot duplicate through massive computation; this reflects the fact that real weather phenomena such as clouds are present at smaller scales than the smallest resolved scale (i.e., the grid box) in the mathematic model.

Physics and chemistry tell us that radiation exchange occurs among molecules in gases, accompanied at atomic scale with actual jumps of electrons between the various energy levels in an atom associated with emissions or absorption of photons. However, because we cannot possibly account for every one of the trillions upon trillions of electron jumps that occur as radiation is being exchanged between the Earth and space, we perform a parameterization instead. We can state that, on average, radiation exchange is some function of temperature and some function of the amount of each gaseous absorber that is present. Because there are many types of parameterizations of processes that occur at a smaller scale than our models can resolve, scientific debate develops over which is best – that is, whether we can partially validate these parameterizations. In effect, are they an accurate representation of the large-scale consequences of processes that occur on smaller scales than we can explicitly treat? (See Chap. 13.)

mathematic models on a computer, which, in turn, need to be tested against a variety of observations – present and paleoclimatic (see *Weather and Climate Forecasting Models*). This is how the scientific method is typically applied. When a model appears plausible, it can be "fed" unprecedented changes (e.g., projected human global change forcings) and then asked to make projections of future climate, ozone levels, forestation, extinction rates, and so forth.

As the "Weather and Climate Forecasting Models" box illustrates, in attempts to forecast climatic change, validation becomes critically important. In fact, we cannot easily know whether our parameterizations are sufficiently accurate. We must test them, and we need a laboratory in which to test them. This is where the study of paleoclimates has proved so valuable (Chap. 15). We can also test parameterizations by undertaking detailed small-scale field or modeling studies aimed at understanding the high-resolution details of a particular parameterized process that the large-scale model has demonstrated to be of crucial importance to the overall model.

Earth systems scientists are now forecasting twenty-first-century global changes, the outcomes of which range from mild to catastrophic. The reason for such wide variation in our climatic change predictions is twofold: (1) uncertainties in the forcing – we do not know, for example, how many people will use how much of what kind of technology; and (2) uncertainties in the internal responses of the Earth system – we do not know the extent to which changes in the evaporation of water might make clouds wider or brighter, and therefore reduce warming (a "negative" or stabilizing feedback mechanism). On the other hand, such changes could make clouds taller, enabling them to trap more heat and become a destabilizing "positive feedback." Similarly, we do not know whether warming in the soils might stimulate bacteria to produce excess carbon dioxide and methane gases, which might increase the amount of greenhouse heat, thereby acting as a positive feedback. Additionally, we do not know the extent to which the addition of carbon dioxide will increase photosynthesis, thereby becoming a negative feedback on atmospheric carbon dioxide buildup.

We can track each one of these processes, individually with perhaps 90, 80, or sometimes only 50 percent confidence in our accuracy. However, when ten different processes interact, with uncertainties of 50 to 200 percent for the influence of each process, it is difficult to have confidence in the accuracy of the predicted outcome.

The best prediction strategy is to posit a reasonable scenario for the probabilities we assign to each plausible outcome envisioned. How can we validate an "experiment" such as this, when the potential rates of change due to human activities are higher than most of the documented, sustained, global changes in the geologic past? We can say with confidence that previous sustained natural ecologic or climatic changes did not occur as rapidly as they are projected to occur in the global change era of the twenty-first century – with the exception of a few sudden "surprise" events (see

Chap. 15). As a result, today we continue to use the past to test our models on nonsurprise, slow changes. We also try to verify theories of rapid change by studying the effects of volcanoes or El Niños. Earth systems scientists search for all the natural experiments that can be found to try to improve the validity of these models. This search is time consuming; it will take decades more to improve the models and gather the vast data sets needed to validate them to a high degree of confidence. Meanwhile, humans continue to create global changes such as habitat destruction and carbon dioxide emissions.

The complexity and scale of issues concerning the accurate prediction of climate change demand a detailed and skillful approach. A built-in lag time adds urgency to the situation: The speed with which human activity is changing the global environment has far outpaced our understanding of that environment as it was prior to our detrimental influence. We need all the help we can get from the next generation of interdisciplinary Earth systems scientists in order to increase our understanding. The tough policy question is: How long do we attempt to clear the glass on the cloudy crystal balls our climatic models have become before we act on what we may be seeing inside? Although the ultimate fate of the Earth may not hang in the balance, the well-being of hundreds of millions of people and the survival of perhaps millions of species most certainly do.

QUESTIONS

1. What about surface water runoff? In North Carolina, more runoff occurs after deforestation because there is less vegetation to keep the soil intact and thereby to hold soil moisture. In addition, because less evapotranspiration occurs, more water is left to run off. (Degradation of this so-called ecosystem service can lead to flooding if soils erode to the extent that runoff increases significantly.) However, if the deforestation proceeds on a large scale, for example, the entire Amazon Basin, this may reduce evapotranspiration to the extent that it will decrease precipitation, and thus runoff. Climate change is one of the controversies associated with large-scale, rapid deforestation. Discuss this problem.

2. How can you tell if the 0.5°C warming trend of the twentieth century is a natural fluctuation or a forced response to human activities?

3. If weather becomes unpredictable a week or two in advance, how can we even contemplate forecasting climatic changes in the twenty-first century?

FURTHER READING

Bruce, J., Hoesung, L., and Haites, E. 1996. *Climate change 1995 – economic and social dimensions of climate change.* Contribution of Working Group III to the Second Assessment Report of the Intergovernmental Panel on Climate Change, 608 pp.

Henderson-Sellers, A., and McGuffie, K. 1987. *A climate modelling primer*. New York, NY. Wiley.

Houghton, J. J., Meiro Filho, L. G., Callander, B. A., Harris, N., Kattenberg, A., and Maskell K. (eds.) 1996. *Climate change 1995 – the science of climate change*. Contribution of Working Group I to the Second Assessment Report of the Intergovernmental Panel on Climate Change, 448 pp.

Schneider, S. H., and Londer, R. 1984. *The coevolution of Climate and life*. San Francisco, CA: Sierra Club Books.

Watson, R. T., Zinyowera, M. C., and Moss, R. H. (eds.) 1996. *Climate change 1995 – impacts, adaptations and mitigation of climate change – scientific-technical analyses*. Contribution of Working Group II to the Second Assessment Report of the Intergovernmental Panel on Climate Change, 880 pp.

THE BIOSPHERE

17 Biodiversity: Result of Speciation and Extinction

CAROL L. BOGGS AND JOAN ROUGHGARDEN

THE RICHNESS OF LIFE ON EARTH

The diversity of species inhabiting the Earth is now the highest in the planet's history. Estimates for the number of species range from 3½ million to over 111 million. The working estimate used by many biologists is 13,620,000 species. These species are not distributed evenly among taxonomic groups. At present, there are approximately 400,000 species of beetles, one of the largest insect groups. Estimates of the number of bacteria species range from 50,000 to 3,000,000. Differences in species diversity among taxonomic groups have occurred throughout the Earth's history. During the Jurassic Period, for example, the species diversity of mollusks was much greater than that of echinoderms (the group to which modern-day sea stars, brittle stars, and sea urchins belong). Nor are species distributed evenly across the globe. Areas on the mainland tend to support higher species diversity than do islands. Tropical areas commonly support higher species diversity than temperate areas. The Amazon rain forest supports 473 tree species per hectare, whereas the 420 million hectares of temperate forest in all of Europe, Asia, and North America combined support only 1166 tree species.

What factors affect speciation? extinction? migration rates? How do speciation, extinction, and migration combine to produce floras and faunas characteristic of particular places? Comprehending how natural processes operate is crucial to understanding the fate of the Earth's inhabitants, and to any efforts to alter human impacts on that fate.

Species diversity at regional and local scales depends not only on speciation and extinction, which generate the global species pool, but also on migration of species into or out of the area. In 1883, the island of Krakatau, located between Sumatra and Java, was completely devastated by a volcanic eruption. Subsequent species migration onto the pumice-covered island has strongly affected the present-day biologic diversity. Processes governing speciation, extinction, and migration determine in large part whether a given locale contains primarily species found nowhere else, or species with wide distributions.

Natural biologic processes of speciation, extinction, and migration often result from interactions between biologic and geologic or physical systems. In general, human activity does not affect rates of speciation. However, human activities, particularly those leading to habitat destruction, have led to a marked increase in extinction rates above the natural baseline. In tropical forests alone, extinction rates are approximately 1000 to 10,000 times the background rate expected in the absence of habitat destruction and other human activities. This loss rate is very much greater than the present rate of species creation through biologic speciation. Global diversity is dropping precipitously as a result. In tropical forests, 5 to 10 percent of the species are expected to become extinct within the next twenty-five years, with essentially no species replacement during that interval.

Humans also transport organisms around the world, both intentionally and unintentionally. Human diseases such as small pox entered the New World with European settlers, *Eucalyptus* trees have been introduced to Mediterranean climate zones throughout the world, introduced rats and pigs in the Galapagos Islands now threaten the persistence of some tortoise populations, and so on. Homogenization of biologic communities will result from such introductions over the long term, as the same species come to dominate similar habitats throughout the world. The San Francisco Bay is a prime example of this process, with new aquatic species arriving via ships from around the world on a regular basis.

Darwin's ideas on evolution form the basis for understanding the generation and distribution of biodiversity. We begin, then, with Darwin, then consider more recent understanding of how speciation, extinction, and migration interact to mold diversity.

DARWIN

Charles Darwin's book, *The Origin of Species*, contains two primary ideas regarding evolution. The first concerns phyletic diversity, or the diversity of organisms at the species level and above. He argued that species are descended from other species, which eventually has resulted in the creation of organized taxonomic relationships among species. The second idea concerns adaptation. Darwin argued that species are mutable; over time, characteristics of a species can change, due to differences in the reproductive success of individuals carrying specific traits in specific environments. In this chapter and Chapter 18, we explore how these processes, in combination with geologic and physical processes, shape the diversity of life on planet Earth. The linkage between geologic and biologic processes is perhaps most dramatically evident in the Galapagos Islands, whose study helped Darwin himself come to an understanding of the origin of species.

VOYAGE OF THE *BEAGLE*. In 1831, at the age of twenty-two, Charles Darwin was offered the job of naturalist on the HMS *Beagle*. The *Beagle* was to sail around the world, with most ports of call along both the Atlantic and Pacific coasts of South America, plus the Pacific islands. Darwin's background prepared him well for the trip. His father sent Darwin to Cambridge to study theology. However, Darwin spent much of his time there collecting beetles, studying natural history, and talking to influential geologists and natural historians. As the *Beagle*'s naturalist, he collected, preserved, and catalogued plants and animals both along the shore and from inland expeditions during the nearly five-year trip.

Darwin returned from the voyage with an extensive collection of plants, invertebrates, and vertebrates, including birds and reptiles from the Galapagos Islands off the coast of Ecuador. He also brought back notebooks filled with observations, and memories of experiences that changed the way he viewed the world. Once back in England, he married, settled south of London, and rarely traveled again, even as far as London. He did, however, carry on a voluminous correspondence throughout his life.

THE ORIGINS OF DARWIN'S IDEAS ON EVOLUTION. Two sets of experiences from the voyage of the *Beagle* conspired to lead Darwin to his ideas on evolution. The first was a devastating earthquake that he experienced while ashore in Chile. The second derived from several of his collections, most notably the animals from the Galapagos Islands and the fossils from Patagonia.

During the *Beagle*'s voyage, Darwin received first-hand verification of the idea – at that time, a new idea – that the Earth is not immobile and immutable. While on a shore expedition in Chile, he experienced a strong earthquake. On returning to the *Beagle* the following day, he observed mass destruction inland and along the coast, along with evidence that the shoreline had risen a foot or more. This experience convinced him that, over geologic time, vast changes in the Earth's landforms are possible. As discussed in the section entitled "Allopatric Speciation," this is a prerequisite for speciation based on geographic isolation.

Darwin also repeatedly noted geographic variation among the species that he collected or observed. For example, he recorded that the governor of the Galapagos Islands could discern which island a giant tortoise came from based on its shell morphology. Variation among mockingbirds found on the different islands making up the Galapagos also impressed Darwin, as did the similarity between birds from the islands and those found on the South American mainland.

These geographic relationships among species clearly extended back in time. Darwin, (1958, republished) wrote:

During the voyage of the Beagle, I had been deeply impressed by discovering in the Pampean formation [in South America] great fossil animals covered with armour like that on the existing armadillos; secondly, by the manner in which closely allied animals replace one another in proceeding southward over the continent; and thirdly, by the South American character of most of the productions of the Galapagos archipelago, and more especially by the manner in which they differ slightly on each island of the group; none of these islands appearing to be very ancient in a geological sense. It was evident that such facts as these, as well as many others, could be explained on the supposition that species gradually become modified; and the subject has haunted me. (pp. 41–42)

LESSONS FROM THE VOYAGE OF THE *BEAGLE*. From Darwin's observations grew a generally accepted understanding of the number and distribution of species. He proposed a mechanism for creating diversity, through the descent of species from other species with modification. Implicit in his proposal are both taxonomic and geographic relationships among species. The ability to classify organisms hierarchically into species, genus, family, class, and phylum reflects taxonomic relationships. Each higher level of organization (e.g., a family) subsumes a number of members of the level below it (in this case, genera). Geographic relationships among species also result directly from descent of species from other species. "Daughter" species are often, by necessity, geographically close to the ancestral species.

Descent with modification further implies a process of adaptation to the local environment. We leave the discussion of this process, and its implications for species diversity and distributions, for Chapter 18.

MECHANISMS OF SPECIATION

What produces the diversity of species that we observe in nature? To address this question, we must first define what we mean by a "species" and by "evolution." We then use the Galapagos Islands as a case study, looking in more detail at the geologic and biologic history of the islands themselves.

DEFINITIONS OF SPECIES AND EVOLUTION. A *species* consists of biologically similar individual organisms. For organisms capable of sexual reproduction, biologists often describe a species as composed of those individuals capable of breeding with each other and producing viable offspring. Thus, all humans belong to one species, and chimpanzees to another species.

How do we apply this definition of species to organisms,

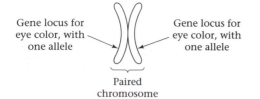

Gene locus for eye color, with one allele

Gene locus for eye color, with one allele

Paired chromosome

Figure 17.1. The relationship among alleles, genes, and chromosomes.

for example, a few species in the genus *Cnemidophorous*, a type of lizard whose females reproduce without mating? The definition is clearly inapplicable to organisms such as these or others that reproduce asexually. Yet, individuals can be identified as members of a species, based on similar morphologic, behavioral, or genetic characteristics.

Changes in the genetic makeup of organisms over time form the basis for *evolution*. A ribbon of deoxyribonucleic acids (DNA) organized into chromosomes within each cell contains the genetic information for a particular trait, such as eye color in the fruit fly *Drosophila melanogaster*. A gene locus is the area of DNA within the chromosome coding for the trait. Not all flies have the same eye color, however. These differences arise from variation among flies in the DNA at the locus for eye color. One variant, or *allele,* may yield brown eyes, whereas another may yield red eyes. Many different alleles at one gene locus can exist within a population, or actually interbreeding group, of individual flies. Finally, the fly's genotype is the sum of all the alleles present within all the gene loci contained within the individual (Fig. 17.1).

Biologic evolution, then, is a change in the allele frequencies of a population over time. These changes in the genetic makeup of the group may result in changes in the phenotype, or observed morphologic, biochemical, or behavioral traits. If these changes are relatively extreme, so that individuals with the old and new genotypes are no longer able to mate and produce offspring, speciation has occurred. (See the box entitled *Allele Frequency Changes and Industrial Melanism.*) However, evolution does not always lead to speciation. For example, a change in the frequency of the alleles coding for eye color in a human population is an example of evolutionary change, but this change does not result in speciation!

ALLOPATRIC SPECIATION. One type of species formation, termed *geographic* or *allopatric* (*allo* means "different," *patris* means "country") *speciation*, results from the geographic isolation of two populations of a given species. The two populations diverge genetically, to the point that successful reproduction is not possible if the two populations come back into contact (Fig. 17.2). We return to the Galapagos Islands to examine how this process works.

The Geology of the Galapagos Islands. The Galapagos Islands are relatively new. They formed as a result of the ascent of molten basalt from a deep mantle "hot spot" break-

ALLELE FREQUENCY CHANGES AND INDUSTRIAL MELANISM

The peppered moth, *Biston betularia*, has a melanic, or dark, form and a white form in Britain. Differences only at one gene produce the two color forms. The allele frequencies have undergone major fluctuations through time in response to environmental changes.

The moth rests on tree trunks during the day. Prior to the Industrial Revolution, the trunks were lichen-covered. The white form, with its specks of black, was well camouflaged against the lichens. Birds, which are visual predators, have difficulty finding the camouflaged moths. With the advent of vast quantities of air pollution, the trunks became covered with soot. The melanic form, *carbonaria*, increased in frequency because it now was protected by camouflage during the day.

In recent years, air pollution has decreased in Britain, particularly with the passage of the Clean Air Acts. The cleanup promoted the return toward mottled coloration of tree trunks; this in turn resulted in a major shift in the frequency of the *carbonaria*. At West Kirby in northwestern England, for example, the *carbonaria* form constituted approximately 90 percent of the population from 1960 through 1970. The frequency of *carbonaria* began dropping around 1970, and by 1990 constituted only approximately 30 percent of the population. Similar patterns occurred elsewhere in Britain. These changes represent a rapid shift in allele frequencies within the population.

ing through the Earth's crust to the surface, as a volcano (Chap. 6). The emerging basaltic flow covers a large area, and can lead to island formation. A string of islands forms over time, as the overlying tectonic plate moves across the hot spot. The youngest islands in the string have ongoing volcanic activity and are in the process of being built; somewhat older islands possess dormant volcanoes. Even older islands in the chain have subsided beneath the sea's surface. (The Hawaiian Islands are another example of islands formed as a result of hot spots.) The geologic origin of the Galapagos means that the islands have been geographically separated from each other over their entire histories. This has maximized the opportunity for speciation to occur.

Darwin's Finches. The fourteen finch species found on the Galapagos Islands, known as *Darwin's Finches*, are all believed to be descended from the same ancestral species on the South American mainland: the blue-black grassquit (*Volatinia jacarina*). The islands are sufficiently isolated from each other that migration among them is relatively rare, even for finches. The initial finches that colonized each island thus gradually changed over time into separate species, through evolution of distinct characters. The finches diverged not only in morphology, but also in ecologic traits such as feeding habits. Today's finches in the Galapagos include seed eaters (*Geospiza* spp.), a bud eater (*Platyspiza* spp.), and insec-

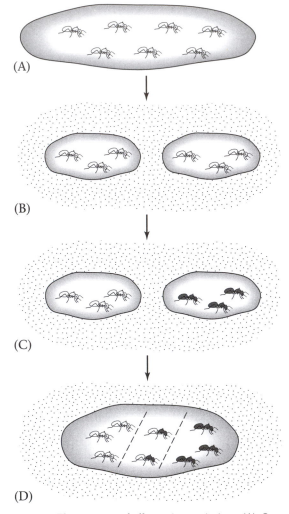

Figure 17.2. The process of allopatric speciation. (A) One species of ant is distributed throughout an island. (B) The island is broken into two islands by rising sea levels. (C) The ants on each island differentiate from each other. (D) Sea levels drop, and the two species of ant come back into contact on the large island. A hybrid zone between the two species may form at the point of contact.

Figure 17.3. Darwin's Finches. (*Source:* D. Lack, *Darwin's Finches*, Cambridge University Press, 1983.)

tivores (*Camarhynchus* and *Certhidea* spp.) (Fig. 17.3). These habits match the environmental characteristics of the islands on which each bird species occurs.

OTHER TYPES OF SPECIATION. Geographic speciation is likely to be the predominant mode of speciation. Other processes can result in species divergence, however. (The box entitled *Reproductive Isolating Mechanisms* presents the general principles of reproductive isolation.) In some cases, an entire species may gradually change through time, to the point that taxonomists judge that the ancestral group of individuals are sufficiently morphologically different from the later group to warrant classifying the two as separate species. The two species do not overlap in time; one is descended from the other. This process is termed *anagenesis*, as distinct from *cladogenesis*, which is the splitting of a parent species into two or more daughter species (Fig. 17.4). Anagenesis is fairly common in the fossil record.

Parapatric Speciation. Occasionally, two incipient species are not separated by a geographic barrier – such as the Pacific Ocean for Darwin's finches – but dwell side by side. If members of the two neighboring groups have limited reproduction with each other, *parapatric* (*para* means "beside") *speciation* may occur. Typically, reproduction is limited because a sharp difference in environmental conditions for each of the two groups leads to adaptation by each population to its own environmental conditions. Offspring from crossbreeding between the two groups do well in neither environment.

Differences in soil or underlying geology often cause these sharp environmental boundaries. For example, the tailings from the Goginian lead mine in Aberystwyth, Wales, contain

REPRODUCTIVE ISOLATING MECHANISMS

Reproductive isolation can take many forms. Isolating mechanisms are typically divided into two categories, depending on whether isolation occurs before or after union of the sperm and egg to form a zygote.

Prezygotic mechanisms:

- Organisms' *behavior* may change, with the result that the two new species never encounter each other when reproductively active. Alternatively, courtship behavior may change, so that individuals of the two species do not recognize each other. For example pintail and mallard ducks have very similar-looking females and may coexist in the same area. Males discriminate between females based on courtship responses.
- *Morphologic changes* may mean that mating between the two new species is impossible. Many insects have genitalia that fit together like a "lock and key." Attempts to mate between two species without matching lock and key morphology are unsuccessful.
- *Chemical incompatibility between the gametes* can prevent formation of zygotes.

Postzygotic mechanisms:

- Individuals may succeeding in mating, only to produce *inviable zygotes*. Problems may arise from differences in chromosome number between the two species, or from other genetic incompatibilities.
- Individuals may mate and produce offspring; however, those *offspring are sterile*. Hybridization between the horse and donkey produces only infertile mules.

The particular type of reproductive isolation has important implications for individuals that happen to mate or try to mate with individuals of a different species. Matings that produce infertile offspring are more costly to the parents in time and energy than are matings that do not produce zygotes. If prezygotic isolating mechanisms occur, the individual has "wasted" less time and effort, and is likely to be able to mate again with a member of its own species. For this reason, selection often favors the development of prezygotic reproductive isolating mechanisms.

high concentrations of lead. The grass *Agrostis tenuis* grows on both tailings and natural soils; however, individuals found on tailings soil are resistant to high lead concentrations. The same is not true of individuals from natural soils. Individuals even a few meters off the tailings cannot survive in tailings soil. The grass is thus highly adapted to particular soil types, which is a necessary prerequisite for parapatric speciation.

Sympatric Speciation. A third type of speciation, termed *sympatric* (*sym* means "with") *speciation*, is controversial. Here, two incipient species live within each others' range; there is no geographic separation at all. Sympatric speciation is associated with *polyploidy*, or the duplication of chromosomes, as well as with highly specialized resource use.

Many organisms are either *haploid* or *diploid* – that is, they contain either one (haploid) or a pair (diploid) of each chromosome. When chromosome duplication is abnormal during reproduction, offspring may have a higher number of each chromosome than do their parents. Those with three of each chromosome are *triploids*, and those with multiple copies of each chromosome are *polyploids*. Triploids may result, for example, from the union of a normal haploid sperm cell with a diploid (instead of haploid) egg cell. Under normal circum-

stances, both cells would be haploid, and the resulting organism would be diploid.

Polyploidy is common in plants. Polyploidy may result either from abnormal chromosome duplication during reproduction, or from hybridization of two closely related plant species. In the latter case, the resulting hybrid may have 4n sets of chromosomes. These hybrids then interbreed in reproductive isolation from the parent species. Although polyploidy is less common in animals, it does occur in some asexually reproducing groups such as sowbugs (*Triochoniscus* sp.), and weevils (*Otiorrhynchus* sp.).

Highly specialized resource use is also associated with sympatric speciation. For example, the true fruit fly, *Rhagoletis pomenella*, lays its eggs in hawthorn fruit. Mating also occurs on the fruit. During the 1800s, a fly race developed that utilizes apples instead of hawthorn. Within the past forty years, another race has begun to utilize cherry. The timing of fruit development differs among hawthorn, apple, and cherry, leading to differences in timing of maturation of the flies. Such differences result in reproductive isolation, and the potential for speciation.

EVIDENCE FOR SPECIATION. Speciation occurs on time scales that are relatively long compared to our observation abilities; these time scales are shorter, however, than geologic time scales (see Chap. 3). What evidence do we have that

Figure 17.4. Species can arise through anagenesis or cladogenesis.

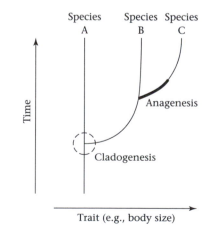

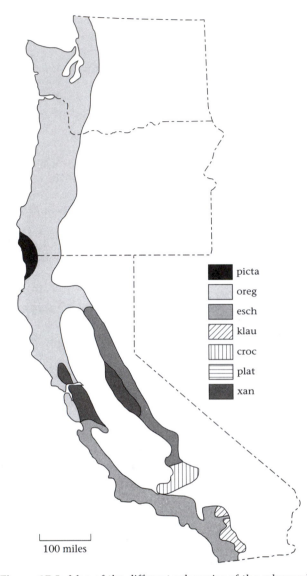

Figure 17.5. Map of the different subspecies of the salamander *Ensatina eschscholtzii*. Also shown are the two southern forms and the northern form of the salamander. picta = *Ensatina eschscholtzii picta*; oreg = *E. e. oregonensis*; esch = *E. e. ecscholtzii*; klau = *E. e. klauberi*; croc = *E. e. corceater*; plat = *E. e. platensis*; xan = *E. e. xanthoptica*. (*Source:* Adapted from C. Moritz, C. J. Schneider, and D. B. Wake, Evolutionary relationships within the *Ensatina eschscholtzii* complex confirm the ring species interpretation *Systematic Biology*, vol. 41, 1992, pp. 273–329.)

that have geographic ranges that form nearly complete rings, with an unoccupied center. For example, the salamander *Ensatina eschscholtzii* lives along the coastal areas of the U.S. Pacific Northwest and through the mountains of the Coast Range and of the Sierra Nevada on either side of California's Central Valley (Fig. 17.5). Morphologic and genetic data indicate that the salamanders spread south through the Pacific Northwest, splitting into two groups in northern California on either side of the Central Valley. As salamander populations continued to spread south, those on opposite sides of the valley differentiated from each other. Thus, at the southern end of the valley, populations that meet from east and west are morphologically distinct and do not interbreed. However, southern populations can interbreed with populations immediately to their north, and so on up the valley, until they reach the northern end, which contains one interbreeding population. The result is one continuous species. However, geographic isolation at the southern end of the ring leads to reproductive incompatibility. Consistent with our understanding of allopatric speciation, this is exactly what we would expect to occur.

EXTINCTION

Species are born through speciation processes and die through extinction. Species can go extinct for many reasons. Change in the physical environment can result in a reduction of available suitable habitat, leading to extinction. For example, changes in sea level at the end of the Ordovician Period may have altered the amount of suitable habitat available for marine organisms, which may have led to extinction events. Alternatively, extinction of one species may lead to the extinction of others. If a prey species goes extinct, its predator will soon follow unless it can evolve to eat a different diet.

Extinction rates have been far from constant throughout geologic history. The geologic record shows periods of high extinction rates. These high rates are sometimes associated with changes in sea level or climate, which could be caused by changes in the relative positions of the continents or by changes in the Earth's orientation toward the Sun (see Chaps. 6 and 15). Other cases are associated with collisions of meteors with the Earth; the dust thrown into the atmosphere by such collisions could have altered global climate for a period of years, leading to mass extinctions (see Chap. 3).

Humans have played an important role in extinction rates. First we significantly alter habitat – through agricultural and forestry activities, dam construction, and the building of urban centers (Chaps. 28 and 29). By increasing the rate of habitat fragmentation or alteration over natural rates, we also increase the extinction rate (Chap. 21). Hunting is a second way in which humans have increased extinction rates. By directly killing animals, we can decrease their numbers significantly. Even if the final individual of the species is not killed by hunters, the population size may have been lowered to the extent that the species could not recover.

The human influence on extinction rates is not new. In

speciation has occurred in the manner we have outlined above? In particular, what evidence is there that geographic isolation, followed by differentiation, is important to the speciation process?

Species Distributions. Present-day distributions of species provide some evidence in support of geographic speciation. New Guinea kingfishers, *Tanysiptera* spp., are closely related birds that are geographically separated by barriers to dispersal on different parts of the island. In some locations, two species may come into contact and form hybrid zones.

Ring Species. Perhaps the most direct evidence for geographic speciation comes from ring species. These are species

100 miles

400 A.D., Great Britain was forested. Timber cutting by humans significantly reduced the forested area even by 1086, and today almost nothing is left of the original forest. Coincident with the destruction of the forest, the brown bear went extinct by 1066, the wild boar, wolf, and the capercaille (European grouse) by the 1700s, and the goshawk by the 1800s (although the latter two species have been reintroduced). Today, the lowland heaths, upland moors, and calcareous grasslands that replaced the forests are themselves under pressure from humans.

PATTERNS OF BIODIVERSITY

The local difference between species generation and loss through speciation and extinction determines local patterns of biodiversity. Colonization is another mechanism by which local, but not global, species diversity may be enhanced. At any one place and point in time, the number of species present is a historic result of past speciation, extinction, and colonization events. The number of species may change over time as well, depending on the relative rates of speciation, extinction, and colonization. Each process may dominate the observed pattern of local biodiversity, depending on the particular circumstance. We examine each process in turn, and the specific patterns that each produces.

LAND-BRIDGE ISLANDS: DOMINANCE OF EXTINCTION. During the last ice age 14,000 ago, sea levels were much lower than they are at present. Land bridges connected many present-day archipelagoes as a result. These include the Caribbean Islands. As the ice cap melted back, sea levels rose, and large expanses of land became fragmented into the present island chains. As a result of this geologic process, once continuous habitat was severely fragmented and isolated.

Species–Area Relationships. What happened to the diversity of species present in the area as a result of island formation? If we assume that 14,000 years ago species were uniformly distributed throughout the area, we can then examine the present species distributions among remaining islands to deduce the effects of fragmentation on diversity. In

particular, what is the relationship between island (or fragment) size and number of species found on the island? If habitat fragmentation has no effect on diversity, all islands should have the same number of species. If extinction is the major force determining diversity, we expect to see lower numbers of species on smaller islands, because a smaller island would have a smaller number of types of habitat.

We restrict this discussion to species that do not disperse well, and hence are not adept at colonizing new areas from afar. (We return to these species later.) For the Caribbean Islands, reptiles (e.g., lizards and snakes) and amphibians (e.g., frogs and salamanders) are groups that might not be successful colonizers because they cannot fly and are not particularly adept seagoing swimmers.

Indeed, the number of reptile and amphibian species on the Caribbean Islands increases as island area increases (Fig. 17.6). This relationship is not linear; however, it is a power function. The number of species increases disproportionately rapidly with increases in area. This pattern is a general one, observed in other faunas on other land-bridge islands. As a rule of thumb, a decrease in area of 50 percent results in a 90 percent decrease in species number.

Nestedness of Species on Islands. If extinction within newly formed fragments determines the current species diversity seen on islands, we might also expect that species assemblages on islands would exhibit "nestedness." That is, all island fragments begin with the same species assemblage. Larger islands, with more habitat and space, lose only a few species – those organisms that are most prone to extinction. Smaller islands lose not only the same extinction-prone species that vanish from larger islands but also some species that can persist on larger islands but not in smaller areas. The result is a pattern of nested subsets of species, with sequentially larger islands containing all the species found on smaller islands, plus a few additional species.

Generality of Land-Bridge Islands. Land-bridge islands are not restricted to oceans. Terrestrial habitats can be isolated fragments of once larger areas, and can exhibit the same species – area relation and the same patterns of nestedness. Mammals in the Great Basin of the western United States provide a dramatic example of terrestrial land-bridge islands.

The Great Basin today contains a series of mountain ranges separated by broad valleys. The valleys contain dry sagebrush communities; the higher mountains contain pinyon-juniper and aspen communities. Yet, as recently as 8000 years ago, the area was cooler and wetter than it is today, with pinyon-juniper communities, meadows, and streams connecting the mountain ranges. At that point in the past, small mammals were fairly evenly distributed throughout the area. As the climate warmed and dried, the pinyon-juniper communities retreated to higher elevations. Today, pinyon-juniper communities and their associated small mammals are restricted to elevations above 2300 meters. For small mammals, the mountain ranges in the Great Basin effectively form land-bridge islands in a sea of dry, inhospitable sagebrush.

Figure 17.6. Species–area curve for reptiles and amphibians on the Caribbean Islands. (*Source:* Adapted from R. H. MacArthur and E. O. Wilson, *The theory of island biography*, Princeton University Press, 1976.)

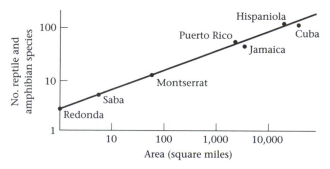

Mountain range

Species	1	2	3	4	5	6	7	8	9	10	11	12	13	14	15	16	17	18	19
a	X	X	X	X	X	X	X	X	X	X	X	X	X	X	X	X	X	X	X
b	X	X	X	X	X	X	X	X	X	X	X	X	X	X	X	X	X	X	
c	X	X	X	X	X	X	X	X	X	X	X	X	X	X	X	X	X		
d	X	X	X	X	X	X	X	X	X	X	X	X	X	X	X	X			
e	X	X	X	X	X	X	X	X	X	X	X	X	X	X	X				
f	X	X	X	X	X	X	X	X	X	X	X	X	X						
g	X	X	X	X	X	X	X	X	X	X	X	X							
h	X	X	X	X	X	X	X	X	X	X									
i	X	X	X	X	X	X	X	X											
j	X	X	X	X	X	X	X												
k	X	X	X	X	X														
l	X	X	X	X															
m	X	X																	
n	X																		

(A)

Mountain range

Species	1	2	3	4	5	6	7	8	9	10	11	12	13	14	15	16	17	18	19
Eutamias umbrinus	X	X	X	X	X	X	X	X	X	X	X	X	X	X	X	X		X	
Neotoma cinera	X	X	X	X	X	X	X	X	X	X	X	X	X	X			X	X	X
Eutamias dorsalis	X		X	X	X	X	X	X	X	X	X	X	X	X	X		X	X	X
Spermophilus lateralis	X	X	X	X	X		X	X		X	X	X	X	X	X		X		
Microtus longicaudus	X	X	X	X	X	X	X	X	X	X	X		X			X			
Sylvilagus nutalli	X	X	X	X	X		X	X		X	X	X	X						X
Marmota flaviventris	X	X	X	X	X	X	X	X	X	X	X								
Sorex vagrans	X	X	X	X	X	X	X	X	X			X							
Sorex palustris	X	X	X	X	X	X			X							X			
Mustella erminea	X		X	X	X	X			X										
Zapus princps	X	X	X			X										X			
Spermophilus beldingi	X	X																	
Lepus townsendi		X				X													

(B)

Figure 17.7. (A) Perfect nestedness for species on mountain ranges. Mountain ranges are numbered from those with the most to least number of species. x = species occurs in the range. (B) Nestedness relationships for nonflying mammals on mountaintop islands in the Great Basin. (*Source:* Adapted from A. Cutler, Nested faunas and extinction in fragmented habitats, *Conservation biology*, vol. 5, 1991, pp. 496–505.)

The small mammals on these habitat islands show the same type of species–area relationships that are observed in the Caribbean Islands. Further, the mammals show significant nestedness, with species dropping out of the community in sequence as island size decreases (Fig. 17.7).

OCEANIC ISLANDS: COLONIZATION PLUS EXTINCTION. Not all organisms found in island habitat fragments are incapable of dispersal. Birds, for example, can fly from island to island. Juvenile spiders can move to new areas by "ballooning," or drifting on silk threads in the wind. Mud-dwelling invertebrates can move from pond to pond on the webbed feet of ducks or other waterfowl. Cottonwood tree seeds are blown in the wind, and seeds of many fruiting plants are transported to new localities inside the intestinal systems of the birds or mammals that eat the fruit. Even some fish can invade new streams during floods.

Once a spider or cottonwood seed has landed on a new habitat island, it must reproduce and establish a viable population in order to be a successful colonist. Such colonization events raise the possibility that species diversity on the habitat island can be determined not only by subtraction of species through extinction, but also by addition of species through colonization.

Colonization and extinction were first used to describe the diversity of areas that are already established, such as oceanic islands, as opposed to newly formed areas, such as land-bridge islands. Although extinction may predominate initially in new habitat fragments, such areas are conducive to colonization within a short time.

Species–Area Relationships. The biologic diversity on an oceanic island is characterizable by a species–area curve. As area of the island increases, the number of species also increases. In this case, the increase is due to not only a smaller number of extinctions on the island, but also an increase in the number of immigrants that actually find the island and become established. The increasing success of colonization with island area is due to two factors. First, a larger island represents a larger target for dispersing individuals, such that more are likely to arrive on the island, increasing possibilities

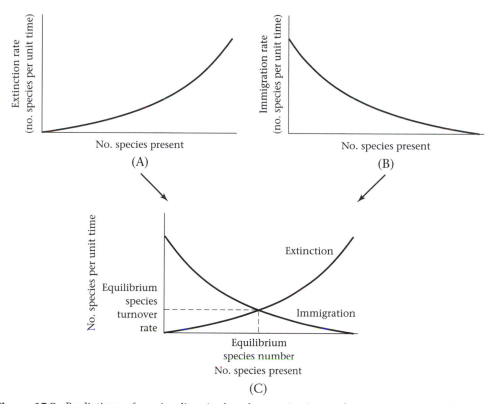

Figure 17.8. Predictions of species diversity based on extinction and immigration rates. (A) Relationship between extinction rates and number of species already present on an island. (B) Relationship between immigration rates and number of species already present on an island. (C) The combination of extinction and immigration rates yields an equilibrium number of species, and a predicted turnover rate of species on the island.

for successful establishment of populations. Second, a larger island has more space and potential habitats, again increasing the likelihood that an immigrant will successfully found a population.

Equilibrium Species Diversity. The interplay between extinction and colonization leads to predictions about the species diversity on any given island. Consider first how the extinction rate, or the number of species going extinct per unit of time, changes with the number of species already present on an island. At low species numbers, the extinction rate is low because there are few species present that could go extinct, and few competitors or predators are present. As species numbers increase, there are more species that can possibly go extinct; also, there is likely to be more crowding and competition for space and other resources. Thus, the extinction rate increases with increasing numbers of species (Fig. 17.8).

The colonization rate, or the number of new species arriving on the island per unit of time, decreases with increasing numbers of species already present (Fig. 17.8). As species numbers increase, the arriving species is less likely to represent a group that is not already present on the island.

Curves for both extinction and colonization should be convex to the number-of-species axis. Why? Extinction rates should rise disproportionately rapidly with increasing numbers of species due to increased crowding and competition under high species diversity. Colonization rates should drop disproportionately rapidly with increasing numbers of species if early colonists are those that can migrate to the island easily, while later colonists are less mobile.

Combining extinction and colonization curves into a single graph allows us to make some predictions about species diversity. First, the point at which extinction rates equal colonization rates is an equilibrium for species diversity (Fig. 17.8). Move to the left of that point, and colonization rates are greater than extinction rates, so species numbers increase over time. Move to the right of that point, and extinction rates are greater than colonization rates, so species numbers decrease with time.

This predicted equilibrium number of species says absolutely nothing about which species are present on the island. It merely predicts the number. However, our second prediction based on this graph is that change in identity of species present will occur through time, because extinction and colonization rates are not zero. The rate of species turnover is expected to be equal to the colonization or extinction rate when the number of species is at equilibrium (see Fig. 17.8).

Effects of Island Area and Distance to the Mainland. Island area and distance from the mainland source of immigrants will affect rates of extinction and colonization,

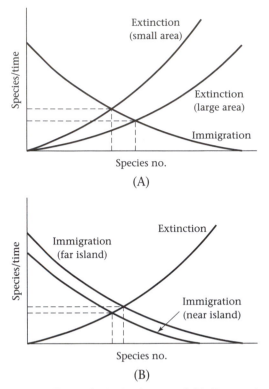

Figure 17.9. Effects of (A) island area and (B) distance from the mainland on extinction and immigration rates, and thence on predicted species equilibrium numbers and turnover rates.

and hence the number and turnover rate of species at equilibrium. As for land-bridge islands, as island area increases, the extinction rates decrease. Thus, the equilibrium number of species increases, and the turnover rate for species present on the island decreases. Distance from the mainland source affects colonization rates – as distance increases, colonization becomes more difficult and colonization rates decrease. Thus, the equilibrium number of species decreases as distance increases, and the turnover rate for species on the island also decreases (Fig. 17.9).

Realism. How realistic is this model of species diversity under extinction and colonization? Records of bird extinctions and colonizations in the Solomon Islands in the Pacific Ocean support the model. Extinction rates increased with number of species present on the islands, and immigration rates decreased. For both extinction and immigration, the relationship was concave (Fig. 17.10).

The model was tested experimentally on islands composed of mangrove trees off the Florida Keys. Insects on several islands were censused. An exterminator's tent was erected over the mangroves, and all the insects present were killed by methyl bromide. Subsequent censuses documented the number of species present at various times after the extinction event. As expected, the closest islands to the mainland had the largest equilibrium number of species, and the farthest island had the fewest species. Species composition on each island differed before and after the extinction event. Finally, the relationship among extinction rates, coloniza-

tion rates, and number of species present accurately predicted the increase in species number during the recovery period.

Bird censuses taken on the Indonesian island of Krakatau provide evidence that species turnover continues after an equilibrium number of species has been reached. In 1883, a volcanic eruption on that island killed all life there. Based on the species–area relationship for neighboring islands, we expect to see approximately twenty-seven bird species at equilibrium on an island of approximately 8 square kilometers. Birds were censused in 1908, 1920, and 1934. By 1920, the number of species had reached twenty-seven. The number remained constant at twenty-seven in 1934; however, five of the original twenty-seven species had gone extinct and five more had colonized the island. Thus, not only was the expected equilibrium species number reached, but species turnover was continuing.

Generality of Oceanic Islands. Like land-bridge islands, "oceanic" islands can be terrestrial. The Great Basin of the western United States again provides an example of island habitats on mountaintops. Just as mammal distributions fit the predictions for land-bridge islands, birds, which disperse more readily, fit the predictions for oceanic islands in this area.

The Role of Habitat Connectedness. Colonization probability for any given species is affected by the ability of the species to migrate and successfully establish a population, which in turn is affected by the presence or absence of corridors between habitat fragments or islands. Such corridors, or connections, between fragments can have diverse spatial and temporal characteristics.

Linear stretches of habitat can function as corridors con-

Figure 17.10. (A) Extinction and (B) colonization rates for birds on the Solomon Islands. E = extinction rate; I = immigration rate; P = constant pool size; S = species number.

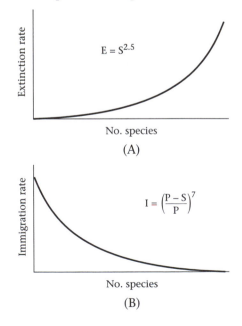

necting habitat fragments. For example, white-footed mice move between woodlots via fencerows. Riparian areas (the habitat along streams or rivers) can also act as corridors through a dry landscape for species needing complex vegetation structure.

Corridors may connect habitat fragments year-round, or only during part of the year. For example, ice cover over Lake Huron connects islands of the Drummond Archipelago to the mainland for at least part of the winter, allowing non-hibernating mammals to migrate to the islands freely. As a result, the species diversity of such mammals is higher on islands of the archipelago than, for example, in the Thousand Island region in the Saint Lawrence River, where rapid currents prevent ice cover from connecting islands with the shore during the winter.

Hotbeds of Species Radiation. Rapid speciation can also increase local species diversity. Cichlid fish living in Lake Victoria are a case in point. Currently, over 300 species of fish in the family Cichlidae live in Lake Victoria. Almost all of them are unique to that lake (Fig. 17.11). The water level of the lake has fluctuated dramatically over time; 12,000 years ago the lake dried up completely. Today's cichlid diversity must have arisen over the past 12,000 years, which is incredibly fast in evolutionary time. By comparison, the fourteen species of Darwin's Finches likely evolved over the past 4 million years on the Galapagos Islands.

Why might these fishes have such a rapid rate of speciation? One possibility is that individuals do not move very far during their lifetimes, and preferences for the appearance of mates change fairly rapidly. Behavioral reproductive isolation could evolve rapidly in a localized area, leading to speciation.

LESSONS FOR CONSERVATION

What are the practical consequences of the forces that shape species diversity? For conservation planners, several lessons stand out. The first lesson is that extinction is far outstripping speciation at this point in time. This adds urgency to conservation work. Restoration biology will also be an increasingly important field in the future, as we attempt to clean up polluted sites and restore some sort of functioning biologic community to damaged or fragmented habitats.

Second, the shape, size, location, and geologic history of habitat fragments strongly influence the number and identity of species present in an area. Conservation and land use planning must take into account the area and connectedness of habitat fragments, and the existence or absence of corridors. These pathways themselves may be semipermeable – allowing the free diffusion of some organisms but not others.

Finally, in this chapter we have used "biodiversity" to refer almost solely to diversity of species present in an area. However, the term is much broader, encompassing genetic diversity, community diversity, and regional diversity. In Chapter 18, we examine in more detail diversity at the community level.

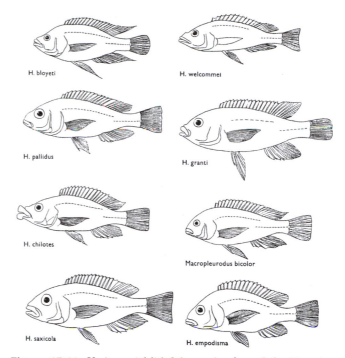

Figure 17.11. Various cichlid fish species from Lake Victoria. (*Source:* P. H. Greenwood, The cichlid fishes of Lake Victoria, East Africa: the biology and evolution of a species flock, *Bulletin of the British Museum (Natural History) Zoology*, London supplement 6, 1974.)

QUESTIONS

1. Geologists and biologists typically think of time on completely different scales (see Chap. 3). For geologists, a short time interval may be 1 million years, whereas for biologists that is a very long time. What are the implications of these different perceptions for interpretation of the fossil record and rates of evolution?

2. Humans fragment the habitat of a variety of plants and animals through construction of urban areas and highways, as well as through forestry and agricultural practices. What are the implications of this fragmentation over the short run for species diversity in the fragments? Over the longer run?

3. Many climatologists currently predict that the global climate will warm approximately 3°C over the next 100 years (see Chap. 16). Associated with this warming, pinyon-juniper associations in the mountains of the Great Basin are predicted to move up in elevation. What is the predicted impact on mammalian species diversity?

4. Biologic suture zones are geographic areas where a number of species ranges come into contact. For example, in central Texas, the flora and fauna found in the southeastern United States end and the flora and fauna that are characteristic of the southwest begin. What processes might form such suture zones?

FURTHER READING

Brown, J. H. 1971. Mammals on mountaintops: nonequilibrium insular biogeography. *American Naturalist* 105:467–478.

Charles Darwin. 1958. *The Autobiography of Charles Darwin & Selected Letters*, ed. Francis Darwin. New York: Dover.

Grant, P. R. 1986. *Ecology and evolution of Darwin's Finches.* Princeton, NJ: Princeton University Press.

Heywood, V. H., and Watson, R. T. 1995. *Global biodiversity assessment.* New York: Cambridge University Press.

MacArthur, R. H., and Wilson, E. O. 1967. *The theory of island biogeography.* Princeton, NJ: Princeton University Press.

Otte, D., and Endler, J. A. 1989. *Speciation and its consequences.* Sunderland, MA: Sinauer Associates, Inc.

18 Evolution: Adaptation and Environmental Change

CAROL L. BOGGS AND JOAN ROUGHGARDEN

THE EVOLVING WEB OF LIFE AND THE ENVIRONMENT

The view out of our window at Stanford University includes oak, buckeye, and *Eucalyptus* trees, various grasses, the occasional ground squirrel, feral cat, and raccoon, and assorted insects. Were we to transport our offices to the Caribbean or to the high elevations of the Rocky Mountains of Colorado, the identity and number of plant and animal species we could observe would be different. Why do particular groups of organisms occur in particular locations? Does the number of species affect the function of the biologic system in that loca-tion? What effects do environmental or geologic changes have on the species present outside our window?

As noted in Chapter 17, Charles Darwin's message contains two primary lessons. One, dealt with here, is that the characteristics of organisms can change over generations in response to environmental pressures. This phenomenon is termed *adaptation*. The sum of extinction, speciation, and adaptation leads to changes in community composition and regional biodiversity. In this chapter, we explore the relationship between organisms and their environment, the effects on community structure and function, and potential impacts of global climate change.

We return for a moment to Darwin's Finches on the Galapagos Islands, whose speciation patterns we discussed in Chapter 17. One of the observations made by Darwin, and elaborated in recent studies by Peter and Rosemary Grant and their students, is that the beak sizes of finch species are correlated with the size of the seed the bird eats. For example, those birds with deeper, larger bills are more able to eat larger, harder seeds. This effect occurs even within a single species. Individual *Geospiza fortis* on Santa Cruz Island forage in different habitats during the dry season, depending on their bill depths. Those birds with slender bills occur in areas with more abundant small seeds, which are easier for them to crack open.

Traits such as beak depth can change through time as the environment changes. Rainfall on the Galapagos dramatically deceases during El Niño weather events. Changes in rainfall lead to significant changes in the seeds available to the birds. Between 1976 and 1977, for example, seed abundance declined from 10 to 3 grams per square meter. *G. fortis* rapidly depleted the supply of small, soft seeds on the island of Daphne Major. Birds with larger body sizes and bill depths were most able to eat the remaining larger, hard seeds. These birds had a survival advantage over the smaller birds. As a result, beak depth increased by approximately 4 percent on average in the population the following year.

This match between beak depth and seed size is an example of adaptation. The changes in beak depth with changes in food availability demonstrate evolution of traits in response to environmental pressure.

A pattern equivalent to that seen in Darwin's Finches also occurs in bees and many other nectivorous insects. The length of the proboscis (or the "tongue" used to extract nectar from the flower) matches the length of the corolla tube of the flower species visited by particular bee species (Fig. 18.1). In this case, location of the food eaten by the bees, rather than size of the meal, affects mouthpart morphology.

Organisms of all kinds, not only animals, can exhibit adaptation. For example, European annual plants invaded California over the past several hundred years. On most soil types, these invaders outcompete the native California plants such as *Plantago erecta*, driving them to extinction. However, many native plants have adapted to be able to grow on serpentine-based soils. Such soils have high concentrations of chromium, magnesium, and nickel, and low concentrations of calcium, molybdenum, nitrogen, and phosphorus. On the other hand, introduced European annual plants have not adapted to the harsh environment of such soils, and hence

W. G. Ernst (ed.), *Earth Systems: Processes and Issues*. Printed in the United States of America. Copyright © 2000 Cambridge University Press. All rights reserved.

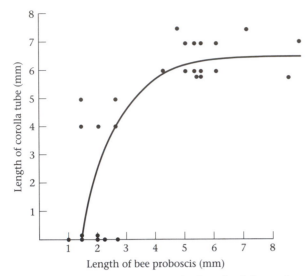

Figure 18.1. Relationship between the length of the proboscis (or tongue) of solitary bees, and the length of the corolla tube of the flowers from which they feed, in interior and arctic Alaska. Most bee species feed on multiple flower species, and each flower may be visited by several bee species. mm = millimeters. (*Source:* Adapted from W. S. Armbruster, and D. A. Guinn, The solitary bee fauna (*Hymenoptera, Apoidea*) of interior and Arctic Alaska: Flower associations, habitat use, and phenology. *Journal of the Kansas Entomological Society*, vol. 62, 1989, figure 2, p. 477.

Figure 18.2. *Colias philodice eriphyle* basking to achieve a body temperature that supports flight. (*Source:* Photograph courtesy of Ward Watt.)

cannot survive in serpentine areas. As a result, many native plants such as *P. erecta* are now restricted to serpentine soils.

Adaptation need not be restricted to cases involving food, such as seeds, nectar, or soil nutrients. The sulfur butterflies, *Colias* spp., exhibit an adaptation associated with thermoregulation. These butterflies must warm up to body temperatures nearly as high as ours in order to be able to fly; however, all warming occurs passively by solar basking. During basking, the butterflies orient perpendicular to the Sun with their wings folded over their backs (Fig. 18.2). Black melanin pigment on the wings enhances absorption of solar heat while the butterfly is warming up. Individuals from high-elevation populations experience relatively cool air temperatures, and thus need to absorb more heat to warm up. These individuals have more melanin on the underside of their wings compared with individuals from lower-elevation populations. Too much melanin at low elevations leads to rapid overheating of the butterfly. Melanization thus increases with elevation, in synchrony with changes in air temperature and requirements for solar basking as part of thermoregulation.

THE NICHE

The size of the seeds eaten by Darwin's Finches is a part of the birds' niche. Similarly, the air temperature and other climatic factors are a part of *Colias* butterflies' niche. An organism's *niche* is defined as that range of environmental factors over which an individual can have positive fitness. For Darwin's Finches, other aspects of the niche might include the location of the nest – for example, within a cactus plant or on the ground.

Because an organism's niche describes the environmental factors to which it has adapted, adaptation must be understood in the context of the niche structure. Niches have two components. The *abiotic* component includes those elements of the physical environment that affect the fitness, or survival and reproduction, of the organism. The *biotic* component includes interactions with other organisms, such as competition, which similarly affect the organism's fitness.

Abiotic factors determine the organism's fundamental niche, or the range of environments within which the organism can potentially survive and reproduce. For *Colias* butterflies, reproduction requires flight to find mates or egg-laying sites. As noted above, flight occurs only at specific body temperatures. Abiotic factors affecting body temperature include air temperature, amount of solar radiation, and wind speed. These factors interact to affect the niche space used by the butterfly. For example, flight may occur at lower air temperatures only if solar radiation is high, so that the butterfly can warm up by basking.

Biotic factors, such as competition with other organisms, are superimposed on the fundamental niche. Typically biotic factors further limit the niche actually occupied by the organism, and result in a realized niche. The barnacle *Chthamalus stellatus*, found in rocky intertidal regions of Britain, tolerates desiccation within specified values at low tide. Its fundamental niche is broader than its realized niche, however. The barnacle's lower limit in the intertidal zone is determined not by desiccation tolerance, but by another barnacle, *Balanus balanoides*. Both species are found in this zone. *B. balanoides* overgrows *C. stellatus*, pushing it off the rocks. The lower limit of *C. stellatus* is the same as the upper limit for desiccation tolerance in *B. balanoides*.

The influence of abiotic and biotic factors on species distributions can lead to *zonation*, or bands of species replacing each other across an environmental gradient. In the case of the barnacles, the combination of biotic and abiotic factors creates two zones: one in the lower intertidal containing *B. balanoides*, and a second one higher in the intertidal containing *C. stellatus*, with a relatively sharp demarcation between them. Other environmental gradients show the same effect. For example, woods dominated by aspen and spruce replace woods dominated by oak and juniper over increasing elevation in the Rocky Mountains of the United States.

NICHE OVERLAP, COMPETITIVE EXCLUSION, AND LIMITING SIMILARITY. Can two species ever share the same niche? How much difference must there be between niches of two species before they can coexist in the same locality? The answer to these questions will tell us how many species can be "packed" into a single environment.

Niche Overlap. To answer these questions, we must examine the degree and type of niche overlap observed among species. In theory, niches of individual species or populations may be disjunct, abutting, overlapping, or included within each other (Fig. 18.3). For example, five amphipod species show overlapping salinity tolerances (Fig. 18.3). In another example, as noted earlier, proboscis length in bees matches the length of the corolla tube of flowers visited by the bees. In the Elk Mountains of central Colorado, we find two species with short proboscides of nearly identical length, flying in the same meadows. These two species, *Bombus occidentalis* and *Bombus bifarius*, appear to share the same niche. However, upon closer observation, we discover that *B. bifarius* obtains nectar through the corolla tube of flowers, whereas *B. occidentalis* does not pollinate flowers at all. *B. occidentalis* is a "nectar robber," biting a hole in the back of the flower and extracting the nectar without pollinating the plant. Thus, *B. occidentalis* can extract nectar from flowers with a wide variety of corolla tube lengths, and its flower use does not overlap strongly with that of *B. bifarius*.

In practice, ecologists have never found an example of one species' niche being completely included within that of another species *and* the two species coexisting in the same locality. Occasionally, species share partial niches, such as food or nesting areas, but only if other, unshared items limit the survival and reproduction of the species. This observation

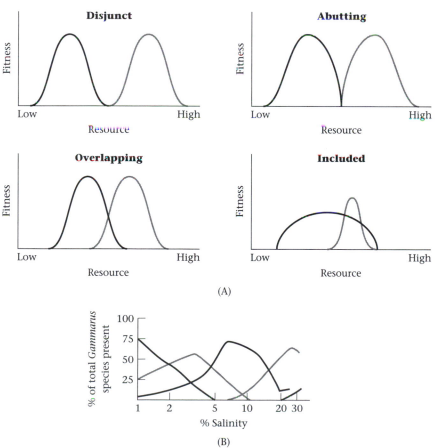

Figure 18.3. (A) Theoretical relationships among unidimensional niches for two species. (B) Salinity niches for five species of amphipods in the genus *Gammarus*. Frequency of individuals is assumed to correlate with fitness. (*Source:* Adapted from T. Fenchel and F. B. Christiansen, Measuring selection in natural populations, *Lecture notes in biomathematics*, vol. 19, Springer-Verlag, 1977.)

is formalized as the *principle of competitive exclusion*, which states that two species cannot occupy exactly the same realized niche.

Competitive Exclusion. Several laboratory and field experiments support the principle of competitive exclusion. Indeed, they indicate that it is competition among species that prevents complete niche overlap. One such set of experiments concerns *Anolis* lizards in the Caribbean Islands, which compete for food.

Individual anoles are generally hungry. Stomach content analyses show that their stomachs are often empty. When the environment is enriched with insects of the correct size, the lizards eat them, and individual lizard growth and reproductive rates increase. Further, on the islands of Bonaire and St. Martin, lizard abundance increases with increasing insect abundance, with evidence for an upper population size limit on St. Martin (Fig. 18.4). These data suggest that food supply limits the populations of individual species, and that intraspecific competition for food occurs.

Does interspecific competition for food affect the ability of *Anolis* species occupying the same niche to coexist on the Caribbean Islands? To answer this question, we must first find an easy way to measure each species' food niche. Anoles eat insects and only insects of a particular size range. For each species, the body size of the lizard, as measured by snout-vent length, dictates the size range of insects that the lizard eats. Larger lizards eat larger insects, because small insects do not provide sufficient sustenance; in contrast, smaller lizards cannot manage to eat larger insects. Thus, body length is an accurate surrogate measure of the food niche.

Observational studies provide strong circumstantial evidence for competition among lizard species. For example, on the Caribbean island of St. Martin the population size of *Anolis gingivinus* is generally lower in areas where it coexists with *Anolis pogus* than in areas where *A. gingivinus* exists alone. Further, the two species tend to occur in different areas on St. Martin, with *A. pogus* restricted to higher, cooler woodlands near the center of the island.

The two species on St. Martin do not differ significantly in body size compared to species pairs found elsewhere in the Caribbean, suggesting that niche overlap is greater on St. Martin. If this is true, we would expect to observe greater overlap in habitat use among species pairs elsewhere – and this, indeed, is the case.

Experimental studies involving manipulations of species numbers within an area provide strong evidence for competition. A species introduction experiment on Anguillita, a tiny cay at the western tip of Anguilla, showed an effect of *A. gingivinus* on *A. pogus*. The cay typically contains only *A. gingivinus*. In two areas, researchers introduced *A. pogus*, without altering the number of *A. gingivinus* present. In two other areas, *A. pogus* were also introduced; however, 35 to 50 percent of the *A. gingivinus* population was removed. In all cases, the survival rate of *A. pogus* over six months was much lower than that of *A. gingivinus*. However, the survival rate of *A. pogus* was much higher in areas where *A. gingivinus* population numbers had been reduced, indicating a significant effect of *A. gingivinus* on *A. pogus*.

The inverse observation, that *A. pogus* has an effect on *A. gingivinus*, is also true. On St. Martin, enclosures, which prevented lizards from entering and leaving an area measuring 12 meters by 12 meters, were stocked with either *A. gingivinus* by itself, or both species together. The growth and reproduction rates of *A. gingivinus* were lower when the two lizard species were together than when *A. gingivinus* occurred alone.

A repeat of the enclosure experiments on another island, St. Eustatius, using the resident species *Anolis bimaculatus* and *Anolis schwartzi* showed no evidence for competition between these two species. Why might these two sets of lizard species show wildly different levels of competition?

Limiting Similarity. A closer look at the degree of niche overlap between anole species within each island provides an answer. On St. Martin, the size ratio between these two species, a measure of niche overlap, is approximately 1.2. On St. Eustatius, the same ratio for the two resident species is 1.7. The two lizard species found on St. Martin are much closer in body size, and hence in size of insect prey eaten, than are the two species from St. Eustatius. We might logically expect that competition would be stronger on St. Martin than on St. Eustatius. Why is it entirely absent from St. Eustatius?

Figure 18.4. Relationship between insect abundance and *Anolis* lizard abundance on the Caribbean island of St. Martin. (A) sites at sea level. (B) Sites in the central hills. (*Source:* Adapted from J. Roughgarden, *Anolis lizards of the Caribbean*, Oxford University Press, 1995.)

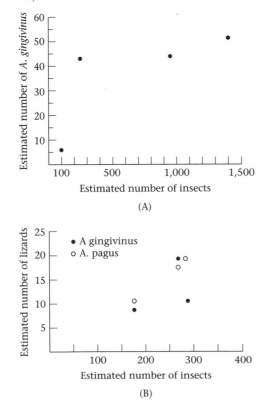

In 1959, G. Evelyn Hutchinson sought to answer the question, "How similar can two species be in feeding habits and still coexist within an area?" In other words, what is the limiting similarity? He surveyed bill lengths of coexisting birds and skull lengths of coexisting mammals as measures of their food niche. Hutchinson concluded that the ratio of the larger to the smaller species must be no smaller than approximately 1.3 in order for the two species to have sufficiently distinct niches to allow coexistence. At much larger values, such as 1.7 observed on St. Eustatius, we might expect that the species would be sufficiently different that competition does not occur.

This value of 1.3 is approximately the cube root of 2. Thus far, we have been considering a linear trait, body length, as a proxy for environmental niche. Body volume or mass will scale as (body length)^3. Therefore, Hutchinson's empirically observed limiting similarity for linear measures, $\sqrt[3]{2}$, suggests that body mass or volume must differ by at least a factor of 2 between two species competing for food in order for them to be able to coexist in the same location.

CHARACTER DISPLACEMENT. Suppose two *Anolis* species coexist on an island but have snout-vent length ratios that are very close to 1.3. Over ecologic time, the species will both persist. What changes might we expect to see over evolutionary time?

Competition between the two species for food will decrease individual fitness for both species. However, individuals of the larger species that are themselves larger than average will experience less competition from the smaller species, and should have a higher fitness (likelihood of survival). Similarly, smaller individuals within the smaller species will have a better chance for survival than other members of their population. Selection due to competition will thus act to increase the size difference between the two species, with the larger species tending to become even larger and the smaller species to become yet smaller. However, the maximum and minimum size attained by these species may be constrained, due to, for example, size diversity of the insect prey.

This phenomenon known as *character displacement*, occurs among *Anolis* species on the Caribbean Islands and in many other locations. Darwin's Finches on the Galapagos Islands provide a particularly striking example (Fig. 18.5). *G. fulginosa* and *G. fortis* occur separately on two islands; their beak depths overlap substantially. Remember that beak depth is an indicator of the size of seed eaten by the bird. Thus the finches' food niche on the two islands is very similar. The birds occur together on Santa Cruz Island. There, beak depths of both species have diverged significantly, with the smaller beak becoming slightly smaller, and the larger beak becoming larger. These data indicate that the niches of the two species have diverged over time on Santa Cruz Island, most likely in response to selection favoring individuals that experience little interspecies competition for food.

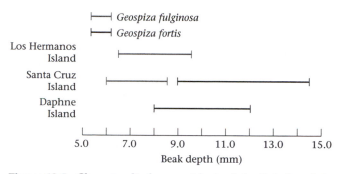

Figure 18.5. Character displacement for beak depth in Darwin's Finches. Beak depth shifts when the two species occur together on the same island. mm = millimeters.

FOOD WEBS

Thus far, we have considered species that eat the same type of food, be it insects or seeds. Species can be grouped by the type of food they eat; within a given area, they form *food webs* (Fig. 18.6). Food webs are composed of several trophic levels. Primary producers, such as plants or algae, capture energy from sunlight and use it to produce living matter. Herbivores eat primary producers. Primary carnivores eat herbivores, and secondary carnivores may eat primary carnivores. Scavengers and decomposers feed on dead biomass. Individuals within a species need not be restricted to feeding at only one trophic level. Humans act as herbivores and as primary carnivores, and hence are omnivores.

Nothing is 100 percent efficient, including the conversion of food into body mass. Most organisms have a conversion rate of food to body mass of approximately 10 percent; a few very efficient organisms exhibit a rate of 15 percent. This means that primary producers have the largest biomass in

Figure 18.6. (A) A food web. (B) a hypothetical biomass pyramid.

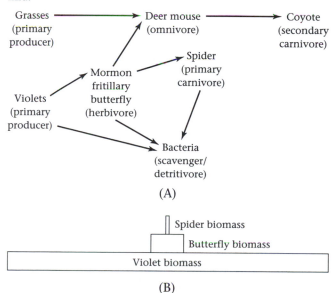

any food web, followed by herbivores, and so on up the food web. The biomass of herbivores is less than 10 percent that of primary producers, because herbivores do not eat all primary producers (Fig. 18.6). The ratio of nutrient to consumer applies at other trophic levels as well.

This rapid attrition of biomass with increasing trophic level raises the question of how many trophic levels can exist within a given food web. The answer has two components. First, the highest trophic level must contain species that maintain positive population sizes. These populations must be sufficiently large that they cannot go extinct due to random chance events such as variation in the physical environment or chance fluctuations in reproduction. Second, the food web must be sufficiently stable to tolerate perturbations. If the numbers of individuals at the lower trophic levels decrease, how are the already smaller numbers of individuals at higher levels affected? Longer food chains have been shown to be less stable, due to a large number of critical connections among trophic levels. The effects of the perturbation at lower trophic levels amplify as they move up the food chain. Thus, stable food webs typically have no more than three or four trophic levels.

COMMUNITIES

In any location, a set of food webs exists, with the species at any trophic level differing at least somewhat in their niches. In effect, this is an interacting set of biologic populations, or a community. Temperate or arctic zone communities are often referred to by the most visible, characteristic species within them. For example, we refer to the pinyon-juniper communities in the American Southwest.

BIODIVERSITY AND COMMUNITY FUNCTION

Communities may be composed of a few or many species. How important is species diversity to community function or stability? This question can be asked at several levels of organization.

COMMUNITY STABILITY AND SPECIES DIVERSITY.
Do communities with a greater number of species exhibit greater resilience to environmental stress than communities with fewer species? We are only just beginning to be able to answer this question.

We might use several measures of community health or stability to assess the effects of species diversity on communities. One measure is the production of biomass over a particular time span. For example, productivity of a community of self-pollinating herbaceous plants from British fields was higher at greater species diversity than was predicted based on the sum of productivities of each species grown alone. These experiments studied up to sixteen species at a time. A field study in Minnesota examined similar questions, but using a larger maximum species diversity. There, fields with higher plant diversity showed less variation in productivity over time than did fields with low plant diversity. The increased stability that accompanies increased diversity was particularly pronounced during periods of drought stress. Thus, the evidence to date suggests that increased species diversity can increase productivity, and stability of productivity – among plant communities, at least.

The stability of the abundances of the species that make up the community provides another measure of community stability. Here the data are different. In the same Minnesota fields, increased species diversity did not result in more constant population sizes for each species. Together with the effects of species diversity on productivity, the data suggest that when some species' population sizes fall due to environmental factors such as drought, other species' population sizes (and productivity) increase. Total productivity is thus stable, whereas the population size of each species fluctuates.

ECOSYSTEM SERVICES AND SPECIES DIVERSITY.
Many of the species within a community perform *ecosystem services*, including nitrogen fixation, water purification, carbon dioxide fixation, and oxygen production, to name only a few. How important are diverse communities to the maintenance of ecosystem services? Can one nitrogen-fixing plant species provide the same services as three nitrogen-fixing plant species? Are two of these species redundant?

The answers are as yet unclear. Some evidence exists for the biogeochemistry of diverse communities, such as the percentage of nitrogen or organic matter in soils. These functions increase initially with increasing species diversity, but then they plateau (Fig. 18.7). We do not yet know whether this pattern applies to other ecosystem services or whether it is replicated in a diversity of community types.

PROCESSES CHANGING COMMUNITY STRUCTURE

Community structure, including both the species present in a given area and their relationship to each other, changes over time. Structure depends on biologic and environmental history, as well as the characteristics of each species present in the area.

Figure 18.7. The relationship between number of species in a plant community in Turrialba, Costa Rica, and percentage soil nitrogen. (*Source:* Adapted from E.-D. Schultz and H. A. Mooney, *Biodiversity and ecosystem function*, Springer Verlag, 1994.)

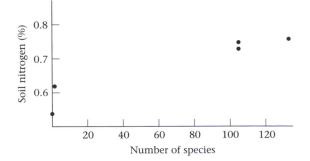

CHANGES IN ECOLOGIC TIME: SUCCESSION. Organisms modify their environments over time. For example, a seedling does not cast a great shadow; however, when it grows into a redwood tree, its shadow is quite large. The presence of that tree changes not only the amount of sunlight striking the ground, but also the soil and air temperature and humidity. Nutrients available in the soil may change as well. Such changes in the environment may alter the outcome of competition between two species or foster the population growth of an invading species. Community composition thus changes over time, in a process known as *succession.*

The process of succession is subdivided into two categories: primary and secondary. Primary succession begins with a disturbance that eliminates the biota of a local area. The disturbance may be a landslide, a volcanic eruption, a flood, or other such event. Secondary succession in terrestrial regions occurs when disturbance eliminates much of the biota but the soil remains more or less intact. Because some seeds and other life forms already exist in the area, and because soils are often well developed, secondary succession can proceed more rapidly than primary succession.

In the typical scenario of primary succession, the first invading species into the area are often weedy. They are good colonizers, having excellent dispersal abilities and a high rate of population growth once individuals settle in an area. Weedy species are often intolerant of competition. These first species may change the physical environment, and may also serve as food for later arrivals. These changes in niche structure allow later invaders to establish populations. These later arrivals often are excellent competitors in the new environment created by the first arrivals, and gradually eliminate the early community. The result is community turnover through time, driven by environmental change and species interactions.

Community succession eventually halts, with the formation of a climax community. Ecologists believed for a long time that a climax community, or an end point toward which community moves as succession occurs, characterizes each area. It is now clear that multiple climax communities are possible in any given area, depending on the type and intensity of disturbance and the pool of invading species.

The classic view of primary succession is thus that each wave of colonists facilitates the arrival of the following wave, through changes in niche structure. At least one alternative view exists. Based on inhibition, it states that early arrivals occupy territory and resist the invasion of other species. It is only when early individuals die that space and resources become available, allowing the invasion of a new set of species to form a new community.

Data from different systems support each view. Marine communities can exhibit the inhibition model of succession. In the rocky low intertidal zone in southern California, boulders are initially colonized by a green alga (*Ulva* spp.). These algae resist invasion by a variety of red algae. Red algae settle only if the green alga is first eaten (and hence removed) by a crab, *Pachygrapsus crassipes*. The classic model of primary suc-

cession fits many terrestrial systems, including the sand dunes of Lake Michigan. A variety of successional pathways are possible, depending on initial conditions and the identity of the first colonizing species. In one pathway, rushes invade a damp depression to form a wet meadow. A red maple swamp later replaces the meadow.

CHANGES IN EVOLUTIONARY TIME: INTERACTION WITH GEOLOGIC PROCESSES. Chapter 17 explains that extinction, speciation, and dispersal processes have determined the biodiversity present in a given area. The relative role of each process has varied among geographic areas and taxonomic groups. These same processes affect community structure over time. Communities occurring on *land-bridge islands* are those that have survived extinction over time, whereas communities on *oceanic islands* are those that result from a combination of extinction and migration. Over the longer term, speciation also plays a role in community structure. Similarly, character displacement and competitive exclusion, as discussed earlier in this chapter, also help determine community structure on long time scales.

Historic events accumulated over very long time scales can also affect community composition. For example, the movement of tectonic plates may separate or bring together communities over time. This is one area in which biologic phenomena may actually help us understand geologic processes in a given region.

An Example: The Caribbean Plate. The origin and history of the Caribbean tectonic plate is one such geologic event that can be better understood by examining the patterns and history of present-day biologic communities within the region. In the process, we find that geologic and biologic phenomena are not independent.

The origin and history of the Caribbean tectonic plate have been problematic. The plate appears to have been squeezed into location between North and South America, having originated someplace else. The age relationships among the Caribbean Islands are also not well understood – which islands are older than others? What is the geologic relationship among islands? We use biologic communities to propose answers to these questions.

The organisms used to generate hypotheses concerning the origin and fate of the Caribbean plate (or any other geologic process) must fulfill three criteria. First, we must be able to generate phylogenetic trees for the organism (see the box entitled *Phylogenetic Trees*). Phylogenetic trees outline the evolutionary relationships among species or other taxa, telling us which species are most closely related to which other species. If we assume that closely related species originate in adjacent areas, we can use these phylogenetic relationships to track land movements. Second, the organism must be a successful colonizer of uninhabited areas but must repel subsequent invasion. If further invasions occurred, we could receive mixed signals as to the age of a particular landform, or its relationship to adjacent landforms. Third, the ideal organisms on which to base these hypotheses are those that have

---

PHYLOGENETIC TREES

Phylogenetic trees are the representation of the evolutionary history of a set of related organisms. Trees generally have a common ancestor at their root; branches represent the evolutionary changes that have resulted in present-day organisms. Systematists use either morphologic traits or various types of DNA sequences to estimate which species are more closely related to each other and to the putative common ancestor. Morphology and DNA do not always tell the same story, however; consequently, some type of consensus tree may be produced.

In constructing phylogenies, we assume that shared morphologic traits or DNA sequences are due to common descent. That is, the ancestors of both species with the trait also possessed the trait. These are called *homologous traits*. However, two species may also share a trait as a result of adaptation to a common environment. That is, traits may converge through time due to selection pressures. These are called *analogous traits*. Wings are an example of analogous traits. The wings of bats and the wings of butterflies have the same function, and very similar shape. Yet, they do not share a common ancestry, having evolved from very different structures over time.

Any set of organisms will all share some traits. We assume, then, that the common ancestor of these organisms also had these traits. As the organisms differentiate through time, some of their traits also differentiate. These are known as *derived traits*, and are useful for building phylogenetic trees. For *Anolis* lizards, an *ancestral trait* is the presence of scales. Derived traits include the number, size, and distribution of scales.

Typically, trees are built by assuming that there are a minimal number of changes between present-day organisms and the ancestral organism. For example, consider the four hypothetical butterfly species, and their ancestral species, shown below:

Species	Wing color	# Growth stages as a larva	Presence of an extra wing vein
1	White	5	No
2	Yellow	6	No
3	Yellow	5	Yes
4	Yellow	5	No
Ancestor	Yellow	5	No

Assuming a minimal number of changes at each branch point leads to the following tree, with trait changes as marked:

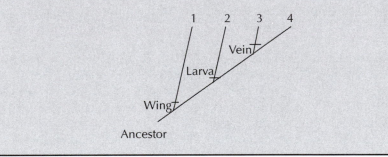

lived in the region for a long time and are thus able to reflect geologic processes over geologic time scales.

Anolis lizards fit these three criteria. First, we have accurate phylogenetic trees for Caribbean anoles. Second, because the lizards on any given island often strongly compete with the other lizard species on the island, there is little niche space for an invader to occupy. Therefore, a small group of invading anoles is unlikely to be able to exclude competitively a resident anole species with a much larger initial population. Third, anoles have probably lived in the Caribbean since its inception. Recognizable *Anolis* species date from at least the Miocene Epoch (20 to 23 million years ago); we have a complete juvenile specimen encased in amber from that period. This lizard is not very different from the present-day species living in Hispaniola. Therefore, *Anolis* must have originated significantly earlier than 23 million years ago. Further, rocks older than approximately 65 to 100 million years are very scarce in the Caribbean. All these attributes taken together suggest that *Anolis* distributions reflect the geologic history of the Caribbean.

The phylogenetic tree of *Anolis* lizards shows that species in the Lesser Antilles and Puerto Rico split into several groups, each of which shares a common ancestor (Fig. 18.8). Note that the northeastern and southeastern Caribbean Islands contain lizards whose common ancestor is very remote in time. Thus, the biologic data suggest that the islands of the northeastern and southeastern Caribbean have different geologic histories.

Did anoles in the Caribbean immigrate from South and North America, or do they have a different origin? We can answer this question by generating phylogenetic trees that now include anoles from those and other localities. We find

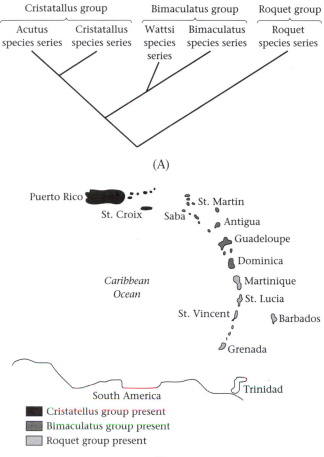

(A)

(B)

Figure 18.8. (A) Phylogenetic tree for *Anolis* lizards from the Lesser Antilles and Puerto Rico. (B) Location of the same lizards throughout the islands of the Lesser Antilles and Puerto Rico. (*Source:* Adapted from J. Roughgarden, *Anolis lizards of the Caribbean*, the Oxford University Press, 1995.)

that the closest non-Caribbean relative of anoles from the northeastern Caribbean is from western Mexico. The closest relative to those from the southeastern Caribbean is from an isolated island south of Panama and west of Colombia. This suggests that the ancestors of present-day Caribbean anoles came from the Pacific coastal regions of Central America and northern South America. The implication then is that the Caribbean plate also originated in this area.

The Caribbean *Anolis* example is perhaps more striking than many, given the uncertainties surrounding the geologic history of the Caribbean Islands. Thus, it demonstrates dramatically the interconnections between biologic and geologic processes inherent in an Earth systems approach to environmental studies.

EFFECTS OF CLIMATE CHANGE ON COMMUNITY STRUCTURE

Niche structure can change as a result of changes in the physical environment, with consequent changes in commu-

nity structure. We have already described this effect in the context of ecologic succession. Changes in the physical environment can arise from other causes, however, and sometimes produce very different results.

Regional or global climate change is one such factor that dramatically affects communities. Such climate change takes place over a range of time scales, from thousands of years for glacial advances to less than a second for asteroids striking the Earth. We know the most about what occurs over geologic time scales (see Chap. 3).

CLIMATE CHANGE OVER GEOLOGIC TIME SCALES. The Earth has experienced a series of expansions and retreats of glaciers, with the latest maximum extent of glaciation in North America occurring 18,000 to 20,000 years ago (Chap. 15). Approximately 16,000 years ago, the temperature warmed dramatically, and the ice sheets melted within a few thousand years. As the ice sheets retreated northward, areas with particular climatic conditions also moved north. Consequently, the niches associated with those climate conditions also moved north (Chap. 21).

Do the communities associated with a particular climate also move as one? Phrased differently, can individual species move at the rate required to keep up with climate change? Pollen deposited in lake beds over geologic time tells us the answer to these questions for tree species in the northeastern United States and Canada. These deposits indicate that species' ranges moved northward as the climate warmed. They moved at a rate whose continent-wide average varies from 100 meters per year for chestnut to 400 meters per year for jack and red pine. Most species showed patterns similar to beech tree, whose range moved northward at a rate of 200 meters per year (Fig. 18.9). These migrations match the rate of climate change in the regions inhabited by the different tree species.

ANTHROPOGENIC CLIMATE CHANGE. Current models of climate change induced by human activity suggest that we will see a doubling of atmospheric carbon dioxide concentrations, with resulting temperature shifts of up to 3°C in some regions over a period of decades to centuries (see Chap. 16). The amount of warming is not predicted to be uniform around the world; some areas will warm more than others, and some may actually cool. Rainfall accumulations will also likely change. These changes are forecast to occur much more rapidly than did the climate shifts experienced by the North American forests after the last glaciation. What is likely to happen to communities experiencing change on this time scale?

Climate change is likely to have the greatest effects on species with narrow thermal niches or near the edge of their ranges. If the range of the species is geographically broad, then only a section of that range may become uninhabitable. The outcome will depend on whether the species can migrate with sufficient speed to keep up with the northward move-

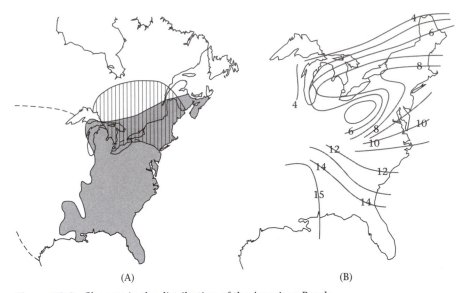

Figure 18.9. Changes in the distribution of the American Beech, *Fagus grandifolia*, in North America. (A) Present range, indicated by stippling. Predicted range under one scenario for doubled atmospheric carbon dioxide concentrations, indicated by horizontal lines. (B) Expansion of American Beech ranges between 15,000 and 4,000 years before present. Lines indicate the leading edge of the range; numbers represent the number of thousands of years before present. (Source: (A) Adapted from Fig. 22.5, p. 301, M. B. Davis and C. Zabinski, 1992, Changes in geographical range resulting from greenhouse warming: effects on biodiversity in forests. In R. L. Peters and T. E. Lovejoy, eds., *Global warming and biological diversity*, Yale University Press, 384p. (B) Adapted from Fig. 10.13, p. 149, M. B. Davis and D. B. Botkin, 1981, Quaternary history and the stability of forest communities. In D. C. West, H. H. Shugart, and D. B. Botkin, *Forest succession concepts and adaptation*, Springer Verlag, New York, 517p.)

ment of suitable habitat. If this occurs, then the entire range will simply be displaced to the north. If this does not occur, we expect to see a shrinkage of the range, with a slower reexpansion northward into the newly available habitat. However, the rate of migration, not the rate of climate change, will control the rate of reexpansion northward. Similarly, if the range is geographically narrow, the entire range may become uninhabitable. If the species is unable to migrate to keep up with the movement of its niche across space, it will go extinct.

For some species, migration rates that match the rate of climate change are unlikely to be a realistic possibility. Beech trees, for example, may need to migrate northward at a rate of 7 to 9 kilometers per year, or forty times faster than the rate observed after the last glaciation (Fig. 18.9).

A second option is adaptation in place to the new climatic conditions. Whether all species within a community will be able to adapt successfully to rapidly changing conditions is unknown, but doubtful. Genetic variation must exist within each population so that selection can produce adaptation to the new environment. The fact that species moved when climate change occurred over geologic time suggests that appropriate genetic variation does not exist in many species. If it did exist, we would have observed fewer range shifts in response to climate change over long time scales.

The most likely scenario is community disassembly. In any given region, climate change will affect some species more intensely than others, simply because some species are closer to the edge of their thermal niche. Consequently, some species will be more susceptible to extinction. Yet, species differ in their ability to disperse: Some species will be able to invade new regions that now contain the old climate, whereas others will not.

Extinction cascades may occur as well, if food webs are torn apart. Herbivores may be better able to move than are the plants they eat. Insects in particular often eat only particular plant species. Hence they may be particularly susceptible to extinction not because of climate, but because of food shortages.

This scenario is bleak, given currently projected rates of climate change. Current models indicate that 30 percent of the Earth's vegetation will be affected, along with untold numbers of animals and other organisms. Yet, crystal balls are notoriously cloudy. Whether widespread community disassembly and disruption will occur depends on the accuracy of climate change predictions and the thoroughness of our understanding of organisms' responses to change – and whether humans continue to alter the environment on a global scale (see Chap. 32).

QUESTIONS

1. Keystone species are those that alter the abiotic or biotic environment in such a way as to allow numerous other species to flourish in the area. The classic example is the starfish *Pisaster*, which preferentially eats a mussel, *Mytilus edulis*. When present, the mussel competitively excludes numerous other invertebrate species from the rocky intertidal. With predation, the mussel population is reduced and these other invertebrates are able to invade and flourish. Keystone species indicate that species identity can affect community composition. Should keystone species be specially targeted for conservation measures?

2. Habitat reserves for endangered species designed today almost never account for the effects of climate change. How might you design reserves to account for such effects? How feasible are your ideas?

3. Ecologic economists have begun considering how to assign a dollar value to species, communities, or ecosystems. Such values would be used in cost/benefit analyses of various proposed environmental policies. Among ecosystem services, for example, nitrogen fixation might be accorded the dollar value of an equivalent amount of nitrogen fertilizer. What technologies and associated costs would be necessary for the replacement of other ecosystem services?

FURTHER READING

Hutchinson, G. E. 1959. Homage to Santa Rosalia or why are there so many kinds of animals? *American Naturalist* 93: 145–159.

Peters, R. L., and Lovejoy, T. E. 1992. Global warming and biological diversity. New Haven, CT: Yale University Press.

Roughgarden, J. 1995. Anolis *lizards of the Caribbean.* Oxford, England: Oxford University Press.

19 Global Biogeochemical Cycles: Carbon, Sulfur, and Nitrogen

ISAAC R. KAPLAN AND JULIE K. BARTLEY

THE CRUCIAL ROLE OF MICROORGANISMS

Industrialization has grown rapidly over the past fifty years, and the environmental consequences have been dire; in fact, many people believe that the recent environmental changes on our planet are caused entirely by anthropogenic excesses. Actually, the Earth's environment has undergone major changes throughout its history, caused by many processes described in this book: Earth–Sun interactions, reactions within the Earth's interior and on its surface, and the metabolic processes of organisms. This chapter describes the diversity of biologic systems, their role in consuming and producing metabolites, how their metabolism affects the Earth's environment, and how humans have benefited from microbes.

The *Webster's Dictionary* definition of a *microorganism* is "any microscopic or ultramicroscopic animal or vegetable organism; especially any of the bacteria, protozoa, virus." Also included are fungi, algae, and other groups of single-celled organisms. The diversity and abundance among micro-organisms is so great in every microcosm of the Earth's surface that they affect nearly all aspects of the Earth's environment, from the atmosphere to the sediment far beneath the ocean floor.

This chapter is devoted chiefly to describing the metabolism of these microbes and the interactions between microbial communities and Earth systems processes. To many, microorganisms are harbingers of disease. Food technologists know that bacteria and yeast are key to the production of many dairy items, baked goods, and alcoholic beverages. Soil scientists study the symbioses between microbes and plants. However, few people realize the importance of microorganisms in the formation of many natural resources, such as petroleum, carbonate and silica minerals, and metal ores. They play an important part in the global cycling of carbon, sulfur, and nitrogen, and in maintaining the chemistry of the ocean and atmosphere.

Six elements are recognized as being of utmost importance to all life on Earth: carbon (C), hydrogen (H_2), oxygen (O_2), nitrogen (N_2), sulfur (S), and phosphorus (P). These elements ("CHONSP") are essential components of the building blocks of all cells, from unicellular bacteria to multicellular mammals. In different combinations and in different ratios, they are present in carbohydrates, lipids, proteins, and nucleic acids. They combine to form skeletal material such as lignin and cellulose in plants, chitin in insects and crustaceans, keratin in mammals, hydroxy apatite in bones, and calcite in invertebrates. A wide spectrum of metabolic systems has evolved to recycle these elements efficiently, such that the products of one set of biochemical processes are used as reactants for another set, thus ensuring that the elements are not irreversibly bound in a form that is unusable by living matter.

The recycling of elements by organisms, in the context of geologic processes, is referred to as *biogeochemical cycling*. In this chapter, we emphasize the great importance of microorganisms in biogeochemical cycling. Various species of microbes participate in every chemical transformation of CHONSP. These tiny organisms, incredibly abundant and diverse on Earth, are the workhorses of biogeochemistry. These microorganisms can degrade presynthesized organic material or synthesize new organic substances by fixing carbon dioxide, both photosynthetically and in the absence of light. In carrying out their metabolic functions, they consume, then release each element, returning it to the biosphere.

EARTH SYSTEMS AND GEOCHEMISTRY

The *biosphere,* the portion of the Earth that is inhabited by organisms, includes the atmosphere, ocean, land surface, and parts of the Earth's crust. The organisms that live on this planet affect the chemical composition of their environment by producing oxygen, synthesizing organic matter, oxidizing or reducing metals, and influencing a host of other processes. Biogeochemistry studies the interactions between living or-

W. G. Ernst (ed.), *Earth Systems: Processes and Issues.* Printed in the United States of America. Copyright © 2000 Cambridge University Press. All rights reserved.

ganisms and the physical processes of the planet. Organisms affect the chemistry of most elements in the Earth's crust; however, in this chapter we survey only the light elements – carbon, hydrogen, oxygen, nitrogen, sulfur, and phosphorus (CHONSP). Carbon, nitrogen, and sulfur deserve special attention because their complex biogeochemical cycles control, in part, the chemistry of the atmosphere, hydrosphere, and lithosphere.

Recycling of elements by living systems occurs within the context of geologic cycling, which is independent of the presence of organisms on the Earth. New rocks are formed by volcanoes and at submarine spreading centers beneath the ocean basins (Chaps. 5 and 6). At the ocean margins, older rocks are drawn under continents by the process of subduction. As rocks are subducted, they are subjected to great heat and eventually partially melt, sometimes forming volcanoes. Crustal rocks interact with the atmosphere and oceans, where they are weathered, eroded, and redeposited as sediments. Elements are neither created nor destroyed in the geologic cycle, merely reworked. In studying the effects of organisms on the cycles of elements, we must remember that geologic processes, driven by solar energy and by the Earth's internal heat and gravity, are the driving forces of the biogeochemical and geologic cycles. The interactions of rocks with the ocean and atmosphere greatly influence the chemistry of seawater (Chap. 12) and the composition of the atmosphere (Chap. 13).

IMPORTANCE OF CHONSP TO LIFE

Living organisms are divided into three large groups, based on their evolutionary history. These three *domains* are *Eucarya, Bacteria,* and *Archaea.* More colloquially, organisms are classified as *eukaryotes* or *prokaryotes.* Eukaryotes, or Eucarya, are organisms that contain a membrane-bound *nucleus,* and often other *organelles,* such as *mitochondria* and *chloroplasts.* The domain Eucarya includes protists, algae, animals, and plants. Prokaryotes, which comprise both Archaea and Bacteria, are organisms that lack a membrane-bound nucleus. Prokaryotes are almost entirely microscopic, and are single-celled or colonial. The differences between Archaea and Bacteria are discussed in more detail in the box entitled "Life's Diversity." In this chapter, we use the term "bacteria" to refer to members of the domain Bacteria (also commonly called Eubacteria), and the term "prokaryote" to refer to all non-nucleated microorganisms (Bacteria and Archaea).

The earliest evidence of life on Earth comes from cherts (see Chap. 5) from western Australia that are approximately 3.5 billion years (Ga) old. Microorganisms resembling modern cyanobacteria (a type of photosynthetic bacteria) occur in these cherts, which – remarkably – preserve the original cellular structure of the organisms. Fossils of multicellular eukaryotes, first found in sediments approximately 1.2 billion years old, are morphologically similar to modern red algae. Multicellular animal-like forms are observed in rocks approximately 570 million years (Ma) old. Animals recogniz-

able as belonging to modern groups first appear in Cambrian age rocks, approximately 550 million years old.

Microorganisms are found almost everywhere on the Earth's surface and in the oceans. Recently, organisms were found in sediments 500 meters below the sea floor of the Pacific Ocean. Bacteria have been found living in salt crusts, within rocks, and under ice packs in the Antarctic. Life on this planet has an absolute requirement for moisture, in order to carry out metabolism and reproduction; anywhere there is moisture, there seem to be organisms. Certain microorganisms can tolerate extreme conditions of temperature, salinity, and pH that make conditions inhospitable for more delicate multicellular organisms. Organisms live in temperatures ranging from the Antarctic's subzero, to the steamy 113°C, of deep-sea hydrothermal vents. Bacteria live in acidic runoff waters of coal mines, at pH values as low as 1.5, and in saline desert lakes where the pH is greater than 10. Some fungi can even thrive in sulfuric acid. Halobacteria, some of which are members of Archaea, require extremely saline environments, growing even in waters where sodium chloride is precipitating, at a salinity of 30% (normal seawater has a salinity of 3.5%).

Life occurs nearly everywhere on the Earth's surface, and each type of organism is adapted to its particular environment. For example, an extreme halophile that grows in the salt-evaporating ponds of a saltworks is unlikely to grow in a swimming pool. This restriction of organisms to ecologic zones can be observed in Grand Prismatic Hot Spring, at Yellowstone National Park (Fig. 19.1). The colors are due to the presence of different types of pigments in organisms, each growing best at different temperatures. In the center of the pond, where the temperature is highest (75°C), extreme thermophiles, primarily Archaea, grow. At the pond's edge, where the temperature is between 59°C and approximately 70°C, a thermophilic cyanobacterium, *Synechococcus*, grows, giving the water a yellow color. Between 30°C and approximately 59°C, a rusty-brown microbial mat is produced by the cyanobacterium *Phormidium*. Farther away from the heat source, at temperatures below 30°C, *Calothrix*, another cyanobacterium, forms a dull-brown colored mat. The ranges of each species of organism are determined in part by the physiologic tolerances of each organism for temperature, and in part by competition with other organisms that live in the hot spring. It is important to remember that although microorganisms exist everywhere, each individual species is generally well adapted to a particular environment and will not grow in all environments.

CHONSP are the major elements that constitute the organic compounds of all these organisms. Cell membranes are made up of carbon, hydrogen, oxygen, and phosphorus; proteins consist of carbon, hydrogen, oxygen, nitrogen, and often small amounts of sulfur; nucleic acids are composed of carbon, hydrogen, oxygen, nitrogen, and phosphorus; and sugars are formed by carbon, hydrogen, and oxygen. Organisms must acquire these elements in order to create their cells, much as an artist needs paint of different colors to

Figure 19.1. Aerial photograph of Lower Geyser Basin, Yellowstone National Park. Four consortia of microorganisms grow in the basin and effluent streams of Grand Prismatic Hot Spring. The distribution of each consortium is controlled primarily by temperature, and each has a distinctive color pattern that can be observed around the spring. This figure also appears in the color plate section. (*Source:* Photograph by Julie K. Bartley.)

portray a landscape. These same elements are used in metabolism, providing the energy sources as well as the building blocks for life on Earth. For example, aerobic respiration requires oxygen (O_2) to convert organic matter to carbon dioxide, producing energy, which is stored in the form of adenosine triphosphate (ATP), itself made of carbon, hydrogen, oxygen, nitrogen, and phosphorus. Organisms transform CHONSP in a variety of ways, some organisms utilizing the by-products of other organisms (Table 19.1). The diversity of metabolic types is so large that despite decades of careful study we are still discovering them (see the box entitled *Life's Diversity*).

ENVIRONMENTAL AND ECOLOGIC ASPECTS OF BIOGEOCHEMISTRY

What is a pollutant? A pollutant is generally considered to be a chemical produced by humans or as a result of human activity that is foreign to the environment, and has a detrimental effect on the environment or ecology. Biogeochemists define pollutants as substances produced by humans that disrupt one or more biogeochemical cycles. By either definition, the list of potential pollutants is virtually endless. A pollutant may be something as dangerous as a toxic chemical, such as a pesticide, or as common as fertilizer, petroleum, or carbon dioxide gas. The effects of pollution may be local or global in scale. Because pollutants disrupt biogeochemical cycles, it is important to understand the mechanisms by which these cycles operate. A better understanding of these cycles can lead to practices that minimize the detrimental effects of human activities on the environment. Without a comprehension of cause and effect, it is difficult to legislate effective public, corporate, and individual policies.

Among processes that disrupt ecosystems locally, eutrophication is one that has received much attention in recent years (see Chap. 9). When excess nitrogen or phosphate is

TABLE 19.1 MICROORGANISM-MEDIATED TRANSFORMATIONS AMONG CHONSP ATOMS, IONS, AND MOLECULES

Species Oxidized	Species Reduced				
	O	C	N	S	H
O		Photosynthesis $CO_2 \rightarrow OC^a$ $H_2O \rightarrow O_2$			
C	Aerobic respiration $O_2 \rightarrow H_2O$ $OC \rightarrow CO_2$		Denitrification $NO_3^- \rightarrow N_2$ $OC \rightarrow CO_2$	Sulfate reduction $SO_4^= \rightarrow H_2S$ $OC \rightarrow CO_2$	Fermentation organic H $\rightarrow H_2$ $OC \rightarrow CO_2$
N	Ammonia oxidation $O_2 \rightarrow H_2O$ $NH_4^+ \rightarrow NO_2$	Nitrification $CO_2 \rightarrow OC$ $NH_4^+ \rightarrow NO_3^-$			
S	Sulfide oxidation $O_2 \rightarrow H_2O$ $HS^- \rightarrow S^0 \rightarrow H_2SO_4$	Anoxygenic photosynthesis; chemoautotrophy $CO_2 \rightarrow OC$ $H_2S \rightarrow S^0 \rightarrow H_2SO_4$	Denitrification/ sulfur oxidation $NO_3^- \rightarrow N_2$ $S^0 \rightarrow SO_4^{2-}$		
H		Methanogenesis $CO_2 \rightarrow CH_4$ $OC \rightarrow CH_4$ $H_2 \rightarrow H_2O$	Denitrification $NO_3^- \rightarrow N_2$ $H_2 \rightarrow H_2O$		

^{a} Organic carbon.

added to a body of water, such as a lake or estuary, plankton blooms commonly occur, using up the excess nutrient as rapidly as they can. When the plankton die, they begin to sink, and are consumed by fish and bacteria in the water column. The proliferation of these organisms uses up the oxygen in the water, causing the water to stagnate, killing fish and other animals that live in deeper water, and releasing foul odors and, occasionally, toxins. Generally, the cause of eutrophication is the runoff of artificially fertilized soils containing high levels of nitrate, or the discharge of phosphate-containing detergents into bodies of water (Chap. 29).

Processes that affect global biogeochemical cycles are familiar as well. One of the most prominent global environmental problems is the release of "greenhouse gases" into the atmosphere (see Chaps. 14 and 16). Gases such as carbon dioxide, methane, and nitrous oxide absorb infrared radiation and act as a heat blanket for the Earth, thus raising the temperature of the atmosphere. The burning of fossil fuels, and other by-products of technology, contribute greatly to the amount of greenhouse gases in the atmosphere.

CHEMICAL AND PHYSICAL PROPERTIES OF CHONSP

This section is not intended as a comprehensive treatment of the chemical and physical properties of the light elements (CHONSP), but rather as a primer. Hopefully, the terminology will be somewhat familiar to you. Table 19.2 shows some of the biogeochemically important compounds formed by carbon, hydrogen, oxygen, nitrogen, sulfur, and phosphorus. Some of the minerals are formed biologically as well as geo-

logically, and many "organic" substances, such as coal, are formed through geologic processes acting upon biologically produced material.

CARBON. Most of the carbon in the crust–atmosphere–ocean system is in the form of carbonate rocks – limestones and dolomites. These minerals, in the modern ocean, are formed largely through the activity of organisms that secrete calcium carbonate ($CaCO_3$) skeletons. The volumetrically most important calcite-secreting organisms are small planktonic organisms, such as coccolithophores (algae) and foraminifera (protozoa). Biologically precipitated calcium carbonate, such as that seen in coral reefs, may form large-scale marine structures. Fossil limestones often serve as reservoirs for oil or water. In the atmosphere and oceans, carbon is present as carbon dioxide (CO_2), methane (CH_4), and bicarbonate (HCO_3^-). Carbon dioxide is a stable gas and presently forms 0.03 percent of the Earth's atmosphere. Methane is not stable over long time periods in the Earth's atmosphere, and therefore is present only in trace amounts. It reacts with oxygen in the atmosphere and stratosphere to form carbon monoxide and, subsequently, carbon dioxide. Most of the carbon in the ocean is present as dissolved bicarbonate (see Chap. 12). It is bicarbonate or dissolved carbon dioxide that is utilized by marine photosynthetic organisms (phytoplankton).

When carbon is used to make cellular material, it becomes organic matter. Organic molecules are found in many forms. *Hydrocarbons* are molecules consisting of only carbon and hydrogen. Compounds containing carbon, hydrogen, and oxygen are called oxygenates (sugars, starches, cellulose, lip-

LIFE'S DIVERSITY

Traditionally, life on Earth has been separated into five familiar kingdoms: Animalia, Plantae, Fungi, Protista, and Monera (animals, plants, fungi, protists, and bacteria). Among these five kingdoms, the first four are eukaryotes, organisms whose genetic material is contained in a membrane-bound nucleus. The fifth kingdom contains all prokaryotes, organisms lacking such a nucleus. More recently, it was discovered that these five kingdoms do not adequately describe the relationships among organisms. When the genetic material (16S ribosomal) ribonucleic acid (RNA) of a diverse array of organisms is sampled, we find that there are three large groups of organisms – not two or five – each distantly related to each other. The three groups of *domains*, that arose from this analysis were termed *Archaea* (recognized as a separate group for the first time), *Bacteria*, and *Eucarya*. The mem-

bers of Archaea are all microscopic, nonnucleated organisms. They perform a wide variety of metabolic functions, some of which are unique to Archaea, and some of which are also performed by Bacteria or by Eucarya. Figure 19.2 depicts the phylogenetic relationships among these three groups. Branch points in the tree indicate divergences. Groups that are closely related share common branch points. Each group of organisms clusters together, sharing a common ancestor distinct from the common ancestor of the other two groups. This indicates that the members of each group are more closely related to each other than they are to the members of any other group. Further, recent analyses of these relationships suggest that Archaea and Eucarya are more closely related to each other than either is to Bacteria (notice

the branch point uniting Archaea and Eucarya).

What does this discovery mean for biogeochemistry and microbial ecology? Prior to the recognition of Archaea as a separate domain, it was thought that most of the diversity of the microbial world had been explored. With this new understanding, new bacteria are discovered frequently. A vast space of "uncharted territory" has been revealed. When the ribosomal RNA from a mixture of organisms is sequenced, Archaea are almost always present, and brand new species are found in many cases. Little is known about the metabolism, the abundance, or even the morphology of these organisms, which makes us wonder how many undiscovered microorganisms are living in environments where they previously went undetected, and how they contribute to the world's ecology.

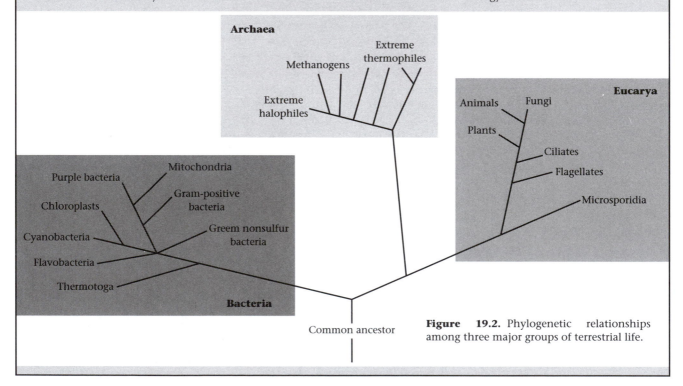

Figure 19.2. Phylogenetic relationships among three major groups of terrestrial life.

ids). Organic matter may also contain carbon, hydrogen, nitrogen, oxygen, and sulfur together, for example, in proteins. All these types of organic matter are found in sediments. Large accumulations of fossil organic matter may be economically important as oil, coal, or natural gas (see Chap. 24).

HYDROGEN. Hydrogen is usually found in compounds with other elements (Table 19.2). Hydrogen gas (H_2), the elemen-

tal form of hydrogen, is present but rare on the Earth's surface. As discussed above, hydrocarbons (e.g., oil and natural gas), are formed when hydrogen combines with carbon. Hydrogen bound to oxygen makes water, one of the most important molecules on Earth. Water provides most of the hydrogen necessary to form organic compounds. It is a key compound for living things, being an absolute requirement for life as we know it. Ammonia (NH_3) is composed of nitro-

TABLE 19.2 IMPORTANT BIOCHEMICAL COMPOUNDS OF CHONSP

	C	H	O	N	S	P
C	Diamond graphite	Organic matter hydrocarbons	Sugars Lipids	Chlorophyll porphyrins amino acids	Dimethyl sulfide Sulfur-rich organic matter	Nucleic acids ATP
H	Methane petroleum coal	Hydrogen gas	Water	Ammonia	Hydrogen sulfide	
O	Carbon dioxide carbonate rocks	Water	Oxygen gas	NO_x compounds nitrous oxide	Sulfur dioxide	Phosphate
N	Geoporphyrins	Ammonia	Nitrate $NaNO_3$	Nitrogen gas		
S	Sulfur-rich organic matter	Hydrogen sulfide	Sulfate gypsum anhydrite		Sulfur (S^0); pyrite	
P			Phosphate nodules apatite			Phosphorus (P^0) (unstable)

gen and hydrogen. Ammonia, as its protonated ion ammonium (NH_4^+), is the form of nitrogen most readily accessible to organisms.

OXYGEN. The most important form of oxygen is oxygen gas (O_2), found in the Earth's atmosphere. This highly reactive gas is produced at ground level by the activity of photosynthetic organisms. High-energy gamma radiation in the upper atmosphere reacts with oxygen gas to form ozone (O_3), the substance that shields the Earth's surface from ultraviolet light (Chap. 13). Oxygen gas is quite toxic, reacting with enzymes and organic matter in cells; accordingly, organisms that live in environments exposed to oxygen have developed complex biochemical mechanisms for minimizing damage resulting from oxygen gas and its by-products (ozone, hydrogen peroxide, and free radicals). Despite the abundance of oxygen in the atmosphere, most of the oxygen on Earth is actually contained in silicate and carbonate rocks. Silicate rocks, the most abundant rock types in the Earth's crust, are produced during the melting of rocks at high temperatures deep in the Earth. The weathering of these rocks is responsible for buffering the pH of the oceans.

NITROGEN. Nitrogen gas (N_2) is an extremely stable compound. The two nitrogen atoms are bound together strongly by a triple covalent bond. For this reason, most of the nitrogen on this planet is contained in the atmosphere as dinitrogen gas (N_2). Some organisms have the ability to break this triple bond, converting dinitrogen gas into "fixed" nitrogen, in the form of ammonia, nitrogen oxide compounds, or organic nitrogen. Nitrogen fixation, carried out exclusively by prokaryotic microorganisms, is a critical aspect of the nitrogen cycle.

Nitrogen combines with oxygen to form several compounds. Nitrous oxide (N_2O) is an important greenhouse gas. Nitrogen oxides (NO and NO_2), are also present in the atmo-

sphere in low concentrations, and are formed primarily by combustive pollution. Nitrate (NO_3^-), the dominant nitrogen species dissolved in the oceans, is an important source of nitrogen for the marine biota. Ammonia and ammonium are also readily assimilated by organisms.

Because nitrogen gas is so unreactive, and because nitrogen oxides are soluble in water and readily assimilated by organisms, nitrogen on Earth does not form many minerals. Ammonia is sometimes found as a minor ion in minerals such as clays and feldspars. The sodium salt of nitrate, $NaNO_3$, is sometimes found as a mineral (soda nitre); large deposits occur in the Atacama Desert of Chile. These deposits have been exploited as economically important sources of nitrate for fertilizer; however, because nitrate is very soluble and is rapidly utilized by organisms, such accumulations are rare.

SULFUR. Elemental (S^0) is a stable mineral, forming a yellow powder, and occurs naturally in this form in some areas in the world, such as the Gulf Coast of the United States, and in Iran. Sulfur has six oxidation states, and each can be formed or used by different organisms. Because sulfur can form many mineral types, a wide variety of sulfur minerals occur in the Earth's crust (Table 19.2). *Sulfides* are the most reduced mineral form of sulfur, and include pyrite and marcasite (FeS_2). Elemental *sulfur* is present in some rocks. The most oxidized form of sulfur is *sulfate*. Sulfate is abundant in the ocean, where it is utilized by organisms. Sulfate minerals are deposited when seawater evaporates, forming the minerals gypsum ($CaSO_4 \cdot 2H_2O$) and anhydrite ($CaSO_4$).

PHOSPHORUS. Elemental phosphorus reacts violently with oxygen or water to form phosphate (PO_4^-), the only geologically stable form of phosphorus (Table 19.2). It is as phosphate that organisms acquire the phosphorus they need to manufacture phosphorus-containing organic compounds such as deoxyribonucleic acid (DNA), ribonucleic acid (RNA), and adenosine triphosphate (ATP). The major phosphorus mineral occurring in rocks is apatite (calcium phosphate hydrate). Organisms incorporate apatite into bones and shells, and biologically formed hydroxy apatite may be found on the sea floor as phosphorite nodules. These phosphorite nodules are presently found on continental shelves such as those off the coasts of southern California, Baja California, and New Zealand. They are also found on land in Morocco, Russia, and China, where they constitute an important resource for fertilizer production.

BIOGEOCHEMICAL CYCLES

Biogeochemical cycles describe the paths followed by elements during their biologic conversion into many different chemical and physical states in the atmosphere, hydrosphere, and lithosphere. In studying the details of biogeochemical cycles, we concentrate on the transformation processes followed by these elements. Especially important are the pathways by which organisms acquire and excrete the elements, and the length of time each resides in the biosphere. All these factors require detailed understanding of the physical, chemical, and biochemical properties of the compounds of interest, combined with an appreciation of the geology and geochemistry that drive the cycles. With so many factors involved, where do we begin to examine biogeochemistry?

MICROORGANISMS AS MODEL SYSTEMS. In order to study the biologic transformations of carbon, nitrogen, and sulfur, we examine the activities of microorganisms. Most of the organisms that we discuss here are prokaryotes (Archaea and Bacteria); however, protists and some fungi may be considered microbes. There are several advantages to studying microbial biogeochemistry as a model for larger cycles. Microorganisms are relatively easy to study; they are found everywhere in great numbers, and they are readily acquired, brought into the laboratory, maintained, and studied (see the box entitled *The Winogradsky Column*). In addition, microbes are responsible for most aspects of biogeochemical cycling. The macrobiota (plants, animals) are quantitatively impor-

---

THE WINOGRADSKY COLUMN

Benthic microbial communities (microbial communities that live at the bottom of a body of water) are stratified: The organisms that live at the top, where light and oxygen (O_2) are plentiful, are different from the organisms that live farther down, where oxygen has been removed from the water.

One can see this stratification in laminated microbial mats in nature. We can also create such a stratified system in the laboratory. This was first done by S. Winogradsky in the late nineteenth century. He used this method to separate microorganisms with different metabolisms. Using this simple system, Winogradsky became the first microbiologist to cultivate many different kinds of organisms with previously unknown metabolisms.

To make a Winogradsky column, place a few inches of garden soil into a cylinder with clear sides (the cylinder must be at least 2 inches in diameter and 12 to 15 inches high – a laboratory graduated cylinder will work). Moisten the soil. Add approximately 1 cup of mud from a pond or other body of water (if you mix it with a little grass or news-

paper, this will provide a long-term carbon source for heterotrophs). Cover the whole thing with 2 to 6 inches of water, and place it in a north-facing window (so it receives light, but not too much). After several days to weeks, a stratified community will begin to form. Figure 19.3 illustrates a typical community. At the top, beginning in the water column and extending down to the sediment–water interface, cyanobacteria and eukaryotic algae typically grow. Below the "algal" layer, one or more layers of pink and green anoxygenic photosynthetic bacteria commonly occur. These layers mark the anaerobic zone. Below these layers, the soil is typically black, evidence of the activities of sulfate-reducing bacteria. If you examine regions of your column under the microscope, you will find that several kinds of microorganisms grow together, but that the zones for most of the organisms are restricted to the parts of the column that are environmentally favorable. The Winogradsky column allows you to see the carbon, sulfur, and nitrogen cycles at work on a miniature scale.

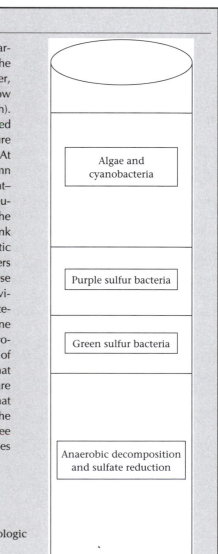

Figure 19.3. Schematic Winogradsky column, showing biologic zonation typically observed.

tant in fixation of elements (especially of carbon dioxide), but are of secondary importance in the cycling of elements. Often, recycling that is attributed to animals is in fact due to the activity of symbiotic microorganisms. For example, methane production by cattle and termites is actually attributable to the methane-producing Archaea that live in the digestive tracts of these animals. Typically, when we talk about recycling and transformation of the biologically important elements, we are referring to, primarily, microbial processes. A notable exception is provided by coals, where the remains of vascular plants constitute most of the organic material. In order to emphasize the importance of the microbial world in biogeochemistry, the next several sections follow a microbial biogeochemical approach. For each element, we first illustrate the microbial transformations and then generalize these results to show each global biogeochemical cycle.

MAJOR METABOLIC PROCESSES. Microorganisms are capable of a vast array of metabolic transformations (Table 19.1). They can manufacture or decompose organic matter, produce or consume oxygen, and oxidize or reduce sulfur, iron, nitrogen, and a variety of other substrates. In the laboratory, we can take advantage of this metabolic diversity by employing a technique called an enrichment culture, which uses the metabolic capabilities of organisms in order to separate them from one another. The box entitled "The Winogradsky Column" describes a particular method of observing ecologic zonation in a microbial community, developed in the late nineteenth century to separate organisms on the basis of their microhabitats. With this technique, microorganisms are separated according to environmental tolerances and metabolic requirements, while fitting comfortably within the confines of a laboratory.

Organisms that use inorganic carbon (carbon dioxide or bicarbonate) to manufacture organic material are called *autotrophs*. Broadly speaking, two types of autotrophy exist: (1) *photoautotrohpy*, in which the source of energy used for organic carbon synthesis is light; and (2) *chemoautotrophy*, in which the source of energy for organic carbon synthesis is chemical energy, derived from changing the oxidation state of an energy-providing molecule, such as ammonia or sulfide (Table 19.1). Autotrophy may be either *aerobic* or *anaerobic*. Aerobic organisms utilize molecular oxygen; anaerobic organisms do not use oxygen in their metabolism. Aerobic photoautotrophs use light energy to manufacture organic matter from inorganic carbon, producing oxygen:

$$CO_2 + H_2O \rightarrow CH_2O + O_2$$

This metabolic type, called *aerobic photosynthesis,* is carried out by land plants, algae, and *cyanobacteria*. Other *photosynthetic bacteria* are capable of photosynthesis but do not produce oxygen. Thus, they perform *anaerobic photosynthesis:*

$$CO_2 + H_2S \rightarrow CH_2O + S$$

Organisms that use organic matter as a food source to manufacture their own cell constituents are called *heterotrophs*. Animals, fungi, many kinds of protists (protozoa), and many prokaryotes are heterotrophs. Heterotrophs recycle the organic matter generated by autotrophs, thus returning carbon and other elements to the biosphere. Similar to autotrophs, heterotrophs may be aerobic or anaerobic, and some may switch between the two modes, depending upon the availability of oxygen. Organisms that require oxygen gas at some stage in their life cycle are called *obligate aerobes*. This group includes nearly all eukaryotes. Even yeast, a fungus used to ferment beer, requires oxygen during its life cycle. *Obligate anaerobes* do not use oxygen gas in their metabolism and are generally inhibited by its presence. Many anaerobic prokaryotes, such as those that live in the cattle rumens, conduct their entire life cycle in the complete absence of oxygen. *Facultative aerobes* use oxygen when it is available but do not require it for survival. Included in this group are many prokaryotes, most of which live in environments where oxygen is only occasionally available for metabolism.

The biochemical pathways involved in heterotrophic metabolism are quite diverse. The ubiquitous pathway, occurring in nearly all organisms, is *glycolysis,* the pathway that converts glucose into pyruvate, generating ATP, a molecule that stores energy in the cell. Under anaerobic conditions, *fermentation* occurs, in which pyruvate is converted into another organic molecule, such as ethanol or lactate. This process is harnessed in the manufacture of wines and cheeses. Under aerobic conditions, the pyruvate is completely oxidized into carbon dioxide and water by another biochemical pathway, called the *citric acid cycle.*

Methanogenesis is the biologic production of methane. Methanogenesis proceeds through two different biochemical pathways, both carried out by methanogenic bacteria, which are all members of Archaea. These obligate anaerobes colonize organic-rich environments as diverse as termite guts and lake sediments. In the absence of an electron source, methanogens can disproportionate small acids or alcohols, forming methane and carbon dioxide, as shown below for acetic acid:

$$CH_3COOH \rightarrow CH_4 + CO_2$$

Alternatively, in environments where an electron source such as hydrogen gas is present, methanogens can reduce carbon dioxide, producing methane and water:

$$CO_2 + 4H_2 \rightarrow CH_4 + 2H_2O$$

In landfills for domestic waste, methane forms in such abundance that it must be collected and removed, or it can endanger the landfill site and surrounding dwellings because of its combustibility. Some landfills produce sufficient amounts of methane to be harvested and burned with natural gas in order to generate electricity.

Sulfate reduction is a heterotrophic process carried out by anaerobic prokaryotes (Bacteria and Archaea). During sulfate reduction, organic matter is oxidized by glycolysis, and sulfate acts as an electron acceptor (oxygen source). The breaking of the bonds in the organic matter yields usable energy in the form of ATP. The sulfides produced in the reduction of sulfate are responsible for the "rotten egg" smell associated with sewage.

Nitrification is the set of biologic processes that oxidizes ammonia (NH_3) to nitrite (NO_2^-) and nitrate (NO_3^-). Nitrification is commonly carried out by chemoautotrophic bacteria; however, some heterotrophs can oxidize ammonia. The oxidation of ammonia provides the nitrifying chemoautotroph with cellular energy (ATP).

Denitrification is the reduction of nitrate (NO_3^-) to nitrous oxide (N_2O) or nitrogen gas (N_2). Most denitrifying bacteria are heterotrophs that obtain energy through nitrate reduction, typically under anaerobic conditions. Many denitrifying bacteria will reduce oxygen gas (aerobic respiration) when it is available, and nitrate when oxygen gas is absent (anaerobic respiration). This ability to switch between aerobic and anaerobic metabolisms is exceptionally useful in environments with rapidly fluctuating oxygen levels, such as the sediment beneath a body of water.

Nitrogen fixation is effectively the reverse of denitrification, in which nitrogen gas is converted to organic nitrogen. Only prokaryotes can fix nitrogen, using an enzyme called nitrogenase. This enzyme has a complex structure that has only recently been elucidated. Iron, sulfur, and molybdenum are critical to the operation of nitrogenase, and the enzyme is easily destroyed by oxygen. For this reason, nitrogen gas fixation must be carried out anaerobically.

THE CARBON CYCLE. Organisms are major contributors to the carbon cycle, which is linked to the other biogeochemical cycles because the synthesis of organic matter drives the biosphere, and thus the biologic alteration of the other elements. Inorganic carbon comes from volcanic sources, primarily in the form of carbon dioxide. Carbon dioxide is then utilized in the chemical weathering of silicate rocks, dissolved in the oceans, or consumed in the formation of carbonate rocks. The biologic carbon cycle begins with the Sun's energy, which is used by photoautotrophs to "fix" carbon dioxide or dissolved inorganic carbon into organic matter. The organic matter is reused and recycled by heterotrophs, and some of it is ultimately buried as sedimentary organic carbon.

The Microbial Carbon Cycle. The energetics of metabolic processes form the basis for the *ecologic succession* of microbiota in a natural community (Fig. 19.4). Interestingly, a vertical succession of metabolic types can be observed in most microbial communities, because the distribution of dominant metabolisms generally depends on the presence or absence of molecular oxygen or oxidizing potential of an environment, referred to as the *redox (reduction–oxidation)* potential. Near the surface, the oxygen content is high, and typically decreases with increasing depth as oxygen is consumed by respiratory organisms. Figure 19.4 illustrates the main features of the microbial carbon cycle, and shows the

spatial arrangement of the various metabolic types. At the top of the diagram, oxygenic photoautotrophs, such as cyanobacteria, use the Sun's energy to fix inorganic carbon into cellular material. This process forms the base of the carbon cycle. In the aerobic (oxygen-containing) zone, near the air – water interface, aerobic respiration is the dominant dissimilatory (energy-yielding) process. Aerobic respiration is the most efficient of the heterotrophic pathways; that is it yields the most energy per molecule of organic matter consumed. The respired organic matter is converted into carbon dioxide and returns to the inorganic carbon pool. When molecular oxygen is no longer available, organisms will use nitrate for respiration, a process that is somewhat less energy-yielding than is aerobic respiration. When nitrate is depleted, sulfate reduction dominates. After sulfate is exhausted, the redox potential of the environment is quite low; at this point, fermentative processes, which do not require an external electron acceptor, take place. Carbon dioxide reduction by methanogens also occurs in the sulfate-depleted anaerobic zone.

The methanogens, obligate anaerobes, are members of the domain Archaea and convert carbon dioxide or small organic molecules into methane (CH_4), which is released into the environment. Methane diffuses upward and is consumed in the aerobic or upper anaerobic zone by methanotrophs (Bacteria), which convert methane into carbon dioxide and organic matter. The methanogenic bacteria are members of the same group that resides in bovine rumens and in the guts of termites, converting organic matter into methane. Archaea play an important role in the world's biogeochemical cycles; however, their diversity and importance has only recently gained widespread appreciation (see the box entitled *Life's Diversity*).

At the bottom of the microbial carbon cycle, unused organic matter is buried in the sediment, becoming part of the rock record. Buried organic matter constitutes a very small fraction of the total carbon fixed in an ecosystem, typically much less than 1 percent. The succession of organisms and metabolic types from most to least energy-yielding, which can be observed in all the microbial cycles discussed below, is perhaps most apparent in the carbon cycle.

The Global Carbon Cycle. Microorganisms perform the bulk of carbon recycling in the global carbon cycle; however, terrestrial biomass (primarily land plants) accounts for approximately 50 percent of the global carbon fixation budget. Fixation in the oceans, primarily due to the activities of microscopic organisms, accounts for the other 50 percent. Thus, the input of macroscopic organisms is quantitatively important to the carbon cycle, particularly in fixation. See Figure 20.3 for an illustration of the global carbon cycle, showing the contributions of both terrestrial fixation and human input.

The global carbon cycle is incompletely understood at present, however; the reservoir sizes and flux rates are only approximately known. For example, we can measure with fair precision the amount of carbon dioxide emitted annually into the atmosphere through fossil fuel burning (greater than

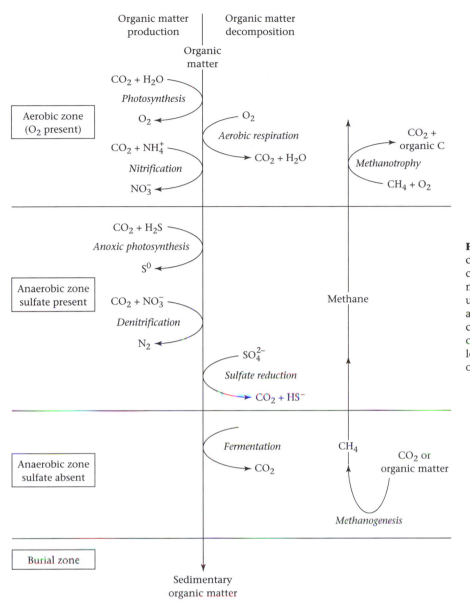

Organic matter
production

Organic matter
decomposition

Organic
matter

$CO_2 + H_2O$

Photosynthesis

Aerobic zone
(O_2 present)

O_2

O_2

Aerobic respiration

$CO_2 +$
organic C

Methanotrophy

$CO_2 + NH_4^+$

Nitrification

$CO_2 + H_2O$

$CH_4 + O_2$

NO_3^-

$CO_2 + H_2S$

Anoxic photosynthesis

S^0

Anaerobic zone
sulfate present

$CO_2 + NO_3^-$

Methane

Denitrification

N_2

SO_4^{2-}

Sulfate reduction

$CO_2 + HS^-$

Fermentation

CH_4

CO_2 or
organic matter

Anaerobic zone
sulfate absent

CO_2

Methanogenesis

Burial zone

Sedimentary
organic matter

Figure 19.4. The microbial carbon cycle. Organisms are spatially restricted according to their environmental requirements. Aerobic organisms thrive in the uppermost zone, where oxygen is available. Anaerobic organisms are zoned according to the availability of electron acceptors. Methane is produced in the lower parts of the anaerobic zone, and is oxidized aerobically.

5.7×10^{15} grams), and we know that this emission is significantly larger than it was in preindustrial times (less than 0.1 $\times 10^{15}$ grams), when less hydrocarbon fuel was burned. The amount of carbon dioxide in the atmosphere has increased, but by a smaller amount than we would have predicted, based on fossil fuel burning. Thus, we know that not all the carbon dioxide produced from fossil fuels remains in the atmosphere. Where does this "extra carbon" go? The obvious answer is that one of the sinks (terrestrial biomass, marine plankton) has been consuming it, and that it is ultimately stored in land plants, soil, the deep oceans, marine sediments, or some combination of these. However, we do not know precisely where it is going. It is extremely important when we evaluate government legislation concerning carbon dioxide emissions that we understand how the biosphere responds to additional carbon dioxide. As our understanding

of the global carbon cycle improves, we may learn the fate of the missing carbon.

THE SULFUR CYCLE. As is true for the carbon cycle, organisms are the most important cycles of sulfur in the near-surface environment. The major participants in the sulfur cycle are prokaryotes. Eukaryotes do use sulfur in biomolecules such as proteins, but not as energy sources or metabolites. The process of using a substance for incorporation into biomolecules is called *assimilation,* and the processes associated with utilizing a substance as an energy or electron source is called *dissimilation.* Dissimilatory processes, conducted by Bacteria and Archaea, are responsible for most sulfur transformations in the biosphere.

The Microbial Sulfur Cycle. As a model for the sulfur cycle, we examine the processes that occur in a *sulfuretum,* a microbial ecosystem that performs the various aspects of the sulfur cycle. Salt marshes typically contain microbial communities that depend upon the oxidation and reduction of sulfur-containing compounds. Figure 19.5 illustrates the biologic processes that occur in such a sulfuretum. Sulfate is the principal species of sulfur in the water column, where it is assimilated primarily by organisms. In the anaerobic zone, where little or no free oxygen remains, sulfate is reduced to hydrogen sulfide (H_2S). Sulfide diffuses upward into the aerobic zone, where it is oxidized through intermediates to sulfate. Hydrogen sulfide may be used by a variety of organisms. It is the likely source of compounds such as thiosulfate ($S_2O_3^{2-}$) and sulfur-containing organic molecules, compounds that are metabolized readily in the presence of oxygen. Hydrogen sulfide itself is utilized directly by anaerobic photosynthetic sulfur bacteria. These photobacteria oxidize hydrogen sulfide to sulfur and fix carbon dioxide photosynthetically. Sulfur cycling through a sulfuretum may be very rapid. Sulfur is removed as sulfides or sulfates when sulfur-containing minerals are deposited. Sulfides may be buried as minerals such as iron-containing pyrite or marcasite, copper-containing chalcopyrite, lead-containing galena, or zinc-containing sphalerite; these minerals constitute some of the Earth's most important metal ores. Sulfates may be buried as evaporite minerals, such as gypsum or anhydrite.

The Global Sulfur Cycle. The biologic component of the global sulfur cycle remains virtually identical to the sulfuretum; plants and animals only assimilate sulfate. Figure 19.6 shows the sulfur inputs to and outflows from the biosphere. Biologic processes are responsible for a significant portion of sulfur transformations in the biosphere; however, physical processes (volcanism, rainfall, river runoff) are responsible for distributing that sulfur over the Earth's surface. This observation highlights the fact that biology and geology act together to produce biogeochemical cycles. For example, dimethyl sulfide (DMS), is biologically produced but is non-biologically oxidized to sulfate in the atmosphere and may be responsible for adding sulfate particles to clouds, a process thought to lower the Earth's temperature.

Human influence on the global sulfur cycle is quantitatively significant. Because many sulfur gases undergo rapid chemical reactions in the atmosphere, most anthropogenic sulfur reacts rapidly and is deposited close to the source of production. Because most sulfur is deposited locally, areas near the sites of sulfur production are commonly hit hardest by the acid rain produced when sulfur emissions descend to Earth as sulfuric acid. Approximately 50 percent of the sulfur emitted on land is deposited on land; the other 50 percent is transported farther and deposited in the ocean.

THE NITROGEN CYCLE As in the sulfur cycle, the major participants in the nitrogen cycle are microorganisms. The most significant part of the microbial nitrogen cycle is the fixation of nitrogen gas into organic nitrogen. Because

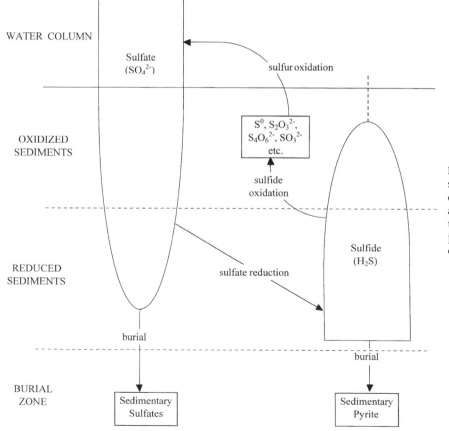

Figure 19.5. Microbial (coastal marine) sulfur cycle. The oxidation and reduction of sulfur takes place according to the availability of oxygen in the sediment. Although sulfate and sulfide are the two major forms, sulfur proceeds through many oxidation states in the microbial system.

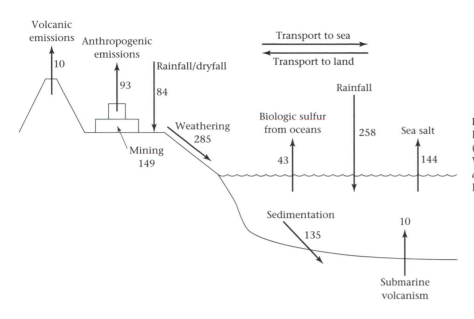

Figure 19.6. The global sulfur cycle. Fluxes in megatons of sulfur per year (10^{12} grams). (*Source:* Adapted from W. H. Schlesinger, *Biogeochemistry: an analysis of global change*, Academic Press, 1991.)

molecular nitrogen is very stable (recall the triple bond between the nitrogen atoms), nitrogen fixation is an energetically costly process. Only prokaryotes are capable of fixing nitrogen; consequently, their presence in the nitrogen cycle is critical. Nitrogen fixers are either free-living or symbionts. Free-living nitrogen-fixing cyanobacteria (e.g., *Trichodesmium*) are important in stimulating primary productivity in nutrient-depleted areas of the ocean. Nitrogen-fixing bacteria that live symbiotically in root nodules of legumes and certain other plants are responsible for fixing atmospheric nitrogen, enabling these plants to live in nitrogen-poor soil.

The Microbial Nitrogen Cycle. The microbial nitrogen cycle consists of five biochemical processes: nitrogen fixation, nitrogen assimilation, ammonification, nitrification, and denitrification (Fig. 19.7). Each nitrogen species can be utilized by organisms. Life has evolved to take advantage of the several oxidation states of nitrogen, and thus to participate in all steps of the nitrogen cycle. Most critical to the microbial cycle, and also important at a global level, is the biologic fixation of nitrogen gas. This process provides usable nitrogen to the biosphere. Without nitrogen fixation, nitrate would be rapidly reduced to nitrogen gas by denitrification. An essentially modern nitrogen cycle arose when all five metabolic transformations had evolved. This event probably occurred after an aerobic atmosphere was established, perhaps by 2100 million years ago. Nitrification requires molecular oxygen; therefore, it is reasonable to assume that an aerobic environment must have been available for these organisms. However, there is some disagreement about when biologic nitrogen fixation arose. Some researchers believe that nitrogen fixation originated early in Earth's history, forming an essential part of the nitrogen cycle. Others hypothesize that nitrogen fixation is such an energetically costly process that it probably did not originate until biologic demand for fixed nitrogen exceeded the supply; therefore,

nitrogen fixation might have been the last of the five metabolic processes to originate.

The Global Nitrogen Cycle. The terrestrial nitrogen cycle is quantitatively important to global nitrogen fluxes. Soil bacteria and symbiotic bacteria play an important part in fixing and recycling nitrogen through the biosphere. Additionally, nitrogen fixation by human activity contributes significantly to the amount of fixed nitrogen in the biosphere. Some researchers estimate that humans are responsible for 50 percent of the world's fixed nitrogen budget through the combustion of molecular nitrogen in the presence of hydrogen, making ammonia for fertilizer, and automobile emission of nitrogen oxides (NO_x). Lightning and ultraviolet radiation in the atmosphere contribute a very small amount of fixed nitrogen to the Earth's surface. The uptake and cycling of this fixed nitrogen is accomplished primarily by microbes, both on land and in the ocean. In environments where nitrogen nutrients are scarce, nitrogen gas fixation is particularly important. Figure 19.8 illustrates the reservoirs and fluxes in the global nitrogen cycle.

Figure 19.7. Microbial transformations of nitrogen. Denitrification is shown in pink; nitrification in red. Assimilatory pathways are indicated by gray arrows.

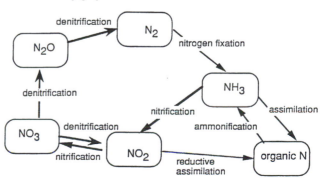

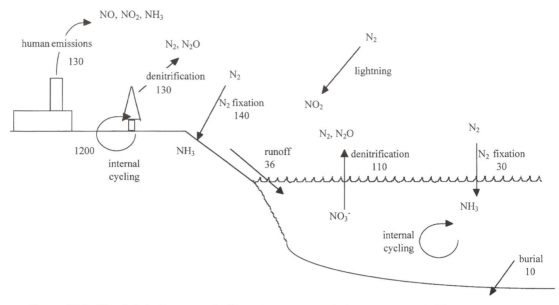

Figure 19.8. The global nitrogen cycle. Fluxes in megatons of nitrogen per year (10^{12} grams). (*Source:* Adapted from W. H. Schlesinger, *Biogeochemistry: an analysis of global change*, Academic Press, 1991.)

ECOLOGY OF THE CARBON, NITROGEN, AND SULFUR CYCLES: SYMBIOSIS AND INTERACTIONS

Close interactions among organisms are important in all the biogeochemical cycles. As we can see from the previous discussion, the by-products of one group of organisms often serve as a food or energy source for another, and organisms have evolved to take advantage of these relationships. When the coupling of these processes takes place in very close association, the relationship is called a *symbiosis*. The following sections discuss the role of symbioses in the biogeochemical cycles. We present three examples of symbiotic relationships, each of which is important in one of the biogeochemical cycles previously discussed.

THE CARBON CYCLE: CORAL REEFS. Coral reefs have formed since the early part of the Paleozoic Era. They are composed primarily of calcium carbonate in the form of calcite or aragonite. Following deposition and submergence beneath the sea floor, the calcite and aragonite in some environments are transformed to the mineral dolomite, $CaMg(CO_3)_2$, which is more stable and less prone to weathering than calcite or aragonite. Ancient reefs, particularly from the Mesozoic and Cenozoic eras, constitute vast areas both below and at ground surface. When they are partially dissolved, these structures give rise to underground caves containing stalagmites and stalactites. Fossil reefs make excellent subsurface reservoirs for petroleum and water.

Modern reefs are formed by a group of corals called *hermatypic* corals. The coral animals live in association with algal species known as zooxanthellae. The symbiosis between coral and zooxanthellae is of benefit to both participants: The algae live in a protected environment within the coral animal; and the algae photosynthesize and provide certain essential nutrients for the animal, which in turn provides other nutrients (e.g., glycol) to the algae. Because the zooxanthellae are photosynthetic, hermatypic corals are restricted to shallow water. In the modern ocean, only hermatypic corals form reefs; corals that lack zooxanthellae are generally smaller and non–reef forming, although they may form carbonate banks. Coral reefs are dominant sedimentary structures located in shallow tropical seas, forming an important part of the global carbon cycle and providing a substrate for complex ecosystems.

THE NITROGEN CYCLE: PLANT–RHIZOBACTERIA SYMBIOSIS. Plants are unable to fix nitrogen; consequently, they depend upon nitrate and ammonia in the soil for nitrogen assimilation. In some plants (e.g., legumes), certain species of Bacteria (e.g., *Rhizobium*) live inside specialized root cells that fix nitrogen and excrete the fixed nitrogen into the plant's roots. In turn, the bacteria assimilate carbon that the plant has fixed and transported to its roots. Thus, the plant receives fixed nitrogen in a nutrient-poor environment, and the bacteria have access to a source of organic carbon and a protective environment to live in. These plant–*Rhizobium* interactions result in significant amounts of fixed nitrogen per year, and enable vascular plants to grow in soils that would otherwise be inhospitable (see Chap. 20).

THE SULFUR CYCLE: SYMBIOSIS IN HYDROTHERMAL VENTS. Relatively recently, a diverse biota of microorganisms and multicellular animals was discovered living in the deep ocean, near very hot (350°C) hydrothermal vents (Chap. 12). Because these communities grow in the complete absence of light, scientists initially could not discern what formed the base of the foodweb for these organisms. How did carbon enter the ecosystem, to be consumed by the heterotrophs living there? What provided the energy source to drive carbon fixation?

It was soon discovered that chemoautotrophs are responsible for carbon fixation in these hydrothermal vent communities. They use sulfides coming from the vents to provide electrons for the fixation of carbon. Further, these chemoautotrophs live in the tissues of pogonophora, large worm-like animals with no mouth and no gut, which act as protective hosts. The bacteria living in the pogonophora fix inorganic carbon, and provide organic carbon (nutrients) for the worms. The hosts may also provide the bacteria with growth stimulants. Other symbiotic bacteria were found to live in association with molluscs in these communities, performing the same task. In the process of fixing carbon, these organisms also form the base of the sulfur cycle in these communities, converting sulfides into sulfate, which is used by other organisms for sulfate reduction or sulfur assimilation. Thus, sulfide supplies the electrons required for the fixation of carbon in these communities, linking the carbon and sulfur cycles closely.

INTERACTIONS OF THE CARBON, NITROGEN, AND SULFUR CYCLES. It should be obvious from the preceding sections that the biogeochemical cycles are intimately interconnected. One cannot discuss the sulfur cycle without knowing how sulfate-reducing bacteria transform carbon. Similarly, we cannot understand the vertical succession of microorganisms in the microbial carbon cycle without a grasp of the oxidation and reduction reactions that occur with sulfur and nitrogen.

Figure 19.9 summarizes these interconnected relationships. Notice the central importance of photoautotrophy. The synthesis of organic matter drives all the biogeochemical cycles on Earth. Without primary producers, the biosphere would rapidly run out of fuel. In fact, many researchers believe that such a shortage of organic carbon occurred very early in the history of life. They hypothesize that the first life forms were heterotrophs subsisting on nonbiologically produced organic matter. Gradually, however, as the conditions on the Earth's surface changed, the environment was no longer favorable for the nonbiologic production of organic matter. The biosphere likely experienced a shortage of organic matter, and autotrophic metabolism became critical as a source of carbon for the biosphere. Evolution may very well have proceeded in this way, with new metabolic types rising to prominence as their biochemical products became critical to the survival of the ecosystem.

THE GEOLOGIC AND GEOCHEMICAL RECORD

GEOLOGIC STRUCTURES. The biogeochemistry of ancient ecosystems is preserved in the geologic record; even small organisms can leave an impressive rock record. Skeletonized plankton, such as foraminifera and diatoms, have an outstanding fossil record. These organisms sink to the sea floor

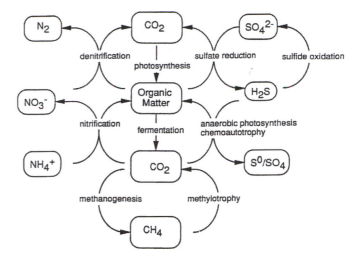

Figure 19.9. Linkage of the carbon, nitrogen, and sulfur cycles. Synthesis of organic matter by autotrophs is of central importance and drives the interdependence of the cycles.

and are compressed by the weight of the ocean and the pressure of burial to form sedimentary deposits of limestone (foraminifera and coccolithophores) or chert (diatoms). The Monterey Formation, exposed in many places along California's coast, is composed of massive chert beds. The spectacular White Cliffs of Dover, in England, is a limestone made of coccolithophore skeletons.

The oldest record of life on Earth is preserved as structures called *stromatolites,* which are layered rocks, formed from the lithification of laminated microbial communities. Some stromatolites are small, perhaps only a few millimeters in height and breadth, whereas others are large, dome-shaped structures several meters in diameter. Even the earliest records of life on Earth indicate that biochemical processes have given rise to prominent geologic structures.

RESOURCES: THE GEOCHEMICAL RECORD OF LIFE

Sulfide Ores. Although sulfide ores are commonly formed by volcanic or other very high-temperature igneous activity (see Chap. 23), certain ores found in numerous locations worldwide appear to have formed at typical surficial or oceanic temperatures. These are known as *stratiform ores.* The name is assigned to them because they are deposited parallel to, or within, sedimentary strata. These ores are typically rich in pyrite as well as in various copper sulfides, galena (lead sulfide) and sphalerite (zinc sulfide) and are often associated with nearby igneous deposits. Examples of stratiform ores can be found in the United States, Australia, Ireland, Spain, and Japan.

Until recently, it was speculated that such deposits must have been emplaced on the sea floor by volcanism or igneous flows, with little or no biologic contribution. However, lead and sulfur isotope measurements on the ores have shown that in many such deposits, the sulfide may have been formed by bacterial sulfate reduction. The metals were likely introduced at the deposition site by magmatic processes, possibly related to subduction zones, where water enriched in metals was squeezed out of a deep sediment column. When the metal-rich brine contacted a marine zone with active sulfate reduction, the metals reacted rapidly with the biologically generated sulfide and were deposited as insoluble metal sulfides.

Coal. Coal and petroleum are formed when organic matter is subjected to burial and decomposition. Three factors determine whether coal or petroleum will be formed from organic matter: (1) the nature of the original organic material; (2) the depth of burial; and (3) the environment of deposition. Whereas the oil-forming process begins in a relatively deep marine or lacustrine environment (see Chap. 24), coal forms following the burial of terrestrial vegetation (e.g., trees, ferns, grasses) that live near a shallow-water environment, typically a lagoon or river delta. During a rise in sea level, a lagoon will become flooded with seawater or fresh water in a river floodplain, inundating vast areas of forest. A deltaic environment may become flooded with fresh water, submerging nearby plants. If the vegetation is buried quickly, decomposition of the wood is slowed. Under such conditions, humic acids form rapidly, in part because the lignin of vascular plants contains a phenolic structure that reacts with sugars released from cellulose, forming polymers. The humic acids are converted into a kerogen following burial, with time, the kerogen is transformed to lignite, then to bituminous coal, and, with extensive heating, to an anthracite coal. "Cracking" of the kerogen to produce crude oil never occurs during coal formation. Methane is released from the residual organic matter, first by biologic processes; then, during later maturation stages, thermal methane gas forms. Because of the phenolic composition of lignin, coals are enriched in aromatic hydrocarbons but are depleted in nitrogen (which derives from proteins). Coals may be sulfur-rich or sulfur-poor, depending on sulfate availability. When the plants are buried in a saline environment, where sulfate is available, sulfate reduction in the organic-rich sediment results in the incorporation of pyrite or marcasite into a sulfur-rich coal. When the coal forms in a river floodplain, sulfate reduction is minimal; therefore, pyrite is typically absent and the coal formed is sulfur-poor.

THE BIOGEOCHEMICAL RECORD: CLUES TO EARTH'S HISTORY

How can we use our understanding of biogeochemistry to reconstruct past environments? By examining the geologic record, we can observe large-scale, long-term change in the biosphere. The evidence available to biogeochemists is threefold. First, the morphology of the organisms preserved in rocks, and of the rocks themselves, may be used to infer aspects of ancient environments. This category of data is broad, encompassing the fields of paleontology and geology. It is essential, however, that the large-scale paleoenvironmental information available from the rock record be utilized to form a rough sketch, and further information can be

added to the picture by utilizing other lines of evidence. The second type of information available is provided by ancient biomolecules themselves (chemical fossils). Individual organic molecules can be extracted from a rock. Although these biomolecules have been subjected to geologic alteration, often they contain significant information about the organisms that formed them and about the environments in which they formed. The third source of evidence is the stable isotopic record. Isotopes of an element are atoms that have the same atomic number (number of protons) but different atomic masses (number of protons plus neutrons). For example, the most abundant isotope of carbon, ^{12}C, contains six protons and six neutrons. ^{12}C constitutes 98.9 percent of naturally occurring carbon. The other stable isotope of carbon is ^{13}C, which contains six protons and seven neutrons and constitutes the remaining 1.1 percent of carbon on the Earth. Stable isotopes do not undergo radioactive decay. All other isotopes of carbon (e.g., ^{14}C) are radioactive, and they decay to other elements over time.

MORPHOLOGY AS A KEY TO EXTINCT ECOSYSTEMS. The oldest fossils currently known come from Western Australia and were found in rocks that are approximately 3.5 billion years old. They are cellularly preserved microscopic fossils that resemble certain modern cyanobacteria in their size, shape, and cellular arrangement. How can we use these fossils to infer aspects of Earth's early biosphere?

First, we must determine that these fossils are actually fossils, and are as old as they seem. Three criteria are used to ascertain this information: (1) The fossil-like objects must be demonstrated to be a part of the rock (not the result of modern contaminants); (2) the geologic source of the rock must be known (i.e., the rocks containing the fossils must be shown to be of the proper age); (3) the microfossil-like objects must be shown to be biologic in nature (not mineral grains or some other nonbiological material). The fossils from Western Australia meet all these criteria and therefore may be considered bona fide microfossils, with an age of approximately 3.5 billion years. That established, what can we learn from them about the early Earth?

If these earliest fossils were indeed cyanobacteria, that suggests that oxygenic photosynthesis may have originated by 3.5 billion years ago. This means that the central ingredient for the accumulation of oxygen in the atmosphere may have been in place quite early in Earth's history. However, other lines of geologic evidence, such as iron-rich rocks called banded iron formations, indicate that significant amounts of oxygen gas did not accumulate until approximately 2.0 billion years ago. What happened in the intervening 1.5 billion years? Something prevented oxygen from rapidly building up in the atmosphere, scavenging the photosynthetically produced oxygen gas. Two main types of oxygen sinks were present. First, large amounts of reduced compounds were present on the Earth. Important sinks included iron (Fe^{2+}), hydrogen sulfide (H_2S), hydrogen (H_2), and methane (CH_4). These substances became oxidized in the presence of oxygen gas, using up oxygen in the process. Second, aerobic respiration is a potent oxygen sink. Figure 19.4 shows that oxygenic photosynthesis and aerobic respiration perform opposite tasks: Photosynthesis produces oxygen gas, and respiration consumes it. Therefore, in order for oxygen to accumulate, organic matter must be buried and made inaccessible to heterotrophs. Even though oxygenic photosynthesis may have arisen early, the development of a modern-style, oxygen-rich atmosphere evolved over a long time; in the end, the interactions of living systems (photoautotrophs) with Earth systems (sediment burial) produced the atmosphere we now observe.

CHEMICAL FOSSILS AS INDICATORS OF EVOLUTIONARY PROCESSES.

All multicellular organisms are eukaryotes (although some eukaryotes are unicellular); therefore, we know that the biosphere prior to the origin of eukaryotes was a microbial one. Eukaryotes contribute significantly to biomass production, and thus are important to biogeochemical cycles. When did eukaryotes first evolve? A partial answer is provided by 1800-million-year-old fossils from Upper Michigan that are possibly eukaryotic in origin. They are coiled filaments, approximately 1 millimeter in diameter and several centimeters long. Based on their size, they have been interpreted as eukaryotes; however, little other morphologic information has been preserved. The morphologic evidence alone does not provide a conclusive answer; a second line of evidence relies on biogeochemistry to reveal the origins of eukaryotes.

Chemical fossils are the molecular remnants of living systems, chemically altered by the heat and pressure of burial. During geologic alteration, called diagenesis or metamorphism, the chemical structure of the original biomolecule is altered. Typically, functional groups, such as alcohol ($-OH$) or amine ($-HN_2$) groups, are lost and double bonds may form or be removed. Figure 19.10 shows the chemical structures of a biomolecule and the chemical species that results following alteration. Many of these chemical fossils provide important clues that enable us to identify the particular types of organisms that formed them. These organism-specific geochemicals are called *biomarkers*. An excellent example of a biomarker is provided by sterols and their geochemical counterparts, the *steranes*. Sterols are manufactured only by eukaryotes; thus, the presence of steranes in the organic matter of an ancient rock indicates that eukaryotes contributed to the formation of that organic matter. Steranes have been found in rocks of Middle Proterozoic age, approximately 1690 million years old, from the McArthur basin in northern Australia. By taking morphologic and chemical evidence together, we can say that eukaryotic organisms likely arose by 1800 to 1690 million years ago.

THE CARBON ISOTOPIC RECORD AS AN INDICATOR OF ENVIRONMENTAL CHANGE.

Carbon isotopes in sedimentary rocks provide an informative record of environmental change through Earth's history. When we study isotope geochemistry, we are interested in changes in the ratios of the stable isotopes over time. Change in the relative proportions of the isotopes provides information about the biologic processes affecting each element. In order to express variation in isotopic composition, the "delta" notation is used. The isotopic composition of a substance is expressed as a ratio of two isotopes of an element, relative to that ratio in a standard, which is assigned a value of zero. For carbon, the standard is the Pee Dee Belemnite, a carbonate rock. The delta notation, expressed in parts per thousand (‰), is determined as follows:

$$\delta E‰ = \frac{R_{sample} - R_{standard}}{R_{standard}} \times 1000$$

In this equation, E is the isotope of interest (e.g., ^{13}C or ^{15}N), and R is the ratio of the heavier isotope to the lighter isotope: $^{13}C:^{12}C$ for the element carbon, $^{15}N:^{14}N$ for the element nitrogen. According to this notation, negative numbers indicate that a substance is relatively enriched in the *lighter* isotope, and positive numbers indicate that a substance is enriched in the heavier isotope.

During biologic carbon fixation, ^{12}C is assimilated faster than ^{13}C. The result is that photosynthetically derived organic matter has an isotopic composition of approximately −25 parts per thousand that is, it is 25 parts per thousand enriched in ^{12}C, relative to a typical carbonate rock (0 parts per thousand). There is some variation in the carbon isotopic composition of organisms; however, overall, marine organic matter has an isotopic composition of approximately −20 to −25 parts per thousand. However, as we examine rocks of increasing geologic age, we find that the average isotopic composition of organic carbon has changed through time, becoming progressively more negative in older rocks.

On average, organic matter of Archean age (2.5 to 3.5 billion years) has an isotopic composition of −34 per mil relative to carbonate carbon. In modern environments, organic mat-

Figure 19.10. Transformation of sterol (biochemical) to sterane (geochemical) as a result of diagenesis. The alcohol (OH) functional group and the double bond are lost during diagenesis; however, the complex, multiringed carbon framework remains intact. The presence of steranes in a rock indicates that eukaryotes gave rise to some of the organic matter contained in the rock.

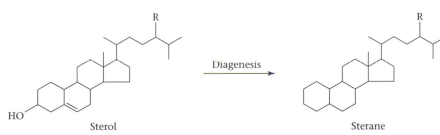

Sterol   Diagenesis →   Sterane

ter rarely has a $\delta^{13}C$ value this low. One of the most significant factors influencing the isotopic composition of organic matter in modern organisms is the amount of carbon dioxide or bicarbonate present in the atmosphere or water. When excess carbon dioxide is present in the atmosphere, the $\delta^{13}C$ of organic matter is lower than that found in organic matter produced under normal amounts of carbon dioxide. In experiments using cyanobacteria-dominated ecosystems, increasing the amount of carbon dioxide in the atmosphere to 5 percent (it is normally 0.03 percent) resulted in organic matter with an isotopic composition of −32 parts per thousand relative to carbonate carbon. These experiments suggest that the Archean atmosphere may have contained much more carbon dioxide than does the current atmosphere. Interestingly, an atmosphere rich in carbon dioxide is indicated by several climate models for the early Earth.

STABLE NITROGEN ISOTOPES AS INDICATORS OF HUMAN HISTORY. The isotopic composition of atmospheric nitrogen gas ($^{15}N{:}^{14}N$) falls within a very narrow range, and because it is used as the international standard for measurement of nitrogen, it has been given a value of 0 parts per thousand. Terrestrial plants, which derive their nitrogen from *Rhizobium* fixation and release of ammonia, have nitrogen 15 isotope ratios, $\delta^{15}N$, similar to those of atmospheric nitrogen. By contrast, most of the nitrogen available in the ocean for primary production is derived from nitrate. This nitrate is enriched in ^{15}N because denitrification converts nitrate characterized by the nitrogen 14 isotope, $^{14}NO_3^-$, to nitrogen gas more rapidly than does $^{15}NO_3^-$. This process leaves the residual nitrate enriched in ^{15}N. Additionally, it has been found that during metabolism, fish, birds, and mammals all concentrate ^{15}N up the food chain (Fig. 19.11); thus, the nitrogen isotopic composition of animals reflects both their food source (terrestrial or marine) and their position in the food web.

Nitrogen isotope ratios have been used by anthropologists and chemists to study the diets and migration habits of prehistoric humans (Fig. 19.11). The $\delta^{15}N$ values of collagen or other protein from bone are measured. When $\delta^{15}N$ values are in the range of approximately 3 to 8 per mil, they indicate that the population subsisted primarily on plants; $\delta^{15}N$ values between 7 and 10 per mil are indicative of a carnivorous diet of land animals, whereas intermediate values indicate mixed diets. $\delta^{15}N$ values of 15 to 20 per mil are indicative of a marine fish, bird, or mammal diet, whereas values in the range of 10 to 15 per mil are suggestive of a marine fish diet supplemented by a plant-derived diet, as would be expected in coastal or island settlements.

THE GAIA HYPOTHESIS: THE BIOSPHERE AS AN AGENT OF GLOBAL CHANGE. It should be clear from the foregoing discussions that the global biogeochemical cycles are com-

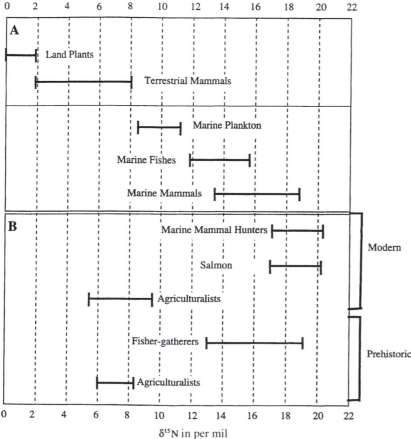

Figure 19.11. Nitrogen isotopes and diet. (A) Distribution of $\delta^{15}N$ values for animals and plants. Terrestrial plants and animals have nitrogen 15 isotope values lower than those of marine plankton and animals. (B) These isotopic differences are reflected in humans with various diets. The top portion of (B) shows the isotopic composition of bone protein in present-day humans. Humans that subsist primarily on marine animals have a much higher isotopic composition than those that eat terrestrial animals and plants. This pattern is preserved in the bone proteins of prehistoric humans: Humans that lived on the seashore have $\delta^{15}N$ values in their bones that reflect a marine diet, whereas those that lived inland show evidence of a terrestrial diet.

plex, interacting cycles with many facets. The cycle of each element is dependent in no small way upon the cycles of the others. The carbon cycle, ultimately, is connected to the cycles of sulfur, nitrogen, phosphorus, oxygen, and hydrogen, because carbon fixation by autotrophs lies at the base of all life on this planet. Thus, all processes occurring in the biosphere must be interrelated, with living processes as their common link. This idea was proposed by J. E. Lovelock in a controversial book entitled *Gaia*. In it, he argues for the interconnectedness of life on this planet, and in favor of the idea that living systems manipulate their environment so that it best suits their needs. In this way, the planet Earth is similar to an organism, and the cycles that occur on it are similar to physiologic systems, carefully controlled by the regulatory properties of the organism. The language that Lovelock uses in his book suggests planned activity by the biosphere. In fact, there is no need to invoke consciousness in the perpetuation of life on Earth. Processes that enhance expansion and diversity of life are favored, and processes that are destructive are naturally self-limiting. In addition, living systems evolve over time to fill ecologic niches provided by other organisms. These processes, which might be described as the interactions of the biogeochemical cycles, do not require that the Earth operate as a single organism.

The "Gaia hypothesis" has spurred great controversy in the fields of geology, environmental science, and biogeochemistry. Whether or not Lovelock's theory is accurate in its description of the effects of life on Earth, it has almost certainly caused scientists to think about the planet in terms of complex, interactive cycles, rather than separable, distinct entities. Feedback in the biogeochemical cycles is a vital part of the continued success of life on Earth. It is important to remember, though, that all of biology operates on the template of the Earth. Mars, which apparently lacks an active tectonic cycle, shows no evidence of recent life, perhaps because elements could not be recycled tectonically once buried. Thus, geochemical cycles could not be sustained. Without volcanoes, plate tectonics, and the light of the Sun, the Earth would be a very different place. All these physical factors influence the nature of life on this planet, and have played an integral part of perpetuating life over 3.5 billion years or more of its history.

SUMMARY

In this chapter, we have discussed the role of organisms in Earth system processes. The various metabolic pathways undertaken by bacteria result in both the fixation and release of various elements to the lithosphere, hydrosphere, and atmosphere. The primary process driving modern biogeochemical cycles is oxygenic photosynthesis, which uses solar energy to produce the Earth's store of molecular oxygen and most of the organic matter in the biosphere. The biogeochemical cycles of carbon, sulfur, and nitrogen illustrate the various oxidation and reduction transformations performed by the biosphere. Without the recycling of biologically important elements, the Earth would be vastly different.

Understanding the importance of organisms in Earth's history permits reconstruction of ancient ecosystems and of their changes through time. Metabolic processes of microscopic algae and bacteria are responsible for deposition of mountains of limestone and chert. They are responsible for the concentration of certain types of ores, and for the production of methane gas and crude oil. The biogeochemical window onto the past provides a framework for examining the modern Earth. Appreciation of the interactions among organisms and element cycles through time leads to better understanding of modern environmental change. Such knowledge about the biosphere can be applied to problems of pressing importance in today's world. Effective public policy must rest on comprehension of various aspects of Earth systems, of which biogeochemistry is an integral part.

QUESTIONS

1. Medieval doctors recommended against sleeping in a closed, dark room with a growing plant. Is there any reason this might be good advice?

2. In the 1950s it was discovered that certain regions were without plant life, although the moisture and nutrient content in the soils seemed sufficient to support growth. It was later discovered that plants flourished when molybdenum (Mo) was added to the soil. Why does molybdenum help?

3. The White Cliffs of Dover in England and the limestones of the Carlsbad Caves in New Mexico are both formed by marine organisms. However, they are very different in their mineralogy. Explain how each formed.

4. While visiting a friend in Minnesota and walking on some of the freshwater swamps (to collect cranberries, no doubt), you notice bubbles rising in the water, but there is no discernible smell. What is the gas in the bubbles, and how is it formed?

5. When you visit your cousin in Maine and walk along the seacoast, you notice a strong smell of rotten eggs. When you move closer, you also notice bubbles on the sediment surface. What is the gas, and how is it formed? Why didn't you smell it in Minnesota?

6. It has been suggested that the Earth's atmosphere originally contained little or no oxygen. How did oxygen accumulate in our atmosphere? Why doesn't it disappear?

7. What is a chemical fossil? What are some scientific or technologic uses for chemical fossils?

8. High nitrate contents in drinking water can be dangerous to your health. Pollution of drinking water by nitrate is thought to originate from three sources: (1) fertilizers synthesized from atmospheric nitrogen; (2) farm wastes such as cattle manure; and (3) leaking septic tanks. How could you differentiate among the possible sources of nitrate in drinking water?

9. Recently, an experiment was conducted in the Southern

Ocean to determine whether addition of iron to the ocean would stimulate primary organic productivity there. Initially, phytoplankton productivity did increase; however, it rapidly returned to prefertilization levels. Given what you know about biogeochemical cycles, how might you explain this result? Do you think a similar result would be obtained if nitrate were added to the ocean?

FURTHER READING

Ehrlich, H. L. 1981. *Geomicrobiology*. New York: Dekker.

Libes, S. M. 1992. *An introduction to marine biogeochemistry*. New York: Wiley.

Schlesinger, W. H. 1991. *Biogeochemistry. An analysis of global change*. San Diego, CA: Academic Press.

20 Global Change and the Terrestrial Carbon Cycle: The Jasper Ridge CO₂ Experiment

CHRISTOPHER B. FIELD AND NONA R. CHIARIELLO

ECOSYSTEMS AND A GREENHOUSE GAS

Much of the discussion concerning carbon dioxide (CO_2) and global change has focused on carbon dioxide and climate. Yet, carbon dioxide has important direct effects on plants. At least in the short term, increased carbon dioxide leads to increased plant growth and decreased use of water. If these effects persist after long-term exposure to increased carbon dioxide, and if they are not suppressed by other factors at the ecosystem scale, they could have dramatic consequences. First, plant growth, including production of food, fiber, and fuel in croplands and forests could increase. In addition to providing more products for human harvest, increased plant production might decrease requirements for cropland and allow increased areas of natural vegetation. Second, decreased water requirements might allow plants to thrive in regions that are currently too dry. This could mean the movement of plants into areas that currently have few or none as well as the replacement of dry land vegetation by ecosystem types that currently need more water. For example, shrublands might replace grasslands or forests might replace shrub lands. Third, increased uptake of carbon dioxide by plants in response to an increase in carbon dioxide concentration in the atmosphere could remove some of the carbon dioxide that is injected into the atmosphere by human activities, especially fossil fuel combustion. If this occurs, the concentration of carbon dioxide in the atmosphere may stabilize, eliminating the need for costly societal changes to reduce carbon dioxide emissions. On the other hand, the concentration of carbon dioxide in the atmosphere has already increased by over 25 percent since the beginning of the Industrial Revolution. Increasingly strong evidence supports the hypothesis that terrestrial ecosystems currently sequester 20 to 30 percent of the carbon released by human activities. If this storage stops in the future, the trajectory of the increase in atmospheric carbon dioxide may steepen substantially, increasing the societal investments necessary to stabilize the atmosphere by a certain date or within a particular carbon dioxide concentration range.

Will the initial effects of increased carbon dioxide on growth and water use penetrate through processes and scales to influence plant production and the total amount of carbon stored in ecosystems? This is one of the central questions in global change research. It is a very difficult question because the direct effects of increased carbon dioxide on plant processes are modified by a host of indirect effects. Some of these indirect effects amplify the direct effects, and others suppress them. In addition, effects of carbon dioxide that are beneficial from some perspectives may be detrimental from others. For example, an increase in plant growth stimulated by carbon dioxide could be a benefit if the stimulated species are valuable economically or from a biodiversity perspective. However, it could be a serious problem if the most stimulated species are weeds. Overall, our understanding is increasing rapidly; however, it remains too incomplete to support accurate predictions across the broad range of ecosystems and over the entire suite of direct and indirect mechanisms.

This chapter surveys the issues that need to be addressed in order to understand ecosystem-scale responses to carbon dioxide (Fig. 20.1), and it details the approach in a single experiment, an ecosystem-scale study in grassland at Stanford University's Jasper Ridge Biological Preserve. Neither the Jasper Ridge CO₂ Experiment nor any other single study can provide all the information necessary to understand the way the world's ecosystems will respond to increased levels of atmospheric carbon dioxide. The Jasper Ridge experiments do, however, add several critical pieces to the carbon dioxide puzzle, with other essential pieces coming from carbon dioxide experiments in other ecosystems, as well as studies with computer models and reconstructions of ecosystem responses to past changes in carbon dioxide levels.

THE FACT OF INCREASED CARBON DIOXIDE

Human actions over the twentieth century have had a dramatic and unequivocal effect on the concentration of carbon

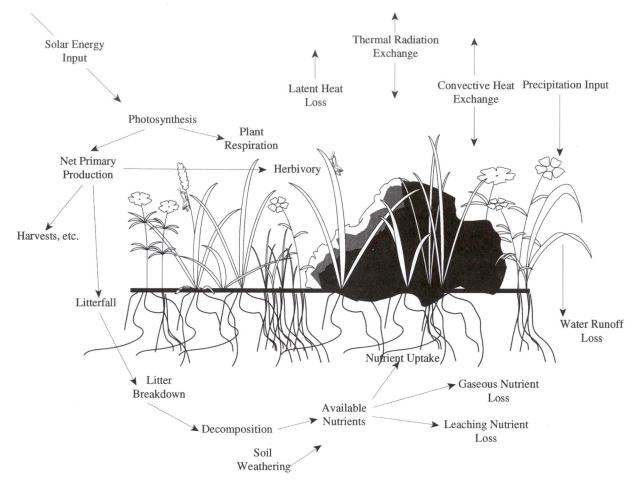

Figure 20.1. The major components of an ecosystem and some of the exchanges of mass and energy among them.

dioxide in the atmosphere. Measurements on air bubbles trapped in glacier ice reveal that atmospheric carbon dioxide concentration ranged between 180 parts per million (180 parts per million equal 0.018 percent of the atmosphere by volume) and 300 parts per million throughout the past 160,000 years (Fig. 20.2). The concentration at the beginning of the Industrial Revolution was approximately 280 parts per million, and it has increased steadily since then, to a current value near 360 parts per million (see Chap. 15). This is 25 percent above the 1750 value and 20 percent above the highest value in the past 160,000 years.

Is there any doubt that much of the recent increase has been driven by human actions? Essentially none. First, our understanding of the global carbon cycle indicates that carbon dioxide released by human action is the only plausible candidate. Second, changes in the atmospheric abundance of the stable carbon isotope, ^{13}C, have the signature of biogenic carbon, characteristic of fossil fuels and plants but not of mineral (inorganic) carbon (Chap. 19). Third, the atmospheric abundance of the radioactive carbon isotope, ^{14}C, indicates that the excess carbon is largely ancient, clearly pointing to an origin from fossil fuels, not modern plants.

Based on the history of per capita consumption of fossil

fuels through economic development, it is very likely that atmospheric carbon dioxide will continue to increase for several decades. The increases are likely to continue until we dramatically decrease consumption of fossil fuels. Depending on assumptions concerning future use of fossil fuels, atmospheric carbon dioxide levels in 2100 may range from approximately 500 to over 800 parts per million. The role of carbon dioxide as a greenhouse gas is dominant in the science of understanding global change. However, the ecosystem effects of increased carbon dioxide are also unequivocal and dramatic, potentially rivaling or even surpassing climate change in overall impact.

APPROACHES TO UNDERSTANDING ECOSYSTEM RESPONSES TO INCREASED CARBON DIOXIDE LEVELS

How can we predict ecosystem- and global-scale consequences of increased carbon dioxide? At least three approaches are promising. First, we can compare today's ecosystems with ecosystems in the past. Atmospheric carbon dioxide has increased by approximately 25 percent during the twentieth century, and in the past 160,000 years it has been as low as half the current level.

One way to build models of future ecosystem responses on the basis of past performance is used by the world's governments through the Intergovernmental Panel on Climate Change (IPCC). In this program, researchers compare the historic record of carbon dioxide emissions from anthropogenic sources (fossil fuel, cement production, and deforestation) with the accumulation of carbon dioxide in the atmosphere. They also use mathematic models to estimate carbon dioxide uptake by the oceans; any residual is tentatively identified as storage in terrestrial ecosystems in response to the carbon dioxide increase. This kind of empirical extrapolation leads to an "uncertainty *factor*," which summarizes the sensitivity of terrestrial carbon storage to atmospheric carbon dioxide. Although these kinds of approaches are invaluable for estimating the long-term trajectory of total carbon in the terrestrial biosphere, they are limited – especially in mechanistic and geographic detail.

A second approach to predicting the ecosystem-scale effects of carbon dioxide involves forward mathematic modeling (see the box entitled *Mathematic Models of Ecosystem Responses to Carbon Dioxide*) The key to the usefulness of these mechanistic models is that they can combine direct and indirect effects at different scales. Because some of the indirect effects occur only over long time intervals or large spatial scales, simulations provide a way to incorporate processes that are difficult to study with experiments. Currently, mathematic models provide the only spatially detailed global estimates of ecosystem responses to increased levels of carbon dioxide. Still, the accuracy of model predictions depends on the accuracy with which the effects and their underlying causes are specified. Some of the mechanisms are understood in detail; others are not.

The third approach to predicting ecosystem-scale effects of carbon dioxide involves experiments in which carbon dioxide, and perhaps other variables, are manipulated on test plots. This approach provides direct measurements of the entire range of ecosystem effects. It offers opportunities to test the hypotheses formalized in mathematic models, and it is a proving ground for identifying and quantifying unexpected mechanisms. Experimental ecosystem studies, however, involve a large array of challenges. Some are technical. For example, it is difficult to alter atmospheric carbon dioxide levels without changing other aspects of the environment. Also, it is difficult to measure the response of plant growth to changing levels of carbon dioxide without disturbing the study plots.

Other challenges are conceptual. For example, how does one determine the number of plants or species or the plot size sufficient to generate the entire range of ecosystem responses and feedbacks? How can one design an experiment that gives the investigator access to infrequent but important processes such as changes in the relative abundance of different species or changes in the frequency of fire or other disturbances?

Each of these three approaches has limitations; however, together they provide a broad menu of tools. The challenge is to unite the tools effectively, building on the strengths of each.

The unreplicated carbon dioxide experiment on the grandest scale – the current experiment involving our entire planet – differs from scientists' purposeful studies in a number of key respects. The scale is vast, and the timing, although rapid in terms of the natural dynamics of the Earth system, is too slow to be applicable to programs designed to help us cope with or manage future change. Perhaps most important, the global experiment is embedded in a matrix of

Figure 20.2. Atmospheric carbon dioxide levels, over three time periods. (Top) Monthly atmospheric carbon dioxide levels at Mauna Loa, Hawaii, based on direct measurements since 1958 by C. D. Keeling. (Middle) Atmospheric carbon dioxide levels since the eighteenth century, showing the Keeling record from 1958 and the record from the Siple ice core (Antarctica) prior to 1958. (Bottom) Atmospheric carbon dioxide levels for the past 160,000 years, based on measurements from the Vostok ice core (Antarctica). ppm = parts per million. (*Source:* Data from T. A. Boden, D. P. Kaiser, R. J. Sepanski, and F. W. Stoss (Eds.), *Trends '93: a compendium of data on global change*, Carbon Dioxide Information Analysis Center, Oak Ridge National Laboratory, Oak Ridge, Tennessee, 1994.)

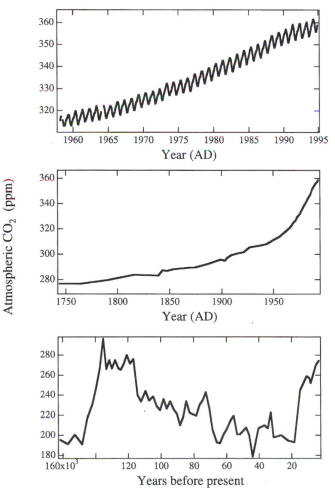

MATHEMATIC MODELS OF ECOSYSTEM RESPONSES TO CARBON DIOXIDE

Mathematic models in science are succinct, quantitative statements of a hypothesis or an array of hypotheses. A model can be a single, simple equation (e.g., $E = mc^2$) or a massive computer program with tens of thousands of lines of code. It can describe a process that is completely understood, or it can suggest a way to understand a poorly known process.

Many of the models in ecology are empirical. These summarize a number of observations, often using regression analysis to fit an equation to data. Empirical models can be very accurate over the range of conditions of the experimental measurements; however, they have little predictive power outside this range. In contrast, mechanistic models predict phenomena at some scale (e.g., the ecosystem), based on processes characterized at a smaller scale (e.g., photosynthesis and respiration at the scale of individual plants). Mechanistic models are potentially applicable beyond the range of calibration data at the output scale; however, their accuracy depends on the quality and completeness of the underlying process models. Often, models are calibrated or adjusted so that one or more aspects of the models response conform to observations at a particular site or over a particular time period. Calibration is a controversial topic: Although it can be essential for fine-tuning a model to address site-to-site or species-to-species differences, it can also confer upon the model the appearance of a level of accuracy that may not exist.

Models of ecosystem or global responses to increased levels of carbon dioxide range from purely empirical to purely mechanistic. The simplest empirical models use the measured sensitivity of plant growth to elevated carbon dioxide levels as an input, and calculate changes in carbon storage or atmospheric carbon dioxide levels as outputs. Although models of this type are unlikely to be accurate, they illustrate one very important point. Changes in terrestrial carbon storage in response to increased carbon dioxide levels do not depend only on the responses of plant growth. Carbon storage is an imbalance between carbon uptake in plant growth and loss through respiration by heterotrophs. Because microbial respiration typically increases with the amount of decomposable material, an increase in plant growth tends to be balanced by increased respiration of a larger pool of soil organic matter. This increased pool represents increased carbon storage; however, the storage is limited by the increasing losses caused by a larger pool. Other empirical models originate not from experimental studies but from an estimated terrestrial carbon sink, determined as the remainder in a calculation involving anthropogenic emissions, changes in the total quantity of atmospheric carbon, and carbon uptake by the oceans. Given a historic record of atmospheric carbon dioxide concentration and a historic record of the magnitude of the terrestrial sink, one can calculate a sensitivity of the strength of the terrestrial sink to increased carbon dioxide levels. The models used by the IPCC assume that an increase in the carbon dioxide level is the only factor causing the terrestrial sink. If other factors are involved (see the box entitled "The Missing Sink"), these models may overestimate the effects of increased carbon dioxide levels.

Mechanistic models of ecosystem responses to carbon dioxide typically include component models for the direct and indirect responses discussed in this chapter. Some of these components are well understood. For example, the short-term response of photosynthesis to increased carbon dioxide levels can be specified with confidence. Other components are more poorly understood. Even the mechanistic models are based on a combination of suspected mechanism and empiricism. Mechanistic models of ecosystem responses to carbon dioxide are very useful for highlighting significant interactions. Examples include the sensitivity of plant root-to-shoot ratio to changes in the relative availability of carbon and nutrients, and changes in nutrient inputs from anthropogenic or biologic sources.

Currently, modeling ecosystem responses to elevated carbon dioxide levels involves two challenges: to significant incorporate potentially significant mechanisms for which good models are not yet available (e.g., changes in plant species composition), and to determine appropriate ways to capture ecosystem-to-ecosystem variation. The current generation of models provides excellent tools for shaping hypotheses and for exploring interactions among component processes. Concrete predictions are most useful when applied to environments that closely resemble the observation or calibration conditions.

---

other changes, many of which are discussed in this book. Without an understanding of the effects of carbon dioxide in combination with altered nutrient deposition, air pollution, ultraviolet-B radiation, intensity of land use, human population pressure, and perhaps natural climate dynamics, we will have dull and unwieldy tools for predicting the future of the Earth system.

THE RANGE OF ECOSYSTEM RESPONSES TO INCREASED CARBON DIOXIDE LEVELS

Ecosystem responses to increased levels of atmospheric carbon dioxide are potentially important components of global change. The impact of increased levels of carbon dioxide and the processes contributing to this impact will be quite different for different aspects of the Earth system. From the agricultural or economic perspective, the key responses are changes in the production and quality of crops for food, fuel, and fiber, and changes in the areas where crop production is possible. Increased carbon dioxide levels may lead to substantial increases in crop production in some climate zones but not in others. Or it could lead to changes in the climatic zones suitable for growing particular crops.

From a hydrologic perspective, the key responses to carbon dioxide levels are related to plant water use: Decreases in plant water use potentially lead to increased runoff and increased water in rivers and reservoirs. On the other hand, water conservation from increased carbon dioxide levels tends to make the lower atmosphere warmer, because any decrease in the use of the Sun's energy for evaporating water (supplying latent heat) results in an increase in the energy that goes into heating the air. Thus, the effects of carbon dioxide on ecosystems potentially amplify greenhouse warming.

From the perspective of the global carbon cycle, increased carbon dioxide levels could lead to an increase in the total amount of carbon in plants and soils, offsetting some anthropogenic emissions and slowing the increase in atmospheric carbon dioxide. However, past increases in carbon dioxide may have already stimulated ecosystem carbon storage. If this capacity for storing carbon saturates in the future, an increased fraction of the anthropogenic carbon will remain in the atmosphere, and the increase in carbon dioxide levels will accelerate with time.

These impacts on diverse aspects of the Earth system can interact with each other and with other components of global change. For example, a decrease in plant water use stimulated by an increase in carbon dioxide levels could partially offset drought caused by climate change. Or, increased runoff could exacerbate flooding in areas already threatened by rising sea level.

THE GLOBAL CARBON CYCLE

On the time scale of tens of millions of years, the total pool of carbon in the atmosphere, oceans, and vegetation is controlled by geologic processes – the degradation of silicate minerals and the precipitation of calcium and magnesium carbonate in the oceans. Although the carbon pools in vegetation are not significant on these long time scales, vegetation and microbes can, largely through physically breaking and scraping at rock surfaces, play a critical role in regulating the rate of weathering of silicate minerals and hence the level of carbon dioxide in the atmosphere.

On a time scale of thousands of years, the oceans exercise the primary control on atmospheric carbon dioxide levels (see Chap. 19). Because the oceans contain – in the form of dissolved carbon dioxide and carbonate ions (CO_3^{--}) – over ten times more carbon than the atmosphere, vegetation, and soils combined (Fig. 20.3), small fractional changes in the

Figure 20.3. The global carbon cycle, showing the major stocks (numbers at the ends of the arrows, in gigatons) and fluxes (numbers in the middle of the arrows, in gigatons per year). This figure summarizes recent estimates of sources and sinks, with an ocean sink of approximately 2 gigatons, a terrestrial source of approximately 1 gigaton due to land use change, and a terrestrial sink of approximately 0.3 gigatons from a number of processes. All stocks and fluxes are in gigatons (1 Gt = 10^{15} g). (*Source:* Adapted from D. S. Schimel, *Global change biology*, 1995.)

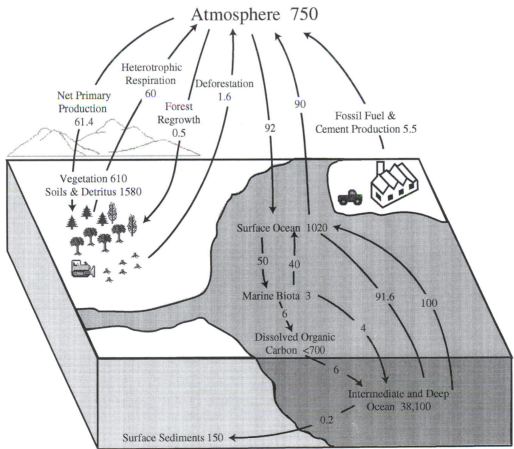

THE MISSING SINK

The global carbon budget is balanced; however, our understanding of it is not, quite. The "missing sink" refers to the fact that known inputs of atmospheric carbon sum to a quantity larger than the annual increment in the atmosphere, plus the known pathways of loss from the atmosphere. There must be one or more additional pathways through which carbon is leaving the atmosphere. Although we can identify candidates, we cannot yet assign quantities. What is missing is not so much the sink as our ability to quantify it.

Records on carbon injected into the atmosphere through direct human activities – primarily fossil fuel combustion and cement production – are quite accurate. Estimates extending as far back as the early days of the Industrial Revolution are reasonably precise, with an accuracy of approximately plus or minus 15 percent. Estimates of human-caused carbon emissions from forest clearing are much less accurate; the uncertainty is greater than plus or minus 50 percent. Deforestation constituted the dominant anthropogenic carbon flux to the atmosphere through the first half of the Industrial Revolution; however, currently it accounts for only approximately 20 percent of the total (Fig. 20.3).

Data on the annual change in atmospheric carbon levels are excellent. The data since 1958 are based on direct measurements, beginning with the pioneering work of C. D. Keeling at Mauna Loa, Hawaii (Fig. 20.2), with many monitoring sites added more recently. Because carbon dioxide is well mixed in the atmosphere, measurements at any single site can be used to quantify long-term changes in the atmospheric content. Data prior to 1958 are based on analysis of air trapped in ice. The oldest records are from deep cores drilled in Antarctica and Greenland (see Chap. 15).

Any time the carbon dioxide concentration in the atmosphere changes, carbon exchange with the ocean tends to moderate that change. Thus, when atmospheric carbon dioxide levels rise, the oceans tend to take up carbon dioxide and maintain equilibrium with the atmosphere. Precise calculation of ocean carbon uptake is a major task, requiring a complete specification of ocean currents, mixing across depths, and buffering by ocean carbonates. The critical information can, however, be obtained by calibrating a much simpler model, using a tracer to determine the value for a parameter specifying ocean carbon dioxide uptake as a function of the effective concentration difference between atmosphere and ocean. Radioactive carbon (^{14}C) produced during atmospheric tests of nuclear weapons is an excellent tracer because its radioactivity makes it relatively easy to quantify, and its brief but intense abundance in the atmosphere provides a suitable tracer signature. However, even with (^{14}C) calibration, the ocean models remain inaccurate to some degree; current ocean carbon dioxide uptake is known to plus or minus approximately 25 percent.

Currently, anthropogenic carbon in carbon dioxide input into the atmosphere is approximately 7.5 gigatons, (including 5.5 gigatons from fossil fuel primarily in the north temperate zone), 0.5 gigatons from cement production, and 1.5 gigatons from deforestation (now primarily in the tropics). The annual increment in the atmosphere varies from year to year but tended to be approximately 3.5 gigatons through the 1980s. Ocean uptake was approximately 2.0 gigatons during that period, leaving a sink of approximately 2 gigatons unaccounted for or missing. Since 1800, anthropogenic emissions from fossil fuels have been 150 to 190 gigatons. Deforestation adds 90 to 120 gigatons. The atmospheric increment has been approximately 150 gigatons. If ocean uptake is 40 to 80 gigatons, the cumulative missing sink is 10 to 120 gigatons.

It is likely that a number of processes contribute to the missing sink. One is the regrowth of temperate forests harvested in decades past. Another is transport of carbon from land to the oceans via rivers. Stimulation plant growth in nutrient-limited areas by the nutrients in acid rain and other pollution is likely to cause a significant sink in some areas. Also, ecosystem responses to gradually increasing levels of carbon dioxide may result in a sink, as discussed in this chapter. In terms of carbon dioxide, two aspects of this multiplicity of sinks are significant. First, some of the sink activity will tend to decrease with time. Regrowth of north temperate forests will slow as they become mature. Stimulation of carbon storage by nutrients deposited in pollution will also tend to slow as ecosystems become nutrient-saturated. Second, if only some of the current terrestrial sink is a consequence of the effects of increased carbon dioxide levels on plant production, then models that use the entire size of the missing sink to estimate the sensitivity of carbon storage to atmospheric carbon dioxide levels will tend to exaggerate ecosystem sensitivity to those levels. Studies such as the Jasper Ridge CO_2 Experiment (discussed later in this chapter) will be critical for quantifying the fraction of the missing sink specifically due to the effects of increased carbon dioxide levels on terrestrial ecosystems.

---

carbon content of the oceans represent large fractional changes in the atmosphere or ecosystems. Transfer of carbon dioxide from the atmosphere to the oceans varies with the concentration gradient between the atmosphere and the surface waters; consequently, ocean uptake tends to increase when the concentration in the atmosphere rises. However, because mixing throughout the oceans occurs on a time scale of thousands of years (Chap. 12), the surface waters transfer carbon to deep and intermediate waters relatively inefficiently. Thus, the short-term ability of the oceans to take up or release carbon dioxide is quite limited, whereas the long-term potential is vast.

Vegetation and soils are dominant players in the control of atmospheric carbon dioxide levels on a time scale of years to centuries. The total quantity of carbon in vegetation and soils is greater than that of the atmosphere by a factor of 2 to

3, and the annual fluxes of carbon into ecosystems are approximately one-sixth the total quantity in the atmosphere. The annual flux of carbon into and back out of ecosystems is approximately 120 gigatons; however, the uncertainty regarding this figure is significant – plus or minus approximately 20 percent. In the mid-1990, the annual flux of carbon released into the atmosphere through fossil fuel combustion was approximately 5.5 gigatons. Other anthropogenic emissions include 0.5 gigatons through cement production and approximately 1.5 gigatons from deforestation, which currently occurs primarily in the tropics. The total anthropogenic emissions of approximately 7.5 gigatons represents a little more than 5 percent of the total carbon fixed into carbohydrate through terrestrial photosynthesis. Of this total, also called *gross primary production*, approximately 50 percent is released through plant respiration and 50 percent is converted into plant growth or net primary production, making total anthropogenic carbon emissions a little more than 10 percent of net primary production (Fig. 20.3).

Of the approximately 7.5 gigatons of carbon released annually through human activities, approximately 50 percent remains in the atmosphere, causing an annual increase of approximately 0.5 percent of the current total of 750 gigatons, or approximately 1.5 parts per million of carbon dioxide. Approximately 4 gigatons of the annual release of anthropogenic carbon does not appear in the atmosphere. The best estimate from ocean models is that ocean carbon uptake is approximately 2 gigatons per year, leaving an additional 2 gigatons, which are commonly characterized as "missing" because their exact fate is not known (see the box entitled *The Missing Sink*).

DIRECT AND INDIRECT COMPONENTS OF ECOSYSTEM RESPONSES TO INCREASED CARBON DIOXIDE LEVELS

An ecosystem consists of all the organisms that occupy an area, plus the soil, the water, and the atmosphere. Many studies address the responses of plants in pots to increased carbon dioxide levels; however, the responses of complete ecosystems are often difficult to predict from plant responses. For example, treatments that increase growth in experiments with single plants might have no effect in an ecosystem where limited nutrient availability or herbivores restrict plant growth. Alternatively, a treatment that has no effect on the growth of most species in single-plant experiments can lead to large changes in ecosystem production if one or a few sensitive species dramatically increase or decrease in abundance.

The difficulty of predicting ecosystem responses from single-plant experiments is even greater for processes other than growth. For example, a treatment that initially stimulates plant growth but decreases the concentration of nitrogen in leaf tissues could decrease plant growth in the longer term, if the plant tissue with the lower nitrogen is less palatable to microbial decomposers and therefore releases only slowly the nutrients required for plant growth in succeeding years.

Most potentially significant responses to carbon dioxide increases at the ecosystem scale – changes in agricultural production, hydrology, biome type, carbon storage, and climate – will reflect the outcome of a number of mechanisms acting in concert. Some of these mechanisms are direct consequences of the effect of carbon dioxide increases on plant growth or water use. Others are indirect. Some indirect mechanisms involve the effects of increased carbon dioxide levels on other groups or organisms (e.g., herbivores, pathogens, or decomposing microbes in the soil) or other resources (e.g., soil moisture or mineral nutrients). An understanding of the mechanisms underlying the major direct and indirect effects of increased carbon dioxide levels is an important prerequisite for interpreting responses to carbon dioxide at the ecosystem scale.

DIRECT EFFECTS OF INCREASED CARBON DIOXIDE LEVELS. Photosynthesis, the light-driven series of reactions that plants use to convert carbon dioxide into carbohydrates, is the engine driving plant growth. Approximately 90 percent of the world's plant species use the C_3 (see the Glossary at the end of this book) photosynthesis pathway (Fig. 20.4); among these species, photosynthesis is not saturated with carbon dioxide at levels characteristic of the current atmosphere. Doubling the carbon dioxide concentration (to approximately 700 parts per million) around leaves of these plants usually increases photosynthesis by 50 percent or more, with larger stimulations at higher temperatures (Fig. 20.5).

Plants with the C_4 (see the Glossary at the end of this book) or crassulacean acid metabolism (CAM) photosynthesis pathways (Fig. 20.4) typically are saturated at current ambient carbon dioxide levels, and photosynthesis is insensitive to higher concentrations. Although these species constitute a small minority of the world's plants, the C_4 group contains a number of very important agricultural crops, including corn, sugar cane, and millet. According to one recent estimate, C_4 plants, including crops and wild species, contribute 21 percent of annual, global photosynthesis. Although photosynthesis in C_4 plants commonly does not increase when levels of carbon dioxide are elevated, the studies to date document a variety of growth responses, including some instances of substantial increases. This diversity of response emphasizes the potential importance of the effects of increased carbon dioxide levels beyond any influence on the biochemistry of carbon dioxide uptake.

Regardless of the photosynthetic pathway, increased carbon dioxide partially closes the stomata (pores) in the leaf surface of many plants, decreasing plant water loss (Figs. 20.5 and 20.6). This effect occurs in essentially all the nonwoody plants studied to date, and in many but not all trees. The physiologic role of stomates is to establish an acceptable compromise between allowing carbon dioxide to diffuse into leaves and preventing excess water vapor from diffusing out. Partial closure of the stomates decreases the movement of both carbon dioxide and water vapor. In the case of carbon

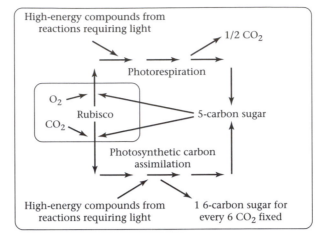

(A)

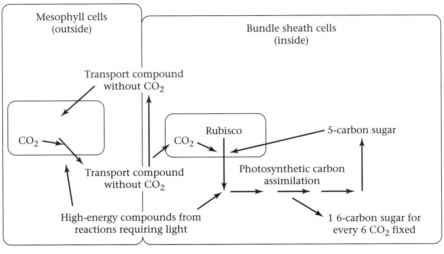

(B)

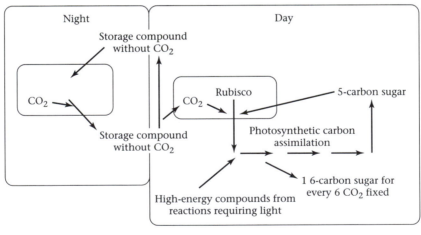

(C)

Figure 20.4. The biochemistry of photosynthesis in plants with (A) the C_3 photosynthesis pathway, (B) the C_4 photosynthesis pathway, and (C) the CAM photosynthesis pathway. In C_3 plants, rubisco catalyzes the combination of carbon dioxide and ribulose-1,5-bisphosphate as the first step in carbohydrate formation. Rubisco also catalyzes the combination of oxygen and ribulose-1,5-bisphosphate in an energy-consuming pathway called photorespiration. In C_4 plants, the reactions in the mesophyll cells, which surround the bundle sheath cells, serve to provide rubisco with a high concentration of carbon dioxide and a low concentration of oxygen. CAM photosynthesis uses the same biochemistry as C_4 photosynthesis; however, all reactions occur in the same cells. In CAM photosynthesis, the stomates open at night.

dioxide, however, the increase in concentration compensates for the decrease in stomatal aperture, and the total absorption of carbon dioxide by leaves typically increases. Therefore, the partial closing of the stomates in response to elevated carbon dioxide levels makes ecologic sense. When carbon dioxide levels increase, plants can satisfy their need for carbon dioxide with the expenditure of less water, allowing an increase in growth per unit of water evaporated from leaves, a ratio commonly called *water use efficiency*. Especially in water-limited habitats, water use efficiency can be a key factor in plant growth and other aspects of ecologic success.

INDIRECT EFFECTS AT THE PLANT LEVEL. Plants grown for weeks or months in elevated levels of carbon dioxide typically had higher rates of photosynthesis than did plants grown in ambient levels of carbon dioxide; however, the difference is often smaller than it was after the onset of elevated carbon dioxide levels. Many C_3 species respond to extended periods of increased carbon dioxide levels by decreasing the biochemical capacity for photosynthesis, especially the concentration of rubisco, the enzyme enabling the initial fixation of carbon dioxide into organic compounds (Fig. 20.4). The most common pattern is that, following this acclimation of photosynthetic capacity, photosynthesis at two times the current ambient level of carbon dioxide (approximately 700 parts per million) increases by 20 to 40 percent instead of the initial 50 to 70 percent (Fig. 20.5). The cause of this acclimation is not well known. Two possibilities are consistent with much of the available data. One is that because available nitrogen, a critical raw material for enzymes such as rubisco, is scarce in the soil, increased carbon dioxide levels lead to lower enzyme levels by spreading a nearly fixed supply of enzymes over a larger plant. Another is that evolution has favored plants that respond to an increase in the abundance of limiting resource (carbon dioxide) by shifting investments in capturing this resource (e.g., nitrogen in rubisco) toward increasing the capture of other scarce resources such as water or available nitrogen.

Plants grown under elevated levels of carbon dioxide typically produce tissues with decreased concentrations of essential mineral elements, especially nitrogen. Nitrogen availability limits plant growth in many ecosystems, and increased growth per unit of nitrogen used (or nitrogen use efficiency) provides an option for growth stimulation by carbon dioxide even in nitrogen-limited sites. The effects of increased carbon dioxide levels on the requirements for other limiting nutrients, of which phosphorus is the most important globally, are less clear. A decrease in tissue nitrogen has little or no impact on the quality of a fiber crop such as cotton; however it can be a major determinant of quality and economic returns in food crops.

Under elevated carbon dioxide levels, the decrease in water loss potentially allows plants to increase growth per unit of precipitation or water added through irrigation. Because water is, globally, one of the environmental factors most limiting to plant growth, an increase in growth per unit of water (or water use efficiency) could lead to increased production in dry areas, perhaps facilitating the establishment of crops and wild species in areas now too dry to support them. Of course, changes in precipitation could either amplify or oppose range changes, depending on altered water use efficiency.

Based on results from a large number of experiments with single plants in pots, stimulation of plant growth in response to increased carbon dioxide levels is generally greatest in species with the highest inherent growth rates. Inherently low-growth potentials, hypothesized by many researchers to help plants cope with stressful environments, appear to restrict sensitivity to carbon dioxide. The low sensitivity of slow-growing plants to elevated levels of carbon dioxide is important because slow-growing plants dominate most of the world's natural ecosystems. Because most of the experimental studies to date have focused on agricultural crops and other fast-growing plants, a link between growth potential and carbon dioxide increases the difficulty of extrapolating from experiments to nature.

INDIRECT EFFECTS AT THE ECOSYSTEM LEVEL. In many natural ecosystems, plant growth is strongly limited by the availability of water or mineral nutrients. In other words, growth continues until all the available nutrients or water are used, and then it stops. With this kind of concrete limitation, only two alternatives can increase growth: Either increase the efficiency of resource use or increase the availability of the limiting resource or resources. The effects of increased carbon dioxide levels on the plant use efficiency of nitrogen and water are discussed above. What are the effects of increased carbon dioxide on the total availability of these potentially limiting resources?

Water enters most ecosystems as precipitation or "run-on" (as from a flooding river) and leaves as runoff and evaporation from plants and soil. A decrease in evaporation through plants could increase runoff, potentially increasing the release of water into streams and reservoirs. Of course, any increases in leaf area will tend to counteract the water savings from decreased water use per unit of leaf area. Decreased water use by plants grown under elevated carbon dioxide levels could also prolong the period of the year when soil moisture is sufficient to support plant growth. In at least one experimental study, the most significant impact of carbon dioxide increases on plant growth was caused by the delayed onset of the late summer drought.

In unfertilized ecosystems, most of the nutrients required for plant growth come from the decomposition of dead plant and animal matter. Typically, the rate of decomposition is an important control on plant growth, because plants cannot grow faster than nutrients become available. As a general rule, the rate of decomposition of dead organic matter depends on nitrogen concentration. Because the decomposing organisms in the soil (principally bacteria, fungi, and nematodes) commonly have higher nitrogen requirements than plants, they grow fastest and process the most organic matter

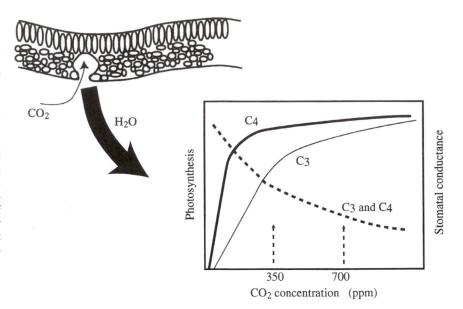

Figure 20.5. A schematic cross section of a leaf with the C_3 photosynthesis pathway and the response of single-leaf photosynthesis (solid lines) and stomatal conductance (dashed line) to carbon dioxide concentration in plants with the C_3 (thin line) and C_4 (thick line) photosynthesis pathways. Because the concentration gradient driving transpiration is much larger than that which drives photosynthesis, the transpiration rate is always much greater than the photosynthesis rate. At current ambient levels of carbon dioxide, photosynthesis is nearly saturated with carbon dioxide in C_4 but not in C_3 plants. ppm = parts per million.

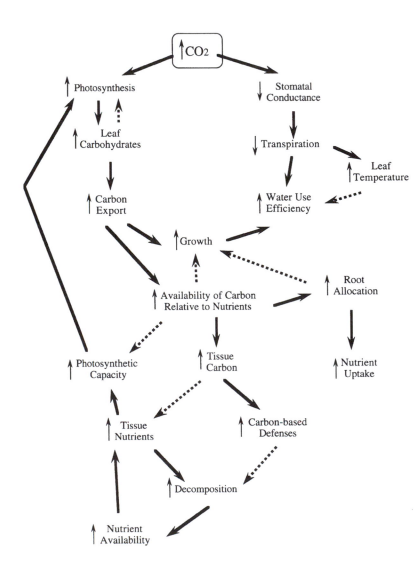

Figure 20.6. Summary of direct and indirect ecosystem responses to increased levels of carbon dioxide. Solid arrows indicate a positive effect and dashed arrows indicate a negative effect. The network of positive and negative effects makes it difficult to predict changes in any component, without experimental studies.

when the organic matter is rich in nitrogen. This implies that organic matter from plants grown under elevated levels of carbon dioxide, which tend to have lower nitrogen contents, may decompose more slowly, decreasing the rate at which nutrients become available for plant growth. This effect could counteract the tendency of elevated carbon dioxide levels to increase plant growth.

Although these indirect effects on nutrient availability are potentially important, they are quite difficult to study experimentally. The quantity of dead organic matter in the soil is often much greater than the total mass of living plants, and the time required to observe the effects of increased carbon dioxide levels on the nitrogen content of the soil organic matter could be many years, except in environments where the pool of soil organic matter is unusually low.

Plant species differ in responsiveness to increased carbon dioxide levels. Differences related to maximum growth potential were discussed earlier. Researchers have recorded higher rates of growth in plants with the C_3 photosynthetic pathway than in those with the C_4 pathway. Although some experiments suggest that trees may be more sensitive than herbaceous plants to increased carbon dioxide levels, other experiments suggest the opposite. Plants also differ in their response to the indirect consequences of increased carbon dioxide level, such as altered soil moisture and nutrient concentrations. If increased carbon dioxide leads to changes in the frequency of fires or in the abundance of particular pests or pathogens, this too could favor some species over others.

Changes in the plant species growing on a site often have large impacts on ecosystem processes. For example, conversion from a grassland to a forest leads not only to increased above-ground carbon storage in biomass but also to decreased nutrient availability related to the slow turnover of nutrients used to construct woody tissues. This kind of conversion typically increases plant access to soil moisture, as a result of the tree's deeper roots. Shrubs invading grasslands may develop islands of increased nutrient availability, and grasses invading woodlands may increase susceptibility to wildfire. Successful establishment of a plant species that is a partner in a symbiosis with a bacterium that fixes atmospheric nitrogen into plant-available form can dramatically change nutrient relations. Few experiments have spanned the amount of time necessary to document this response. Yet, it is likely that, at the ecosystem scale, effects of increased carbon dioxide levels on species composition will be among the most influential factors to alter the budgets of carbon water and nutrients. Because human use of ecosystems is often strongly determined by the plant species present, species changes resulting from increased carbon dioxide levels are, in turn, likely to have major impacts on human uses of ecosystem products – ranging from fuel and fiber to food and medicines.

APPROACHES TO PREDICTING ECOSYSTEM RESPONSES TO INCREASED CARBON DIOXIDE LEVELS

Results from the few ecosystem-scale carbon dioxide experiments to date are diverse. Photosynthesis, plant growth, and the total quantity of carbon in plants and soil all increase in salt marsh vegetation dominated by plants with the C_3 photosynthesis pathway, but not in plots dominated by C_4 plants. Arctic tundra responds to increased carbon dioxide with an increase in carbon uptake and sequestration; however, this effect is transient, disappearing within a few years. In Arctic tundra exposed to a combination of increased carbon dioxide and increased temperature, however, increased carbon sequestration persisted throughout a three-year study. In tallgrass prairie dominated by C_4 plants, elevated carbon dioxide levels lead to increased plant growth in dry but not in wet years.

Based on these limited ecosystems results, plus results from the thousands of studies on responses of single plants to increased levels of carbon dioxide, it is still very difficult to generalize across climate zones and ecosystem types, especially over extended time periods. Several hypotheses have been proposed, some of which have been formalized in mathematic models. Two hypotheses are consistent with much of the experimental evidence.

1. Responses of plant production and carbon storage to increased carbon dioxide levels will be most significant in sites with abundant nutrients, where plants tend to have high growth potential and nutrient limitation is unlikely to constrain responses to carbon dioxide.
2. Responses of plant production to increased carbon dioxide levels will be significant in sites with limited water availability, based on the role of increased carbon dioxide levels in increasing growth per unit of water transpired.

To test these hypotheses and to separate the role of resources from the role of species, we are working with colleagues on experimental studies on ecosystems selected to allow reasonable access to the full array of direct and indirect mechanisms that potentially play an important role in responses to carbon dioxide at the ecosystem scale.

THE JASPER RIDGE CO$_2$ EXPERIMENT

Experimental ecosystem studies ideally should provide access to the entire range of relevant processes, on a politically acceptable time frame. It is important for the experiments to emphasize underlying mechanisms, inasmuch as these are the elements most likely to be generalizable across ecosystems. It is also important for the experiments to be integrated, allowing the investigators to explore relationships among mechanisms that influence different parts of the ecosystem on different time scales. Finally, ecosystem studies should be comparative, providing a framework for testing the generality of the important mechanisms on a single site.

The Jasper Ridge CO_2 Experiment, initiated in early 1992, attempts to identify and quantify general principles that regulate ecosystem responses to carbon dioxide. We chose to study the Jasper Ridge annual grasslands because they have a number of features that increase the prospects for results that are mechanistic, integrated, and comparative. Some of these features also make them somewhat unrepresentative of other terrestrial ecosystems or even other grasslands. Our working hypothesis, however, is that the underlying principles we uncover by studying and manipulating this ecosystem will provide a sound basis for predicting the response of other ecosystems to increasing levels of atmospheric carbon dioxide. The experimental sites are ridgetop grasslands within the Jasper Ridge Biological Preserve of Stanford University (Fig. 20.7). The area has been a site for ecologic studies for many years, including significant studies of population structure in insects, landscape dynamics, plant–animal interactions, plant carbon balance, and remote sensing.

We view the annual grasslands at Jasper Ridge as a model ecosystem for exploring the effects of resource availability, species characteristics, and community composition in controlling ecosystem responses to increasing carbon dioxide levels. Model systems have long played a central role in biologic research. The fruit fly, *Drosophila melanogaster*, remains a dominant subject of experiments in population and developmental genetics. The gut bacterium, *Escherichia coli*, is the standard for research in biochemistry and molecular genetics. Recently, a small weed in the mustard family, *Arabidopsis thaliana*, has been used extensively in studies of plant bio-

chemistry molecular biology. Key characteristics of these and other experimental model systems are that they operate with very general mechanisms, and that experiments based on them are efficient in terms of space, time, and expense. Typically, the utility of a model is dramatically increased by the presence of an unusual feature (e.g., the polytene chromosomes in *Drosophila*), or through the absence of a major complication (e.g., the absence of intracellular compartmentation and multiple copies of the genome in *E. coli*). Each new discovery on a given model system expands its potential for use as a foundation in new types of experiments.

Model systems have not been widely explored in ecology, probably because the objectives of most studies have been to find the unique aspects of a species, community, or ecosystem, or to characterize broad patterns of diversity or function. Ecosystem studies have tended to emphasize "representative" ecosystems; that is, examples of ecosystem types that are widespread or economically important, rather than model ecosystems. With the emergence of global ecology and global change research, the community of researchers must address new needs. One is to understand the extent to which general features of organism and ecosystem function allow global extrapolations, as well as the limits to useful extrapolation. A second need is to design experiments in which it is feasible to subject complete functional units, including complete ecosystems, to a range of novel climates, atmospheres, and disturbance regimes. Similar to the manner in which *D. melanogaster* and *E. coli* have been useful as experimental models in other fields, California annual grass-

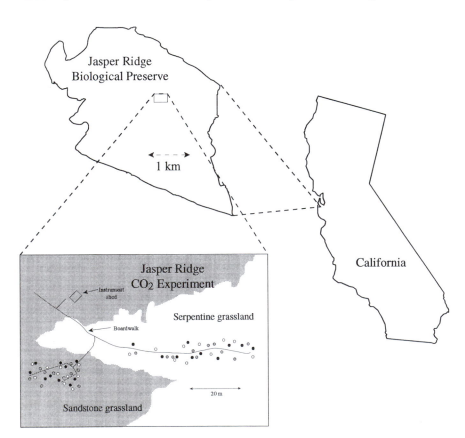

Figure 20.7. Map of the field component of the Jasper Ridge CO_2 Experiment, showing its location in the Jasper Ridge Biological Preserve, and the location of Jasper Ridge in California. The 1200-hectare Jasper Ridge Biological Preserve contains areas with most of the ecosystems from this region. The field plots for the Jasper Ridge CO_2 Experiment span the boundary between sandstone and serpentine grassland. Plots with high carbon dioxide chambers (Hi) are shown as solid black, plots with chambers but ambient (normal atmospheric levels) carbon dioxide (Lo) are shown as gray circles, and no-chamber control plots (X) are shown as open circles. km = kilometer; m = meter.

lands have several features that make them useful as a model system, even though they are not important in terms of global extent.

JASPER RIDGE ANNUAL GRASSLANDS. Jasper Ridge is located on the San Francisco Peninsula in the foothills that flank the outer Coast Ranges and include the San Andreas fault zone. The high degree of geologic activity has resulted in diverse topography, soils, vegetation, and flora. Some of the minor faults within the preserve have caused sharp boundaries between very distinct soil types (Figs. 20.7 and 20.8). These contacts provide a natural setting for cross-system comparisons with constant climate but contrasting substrate and vegetation. One of the most striking examples is the contrast between shallow, nutrient-poor, serpentine-derived soils supporting native forbs and grasses and more fertile sandstone soils dominated by Eurasian annual grasses (Fig. 20.7).

The Jasper Ridge climate is mediterranean, with cool, wet winters and very dry summers (Fig. 20.9). Year-to-year variation in total precipitation is high. Over the past twenty years, annual precipitation averaged 582 millimeters and ranged from less than 200 millimeters to more than 1200 millimeters. During the past four years, it varied from less than 500 to nearly 1200 millimeters. Year-to-year climate differences can provide another axis of environmental variation, allowing us to add rainfall and water budget to the list of "manipulated" factors.

The vegetation of both the serpentine and sandstone grasslands is dominated by annuals, which germinate in the fall and have life-spans ranging from a few to fifteen months. The annual life-form provides three major advantages in terms of experimental tractability. First, researchers need not be concerned about the condition of the plants prior to initiation of the experiment. Second, all stages of the life cycle can be studied, including reproduction, which is an essential control on long-term dynamics. Third, we can, in only a few years, track changes in species composition. This is important because much of the variation in plant properties that affect ecosystem processes results from species differences.

The potential importance of changes in plant community composition is highlighted by the history of California's grasslands. Prior to European settlement and the introduction of domestic grazing stock, the more fertile grasslands likely were dominated by native perennial bunchgrasses, which were then outcompeted by Eurasian annual grasses. This replacement has been reversed in some areas removed from grazing. It is possible that climatic factors play a role in limiting the current success of the perennial bunchgrasses and that increasing carbon dioxide levels may alter the competitive balance between native and exotic species.

Although both the serpentine and sandstone grasslands are dominated by annuals, the diversity of plant properties such as phenology (developmental timing) is still quite large. Both grasslands include perennial species, as well as annuals that vary in flowering time from February to September. This

Figure 20.8. Part of the field component of the Jasper Ridge CO_2 Experiment. The foreground chambers (in the area with wildflowers) are on serpentine soil. The background chambers are on sandstone soil. Individual open-top chambers are approximately 0.65 meter in diameter and 1 meter tall. See color section.

variation in flowering date is paralleled by differences in rooting depth and other properties that influence ecosystem processes. In addition, both grasslands contain legumes with symbiotic nitrogen fixation, as well as species with and without symbiotic associations with fungi called mycorrhizas, which facilitate nutrient uptake. This range of functional diversity is small relative to the variation across biomes; however, the Jasper Ridge grasslands provide sufficient variation to assess whether significant differences in functional properties have implications for ecosystem responses to elevated levels of carbon dioxide.

Because the Jasper Ridge grassland species are small, even very small parcels of ecosystem include many individuals

Figure 20.9. Mean climate at Jasper Ridge, showing temperature (dashed line) and rainfall (solid line). As is typical in mediterranean-type climates, Jasper Ridge receives almost no rainfall in the summer months. mm = millimeters. (*Source:* Data from C. B. Field, F. S. Chapin, III, N. R. Chiariello, E. A. Holland, and H. A. Mooney, The Jasper Ridge CO_2 experiment: design and motivation, in G. W. Koch and H. A. Mooney (eds.), *Carbon dioxide and terrestrial ecosystems*, Academic Press, 1996, pp. 121–145.)

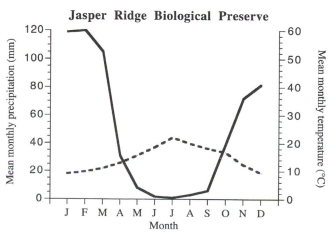

Jasper Ridge Biological Preserve

and a substantial number of species. A field chamber 0.65 meters in diameter contains 1000 to 3000 individual higher plants on serpentine soil and 300 to 1000 plants on sandstone soil. With small plants, individual experimental units can be small, making it possible to design experiments with large numbers of treatments and/or replicates.

EXPERIMENTS AND RESULTS

The general goal of the Jasper Ridge CO_2 Experiment is to link measurements on many individual processes with ecosystem-scale studies under field conditions that are as natural as possible. We accomplish this by employing a combination of field and microcosm studies. The microcosms are artificial ecosystems in which we control soil type, species abundances, and nutrient and water availability. The level of control in the microcosm experiments is critical for extending the range of manipulated variables and for quantifying subtle responses, which tend to be obscured by the natural heterogeneity in the field. Microcosms also provide much greater access to below-ground processes.

FIELD EXPERIMENTS. The field experiments serve as the primary site for testing some hypotheses and as a reference point for interpreting other experiments. The experiments focus on thirty circular study plots, each 0.65 meter in diameter, on both serpentine and sandstone substrates (Figs. 20.7 and 20.8). There are ten replicates of three treatments: "Hi" represents chambers plus carbon dioxide, "Lo" represents chambers without carbon dioxide, and "X" represents no chambers. The individual chambers are polyethylene chimneys with the bottom attached to the soil surface and the top open. A blower attached to each maintains a continuous flow of air through the chamber. The chambers with elevated carbon dioxide levels receive ambient air enriched with an additional 360 parts per million of carbon dioxide. The chambers impose effects on temperature, light, humidity, and access by pollinators and herbivores. The best test for the effects of carbon dioxide is to compare responses among the Hi and Lo plots, where both have the same kind of chamber; the increased carbon dioxide levels in some should be the only difference. By comparing responses in the Lo and X plots, we can quantify the impact of the chambers and use this information to help clarify predictions about future changes in the grassland.

A technology called free air CO_2 exchange (FACE) enables researchers to provide controlled levels of increased carbon dioxide without enclosing plants in a chamber. However, the large scale of the FACE facilities makes it difficult to explore responses to a large range of treatments. FACE experiments in other grassland sites will be invaluable for relating our chamber results to the real world.

As in most other experiments with increased carbon dioxide levels, we find increased photosynthetic carbon dioxide uptake in the dominant species in the plots with increased carbon dioxide. The increase varies with time of the

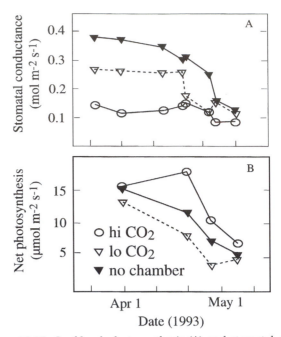

Figure 20.10. Leaf-level photosynthesis (A) and stomatal conductance (B) in *Avena barbata*, the dominant species in the sandstone field plots. Open circles represent data for plants in plots with approximately two times the normal levels of carbon dioxide. Open triangles represent data from plots with chambers at ambient levels of carbon dioxide. Closed triangles represent data from plots with no chambers. The effects of carbon dioxide on photosynthesis are proportionally more significant later in the year, when the effects on stomatal conductance are proportionally less significant. μmol m^{-2} s^{-1} = micromoles per square meter per second; mol m^{-2} s^{-1} = moles per square meter per second. (*Source:* Data from R. B. Jackson, O. E. Sala, C. B. Field, and H. A. Mooney, CO_2 alters water use carbon gain, and yield in a natural grassland, *Oecologia*, vol. 98, 1994, pp. 257–262.)

year. During the period of most active growth, the increase is greater than 50 percent (Fig. 20.10). A small downregulation of photosynthetic capacity occurs; however, it is not large enough to counterbalance the effects of the greater substrate availability (Fig. 20.4). These relatively large effects of increased carbon dioxide levels on photosynthesis do not generally translate to large effects on growth. In two of the first three years of results from the sandstone (moderate fertility) plots, doubled carbon dioxide levels had no effect on the biomass of the annual grasses that dominate the plots. In the third year, which was relatively dry, biomass of the annual grasses in the plots with high levels of carbon dioxide was approximately 40 percent greater than in the plots receiving ambient carbon dioxide levels. The pattern on the serpentine plots is different, with small, statistically insignificant, effects of increased carbon dioxide on the biomass of early-season annuals in each of the first three years.

What prevents the large effects of carbon dioxide on photosynthesis from appearing as large effects on the growth of the early-season species? We do not yet have a definite answer. The list of possible explanations includes intrinsically low growth capacity in some of the Jasper Ridge species,

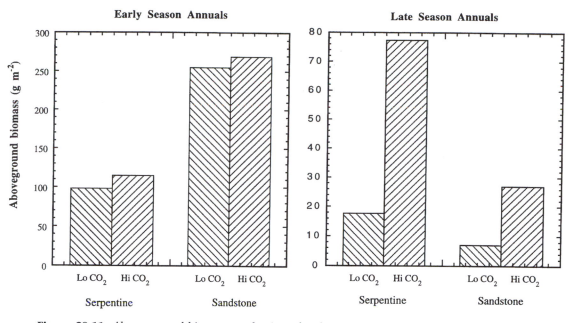

Figure 20.11. Above-ground biomass production of early season annuals and late season annuals in the field plots. The data are mean production over the first three years of the experiment. g m^{-2} = grams per square meter. (*Source:* Data from C. B. Field, F. S. Chapin, III, N. R. Chiariello, E. A. Holland, and H. A. Mooney, in G. W. Koch and H. A. Mooney (eds.), *Carbon dioxide and terrestrial ecosystems*, Academic Press, 1996.)

growth restriction due to limited nutrient availability, and growth restriction resulting from low allocation of new biomass to leaf production.

Though increased CO_2 has no or modest effects on the growth of the early-season species, its effect on growth of the late-season annuals is quite dramatic (Fig. 20.11). Especially in the serpentine grassland, increased carbon dioxide changes the late-season annuals from a minor to a major component of the ecosystem, perhaps precipitating a major change in the structure and function of this ecosystem. The

dramatic growth response of the late annuals in ecosystems exposed to increased carbon dioxide levels does not appear to be a consequence of unusual carbon dioxide sensitivity of these species. Instead, it is an indirect effect of decreased stomatal conductance in the other species (Figs. 20.6 and 20.10). Decreased stomatal conductance in the early-season species results in decreased water loss by transpiration, which leads to increased soil moisture, especially at the end of the activity period for the early-season annuals (Fig. 20.12). Because the length of the growing season for these species is

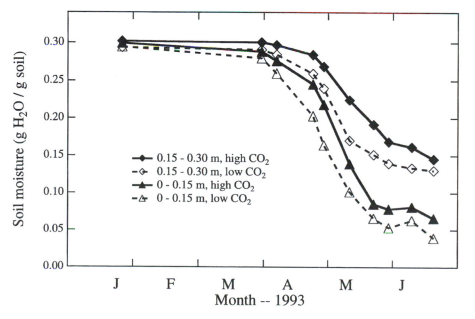

Figure 20.12. Seasonal course of soil moisture in the sandstone field plots in 1993. Late in the growing season, moisture is consistently higher in the plots receiving elevated (Hi-solid lines) than in the plots with chambers but ambient (Lo-dashed lines). g H_2O/g = grams of water per grams of soil; m = meters. (*Source:* Data from C. B. Field, F. S. Chapin, III, N. R. Chiariello, E. A. Holland, and H. A. Mooney, The Jasper Ridge CO_2 experiment: design and motivation, in G. W. Koch and H. A. Mooney (eds.), *Carbon dioxide and terrestrial ecosystems*, Academic Press, 1996, pp. 121–145.)

regulated by day length, they are unable to take advantage of water available late in their growing season. This additional water is, however, available at a critical time in the life cycle of the late-season annuals. These plants survive to reproduce only if their roots reach water deep in the soil profile; by extending the time when water is available shallow in the soil profile, increased carbon dioxide levels increase the opportunity for roots to reach the deep moisture reserves.

Increased soil moisture in the Hi plots has other consequences as well. It appears to stimulate soil respiration, the process through which soil microbes consume dead organic matter, releasing carbon dioxide. This increase in soil respiration tends to counteract the tendency of elevated carbon dioxide levels to cause major changes in carbon storage in the soil. However, increased soil respiration also increases the rate of nutrient release, possibly allowing greater nutrient uptake by plants.

Overall, the Jasper Ridge grasslands are very sensitive to elevated carbon dioxide levels; however, much of this sensitivity is caused by mechanisms not necessarily associated with increased photosynthesis. Total plant growth increases by up to 50 percent; however, most of this increase is in the late annuals, which are, from a grazing or an aesthetic perspective, the most undesirable members of the plant community. It is still too early to say whether the increased importance of the late annuals will lead to the eventual elimination of the early-season species – that does remain a possibility. The increased soil moisture so important in facilitating the success of the late-season annuals also acts to limit the potential for these grasslands to respond to elevated carbon dioxide levels with increased storage of carbon in the soil.

MICROCOSM EXPERIMENTS. The microcosms or *micro-ecosystems for climate change analysis* (MECCAs) are the focal point of experiments designed to quantify particular components of ecosystem responses and to test hypotheses concerning the roles of species characteristics and resources. The individual units are sufficiently small that factorial experiments can be performed with many combinations of carbon dioxide, nutrients, soils, species, and water. With these small units we can, for example, look at sufficient numbers of species to test the hypothesis that responsiveness of monocultures to carbon dioxide is related to maximum growth potential, or that the most responsive ecosystems are those with limited water but abundant nutrients, or that the abundance of nitrogen fixers is likely to increase in response to elevated levels of carbon dioxide.

The MECCA facility consists of twenty enclosures that contain microcosms with a soil column 0.95 meter deep. The soil column contains an artificial sandstone or serpentine soil profile (Fig. 20.13). Some of the microcosms are used for one-year experiments that focus on the initial effects of variation in resources or species characteristics and on the mechanisms driving that variation. Others are used for long-term treatments with single plant species, in experiments focusing on

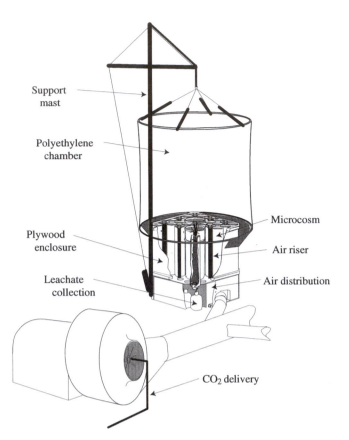

Figure 20.13. Schematic of a microcosm unit, each of which houses approximately thirty microcosms with a number of experimental treatments. These microcosm units are located on the Jasper Ridge Biological Preserve, approximately 1 kilometer from the field plots.

the cumulative ecosystem effects of changes in plant growth, tissue chemistry, and plant water use. Finally, some microcosms are used for long-term community experiments, focusing on the impacts of elevated carbon dioxide levels on plant community composition and on the implications of plant community structure for ecosystem processes.

Results from the microcosm experiments generally parallel results in the field. The sensitivity of plant growth to elevated levels of carbon dioxide is low in most species, especially at low nutrient levels. Yet, when nutrients are abundant, the growth responses of some species are dramatic, with doubled carbon dioxide leading to growth increases greater than 100 percent (Fig. 20.14). In general, stimulation of plant growth caused by elevated carbon dioxide levels is greatest in the plants that grow most rapidly, independent of whether the rapid growth is an intrinsic species characteristic or a response to increased nutrients. The one exception to this response involves water. In the Jasper Ridge climate, increased water stimulates plant growth; however, sensitivity to carbon dioxide is greatest when water, being in short supply, represents the critical limit to growth.

Increased dominance by late annuals is sufficient in community MECCAs to drive increased plant growth in response to increased carbon dioxide levels. Overall, the MECCA results also support the interpretation that in this ecosystem, many of the significant consequences of increased carbon dioxide reflect indirect effects mediated through changes in the water budget, not effects resulting primarily from changes in photosynthesis.

If we had focused our experiments at the scale of the individual plant, rather than the scale of the complete ecosystem, we could have observed many of the initial effects of increased carbon dioxide. Yet, our ability to assess the importance of each effect and the consequences of combining them would have been limited. Some responses to carbon dioxide – for example, changes in species composition – are impossible to observe in studies on single plants. We could observe decreased stomatal conductance in single-plant studies; however, it would be difficult to extend this observation to increased soil moisture, increased soil respiration, increased nutrient release, or increased growth of late-season annuals. This experiment provides a number of spectacular examples of important indirect effects, where the outcome at the ecosystem scale is separated from the physiologic mechanism by a number of steps. Although we are still developing the tools to predict the outcome of these multistep mechanisms in other ecosystems, we can make an increasing range of predictions with reasonable confidence based on the results of these experiments.

Figure 20.14. Aboveground biomass production in the dominant species in the sandstone grassland (*Avena barbata*) in a microcosm experiment, performed in 1994, testing effects of added nutrients and elevated carbon dioxide levels, both alone and together. Whereas, either doubled carbon dioxide or a heavy fertilizer application increased growth by about approximately 20 percent, the two in combination increased growth approximately 2.5-fold. g per MECCA = grams per micro-ecosystems for climate change analysis.

Jasper Ridge CO₂ Experiment - Avena barbata (1994)

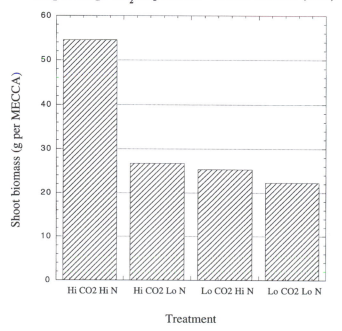

THE GLOBAL SCALE

The Jasper Ridge CO$_2$ Experiment supports the hypothesis that responses to increased levels of carbon dioxide are significant in species with the intrinsic capacity to grow rapidly and in settings where mineral nutrients, especially nitrogen, are abundant. This combination – rapid growth potential and high nutrient availability – is characteristic of managed agriculture, especially in developed regions where farmers have access to fertilizer. Thus, our results are consistent with relatively significant effects of elevated carbon dioxide levels on agricultural plant production. When fixed carbon is the desired product, increased carbon dioxide levels may lead to increased crop value. Agricultural production is always, however, the balance between enhancing the growth of selected species and suppressing the growth of weeds as well as losses from pests. Our experience with increased late annuals argues for careful attention to weed success in response to elevated levels of carbon dioxide.

Because fast-growing species with abundant nutrients respond significantly to increased carbon dioxide levels, we can infer that slow-growing plants or plants growing in nutrient-limited sites will respond to these levels less dramatically. These responses are characteristic of most of the world's natural ecosystems, including forest and the high-latitude ecosystems with large carbon stores in soils and peat. In these ecosystems, seemingly insignificant responses of plant growth to elevated carbon dioxide levels will tend to restrict future increases in carbon storage, especially if nutrients become even more limiting. The ultimate capacity for carbon storage is set by the balance between carbon inputs through photosynthesis and carbon losses through respiration by plants and heterotrophs; therefore, it is difficult to predict the resultant total carbon fixation based on the sensitivity of plant growth to increased levels of carbon dioxide. In areas where increased soil moisture increases rates of respiration by heterotrophs, the response of soil carbon is likely to be small or even negative. Therefore, it is unlikely that natural ecosystems will act as a strongly increasing sink for atmospheric carbon dioxide.

At Jasper Ridge, the effects of elevated carbon dioxide levels mediated through the water budget are at least as significant as the effects resulting from altered photosynthesis. This is likely to be a general result in water-limited ecosystems. Typically, the driest ecosystems are deserts, and sites of increasing water availability are occupied by grasslands, shrublands, and forests. Elevated carbon dioxide levels are likely to facilitate establishment of vegetation types with greater water requirements, unless climatic changes related to these elevated levels lead to increased drought. The extent and pace of vegetation changes will almost certainly vary from site to site, depending on the availability of plant colonists, the frequency of disturbance, and the level of human influences.

At Jasper Ridge, elevated carbon dioxide levels lead to both positive and negative consequences. These mixed results are probably applicable on a global scale. Agricultural production will increase, at least in some areas; however, the quality of some crops will decrease and problems with weeds will increase. In natural ecosystems, carbon storage will tend to increase; however, some desirable species will be replaced by undesirable ones (and vice versa), and some sites may change in character fundamentally, from grasslands to shrublands or from shrublands to forests.

QUESTIONS

1. Should plants with the C_4 photosynthesis pathway be more or less sensitive to increased carbon dioxide levels than plants with the C_3 pathway? Would the sensitivities be different for effects on photosynthesis and water use?

2. Plants grown at increased carbon dioxide levels tend to produce leaves with decreased nitrogen per unit of leaf mass. What would you expect to be the implications of this tendency for photosynthesis? For decomposition? For carbon storage in soil organic matter?

3. Some experiments, show a close correlation between increased leaf area and decreased stomatal conductance stimulated by elevated carbon dioxide levels. In these cases, would you expect elevated carbon dioxide levels to lead to increased soil moisture?

4. In deserts, much of the soil moisture is lost by evaporation from the soil surface, not by transpiration through plants. Will the effect of elevated carbon dioxide levels on soil moisture be amplified or suppressed in a desert? Why?

5. Much of the Sun's energy absorbed by an ecosystem is used to evaporate water. A significant fraction of the remainder is used to heat the air. When stomatal closure causes evaporation plus transpiration to decrease, heat in the air increases, leading to a physiologic effect of carbon dioxide on air temperature. In what kinds of ecosystems would you expect these physiologic effects on air temperature to be most significant? How do these ecosystems compare with the sites where warming caused by the greenhouse effects of elevated carbon dioxide levels is likely to be greatest?

FURTHER READING

Bazzaz, F. A., and Fajer, E. D. 1992. Plant life in a CO_2-rich world. *Scientific American* 266:68–74.

Field, C. B., Jackson, R. B., and Mooney, H. A. 1995. Stomatal responses to increased CO_2: implications from the plant to the global scale. *Plant, Cell and Environment* 18:1214–1225.

Jackson, R. B., Sala, O. E., Field, C. B., and Mooney, H. A. 1994. CO_2 alters water use carbon gain, and yield in a natural grassland. *Oecologia* 98:257–262.

Mooney, H. A., Drake, B. G., Luxmoore, R. J., Oechel, W. C., and Pitelka L. F. 1991. Predicting ecosystem responses to elevated CO_2 concentrations. *BioScience* 41:96–104.

Sellers, P. J., Bounoua, L., Collatz, G. J., Randall, D. A., Dazlich, D. A., Los, S., Berry, J. A., Fung, I., Tucker, C. J., Field, C. B., and Jenson, T. G., 1996. A comparison of the radiative and physiological effects of doubted CO_2 on the global climate. *Science* 271:1402–1405.

21 Ecology: Possible Consequences of Rapid Global Change

TERRY L. ROOT

SURVIVAL OF THE FITTEST AND ENVIRONMENTAL CHANGE

Plants and animals participate in beneficial, competitive, and predator–prey interactions that control their access to life-sustaining resources – food, energy, and space. Many factors influence the waxing and waning of individual species; however, physical environmental limitations and species interactions play the dominant roles controlling the degree of success or failure of all organisms in the attempt to perpetuate their kind. Environmental conditions and biotic interaction also limit the carrying capacity of the Earth's dominant species – humans. How do changes in climate affect the intricate web of life? How are human activities impinging on habitats, biodiversity, and the density and ranges of species? This chapter looks at this complex problem, using as its primary example North American birds. Some of these birds are likely to shift their densities and ranges extremely quickly in response to rapidly changing biologic and abiologic conditions. Other birds are expected to be constrained in their responses, having to wait until other, slower factors, such as shifts in the ranges of vegetation, change. This differential response could change the composition of biologic communities, affecting such "ecologic services" as natural pest control.

cology has been defined as the study of interactions that determine the densities and ranges of organisms. There are two basic types of interactions: those we call biotic, which occur among species, such as competition, predation, and mutualism (interactions that are mutually beneficial); and those between species and their abiotic environment.

SPECIES DENSITIES

The density of species varies dramatically both among species and within species. For example, the spotted owl is rare, whereas the great horned owl is fairly common, and the bald eagle is relatively rare in the continental United States but occurs in rather large numbers in Alaska. Why is there so much variation between such relatively similar species of owls, and why are bald eagle densities so different in different locations?

Biotic interactions are known to shape the densities of species. The interdependence of the populations of lynx and snowshoe rabbit is one of the classic examples of how biotic interactions, in this case a predator and its prey, help regulate the densities of both species (Fig. 21.1). An increase in the density of the snowshoe rabbit allows the lynx population to increase; however, the increase in the number of lynx precipitates a decrease in the hare population, which in turn results in a decrease of the lynx population. This last decrease, however, allows the hare population to rebound and the cycle begins again. Such cycling occurs to a greater or lesser extent for most, if not all, predators and their prey, including between herbivores and the plants they eat.

Various environmental factors can also strongly influence the density of species. For example, in the southern and central regions of the United States the scratching by dogs and cats is endless when the winters remain fairly warm and the spring is relatively warm. Those conditions are just right for fleas, which allows for the density of fleas to be quite high. Dogs and cats in those same regions scratch relatively less frequently when the winter is unusually cold or the spring relatively dry. Fleas do not survive well under such cold and dry conditions, which means the density of fleas is substantially lower.

Humans influence the density of species in many significant ways. When natural habitat is destroyed to make room for farms, factories, and housing developments, the densities of most species in the natural habitat will decline. However, the numbers of "weedy" species, such as roaches or dandelions, typically increase. In addition to such direct effects, hu-

W. G. Ernst (ed.), *Earth Systems: Processes and Issues*. Printed in the United States of America. Copyright © 2000 Cambridge University Press. All rights reserved.

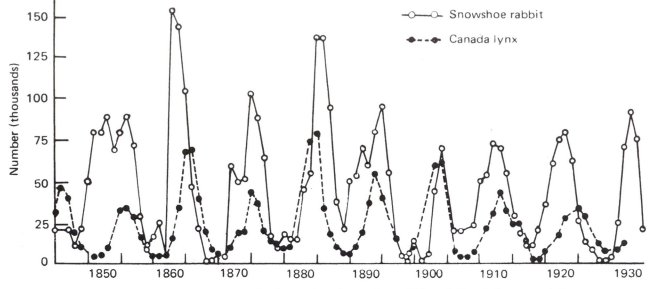

Figure 21.1. Nine- to ten-year cyclic fluctuation of snowshoe rabbit and Canada lynx population in northern North America. This cycle is illustrated in the fur returns from both snowshoe rabbit and Canada lynx. (*Source:* Adapted from C. A. Viller, *Biology*, 7th edition, W. B. Saunders, 1977.)

mans also influence the density of species indirectly, for example, by introducing nonnative species. One such species that people are particularly fond of is the domestic house cat. The problem arises not when cats remain in the house, but when they are let out to range free or when they become feral. A study focusing on well-fed pet cats in one small town in England found that roughly seventy pet cats killed approximately 2180 animals in one year. Free-ranging cats are particularly destructive to ground-nesting birds. Indeed, the lighthouse keeper's cat on Stephen Island is believed to have caused the extinction of that island's flightless wren. Accidental introductions of species have been equally as devastating. For instance, the brown rat snake was an unwelcome stowaway on a cargo ship that docked in Guam. As it became established and its population grew, all the ground-nesting birds, and many tree-nesting birds (the snake can climb trees), were decimated by this snake.

Changes in the densities of some species actually have been orchestrated by people. For example, in the late 1800s, the gray wolf was purposely eradicated from the forty-eight contiguous states because ranchers, farmers, and other settlers viewed this predator as a threat to their livelihood. They did not realize that the presence of the wolves was controlling the density of the coyote population, which in turn was controlling the population size of red foxes. With the eradication of the wolf, the coyote population increased almost fivefold in only thirty years (Fig. 21.2). Consequently, the U.S. government began an extensive and expensive campaign to control the coyote population, because it then was perceived as a threat. The resulting decline was met with a dramatic tenfold increase in the red fox population. Recent findings have indicated that despite the millions of dollars private groups and the government have spent trying to maintain and increase waterfowl populations, the persistent

decline in the population of many dabbling ducks, such as the mallard and northern pintail, is probably significantly influenced by the increase in the red fox population. Red foxes are much more efficient predators on the nests of these ducks than are coyotes. Therefore, the act of eradicating wolves around the turn of the century carried with it unanticipated consequences 100 years later. The close links among species indicate that actions by people, which either directly or indirectly affect the density of species, can have unforeseen cascading effects on many untargeted species. Many other examples of such cascading effects abound. For example, the introduction and expansion of the house sparrow and the European starling in North America pushed native birds such as the house wren and the eastern bluebird out of their preferred habitats.

THEORY

Understanding some of the theory behind the dynamics of single populations helps us to quantify some of the interactions that directly affect the densities of populations. The population size at a given time depends upon the population size at a previous time. For example, the population size during your generation depends on the population size during your parents' generation. For a single species, birth and death rates in one generation are the only factors that determine the change in the density of a population from one generation to the next (assuming, of course, that there is no immigration or emigration). Per capita growth rate is equal to the per capita birth rate minus the per capita death rate. When the per capita birth rate is larger than the death rate, then the number of individuals added to each generation will be more than those added to the previous generation.

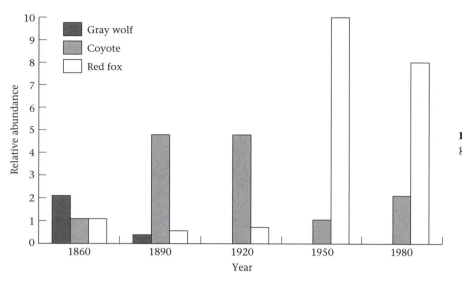

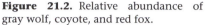

Figure 21.2. Relative abundance of gray wolf, coyote, and red fox.

Over time, such populations will grow exponentially (Fig. 21.3). This occurs when a population grows without any limits placed on it.

Sustained exponential growth is not possible in the real world because eventually a critical resource, such as food or space, will become rarer and then finally will be totally utilized. As it becomes rarer, the number of individuals added to the population will decrease due to either the increased death rate or decreased birth rate, because the resources are not available to support additional individuals. For a sustainable population the birth rate must equal the death rate so that all individuals have their required amount of critical resources. This, in turn, limits the size of the population. The maximum size of the population that the critical resource(s) can support is called the *carrying capacity* of the given area. The plot of population size over time is called an S-*shaped, sigmoid,* or *logistic curve* (Fig. 21.4). The population has the

fastest growth rate when it is halfway to its carrying capacity; at that point, the resources are not yet rare and restrictive, and a relatively large number of adults are present in order to produce young. Numerous examples exist of populations, both in the laboratory and in the wild, that exhibit logistic growth.

The factors limiting the density of a population can be either biotic or abiotic, or both. For example, the carrying capacity of a seed-eating bird, such as a sparrow, can be set by the co-occurrence of a population of a seed-eating mammal, such as a mouse, and vice versa. Both species need to eat seeds in order to survive; but the number of seeds is limited. Individuals, both within and between the species,

Figure 21.4. Logistic growth curve. In logistic population growth, the total size of the population (N) grows in an S shape, first growing slowly, then more rapidly, then more slowly, leveling off near the carrying capacity (K). Growth rate increases, reaches a peak in the middle of the total population curve (the point of inflection, where the curve stops bending to the left and starts bending to the right), and then decreases toward zero as population numbers approach the carrying capacity.

Figure 21.3. Global population (N) will increase exponentially over time (t) according to the equation $N_t = e^{rt}$, where r is the growth rate and e is the base of the natural logarithms ($e = 2.718$).

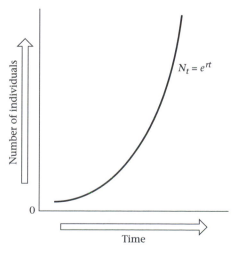

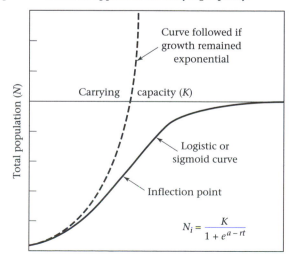

will compete for the seeds. Those that win will survive. Those that lose completely will die, and those that lose partially will be in poor physical condition and thereby might produce fewer offspring. Hence, the biotic interaction of competition can limit the density of both populations.

Environmental factors can also set the carrying capacity of a population. Imagine a population of parrots living on an island in the Atlantic Ocean where they nest in cavities carved in large, dying trees. If a hurricane came through and blew down most of the large, dying trees, the nest sites could be so limited the following nesting season that the parrots' birth rate would drop dramatically. Therefore, the carrying capacity, which is limited by nest-site availability, is decreased and remains low for numerous generations – at least until more trees become available for nesting sites.

An analogous situation occurs when humans destroy, whether purposely or not, the habitat of species: The carrying capacity of the region drops. A politically controversial example is the spotted owl, which lives in the ancient forests of the northwestern United States. As these forests are logged, the number of breeding sites for the owls declines, thereby decreasing the carrying capacity of the area for the owl. If carrying capacity drops too low, then extinction may result. The question becomes, what is too low? Small populations are much more vulnerable to extinction than are large populations. For example, a natural disaster, such as a severe cold snap in the region accompanied by snow that could make foraging difficult, could wipe out an entire small population.

Much concern exists throughout the world about the growth rate of the human species. The United Nations Population Convention held in Cairo in 1994, attended by diplomats from nearly all of the countries around the world, attests to the grave concern about overpopulation and overconsumption by humans. In short, the growth rate of humans in many regions has not declined sufficiently to keep pace with availability of many critical, nonrenewable resources (see Chap. 22 for a more detailed explanation). Consequently, the quality of life for humans in these regions is painfully low because in these locations (perhaps throughout the world) we are exceeding our carrying capacity. Natural corrections to such overshoots of carrying capacity necessitates that the death rate be higher than the birth rate (Fig. 21.5). Unsustainable, excessive utilization of nonrenewable resources may be providing temporary relief, thereby allowing the maintenance of a higher quality of life for these populations. Such "borrowing from the future," however, reduces the long-term carrying capacity of the planet, unless we can devise *sustainable* technical and organizational means to effectively increase our carrying capacity.

Concerns about human overpopulation are not new. In 1960, von Foerster, Mora, and Amiot wrote an article in the journal *Science* on human population size with the intriguing tongue-in-cheek title "Doomsday: Friday, 13 Nov. AD 2026." In this article, they plotted all the independent estimates of human population size they could find (a total of twenty-five) from 0 A.D. to 1960 A.D. (Fig. 21.6) and determined that

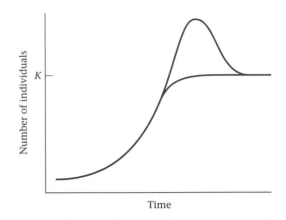

Figure 21.5. Often, population numbers seem to overshoot the carrying capacity of the ecosystem and then drop back to (or below) it. K = equilibrium carrying capacity.

the points could be well approximated by a line. The equation of the line revealed that when time, in years, was equal to 2026.87, the human population size would equal infinity. Because November 13 falls on a Friday in 2026, and because 0.87 of a year is roughly equivalent to mid-November, they devised the attention-getting title. Obviously, the human population size cannot become infinite, but the real message is that the human population growth rate is unsustainably high, and to avoid low quality of life for many people and high death rates, we must decrease our birth rate.

In the decades since this article was published, we might expect that human growth rate would have slowed dramatically – after all, the year 2026 is not that far off. Indeed, the growth rate has dropped significantly; however, it is still quite high. Using the equation they derived for the line, the estimated population size for 1975 is 3.7 billion people, whereas the actual number was close to 4.0 billion. For 1980, the estimate is 4.0 billion, whereas the actual number was near 4.4 billion, and for 1993, the numbers are 5.5 billion and approximately 5.6 billion, respectively. Evidently, we remain on track toward "doomsday." In reality, of course, we will not ever have an infinite number of people on the Earth, because the limitation of critical resources will prevent it.

RANGES OF SPECIES

Besides investigating how interactions determine the density of species, ecology also includes the study of biotic and abiotic interactions that shape the geographic ranges of species. This means examining what factors at a large spatial scale limit species ranges. An example of how biotic interactions shape ranges is that of the North American breeding ranges of the black-capped chickadee and the Carolina chickadee. Competition seems to have forced the black-capped chickadee into areas farther north than those occupied by the Carolina chickadee. Although many studies have investigated the importance of biotic interactions, most have been done

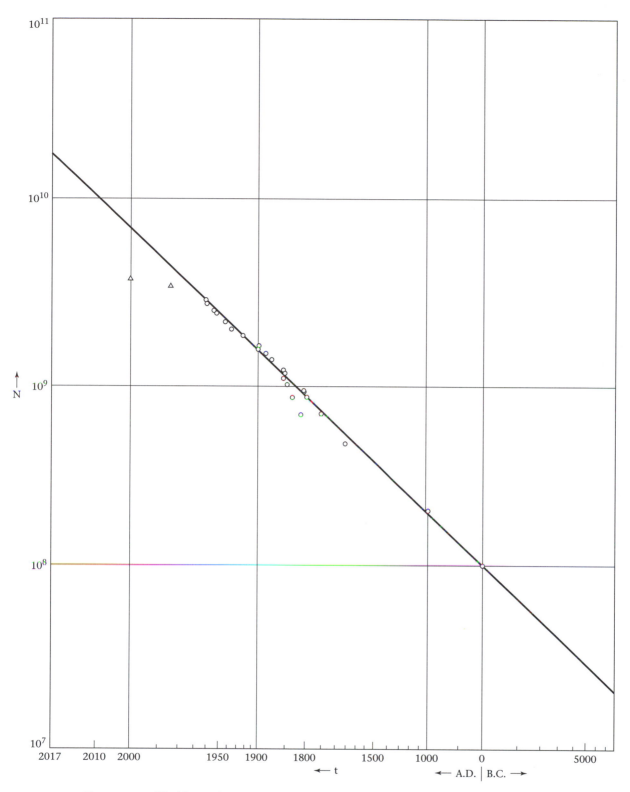

Figure 21.6. World population *N*, observed (circles), and projected (triangles) as a function of time *t*. (*Source:* Adapted from H. Von Foerster, P. M. Mora, and L. W. Amid, *Doomsday: Friday, 13, November A.D. 2026, Science*, vol. 132, 1960, pp. 1291–1293.)

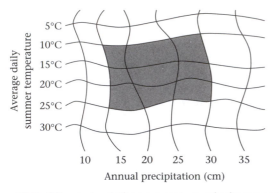

Figure 21.7. Climate isopleths on a map reveal where a species could live.

on study plots roughly the size of tennis courts. Several attempts have been made to extrapolate these small-scale findings to larger spatial scales. Most people, however, now agree that the primary factors limiting species ranges at a large spatial scale are environmental.

Many different environmental or abiotic factors can act as limiting agents. For example, the compactness of soil can determine the type of organisms that live in an area. Many more nematodes live in relatively open and airy humus than in dense, compact clay soil. Climate also has a striking effect on the presence of species. Indeed, temperature and precipitation data can be used to approximate the type of vegetation present. A classic example is the saguaro cactus. The northern range limit of this succulent plant is dictated by temperature dropping below −7°C. In such cold weather, the water in their tissues freezes, which causes their cells to rupture, which in turn causes the plants to die. Therefore, if any young plants become established beyond this −7°C isotherm, they will not survive the winter.

Graphic theory can help us understand how environmen-

tal factors may determine the shapes of species' ranges (Fig. 21.7). For simplicity, assume that a species' range is shaped by only two abiotic factors, for example, temperature and precipitation. Further assume that our species can tolerate daily summer average temperatures from 0 to 30°C, but outside this range (both lower and higher) individuals will die. Similarly for precipitation, our species must have 10 to 50 centimeters of annual precipitation, and any more or less will cause death. Plotting these climate isopleths on a map reveals where our species could live (Fig 21.7). Other factors, such as overharvesting, lack of nutrients, habitat fragmentation, or diseases, could prevent our species or any other from filling this entire potential range. By understanding the mechanisms shaping species ranges, such as physiology or habitat destruction, we can better understand how global change could potentially shift and reshape the ranges of different types of species. An example for North American birds is provided below.

NORTH AMERICAN BIRD CASE STUDY

Research on birds wintering in North America shows that for the vast majority of species, range boundaries are influenced by environmental factors. The isopleths of six different factors were compared to the north-, east-, and west-facing range boundaries for virtually all birds that winter in North America. South-facing boundaries were not included because most of the species' ranges extend into Mexico, where data are not available. The associations were striking, as shown in Figure 21.8, and quite common, as shown in Figure 21.9. For example, certain birds have northern range boundaries associated with average minimum January temperature isotherms, other species have boundaries associated with vegetation ranges, and still others have boundaries associated with both.

These facts could have important consequences for a com-

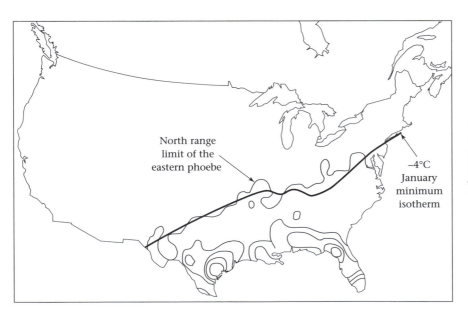

Figure 21.8. The distribution and abundance of the winter range of the eastern phoebe. The northern boundary lies very close to the −4°C isotherm of January minimum temperature (heavy solid line).

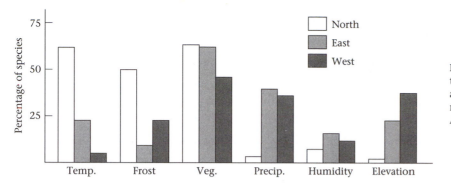

Figure 21.9. Isopleths of six different factors were compared to the north-, east-, and west-facing range boundaries for roughly all birds that winter in North America.

munity of birds if the Earth begins to warm rapidly. Those species limited only by winter temperature could expand their ranges north immediately as the winter temperature warms (Fig. 21.10; range of species A). Those birds limited by vegetation will also move north, but at a much slower rate, because they will have to wait until the vegetation they require moves north with the rising temperatures (Fig. 21.10; range of species B). This expansion assumes that the vegetation can cross habitats that have been highly disturbed and fragmented by human use. The differential speed of range expansion could result in a disaggregation of the bird communities as we know them.

Will such fragmentation of communities be catastrophic,

Figure 21.10. Schematic diagram of the ranges of species A, limited by temperature, and species B, limited by vegetation. (A) Present-day range locations; (B) Future range locations with global warming.

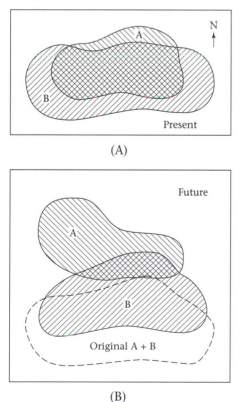

benign, or something in between? That will depend on the location, timing, and species involved. For example, if a prey species moves away from a predator that specializes on it, then the density of the predator could decrease in the areas the prey deserts. In the areas where the prey is no longer hunted by this particular predator, then the prey population could increase dramatically. If this prey species is, for example, an agricultural pest and it resides or moves into an agricultural area, then from a human point of view this fragmentation of communities would be catastrophic, at least at the local level. Another example is the possible disruption of a predatory regulation of spruce bud worm. Studies have shown that both warbler predators that specialize on bud worm caterpillars will most likely shift their ranges in response to global warming and decline in density due to human land use changes. The spruce bud worm, however, will likely change its range as the forest range shifts, which is expected to occur at a much slower rate than that of the warblers. The consequences of global warming then could be that for a time, until another predator learns to specialize on spruce bud worms or technology develops a species-specific insecticide, the population of spruce bud worm could be high, possibly causing significant damage to the forest and, in turn, dramatic economic hardships to timber companies. Admittedly, considerable speculation is a part of the above scenario. For some the probability of occurrence and the potential costs are both sufficiently high to warrant investigation of possible mitigating actions.

Policy makers need to know the probability of the occurrence of catastrophic events, such as the one hypothesized above, in order to help them determine what action(s) to undertake. Knowledge concerning the probability of the occurrence of benign events is not needed as urgently, because policy interventions are not as critical, given such a possibility. An allegory of two fifty-year-olds who consider taking an aerobics class may help clarify this idea. One person is in good physical condition and healthy; her physician recommends the appropriate level of exercise. The other person has recently had a mild heart attack; his physician wants to do tests to help define the type and extent of exercise he can safely perform. The probability of catastrophic damage is so low in the healthy, fit individual that no modifying policy is needed. The probability of a devastating catastrophe, such as death, is so high in the individual with heart disease that intervention is prudent. Of course, even in the example of the healthy person, detailed knowledge of the person's physical condition could help in tailoring an optimal exercise program. The urgency in acquiring such knowledge, however, is greatly diminished when the potential impact is not negative.

Currently, scientists are attempting to determine the probability that biologically catastrophic events could occur with rapid global warming. Their first step is to quantify the importance of temperature in shaping the range boundaries of many different species. The winter physiology has been studied for several species whose northern range limits are dictated by temperature. These studies provide information about three parameters: basal metabolic rate (BMR), lower critical temperature (TCRIT), and conductance (COND). BMR is the metabolic rate of sleeping, nondigesting, nongrowing, nonreproducing individuals – it is analogous in function to the idle of a car (Fig. 21.11). TCRIT is the temperature at which, if it becomes any colder, the animal must increase its metabolism to maintain body heat. COND is the rate of increase in metabolism as temperatures decrease below TCRIT. These values plus the temperature at the northern edge of the ranges (TDIST) can be used to calculate the northern-boundary metabolic rate (NBMR) in the following way:

$$NBMR = COND \ (TCRIT - TDIST) + BMR$$

Plotting body mass against BMR and NBMR on a log-log plot shows that the slopes of the two lines are not significantly different, but the *y*-intercepts are (Fig. 21.12). Therefore, NBMR can be rewritten in terms of BMR:

$$NBMR = 2.5 \times BMR$$

Consequently, the northern boundaries of these species ranges occur near regions in which it is just cold enough on average that the individual birds along the species' northern range boundary need to raise their metabolic rate to 2.5 times their basal rate in order to compensate for the decrease in temperature. Hence, as the Earth warms, these species will likely be able to expand their ranges northward, tracking the same isotherms that currently limit their ranges (TDIST).

In cold environments, more food is needed to fuel higher metabolic rates. Generally, birds do not store fat the way mammals do – by putting on fat in the fall and using it up throughout the winter. Instead, birds eat during the day and store fat. At night, when they do not forage and it is usually colder than during the day, they use most of the fat that they stored during the previous day to fuel their metabolism. The next morning they must forage again to store enough fat to survive the day and the next night. For example, a northern cardinal living in Michigan, which is near the northern boundary for this species, must sustain a metabolic rate approximately 2.5 times BMR from dusk to dawn. The average amount of fat remaining at dawn would fuel their metabolism and thereby allow them to survive without eating for only three hours. Therefore, the individuals at the northern

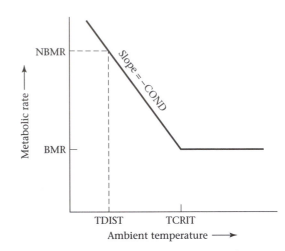

Figure 21.11. Plot of metabolic rate against ambient temperature (see text for explanation). BMR = basal metabolic rate; COND = conductance; NBMR = northern-boundary metabolic rate; TCRIT = critical temperature; TDIST = temperature distribution.

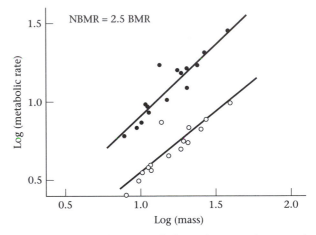

Figure 21.12. Plot of log (metabolic rate) against log mass (see text for explanation). BMR = basal metabolic rate; NBMR = northern-boundary metabolic rate.

edge of the cardinal's range must begin feeding almost immediately at dawn, even in severe storms. Hence, individuals could not survive much farther north because they could not survive the colder, longer nights. As the Earth warms, however, they may be able to move farther north.

What will occur with rapid global warming? Will the birds be able to move farther north quickly and track the changing temperatures? This will occur only if the ranges of birds are dynamic and not static. Certainly, this characteristic will vary with species. For example, the wrentit, which lives in the vegetation of the California chaparral, will probably not expand its range until the chaparral vegetation shifts north. In contrast, the Baltimore oriole, which occurs in many different types of habitats, could, and probably is already, expanding its range northward, perhaps in response to the warmer climate of the past thirty years.

Species ranges and densities are dynamic and can change

Figure 21.13. Winter range boundary for the myrtle variety of the yellow-rumped warbler (*Dendroica coronata*) for the winters of 1969–70 (solid line), 1970–1 (dashed line), and 1971–2 (dotted line). (*Source:* Adapted from J. Kingsolver, P. Kareiva, and R. Huey (eds.), *Biotic interactions and global change*, Sinauer Associates, Inc., 1993.)

on an annual basis, at least. An example is the yellow-rumped myrtle variety of the warbler. Figure 21.13 shows significant annual shifts in the location of the northern boundary of this warbler's winter range for three consecutive years. Also, the density of the eastern phoebe, an insect-eating bird, is higher at more northerly locations in warmer years than in colder years, and vice versa (Fig. 21.14). These patterns suggest that individuals expand north in warm years and retreat south during cold ones. Consequently, if global temperatures rise rapidly, some species of birds are likely to be able to track the increasing temperature and quickly expand their ranges northward. Other species, however, such as those limited by the availability of vegetation, probably will not be able to move quickly.

If the species that moves northward is a predator and the species that does not move is one of its prey species, then the predator may no longer hold the prey population size in check. For instance, if the Cape May warbler, which is a specialist on gypsy moth caterpillars, moves north and the moth does not move, then an outbreak, or high density, of caterpillars could defoliate and damage vast tracks of forest. This could be economically damaging to the forest ecosystem.

Ideally, an understanding of the possible consequence of global warming will help policy makers to fashion policies designed to help mitigate potential ecosystem disruptions. Thus it is important that all sectors of society develop heightened awareness of these matters and their implications for the biosphere.

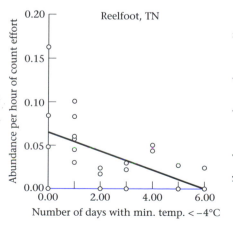

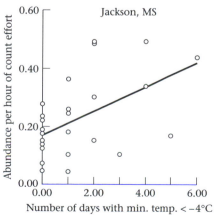

Figure 21.14. Density of the eastern phoebe, an insect-eating bird, is higher at more northerly locations in warmer years than in colder years, and vice versa.

CONCLUSION

Increased knowledge of the interactions among species, and between species and their abiotic environment, will lead to a better understanding of what shapes the ranges and densities of organisms. The factors (biotic or abiotic) that limit the populations of many species will determine the response of ecosystems to global change disturbances. The result could be a rapid change in the density and/or range of species, which in turn could significantly change the density and/or range of other species. An example is the extermination of the gray wolf in the continental United States, which resulted in an expansion of the coyote population. Subsequent hunting and trapping of the coyote caused its population density to drop, allowing the population of the red fox to increase. This increase is probably a significant factor in the decline in dabbling ducks. An old cliché in the field of ecology applies: In nature, you cannot change just one thing. Providing information to policy makers hopefully will allow them to determine the appropriate mitigation strategies to help ensure that the fewest irrevocable extinctions occur.

QUESTIONS

1. What kinds of technologies or organizational arrangements have been or could be used to increase the carrying capacity for humans? Are these techniques sustainable, or could they be made sustainable?

2. Why is the rate of climate change or scale of habitat destruction significant for the viability of species?

3. Name three ways that altering the structure of biologic communities might affect how ecosystems work.

4. Under what circumstances are abiotic factors likely to control the ranges or densities of certain species? What circumstances are most dependent on biotic factors?

FURTHER READING

Kingsolver, J., Kareiva, P., and Huey, R. (eds.) 1993. *Biotic interactions and global change*. Sunderland, MA: Sinauer Associates, Inc.

Mac, M. J., Opler, P. A., Puckett Haecker, C. E., and Doran, P. D. (eds.) 1997. *Status and trends of the nation's biological resources*. Washington, DC: U.S. Department of the Interior, U.S. Geological Survey.

Peters, R. L., and Lovejoy T. E., (eds.) 1992. *Global warming and biological diversity*. New Haven, CT: Yale University Press.

Pimm, S. L. 1991. *The balance of nature?* Chicago, IL: University of Chicago Press.

SOCIETAL AND POLICY IMPLICATIONS

RESOURCE USE AND
ENVIRONMENTAL TECHNOLOGY

22 Population and the Environment

SUSAN E. ALEXANDER AND PAUL R. EHRLICH

EVER-INTENSIFYING ANTHROPOGENIC ENVIRONMENTAL CHANGE

At the dawn of the new millennium the news on world population is paradoxical. With over 6 billion people on earth, and almost a quarter million added each day, the scale of human activity and environmental impact is unprecedented. At the same time, this tremendous growth in population has been accompanied by major improvements in many areas of human life. Recent gains in average life expectancy, declines in fertility rates, reductions in child mortality rates, greater access to health care, and increases in the world food harvest have led to substantial improvements in the standard of living of hundreds of millions of people. Yet, this incredible progress has come at the expense of the environment and its future capacity to sustain life.

Soaring human population growth over the past century has created a visible challenge to Earth's life support systems. Critical natural resources such as fertile topsoil, clean groundwater, and *biodiversity* are diminishing at an exponential rate, orders of magnitude above that at which they can be regenerated. In addition, the world faces an onslaught of other environmental threats including deforestation, global climate change, intensified acid rain, stratospheric ozone depletion, and health-threatening pollution. Overpopulation, high levels of per capita consumption, and the use of deleterious technologies combine to increase the scale of human activities to a level that underlies all of these problems. Essentially, to support its growing population, humanity is depleting a one-time resource bonanza, and living off earth's capital. This capital includes such things as fossil fuels, metals, deep soils, underground aquifers, as well as the millions of other species with which we share the Earth. These intensifying trends cannot continue indefinitely. Hopefully, through increased understanding, appreciation, and valuation of *ecosystems* and their services, earth's basic life-support systems will be better protected for the future.

As the unprecedented growth in human numbers continues, the ability of human beings to shape and manipulate the physical world and all its living inhabitants far exceeds that of any other species on Earth. Throughout most of our existence, however, the human population was not capable of radically altering the environment on a planetary scale. Only within the past few centuries, and especially in recent decades, has humanity caused environmental changes comparable to those that took centuries and millennia to occur through natural processes. Although humankind has long affected the local environment, it has now become an agent of unparalled global change in the biosphere, atmosphere, and hydrosphere. How have we reached this position, and what role do humans play in the future of the environment and its capacity to sustain life?

Human populations, natural resources, and environmental issues are linked in numerous and complex ways. Life depends upon a wide variety of services provided by natural ecosystems that are basically free of charge and largely taken for granted. These *ecosystem services* include water purification, soil generation and preservation, microbial recycling of nutrients, climate stabilization through the regulation of atmospheric gases, and maintenance of a vast genetic library of plants, animals, and microorganisms. Without these services, life in its present form would not exist. For example, from the diversity of wild species, humanity has extracted the very basis of civilization in the form of crops and domestic animals, to say nothing of providing a substantial proportion of medicines used today to combat disease.

The impact of humanity on the environment is linked strongly not only to population growth, but also to patterns of production, consumption, and activity levels associated with varying lifestyles, values, and economic status. In a comparison of industrialized nations and the developing world, the differences in these factors are very large. Understanding and raising awareness of the processes and issues involved are only a beginning. The achievement of *sustainable development*, while closing the gap between rich and poor nations, is necessary as the global community struggles to solve these critical, environmental problems. This chapter provides a framework for exploring the link between population, environmental processes, and ecosystem services. We discuss basic *demography* and the current population situation, then examine specific global environmental issues and contributions of population growth to these problems.

DEMOGRAPHY

Basic demographic principles provide a framework for understanding how populations grow and the effects they have on their environment. Three main processes control the size, composition, and distribution of any population: birth, death, and *migration*. Ignoring migration for the moment, when births per unit time exceed deaths, a population grows. Conversely, when deaths exceed births, a population declines. In practice, the crude *birth rate*, the number of births per 1000 individuals in a population per year, is calculated by dividing the total number of births by the mid-year population size and multiplying that result by 1000. Similarly, the crude *death rate* is the number of deaths per 1000 individuals in a population per year, and is calculated in a similar manner. The difference between the birth rate and the death rate determines the rate of *natural increase* in a population. Throughout most of human history the annual population growth rate was very small because the relatively high birth rate offset the relatively high death rate. By modern times, however, this was no longer the case. In 1999, for example, almost 2.5 times as many births (23) than deaths (9) per 1000 individuals occurred, creating a net increase of more than 225,000 people each day.

In theory, when the birth rate equals the death rate, natural increase stops. However, calculations of the actual growth rate of a population must take into account immigration and emigration as well as the birth rate and death rate: It is the sum of the natural population increase or decrease and the net migration. Birth rates and death rates often are given as crude rates incorporating all age groups; however, demographers typically refer to age-specific rates, such as those for individuals between the ages of eleven and fifteen. In contrast to birth and death rates, which are given per 1000 individuals, population growth and natural increase rates are usually expressed as percentages or the average increases per 100 individuals per year. Ultimately, population size is said to be stationary when the rate of natural change reaches zero (birth and death rates equal) with no net migration. This scenario is known as *zero population growth* (ZPG).

Alternatively, when the growth rate of a population exceeds zero for a period of time, as has the human population in recent history, the population experiences a phenomenon known as exponential growth (Fig. 22.1). Exponential growth means that the population increases over a period of time proportional to the number of individuals present. This concept is similar to the compounding of interest in a bank account. Over time, the initial deposit of money will earn interest, and, in turn, the new money will earn interest; that is, interest is earned on interest. Similarly, children eventually grow older and have children themselves. Therefore, the greater the population size, the greater the absolute increase at any given growth rate. It is this same phenomenon that has occurred within the past few centuries as the human population has grown exponentially at a rate of approximately 1 to 2 percent each year.

Birth rate, death rate, and migration also determine the composition of a population, particularly its age distribution and sex ratio. Age composition diagrams show the relative percentage or number of males and females of each age group, and relate directly to growth rate and future population size (Fig. 22.2). Populations with high birth rates and relatively low death rates, such as those of many less developed countries, will have far more people in the younger age groups than in the older age groups, and typically high

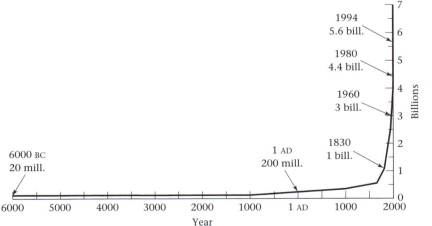

Figure 22.1. World population rose from a level of approximately 20 million around 6000 B.C. to 1 billion by the year 1830. By the end of the twentieth century, the world's population had reached 6 billion, an increase of 4.4 billion in the previous one hundred years.

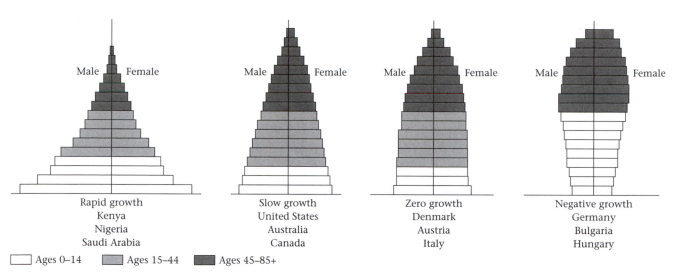

Figure 22.2. Population age-structure diagrams for countries with rapid, slow, zero, and negative population growth rates. Bottom portions represent prereproductive years (ages 0–14), middle portions represent reproductive years (ages 15–44), and top portions represent postreproductive years (ages 45–85+). (*Source:* G. Tyler Miller, *Sustaining the Earth: an integrated approach*, Wadsworth Publishing, 1994.)

growth rates. In a rapidly growing country such as Kenya, 46 percent of the population is under the age of fifteen, whereas only 3 percent of the population is over age sixty-five. The rate of natural increase in Kenya is 2.1 percent per annum. In more developed regions, populations tend to have low birth and death rates, which are reflected in their *age structure*: The numbers of people in each age group are more evenly distributed, and overall growth rates are lower. In a slowly growing country such as the United States, with a rate of natural increase of 0.9 percent, (rate includes immigration) 21 percent of the population is under the age of fifteen and 13 percent over the age of sixty-five. Some countries have already reached ZPG, and in Germany, the rate of natural increase is actually negative, −0.1 percent per year, with approximately 16 percent of the population under the age of fifteen and 16 percent over the age of sixty-five. Changing demographics in countries making the transition to ZPG and population decline initially leads to aging populations and, often, new problems. In the United States a "baby boom" occurred when 80 million people were born between 1946 and 1964. On the tail of this boom is a group of 47 million people born between 1970 and 1985, called the baby bust generation. As the baby boom generation ages, declining job markets and stresses on health insurance programs and pension plans may create an economic burden on the baby busters. On the other hand, positive social and economic changes such as less competition for jobs and greater opportunity for education may also occur.

The theory of *demographic transition* is often used to explain the decline in population growth that accompanied the industrialization of various countries. This theory contends that populations progress through three distinct phases. First,

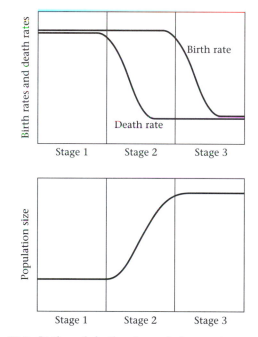

Figure 22.3. Birth and death rates and changes in population size are illustrated for the three stages of the demographic transition in industrializing nations. In Stage 1, both birth and death rates are high; stage 2 birth rates remain high and death rates decline; stage 3 birth and death rates are both low. In stage 2, when the birth rate remains high while the death rate is declining, population growth leaps. Population growth ends in stage 3.

populations are characterized by high birth and death rates, then by high birth rates and declining death rates, and finally by low birth and death rates (Fig. 22.3). Prior to industrialization, most nations had high birth and death rates, which kept the population from growing rapidly. In colonial America, for example, it was not uncommon for a woman to give birth to twelve or thirteen children in her lifetime. With industrialization, however, came health improvements in sanitation and nutrition that contributed to decreasing death rates, and more women survived their childbearing years. Initially, women continued to have many children even as infant mortality rates dropped. Declining death rates combined with continued high birth rates led to rapid population growth. For example, during industrialization in Great Britain, the population quadrupled. Eventually, with the higher standards of living and the higher education levels that accompanied economic growth, birth rates dropped and populations entered the final phase of the demographic transition during which both birth and death rates were low.

Some form of this demographic transition has occurred in most of the *developed countries* of the world. In *developing countries*, however, death rates dropped significantly following World War II, while birth rates remained relatively constant. Western intervention introduced the means of lowering death rates but did not adequately address the resulting need to lower birth rates. For example, by the early 1990s, the population dynamics of sub-Saharan Africa posted the highest crude birth rates and the highest rates of natural increase of any major world region. Continued growth in this region can be attributed to the general trend of decreasing death rates and relatively constant birth rates. However, some countries in this region, have been able to reduce their *fertility*. In Kenya, Zimbabwe, and Botswana, birth rates fell by 15 to 25 percent as education rates rose, contraceptive use increased, and infant mortality rates decreased. This trend potentially indicates the transition toward substantially lower population growth rates. However, great social and economic challenges continue to exist in making a rapid demographic transition in most sub-Saharan countries, where children are economic assets and important sources of security, especially for the elderly.

Populations including large percentages of young people have a powerful built-in *demographic momentum*, a term used to explain the continued growth of populations even after family sizes have stabilized near replacement level (the number of children a couple must have in order to replace themselves in the next generation). The actual replacement rate is slightly over two children to replace two people, because some females die before they reproduce. When family sizes decline in the developing regions, the skewed age structure, with many young women still approaching reproductive age, will continue to elevate birth rates for a long time. For example, in a typical, rapidly growing population in a poor nation, even after fertility rates decrease and the average number of births per woman in the population reaches replacement rate, the population will continue to grow for approximately 60 to 70 years (roughly a human lifetime) due to demographic momentum. There will be fewer children per female, but the absolute number of females giving birth will grow due to the high proportion of young adults moving through their reproductive years. ZPG will not be reached until that young group reaches old age, substantially raising the population's death rate.

The Earth's carrying capacity ultimately limits unchecked growth. *Carrying capacity* is defined as the maximum number of individuals that can be sustained by a given environment and its resources, without reducing that region's capacity to support populations in the future. Overpopulation occurs when the impact on local resources and the environment is unsustainable. When the population size exceeds this limit and diminishes its resource base, population size begins to decline as well. Although carrying capacity is easily defined in theory, it is far more difficult to identify and understand in practice, and can easily vary between similar resource bases depending upon human use and activity patterns. On a regional or national level, carrying capacity is a function of many factors. These include food and energy supplies, freshwater availability and other ecosystem services, technologic innovations, human lifestyles and values, and the social, political, and cultural climate of that area. In many parts of the world, the increase in human numbers combined with their patterns of resource consumption and associated technology use already exceeds the local carrying capacity. In these areas, depletion of renewable resources such as soil, groundwater, forests, and fisheries, occurs more rapidly than replenishment.

The world's current population size already exceeds some estimates of the optimal number of people that Earth can sustain in the long term. An optimal population size, by most social standards, would allow for all human beings to live decently and have access to essential resources. Economic development would occur while safeguarding the common environment, thereby providing for the needs of the present generation without compromising the needs of future generations. Among other things, the optimum population size would be small enough to preserve and sustain large tracts of natural habitat and an abundance of resources. It would not in itself directly foster unsustainable environmental degradation. The population would, however, be sufficiently large throughout the world to form vibrant cities that stimulate intellectual, artistic, and technologic creativity, and enhance the quality of life. Using these criteria, one study has estimated an optimal population size, given present and foreseeable technologies, in the range of 1.5 to 2.0 billion people. This conservative estimate, far below the present population size, incorporates a large margin of error to provide a buffer from unforeseen environmental impacts. Although the above number may underestimate the optimal population size, it is clear that humanity must work toward decreasing current growth rates in order to achieve sustainable development. The following sections review the current population situation, outlining how human carrying capacity and environmental impact vary depending not only upon population size and location, but also upon patterns of consumption and technology use.

POPULATION EXPLOSION

The world's population is growing at an alarming rate. The human population has quadrupled in size over the past 100 years. This is a remarkable feat, considering that only 10,000 years ago fewer than 10 million human beings lived on Earth. It took 8000 years for the human population to reach 100 million, and not until around 1850 did we reach the 1 billion level. In the past 200 years, with the advent of the Industrial Revolution, the growth of humanity has been unprecedented. The current world population of 6 billion is expanding at a rate of approximately 1.4 percent per year. Demographers project that human numbers will climb to over 8 billion within the first half of the twenty-first century, with a peak population of 8 to 12 billion, before declining to a sustainable number. They do not, however, eliminate the possibility that rises in death rates caused in part by environmental deterioration may limit numbers before 8 billion is reached. As noted earlier, population growth rates differ from region to region and are especially disparate between less developed and developed countries. For example, in 1999, more than 80 percent of the world's population lived in less developed countries, whereas less than 20 percent lived in developed countries. The global population increases by nearly 90 million people each year (Fig. 22.4). More than 95 percent of this increase occurs in developing countries of Africa, Asia, and Latin America.

The combined population growth rate of less developed nations is approximately 1.7 percent per year. Barring significant changes in fertility or mortality rates, the population of those nations will double in approximately 40 years. When China, which in past decades has significantly decreased its rate of natural increase, is excluded from this group, the growth rate increases to 2.0 percent per year and the doubling time is only 35 years. In comparison, the more developed nations have a combined growth rate of 0.1 percent and a doubling time of 543 years. With the current age structure in these countries, the difference in population numbers will continue to widen, even as less developed countries reduce their population growth rates. Encouragingly many countries have succeeded in reducing their growth rates over the past several decades, especially in Asia and Latin America. However, the momentum inherent to the system remains strong. Although the global rate of population growth has actually declined significantly from a high of 2.1 percent in the late 1960s, the absolute number of people added to the world each year remains high. Compounding problems, the areas of fastest expansion are also areas where food supplies and natural resources are in shortest supply. Overpopulation already exists in many nations, and continued population growth will ultimately threaten the lives of all human beings through nonrenewable resource depletion and environmental degradation. If death rates continue to fall faster than birth rates, the world may experience population crashes involving millions of premature deaths.

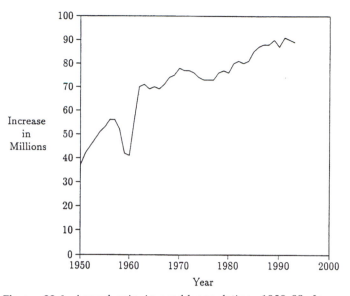

Figure 22.4. Annual gain in world population, 1950–93. In 1950, births exceeded deaths by 36 million. This rate has increased and, with the exception of a decrease in numbers the late 1950s, is currently growing by approximately 90 million annually. (*Source*: Courtesy of the Population Reference Bureau, *World population data sheet*, Washington, DC, 1993.)

The impacts of human activities on Earth, however, are not solely a matter of sheer numbers. They also reflect the manner in which populations use natural resources and interact with the environment. For example, all people do not exert the same pressure on the environment. Social values and norms, as well as the economic and political structure of society, influence people's impact. Total environmental impact can be expressed as a combination of three separate, but interacting, factors. In a simple equation, $I = PAT$: the impact (I) of a population is the product of population size (P); per capital consumption or level of affluence (A); and a technology factor (T) which representing an index of environmental damage due to the technologies used to support consumption. In practice, however, A times T can be difficult to measure, and per capita energy consumption is often used as an imperfect substitute for these terms. This simple yet powerful identity demonstrates how alterations in any of the three factors, P, A, and T, can change a society's impact on the environment.

The $I = PAT$ identity makes it clear that environmental trends reflect life-styles (the A and T factors) as well as sheer numbers (the P factor). Although less developed countries together contain nearly 4 times the combined population of developed countries, they also have a per capita rate of energy consumption that averages approximately one-seventh of that in developed countries. Indeed, the total environmental impact is less than half that of the citizens of industrialized nations, where the rates of consumption and use of environmentally damaging technologies are very high. For example, the current population growth rate in the United States is approximately 0.9 percent. This rate is far lower than that in the majority of African nations such as Malawi, which has a growth rate of 2.8 percent. Yet, per capita commercial energy use is 300 times greater in the United States than in Malawi. Because reductions in any of the three factors, P, A, or T, can reduce population impact, it is likely that the exploding populations of developing countries may, for the time being, have less of an environmental impact than the nearly stationary populations of developed countries. Similar to demographic momentum, which is built into population growth, social momentum is built into the A and T factors. It takes decades, for example, to develop and deploy new energy technologies. Developing nations that follow the example of the now-industrialized nations in basing their energy economies on fossil fuels will have long-term repercussions.

Many developing countries experience problems due to land shortages. The majority of people in these countries make their living directly from the land, yet almost all fertile and accessible land is already in use (see the box entitled *The Population–Environment Link: The Philippines*). This problem is certainly exacerbated by continued population growth. In the 1970s, land under cultivation was expanding at approximately 0.5 percent per year. This rate dropped by almost 50 percent in the 1980s as expansion moved into semiarid and montane zones, highly susceptible to degradation (Chap. 29). Even as more land was cultivated, the amount of per capita arable land in the 1990s declined by approximately 1.9 percent per year. Clearly, population growth exceeds the rate of expansion of potential agricultural land usage. Renewable natural resources at times exhibit a nonlinear response to exploitation and population growth. These resources may sustain gradual population growth over many years prior to becoming nonrenewable. Then, especially during periods of very rapid growth, sustainable yields may be exceeded and the resource may become, in essence, nonrenewable – for example, when overpumping of groundwater leads to collapse or salinization of the aquifer. Three key natural resources – soils, groundwater, and biodiversity – are approaching or already at this point in many parts of the world (including the United States).

Based on this understanding of the main forces driving change in the global environment, the following sections explore the implications of population impacts on specific environmental problems and issues encountered throughout the world.

ENVIRONMENTAL IMPLICATIONS OF POPULATION GROWTH

Every day – when we read the newspaper or watch television, see litter on the ground, or notice the smog in the air – we are made aware of the fact that human activities are straining the health of our global ecosystems. The depletion of the Earth's natural resources is not a new phenomenon. However, the rate, scale, and complexity of resource use now

---

THE POPULATION–ENVIRONMENT LINK: THE PHILIPPINES

To explore this population–environment link, consider the specific example of the Philippines. With over half of the country consisting of hilly and mountainous terrain, and heavy tropical rainfall common, soil erosion and other forms of environmental decline are particularly problematic. The Philippine economy is based primarily on its natural resources, including fertile soil for agriculture, mature forests for timber, and unpolluted water for agriculture and fisheries. Unfortunately, the environmental problems already are severe, caused, in part, by population growth, poverty, unwise development, unequal land distribution, and a struggling economy. A shortage of agricultural land for the country's growing population makes loss of productivity from ongoing soil erosion extremely serious. As remaining agricultural settlements, populated primarily by subsistence farmers, shift onto the steeper, unsuitable slopes of the upland forests, this vulnerable land is degraded; consequently, it becomes unsuitable for even subsistence agriculture and timber production.

The removal of forest cover disrupts critical watersheds, causing flooding, erosion, and silting of the reservoirs essential to rice production. Water deficits for agriculture already exist in the Philippines. Along this country's extensive coast, fisheries are negatively influenced by development, fishponds, and lowland and mangrove forest destruction. In areas where the fish catch had previously always increased, the catch is now steady year to year. In a country that depends on agriculture, little new land is available and traditionally farmed land shows signs of strain and overuse. Necessary watersheds are being degraded, yet the Philippine's population growth rate remains steady at 2.5 percent.

Clearly, these environmental problems are strongly linked to population growth and the need to feed more and more people; reduced population growth rates are necessary for the health and well-being of the country's people. Sensitive ecologic zones such as the upland forests, mangroves, and estuaries are of particular importance and must be managed carefully, with population concerns incorporated into development projects.

---

differs from that in earlier times; humanity is now a global force.

Both directly and indirectly, those natural ecosystems that provide essential services to human populations are becoming irrevocably stressed. The environmental link to population growth is seen clearly in such ecosystem services ranging from the food supply derived from marine and terrestrial habitats, to the control of atmospheric gases, to the removal of wastes from our environment. The environment has, to date, provided us with these essential services. At some point, however, the increasing scale of the human enterprise may destroy its ability to do so. Unfortunately, many activities that are a part of our everyday lives and often lead to economic gains also degrade the environment and the essential resources upon which humanity depends. Deforestation, acid rain, land degradation, and water pollution are now present throughout the world, and there are hints of ominous changes in the global climate.

GLOBAL CLIMATE CHANGE

Life on Earth became possible with the formation of the atmosphere. Among other things, the atmosphere provides us with a stable climate, fresh air to breathe, and protection from the Sun's harmful ultraviolet rays. The natural greenhouse effect keeps the Earth's surface habitable by trapping infrared radiation using natural gases in the atmosphere such as water vapor, carbon dioxide, methane, nitrous oxide, ozone, and many others. The relative and absolute amount of these gases determines whether the Earth's surface will be too hot, too cold, or just right (see Chaps. 14, 15, and 16). The alteration of the Earth's atmosphere through increasing emissions of carbon dioxide from fossil fuel burning and changing land use patterns, and the release of other greenhouse gases such as methane and chloroflourocarbons (CFCs), is arguably the most serious environmental problem induced by human activity today. This alteration in the composition of the atmosphere due to global air pollution could bring about climate change at a rate unprecedented in human history.

For approximately the past 200 years, since the beginning of the Industrial Revolution, human beings have dramatically increased the concentration of otherwise naturally occurring greenhouse gases. These increased levels of greenhouse gases in the atmosphere ultimately have the potential to raise the average surface temperature of the planet by several degrees Celsius. Such an increase could trigger dramatic changes in climate, altering patterns of rain, increasing the frequency of violent storms, raising sea level to cause extensive coastal flooding, and severely damaging the agricultural system in the process. Although industrialized nations support less than 25 percent of the global human population, because of their early dependence upon coal, oil, and natural gas, they have contributed approximately 75 percent of the total anthropogenic carbon dioxide released into the atmosphere through fossil fuel consumption. The United States has been the largest emitter, followed by western European nations and the former Soviet Union. Today, developing countries produce approximately 30 percent of carbon dioxide emissions.

Coal releases the most carbon dioxide per unit of energy generated; natural gas releases the least, and petroleum is in between. Many industrialized countries have reduced their use of coal and are experimenting with new sources of energy, which bodes well for the environment. In developing countries, however, industrialization creates a high demand for cheaper energy sources, especially coal. Even modest and very legitimate economic gains carry environmental prices. If China exploits its vast, untapped coal reserves as it contin-

ues to develop, increased carbon dioxide emissions could easily offset the recent emission reductions made by all of the western industrialized nations. Therefore, while developing nations now contribute relatively smaller amounts of carbon dioxide through fossil fuel use, they will contribute orders of magnitude more as they modernize and their populations continue to grow. It is in the self-interest of all nations to help provide both the technology and economic means to use more efficient and cleaner energy sources.

Fossil fuel use is not the only contributor of carbon dioxide to the atmosphere. The amount of carbon dioxide being added to the atmosphere through deforestation and changing land use patterns, while considerably less than that from fossil fuel combustion, is nonetheless significant. Trees and other plants take up carbon (as carbon dioxide) during photosynthesis. When trees are cut down and decay or are burned, they release their stored carbon into the atmosphere. In the tropics especially, large tracts of forest are destroyed to make land available for food production. In many nations, such as Brazil, China, and India, the pattern of deforestation correlates strongly with high rates of population growth.

Biomass burning, the burning of vegetation, is recognized as a widespread phenomenon that significantly affects regional and global climate. In developing countries in the tropics, fire is used extensively following the cutting of forests for shifting agriculture and permanent conversion to ranching and croplands, for extracting energy for cooking from fuel wood and charcoal, for maintaining agricultural lands, and for eliminating agricultural waste. Ranching, timber production, and clearing land for agricultural use are common temporary solutions to economic hardship and overpopulation. These countries are repeating the actions of industrialized countries – such as Canada and the United States – which, through unwise forestry management, have already severely depleted forests to build homes and provide paper products, among other things (Fig. 22.5). Decreasing the rate of expansion of the human population will help stem carbon dioxide emissions; however, the affluence and technology factors represented in the $I = PAT$ equation cannot be overlooked. Reduction in use of fossil fuels, substitution of alternative sources of energy, and increasing technologic efficiency are also required if carbon dioxide emissions are to be significantly reduced.

Several other trace gases contribute to the natural greenhouse effect. Methane, also known as natural gas, actually traps heat more efficiently in the atmosphere than carbon dioxide, but is relatively short-lived. The population link to increased methane emissions is very strong. Sources of methane include rice paddies, cattle, termites, coal mines, garbage dumps, and soils from recently deforested land, all of which relate directly to human activity and population growth. Nitrous oxides are similarly significant greenhouse gases. The anthropogenic sources of nitrous oxides include nitrogen fertilizers, deforestation, coal burning, and auto emissions. Not surprisingly, the most potent and long-lived greenhouse gases are synthetic CFCs. These industrial chemicals, in-

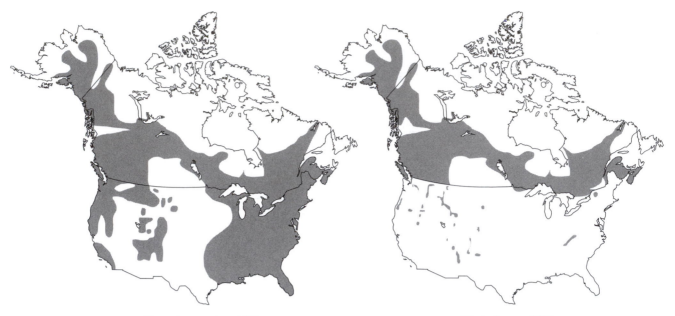

Virgin forests, circa 1600   Virgin forests, 1993

Figure 22.5. Vanishing old-growth forests in the United States and Canada. Since 1620, 90 to 95 percent of the virgin forests that once covered much of the lower forty-eight states have been cleared away. Most of the remaining old-growth forests in the lower forty-eight states and Alaska are on public lands. Approximately 60 percent of the old-growth forests in western Canada have been cleared, and much of what remains is slated for cutting. (*Source:* Adapted from, G. Tyler Miller, *Sustaining the Earth: an integrated approach*, Wadsworth Publishing, 1994; data from Wilderness Society, U.S. Forest Service, and Atlas Historique du Canada.)

vented around 1930, are released solely through human activity and are used in air conditioners, refrigerators, and aerosol cans. Although their effects will be felt for years to come, the rate of CFC emissions has decreased significantly in many areas of the world through successful phase-out programs such as the Montreal Protocol (see Chap. 3).

The most significant impact of global climate change is likely to be on agricultural production. One possible result is a drying up of the world's grain belts and dwindling of the world grain harvest, which could result in the starvation of hundreds of million of people. Also, possible rises in sea level and the ensuing loss of cropland could produce many environmental refugees, especially in lower-elevation countries with dense populations such as Bangladesh. (The general topic of food production is addressed in the section entitled "Food and Agriculture.")

OZONE

The depletion of ozone in the upper atmosphere constitutes another threat to the global atmosphere. In addition to contributing to the greenhouse effect in the lower atmosphere, CFCs rise into the stratosphere and cause a thinning of the ozone shield (see Chaps. 13 and 16). The ozone shield is a protective layer of ozone gas located approximately 20 to 40 kilometers above the Earth's surface. It filters out the Sun's damaging ultraviolet-B (UV-B) rays before they can reach hu-

man beings and other life forms. Stratospheric ozone is being depleted by the decomposition of CFCs and other related compounds. Approximately 3 percent of the protective ozone layer over the United States and other temperate countries has been depleted since the invention of CFCs nearly 70 years ago. However, the best-known ozone decrease lies over the Antarctic continent (the so-called ozone hole), where more than a 60 percent loss of ozone has been measured during the southern spring.

With less protective ozone, increases in ultraviolet radiation are likely to affect human health directly through an increase in the incidence of skin cancer, cataracts, and depression of the immune system, and indirectly through damage to both fisheries and crops sensitive to this higher level of radiation. UV-B rays are toxic and can inhibit photosynthesis in large numbers of forest, crop, and marine species of plants. The United States estimated a 5 to 10 percent loss in potential crop yield due to ozone exposure. Certain marine fisheries may also be at risk due to the sensitivity of marine algae to UV-B. Studies of phytoplankton in Antarctica show declines in production during the period of time when the ozone hole over this region is largest. These Antarctic phytoplankton form the basis for aquatic food chains throughout the world's major oceans, and losses may be felt in fisheries many thousands of miles away. Additionally, there are indications of a worldwide decrease in numbers of frogs and other amphibians that may be related to thinning of the ozone layer. However, it appears that recent technologic innovations leading to CFC elimination and substitution have been successful in slowing down the rate of ozone-depleting emissions, especially in the developed world. CFCs, however, take many years to reach the stratosphere and, once there, can remain active for up to 100 years. Therefore, even after CFC production is halted, they will remain with us for a very long time.

The developing world produces only approximately 17 percent of the global CFC total; however, this level is expected to increase with modernization. In developing countries, the largest growing use of CFCs is for refrigeration. Both India and China, the two most populated nations, expect to increase their stocks of refrigerators significantly over the next decade. In Beijing alone, the proportion of households with refrigerators rose from 3 percent to 60 percent in the 1980s. This same amount of increase is expected in other large cities by the year 2000, however. Per capita output would still be far below that of the United States; however, there are well over 1.2 billion people in China – the sheer number of people there could be a critical factor in CFC release. On a positive note, both China and India are working in accord with the Montreal Protocol to reduce CFC production and address the problem of ozone depletion.

Interestingly, ozone behaves in a different manner in the troposphere and in the lower atmosphere. Unlike stratospheric ozone, tropospheric ozone is a component of photochemical smog. It can cause eye and lung irritation, as well as damage to trees and crops. Legumes, especially soybeans, are particularly at risk. Increased ozone levels in North America and Europe have already caused measurable losses in agricultural productivity. This harmful ozone is formed when the Sun's rays stimulate the reaction of hydrocarbons, nitric oxide, and nitrogen dioxide emitted primarily from vehicle exhaust and biomass burning. Thus, the same gas that protects us from harmful solar radiation is itself harmful when it accumulates at ground level. This type of air pollution forms a third human-induced change in the atmosphere – one that leads to other problems such as acid rain.

ACID DEPOSITION

Anthropogenic air pollution has already caused extensive damage worldwide by altering the natural chemical balance in a number of lakes, waterways, and forests. Acid deposition occurs when sulfur and nitrogen oxides are released as gases into the atmosphere and converted into sulfur and nitric acids. These acids dissolve in water droplets and eventually fall to the ground in the form of acid rain, snow, and fog, and also as dry particulates. Numerous lakes, rivers, and streams in eastern North America and northern Europe have become heavily acidified and are no longer able to support established populations of fish and amphibians. In addition, the evidence is clear that forests have already deteriorated due to the slow buildup of acid in the soil. Forests in Germany were among the first to show signs of damage, and

now forests in central Europe and some high-altitude forests in the eastern United States and Quebec appear to be dying off as well. The damage is not confined to living organisms. Acidification also accelerates corrosion and aging of a number of buildings and historic monuments such as the Acropolis in Athens, Mayan artifacts in Mexico, and the Jefferson Memorial in Washington, D.C.

High levels of sulfur and nitrogen oxides in the atmosphere occur primarily as a result of the combustion of fossil fuels, especially coal. Power plants, ore smelters, and industrial boilers are the principal sources of most sulfur oxide and some nitrogen oxide emissions from human activity. Automobile exhaust fumes constitute another substantial source of nitrogen oxides. Other anthropogenic sources include biomass burning in forests, savannas, and grasslands. The industrialized countries of Europe and North America are experiencing the highest rates of acid deposition. Other industrialized countries such as Japan, and less developed countries such as China, also experience these problems. All these sources of pollution have obvious links to population growth. The effects of pollutants produced in one region of the world can be felt in other areas – even thousands of miles away. For example, coal burning in China for heating and cooking causes acid deposition in regions far from urban industry; acids originating in Europe or the United Kingdom may be deposited in Scandinavia; and the United States most likely produces half the acid precipitation that falls in Canada.

OTHER FORMS OF POLLUTION

Global warming and acid precipitation are not the only manifestations of pollution from the industrial, agricultural, and human waste produced by growing and consuming populations. Humanity faces health-threatening degradation of water and land due to emissions from industrial processes, transportation, and domestic energy consumption. In urban areas of both developing and developed countries, industrial pollutants contaminate the air, water, soil, and food that people depend upon to survive, and they impose serious health and productivity costs.

Agriculture, industry, mining, and human settlements contribute significantly to freshwater pollution. Untreated sewage, heavy metals, toxic chemicals, sediment, and organic matter are only some of the pollutants regularly entering our surface freshwater and groundwater sources. In much of the developing world – and in too much of the developed world – sewage is dumped untreated into surface waters, where excess nutrients can cause algal blooms and oxygen depletion. In such conditions, pathogens can easily enter into household water. This deterioration of water quality is a major health threat. Many people in these areas lack access to clean drinking water, and even more lack adequate sanitation services. For example, in Bangkok an estimated 10,000 metric tons of raw sewage are dumped daily into nearby rivers and canals. Drinking and bathing in polluted waters allows for rapid transmission of infectious diseases such as cholera and schistosomiases. In addition to being a direct threat to human health, water pollution commonly degrades entire ecosystems. In the developed world, regions such as the Great Lakes in North America have been severely damaged by human activity. The fish population fell sharply in Lake Erie, which was for a long time the recipient of toxic effluents. Now, even though Lake Erie is viewed by many as a fishing miracle due to its successful stocking programs, many consider the fish too contaminated to eat (Chap. 9).

Despite their vast size, Earth's oceans are also seriously threatened. A significant increase in coastal marine pollution accompanied the urban, commercial, and industrial development of coastlines and destruction of natural coastal habitat. Over half the world's population lives in coastal areas. The oceans' waters have long been a dumping ground for oceangoing vessels, and growing populations have increased land-based runoff from sewage, agriculture, and land erosion. The excessive nutrients from these sources, especially nitrogen and phosphorus, can cause algal blooms that may block sunlight and release deadly toxins leading to the demise of many fish and invertebrate populations. Almost 80 percent of marine pollution comes from land-based sources (Fig. 22.6). The most serious threats to the marine environment are from sewage, toxic chemicals, radioactive wastes, plastics, sediments, metals, and oil. Oil is one of the most visible contaminants of the oceans. Large spills such as the 1989 *Exxon Valdez* catastrophe in Alaska and those occurring during the Persian Gulf crisis of 1990 to 1991 can instantly alter ecosystems for years, and possibly decades, to come. Smaller spills that occur more frequently may have a much greater cummulative effect than individual large spills. Estimates indicate that shipping activities release some 600,000 tons of oil into the ocean annually. All forms of ocean life, ranging from phytoplankton to fish to coral reefs, are in danger as the pollution of global water resources steadily grows.

The same processes that pollute water resources also pollute the land. The disposal of urban and industrial wastes overtaxes landfills and creates a serious space problem. In many large cities, landfills are completely full and an urgent need exists to dispose of garbage by some other means. A dramatic example of this problem occurred in the late 1980s, when a barge piled high with garbage floated off the eastern coast of the United States for many weeks with nowhere to unload its unwanted cargo. Hazardous waste disposal presents a similar dilemma; few communities want to serve as dumping sites for unsafe, toxic materials. Advances in technologies that produce waste as a by-product far outnumber advances in waste removal technology. Governments and private industries have disposed of radioactive and toxic wastes and allowed them to infiltrate the groundwater. Even where contained waste was thought to be safely stored in underground canisters, leaks into the surrounding soil have occurred and are threatening human health. It is estimated that cleaning up wastes from the U.S. nuclear weapons-making program will cost at least $300 billion.

Because the amount of solid waste generated by a population increases with income levels, it is not surprising that industrial countries produce far more waste than do developing countries. Clearly, as nations improve their economic status and as populations continue to grow, the only way to avoid being overcome by pollution is through a revolutionary change in values and life-styles. These changes demand a decrease in waste production, as well as technologic innovations in the processing and removal of waste.

LAND DEGRADATION AND WATER SUPPLY

Nearly every landscape on Earth has been altered as a result of human activities such as agricultural usage, grazing, deforestation, and urbanization (see Chap. 28). A large portion of this alteration has led to some form of land-surface damage. In these areas, productivity has declined and, in some cases, cannot be restored with current technology. In particular, forested areas have been destroyed as the need for agricultural lands and fuel wood supplies has grown. Indeed, much European and North American farmland was covered with forests only a few centuries ago. The destruction of tropical forests, some of the last remaining relatively untouched areas, has accelerated dramatically since the mid-1900s. Pressure to feed growing populations in developing countries has led to unsustainable slash-and-burn agriculture and direct forest destruction for settled farming, grazing, and logging. As mentioned earlier, clearing and burning leads to the release of greenhouse gases in the atmosphere, and can result in loss of fertility in soils.

Many countries show visible signs of degradation in the form of erosion, waterlogging, and salinization of soils. Since 1950, the world's population has more than doubled. With that increase, approximately 11 percent of the Earth's land

Figure 22.6. Sources of marine pollution. Land-based runoff and emissions account for the majority of marine pollution. (*Source:* Adapted from L. Brown, A. Durning, C. Flavin, H. French, N. Lenssen, M. Lowe, A. Misch, S. Postel, M. Renner, L. Starke, P. Weber, and J. Young, *State of the world 1994: a Worldwatch Institute report on progress toward a sustainable society*, Norton, 1994.)

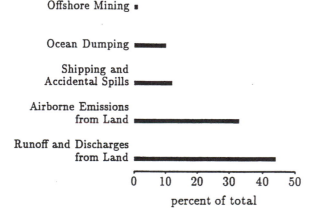

surface, over 1.2 billion hectares, has undergone some degree of degradation. Of this total area, the largest contiguous degraded areas occur in Asia and Africa, the same areas that are home to many of the world's poor. Central America shows the highest percentage of vegetated land that has been degraded to some extent (25 percent), followed by Europe (23 percent), Africa (22 percent), and Asia (20 percent). Suitable agricultural land is increasingly scarce in these regions and, commonly, already under cultivation in populated regions of the world. The overuse of fertilizers and pesticides (although not in the poorest areas), deep plowing, improper irrigation, and inadequate drainage lead to a slow deterioration of the land and potentially to the irrevocable loss of one of any country's most valuable assets: its soil. Even in areas with prolific production, such as the central corn belt in the United States, crop yield potentials have declined and a significant amount of topsoil is lost every year. Many of the methods used to increase productivity are quick-fix solutions focusing on current crops rather than on sustainable harvests. In a recent study assessing global soil degradation (classifying degradation as light, moderate, severe, and extreme), poor agricultural practices were directly linked to the degradation of 28 percent of agricultural land worldwide (Table 22.1) As the human population grows, this loss of high-quality agricultural land will seriously threaten food production worldwide.

Overgrazing, as well as poor agricultural practices, can lead to desertification. Rates of desertification are increasing not only in many arid and semiarid regions, such as in Africa and Australia, but also in the United States. Severe overgrazing by cattle decreases both vegetation and soil compaction, which leads to erosion, soil degradation, and severe disruption of ecosystems. Overgrazing is the leading offender in soil degradation, followed closely by deforestation, then by other agricultural activities. Currently, most deforestation is taking place in the developing regions of the tropics. In the 1980s, close to 17 million hectares of tropical forest were lost each year to deforestation. The serious soil degradation that occurs may leave the land unfit for most agricultural use after only a few years. One of the most serious impacts of tropical deforestation, though, is the loss of species diversity, a topic that will be addressed in the section entitled "Biodiversity."

The degradation of groundwater occurs in much the same manner as the degration of soil – through pollution as well as continued exploitation. Freshwater aquifers are being drained at a rate far exceeding recharge. In certain locations, groundwater levels have fallen by tens of meters. The vast majority of this water is being withdrawn for agricultural and industrial use, and a small amount is used for domestic purposes. Much of the irrigation water in the world is obtained from underground aquifers and lakes. The southern portion of the Ogallala Aquifer in the Great Plains region of the United States is an excellent example of water withdrawn primarily for irrigation purposes. This aquifer has been over-pumped for years, and portions of it are expected to be exhausted by the early part of the twenty-first century. The outlook for grain crops that produced high yields under irrigation in this region over the past thirty years is today much less certain as reversion to rain-fed agriculture looms.

Renewable water supplies are also affected by the changes in land-use patterns brought about by urbanization, especially in developing countries. Deforestation, the alteration of natural runoff patterns, eroding soil, and evaporation, as well as toxic substances from urban runoff, are severely degrading water quality. The world's freshwater supply is unevenly distributed, as is the demand for water. Already, many nations in the Middle East and Africa face severe water shortages. The large populations of Beijing and other cities in the semiarid regions of China are depleting their freshwater sources rapidly. Worldwide, human water use is currently increasing at a rate of 4 to 8 percent per year, much higher than the present rate of population growth. The highest per capita water use occurs in the industrialized countries of North America and in Central America. Most of the expected increase, however, will take place in developing countries as industrial and agricultural use rises. Water use in developed countries is actually expected to decline slightly. The impact of the overall rise can be lessened through improved water resource management such as changes in crop rotation, irrigation practices, and runoff control. The implementation of conservation measures and reallocation of agricultural water to more efficient uses is essential.

BIODIVERSITY

The environmental issues discussed thus far together contribute to the loss of plants, animals, and microorganisms on our planet at a rate approaching the magnitude of mass species extinctions millions of years ago (see

TABLE 22.1. ESTIMATES OF LAND DEGRADATION WORLDWIDE, 1945–1990 (VALUES GIVEN IN MILLIONS OF HECTARES)

Source	Asia	Africa	South America	Europe	North/Central America	Oceania	World
Overgrazing	197	243	68	50	38	83	679
Deforestation	298	67	100	84	18	12	579
Agricultural activities	204	121	64	64	91	8	552
Other	47	63	12	22	11	0	155
Total	746	494	244	220	158	103	1965
% of Total Vegetated land	20	22	14	23	8	13	

Source: L. Brown, A. Durning, C. Flavin, H. Freuch, N. Lenssen, M. Lowe, A Misch, S. Postel, M. Renner, L. Starke, P. Weber, and J. Young, *State of the World 1994: A Worldwatch Institute Report on Progress Toward a Sustainable Society*, New York, Norton, 1994.

Chap. 3). Population and species extinctions are accelerating rapidly through tropical deforestation, and through the destruction of temperate forests, wetlands, coral reefs, and other unique habitats supporting the world's growing human populations. The Earth contains billions of distinct populations comprising 5 to 30 million species, of which only 1.4 million have been described and catalogued, leaving the vast majority of species and their potential genetic resources unknown. Evidence from the fossil record indicates that, on average, species have persisted between 1 and 10 million years before factors such as climate change, severe drought, natural catastrophes, new predators, or genetic mutations cause them to go extinct. Similarly, the fossil record also suggests that for most of the history of life on Earth, rates of species extinction and rates of species evolution have been approximately the same. However, recent estimates indicate that contemporary rates of extinction far surpass rates of speciation. Some experts project that 50 percent of all modern species will go extinct within the next 100-year period if current rates of deforestation continue. Most of these extinctions will take place in the tropics, where it is estimated that an overwhelming 50 to 90 percent of all species live. If this projection proves to be true, and assuming that 50 percent of all existing species evolved within the past 50 to 100 million years, then today's rates of extinction are on the order of 1 million times faster than rates of speciation.

This major episode of species extinction would rival five other massive species die-offs over the past 500 million years that resulted in the loss of a significant portion of global flora and fauna. The most recent event took place 65 million years ago when the dinosaurs disappeared. Following each of these events, it took approximately 10 million years for the number of species on Earth to return to the previous level. Unlike these other massive extinction events, however, the current die-off is driven by human factors: especially land-use change and habitat loss, species introductions, overharvesting of plants and animals, pollution, and other activities that degrade or destroy natural ecosystems and the species within them.

The development of agriculture 10,000 years ago changed land use patterns significantly through the conversion of wildlife habitat to farmland, rangeland, and human settlements. These trends rose sharply in the twentieth century as human populations tripled and resource consumption and waste production increased. Through loss of species, humanity erodes genetic variability, which is the base of agriculture, a source of new medicines, and a contributor to countless ecosystem services. These services include climate control through regulation of atmospheric gases, water recycling and the regulation of the hydrologic cycle, maintenance of soils and soil fertility, recycling of nutrients, as well as waste and pollution disposal. In addition, biologic resources provide clothing and housing. The socioeconomic benefits of biodiversity include a $40 billion worldwide market for medicines from wild sources, and annual economic gains worldwide in the use of wild species to improve various aspects of agricul-

ture. For example, approximately 25 percent of the pharmaceuticals in the United States contain active ingredients derived from wild plant species. Preserving biodiversity will benefit humans by ensuring genetic variability, ecosystem services, and direct economic benefits, as well as providing aesthetic and ethical pleasure.

FOOD AND AGRICULTURE

The world's food supply is tied directly to loss of biodiversity. Biodiversity is the basis of agriculture and has provided all the crops and domestic animals used by humanity. From the many thousands of wild plant species across the world that people have attempted to cultivate, approximately 100 species provide most of the world's food today. Indeed, more than half of the calories consumed by humanity are supplied by only three grass species (Chap. 29): rice, wheat, and corn (maize). These crops are commonly threatened by pests and diseases, and genetic variability is necessary in order to continue to develop new, productive varieties of these crops. Although global food production has risen dramatically over the past twenty-five years, the need to produce more food to keep pace with a growing population remains a serious problem. In parts of North America, Asia, and Australia, food production has increased more rapidly than population levels; however, in parts of Africa, Latin America, and South Asia, food production lags significantly behind population growth. This is complicated by the fact that available food is not evenly distributed; consequently even in times of surplus stock, many people go hungry. The proportion of chronically undernourished people in the developing world has decreased substantially, from an estimated 36 percent in 1969 to 1971 to 20 percent in 1990 to 1992. Yet, even with this encouraging trend, over 800 million people were undernourished in 1990 to 1992.

The combined pressures of population growth and degradation of the environment dim the prospect of continued production increases at the current rate. Some areas can still provide new farming and grazing land; however, Green Revolution technology and the heavy use of fertilizers, pesticides, and intensive cropping, not expansion of cropland, have led to increased food production in the past few decades. This improved technology has exacted an environmental cost in the form of increased erosion, nutrient depletion, waterlogging, salinization, loss of genetic resources, and vulnerability to pests, which will eventually decrease yields. Today, after more than three decades of increased per capita food output, crop yields have leveled off due to this degradation of agricultural lands. In India, a country generally viewed as a Green Revolution success, at least 30 percent of irrigated croplands have lost much of their productivity due to salinization. Similarly, Pakistan has lost approximately 20 percent to the same problem. Estimates indicate that the world total of irrigated lands lost to salinization is over 20 percent. Additionally, as populations expand, it is the marginal land that is incrementally cultivated, areas that are not sustainable for the future. Conventional farming practices must evolve toward more sustainable agricultural strategies that will reduce environmental damage while continuing to provide an adequate food supply in the future. Some examples of these strategies include decreasing the use of pesticides in favor of crop rotation, early or delayed planting, planting more than one crop, and using natural predators, parasites, and diseases to maintain pest populations below damaging levels. Modern biotechnology techniques such as genetic engineering also offer future hope for expanding the global food supply through enhanced yields, improved pest and disease resistance, and improved nutritional value, among other things.

The oceans are another important supplier of food, and a major source of protein for the world's poor. Marine fisheries experienced a sixfold increase in harvest catch size from 1950 to 1997 as fishing efforts intensified, more efficient equipment was developed, and new navigation methods were used (Fig. 22.7). In recent years, however, this increase has resulted in overexploitation of traditional stocks, decline in catches of top commercial fish, and changes in the composition of marine ecosystems. The world fish catch is in jeopardy. On Georges Bank in the northwest Atlantic, overfishing has caused a precipitous decline in catches of high-value fish such as cod, haddock, and flounder, and an increase in less commercially desirable species such as skates and dogfish. At low population levels, the flounder and other groundfish are reportedly outcompeted for food or space and preyed upon by the lower-valued fish, thus altering the original species proportions. This is a common and potentially destructive trend, because many traditional marine stocks may have

reached their maximum sustainable yield, and further exploitation could cause significant declines for years to come. International ocean management programs, quota-based fisheries management, and fishing agreements such as the regional seas programs of the United Nations Environment Programme are developing in order to help maintain productive fisheries and sustainable harvest of the oceans (see the box entitled *Increasing Food Demands: The Crisis in Peru*).

HEALTH AND EPIDEMIOLOGIC ENVIRONMENT

Human health has strong links to environmental processes and the health of the planet as a whole. Population growth can affect human health indirectly through alteration of the environment, especially in the form of air, land, and water pollution, malnutrition due to inadequate food supplies and distribution, and declining water availability. The spread of disease and threat of epidemics is also more likely as human communities become more crowded, transmission barriers break down through improved transportation, malnutrition compromises immune systems, and exposure of sizable human populations to disease-causing organisms in the environment increases. Although life expectancy has improved in all developing regions over the past few decades (yet not necessarily in every country), discrepancies remain between the health prospects of developed countries and those of developing nations.

In the developing countries, health risks are linked closely to poor sanitation, overcrowding, and poor housing conditions, indoor air pollution from wood and dung burning, and poor water quality. Deteriorating water quality is a particular threat in these countries, where in 1990, despite the fact that many additional people were supplied with water and sanitation throughout the 1980s, 1.2 billion people still lacked access to clean drinking water, and 1.7 billion lacked access to sanitation (Fig. 22.8). When untreated sewage is discharged directly into surface waters, infectious diseases spread through the acts of washing and drinking from these waters. Cholera epidemics in Africa in the 1970s, and in both Africa and Latin America in the 1990s, have been attributed in part to unsafe drinking water and poor sanitation. Diseases in which other organisms are involved in the life cycle, such as malaria, giardia, and schistosomiases, are also easily spread and rampant in many developing regions. Cases of malaria have been reported in more than 100 countries, and the number appears to be growing as malaria-carrying mosquitoes become more resistant to insecticides, and as strains of the malarial organisms (such as *Plasmodium* species) become resistant to antimalaria drugs. It is estimated that worldwide, 1 million people die of malaria each year.

Industrial pollution and urban development, for years highly destructive of the environment in industrialized countries, are now of growing concern in developing countries as well. Exposure to toxic chemicals, smog and urban fumes, hormone-mimicking compounds, hazardous wastes, and myriad other industrial substances potentially causes rises in

Figure 22.7. World fish production, 1950–96. Total world fish production has slowed since the 1970s, but has not yet declined. Unprecedented growth in aquaculture has played an important role in maintaining this trend. In 1995, almost 20% of all fish and shellfish production was attributable to aquaculture. (*Source:* Adapted from L. Brown, C. Flavin, H. French, J. Abramovitz, C. Bright, S. Dunn, G. Gardner, A. McGinn, J. Mitchell, M. Renner, D. Roodman, L. Starke, and J. Tuxill, *State of the world 1998: a Worldwatch Institute report on progress towards a sustainable society,* Norton, 1998).

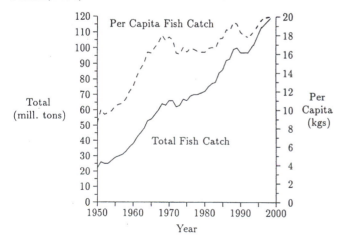

INCREASING FOOD DEMANDS: THE CRISIS IN PERU

As the human population increases on Earth, the demand for food also increases. Pressure on marine fisheries to help meet this need through increasing yields places excess stress on many historically productive areas. The North Atlantic, Central Pacific, and South Pacific fisheries have been especially hard hit. The Peruvian anchovy fishery is an example of a large-scale marine fishery that collapsed, in part, as a result of exploitation and overuse. The Peruvian coast is characterized by upwelling events in which cold, nutrient-rich water is brought to the surface from depth. The abundant nutrients in these waters support the growth of phytoplankton and create areas of high biologic productivity. Upwelling systems throughout the world occupy only 0.1 percent of the oceans yet contribute approximately 50 percent of the world's fish catch. However, between 1959 and 1989 the species composition in the Peruvian system changed dramatically, and the total biomass of six major species decreased by close to 50 percent.

In 1950 Peruvians began to harvest their large population of anchovies for oil and fish meal export. Previous to this time, anchovies were not harvested on a large scale, and other fish were harvested only for human consumption. However, the growing demand for fish meal by processing plants over the next few years made Peru the largest fishing nation worldwide, with anchovies constituting almost 95 percent of the total catch. Throughout the 1960s, Peruvian harvests approached and exceeded 12 million metric tons, despite warning by biologists that the fishery could not sustain harvests over 9.5 million metric tons, including those taken by seabirds. The increase in annual anchovy catches continued until the spring of 1972, when record numbers of anchovies were found near the shore in March followed by a severe decline in numbers for the remainder of the season. The El Niño/Southern Oscillation, an atmospheric–oceanic phenomenon that occurs every three to seven years in varying intensity, arrived and brought with it warm surface waters that disrupted the typical coastal upwelling pattern. The El Niño event can last for up to 18 months, and the fish catch over the next four years plummeted to between 1 and 3 million metric tons annually. Throughout the 1980s, as more El Niño episodes occurred, the annual average catch was only 1.2 million metric tons, or approximately 10 percent of the harvest during the 1960s.

Although El Niño events contributed to the collapse of the Peruvian anchovy fishery, overfishing prior to the 1972 event placed the population in jeopardy. With lower numbers and continued overfishing, the population could not naturally protect and restock itself.

The continued success of many fisheries in the presence of population growth and environmental change will require better understanding of specific marine ecosystems, such as the Humboldt Current ecosystems off Peru, and also rethinking of traditional fishery management practices.

One system worth watching closely is that of New Zealand. That country has implemented a quota-based management system that places limits on catches and was created to reduce overfishing and protect the fishing industry. Fishermen must purchase an individual transferable quota (ITQ) allowing them to harvest a given amount of fish. The total amount of fish harvested is determined prior to the beginning of the season through assessment of the stock. By selling fisherman exclusive rights to a set amount of the resource, larger companies cannot outcompete smaller companies, sustainable stock yields are maintained, and more centralized management exists. However, the downside of limited-access systems includes economic rigidity, less competition, and problems associated with inaccurate stock assessments. Many countries with limited ITQ systems – including the United States, Canada, Iceland, and Australia – are monitoring the New Zealand fishery for signs of success or failure.

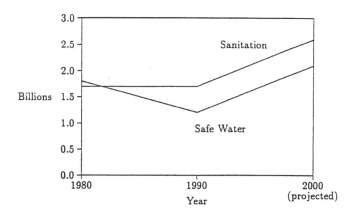

Figure 22.8. Population without access to safe water and sanitation in developing countries. Although 1.3 billion people were supplied with new water sources and 750 million people with sanitation in the 1980s, population growth prevented substantial gains. In the 1990s, the number of people lacking access to safe drinking water and sanitation continued to increase. (*Source:* Adapted from P. H. Gleick, *Water in crisis: a guide to the world's fresh water resources*, Oxford University Press, 1993.)

the incidence of cancer and other serious diseases. Air pollution is linked to the increased occurrence of respiratory and pulmonary diseases, especially in children. Humanity's reproductive health is also threatened due to the increased presence of synthetic chemicals in the environment. Some of these chemicals interfere with the endocrine system – a system of chemical messengers that controls many of the functions of our bodies – early in the developmental process. Disruption of this system through hormone-mimicking compounds may already be lowering sperm counts, causing malformed reproductive organs, and otherwise harming developmental processes in human beings and other species. Overall, the human health and epidemiologic environment faces increased risks as populations expand and the natural environment is more intensely altered.

ENERGY USE

A reduction in the impact of growth on the environment through energy conservation is absolutely imperative. Energy use is the central component of human activity contributing to the degradation of environmental quality, and global energy consumption has increased by over 50 percent since 1970 (see Chap. 24) Reductions in commercial fossil fuel use alone would greatly reduce the ongoing environmental impact of the human enterprise. Generally, people in developed countries use approximately seven times more energy than people in developing countries. The average American uses almost a dozen times more energy than the average citizen of a poor country. Coal, oil, and natural gas are the primary energy sources in developed countries. In developing regions, however, a number of countries continue to depend heavily on noncommercial fuels such as wood, crop waste, and animal dung. Energy consumption in these countries tends to be small, but very inefficient. In many houses, people commonly burn raw coal and wood for heating and cooking in open fireplaces with inefficient stoves and heaters that release toxic pollutants, in turn wasting potential energy.

Industrialized countries have been responsible for the majority of the growth in commercial energy use worldwide. Although energy consumption has increased in developing countries as well, per capita differences remain large. The excessive consumption of energy must decrease in the industrialized world in order to bring about a more equitable distribution of commercial energy. As the developing world modernizes, there will be a rising demand for oil and other commercial energy sources. Because this rapid growth inevitably will have substantial and deleterious effects on the environment, the use of environmentally sustainable technologies has become imperative. Advances in the development of alternatives energy sources such as wind, solar, small-scale hydro, geothermal, and nontraditional biomass are promising; these sources are far less environmentally destructive than today's major commercial sources.

Reductions in energy use and conversion to renewable energy sources are needed in both developing and industrialized countries. The Holdren scenario, an optimistic scheme for minimizing total energy use, involves per capita energy reduction for industrialized countries (from 7.5 kilowatts in 1990 to 3.8 kilowatts by 2025) and per capita increase in energy use by developing countries (from 1.0 kilowatts in 1990 to 2.0 kilowatts by 2025). Based on an eventual population size of 10 billion people, per capita consumption worldwide will converge on 3.0 kilowatts by the year 2100 (Table 22.2). Under this scenario, total energy use is expected to more than double from 13.0 terawatts (trillion watts) in 1990 to 30 terawatts in 2100. Increased energy efficiency through continually improving technology can allow for a higher standard of living for more people. Clearly, environmental impacts will increase; however, conversion to new energy technologies, ideally, will lessen the projected damage.

MANAGING THE ENVIRONMENT: SOLUTIONS AND CONCLUSION

Growth in human numbers and per capita consumption has multiple and diverse impacts on the environment. Reducing fertility, slowing or even halting population growth, will go a long way toward stemming and ultimately reversing the rate of increasing environmental degradation. This reduction is necessary, but not sufficient. At the same time, growing awareness of the link between population and the environment will encourage the implementation of sustainable practices. Many of the strategies that would lower birth rates are "no-regrets" strategies, ones we would consider important to use (even if they did not lower fertility rates) because they have other benefits as well. Outstanding among these is empowering women by providing them with education, health care, and access to safe contraception.

The 1980s and early 1990s marked an increase in an international attempt to promote sustainable development. The United Nations Conference on Environment and Development (UNCED) held in Rio de Janeiro in 1992 was one of the largest international conferences ever held. Cooperative agreements were signed at the meeting to strengthen and enhance international support for sustainable development.

TABLE 22.2. HOLDREN SCENARIO

Year	Population (billions)	Energy/Person (kilowatts = kw)	Total Energy Use (terawatts = tw)
1990 Rich	1.2	7.5	9.0
1990 Poor	4.1	1.0	4.1
Total	5.3		13.1
2025 Rich	1.4	3.8	5.3
2025 Poor	6.8	2.0	13.6
Total	8.2		18.9
2100	10	3	30

Source: J. Holdren, Population and the energy problem, *Population and Environment*, vol. 12, no. 13, 1991, pp. 131–255.

The meeting outlined the responsibilities of countries toward the global environment. Population issues received special attention two years later at the United Nations 1994 Conference on Population and Development, held in Cairo. Progress was made on many specific issues, including commitment toward reducing infant and maternal mortality rates, empowering and educating women, and the recognition of overpopulation. Other issues were hotly contested, such as promoting the use of contraceptives and making available the option to terminate health-threatening or unwanted pregnancies. Unfortunately, developed nations have yet to demonstrate a firm commitment to reduce their consumption and to deploy more efficient technologies.

International conferences and agreements are only part of the effort to address and solve these problems. Scientists from different countries and different disciplines are also working closely together to gain a better understanding of the processes and mechanisms at work in the biosphere, atmosphere, and geosphere. For example, the International Geosphere-Biosphere Program (IGBP) is a cooperative venture to describe and understand the interacting physical, chemical, and biologic processes regulating the total Earth system, and the global change processes influencing this system. The Intergovernmental Panel on Climate Change (IPCC) has brought together the world's leading atmospheric scientists to help reach a scientific consensus on global climate change. These programs help provide guidelines for the direction and actions that we must take to improve our global environment. However, it is principally at the national level that people are responsible for managing their own resources and protecting the environment. In the United States, legislation, such as the Clean Water Act and the Clean Air Act, is reducing environmental pollution. However, this kind of legislation is under pressure from various industries and other groups who seek to weaken or abolish it because it typically requires implementation of new, more costly technology. The Endangered Species Act, another U.S. policy, helps to protect a number of plants and animals that are believed to be in danger of extinction; however, this act does not adequately address the connection between saving species and the need to protect whole ecosystems.

Encouraging signs abound, and there have been a number of environmental successes, both nationally and internationally, indicating that change is possible. The dramatic decrease in the world's crude birth rate over the past three decades, including arrival at ZPG in many western European nations, China's reduction in birth rate (albeit with the creation of some social problems), and the general growth in overall population awareness is encouraging. There has also been a reduction in water pollution with the advent of the U.S. Clean Water Act over two decades ago. Some rivers and lakes have been cleaned up and are now showing dramatic signs of recovery, such as Lake Erie, the Nashua River in Massachusetts, and Maine's Penobscot River (also London's Thames River). Highly destructive pesticides such as PCB and DDT are being banned. Political leaders in many countries are currently campaigning under environmental programs and platforms. Many possibilities for change exist; they inevitably involve profound alterations in life-styles and values, as well as technologic innovations.

Achievement of sustainable development is a great challenge facing humanity today and will ultimately require action on all components of the $I = PAT$ equation: population, affluence, and technology. Changes in consumption patterns as well as reduction in environmental impact through advances in technology are especially promising as a result of actions such as energy conservation, changes in agricultural practices, education, and health improvement. Yet even with drastic technologic changes and increased affluence, increasing population numbers will continue to intensify the environmental impact. Inasmuch as the very life of the planet continues to be challenged, the time for us to act is now.

QUESTIONS

1. What are important factors to consider when developing plans to ensure a sustainable environment?

2. Describe specific links between population trends, economic development, and environmental protection.

3. How can humanity best meet the challenge of population growth and distribution, human needs, resource preservation, and management?

4. Propose some possible population policy options for the developing world.

5. Propose some possible population policy options for the developed world.

6. List specific human needs and qualities of life that you consider essential to sustainable development (e.g., good health, food, security).

7. Roughly estimate your national population carrying capacity based upon these essential needs.

8. Identify the various uses of fresh water in the world today. For what purpose is the majority of fresh water used?

9. How does fossil fuel use lead to serious environmental problems?

FURTHER READING

Brown, L., Flavin, C., French, H., Abramovitz, J., Bright, C., Dunn, S., Gardner, G., McGinn, A., Mitchell, J., Renner, M., Roodman, D., Starke, L., and Tuxill, J. 1998. *State of the world 1998: a Worldwatch Institute report on progress towards a sustainable society.* New York: Norton.

Daily, G. C., and Ehrlich, P. 1996. Socioeconomic equity, sustainability, and carrying capacity. *Ecological Applications,* vol. 6, pp. 991–1001.

Daily, G. C., Ehrlich, P., and Ehrlich, A. 1994. Optimum human population size. *Population and Environment* 15(6):468–475.

Ehrlich, P., and Ehrlich, A. 1990. *The population explosion.* New York: Simon and Schuster.

Ehrlich, P., and Ehrlich, A. 1991. *Healing the planet: strategies for resolving the environmental crisis.* Reading, MA: Addison-Wesley.

Gleick, P. H. (ed.) 1993. *Water in crisis: a guide to the world's fresh water resources.* New York: Oxford University Press.

Holdren, J. 1991. Population and the energy problem. *Population and Environment* 12(3):231–255.

Population Reference Bureau (PRB). 1999. *World population data sheet.* Washington, DC: PRB.

Schneider, S. 1989. *Global warming: are we entering the greenhouse century?* San Francisco, CA: Sierra Club Books.

Scientific American. 1990. *Managing planet Earth: readings from Scientific American.* New York: W. H. Freeman.

Shiva, V. 1991. *The violence of the green revolution: third world agriculture, ecology, and politics.* London: Zed Books.

Turner, B. L., Clark, W. C., Kates, R. W., Richards, J. F., Mathews, J. T., and Meyer, W. B.(eds.) 1990. *The Earth as transformed by human action: global and regional changes in the biosphere over the past 300 years.* New York: Cambridge University Press.

United Nations Population Fund. 1991. *Population, resources, and the environment.* London: Banson.

Wilson, E. O. (ed.) 1988. *Biodiversity.* Washington, DC: National Academy Press.

World Resources Institute. 1998. *World resources 1998–99: a guide to the global environment.* New York: Oxford University Press.

23 Mineral Resources: Assets and Liabilities

MARCO T. EINAUDI

M aintaining an adequate supply of minerals for a growing world population requires continued discovery and extraction of mineral resources. At the same time, we must ensure that the consequences of mining are not deleterious to the environment and that we do not deplete our resources to the point where future generations cannot meet their needs. This challenge requires a multidisciplinary approach, one that spans the fields of geology, engineering, chemistry, economics, public policy, and international relations. In the near future, we need to bring about the effective joining of these disciplines to address the problems of resource depletion and related environmental degradation. Our success in doing so will determine whether *sustainable development* will amount to anything more than a concept.

This chapter lays out some of the fundamentals of understanding, and also the gaps in our understanding, that serve as a basis for discussion and future study of the sustainable development of mineral resources. The core of the chapter deals with the geology of a subset of mineral deposits chosen to illustrate processes that form mineral resources at the interface between the solid Earth and its fluid envelopes. The end of the chapter examines the meaning of mineral reserves, their uneven world distribution, and their long-term availability, and it proposes a method for approximating the potential environmental hazards from mining the Earth's mineral resources.

WHAT ARE MINERAL RESOURCES?

Mineral resources consist of the naturally occurring materials used for industrial, agricultural, and other purposes. Some authors include fluids (e.g., air, water, petroleum, natural gas) as mineral resources. Air could be included not because life is dependent on its oxygen content, but because most of the world's ammonia and nitrate (the most important fertilizer) are "mined" from the atmosphere by reacting hydrogen in natural gas with nitrogen in air. Water could be included not because life is dependent on it, but because much of the world's halite (salt), bromine (used in fire-retardant and agricultural chemicals), and magnesium (used in refractories and aluminum alloys) are "mined" from seawater and salt lakes. Most authors restrict the term *mineral resource* to those materials composed of *minerals* (Chap. 5). This more restrictive, mineralogic definition is adopted here for the most part; coal, peat, oil shale, and oil and gas are excluded as mineral resources. Fossil fuels are treated in depth in Chapter 24.

Mineral resources include a broad range of Earth materials ranging from diamonds to gravels and from antimony to zinc. Some mineral resources are rocks that are mined in their natural form with minimal processing, such as building stone, sand, and gravel. Other mineral resources are mined to extract a given mineral that is itself the sought-after commodity; examples include asbestos, diamonds, and native sulfur. Finally, some mining operations extract rocks containing an abundance of useful elements that are concentrated in specific *ore minerals*. In the latter case, rocks containing the ore mineral are mined, the ore mineral is separated from waste by crushing and concentrating, and then the ore minerals undergo specialized treatment to remove the sought-after elements. Typically, it is the latter type of resource extraction that produces toxic wastes.

Most books on mineral resources commence with a presentation of various classifications of mineral deposits. Although seemingly a "dry" subject, classification is important because it reflects and directs the way we think. If our main concern is the *origin* of mineral deposits, we classify according to genetic processes. If our main concern is *mineral exploration*, we classify according to geologic environment. If our main concern is supply and distribution, we classify according to commodity. In the classification presented in Table 23.2, mineral deposits are grouped by commodity and use

TABLE 23.1. USES OF NATURALLY OCCURRING ELEMENTS IN EVERYDAY LIFE.

Element	Occurrence in Nature	Used as: Elemental	Alloy or mixture	Compound
hydrogen	compound, gas	rocket fuel		
helium	elemental, gas	balloons, lasers		
lithium	compound, solid	pacemaker batteries		lubricant additive, glass
beryllium	compound, solid	X-ray tube windows	watch springs	
boron	compound, solid			tennis rackets, eye disinfectant
carbon	elemental and compound, solid	diamond, pencils, tire colorant		plastics
nitrogen	elemental and compound, gas	cryogenic surgery		rocket fuels
oxygen	elemental and compound, gas	combustion, water purification		
fluorine	compound, gas			toothpaste additives, refrigerator coolants, teflon
neon	elemental, gas	neon lights, fog lights, TV tubes, lasers		
sodium	compound, solid	street lights	batteries	kitchen salt, soda, glass
magnesium	compound, solid		racing bikes, airplanes	bricks, pigments
aluminum	compound, solid	window frames, doorknobs, foil, fireworks, flash bulbs, power cables	planes, cars	
silicon	compound, solid	micro chips, solar cells		glass, cement, silicone rubbers

(continued)

TABLE 23.1. (*cont.*)

Element	Occurrence in Nature	Used as:		
		Elemental	Alloy or mixture	Compound
phosphorous	compound, solid	matches, fireworks		fertilizers, detergents, toothpaste, pesticides
sulfur	sulfide, sulfate, & elemental, solid & aqueous	matches, firework, batteries		permanent wave lotion
chlorine	compound, gas	water purification		bleach, PVC plastics
argon	elemental, gas	light bulbs, lasers		
potassium	compound, solid			fertilizer, matches, gun powder
calcium	compound, solid		cable insulation, batteries	fertilizer, plaster of Paris
scandium	compound, solid			large-screen TVs, stadium lighting
titanium	oxide, solid	heat exchangers	bone pins, airplane motors	white pigments for paper and paints
vanadium	oxide, solid		tools, springs, jet engines	
chromium	oxide, solid	chrome plating	tools, knives	lasers, stereo, video tapes, green and yellow pigments
manganese	oxide, solid		steel	batteries, violet pigments
iron	oxide, sulfide, solid		bikes, cars, bridges, nails	yellow, red, and brown pigments
cobalt	sulfide, solid	gamma radiation source	razor blades, permanent magnets	catalytic converters, blue pigments
nickel	sulfide, solid	coins	knives, forks, spoons	rechargeable batteries
copper	sulfide, oxide and elemental, solid	water pipe, power cables	pennies, Statue of Liberty, bells	
zinc	sulfide, solid	galvanized platings, batteries	water valves	white pigments in rubber
gallium	sulfide, solid	quartz thermometers		computer memory, transistors, used to locate tumors
germanium	sulfide, solid	wide-angle lenses, infrared night vision	dentistry	fibre optics
arsenic	sulfide, oxide; solid	shotgun pellets	metal for mirrors	light-emitting diodes (LED displays), glass, lasers
selenium	elemental and sulfide, solid	light meters, copy machines, solar cells		dandruff shampoos
bromine	compound, liquid			photographic films, disinfectants, tear gas
krypton	elemental, gas	fluorescent bulbs, flash bulbs, UV lasers		
rubidium	comp0ound, solid	vacuum tubes, photoelectric cells		

(continued)

TABLE 23.1. (*cont.*)

Element	Occurrence in Nature	Used as:		
		Elemental	Alloy or mixture	Compound
strontium	compound, solid			phosphorescent paint, crimson fireworks
yttrium	compound, solid			color TV screens, superconductors, radar, lasers
zirconium	compound, solid	nuclear fuel rods, catalytic converters		zircon gemstones
niobium	compound, solid		cutting tools, welding rods, pipelines	
molybdenum	sulfide, solid	filaments in electric heaters	rocket motors, high-speed tools	lubricants
ruthenium	elemental, solid	eye treatment, thickness meters for eggshells	fountain pen points, electrical contacts	
rhodium	elemental, solid	headlight reflectors, catalytic convertors, telephone relays	airplane spark plugs	
palladium	elemental, solid	catalytic converters, hydrogen separation	dental crowns, anti-tumor agents	
silver	elemental and sulfide, solid	mirrors, batteries	silverware	photograpic film and paper
cadmium	sulfide, solid	rechargeable batteries, screw and bolt platings	regulator in nuclear reactors	read and yellow pigments
indium	compound, solid	solar cells, mirrors	regulator in nuclear power	transistors, photo cells
tin	oxide, sulfide; solid		organ pipes, cups, plates, coins	opalescent glass, enamel, weather-resistant vinyl
antimony	sulfide, solid		sodder, lead batteries	fire retardant, ceramic glazes
tellurium	sulfide, solid	percussion caps, vulcaniation of rubber	electrical resistors	
iodine	compound, solid	disinfectant, halogen lamps		ink pigments, salt additive
xenon	elemental, gas	UV lamps, projection lamps, electronic flashes (sun lamps)		
cesium	compound, solid	photoelectric cell, gamma radiation source		scintillation counters
barium	sulfate, silicate; solid		spark plugs, vacuum tubes	fluorescent lamp, green fireworks, yellow and red pigments
lanthanum	compound, solid		lighter flints	battery electrodes, catalytic converters, camera lenses
Lanthanum series, *14 elements from cerium to lutetium*	compound, solids	self-cleaning ovens	catalytic converters, disc drives, magneto-optic alloys for CD's	color TV tubes, X-ray screens, fluorescent lamps, glass colorings

(*continued*)

TABLE 23.1. (*cont.*)

Element	Occurrence in Nature	Used as: Elemental	Used as: Alloy or mixture	Used as: Compound
hafnium	compound, solid	nuclear submarines, vacuum tubes		
tantalum	compound, solid	capacitors	vacuum tube filaments, cutting tools, weights	
tungsten	tungstate, solid	lamp filaments, welding electrodes	rocket nozzles	cutting and boring tools
rhenium	compound, solid		oven filaments, jewelry plating, electrodes, thermocouples	
osmium	elemental, solid		fountain pen points, compass needles, clock bearings	
iridium	elemental, solid	satillite thruster engines	hypodermic needles, helicopter spark plugs	
platinum	elemental, sulfide; solid	crucibles, jewelry	dental crowns, jewelry	anti-tumor agents
gold	elemental, sulfide, teluride; solid	hoarding, jewelry	jewelry, electrical contacts, dental crowns	
mercury	elemental, liquid; sulfide, solid	thermometers, barometers, streetlights	dental fillings	seed protection
thallium	compound, solid		thermometer filling	insecticides, infrared detectors, heart research
lead	sulfide, solid	radiation protection	batteries, roof coverings, solders, ammunition	(gasoline additives) (white pigments)
bismuth	sulfide, elemental, solid	catalyst in rubber production	fuses, fire sprinklers	antiacids, anti-diarrheals
polonium	compound, solid	nuclear batteries, neutron source, antistatic agents		
radon	elemental, gas	earthquake prediction, health threat in homes built on granite		
radium	compound, solid			neutrn source, glow-in-the-dark paint
actinide series, *3 elements from actinium to uranium*	compounds, solids	neutron sources, nuclear and breeder reactor fuels (*thorium* and *uranium*), coating on filament wire (*thorium*)	gyro compasses	gaslight mantles (*thorium*), glass coloring (*uranium*)

TABLE 23.2. MINERAL RESOURCES: TYPES, ENVIRONMENTAL ISSUES, AND WORLD SUPPLY

Name of commodity (chemical symbol if elemental) Annual value in millions of US$	Important ore minerals or materials: rock [rx], laterite [lt], mineral [m], liquid [lq], brine [b], seawater [sw], minor element in other ore minerals [tr]	Typical ore grades (grade = wt % of desired constituent), or by-product [bp] (Future resource in parentheses)	Major Environmental Issues associated with: (1) mining; (2) concentrating; (3) extraction of final product. BFF = burning of fossil fuels (rather than mining) is major source of pollution involving given element	World Reserve Life Index: world reserves divided by world production (1992)	Countries with Major Reserves [% of world reserve] or [% of world production]	U.S. Imports as % of consumption (1992)
ENERGY RESOURCES						
Enriched uranium with 2–3% of ^{235}U $1,000 mill	(m) uranium oxide, U_3O_8	0.1–0.2 % U_3O_8	(1) radioactive aerosol in underground mines; (2) radiation hazard from mine wastes and concentrator tailings	65 years	(no data on USSR & China) Australia 23%, USA 17%, Canada 16%, S. Africa 15%	
METAL RESOURCES						
Iron and Ferroalloy Iron, Fe carbon steel, $375,000 mill[1]	(steelmaking) (m) Fe oxides	30–65% Fe	(1,2) **large volume** of tailings (3) **gases** from electric-arc furnaces enriched in **toxic metals** Cd, Cr, Pb *Elemental Fe is not a major polluter*	178 years	USSR 42%, Australia 15%, Brazil 7%, China 6%	12%
Manganese, Mn $4,000 mill	(m) Mn oxides, hydroxides, carbonate	25–50% Mn. (*Mn-nodules: 25% Mn*)	(1) large volume open-pit mining *Mn is an essential life-nutrient*	43 years	S. Africa 46%, USSR 38%, Gabon 6.5%, Brazil 2.6%	100%
Nickel, Ni $7,000 mill	(m) Fe-Ni **sulfide**, (lt) Ni-Mg hydrosilicate	1.5% Ni	(3) **SO$_2$ emissions** (sulfide ores)	51 years	Russia 14%, Canada 13%, New Caledonia 10%	64%
Chromium, Cr $5,000 mill[2]	(m) complex oxide	33–55% Cr_2O_3	no major environmental problems; *chromite ion an essential nutrient*	109 years	S. Africa 69%, USSR 9%, India 4%, Turkey 0.6% (*production: USSR 25%, China 19%, USA 10%*)	74%
Silicon, Si $5,000 mill	(m) quartz, SiO_2	95–99% SiO_2	(1,2,3) Silica **dust** linked to silicosis (respiratory disease)	unlimited		36%
Cobalt, Co $1,000 mill	(m) Co **sulfide**, (tr) in ~ 20 other **sulfide** minerals	co-product, 0.4% Co; **bp** of Cu, Ni, and Ag mining	see Cu, Ni, and Ag production. *Co is a component of vitamin B$_{12}$ but alone may be carcinogenic*	161 years	Zaire 50%, Cuba 26%, Zambia 9%, USSR 4%	76%
Molybdenum, Mo $900 mill	(m) Mo **sulfide**, molybdite hydrate, calcium molybdate	0.1–0.5% MoS2; **bp** of Cu mining	(1,2) large volume of tailings (3) **SO$_2$ emissions** must be contained. *Mo is essential to human metabolism*	51 years	USA 49%, Chile 21%, China 9%, USSR 8%	low
Tungsten, W $400 mill	(m) tungstate ((b) *in brine of Searles Lake*)	1.0% WO_3 (*b. 0.007% WO_3*)	no significant environmental problems	58 years	China 46%, USSR 12%, Bolivia 2.5%, Portugal 1%	85%
Light Metals						
Aluminum, AL $25,000 mill	(lt) bauxite, mixture of Al-hydroxides (rx) *alunite, nepheline*	35–50% Al_2O_3 (*acid solutions from leaching Cu ores*)	(1) no major problems, (2) caustic leach solutions must be contained, (3) reduction emits Co$_2$ & **fluorocarbons** to atmosphere	bauxite, 219 years	bauxite: Australia 24%, Guinea 24%, Brazil 12%, Jamaica 9%	100%
Magnesium, Mg $1,000 mill	(m) Mg carbonate (b) subsurface brine (sw) in seawater	ore: 70–95% $MgCO_3$ brine: 3% Mg seawater: 0.13% Mg	(3) metal-producing plant emissions of local concern. *Mg essential for human enzymes and proteins*	unlimited	Mg compounds: China 30%, USSR 26%, North Korea 18%, Greece 1%	23%
Titanium, Ti $200 mill	(m) Ti & Fe-Ti oxides	20–30% TiO2	(1) placers disrupt beach systems *Ti is not toxic*	79 years	Australia, India, South Africa, Sierra Leone	46%

(continued)

TABLE 23.2. *(cont.)*

Name of commodity (chemical symbol if elemental) Annual value in millions of US$	Important ore minerals or materials: rock [rx], laterite [lt], mineral [m], liquid [lq], brine [b], seawater [sw], minor element in other ore minerals [tr]	Typical ore grades (grade = wt % of desired constituent), or by-product [bp] [Future resource in parentheses]	Major Environmental Issues associated with: (1) mining; (2) concentrating; (3) extraction of final product. BFF = burning of fossil fuels [rather than mining] is major source of pollution involving given element	World Reserve Life Index: world reserves divided by world production (1992)	Countries with Major Reserves [% of world reserve] or [% of world production]	U.S. Imports as % of consumption (1992)
Base Metals						
Copper, Cu $20,000 mill	(m) Cu-Fe sulfides, Cu-Fe hydrosilicate, Cu-carbonate-hydrate	0.2–5.0% Cu	(1,2) large volume of waste rock and tailings, (3) SO2 and As, Cd, & Hg emissions greatly reduced in modern smelters *copper essential element for life*	35 years	Chile 28%, USA 15%, USSR 12%, Poland 6%, Canada 4%	13%
Lead, Pb $3,000 mill	(m) lead sulfide	4–8% Pb	(BFF) (3) **Pb oxide/sulfate smelter emissions do not meet** 1978 EPA standards. **Pb is toxic in small amts**	20 years	USA 16%, Australia 16%, China 11%, Canada 10%, Peru 3%	8%
Zinc, Zn $8,000 mill	(m) zinc sulfide	2–4% Zn	zinc occurs & is smelted with lead. *zinc is essential in many life processes*	19 years	Canada 15%, Australia 12% USA 11%, Peru 5%, China 3.6%	34%
Tin, Sn $1,800 mill	(m) tin oxide	0.5% Sn	no major environmental problems	40 years	China 20%, Brazil 15%, Malaysia 15%, Indonesia 9%	73%
Chemical and Industrial Metals						
Rare Earths, REEs $150 mill	(m) fluorocarbonate, phosphate	bp from gold, titanium, & tin placers	toxic effects from large doses of pure REE metals; some minerals require special handling (radioactive)	2,000 years	China 43%, USSR 19%, U.S. 13%, Australia 5.2%, India 1.1%	low
Cadmium, Cd $100 mill	(tr) in Zn sulfide	bp of zinc smelting, up to 1.3% in Zn sulfide	(BFF) (3) environmental issues linked to Pb-Zn smelting. **Cd highly toxic.**	27 years	Canada 15%, U.S. 13%, Mexico 6.5%, Japan 2%	49%
Antimony, Sb $100 mill	(m) Sb-sulfide (tr) in Cu-Pb-As-Sb-sulfides	bp from Pb-Zn-Ag ores	(BFF) (3) environmental issues linked to Pb-Zn smelting. **Sb highly toxic.**	69 years	China v. large, Bolivia 7%, S. Africa 5.6%, Mexico 4.3%	58%
Arsenic, As $60 mill	(m) numerous sulfides	bp from gold & base-metal ores	(BFF) (3) environmental issues linked to smelting: **highly toxic.**	not known	not known	100%
Mercury, Hg $20 mill	(lq) native element (m) mercury sulfide	0.2% Hg bp of gold ores	(BFF) (1,2,3) **toxic** mercury vapors in underground mines, & in reduction of sulfide ores.	27 years	Spain 58%, USSR 8%, Mexico 3.8%, Turkey 2.3%	62%
Precious Metals						
Gold, Au $25,000 mill	(m) native gold, gold-silver alloy, gold tellurides	0.0002–0.001% Au	(1) large volumes of waste dumps (2,3) **amalgamation** still a problem in some areas (e.g., Brazil); **cyanide** leaching requires special handling; **arsenic** a common companion.	20 years	S. Africa 45%, USSR 14%, U.S. 11%, Australia 4.9%, Canada 4%	56%
Silver, Ag $2,000 mill	(m) gold-silver alloy, silver sulfide, Pb-Cu-Sb-sulfide (tr) in Cu-Fe-sulfides	0.006% Ag bp of gold & base-metal	(1–3) linked to gold and base-metal production	20 years	USSR 16%, Mexico 13%, Canada 13%, U.S. 11%, Peru 9%	45%
Platinum Group Elements (PGE) $3,500 mill	(m) various sulfides, alloys of platinum, palladium, etc.	0.0003–0.002% PGE bp of Ni-sulfide ores	(1–3) same environmental problems as base-metal mining/ smelting.	190 years	S. Africa 89.3%, USSR 10.5%	94%

(continued)

GEM RESOURCES

Commodity	Source type	Composition / process	Resource life	Production / Exports	Environmental issues	%
Diamond $5,000 mill	(m) diamond (carbon)	in S. Africa: 0.01 cm³/tonne	6 years (unreliable)	*gem production: S. Africa, Botswana, Russia*	no important environmental issues	100%

FERTILIZER AND CHEMICAL RESOURCES

Commodity	Source type	Composition / process	Resource life	Production / Exports	Environmental issues	%
Lime, Flux (CaO) $7,500 mill	(rx) limestone, dolostone	>95% total carbonate, <5% SiO2, <2% Al2O3	resources are large	*production: USSR 19%, china 14%, USA 12%*	(1) competition w/ urban land-use; (3) high energy costs	none
Phosphate $4,000 mill	(m) calcium phosphate-fluoride	20–35% P_2O_5	85 years	Morocco/W. Shara 50%, USSR 11%, USA 10%	(1) strip mining in urban areas, (3) radioactive tailings + radon	low
Salt $3,900 mill	(r) evaporite (sodium chloride) (b) brines	~100% NaCl (salt beds) (solution mining)	**unlimited**	*production: USA 19%, China 15%, Germany 8%*	(3) wastewater disposal problem	13%
Potash $4,000 mill	(m) potassim chloride	(solution mining)	375 years	Canada 47%, USSR 38%, Germany 8%, Israel 0.6%	no important environmental issues	67%
Sulfur $5,000 mill	(m) native sulfur, S (m) iron sulfide, FeS_2	(solution mining) **bp** of smelting, roasting	native S: 27 yrs	USSR 18%, Canada 11% U.S. 19%, Poland 9%	product of smelter flue gas clean-up	18%
Nitrogen $13,000 mill	(g) nitrogen in air (m) sodium nitrate (niter)	air+H_2gas = **ammonia** niter deposits: 7% nitrate	**unlimited** niter res = 1/2 of annual world prod'n	*(prod: USSR, China, U.S.)* niter: Atacama Desert, Chile 100%	no environmental concerns	15%

CONSTRUCTION AND MANUFACTURING RESOURCES

Construction

Commodity	Source type	Composition / process	Resource life	Production / Exports	Environmental issues	%
Cement $65,000 mill	(rx) clayey limestone	25% clay + sand, <2.5% dolostone, 75–72% limestone	**unlimited resource**	*production: China 22%, USSR 10%, Japan 7.5%, U.S. 5.7%, India 4.3%*	(3) 2nd in **CO_2 emissions** to BFF; **SO_2 emissions** rank behind BFF, petroleum, & metal refining.	7%
Aggregate $20,000 mill	(sed) sand & gravel	sources must be near consumer	very large	local sources	(1–3) competition for land use in urban areas; interference with groundwater; particulate emissions	none
Dimension Stone $700 mill	(rx) limestone, granite, basalt	source rock must be free of fractures, pyrite.	unlimited	mostly local sources	no major environmental issues	low
Gypsum (plaster) $700 mill	(m) marine evaporite, gypsum	essentially 100% gypsum	very large	*production: U.S. 15%, China 8.4%, Canada 8.4%, Iran 8%*	no major environmental problems	35%

Fillers, Extenders

Commodity	Source type	Composition / process	Resource life	Production / Exports	Environmental issues	%
Clays $5,500 mill	(m) kaolinite, smectite		very large	*major exporters: U.S.A., U.K., Brazil*	no major environmental issues	none
Asbestos $1,000 mill	(m) chrysotile (m) amphibole (crocidolite, amosite)		very large	*production: Canada, China, S. Africa, Brazil, Zimbabwe*	(1,2) **inhalation of fibers** causes asbestosis, and lung cancer (amphibole, **rather than** chrysotile)	95%

Notes: [1 value of iron production is presented as the simplest form of steel, carbon steel. [2 value of chromium production represented by ferrochromium, the most important form in which chromium is sold. Table based largely on Kesler (1994), with additional data from U.S. Bureau of Mines (1993) and Brobst and Pratt (1973). See Glossary for definition of commodities listed and their major uses.

into energy metals (uranium), metals used in steel-making (the ferrous metals), metals used for other industrial purposes (light metals and base metals), precious metals and gems, and a variety of nonmetals (primarily compounds and minerals) used in the construction, fertilizer, and chemical industries. This table can serve as a data base for those readers interested in the occurrence and environmental hazards associated with a particular type of mineral resource or commodity.

Crustal abundance is a general measure of the availability of a given resource. When we classify according to crustal abundance, we find that nine *geochemically abundant* elements account for 99 percent of the weight of the Earth's crust. Among these nine elements are the commonly used metals aluminum (8.0 weight percent), iron (5.8 percent), magnesium (2.8 percent), and titanium (0.9 percent). The remaining ninety-two elements account for less than 1 weight percent of the crust, and each individually constitutes less than 0.1 weight percent. These are termed *geochemically scarce* elements and include such metals as nickel, tungsten, copper, lead, and gold. In terms of future resources, as first proposed by Brian Skinner in 1976, it is the latter group that will be in short supply within the next twenty to fifty years because we are extracting these at a rate that, in the context of their crustal abundance, is proportionally faster than that for the abundant metals. Figure 23.1 illustrates this concept. If we use iron as a frame of reference for rate of extraction relative to crustal abundance, we can assign a multiplication factor (here termed the *depletion factor*) to each metal that reflects the rate of production relative to iron. For example, elements with depletion factors between 50 and 100 are being extracted at 50 to 100 times the rate of iron (e.g., gold, mercury, and lead). As pointed out by Skinner in 1976, we should be using abundant metals more and scarce metals less. In the last 20 years we have been proceeding in the

opposite direction, a fact that will force us to reevaluate our use patterns as resources become increasingly scarce.

SURFACE DEPOSITS: LINKS BETWEEN THE FLUID ENVELOPES AND THE LITHOSPHERE

Minerals of importance to us occur throughout the Earth; however, unless they are concentrated by natural processes, they cannot be economically extracted. The key in the chain of natural processes that form mineral deposits is the presence of fluids such as magma and water. Fluids transport metals from mantle to crust, from crust to surface, and from place to place on the surface. The Moon, which lacks fluids, also lacks most of the types of mineral deposits known on Earth.

The sequence of processes that form ore deposits can be thought of in terms of cycles of (1) release of components from a *source* region to the fluid, (2) *transport* of components by the fluid from the source region to the site of deposition, and (3) *deposition* of minerals from the fluid in concentrated form. A common example is the formation of salt deposits: Rain dissolves minerals at the Earth's surface, streams transport sodium chloride in solution, and evaporation deposits salt crystals. Some ores (especially those of the abundant metals) are formed by only one such cycle, whereas others (especially those of the scarce metals) require numerous cycles of enrichment. Within this framework, we can classify all mineral deposits into one of three types, based on process: (1) *surface deposits*, formed by surface waters and groundwater; (2) *hydrothermal deposits*, formed by hot waters in the subsurface; and (3) *magmatic deposits*, formed directly from magmas in the subsurface. Most large mineral deposits form within the upper 0 to 10 kilometers of the crust, where fluids, both magma and water, undergo the largest changes in tem-

Figure 23.1. Annual world production of various elements in 1992 versus their crustal abundance. The heavy line drawn through iron is used as a baseline: Elements lying below the line are being mined more slowly than iron, and elements above the baseline are being mined more rapidly. Ag = silver; Al = aluminum; As = arsenic; Au = gold; Bi = bismuth; Ca = calcium; Cd = cadmium; Co = cobalt; Cr = chromium; Fe = iron; Hg = mercury; Mg = magnesium; Mn = manganese; Mo = molybdenum; Ni = nickel; Pb = lead; S = sulfur; Sb = antimony; Se = selenium; Sn = tin; Ti = titanium; W = tungsten; Zn = zinc. (*Source:* Drawing based on the relations first pointed out by B. J. Skinner in 1976.)

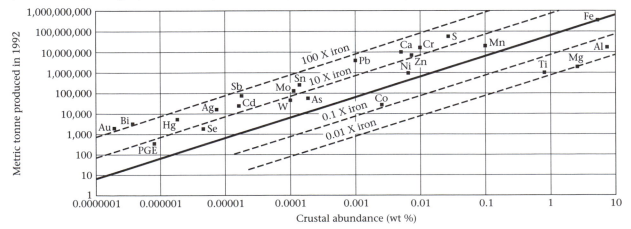

perature and pressure as they meet the cold, brittle, fractured rocks of the upper crust. Mineral deposits do form at depths greater than 10 kilometers but are smaller and less abundant.

For two reasons, we focus first on the geology of mineral deposits formed at the surface: (1) We can directly observe surface processes, and most people have some personal experience with surface phenomena such as stream erosion, longshore currents, hot springs, geysers, and volcanic eruptions (all of which have formed mineral deposits in the geologic past). (2) Surface processes that form ore deposits are analogous to surface processes involved in mining-related environmental hazards such as acid mine drainage. Because the study of the origin of surface deposits finds direct application in environmental geology, the discussion that follows emphasizes these cross-links.

WEATHERING: ACIDITY AND OXIDATION STATE. Acidic solutons tend to transport more metals in solution than do neutral solutions. Therefore, the balance between acid consumption and acid generation as waters react with the air, soil, and rocks during weathering is critical in determining the mobility of metals on the surface.

Mechanical and chemical weathering are processes that set the stage for the formation of a broad variety of useful mineral concentrations. Mechanical weathering cracks rocks and crushes and abrades rock particles, allowing rain and groundwaters greater access for chemical attack. Rain and surface waters are naturally acidic because carbon dioxide gas in air and organic-rich soils combines with water to form carbonic acid (H_2CO_3):

$$CO_2 + H_2O = H_2CO_3 \qquad (1)$$

Carbonic acid is the chief natural acid involved in chemical degradation of rocks exposed at the surface (Chap. 8). Common rock-forming minerals, such as calcite and feldspar, react with acidic solutions and release various components to solution (in the following and subsequent reactions, chemical symbols are presented on the first line and names of minerals (underlined) and aqueous species are given on the second line):

$$CaCO_3 + \quad H_2CO_3 \quad = \quad Ca^{2+} \quad + \quad 2HCO_3^- \qquad (2)$$
calcite + carbonic acid = calcium ion + bicarbonate ion

$$2NaAlSi_3O_8 + H_2CO_3 = Al_2Si_2O_5(OH)_4 \qquad (3)$$
$$+ 4H_4SiO_4 + 2Na^+ + 2HCO_3^-$$
Na-feldspar + carbonic acid = kaolinite (clay) + silica + sodium ion + bicarbonate ion.

Reactions such as (2) and (3) consume acid (acidity declines as the reaction proceeds from left to right) and cations are released to solution. Therefore, these reactions tend to decrease the mobility of metals and cause metals to precipitate out of solution. In competition with reactions (2) and (3) are acid-generating reactions involving surface waters and sulfide minerals. For example, the common sulfide mineral pyrite (FeS_2) yields acid as it reacts with rainwater,

$$4FeS_2 + 14H_2O + 15O_2 = 4Fe(OH)_3 + \qquad 8H_2SO_4 \quad (4)$$
pyrite + rainwater = iron hydroxide + sulfuric acid

Surface waters in contact with the atmosphere contain abundant dissolved oxygen. Rain is naturally acidic and oxidized; however, oxygen is rapidly consumed as surface waters percolate downward toward the water table and react with organic material and ferrous minerals in soils and rocks. When the bedrock contains calcite or feldspar, the acidity also shifts toward neutral. On the other hand, when the bedrock is rich in minerals that do not consume acid, such as quartz and clays, then the dilute aqueous solutions of streams and groundwaters remain on the acid side of neutrality. Thus, a knowledge of minerals present in soils and bedrock can be used to predict whether a particular drainage or groundwater system, if it becomes acid by natural or anthropogenic causes, is likely to remain acid and transport toxic metals or be neutralized and deposit metals. Similar reasoning can be applied to understanding the likelihood of damage to ecosystems as the result of acid rain and acidification of streams (see Chap. 9).

CHEMICAL ENRICHMENT. Weathering is most rapid in tropical and subtropical climates, where abundant rainfall and relatively higher temperatures increase rates of reaction between waters and minerals (see Chap. 8). Except in the case of some rocks such as pure limestone [e.g., reaction (2)], most weathering reactions do not lead to complete dissolution of minerals. Instead, some elements go into solution, and relatively insoluble minerals such as clays and iron hydroxides are left behind [e.g., reactions (3) and (4)].

Laterites: The World's Source of Aluminum. Iron hydroxides (limonite) and aluminum hydroxides (clays) are typical insoluble minerals formed in situ by weathering of igneous rocks. Soils that are highly enriched in, and cemented by, insoluble hydroxides are known as *laterites*. Iron-rich laterites from the weathering of iron-bearing silicate minerals are a minor source of iron, and nickel-rich laterites from the weathering of nickel-bearing olivine in ultramafic igneous rocks serve as a source of nickel. Aluminum-rich laterites, from the weathering of feldspars in granitic rocks and known as *bauxite*, are the world's dominant source of aluminum. Favorable conditions for the formation of laterites include not only a subtropical to tropical climate, but also good drainage in areas of low relief such that erosion does not compete with the enrichment process. Laterites are important in the study of paleoclimates and plate tectonic reconstructions (see Chaps. 6 and 15). For example, the bauxite deposits of southern France and Arkansas indicate that deposits in each of these regions must have been formed in a tropical climate prior to their transport to temperate latitudes by continental drift.

Supergene Enrichment: Major Source of Copper. In contrast with laterites, which form in situ from normal crustal rocks (i.e., rocks in which the ore element is not present in anomalous concentration), *supergene enrichment* deposits form in or near rocks that already contain anomalous concentrations of metals in the form of sulfide minerals. The process involves weathering, acid-leaching, and transport of metals in acidic groundwater followed by reprecipitation of metals in concentrated form. The greater the original concentration of pyrite in the weathering source region, the greater the production of acid – reaction (4) – and the greater the leaching and transport of metals, especially in those cases where the enclosing rock is incapable of buffering the acidity of the groundwater to near-neutral conditions. Under such conditions, weathering of sulfide-bearing rocks leads to total dissolution of sulfide minerals and only the insoluble clays and iron hydroxides are left behind. It follows that outcropping ore deposits hosted by limestone or relatively fresh granitic rocks would be the least likely to generate abundant acid during mining.

The site of supergene enrichment represents the point at which the metal-laden leach solution encounters geochemical conditions that cause the metals to precipitate from solution. These new conditions may be either acid neutralizers in the near-surface (e.g., limestone or granite) or reductants at depth (e.g., unweathered pyrite at the water table). The reaction involving limestone is identical to that used by humans in "liming" acid mine drainages to reduce acidity and remove toxic metals in streams and lakes. Reduction at depth leads to precipitation of metals as metal sulfides, referred to as *supergene sulfides*.

Although supergene sulfide enrichment can occur for numerous metals, the most important types of supergene metal deposits are those of copper and silver. This is because copper sulfides, and to a lesser extent silver sulfides, are less soluble than most other common sulfides, and the copper oxides (e.g., cuprite) and hydroxycarbonates (e.g., malachite) are highly soluble in the acid solutions that predominate in oxidizing ores. A major proportion of the world's copper production has come from supergene copper deposits in which copper sulfides (e.g., chalcocite, Cu_2S) form thick, blanket-like deposits.

Supergene (weathering) profiles, if documented in three dimensions (as can be done in large mines), are sensitive indicators of the pathways of rainwater percolation, oxygenated groundwater flow in the subsurface, and morphology of the groundwater table. Supergene profiles can be used to study rates of erosion, timing of continental uplift, and past climates; therefore, they are of interest not only to exploration geologists but also to the larger scientific community.

ACID MINE DRAINAGE AND TOXIC METALS. We have seen that the natural process of weathering of sulfide-bearing rocks at the Earth's surface generates acid and transfers metals and sulfur into solution. The effect of mining is to increase significantly the efficiency of this process, because mine openings and waste dumps allow oxygenated groundwater far greater access to sulfide-bearing material. The resulting acidification of streams and lakes and dispersion of toxic elements such as arsenic, lead, cadmium, and mercury are of major environmental concern. Numerous *Superfund sites* in the United States (sites where the cost of remediation is funded partially through federal sources) are located in areas of past mining. At Butte, Montana, prior to 1950, tailings and other mine wastes were dumped into the local rivers; today, anomalous metal values are found hundreds of kilometers downstream. Present-day mining practice in the United States is far cleaner; it must meet strictly enforced federal standards on clean air and water, ground reclamation, and other pollution abatement guidelines. *Environmental impact statements* are required before any large mining operation is initiated, and bonds must be posted to assure cleanup after mining terminates. In spite of these precautions, the public's continuing fear of environmental damage due to mining reflects the press coverage given to relatively rare environmental accidents (see the box entitled *The Summitville Affair*).

MECHANICAL ENRICHMENT: PLACERS. We have seen that reactions involving dissolution of sulfide minerals are important in creating both mineral deposits and polluted groundwater. Here we turn to surface processes involving ore minerals that are chemically and physically stable under Earth surface conditions. For these processes, natural and anthropogenic chemical pollution is not an issue, although disruption of the Earth's surface during mining can be.

Ore minerals that are resistant to abrasion and dissolution under Earth surface conditions may be transported from the site of weathering by running water and other physical processes. Running water sorts mineral grains according to size and specific gravity (S.G., defined as grams per cubic centimeter; also termed *density*). All else being equal, relatively dense ore minerals (e.g., for native gold, the S.G. is 19; for native platinum, the S.G. is 19) will be concentrated closer to their source than less dense ore minerals (e.g., for rutile, titanium oxide, the S.G. is 4.2; for diamond, the S.G. is 3.5), and may be separated from the common rock-forming minerals that constitute the bulk of a stream's particulate load (e.g., for quartz, the S.G. is 2.7; for feldspar, the S.G. is 2.6). Such concentrations, known as *placers*, form in areas of sharp gradients in current velocity, as at the base of waterfalls or mountain fronts, on the inner side of stream meanders (*stream placers*), and where streams enter bodies of realtively calm water (*delta placers*). Longshore currents of coastal regions transport mineral grains from alluvial sources, and wave action on beach fronts is very effective at sorting mineral grains in sand. Thus, nature's own "gold pan" creates *beach placers*. Examples of these different placer environments include: gold in stream placers of the western Sierra Nevada Foothills, which led to the California Gold Rush of 1849; the Witwatersrand goldfields in South Africa, whose ancient placers contain 45 percent of the world's known gold reserves; and beach placers containing gold at Nome, Alaska. Beach placers contain diamonds in East Africa, and tin (cassiterite, SnO_2) in Indonesia. On a very small scale, the process can be observed along present-day granitic coastlines

THE SUMMITVILLE AFFAIR

At Summitville, Colorado, a gold deposit was mined from 1985 until 1992 by open pit and cyanide heap-leach techniques. In 1992, the mining company operating at Summitville declared bankruptcy and abandoned the site. Uncontrolled, the acid mine waters and cyanide-processing solutions began to infiltrate the local stream drainages. Summitville mine drainage one year following the mine closure ranged in pH from 1.5 to 3 and in heavy metal content from 40 to 1500 parts per million, far more acidic and concentrated in metals than other Colorado mine waters. These waters are draining into the Wightman Fork of the Alamosa River (Fig. 23.2). Although the thought of cyanide-contaminated water is terrifying, in reality cyanide is relatively benign because it degrades quickly in oxidizing (and especially in acidic) conditions on the Earth's surface. Acidic waters and toxic metals present us with a greater problem.

Summitville is an unusually toxic site primarily because during the formation of the ore body 40 million years ago, the enclosing rocks were hydrothermally altered to a mixture of clays, pyrite, and quartz. Hydrothermal alteration is the product of water–rock reaction at temperatures ranging from 600°C to less than 100°C. At the lower range of hydrothermal temperatures (less than 250°C), fluids tend to become acidic. The result of reactions involving sulfuric acid and volcanic rocks is a mixture of clay minerals and pyrite. Following uplift, erosion, and exposure of the mineral deposit at the Earth's surface, this mixture of clay and pyrite becomes an extremely efficient acid generator not only because of the abundance of pyrite, but also because of the lack of minerals such as calcite and feldspar that could serve to mitigate the acidity and decrease metal mobility. At Summitville, the U.S. Geological Survey (USGS) has shown that concentrations of copper in mine drainage increased from 25 to 100 milligrams per liter during the production period, as mining exposed increasing volumes of unoxidized pyrite. Following abandonment of the site, copper concentrations continued to increase (approximately 300 milligrams per liter in 1993), indicating the potential for a long-lived and worsening environmental problem.

As is typically the case in mining districts, the actual ore deposit mined at Summitville occupies a small fraction of a broad area of lower-grade material, untouched by mining, that generates natural acid drainage and toxic metals through weathering processes. For example, as shown in Fig. 23.2, the Alamosa River above Wightman Fork (up-stream from the point where Summitville mine drainage joins the river) drains an area affected by natural acid generation and contains as much dissolved aluminum as it does downstream from Summitville. The problem, if viewed from a broader perspective, becomes one of isolating anthropogenic sources of pollution from natural ones. Should mining companies be required to clean up any water leaving their properties to drinking-

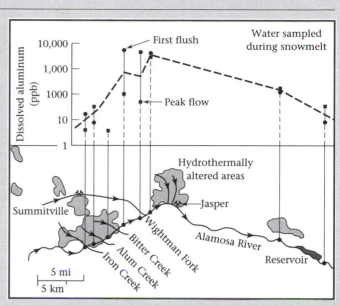

Figure 23.2. Map of hydrothermally altered areas (lower panel) and plot of dissolved aluminum in stream water (upper panel) for the Summitville area, Colorado. Illustrates the broad areas of hydrothermally altered rocks, extending well beyond the actual Summitville mine area, which serve as a source of both natural and anthropogenic acid and metal pollution. Dissolved aluminum in stream water can be used as an indicator of acidity. Significant amounts of aluminum occur in the river water upstream from the Summitville drainage. ppb = parts per billion. (*Source:* Adapted from U.S. Geological Survey public information leaflet, 1994.)

water quality? Or, should they be required to clean up to a premining baseline (*prior baseline*)? In many cases prior baselines are unknown within districts with a long history of mining. In the context of public policy, this fact underscores the urgent need to acquire more data on the environmental impact of both mining and natural pollution.

The Summitville experience suggests that perhaps some types of deposits are simply too environmentally dangerous to mine. To address this question, the USGS is developing "environmental ore deposit models" for a diversity of deposit types. The goal is to combine the geologic characteristics of ore deposits with quantitative models of groundwater flow, natural acid generation and consumption, and metal dispersion in the near-surface. The models can be used both in remediation of past mining and in predicting the effects of future mining. In remediation, the models serve to establish the environmental conditions that existed prior to mining so that remediation is aimed at returning the site to a prior baseline rather than to an arbitrary pristine condition. In prediction, the models would help to identify realistic remediation goals for future mining projects, and to identify those prospects that should not be developed because they would present too great an environmental hazard.

where the accessory mineral magnetite (with an S.G. of 5.2), released by weathering, is transported by streams to the nearby beach front and sorted into thin, black beds interlayered with quartz sand.

In many cases, multiple episodes of erosion, transportation, and deposition are required before a financially attractive placer is formed (see *How Many Concentration Cycles are Needed to Form Placers?*).

MINING THE HYDROSPHERE

Seawater and Brines. On an annual basis, streams transport 4000 million metric tons of dissolved material to the oceans. This influx rate of dissolved constituents to the oceans is matched by their removal rate from ocean water through biogenic and abiogenic precipitation of minerals to form deep-sea sediments and through reactions between seawater and oceanic crust along mid-ocean ridges. Thus, the composition of seawater can be considered to be in a steady state. Dissolved components are primarily sodium chloride (halite) and lesser amounts of magnesium salts and calcium sulfate, the total equal to 3.5 weight percent dissolved solids. Streams also transport dissolved constituents to closed basins, the realm of salt lakes or playas. Dissolved components in salt lakes are similar to those in the oceans but are more concentrated (up to 25.5 weight percent dissolved solids). In contrast with the oceans, the composition of which is controlled by global-scale processes, compositions of brines in salt lakes are controlled largely by chemical weathering of rocks in the local watershed combined with evaporation of lake water. Thus, considerable variation occurs in the composition of salt lakes as a function of their geologic and climatic setting.

Magnesium is the only metal presently extracted from surface water; production of magnesium and magnesium compounds from these sources accounts for some 40 percent of total production in the United States. Halite and bromine also are commercially obtained from seawater; however, attempts to recover other commodities such as potash, uranium, and gold have not been commercially successful. Given the higher concentrations of salts in salt lakes, these serve as a source for a far greater range of commodities than does seawater. Potash and numerous other nonmetals [boron, bromine, chlorine, iodine, lithium, sodium carbonate (trona), sodium sulfate, and sodium chloride (salt)] are extracted from brines.

Will seawater someday become a major source of minerals for human development? Ulrich Petersen of Harvard University believes that this will occur as technologies improve and costs decline, and as mining on land becomes increasingly costly due to environmental hazards and depletion of near-surface deposits. With a volume of 1.35 billion cubic kilometers, the oceans contain 4.9×10^{10} million metric tons of dissolved solids, including many useful salts and metals. Production of minerals from seawater may become linked to production of fresh water in desalinization plants, where by-products potentially include metals such as lead, zinc, and silver and nonmetals such as potash, sodium sulfate, and calcium sulfate (gypsum).

Evaporites. Evaporite deposits are sedimentary rocks that are formed as bodies of water evaporate. Thick deposits form

---

HOW MANY CONCENTRATION CYCLES ARE NEEDED TO FORM PLACERS?

Although a single cycle of concentration may have been responsible for many of the world's placers, there are examples where numerous cycles, each separated by a significant period of geologic time, were involved. The first gold placers to be worked by the California Forty-niners were recent placers, formed in present-day streams draining the western front of the Sierra Nevada. As these stream placers were worked out, prospectors moved upstream and discovered that the immediate source of the gold was older placers of Early Tertiary age in uplifted gravels (the gold in the Tertiary gravels had its source in gold-quartz veins formed during the emplacement of Sierran granite during Mesozoic time, approximately 150 to 100 million years ago). The gold miners drove tunnels into these Tertiary gravels, following the meanders of ancient streams in the subsurface. Their tunnels present us with detailed maps of the stream channels that existed prior to the uplift of the Sierra Nevada range 60 million years ago. The labors of the Gold Rush miners of 1849 contributed significantly to our knowledge of the paleogeography and uplift history of the Sierra Nevada.

Diamond placers along the west coast of Namibia and South Africa also are the result of numerous cycles of concentration. The diamonds originated 800 kilometers inland in diamond-bearing *kimberlite pipes* formed over 1000 million years ago. One hundred million years ago (Mid to Late Cretaceous), weathering released the diamonds to streams draining westward toward the proto–South Atlantic Ocean during the breakup of Pangea. Conditions along the Late Cretaceous shoreline were inappropriate for the formation of high-grade diamond placers; only low-grade, diamondiferous sediments formed in shallow water offshore and were lithified. Forty million years later, the diamondiferous sedimentary rocks were eroded during uplift of the coast in Tertiary time to form the high-grade beach placers mined today: from mantle source to emplacement in the shallow crust as kimberlite; from kimberlite via erosion and transport to sedimentation, burial, and lithification; from shallow-marine sedimentary rocks via uplift, erosion, and sorting by surf action to placers. Because of the uniqueness of diamonds, this may be the only case in which a "family" of mineral grains can actually be followed from mantle to surface through two *rock cycles* spanning 1000 million years (see Fig. 5.19).

only in particular environments, such as an inland sea connected to the ocean by a shallow sill. As evaporation proceeds, salts are replenished through influx from the open ocean. Evaporite deposits that are formed from seawater (*marine evaporites*) are relatively predictable in composition due to the uniform composition of seawater. In contrast, evaporites formed from salt lakes (*nonmarine evaporites*) vary considerably and are compositionally distinct from marine evaporites. Marine evaporites are the world's major source of salt (halite, NaCl), calcium sulfate (gypsum, $CaSO_4 2H_2O$, used for plaster), and potassium salts (carnalite, $KClMgCl_2 6H_2O$, used for fertilizers). Nonmarine evaporites are important sources of sodium sulfate, sodium carbonate (soda ash or trona), and boron (sodium borate minerals such as borax, $Na_2B_4O_7 10H_2O$) used in glassmaking. Nonmarine evaporites in the Green River Formation of Wyoming contain 96 percent of the world's reserve of sodium carbonate, and nonmarine evaporites near the town of Boron and Death Valley in southern California contain the world's largest deposits of borax.

Bacterial Sulfate Reduction and Metal Precipitation. In aerated surface waters, oxygen is consumed by both biogenic and abiogenic reactions. Biogenic reactions are dominated by microbial decomposition of organic matter, whereas abiogenic reactions are dominated by oxidation of ferrous iron. When all the free oxygen has been consumed, the water (or environment) is termed *anaerobic*; however, even at this point the oxidation state of water can continue to decrease. One of the chief agents that drives water toward the lower oxidation state is again biogenic, involving the reduction of aqueous sulfate to hydrogen sulfide by anaerobic bacteria, such as *Desulfovibrio desulfricans*. The importance of the oxidation state in the context of metals is twofold. It affects (1) metal solubilities and (2) the availability of reduced sulfur (H_2S) that is necessary for precipitation of most metals as sulfide minerals. The greater the production of hydrogen sulfide in a given environment, the greater the precipitation of metals out of solutions that pass through that environment. The metals precipitate as metal sulfide minerals. At Earth surface temperatures, anaerobic bacteria play a key role in this process.

Sulfate-reducing bacteria inhabit stagnant tidal flats and marshes, oxygen-deficient bottom waters of oceans, oil field brines, and petroleum. The hydrogen sulfide gas they produce can be retained in rocks or leak into the atmosphere. Commonly, the hydrogen sulfide gas reacts to form organic sulfur molecules in petroleum and coal; it also readily reacts with ferrous ions or iron oxide minerals to form iron sulfide minerals (pyrrhotite, FeS; and pyrite and marcasite, FeS_2). In addition to precipitating iron sulfide, biogenic hydrogen sulfide precipitates small quantities of copper, cobalt, zinc, uranium, and other metals, forming vast, low-grade repositories of these metals in black, organic-rich shales. In the distant future, once we have exhausted high-grade ores, such shales may become an important source of metals.

HYDROTHERMAL DEPOSITS: CRUSTAL INFLUENCE

Most of the metal ores of the world are of hydrothermal origin, formed by heated waters (*hydrothermal solutions*) of various origins as they circulate through the crust. In order to understand the distribution of hydrothermal deposits in the Earth's crust, we need to know the dominant sites of hydrothermal fluids, the forces that drive fluid flow, and the causes of mineral deposition.

HYDROTHERMAL FLUID FLOW AND MINERAL DEPOSITION

■ **Permeability.** We can predict that water is concentrated where the crust is most permeable. *Permeability* can be defined as the volume percent of interconnected openings in rocks (as contrasted with *porosity*, which refers to the volume percent of openings (pore spaces) between mineral grains, with pore spaces not necessarily interconnected). From the study of ancient mineral deposits, exposed at the surface by uplift and erosion, we can conclude that permeability and fluid flow are largely a consequence of fractures; highly fractured rocks allow greater fluid flow than do unfractured rocks. Fractures form in brittle rocks. As pressure and temperature increase with depth in the crust, rocks pass through a transition from brittle to ductile behavior. Below this transition zone, rocks behave as plastic solids, openings are not sustained, and hydrothermal fluid flow is negligible. Although the transition from brittle to ductile is a function of the local geothermal gradient and of the rock type involved, we can estimate with confidence that this transition occurs approximately 10 kilometers below the Earth's surface.

■ **Fluid Density Contrasts.** The second part of our inquiry regards the driving force behind fluid flow. Here we can appeal to everyday experience. If you pour water into a glass filled with vegetable oil, the vegetable oil rises buoyantly to the surface: Density contrasts between two different fluids at the same pressure and temperature cause fluid flow. With water alone, density contrasts are evident when waters (see Chap. 11) of different salinity (and, consequently, of different density) and/or temperature are mixed. For waters of the same salinity, if you heat a pot of water on the stove, the heated water rises over the hot spot in the center and sinks down the sides, setting up numerous convection cells. We conclude that in hydrothermal systems, *fluid flow is the result of contrasts in fluid density caused by thermal gradients*. Given this conclusion, where are the steepest thermal gradients in the Earth's upper crust?

■ **Heat Sources.** The flux of hydrothermal fluids in the Earth's upper crust is largely the result of thermal input by magmatic intrusions. Hydrothermal flow caused by magma emplacement convects a large proportion of the Earth's heat to the surface. Hydrothermal flow also occurs during compactive expulsion of water from large sedimentary basins on the continents and during regional metamorphism; however

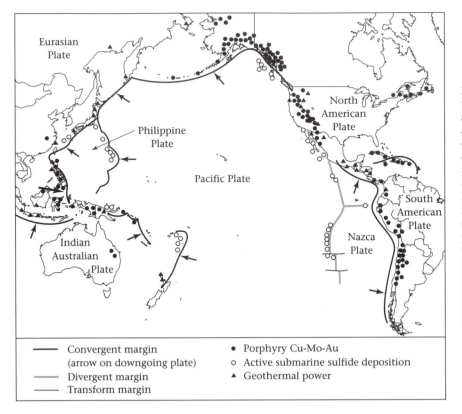

Figure 23.3. Relation between plate tectonic features of the circum-Pacific region and distribution of some types of active and ancient hydrothermal systems, including (1) porphyry-copper (Cu-Mo-Au) deposits (black dots), (2) geothermal power stations (black and red triangles), and (3) sea-floor hydrothermal systems known in 1993 that are actively depositing iron, copper, and zinc sulfides (open circles). The close spatial relation between plate boundaries and hydrothermal systems is striking, although the absence of known porphyry-copper deposits along portions of the circum-Pacific arc (e.g., Japan) suggests a lack of sufficiency in the relation between subduction and porphyry-copper deposits.

In figure:

Eurasian Plate

Philippine Plate

North American Plate

Pacific Plate

South American Plate

Nazca Plate

Indian Australian Plate

— Convergent margin
(arrow on downgoing plate)
═ Divergent margin
─ Transform margin

• Porphyry Cu-Mo-Au
○ Active submarine sulfide deposition
▲ Geothermal power

fluid flux in these environments is minor compared to that caused by magmatic intrusions. It follows that hydrothermal ores are most likely to be found in areas of abundant magmatism: the mid-ocean spreading centers, and the subduction-related magmatic arcs and back arcs of oceanic and continental settings (see Chap. 6, and Fig. 23.3). Significant occurrences are also likely to be found on the margins of large continental sedimentary basins, especially those that have been compacted by collisional tectonics.

■ **Metal Transport and Deposition.** The origin of hydrothermal ores can be placed in the context of the source-transport-deposition cycle introduced above. Hydrothermal fluids leach metals from a source region at depth in the crust, transport the metals in solution as they buoyantly rise toward the surface, and then precipitate the metals in concentrated form as they cool. This conceptual model seems relatively straightforward, and we might therefore ask: Given the abundance of hydrothermal fluids in the Earth's crust, why are ore deposits so scarce? The answer is that there are numerous different paths (both physical and chemical) for the potential ore-forming solution to travel, and many of these paths do not lead to the formation of ore deposits. Therefore, it is important that those who study ore deposits focus not on the sources of metals and fluids but on the *causes of deposition of ore minerals in concentrated form.* This is the emphasis of the sections that follow.

**HYDROTHERMAL FLUIDS AND CONTINENTAL VOLCA-
NOES.** On the exposed surface of the continental crust, hydrothermal ore-forming fluids are rarely present. This is be-

cause during their rise from the middle crust, these fluids lose heat to their surroundings and become depressurized. Both of these effects may cause deposition of ore minerals in the subsurface. Closer to the surface, the fluids boil and/or are increasingly diluted with local groundwater. Both of these processes also can cause ore deposition. By the time these fluids arrive at the Earth's surface, they are low-concentration analogs of their former selves, consisting of *dilute hot springs* and geysers such as those encountered in *geothermal* areas (e.g., Yellowstone, Wyoming; see Fig. 19.1). The bulk of the metal has been precipitated in the subsurface, in dispersed or concentrated form.

In some cases, where the rise of hydrothermal fluids is so rapid that little cooling, mixing, and dilution by groundwater occurs, potential ore-forming fluids may arrive at the surface. Such surface discharges are recognized in many continental and island-arc volcanoes where high-temperature fumaroles and acid springs are surface expressions of degassing magma bodies beneath the volcano. Violent eruptions carry the majority of metals, sulfur, and other volatile components (carbon dioxide, hydrochloric acid) into the atmosphere as *volcanic gases, aerosols,* and *particulates.* Volcanic fumaroles commonly are encrusted with native sulfur, arsenic, and metal hydroxides, indicating the presence of these ore components in the gases. Some of the sulfur emitted as hydrogen sulfide gas reacts with atmospheric oxygen or sulfate gas and forms deposits of native sulfur inside the volcanic craters. Such deposits are local sources for elemental sulfur in numerous Circum-Pacific countries.

■ **Gold Flakes and Sulfur Dioxide in the Atmo-**

sphere. Recent studies have shown that long-term fluxes of gold, copper, and gaseous SO_2 to the atmosphere from individual volcanoes may be equivalent to the content of metals and sulfur in some of the world's major ore deposits. For example, metal fluxes to the atmosphere in Augustine, Alaska, and Mount Etna, Italy, calculated over the lifetime (millions of years) of these volcanoes, are equivalent to a moderate-sized copper-gold deposit. At Mount Etna, gold is emitted to the atmosphere at the rate of 0.1 to 1.0 metric ton per year! Fluxes to the atmosphere and hydrosphere at White Island volcano, New Zealand, have been estimated to be 10 million metric tons of copper and 45 metric ton of gold over the 10,000-year life of the hydrothermal system. These numbers are equivalent to the metal content of a large copper deposit.

Erupting volcanoes, therefore, can be thought of as "failed" ore-forming systems: A cap (or seal) must be kept on the fluids (both magma and hydrothermal) to prevent them from breaching the surface and losing metals and sulfur to the atmosphere. The "cap" can be physical – such as an unfractured rock mass over the magma chamber – or chemical. A chemical cap could involve the conversion of sulfur dioxide gas to aqueous hydrogen sulfide by reaction with water circulating in the margins of the magma chamber. The hydrogen sulfide produced by this reaction can remain in solution in deeply circulating groundwaters and/or can combine with metals and precipitate metal sulfide minerals below the volcanic edifice. Some sulfur dioxide combines with calcium (supplied by the melt, calcium-bearing minerals, or hydrothermal fluids) to form the calcium sulfate mineral anhydrite, common in some copper deposits and in rocks ejected from erupting volcanoes. In either event, sulfur dioxide is fixed as a solid by a natural "scrubbing" process prior to escaping into the atmosphere.

■ **Porphyry-Copper Deposits: The Volcanic Connection.** The link between certain types of ore deposits – especially those of copper, gold, and molybdenum, known as *porphyry-type* deposits – and andesitic volcanoes is well established from studies in the Circum-Pacific region (the "Rim of Fire"; see Chap. 25). As shown in Figure 23.3, porphyry-type deposits of copper and gold occur in linear arrays within the Late Tertiary to Pleistocene volcanic arcs of Chile, Papua New Guinea, Indonesia, the Philippines, and the Aleutian Islands of Alaska. The deposits occur in the roots of volcanic arcs, closely associated with small granitic stocks emplaced at depths of 1 to 5 kilometers below the volcanoes. Sulfide ore minerals occur in thin quartz veinlets and are disseminated along microfractures throughout large volumes of hydrothermally altered rocks. The geologic and geochemical characteristics of porphyry-type deposits indicate that they form from saline aqueous solutions that are produced during the end stages of crystallization of granitic magmas at 600°C (see Chap. 5). Thus, the metals and sulfur in porphyry-type deposits come from the same source as the magmas. Because island-arc and continental-arc magmatism are closely tied to subduction at convergent plate boundaries (see Chap. 6, and Fig. 23.3), the magmas and ore components likely originate

through partial melting of basaltic ocean crust and/or overlying mantle peridotite in the subduction zone at depths of approximately 80 to 100 kilometers and interaction of these magmas with continental crust. Porphyry deposits contain an estimated 65 percent of the world's reserves of copper and also serve as important sources of molybdenum, rhenium, zinc, cadmium, gold, silver, and other metals. Sulfur is present in the ores both as metal sulfides and as calcium sulfate (anhydrite). A moderate-size porphyry-copper deposit contains 2 to 3 million metric tons of copper and 6 million metric tons of sulfur; megadeposits contain 10 to 50 million metric tons of copper and 20 to 30 million metric tons of sulfur.

HYDROTHERMAL FLUIDS AND OCEANIC VOLCANOES. In contrast to the surface of the continents, where hydrothermal ore-forming fluids are absent, the surface of oceanic crust or the submarine portion of island arcs is a relatively common site of ore fluids and ore deposition. Here, hydrothermal ore fluids can reach the surface (the sea floor) because the high pressures maintained by the overlying 3 kilometers of seawater suppress boiling, one of the major causes of ore deposition. Further, if some mixing occurs in the fractured basalts below the seafloor, the ambient water is salty, not dilute, and this saltiness helps to keep metals in solution. *Submarine hot springs*, unlike continental hot springs, are major ore formers. Such hot springs are concentrated in fracture zones in oceanic crust, especially mid-ocean and back-arc spreading centers, and in the roots of island arcs, where ocean water penetrates deeply, is heated, leaches metals from basalt, and then rises buoyantly toward the surface.

■ **Volcanogenic Massive Sulfide Deposits.** A significant portion of the supply of metals such as copper, lead, and zinc has come from *volcanogenic massive sulfide deposits*, found in volcanic terranes exposed on the continents. Geologic studies of these deposits have shown that they originally formed in submarine volcanic successions that later were uplifted and became part of continental blocks. The volcanic rocks hosting massive sulfide deposits are predominantly oceanic pillow basalts or submarine portions of island-arc volcanoes. The massive, sheet-like morphology of the ore (interlayered with the enclosing volcanic rock), the presence of underlying mineralized fracture zones thought to be feeders for hydrothermal solutions, and the general absence of wall-rock alternation in overlying (younger) volcanic rocks – all point to an origin directly linked in time to the emplacement of the underlying volcanic rocks. This model for the origin of massive sulfides has recently been reinforced by direct observation of ore-forming processes in the ocean basins.

■ **Black Smokers: Submarine Hot Springs and Life.** In 1979, exploration of the sea floor aided by deep-diving submersibles led to the first discovery of active hot springs depositing sulfide minerals in the East Pacific Rise at 21° N latitude. During the period 1979–94, some fifty separate, sulfide-depositing vent fields were discovered in the Pacific Basin alone, in environments ranging from mid-ocean ridges on the east to submarine island arcs and back-arc basins in the west and southwest (Fig. 23.3). Venting fluids at 350°C are charged with hydrogen sulfide and metals (such as iron, copper, and zinc) and have salt contents ranging up to several times that of seawater. These fluids originate from seawater that circulates deeply through fractured oceanic crust on the margins of mid-ocean ridges or through the roots of submarine island-arc volcanoes. Seawater convection is driven by magmatic heat. Cold seawater penetrates close to the margins of magma chambers, where it is heated. Water–rock interactions lead to exchange of components between heated seawater and volcanic rocks: Basalt is converted to *greenstone* as seawater is converted to a hydrothermal fluid containing metals in solution. Metals remain in solution until the hydrothermal fluid vents into cold seawater, where the sharp drop in temperature leads to precipitation of minute metal-sulfide particles in plumes (*black smokers*) that rise tens of meters above the sea floor (see Figs. 12.4 and 12.5). With time, the sulfide minerals precipitating around the vent build chimneys up to several meters high, which then collapse as new chimneys are born. Collapsed chimneys and sulfide particles that rain down from the black smoker plumes form mounds of iron, zinc, and copper sulfide minerals. To date, although the majority of active hydrothermal vents explored on the sea floor are not presently economic, some of the largest mounds contain 10 million metric tons of sulfide, equivalent to small volcanogenic massive sulfide bodies mined on land. At greater distances from the vents, manganese and iron oxides are precipitated in deep-sea sediments, forming layers of ocherous muds and manganese nodules. Such iron-and/or manganese-rich deposits commonly are also enriched in other metals and, although not mineable today because of the high costs of recovery, represent a significant resource for the future.

Study of samples of both vent fluids and minerals precipitated from the fluids has led to a more detailed understanding of the processes involved in formation of ancient ore deposits mined on the continents, and in the cycling of chemical components between seawater and underlying crust. Exploration of hydrothermal vents on the sea floor amplified prior models of ore-forming processes – and it also brought some surprises. The submarine vents are colonized by unexpected animal life that is unique to the submarine hot spring environment, an environment that most forms of life would find toxic. Studies of vent fauna have identified 300 new species in 90 new genera, 20 new families, and 1 new phylum. Some vent species are colorful: giant white tube worms with red plumes (*Riftia pachyptila*), yellow siphonophores (nicknamed dandelions), pinkish eelpouts of the Zoarcidae family. Others are chalky white: filter-feeding clams, brachyuran crabs, and alvinellid worms. What do these vent species feed on? We now know that vents are colonized by *chemosynthetic bacteria* that compete with the metals for access to the hydrogen sulfide. These bacteria use oxygen and carbon dioxide from seawater to oxidize hydrogen sulfide discharged from the vents and to produce carbohydrate and native sulfur. Geologists, chemists, and biolo-

gists are collaborating in multidisciplinary teams to understand the interrelation between hydrothermal processes and life in the vent ecosystems. Are these the only ecosystems on Earth that exist without photosynthesis? Such systems were probably abundant in the early Archean oceans. Are these vents the cradle of life?

MINERAL RESERVES, DISTRIBUTION, AND AVAILABILITY

ESTIMATION OF QUANTITIES. We have been using the terms *mineral resources* and *mineral deposits* interchangeably to refer to a variety of anomalous concentrations of potentially useful minerals on and in the Earth's crust. In order to address issues of present and future vailability, we need to know the relative proportion of economic and subeconomic mineral deposits. Which deposits can be mined at a profit, and how much metal do they contain?

Resources. A *resource* is the amount of a given commodity contained in all deposits – discovered and undiscovered, economic and subeconomic (Fig. 23.4). By its definition, the term is restricted to national or global scales and is intended as a measure of the crust's mineral endowment (e.g., the "U.S. copper resource" or the "world copper resource") and cannot be applied as defined above to holdings of individual mining companies (imagine a stockholders' meeting where the CEO declared that the company had just acquired new undiscovered resources!). Assessing the amount of a given commodity in undiscovered resources is a conceptually difficult procedure based on geologic extrapolation and statistical computations. In the United States, such resource assessments have become a highly charged issue because they form part of the basis for federal decisions on *land withdrawal*, the prohibition of mining and other private activities on federal land (amounting to 25 percent of the United States). Resource assessments are not testable other than by drilling, and most of the land that has been withdrawn from exploration has not been adequately explored.

Reserves. *Reserves* include only that portion of the identified resource that is presently economic to recover and is demonstrated by detailed geologic understanding. *Reserve base* includes reserves and marginal reserves (Fig. 23.4) and commonly also includes an unspecified portion of subeconomic (but demonstrated) deposits. The vagueness of the definition of "reserve base" stems from the rapidity with which portions of mineral deposits can switch from marginal to subeconomic. Critical to an understanding of these terms (and to the proper use of related numbers) is the phrase "presently economic": the metal reserve or reserve base of a company, a nation, or the world will vary from year to year due to changes in geologic information, technology, costs of extraction and production, prices of mined products, and environmental factors. All these variations are a function of deposit type and locality. Therefore, the reserve of a given commodity can change independently of either reserve de-

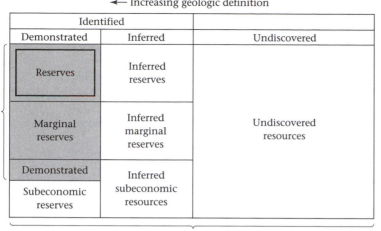

← Increasing geologic definition

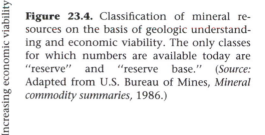

Figure 23.4. Classification of mineral resources on the basis of geologic understanding and economic viability. The only classes for which numbers are available today are "reserve" and "reserve base." (*Source:* Adapted from U.S. Bureau of Mines, *Mineral commodity summaries*, 1986.)

pletion through mining or reserve increase through discovery of new deposits.

Whether or not a portion of a given mineral deposit constitutes a reserve depends first on a detailed understanding of the morphology, size, and mineral content of the deposit through geologic study, sampling, and analysis of drill holes and of surface and underground exposures. Engineering, economic, and environmental factors are then assessed, and if the deposit is feasible to mine, it becomes a reserve (also known as an *ore reserve*). Ore reserves are in some cases reported for a given ore deposit or mining district in the annual reports by mining companies to their stockholders. On an annual basis, the recently eliminated U.S. Bureau of Mines

(USBM) submitted requests to mining companies for data on production and reserves. Because the companies are not required by law to respond, the response rate is rarely above 75 percent (the USBM estimated the nonresponses on the basis of prior reporting and other information). Other countries have similar, though in many cases less reliable, accounting methods. Domestic and foreign reserve numbers, available from the USBM, form the basis for discussion in the following sections. Here we choose reserve data for ninety-seven countries and twelve widely used metals (both geochemically abundant and scarce) to illustrate issues ranging from geographic distribution of mineral reserves to potential for environmental hazards related to mining these reserves. The twelve metals chosen (listed in the caption to Fig. 23.5) represent 81 percent of the dollar value of all metals mined in 1992.

GEOGRAPHIC DISTRIBUTION OF METAL RESERVES. The distribution of mineral reserves among the countries of the world is highly uneven. One might expect that the larger the country, the larger the mineral reserve; however, this is true only in the extreme (Fig. 23.5): the former Soviet Union, once the world's largest country, had the largest percentage (12 percent) of the world metal reserve. At present, the top five countries in terms of land area (Canada, China, United States, Brazil, and Australia) contain the highest metal reserve. Other than these top five countries, the world's nations display no correlation between size of country and size of metal reserve (Fig. 23.5).

What are the factors that determine the distribution or known metal reserves among countries? Factors that are more important than size of country include the geologic setting, the degree of soil and sand cover, and the degree

Figure 23.5. Percent of metal reserve plotted against percent of land area for each of ninety-seven countries in 1990. The percentage of metal reserve for each country represents the mean of its global shares of each of the following twelve metals: iron ore, nickel, manganese, molybdenum, chromium, tungsten, bauxite, titanium, copper, zinc, lead, and tin. These metals represent 81 percent of the global value of all metal reserves. Plot illustrates the general lack of a relationship between land area and metal reserves. (*Source:* Data on reserves from World Resource Institute, 1992.)

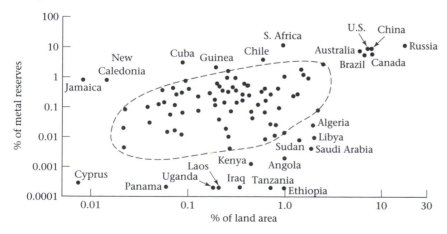

to which the country has been explored and/or "mined out." The geologic setting of a given country has an important influence on its reserves of different metals. For example, we know that ore-forming processes have evolved through geologic time: Certain metals such as nickel, chromium, and titanium were concentrated as ore deposits predominantly prior to 1500 million years ago, whereas others such as tin, tungsten, and molybdenum were concentrated primarily since 500 million years ago. Thus, there is likely to be a correlation between the proportion of crustal age within a given country and the reserves of given metals (Table 23.3).

There are twelve outliers on Figure 23.5 that illustrate some additional factors that govern the distribution of metal reserves. For example, the medium to large countries with very low reserves are the east and north African nations of Kenya, Tanzania, Ethiopia, Angola, Saudi Arabia, Libya, Sudan, and Algeria. Reserves are low in these countries because of a relative lack of exploration and the high proportion (30 to 75 percent) of land area that is deeply covered by young sediments. In contrast, very small to medium-size countries with very high reserves are Jamaica, New Caledonia, Cuba, and Guinea. Reserves are relatively high in these countries because the combined factors of rock type and climate (tropical to subtropical) have led to extensive development of laterites, mined for nickel (New Caledonia and Cuba) and bauxite (Jamaica and Guinea). Chile has an anomolously high reserve for its territorial size because it encompasses some 50 percent of the Andean Cordillera, a continental margin magmatic arc that is uncommonly rich in porphyry copper and related metal deposits. South Africa also is anomolously rich in minerals. South Africa not only qualifies as the highest-rated country in terms of reserves of the twelve metals considered in Figure 23.5, but also contains the bulk of world reserves of gold and gem-quality diamonds. The important point to remember is that mineral reserves are not evenly distributed; this fact has important consequences for the future of both the developed and the less developed nations of the world.

METAL AVAILABILITY. During the 1970s, the specter of increasing scarcity and insufficient supply dominated discussions of mineral resources. Many mineral economists felt that most of the great ore bodies of the world had been found, and that political and environmental restrictions were severely curtailing exploration for new deposits. During the 1980s and early 1990s, a turnabout in perceptions of resource scarcity took place. It is now widely believed that, at least for the foreseeable future, there is an abundance of mineral resources globally. Is there a way to validate this change in perception, or is it largely based on anecdotal evidence?

Physical Measures: The Short-term View. An empirical approach to assessment of the short-term, nonrecycled mineral supply situation is to divide present world reserves by present world production for each commodity of interest. This yields the period of years for which current reserves are adequate, known as the *world reserve life index (WRLI):*

WRLI = (world reserve)/(world production)

Changes in world reserves and/or production rates will change the WRLI. As shown in Table 23.2, which summarizes the WRLI for twenty-seven commodities, the WRLI for 1992 is: thousands of years for soda ash, rare earths, and magnesium; 190 to 160 years for iron ore, cobalt, and platinum; 100 to 50 years for the ferroalloy metals chromium, tungsten, nickel, and molybdenum; and 35 to 20 years for copper, sulfur, gold, mercury, lead, zinc, and silver. Some mineral economists believe WRLIs yield overly optimistic lifetimes because production rates may increase due to increasing population and demand. A contrasting view is that the life index

TABLE 23.3. RESERVES (IN MILLIONS OF METRIC TONS) OF METALS CHARACTERISTIC OF DIFFERENT GEOLOGIC TIME PERIODS, ILLUSTRATING THE UNEVEN DISTRIBUTION OF RESERVES

Archean & Early Proterozoic[a]		Early Proterozoic[a]		Early–Middle Proterozoic & Paleozoic[b]		Paleozoic & Cenozoic[b]	
	NiCrTi		Fe		PbZn		SnWMo
South Africa	337	FSU	23,500	Australia	33.0	United States	3.17
Brazil	71	Australia	10,200	United States	31.0	China	3.10
FSU	55	Brazil	6,500	Canada	28.0	Chile	1.25
India	54	Canada	4,600	FSU	19.0	Brazil	1.22
Zimbabwe	44	United States	3,800	China	11.0	Malaysia	1.12
Canada	35	China	3,500	Mexico	9.0	FSU	1.08
Norway	32	India	3,300	Peru	9.0	Canada	0.82
China	31	South Africa	2,500	Spain	6.7	Indonesia	0.68
Australia	31	Sweden	1,600	North Korea	6.0	Thailand	0.30
Cuba	19	Venezuela	1,200	Ireland	5.7	Australia	0.26

[a] The first two columns represent metals that were concentrated dominantly in Archean and Early Proterozoic time (3000 to 1500 million years ago).

[b] The third and fourth columns represent metals that were concentrated dominantly in Middle Proterozoic to Cenozoic time (1500 to 0 million years ago).

Note: Countries with a large proportion of Precambrian crust (South Africa, Brazil, FSU) contrast with countries with a large proportion of Paleozoic and Cenozoic crust (United States, China) in their share of different metals.

should be calculated using the world reserve base, yielding lifetimes that are two five times longer, depending on the commodity. Lifetimes also could increase due to technologic breakthroughs and/or economic changes. Where does reality lie? I suggest that it lies in the longer time periods, based on one simple observation: Mineral exploration continues to discover new reserves in both operating mines and new districts.

We examine operating mines first. Historically, the charge to mine geologists was to find a new metric ton of ore for every metric ton mined. During the early part of the twentieth century, when mines were young, this was not a difficult task. Potential reserves in any given deposit were largely unknown, and miners developed their reserves from year to year. For the most part, they were successful: The ore reserve at any given mine remained constant through the years in spite of ongoing production. What about today? Mining has changed significantly because increases in scale require increases in investment needed to initiate production, hence greater care in estimating the potential reserve of a given mineral deposit. Geologic reserves are calculated during the exploration phase on the basis of results from drilling. Reserves are modified as drilling progresses and eventually form the basis of mine feasibility studies. This might suggest that today the full potential of any given deposit is known before mining commences, and that once a given deposit is added to the USBM world reserve data base, its reserves are depleted directly as a function of production. However, this conclusion is valid only in limited cases. The norm is continued discovery of new reserves within individual deposits. For example, during a fourteen-year period (1978 to 1992) the twenty largest copper mines in the United States mined 33 percent of their original reserve but added 21 percent of new reserve. These twenty deposits as a group have extended their lives well beyond their 1994 life prediction, as most of them have done every year since they went into production. Should this case history make us complacent about our mineral reserves? No, because eventually all mines are depleted of reserves. Rather, this history illustrates the difficulty of forecasting future supply and makes us aware of the inadequacy of the WRLI for long-term prediction.

One might expect that the value or number of new mineral deposits discovered would decline with time as the number of easy-to-find deposits declines. However, numerous factors can combine to maintain high exploration success, including: (1) advances in geologic understanding and identification of new types of deposits, (2) improvements in exploration and extraction technology, (3) increases in real prices of some metals that make reserves out of marginal reserves and subeconomic deposits, and (4) political changes that open up largely unexplored regions. It appears that exploration productivity in industrialized nations with a long history of domestic mining either is not declining or is declining at a low rate. For example, a recent study concluded that the value (adjusted for inflation) of mineral discoveries in the United States from 1955 to 1980 averaged approxi-

mately $30 billion per five-year period, with no clear upward or downward trend.

These successes are cyclic and dependent on the commodity. The 500 percent increase in real price of molybdenum between 1975 and 1980 led to a boom in molybdenum exploration, extraordinary exploration success, a saturated market, and termination of molybdenum exploration in 1980. In the United States, the abrupt decline in copper and molybdenum exploration in the early 1980s was replaced by a surge in gold and silver exploration. This surge was brought on by the rise in relative price of precious metals between 1970 and 1990 and by improvements in extractive technology. The result was discovery of dozens of large, low-grade disseminated gold deposits in Nevada that doubled the U.S. gold reserve in ten years (1980–90) and placed Nevada second (behind Arizona) in the United States annual dollar value of nonfuel mineral production ($2.6 million). In 1976, the discovery of the Olympic Dam copper-gold deposit in Australia, containing a copper and gold reserve valued at $92 billion, increased world copper reserves by 10 percent. In 1988, the discovery of the Grasberg copper-gold deposit in Indonesia, containing a copper and gold reserve valued at $39 billion, increased world reserves of both copper and gold by 3 percent. These are the types of discoveries that lead to turnarounds in people's perceptions of resources availability. In practice, depletion of mineral reserves has not proved a serious threat to the availability of mineral supplies.

In spite of the anecdotal evidence discussed above, inspection of historic trends in WRLIs over the period 1975 to 1992 suggests that discoveries are gradually losing ground to production (Fig. 23.6). Of the twelve metals considered here, only reserves of tin and zinc are outliving predictions based on the 1975 WRLI. Nickel, iron, and aluminum reserves are following predictions, copper and lead are losing ground slightly, and manganese, chromium, and titanium are being depleted at rates that are two to four times those predicted in 1975.

Economic Measures. Although the physical measures used above yield useful results concerning supply, these measures alone can be misleading because economic factors are considered only insofar as the definition of "reserve" has an economic basis. Also, as the WRLI measure has demonstrated, some physical measures may understate mineral resource availability. An alternative approach is to examine the trends in labor and capital costs per unit of extractive output. When this measure is applied to a variety of resource industries, the results suggest that resources in general, and especially fuel and nonfuel mineral resources, have become increasingly available. The main factors in this increase are the improvement of extractive technology and efficiency. Production costs have continued to decrease. For example, the fact that there has been an increase in reserves through time in some producing copper districts hides the fact that grades of copper in these mines have declined over the years. In other words, increases in metal content in reserves in many cases are the result of increasing tonnages of lower-grade

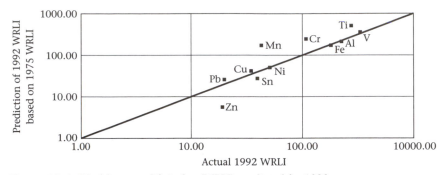

Figure 23.6. World reserve life index (WRLI) predicted for 1992 plotted against the actual WRLI for 1992 for eleven widely used metals: aluminum (AC), chromium (Cr), copper (Cu), iron (Fe), lead (Pb), manganese (Mn), nickel (Ni), tin (Sn), titanium (Ti), vanadium (V), and zinc (Zn). The predicted WRLI is based on the 1975 WRLI discounted through time to 1992 at constant 1975 reserves and 1975 production rates. The plot tests the validity of the WRLI as a measure of future availability of metals. The WRLI is a valid indicator of availability over this seventeen-year period only for nickel, iron, and aluminum. Metals plotting below the line are more available than the WRLI would predict, whereas metals plotting above the line are less available than the WRLI would predict. (*Source:* Data from D. J. Weisser, M. Dalheimer, D. Köhler, and B. Sarbas, Sources and production of widely used metals: an assessment of world reserves and resources, in D. J. McLaren and B. J. Skinner (eds.), *Resources and world development*, Wiley, 1987; pp. 295–303; the World Resource Institute, 1992; and S. E. Kesler, *Mineral resources, economics, and the environment*; Macmillan, 1994.)

material that can be mined. Mining of lower-grade ores is possible because of new extractive technology.

Our conclusion is that mineral resources have become increasingly available over the past century on a worldwide basis. The short-term future – the next twenty to thirty years – looks rosy. However, what can we say about the long term? All the measures of resource availability discussed to this point have been empirical. We have examined the empirical record of the past (and the present) and have tried to predict the future. Scientists are more comfortable with theoretic approaches, in which models are built to be predictive on the basis of fundamental physical and chemical laws. The question of future mineral resource supply is certainly subject to economic and geologic principles that can be cast into mathematic formulations. However, some of the factors influencing long-term future supply of resources are largely unpredictable, including: (1) rates of discovery of presently known types and of presently unknown types of mineral deposits; (2) rates of population growth; (3) changes in consumer demands, effectiveness of recycling, and degree of conservation; (4) and rate of technologic change.

Geologic Research and Future Supply. Present-day empirical and theoretic approaches to questions of future resource supply do not yield precise answers. Most experts would agree, however, that there will be severe shortages of many metals in a century or less. It is not too soon to start planning for this future. To continue to discover and safely

extract minerals requires continuing education and research in geology, mining engineering, and environmental science supported by enlightened political institutions.

Given adequate educational and research facilities, there are numerous research directions related to future mineral supply that geologists should pursue. Two research areas in need of further work relate to: (1) better characterization of marginal reserves, and (2) investigation of the mineralogic sites of metals. What is the size of our marginal reserve of metals? Where are these low-grade mineral deposits located, and what are their characteristics? These questions are difficult or impossible to answer at present because mining companies typically do not publish information on marginal reserves and spend little time or effort characterizing them either geologically or in terms of metric ton and grade. Increases in reserves within individual mining districts are the result either of reclassification from marginal reserves or of new discoveries. In attempting to predict supply, it would be useful to know the relative importance of these two factors. In order to guide research aimed at new extractive technologies, we need to understand the geologic and mineralogic characteristics of marginal reserves. What are the changing mineralogic sites of metals (e.g., copper in sulfide minerals as opposed to copper in silicate minerals) on passing from high-grade ore to low-grade ore and eventually to common rocks? Data required to answer these questions would also have direct application in mineral exploration and environmental geochemistry. For example, the data would yield information on geochemical halos of broadly dispersed elements surrounding ore, which could be used to enlarge the size of exploration targets. Knowledge of distribution of minerals and toxic elements in and around mineralized bodies would aid in establishing geochemical (environmental) baselines prior to mining, and in understanding the natural medium through which potentially toxic mine waters and groundwaters travel to the accessible environment.

The geochemically scarce metals are unevenly distributed in the minerals of the Earth's crust. For example, the copper content of the uncommon sulfide ore mineral chalcopyrite is 33 weight percent whereas the copper content of the common silicate (non-ore) mineral biotite may be near 0.02 weight percent. The average grade of copper ore mined today is approximately 1 weight percent, whereas the average grade of copper in granite is 0.002 weight percent. These examples represent extremes of distribution; nevertheless, the numbers suggest that a gap exists in both grade and mineralogic site of geochemically scarce metals on passing from ore to common rock. Because the energy required per unit mass of metal extracted is several orders of magnitude less for sulfides than for silicates, this pattern of distribution leads to an extractive barrier, referred to as the *mineralogic barrier*. The mineralogic barrier, depicted for copper in Figure 23.7, is un-

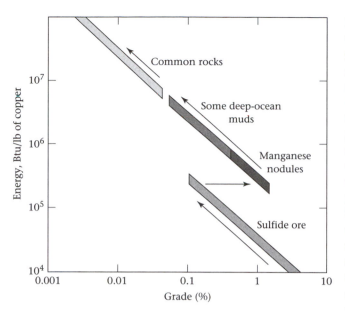

Figure 23.7. Energy input per unit mass of metal recovered plotted against grade of various copper-bearing rocks. Present-day copper-sulfide ores require increasing energy input as grades decline; however common rocks will require more than order of magnitude more energy to extract copper from silicate minerals. This discontinuity has been termed the *mineralogic barrier* (Skinner, *American Scientist*, vol. 64, 1976). After conventional sulfide ores are depleted, but before the transition to common rocks is pursued, it is likely that other copper-rich rocks such as manganese nodules and metalliferous deep-ocean muds will be mined. Btu/lb = British thermal units per pound. (*Source:* Adapted from J. R. Craig, D. J. Vaughan, and B. J. *Resources of the Earth*, Prentice-Hall, 1988.)

TABLE 23.4. CHANGE IN EXTRACTION (DEPLETION FACTORS, SEE FIG. 23.1) RELATIVE TO CRUSTAL ABUNDANCE DURING PERIOD 1976 TO 1992 (NORMALIZED TO IRON)

Element	1976	1992	% Change
Gold	110.00	160.00	45%
Mercury	110.00	55.00	−50%
Silver	20.00	35.00	75%
Lead	20.00	55.00	175%
Copper	18.00	25.00	39%
Chromium	8.50	20.00	135%
Zinc	8.00	15.00	88%
Molybdenum	4.00	15.00	275%
Tungsten	2.00	7.00	250%
Manganese	1.40	3.50	150%
Nickel	1.05	2.00	90%
Platinum	1.05	7.00	567%
Iron	1.00	1.00	Reference
Titanium	0.04	0.02	−63%
Aluminum	0.04	0.04	0%
Magnesium	0.02	0.01	−44%

Note: As examples of depletion factors, in 1992 gold was extracted 160 times faster than iron relative to their crustal abundances (see text).

Source: 1976 data from Brian Skinner, *American Scientist*, vol. 64, p. 258.

---

known for most metals. Also unknown is when we will reach that barrier, although some predictions can be made regarding which metals will arrive at the barrier first. At the beginning of this chapter we used the fact that the crustal abundance of metals is directly reflected by annual production (Fig. 23.1) in order to identify those metals that are being mined "faster" than iron. We introduced the term *depletion factor* to describe the distance of different metals from the iron reference line. Some sense of the rate of movement of different metals toward their mineralogic barrier (away from the iron reference line of Fig. 23.1) can be achieved by comparing the depletion factors in 1976 with those in 1992 (Table 23.4). We find that over the past sixteen years, the depletion factor has increased for all the geochemically scarce metals except mercury, and has decreased or remained constant for all the geochemically abundant metals. Those metals with the highest rate of increase in their depletion factor are platinum, molybdenum, tungsten, and lead. In 1976 two metals had factors greater than 50; in 1992 there were three such metals. In other words, we are increasingly using the scarce metals more and the abundant metals less, a trend that will inexorably draw us into a "second iron age," as suggested by Brian Skinner.

Recycling and Substitution. In parts of the United States, the recycling of paper, aluminum cans, plastic containers, and glass bottles has become second nature. To date, much of the steel and rubber in automobiles is being recycled; however, most of the remaining materials are not recovered in the process. The benefits of recycling are clear. To produce steel from scrap metal rather than from iron ore requires 60 percent less energy, and in the process air and water pollution are cut by 70 percent. To produce aluminum from scrap rather than by mining and processing aluminum ore requires 95 percent less energy. But how effective is recycling in meeting present demands?

Over the past three decades, recycling of metals has been significant on a world basis only in the case of a few of the widely used metals. Recycled lead and copper represent 35 to 40 percent of world consumption, because both find dominant uses in easy-to-recycle concentrated (nonalloyed) form (lead in batteries, copper in cables and pipes). Recycled aluminum, iron, titanium, nickel, and tin each represent some 15 to 30 percent of their world consumption. Recycled zinc represents only 7 percent of world consumption because its major uses as coatings on iron and steel, as an alloying element and as a pigment, lead to dispersal and loss to recycling. As is the case for zinc, recoveries of elemental manganese, vanadium, and chromium are impractical; these metals are recycled as components in iron and steel scrap. Owing to present costs, technology, and manufacturing methods, most of the remaining metals cannot be recycled. The situation is even worse for the industrial minerals. Cement and sand and gravel used in concrete end up in landfills. Reflecting their high solubilities, fertilizers and certain other chemicals manufactured from minerals end up polluting groundwater.

Some argue that we should find substitutes that are less environmentally hazardous to produce or consume, are less costly, and/or are more abundant. In some industries and applications, this makes sense. For example, asbestos cement was replaced by fiber cement in the 1980s. Lead in gasoline has been replaced in most developed countries by alternate octane-enhancing compounds in order to avoid toxic gas emissions. Glass fiber optics are increasingly substituted for copper conductors in communications technology. Many steel parts in automobiles are being replaced by plastic and aluminum to decrease weight and increase gas mileage. Polymers and other composite materials are being used increasingly in construction of automobiles and airplanes. The cast-iron engine block may be replaced in the future by ceramics. The incentive to find substitutes is clear. Certain minerals will always be needed; however, there is a point at which the price (in dollars or health hazards) of supplying a product is higher than the price of developing technology to render that product obsolete.

Many metals and industrial minerals presently cannot be substituted for in their major uses, because of either cost or technical difficulties. For example, there is no known substitute for titanium in most aircraft and space use, and no cost-effective substitute for titanium oxide pigment. Substitutes for the steel-alloying metals (manganese, vanadium, and chromium) and for nickel alloys presently are limited by higher costs and technical drawbacks. Platinum is unchallenged as a catalyst. Gold is gold. For the industrial minerals, there is no substitute for the limestone used to manufacture cement and lime (for steel-making), no substitute for the sand and gravel used in concrete, and no substitute for the potash and phosphate minerals used in fertilizers. Substitution is not yet the solution to the problems of resource depletion and environmental degradation.

SUSTAINABLE DEVELOPMENT: VALUES AND ENVIRONMENT

The concept of sustainable development is commonly defined as a path that "fulfills the needs of the present without compromising the ability of future generations to meet their own needs" (p. 6, Commission on Geosciences, Environment, and Resources, World Commission on Environment and Development, 1987, National Research Council, 1994, "Assigning economic value to natural resources," National Academy Press, 185p.). These needs can be thought of as the well-being of society as a whole, incorporating such aspects as housing, food, clean air and water, health and education, and the preservation of ecosystems. Because mineral resources are nonrenewable and are depleting through time, society is reducing its future productive capacity with regard to that stock of resources. Although sustainability need not require conservation of mineral reserves, as would be the case for renewable resources such as water and forests, some form of substitution is necessary so that per capita well-being in-

creases through time. Therefore, depletion of mineral resources is consistent with sustainability as long as it is conducted in an environmentally sound manner and the proceeds from mining are invested in some measure into education and technology (substitution among various types of capital). The difficulty in further defining the concept of sustainability is obvious: How do we assign appropriate measures of capital substitution? How do we value an ecosystem? However, the statement itself has presented a challenge to scientists and public policy makers alike.

At present we have no generally accepted measure of the "well-being of society." National economic measures, such as gross domestic product (GDP) or gross national product (GNP), fall short of measuring well-being. These economic measures, although reflecting natural resource use to some extent, fail to account for resource depletion and environmental degradation. In fact, one could argue that GDP is a negative measure of environmental quality (and therefore of well-being), especially in those countries where the GDP is dependent principally on production of nonrenewable natural resources.

PER CAPITA DISTRIBUTION OF METAL VALUE. Above, we examined reserves of twelve metals held by ninety-seven nations from a geologic perspective and we found that there exist significant disparities in distribution. How do these disparities affect the potential for well-being of the peoples of the world? We can examine the dollar value of national metal reserves as a measure of potential national development and well-being (assuming that nations invest the income from mineral production in education and environmental protection). The dollar value of national metal reserves is given by

$$Me\$ = {}_i[(tMe^i) \cdot (\$/t^i)]$$

where tMe^i represents millions of metric tons of metal i in reserve, and $\$/t^i$ represents the dollar value of metal i per metric ton in November of 1994. This gross value is not equivalent to "net value," which is determined by commodity price less all costs incurred in extraction. The countries with the highest per capita value of metal reserves (Me$ per capita) are Australia, Surinam, Guyana, Jamaica, and Guinea, in descending order. Contrasting this result with the ranking of nations in terms of their percent share of global mineral reserves (see Table 23.3), we find that only Australia repeats in the top five.

To examine the relation between mineral wealth and national development, we can recast the value of metal reserves (Me$) to a measure comparable to gross national product (GNP), which is measured on an annual dollar basis. What is the potential annual value of each nation's reserves? One way to do this is to define the potential gross domestic metal product (GMeP), in dollars per year, by the following relation:

$$GMeP = {}_i\{[(tMe^i)/(WLRI^i)] \cdot [\$/t^i]\}$$

where $WLRI^i$ represents the world reserve life index for metal i. GMeP is a measure of the *potential* dollar value per year *if* metal production rates in each country were proportional to the world production rates. Using the value of $(tMe^i)/(WRLI^i)$ as an index of future mining rates, instead of the actual published production rates for each country, is both a convenience and a way of smoothing out fluctuations in national production brought on by transient political and social factors.

The relation between GNP and GMeP on a per capita basis for the world's nations is shown in Figure 23.8. Six of thirteen nations with a GNP per capita greater than $10,000 (highly developed countries) have a GMeP per capita of greater than $100 (Norway, New Zealand, Finland, Sweden, Canada, and Australia, in order of increasing GMeP per capita). For these countries, the hypothetical GMeP constitutes 1 to 15 percent of the GNP, and mineral wealth could contribute in a substantial way to national well-being. In contrast, only one of thirteen nations with a GNP per capita less than $300 has a GMeP per capita greater than $100 (Jamaica), and nine have a GMeP per capita less than $1 (Mozambique, Burkina Faso, Niger, Laos, Vietnam, Nigeria, Uganda, Tanzania, and Ethiopia, in order of decreasing GMeP). If present reserves are indicative of future mineral discoveries, then these latter nations will have to rely on other sources of capital for continued development. The fact that history is not necessarily the key to the future is illustrated by Laos and Vietnam: Both countries are presently considered by exploration geologists to be highly prospective for base and precious metals. Their pres-

ent reserves are not thought to reflect their mineral potential.

Of the thirteen highly developed countries, the United States occupies an intermediate position between the high GMeP per capita nations listed above and the lower GMeP per capita nations of France, Spain, Austria, Japan, Italy, Germany, and the United Kingdom (Fig. 23.8.). These highly developed countries with presently low GMeP per capita will remain at low GMeP for the foreseeable future. They are in a relatively mature stage of development of their mineral resources. Less developed countries with an unusually high potential GMeP per capita (ranging from 15 to greater than 100 percent of the GNP per capita) include Papua New Guinea, Solomon Islands, Mauritania, Liberia, Zimbabwe, Guinea, Jamaica, and Guyana. These nations have the opportunity to increase their standard of living based on mineral resources; however, uncontrolled development and a poorly developed technical infrastructure could lead to environmental disaster. China and India, the two countries with the highest population and among the lowest of GNPs, are facing the greatest social need to develop their standard of living. Both occupy the lower middle of the pack in Figure 23.8, with a potential GMeP per capita that is only 4 to 8 percent of the GNP per capita. Unless further exploration yields significant new discoveries in these countries (which is likely in both cases), China and India will have to develop their standard of living by means other than metal mining.

BALANCE BETWEEN METAL WEALTH AND ENVIRONMENTAL DEGRADATION. The problem of surface disruption and groundwater contamination caused by mining is a legacy of the past but need not be a promise of the future. One way to prepare for the future is to attempt to predict the potential for environmental damage caused by future mining around the world. Such a prediction cannot be made in detail because of the numerous factors involved, each of which should be assessed for each deposit: rate of mining of reserves, mining method, topography and climate, composition of ores, and environmental practice. Rather than attempt a detailed approach, we will look only at the broad picture and consider two additive factors: physical disruption of the land surface, and chemical contamination of surface water and groundwater.

Surface Disruption. In order to extract and concentrate metals from the Earth, we need to create underground workings, open pits, waste dumps, and tailings dumps. Dumps, whether chemically toxic or not, can cause silt buildup in streams and airborne dust pollution. Dumps and open pits effectively remove land from most alternate uses. However,

Figure 23.8. Plot of gross national product (GNP) per capita versus potential gross metal product (GMeP) per capita in 1992 dollars for ninety-seven countries of the world. GMeP is based on twelve metals (listed in the caption to Fig. 23.5). Named countries include: (1) highly developed countries with GNP per capita of $10,000 (small black squares); (2) India and China, containing the world's largest populations (large black squares); (3) countries with the highest population density (200 people per square kilometer, highlighted in red); and (4) countries with the highest ratio of GMeP to GNP (group at bottom right). Dashed lines represent the potential contribution of the GMeP to the GNP (in percent).

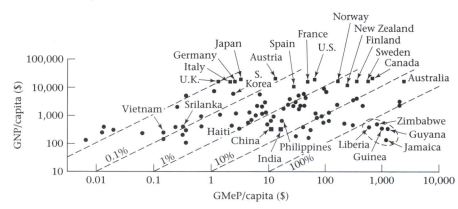

just as in the case of metal smelters, the environmental effects of mining are highly localized and are not toxic at the global scale. Surface mines, and waste dumps from surface and underground mines and related beneficiation (concentration) facilities, occupy only 0.25 percent of the land area of the United States. Nevertheless, surface disruption due to mining becomes a social and environmental issue when it occurs in or near densely populated areas or near sites of historic or religious importance, and when it is potentially disruptive to delicate ecosystems and threatens endangered species.

The physical surface effects of mining are a direct function of mining rates per square kilometers (although surface mining leads to greater modification of land surface than does underground mining). As a generalization, then, we can establish a measure of a given country's potential for surface disruption due to mining, the national *surface disruption index* (SDI),

$$SDI = {}_i[tO^i/WRLI^i] \ [PD]$$

where tO^i represents metric tons of of ore in reserve of metal i, and PD represents population per square kilometer. SDI sums all the present ore reserves of a given country over the average mining rate (WRLI) and thus is time-dependent (similar to GNP and GMeP); however, this measure accounts neither for past mining nor for future mining of unknown deposits outside the present reserve. The potential for environmental risk to human populations is accounted for by multiplying the sums of mining rate by population density. Some countries, such as Chile, are mining their reserves at a rate that is likely higher than the world average: for such countries, the SDI (and CPI, see below) represents their minimum position on the world scene. Other countries, such as the Philippines, are mining their reserves at a rate that is lower than the world average; for such countries, affected by political instability in remote regions where ores are located, the SDI represents a maximum for the next decade.

■ **Chemical Pollution.** Humans move a significant amount of minerals and fluids from deep crustal sources to the surface by extracting petroleum (Chap. 24), and by mining industrial minerals and metal sulfides. All these resources contain toxic substances that are released to varying degrees to the biosphere when they are brought to the surface. For example, the total anthropogenic sulfur flux to the surface is close to 400 million metric tons per year, or approximately twenty times the natural sulfur flux from deep sources to the continental surface and atmosphere! Sixty-two percent of this flux comes from sulfur compounds in fossil fuels, and some 30 percent from native sulfur, metallic sulfides, and sulfates mined for industrial purposes. Approximately 27 percent of this anthropogenic sulfur (equivalent to 130 million metric tons per year) is not contained; it is released to the surface environment (30 million metric tons per year) or to the atmosphere (100 million metric tons per year). One-half of the anthropogenic sulfur flux to the atmosphere returns to the surface as acid rain (sulfuric acid). Coal-burning elec-

tric power plants contribute the majority of sulfur to the atmosphere, estimated by the Environmental Protection Agency (EPA) at 82 million metric tons of sulfur per year globally. Metal smelting, petroleum refining, and cement manufacturing each contribute an approximately equal share of the remaining 18 million metric tons of sulfur per year to the atmosphere. In cement manufacturing, the sulfur emissions are the result of coal-fueled power generation, whereas the metal and petroleum industries emit sulfur from the raw material during processing. In addition to sulfur emissions, the burning of fossil fuels contributes the majority of anthropogenic carbon dioxide to the atmosphere and is responsible for the bulk of the toxic elements – lead, cadmium, antimony, arsenic, and mercury – that are released to the environment (see Table 23.2).

"Scrubbing" refers to removal of particulates and toxic gases generated during industrial processes. Approximately one-fourth of coal-burning power plants in the United States are 90 percent efficient in sulfur removal; however, globally this figure is closer to 60 percent. Sulfur dioxide scrubbing is achieved by reacting powdered lime with flue gas; calcium reacts with sulfur dioxide to form anhydrite ($CaSO_4$), the same mineral formed during natural scrubbing of volcanic eruptions, as previously described. The anhydrite can be sold or dumped in landfill. A more effective scrubbing technique that may become feasible for coal-fired power plants is to add calcium to form anhydrite *during combustion*. End-of-pipeline methods, such as addition of lime and crushed limestone (reaction (2)) to agricultural lands, rivers, and lakes affected by acid rain, are neither economic nor effective over the long term.

The global efficiency of sulfur removal during processing of minerals is approximately 90 percent. Metal smelters are the most efficient because of the higher concentration of sulfur dioxide in the flue gases. In the United States, 95 percent of the sulfur is recovered during smelting as a result of the new technology of *continuous smelting*. The sulfur is recovered as sulfuruc acid and recycled for use in acid leaching of soluble copper oxide ores. Sulfur emissions remain high, however, in smelters that are remote from population centers or in older smelters in developing countries.

An important factor in assessing potential environmental problems resulting from processing a given country's metal reserve is that many countries export concentrates of ore minerals for smelting in other countries. Therefore, there is no direct correlation between a nation's production of a given metallic ore and that nation's emission of environmentally toxic substances. Inspecting the balance between concentrates imported and concentrates smelted reveals that the United States, Chile, and Canada are effectively exporting their pollution to Japan, Germany, and the former Soviet Union. In the following, therefore, we limit our analysis to mining-related chemical pollution.

As discussed above, chemical contamination of groundwaters and soils results from leaching of toxic elements from mine dumps and mine workings and their incorporation into streams and groundwater. The potential for chemical pollution varies significantly across the ore types. Because heavy metals are most easily dissolved and transported in acid solutions, those ores that are rich in acid-generating minerals are the most hazardous. Because acid is generated by oxidation of sulfides, sulfide-bearing ores are more hazardous than nonsulfide ores. As a first approximation in assessing potential chemical contamination for different countries, we separate sulfide-bearing metal reserves (nickel, copper, molybdenum, lead, zinc, tin, tungsten) from nonsulfide metal reserves (chromium, iron, manganese, aluminum) and incorporate the former into a measure of the national potential for chemical pollution from mining, the *chemical pollution index (CPI)*,

$$CPI = [_i(tO_s{}^i/WRLI^i)] \cdot [PD]$$

where $tO_s{}^i$ designates "metric tons of sulfide ore of metal i." The CPI cannot, however, be applied to all metals in all countries and all sulfide-bearing metal reserves. We need to assess, in particular, the degree to which nature's own toxic superfund program has been effective. For example, recent and past lateritic weathering of nickel-sulfide ores and other nickel-bearing rocks has effectively removed the acid-generating capacity of many nickel ores. Therefore, in the computation of national CPIs, we remove the metric tons of sulfide ore of nickel from the CPI of nations that contain all or most of their nickel reserves in nickel laterites, not in nickel sulfide ores.

■ **Combined Physical and Chemical Hazards.** The potential for overall environmental degradation due to mining requires an additive function that includes both the CPI and SDI. Because chemical pollution of air and water is a more serious threat to ecosystems than is the relatively localized surface disruption, the CPI should be given more weight than the SDI. Here we arbitrarily give chemical pollution three times the weight of physical pollution in order to calculate a nation's *environmental hazard index (EHI)* related to mining:

$$EHI = (SDI) + 3(CPI)$$

EHI, computed for ninety-seven countries and twelve metals, identifies Chile, the United States, the Philippines, and India as the countries with the greatest potential for negative environmental impact on their populations as a consequence of mining its reserves. However, these more developed nations should be the most capable of installing environmentally sound mining practices. If this is generally true, then plotting EHI against GNP/per capita yields a measure of the capability of a given nation to deal with the environmental costs of mining. Figure 23.9 suggests that, according to this measure, the United States, Japan, Poland, the former Soviet Union, and Chile should be capable of dealing effectively with their high EHIs, whereas the Philippines, Indonesia, China, India, Zaire, and Jamaica currently may be incapable of controlling their high EHIs. This analysis assumes that the economic burden of sound environmental practice

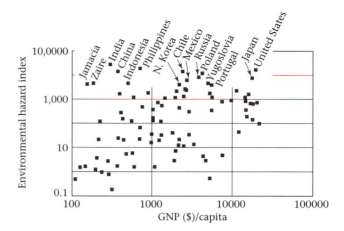

Figure 23.9. Plot of environmental hazard index (EHI) versus gross national product (GNP) on a per capita basis for 1992 for ninety-seven countries, illustrating potential balance or inbalance between these two factors for the nations of the world based on twelve commonly used mineral commodities (listed in the caption to Fig. 23.5. The fifteen countries with highest EHI are identified.

is borne in part by federal funds. We have seen this to be the case even in the United States (recall Summitville, Colorado). The analysis also must be modified by the degree to which mining in less developed countries is conducted by multinational corporations that export sound mining practices.

CONCLUSIONS

Based on what we know about existing mineral reserves, success in exploration for known types of mineral deposits, and present uses and rates of consumption of minerals, few, if any, mineral resources will be in short supply for the next twenty to thirty years. Although this conclusion may suggest an impending downturn in industrial development, this is unlikely to occur because of additional factors that contribute to the sustainability of mineral-based industrial development: recycling will become increasingly important as a source of metals, scarce resources will be replaced by more abundant resources, and new types of mineral deposits (*unconventional deposits*) will be discovered. In the coming decade, mining will continue to focus on conventional deposits exposed on the continents. For many of the developing countries, which are in an early stage of mineral exploration and development, as new resources become available, their economic growth will be facilitated. Mining on the continents will proceed toward increasingly lower-grade deposits located close to the surface. Possibly by the year 2025, the search for conventional high-grade deposits will extend to depths of up to 5 kilometers below the surface; such deep exploration will require the technology and scientific understanding of the developed nations. Through time, the ratio of abundant-to-scarce metals produced will have to increase as the mineralogic barriers for geochemically scarce metals are reached. As the mineralogic barriers are approached, non-conventional metal deposits such as metal-bearing brines, metalliferous black shales, and others whose existence is unknown today will become the targets of exploitation. Saline lakes and saline groundwaters will become increasingly important sources of chemical and fertilizer compounds. Mining in the oceanic realm will focus on manganese nodules and to a lesser extent on deep-sea massive sulfide deposits and their hydrothermal vents. The metals of the future will be iron and the abundant light metals, such as aluminum and magnesium. Glass and ceramics will take on increasing importance, as will recycling and conservation.

Revising national economic measures to incorporate natural resource depletion and environmental quality is a challenge that recently has been taken up by many institutions, including the U.S. Congressional Budget Office, the United Nations, and the World Bank. Such assessments are needed to prepare for an uncertain future that will require national and international policies for resource use and environmental protection. Among the many technical and philosophic issues involved are: the assessment of costs that arise from environmental damage (these should be charged against the GDP); the characterization of natural environments (such as water quality and ecosystems) that would be adversely affected by resource production; and the assignment of economic value to environmental quality. Humanity is about to face its greatest challenge. How do we meet the future needs of a population that will grow to 10 billion or more (see Chap. 22) by the year 2050? This challenge comes at a time when resources are dwindling and environmental degradation is increasing. Human activity has already outpaced the rate of natural processes in transferring mass from the Earth's crust to its surface. As the population continues to grow, it is likely that the rate of mass transfer and environmental degradation also will grow, especially if technologic change and conservation do not proceed apace. In the United States and other highly developed nations, the rate at which we use our mineral resources is greater than the rate of population growth. In the less developed nations, the reverse is true. If this pattern remains in place, and given the unequal distribution of minerals, it is likely that international relations will be deeply affected by struggles to control resources. Solutions must be found before this point is reached. The trend toward exponential population growth and use of minerals and other natural resources must be reversed.

QUESTIONS

1. Examine Table 23.1 and list those elements that you use in your everyday life. How many of these elements do you attempt to recycle?

2. Use your understanding of the geologic setting of porphyry copper deposits, combined with the data presented in Figures 23.3 and 6.12, to indicate where, outside of the circum-Pacific region, you might find additional deposits of this type.

3. Table 23.4 summarizes rates of extraction relative to crustal abundance (normalized to iron) for the geochemically scarce metals. In the period from 1976 to 1991, mercury is the only metal listed that has undergone a decrease in its rate of extraction. In contrast, platinum displays the highest percent increase in rate of extraction. Explain these two changes.

4. In 1976, Brian Skinner of Yale University wrote: "Whichever way we turn, we are forced back to the realization that one day soon we will have to come to grips with the way in which earth offers us its riches. That day is less than a century away, perhaps less than half a century. When it dawns, we will have to learn to use iron and other abundant metals for all our needs. The dawn of the second iron age is much closer than most of us suspect" (p. 269). Explain what Skinner means by the "second iron age."

5. In February 1995, the U.S. House of Representatives voted to change the means of arriving at environmental regulations from health-based decisions to cost/benefit analysis (are costs justified by the anticipated benefits to the public?). Regulated industries would have the right to challenge in court the adequacy of the risk assessments. Proponents state that these changes will ensure fairness to industries that must spend billions of dollars each year to comply with federal regulations. Opponents state that this legislation would remove a generation of laws that protect public health and the environment. What do you think?

6. Can we ensure continued economic growth in all the nations of the world? Some believe that if the less developed countries were to achieve the same standard of living as the highly developed countries, we would be facing an irreversible environmental disaster. Why might this be true? If it is true, should the developed nations reduce their use of mineral resources to (1) allow the developing nations to increase their standard of living, and (2) ensure that future generations can meet their own needs?

FURTHER READING

Craig, J. R., Vaughan, D. J., and Skinner, B. J. 1988. *Resources of the Earth*. Englewood Cliffs, NJ: Prentice Hall.

Holland, H. D., and Petersen, U. 1995. *Living dangerously: The Earth, its resources, and the environment*. Princeton, NJ: Princeton University Press.

Kesler, S. E. 1994. *Mineral resources, economics, and the environment*. New York: Macmillan.

Skinner, B. J. 1976. A second iron age ahead? *Scientific American* 64:258–269.

World Resource Institute. 1992. *World Resources, 1992–93*. New York: Oxford University Press.

24 Energy Resources and the Environment

JANE WOODWARD, CHRISTOPHER PLACE, AND KATHRYN ARBEIT

ENERGY USE AND HABITABILITY OF THE PLANET

The patterns of energy use that have evolved over the past century have produced an extraordinary standard of living for industrialized countries while allowing the people of less developed countries (LDCs) to subsist. Unfortunately, the development and use of our energy resources has been one of the leading contributors to environmental degradation, affecting all Earth systems.

In the 1980s, growing awareness of possible climate change and measured increases in greenhouse gas emissions focused attention on fossil fuel combustion in power plants, automobiles, and other energy conversion devices. The environmental impacts of energy production and use had previously been considered site-specific, such as oil spills; however, the combined effect of fossil fuel consumption and increased deforestation primarily due to fuel wood consumption have increased global atmospheric greenhouse gas concentrations, causing society to question the sustainability of its energy practices and the stability of the planet's climatic balance.

Today, the three fossil fuels – oil, coal, and natural gas – combine to satisfy approximately 90 percent of the industrialized world's energy needs, whereas biomass (e.g., fuel wood, animal and crop wastes) and physical labor are still relied heavily upon by approximately 75 percent of the world's population in LDCs. Despite continuing efficiency improvements in the industrialized world, overall demand for energy continues to grow and the consumption of these depletable resources continues to increase. Although the estimated proven reserves of these resources range from 50 to 300 years at current consumption rates, we must question the cost of using these resources. Society has already developed the easiest and cheapest reserves, consequently, costs are likely to rise in the future, although technology may moderate this trend in the short term. Traditionally, the price consumers pay for a depletable resource includes only its private costs, understating the true social costs of the environmental and public welfare harms borne by society as a result of resource consumption. Many of the costs associated with the combustion of fossil fuels are obvious but unquantifiable, and still others we have yet to understand. Market forces and growing environmental awareness in the developed world has already begun to drive more efficient use and recovery of fossil fuels. Measures are gradually being imposed to mitigate and control current environmental impacts of exploration, production, transportation, refining, conversion, and end use while lawyers battle over who should pay for the unregulated practices of the past.

Our patterns of energy use today cannot be sustained for the world's current population of six billion people, and current practices certainly will not be able to meet the needs of the planet's burgeoning population into the twenty-first century. Change is in progress. Traditional observers of the energy economies forecast a future consisting of the status quo, with modest efficiency improvements in the way fossil fuels are produced and used, coupled with limited growth in renewable energy technologies. However, evidence increasingly suggests a more hopeful future. Even in the ranks of major oil companies, some energy experts suggest that current technologies will make fossil fuel use and renewable resources significantly more efficient and that these technologies will commercialize rapidly in the first half of the twenty-first century. Competitive market forces and policies that price energy services at their true social costs will help drive this change. Some options include removing subsidies from the most polluting energy sources, introducing taxes to reflect the social costs of pollution, creating environmental standards to reduce environmental damage, raising the price of energy produced from harmful sources, and providing market incentives such as the "feebate" – a self-funding mechanism that taxes those who buy products (such as automobiles) that are inefficient and polluting and then uses these revenues to offer rebates to those who purchase clean, efficient products.

Knowledge of energy resources is essential to understand human and Earth systems and to design policies that will lead to the optimum balance between humanity and the Earth for their long-term coexistence and health. Many questions must be addressed:

- How do we repair the accumulated damage our energy use has done to the Earth?
- How do we alter our patterns of energy consumption to make optimal use of the planet's resources, especially those that are depletable and polluting?
- How do we phase out existing automobiles, buildings, mo-

tors, lights, and so forth, and allow rapid adoption of the more environmentally benign technologies that are currently available?

- How do we provide energy services today for over half the world's 6 billion people who do not have electricity or automated transportation without crippling the world economy or devastating the environment?
- If these issues are challenging us now, how do we address them in 2050, when there may be 10 billion people on the planet?

The technologies and policy tools now exist to answer these questions and achieve these goals. The challenge is how to adopt policies and pursue development in an appropriate and sustainable manner. As educated and motivated students of the Earth's systems, we can help produce and implement sound answers to these questions.

Energy is the ability to do work, such as lifting an object or heating a substance. Harnessing energy allows life to exist on a fundamental level and allows us to enjoy our current standard of living as members of a larger society. Energy provides a wide variety of services to society including heating, cooling, cooking, illumination, motor drives for industrial and agricultural processes, and transportation. This chapter examines the nature of energy resources and humanity's use of energy. It also examine the significance, origin, distribution, and environmental impact of the world's depletable and renewable energy resources.

Although energy is all around us, it is rarely in the form and location that we desire it. Energy can be found in many forms: potential, such as energy contained in water behind a dam; mechanical, such as the energy contained in rotating gears; kinetic, such as the energy found in moving media (e.g., wind or water); chemical, such as the energy contained in fossil fuels and biomass; thermal, such as the energy contained in the thermal gradients beneath the Earth's crust; electrical, such as the potential across a battery; radiative, the energy in light; and nuclear, such as the energy released when an unstable atom is bombarded by a neutron. Usually, we must convert energy from one form to another for use and distribution. Table 24.1 shows many of the possible energy forms and conversions. The laws of thermodynamics, specifically the laws of entropy and the conservation of energy, control the outcomes of energy conversions. Each conversion neither creates nor destroys energy but converts some of it to a useless form. In other words, every time a conversion is made, the total amount of energy remains unchanged, however, a portion of the energy involved in the transfer process is effectively lost through conversion to an unrecoverable form. Typically, energy ''losses'' are in the form of low-temperature heat that is difficult or impossible to recapture. The efficiencies of energy conversions determine the quantity of a resource needed to provide a particular energy service.

When discussing energy resources, *power* must be distinguished from energy. As previously defined, energy is the ability to do work, whereas power is the rate at which work is performed. Energy is most commonly measured in *British thermal units* (Btu) or *kilowatthours* (kWh), and power is most commonly measured in *kilowatts* (kW). A British thermal unit is the amount of energy required to raise 1 pound of water 1°F (Fahrenheit), a kilowatthour is a kilowatt exerted over one hour, and a kilowatt is a rate of energy use or consumption equal to 1000 joules/per second. One kilowatthour is the equivalent of 3416 British thermal units. On the world

TABLE 24.1. ENERGY CONVERSIONS				
Conversion From	To Chemical	To Electrical	To Heat	To Light	To Mechanical
Chemical	Food	Battery	Fire	Candle	Rocket
	Plants	Fuel cell	Food	Phosphorescence	Animal muscle
Electrical	Battery	Toaster	Fluorescent	Electric motor	
	Electrolysis	Heat lamp	lamp	Relay	
	Electroplating	Sparkplug	Light-emitting		
	Transistor		diode		
	Transformer				
Heat	Gasification	Thermocouple	Heat pump	Fire	Turbine
	Vaporization		Heat exchanger		Gas engine
					Steam engine
Light	Plant photosynthesis	Solar cell	Heat lamp	Laser	Photoelectric door
	Camera film		Radiant solar		opener
Mechanical	Heat cell	Generator	Friction brake	Flint	Flywheel
	(crystallization)	Alternator		Spark	Pendulum
					Waterwheel

Source: Roger A. Hinrichs, *Energy*, Harcourt Brace Jovanovich, 1991, p. 32.

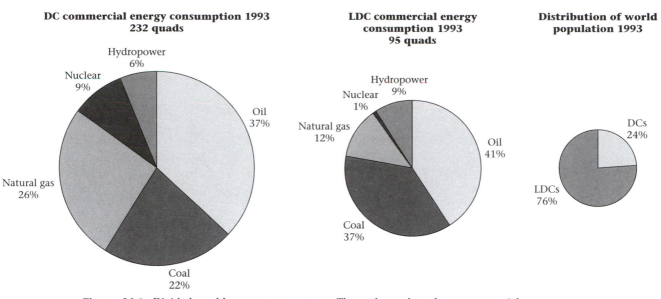

**DC commercial energy consumption 1993
232 quads**

Hydropower
6%

Nuclear
9%

Oil
37%

Natural gas
26%

Coal
22%

**LDC commercial energy
consumption 1993
95 quads**

Hydropower
9%

Nuclear
1%

Natural gas
12%

Oil
41%

Coal
37%

**Distribution of world
population 1993**

DCs
24%

LDCs
76%

Figure 24.1. Divided world energy use patterns. These charts show how commercial energy use patterns differ between the developed and less developed nations. The less developed countries (LDCs) are far more dependent on coal, the dirtiest fossil fuel, and have a much larger portion of the world population than developed countries (DCs), raising the risk of serious environmental damage from coal use. 1 quad = 10^{15} British thermal units. (*Source:* Data from *BP statistical review of world energy, Worldwatch, and world resources*, 1996.)

or national level, it is often more convenient to measure energy with much larger units called *quads*, with 1 quad equaling 1 quadrillion (10^{15}) British thermal units.

ENERGY USE

The world population's divided energy use patterns (Fig. 24.1) have profound implications for the future of our planet. Of the almost 80 percent of the world's 6 billion people living in LDCs, over 50 percent rely exclusively on noncommercial biomass (e.g., crop and animal waste, bushes, trees, charcoal) and on physical labor to satisfy basic energy needs such as warmth and food production and preparation. The remainder of the world's population, which consumes far more energy per capita than do those living in LDCs, relies almost exclusively on *commercial energy sources* (not biomass and labor) and tradable *energy currencies* such as electricity and transportation fuels. As LDC populations grow (Chap. 22) and more of the world's nations industrialize, the world faces the challenge of providing the rapidly growing developing nations with the energy services that they will demand and that only the developed countries currently possess. How we meet humanity's current and future energy demands will greatly influence the health of the planet and the standard of living of future generations. The remainder of this chapter examines many of the energy resources and related technologies available to meet this challenge.

DEPLETABLE RESOURCES. Although a wide variety of energy resources are currently available, society meets the ma-

jority of its energy needs through the utilization of four depletable resources: oil, coal, natural gas, and uranium. In 1992, these resources accounted for over 80 percent of the world's total energy consumption and approximately 95 percent of the world's commercial energy consumption.

Fulfilling almost 90 percent of the world's commercial energy needs, oil, coal, and natural gas are all chemical stores of energy with similar origins. Although there are distinctions among the types of organic material and depositional environments that lead to their formation, oil, natural gas, and coal all derive from the remains of ancient plants and animals and therefore are commonly referred to as *fossil fuels*. Fossil fuels are also all characterized as *depletable resources* because their rate of formation in the Earth's subsurface is far surpassed by the rapid rate of human consumption. Nuclear energy is also depletable because new uranium deposits are not being created inside of the Earth and existing deposits are actually slowly diminishing through their own natural radioactive decay. (For further discussion on this topic, see the box entitled *Determining How Long Depletable Resources Will Last.*)

Chemical bonds between carbon and hydrogen atoms store the energy that humans harness from fossil fuels. Combustion, the burning of materials in the presence of oxygen, releases the chemical energy stored in the carbon–hydrogen bonds in the form of heat. The thermal energy released from hydrocarbons can be used for a variety of purposes: space heating for residential, commercial, and industrial purposes; electric power generation; and driving the jet and internal combustion engines of transportation vehicles. The propor-

Two important terms are used to quantify the abundance of depletable resources such as oil, coal, natural gas, and uranium (see also Chap. 23): resources and reserves. The McKelvey diagram, developed by the U.S. Geological Survey (Fig. 24.2), is a useful way to illustrate the difference between resources and reserves. Although developed to differentiate the categories of oil resources, the McKelvey diagram can be applied to other resources as well. The large rectangle represents the *resource base,* the amount of the resource that experts believe exists. Scientists define the size of the resource base by evaluating geology around the world and then estimating through extrapolation and analogy to known oil, gas, coal, and uranium deposits. The horizontal axis of the graph represents the level of uncertainty, and the vertical axis represents the relative recovery cost. The right-hand portion of the diagram represents undiscovered resources, whereas the lower left portion of the diagram depicts the resources that are known to exist but are too expensive to extract under current conditions. The smaller box in the upper left-hand corner represents *reserves;* reserves represent those resources that are already discovered or are highly likely to exist and are economically recoverable at current prices with current technology. *Proved reserves* have been physically demonstrated to exist, and represent a subset of reserves that includes only those resources already discovered.

The size of the reserve box can increase as a result of change in a variety of factors. An increase in the market price of the resource would extend the reserve box down and eventually to the right as exploration increases; the discovery of additional reserves extends reserves to the right, and technologic advances can extend the box in both directions. For example, the *recovery factor,* the amount of a resource that producers can economically recover from a deposit is approximately 30 to 35 percent for an average oil field. If oil prices rise or recovery technology improves, then the recovery factor will also rise because it becomes economical to extract more of the resource. Reserves can thereby be increased without the discovery of new fields.

Behavioristic models such as the one

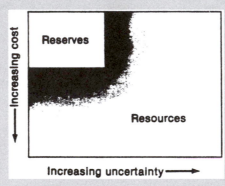

Figure 24.2. The McKelvey diagram for categorizing resources and reserves relates the variables of geologic certainty and economic feasibility. (*Source:* Adapted from R. A. Hinrichs, *Energy,* Harcourt Brace Jovanovich, 1991.)

developed by M. King Hubbert in 1969 provide an alternative method for estimating the size of the resource base. Behavioristic models assume that as a resource is developed, rational economic behavior will cause society to use it in certain predictable ways. When the resource is first tapped, technologies are young and expensive, consequently, growth is slow. As use becomes widespread and technology improves, prices diminish and use of the resource escalates. Eventually, scarcity becomes the dominant factor and prices increase, encouraging conservation, efficiency improvements, and resource substitution. By fitting curves derived from the use patterns of well-developed resources to less developed resources, we can estimate how long our resources will last. Unquestionably, this type of model is not perfect; however, it has some validity because it is based upon human experience and economic theory. Figure 24.3 shows *Hubbert curves* constructed for U.S. and world oil supplies.

It is helpful to relate current proved reserves to the current annual production rate. Typically when we divide the existing proved reserves by the annual production rate, the result is the *ratio of reserves to production,* a measure that suggests how many years current reserves will last. However, the ratio of reserves to production does not account for growth in demand, new technology, price changes, or discoveries. Still, it serves as an easy-to-calculate tool for comparing the expected supply of our energy resources with present consumption, technology, and prices.

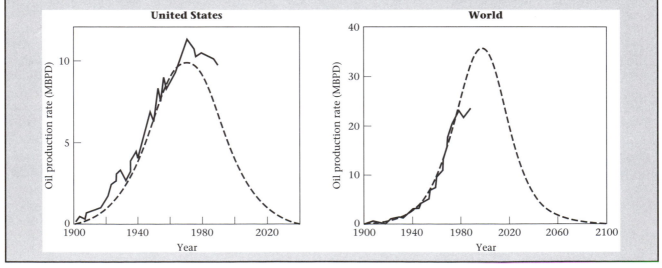

tions of carbon and hydrogen atoms vary in each fuel and directly determine the amount of carbon dioxide emitted during combustion. A growing number of scientists believe that the accumulation of carbon dioxide in the atmosphere is causing global warming; the carbon content of fossil fuels carries a particular environmental consequence (Fig. 24.4).

Combustion of fossil fuels currently accounts for approximately 5.3 trillion kilograms, or approximately 10 percent of all emissions, anthropogenic and natural, of carbon emission to the atmosphere each year. As a result of human interference in the global carbon cycle, concentrations of carbon dioxide in the atmosphere have risen 26 percent since the beginning of the Industrial Revolution in the early 1800s. Scientists expect future carbon dioxide levels to rise by half a percent per year as a result of human activities (see Chap. 19 for more information on the carbon cycle). The combustion of fossil fuels also releases other pollutants. The incomplete combustion of carbon-based fuels emits carbon monoxide (CO), a poisonous gas. Nitrous oxides (NO_x) form whenever anything is combusted in air because high combustion temperatures cause nitrogen and oxygen in the exhaust to form various oxides of nitrogen. Nitrous oxides cause concern because they are a precursor to ground-level ozone and smog formation. Fossil fuels containing sulfur emit sulfur oxides (SOx). Sulfur oxides combine with water droplets in the air to form sulfuric acid droplets, which fall to the Earth as acid rain. Because combustion is seldom complete, it releases various hydrocarbons (HCs) from carbon-based fuels. Hydrocarbons contribute to the chemical mix in the atmosphere that forms smog. Combustion of carbon-based fuels also produces small airborne particles, or particulate matter (PM). Particulate matter contributes to respiratory ailments and has a negatively impact on visibility.

■ **Oil.** Oil makes the single largest contribution to both global and U.S. energy demand. It accounted for 40 percent of both developed nations worldwide, and U.S. commercial energy demands, and 31 percent of the total world energy demand (including the LDCs) in 1995. Oil is typically measured and traded in units of barrels; one barrel is equal to 42 U.S. gallons. World oil production has leveled off since the late 1980s, remaining slightly below its all-time high of 63 million barrels per day in 1979. The United States accounts for 25 percent of the world's oil consumption, approximately 16.5 million barrels per day! Of this total, over half is currently imported, and the United States will be even less able to satisfy its own oil needs in the future. U.S. demand for oil is expected to increase at a modest rate of 1 percent per year for the next two decades, whereas world demand for oil is expected to grow at an annual rate of 1 to 2 percent, largely reflecting rapid population growth and industrialization

Figure 24.3 (opposite). Comparison of Hubbert production curve estimations (dashed lines) and actual U.S. and world oil production (solid lines). MBPD = million barrels per day (*Source:* R. A. Hinrichs; *Energy*, Harcourt Brace Jovanovich, 1991.)

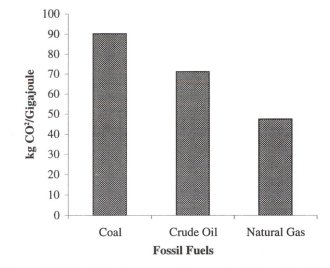

Figure 24.4. Fossil fuels vary in carbon content: Natural gas has the lowest level of carbon per energy unit, and coal has the highest. kg CO_2/gigajoule = kilogram of carbon dioxide per gigajoule. (*Source:* Data from Anonymous, *The environmental costs of electricity*, Pace University, 1991.)

in the LDCs. Figure 24.5 illustrates world oil production and consumption in 1995 by region, providing a useful context for understanding the political, economic, and environmental importance of oil.

In the late 1800s and early 1900s, crude oil was predominately used to produce kerosene for illumination; today, oil is primarily used for producing transportation fuels. Other major uses include electric power generation, heating, and nonenergy applications such as the formation of chemical feed stocks, plastics, fertilizers, asphalt, and coke (a carbon-rich product used in steel manufacturing). A typical barrel of crude oil in the United States is refined into a wide variety of end products, as shown in Figure 24.6, with transportation fuels representing two-thirds of the average barrel. Oil has played a critical role in world trade and politics to date and shows no sign of diminishing in importance. The price of crude oil has reflected many significant economic and political events (Fig. 24.7), the most dramatic being the 1973 Arab oil embargo, when Arab oil-exporting countries embargoed Western nations, many of which (including the United States) were becoming dependent on imports.

Oil is a complex mixture of hydrogen and carbon atoms (hydrocarbons), with an average ratio of hydrogen to carbon of approximately 1:7 by mass. Molecules of crude oil vary greatly in composition depending on the source of the organic material from which the oil formed, the temperatures and pressures to which it was exposed, and length of time the material was buried by overlying sediment. Refining processes break up, reform, and separate these complex molecules into desirable products such as gasoline.

The skeletal remains of single-celled microorganisms such as algae and diatoms provide the source materials for oil.

**Production
130 quads**

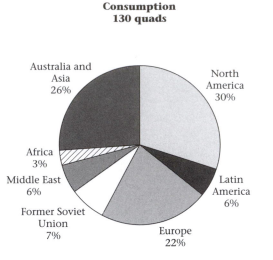

Australia and
Asia
11%

North
America
20%

Africa
10%

Latin
America
9%

Europe
10%

Middle East
29%

Former Soviet
Union
11%

**Consumption
130 quads**

Australia and
Asia
26%

North
America
30%

Africa
3%

Middle East
6%

Latin
America
6%

Former Soviet
Union
7%

Europe
22%

Figure 24.5 (above). 1995 world oil production and consumption. Discrepancies exist between where the oil is produced and where it is consumed, indicating the movement of money and the oil transport that must take place as a result of this concentration in production. 1 quad = 0.17 billion barrels of oil; therefore, 1995 consumption equals 22.5 billion barrels of oil. (*Source:* Data from *BP statistical review of world energy*, 1996.)

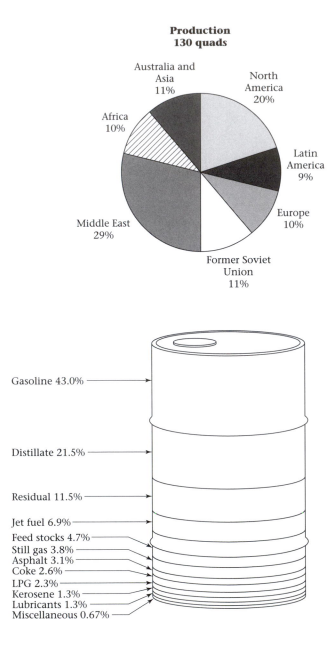

Gasoline 43.0%

Distillate 21.5%

Residual 11.5%

Jet fuel 6.9%

Feed stocks 4.7%
Still gas 3.8%
Asphalt 3.1%
Coke 2.6%
LPG 2.3%
Kerosene 1.3%
Lubricants 1.3%
Miscellaneous 0.67%

Figure 24.6 (left). Crude oil refining products, as illustrated for 1 barrel of oil. Gasoline is the main product of the refining of crude oil; however, refining yields many other products with a variety of uses. (*Source:* Adapted from Robert O. Anderson, *Fundamentals of the petroleum industry*, University of Oklahoma Press, 1984.)

Figure 24.7. (below). Crude oil prices since 1861. Oil prices remained fairly constant until the early 1970s, when U.S. dependence on imports grew and prices rose substantially. A series of oil shocks helped to push prices higher. (*Source:* Data from *BP statistical review of world energy*, 1996.)

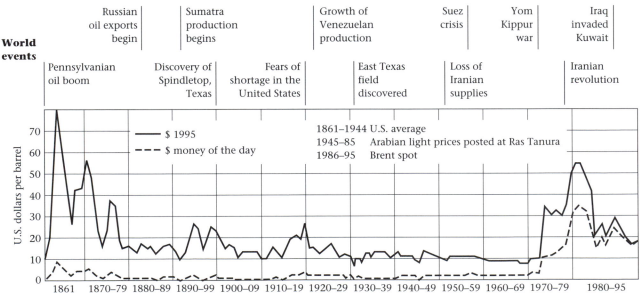

World events

| Russian oil exports begin | Sumatra production begins | Growth of Venezuelan production | Suez crisis | Yom Kippur war | Iraq invaded Kuwait |

| Pennsylvanian oil boom | Discovery of Spindletop, Texas | Fears of shortage in the United States | East Texas field discovered | Loss of Iranian supplies | Iranian revolution |

— $ 1995
--- $ money of the day

1861–1944 U.S. average
1945–85 Arabian light prices posted at Ras Tanura
1986–95 Brent spot

U.S. dollars per barrel

70
60
50
40
30
20
10
0

1861 1870–79 1880–89 1890–99 1900–09 1910–19 1920–29 1930–39 1940–49 1950–59 1960–69 1970–79 1980–95

Normal cycles of upwelling and dramatic climatic events sometimes lead to the development of oxygen-starved zones in oceans and lakes, which cause these microorganisms to die off in large numbers. Over millions of years, sediment accumulates and deeply buries this organic material. The combination of long periods of time and the extreme temperature and pressure caused by deep burial leads the organic material within the sedimentary rock to convert into crude oil, residing in the pores of the rock. The stratum in which the oil forms is referred to as the *source rock*. Over time, high pressures may expel the oil, associated water, and gas from the source rock and slowly force the fluids upward through permeable rocks. The fluids stop when their ascent becomes blocked by an impermeable rock layer, or trap. Migrating oil accumulates beneath the trap, commonly an anticlinal fold or warping of the rock strata, to form an oil field. The rock beneath an impermeable layer where oil, associated water, and natural gas accumulate is referred to as a *reservoir rock*. Economically attractive reservoir rocks (Fig. 24.8) are formations such as sandstones that have both high *porosity* (a large

Non-OPEC Nations

OPEC Nations

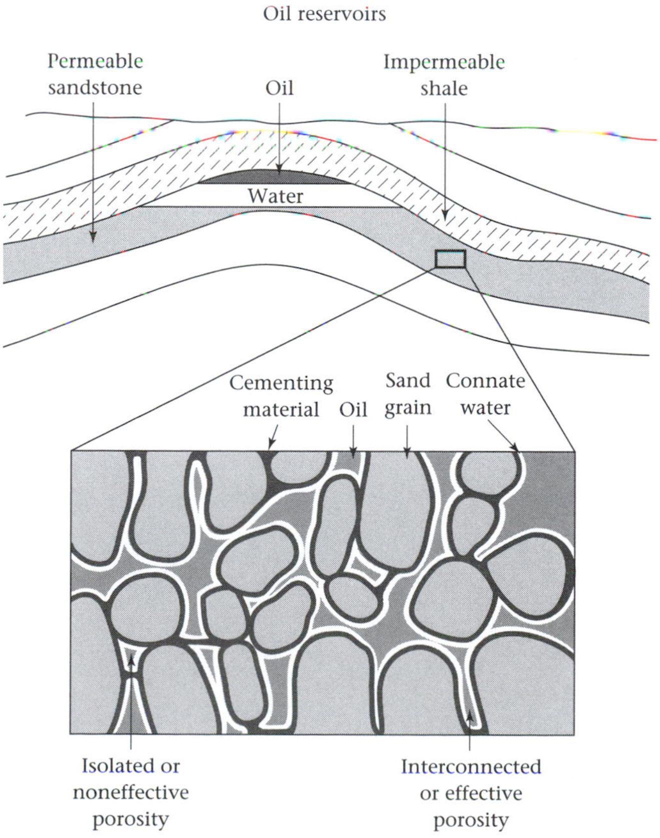

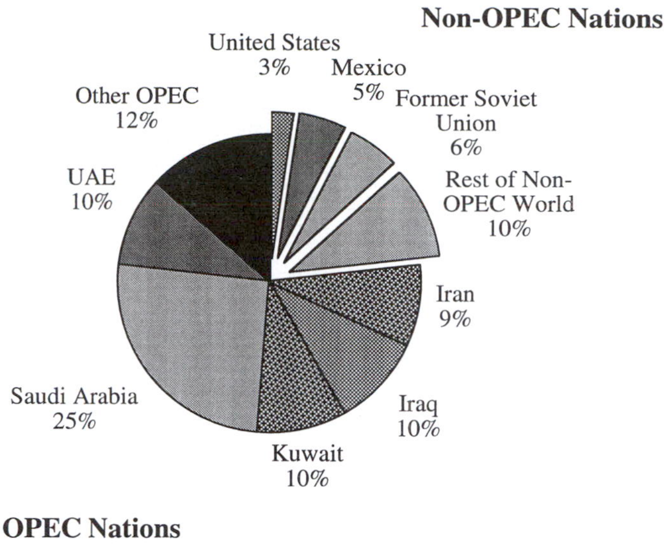

Figure 24.9. Distribution of world oil proved reserves (over 1 trillion barrels). World oil reserves are not declining, because production is being replaced with new reserves; however, the proved reserve distribution has become increasingly concentrated among OPEC nations, particularly Saudi Arabia. (*Source:* Data from *BP statistical review of world energy*, 1996.)

Figure 24.8. An anticlinal oil trap. An oil accumulation must have a reservoir rock with high porosity and permeability, typically sandstone, and a trap, often a layer of impermeable shale over the reservoir rock. Porosity is the volume of pore space in a rock relative to the total volume the rock occupies; permeability measures how easily a fluid can move through the interconnected pores of rock. (*Source:* Adapted from Douglas G. Brookins, *Mineral and energy resources*, Merrill Publishing, 1990.)

amount of void space) and high *permeability* (fluid flows through it easily).

Because of its global importance, oil is the best understood of the fossil fuels in terms of the size and distribution of the resource base and reserves. World reserves total slightly over 1 trillion barrels, with the majority located in the Middle East (Fig. 24.9). The current world production rate of approximately 25 billion barrels per year results in a ratio of reserves to production of 43. To date, approximately 700 billion barrels of oil have been produced worldwide since the first oil well was drilled in Titusville, Pennsylvania, in 1859. The United States has produced 25 percent of this oil and is by far the most thoroughly explored country in the world. Enhanced oil recovery (EOR) may extend the world supply to 125 years at current consumption rates. Exploitation of unconventional hydrocarbon reserves such as extra heavy oil, bitumen, and shale oil could extend the supply to over 300 years, but at significantly higher environmental and economic costs (Fig. 24.10).

From exploration to end use, each phase of oil use has associated environmental impacts. During exploration and production, seismic data collection devices damage habitats, oil wells release hydrogen sulfide and methane gases and can leak oil, and drilling creates fluid wastes such as drilling mud (the drill bit coolant and lubricant) and saline formation water produced with the oil. In the past, groundwater contamination has also been a problem. When wells are not properly sealed following abandonment, oil and saline formation water can leak through the well bore into nearby aquifers, contaminating freshwater supplies. As the most thoroughly explored oil-producing country, the United

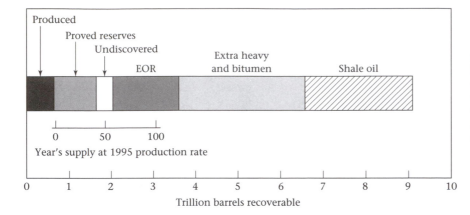

Figure 24.10. Worldwide oil resource base. If all conventional and unconventional oil resources were utilized, the world could have a 350-year supply of oil at current production rates. (*Source:* Adapted from Kenneth W. Haley, *Outlook for oil markets in the 1990s*, Chevron, 1994.)

States has the largest number of abandoned wells, many of which were sealed improperly.

The most publicized environmental impacts of oil use are oil spills that occur during transportation. Oil can be transported by tanker, rail, truck, and pipeline, all of which can result in spills. The world's largest spill since World War II occurred in 1991 during the Gulf War, offshore of Kuwait. The contents of five tankers and a sea island terminal, approximately 300 million gallons of oil, spilled into the Persian Gulf. The *Exxon Valdez* spill, the most publicized and largest spill in U.S. waters, released more than 10 million gallons into Alaska's Prince William Sound in 1989. Although highly publicized spills occur every couple of years, more oil is lost through leaking pipelines than through tanker spills. Russia has had an extremely poor record in oil transportation: Of all the oil Russia has produced at the well head, 15 percent does not reach the tankers because of leaking pipelines. Russia now faces a huge environmental cleanup from decades of uncontrolled spills. The refining process also has negative environmental impacts far from the site of oil production, including stack emissions (which are now highly regulated in developed nations), spent chemical catalysts, and accidental releases (explosions, fires, and spills). The distribution of gasoline and other refined oil products also have environmental impacts. Leaks from underground storage tanks, vapor releases from gas tanks, and small spills all occur at local gas stations. The end uses of oil and oil products have a variety of serious environmental impacts ranging from solid waste disposal issues to the atmospheric emissions of carbon dioxide, carbon monoxide, sulfur oxides, nitrous oxides, hydrocarbons, and particulates.

■ **Coal.** Coal is the second largest source of world energy, satisfying 27 percent of world commercial energy demand and approximately 23 percent of world total energy demand in 1995. In the United States, coal meets 23 percent of the energy demand and is the nation's third most important energy resource behind oil and natural gas. Coal is the most important resource for electric power generation in the United States and the world today. A short ton (910 kilograms or 2000 pounds) is the standard unit for measuring coal. The world produced 89 quads of coal in 1995, over half

of which was produced in the United States and China (Fig. 24.11). The United States does not import any coal and actually exports approximately 10 percent of its annual production to Europe and Asia. Over the next few decades, U.S. demand for coal is expected to grow modestly; however, world demand for coal is expected to increase rapidly, reflecting industrialization and population growth in countries such as the former Soviet Union (FSU) and China, which have large domestic coal reserves. In the United States and other industrialized countries, coal is used primarily for electric power production and to a modest degree for residential and industrial heating and non-energy-related functions such as coke production for steel manufacturing. A labor-intensive resource, coal has had a particularly strong economic and environmental impact on the communities in which it is mined. Curtailing its use may pose serious economic dislocations and lead to powerful political battles to protect jobs, corporate capital investments, and the perceived economic health of the local communities.

Figure 24.11. World coal production by country in 1995: 89 quads, or 3.7 billion short tons. Over half the world's 1995 coal production takes place in China, which is expected to see rapid growth in coal use, and the United States. 1 quad = 41.6 million short tons. (*Source:* Data from *BP statistical review of world energy*, 1996.)

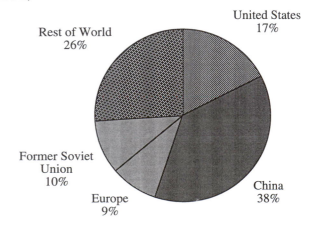

TABLE 24.2. RANKS OF COAL

Rank	Carbon (%)	Energy Content (Btu/lb)
Lignite	30	5000–7500
Bituminous	50–70	11,000–15,000
Anthracite	90	14,000

Btu/lb = British thermal units per pound.

Source: Roger A. Hinrichs, *Energy*, Harcourt Brace Jovanovich, 1991, p. 458.

Coal forms from decomposing woody plant material that accumulated in ancient swamps and deltas many millions of years ago when large regions of the Earth were covered with densely forested swamps. Plants grew and decayed in the same place for thousands of years, resulting in huge beds of decomposing plant material. These accumulations, alternating with periods of sand, silt, and mud deposition, left layer upon layer of sediment and decaying plant material. Burial of these accumulations over millions of years produced elevated temperatures and pressures that brought about the physical and chemical changes in the beds of decaying plant material to form coal deposits. Huge volumes of plant material formed the coal seams mined today; 20 meters of decaying plant material are needed to produce a 1-meter-thick coal seam.

Depending upon the relative amounts of carbon, hydrogen, and oxygen in a coal, it can be categorized into one of three ranks or types: lignite, bituminous, or anthracite. These ranks are differentiated by their carbon and energy contents, as shown in Table 24.2. The plants from which the coal formed and the amount of change the organic material has undergone as a result of heat, pressure, and time control the ratios of carbon, hydrogen, and oxygen in the coal and, consequently, its rank. Lignite and bituminous coal are used

primarily for electric power generation, and anthracite typically is used for heating buildings and homes. Bituminous coal is the most abundant type of coal in the United States and the world.

According to current knowledge, coal is the world's most abundant fossil fuel. At today's annual production levels, the world has approximately 230 years worth of proved reserves of coal remaining, and the United States has approximately a 260-year supply. The United States, the former Soviet Union, and China hold the largest of the world's coal reserves, each with over 200 billion tons (Fig. 24.12). Because of such large proved reserves of coal, scientists concerned with resource scarcity have not studied the coal resource base to the extent that they have studied oil; the world will clearly have a reliable coal resource for centuries to come.

Coal as currently used causes the most environmental damage of the fossil fuels. This is unfortunate because of its abundance and the increasingly important role it will play, especially as nations with vast coal reserves such as China, India, and the former Soviet Union industrialize. Coal has serious environmental impacts at all stages of mining, production, and use (Fig. 24.13).

Coal can be mined either in subsurface or surface mines. Underground mines have harmful environmental impacts, ranging from surface subsidence to subterranean fires to methane releases. Surface or strip mining, which involves removing the soil and rock covering the coal deposit, leads to erosion, acidic runoff, groundwater contamination, deforestation, and complete habitat destruction. Coal emits the largest amounts of airborne pollutants of all the fossil fuels, producing more sulfur oxides, nitrous oxides, carbon dioxide and particulates per unit of energy during combustion than either oil or natural gas. The environmental impact of coal burning actually varies little between the various ranks. This is because the major impact from coal relates to the coal's sulfur content, a factor not necessarily related to rank. Sulfur content has particular significance because sulfur oxide emissions from coal are the leading contributor to acid rain. Coal used extensively in power plants in the central and eastern United States has caused a severe acid rain problem in the northeastern United States and Canada. Coal use also produces a large volume of ash, which commonly contains high levels of toxic metals, creating a serious disposal problem. Workers who mine and process coal typically work under substandard conditions (historically in the United States and in China and other LDCs today) that have subjected them to accidents, underground fires, mine collapses, and high levels of air pollutants. Poor air circulation and high levels of coal particulates and debris within mines produce respiratory conditions that are clearly linked to heart and lung disease. Although mining conditions have improved in the industrialized world, coal mining remains one of the most dangerous occupations globally.

The U.S. government and the coal industry have tried to lessen the environmental impacts of current coal use through policies such as land reclamation, mine safety standards, and

Figure 24.12. World coal proved reserves by region for 1995: over 1 trillion tons. North America, the former Soviet Union, and Asia contain most of the world's abundant coal reserves. (*Source:* Data from *BP statistical review of world energy*, 1996.)

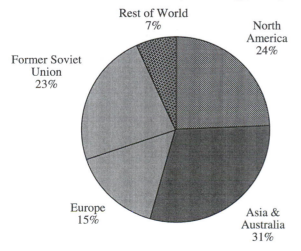

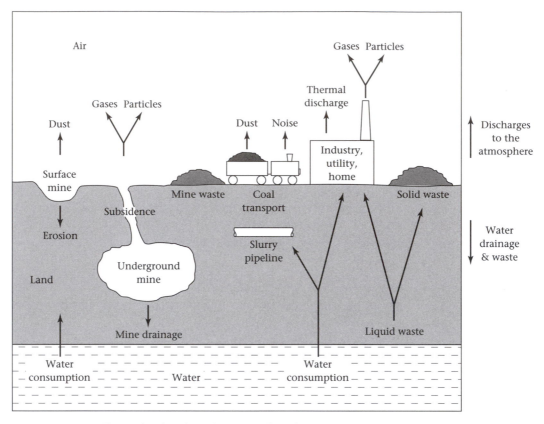

Figure 24.13. The cycle of coal production and use has many environmental impacts, stemming from mining, transport, and use. (*Source:* Adapted from Edward Cassedy and Peter Z. Grossman, *Introduction to energy*, Cambridge University Press, 1990.)

emission controls. However, even with current mitigations in place, coal remains the dirtiest of the fossil fuels. The coal industry is working on methods to produce and consume coal more cleanly in order to ensure that coal has a continuing role as the world's most important source of electric power.

■ **Natural Gas.** As the third largest component of world energy use, natural gas satisfied 23 percent of world commercial energy demand and approximately 21 percent of world total energy demand in 1995. In the United States, natural gas plays a slightly more important role, satisfying 26 percent of U.S. energy demand, second in importance only to oil. Globally, natural gas use has grown dramatically over the past four decades. The United States and the former Soviet Union are by far the largest producers of natural gas, and the former Soviet Union is the largest holder of natural gas reserves (Fig. 24.14). The standard unit of measurement for natural gas is 1000 cubic feet (or 1 mcf).

Growth in the use of natural gas is being spurred by several factors. Until recently, natural gas was considered a secondary fuel to oil largely because of its lower energy density; a volume of 6000 cubic feet (170 cubic meters) of natural gas at standard temperature (20°C) and pressure (1 atmosphere) is necessary to provide the same energy content as one 42 U.S. gallon barrel of oil (5.6 cubic feet, or 0.16 cubic meter).

Figure 24.14. World natural gas production for 1995. 75 quads. The distribution of natural gas consumption is dominated by North America and the former Soviet Union; however, its use is growing rapidly in many other countries. Total world production will grow in the coming decade, and the distribution of production will diversify further. 1 quad = 27.8 trillion cubic feet of natural gas; therefore, 1995 consumption equals 2113 trillion cubic feet. (*Source:* Data from *BP statistical review of world energy*, 1996.)

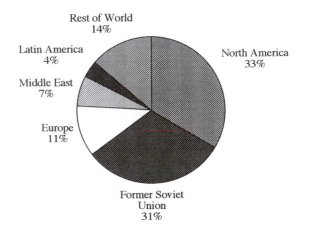

Japan and several other countries compress or liquefy natural gas for storage and transport, although the cost of doing so tends to be prohibitive. Even in a liquid state, natural gas does not rival the energy density of the same volume of oil. This disadvantage is largely compensated for by the fact that the environmental impacts of natural gas, from production to end use, are far less than those associated with oil or coal. Furthermore, natural gas has a more global distribution than oil; consequently many countries, including the United States, are less likely to require natural gas imports in the near term.

Historically, natural gas consumption in the United States has been relatively balanced among electric power generation and industrial, residential, and commercial use. The largest growth in the near future is expected from the electric power industry because of deregulation of the industry, the fact that natural gas is attractive economically, and increasing concern for the environment. The transportation sector is also likely to become a significant user of natural gas within the next twenty years as substitutes for gasoline are sought. Natural gas use is growing rapidly throughout the world, with industrial uses and electric power generation leading demand.

The origin of natural gas is similar to that of coal. Scientists have come to understand that coal seams, especially those buried too deep to mine, are both source rocks and reservoir rocks for natural gas that can be tapped. Natural gas source rocks can also be dark shales rich in organics such as woody plant material or skeletons of single-celled microorganisms.

Natural gas is largely composed of methane (CH_4), the simplest hydrocarbon molecule. Like oil, it can be described as sweet or sour if its sulfur content is low or high. It can also be described as wet or dry depending on the presence of natural gas liquids and other energy gases. If natural gas is greater than 90 percent methane, it is referred to as dry. Natural gas is also described as associated or nonassociated, depending upon whether or not it is physically associated with a significant amount of oil. Contrary to popular belief, most gas is not associated with oil; only 20 percent of natural gas reserves in the United States is estimated to be associated. Exploration efforts explicitly focused on finding and developing these nonassociated reserves is a recent phenomenon that began in the 1980s and continues to grow.

World proved reserves of natural gas currently stand at approximately 4900 trillion cubic feet, with 72 percent of these reserves concentrated in the Middle East and the former Soviet Union (Fig. 24.15). At current production levels, world and U.S. ratios of reserves to production stand at approximately 65 and 9, respectively. The size of the natural gas resource base has been growing in recent years as scientists' understanding of its origin and occurrence has improved and as companies are actively beginning to look for gas resources in response to rising demand.

Natural gas and oil have similar categories of environmental impacts in terms of the exploration and production of

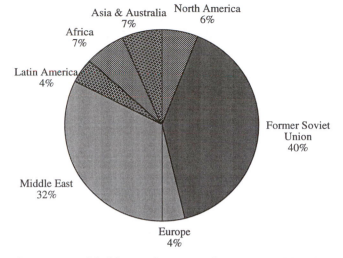

Figure 24.15. World natural gas proved reserves in 1995: 4900 trillion cubic feet. The Middle East and the former Soviet Union hold a majority of the world's proved reserves. (*Source:* Data from *BP statistical review of world energy*, 1996.)

each resource. However, natural gas commonly undergoes only simple, in-field refining to remove other gases and liquids, and almost all transportation is by pipeline. The most significant environmental impacts of production are releases of natural gas through leaks and flaring (combustion) of natural gas to the atmosphere. Scientists estimate that methane, a greenhouse gas, accounts for 15 percent of global warming; however, only approximately 15 percent of the methane emissions to the atmosphere are anthropogenic, and only 10 percent of these come from the natural gas infrastructure. The remainder of the methane released to the atmosphere comes from natural processes such as the decomposition of wood by termites, rice paddies, and the flatulence of cows (all surprisingly significant methane producers). Methane creates a serious concern for global warming because 1 gram of methane produces the same amount of warming as 69 grams of carbon dioxide. Leaks in pipelines, accidents in distribution, leaks during drilling, and incomplete combustion in power plants or other end uses can all release natural gas to the atmosphere. Natural gas–fired electric power plants pollute the least of all the fossil fuels, producing virtually no sulfur oxide emissions and lower carbon dioxide and nitrous oxide emissions than oil or coal (Table 24.3).

Nuclear Fission of Uranium. Energy can be produced from the nuclei of atoms in two ways: nuclear fission and nuclear fusion (Fig. 24.16). Nuclear fission, the only commercial nuclear energy source available today, most often involves the bombardment of a ^{235}U atom by a neutron. This causes the splitting or fissioning of the uranium atom into two fragments and the release of several more neutrons, which can proceed to bombard more ^{235}U, creating a chain reaction as well as releasing energy. In commercial applications, this energy heats water in a reactor to make steam that turns a turbine generator to produce electricity.

TABLE 24.3. CONVERSION EFFICIENCIES AND AIR POLLUTANTS, VARIOUS ELECTRICITY-GENERATING TECHNOLOGIES

Technology[a]	Conversion Efficiency[b] (percent)	Emissions (grams per kilowatthour)		
		NO_x	SO_2	CO_2
Pulverized coal-fired steam plant (without scrubbers)	36	1.29	17.2	884
Pulverized coal-fired steam plant (with scrubbers)	36	1.29	0.86	884
Fluidized bed coal-fired steam plant	37	0.42	0.84	861
Integrated gasification combined-cycle plant (coal gasification)	42	0.11	0.30	758
Phosphoric acid fuel cell (using hydrogen reformed from natural gas)	36	0.04	0.00	509
Aeroderivative gas turbine	39	0.23	0.00	470
Combined-cycle gas turbine	53	0.10	0.00	345

Note: Data are for particular plants that are representative of ones in operation or under development. NO_x = nitrous oxides, SO_2 = sulfur dioxide, CO_2 = carbon dioxide.

[a] Coal plants are burning coal with 2.2 percent sulfur content.

[b] Higher heating values, which give lower efficiency levels, are used throughout.

Source: Christopher Flavin and Nicholas Leussen, *Power surge: guide to the coming energy revolution*, New York, Norton, 1994, Table 5-1, p. 101.

of the nuclear power industry quickly followed. Strong encouragement from the U.S. government and a close association with the nuclear weapons program drove the rapid development of the U.S. nuclear power program. Over the past thirty years, the U.S. nuclear power industry has received by far the most federal subsidies of the energy industries, receiving substantial grants or discounts on liability insurance, research and development funds, uranium fuel, and waste disposal. Less than forty years after the first commercial plant was completed, the world now has over 400 plants. In 1995 nuclear fission met 7 percent of world commercial energy demand and 6 percent of world total energy demand. The United States, which has 30 percent of the world's nuclear power facilities, currently meets only 8 percent of its energy demand with nuclear energy but satisfies approximately 20 percent of its electric power demand with nuclear energy. Although many countries have nuclear power facilities, the United States, France, Japan, and the former Soviet Union account for almost 70 percent of world nuclear energy production (Fig. 24.17).

Originally pitched to the American public as a source of electricity "too cheap to meter," nuclear-generated electricity has proved to be more expensive than many of the alternatives existing today. The cost of installing other sources of new electricity-generating capacity have continued to decline while the cost of nuclear energy increases annually. The costs of building and operating nuclear plants have been routinely underestimated, and the costs of waste disposal and decommissioning are thus far undefined and growing. Public concerns about power plant and waste disposal sites, high variability in U.S. plant design, and evolving safety and environmental standards have created uncertainty for the owners and financiers of nuclear facilities. The world's ability to manage the nuclear material it already has produced is seriously in question. Most nations do not yet have a means of safe permanent waste disposal. For these reasons, analysts expect nuclear fission energy production and installed nu-

Aside from the energy systems of a few ships and submarines, the sole end use of nuclear energy is electric power generation; consequently nuclear energy commonly is measured in terms of the most common electrical unit, kilowatt hours. The world's first commercial nuclear power facility came on line in 1957 in Pennsylvania, and the development

Figure 24.16. The deuterium-tritium fusion reaction and the ^{235}U fission reaction. As seen in the fusion diagram, nuclear reactions lose mass as it is converted into energy according to the equation $E = mc^2$. Fusion reactions lose more mass (relative to total mass) than do fission reactions; consequently, they release much more energy per unit of mass than do fission reactions. (*Source:* Courtesy of Lawrence Livermore National Laboratory.)

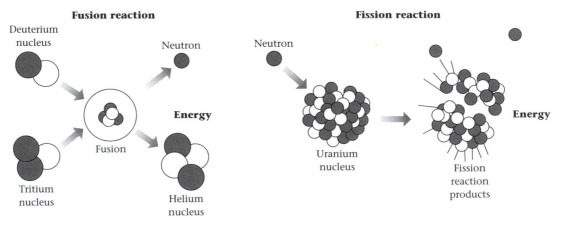

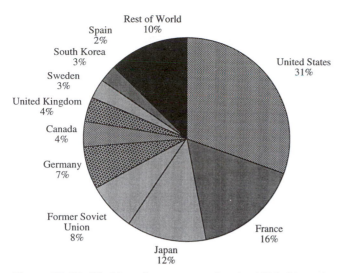

Figure 24.17. World nuclear consumption for 1995: 24 quads. Nuclear consumption is concentrated among a group of developed nations including the United States, France, Japan, and the former Soviet Union. (*Source:* Data from *BP statistical review of world energy*, 1996.)

clear capacity to decline globally and in the United States as aging plants are decommissioned. However, growth is still expected in some countries (e.g., France, Japan, Taiwan, the Koreas). Nuclear waste will have a long and toxic legacy, undoubtedly affecting generations to come.

The fuel for most fission nuclear reactors is ^{235}U, a naturally occurring isotope found in uranium ore at a concentration of approximately 0.7 percent. Uranium-bearing ores can be minded in either open pit or subsurface mines. On-site uranium mills extract the uranium with chemical methods, converting the uranium to an oxide form called yellow cake, largely U_3O_8, which is then shipped to a plant for isotopic enrichment to raise the ^{235}U concentration to approximately 3 to 4 percent. Once enriched, this material can be used to make fuel pellets, rods, and assemblies for a conventional light-water nuclear reactor.

The United States, Canada, Australia, Brazil, sub-Saharan Africa, and the former Soviet Union all have sizable uranium deposits. World proved reserves total approximately 4 million short tons. Assuming conventional thermal light-water nuclear reactors are used, the world ratio of reserves to production is approximately 65 at current annual production levels. The size of the resource base can be dramatically extended by reprocessing the spent fuel from nuclear reactors and using breeder reactors instead of light-water reactors. Breeder reactors actually make more fuel (for a different type of nuclear reactor) than they consume. Utilizing breeder and reprocessing technology might increase the ratio of reserves to production to 3300. The Earth's ultimate uranium resources could last 10,000 years or more at current production levels if the best available technology is utilized. Unfortunately, plutonium, an important component of nuclear weapons manufacture, is an intermediate product of breeder

reactors. Concerns over nuclear weapon proliferation and the increased costs of fuel reprocessing have been the major factors in the refusal of the United States and several other countries to adopt breeder technology.

An argument can be made that nuclear energy is more environmentally friendly than fossil fuels. Fission has advantages over the combustion of hydrocarbons, including the high energy density of uranium and a lack of the atmospheric emissions produced from the burning of fossil fuels. Uranium has an incredibly high energy density; 1 gram of uranium is the equivalent of 90 kilograms of coal. This high energy density means that per unit of energy, the environmental consequences of physically mining, processing, and transporting uranium are arguably much lower than those of coal. Fission energy also produces no carbon dioxide, sulfur oxides, nitrous oxides, carbon monoxide, hydrocarbons, or particulates, because nothing is combusted in the process of producing energy.

Uranium mining shares some of the same negative environmental impacts as coal mining. They both can lead to increased erosion, land subsidence, and chemical contamination of ground water and surface waters. Uranium tailings (mining wastes), however, are far more dangerous than coal tailings. These tailings often contain high levels of radioactive materials such as uranium and radon. Communities near former uranium mines often experience exposure to elevated levels of radioactivity in groundwater supplies and are subjected to radioactive tailings dust, the long-term impacts of which are yet to be understood. For most people, the greatest environmental concerns regarding nuclear energy are waste disposal and the possibility of nuclear catastrophes such as the 1986 Chernobyl incident resulting from poor containment of a runaway reaction (meltdown) in the reactor core. The Chernobyl accident released ten times the amount of nuclear material to the atmosphere as the bomb dropped on Hiroshima.

Many of the radioactive isotopes found in nuclear waste have extremely long half-lives, raising ethical and technical issues about where and how to store a potentially lethal waste that will endure for thousands of years. No country has yet constructed a permanent disposal site for high-level radioactive waste; however, intensive feasibility studies such as that by the United States on a potential site at Yucca Mountain in Nevada, continue at great expense. Meanwhile, waste products continue to accumulate in temporary storage facilities, primarily on-site at nuclear plants, while governments wrestle with the scientific and political problems of waste disposal.

RENEWABLE RESOURCES. Today, use of renewable energy resources accounts for less than 10 percent of all commercial energy consumption and has been largely limited to biomass and hydroelectric resources. Unlike depletable fossil fuels and uranium, the remaining energy resources discussed in this chapter are renewable. *Renewable resources* are so called because society may sustain their use indefinitely (at least on human time scales) with proper management. There are three main categories of renewable resources: those whose rate of renewal approximates that of reasonable use so that, if properly managed, use of the resource is sustainable; those whose use has no bearing upon future availability; and those that improve energy efficiency. Of the resources discussed in this chapter, geothermal and biomass are renewable of resources of the first type, whereas wind, solar energy, ocean power, and hydroelectricity are of the second. Both of these resource types typically have minimal environmental impacts that tend to be localized, unlike the global and regional impacts of conventional depletable resources. Energy efficiency improvements rarely have any negative environmental impacts and tend to produce significant environmental benefits relative to less efficient practices.

Wind, solar radiation, tides, and waves are called *intermittent resources* because they are not always available. They are also *nondispatchable resources*, which means that they cannot be summoned to produce energy on demand. A gas turbine power plant can generate electricity within minutes of the need for energy; however, wind turbines cannot dispatch energy when there is no wind. The nondispatchable nature of many renewable resources has hindered their widespread deployment. For the time being, intermittent renewable will be used most frequently when supplemented by a dispatchable energy source to ensure reliability. Advances in energy storage technologies, which to date have been limited primarily to lead-acid batteries, will make intermittent resources more self-sufficient and widely utilized because energy could be stored when produced and dispatched when needed.

Most renewable resources are also *site-specific resources* because their use is not possible or practical everywhere. For example, ocean thermal energy can be feasibly supported only in coastal areas with large thermal gradients in the first 1000 meters of water depth, and wind power production requires areas with frequent, predictable winds in excess of 16 kilometers per hour (10 miles per hour). Power producers must place hydropower facilities along rivers and geothermal plants in areas with high thermal gradients in the Earth's upper crust. Photovoltaics cells, although functional anywhere in the world, operate more efficiently and therefore more economically in areas of high *solar insolation* such as low-latitude deserts. Biomass energy can be transported to the point of use but becomes prohibitively expensive far from its source because of its modest energy density.

All renewable resources will likely make gradual gains in use globally; of these, solar and wind energy hold the largest near-term promise, and their short-term growth will be the most dramatic. These resources can be converted to electricity by technologies that are rapidly evolving and approaching cost competitiveness with more conventional fuels. Growing markets for these resources will in turn create greater cost competitiveness as manufacturing volumes grow. Photovoltaic cells and wind turbines can be developed for large-scale or small-scale applications, and their use has no bearing on future availability of the resource, making them

highly attractive energy technologies. Renewable sources of energy must play an increased role in meeting world energy needs if a sustainable energy economy is to be achieved.

■ **Hydroelectric Energy.** Over 2000 years ago, ancient peoples recognized that running water could be harnessed as an energy source, and they utilized it for services such as milling grains. Later, the kinetic and potential energy of water was also transformed into mechanical energy to pump water and power simple machines such as bellows. In 1882, the first hydroelectric plant was built in Appleton, Wisconsin. By the time of the Great Depression in the 1930s, hydroelectric power provided 40 percent of the electricity in the United States. Today hydroelectric power meets 7 percent of world commercial energy demand and 6 percent of world total energy demand. In the United States hydroelectric installations satisfy 4 percent of U.S. energy demand, all in the form of electricity generation.

Hydroelectric facilities are commonly installed in dams. Less than 30 percent of all dams worldwide have hydroelectric facilities, and most of those that do were not built solely for the purpose of electric power production. In addition to the major benefit of electric power production, dams provide reservoirs for recreation, aid in flood control, improve transportation, and increase the local water supply. Large-scale hydroelectric facilities are typically not economic for energy production alone; other benefits need to be considered. Therefore, federal, state, and municipal agencies own over 80 percent of the hydroelectric generating capacity in the United States.

The Earth's hydrologic cycle (see Fig. 9.2) makes hydroelectric power available. This cycle is powered by the Sun; thus, in a broad sense, hydroelectric power is also a form of solar energy. The Sun's radiative energy causes evaporation on the surface of the Earth. Water vapor that has accumulated in the air may fall as rain or snow on land, flowing downhill toward lakes and oceans. The gravitational potential energy from the vertical drop of this water is converted to kinetic energy as the water flows down to sea level. Dams constructed in the path of rivers can capture some of this gravitational potential energy through the use of turbines and generators.

Hydroelectric power plant sites are described by head (the vertical distance between the top of a reservoir and the power plant), power rating (generating capacity in megawatts), and layout (type of facility). Table 24.4 shows the extent to which regions have developed their hydropower resources. Current global hydroelectric power generation is only approximately 15 percent of what appears to be technically feasible. Some believe that hydropower could provide 50 percent of the world's energy needs if fully developed. The future development of hydropower will likely be limited, however, by its environmental impacts and the lack of nearby markets for hydroelectricity. These factors are expected to limit hydroelectric generation to approximately 25 percent of its technically achievable potential.

Unlike energy production from fossil fuels, hydroelectric

TABLE 24.4. HYDROELECTRIC POTENTIAL

Region	Possible	Installed	Developed (percent)
Africa	1150.00	36.00	3
South Asia and Middle East	2280.00	171.00	8
China	1920.00	109.00	6
Former Soviet Union	3830.00	220.00	6
Japan	130.00	87.00	67
North America	970	536.00	55
Latin America	3540.00	367.00	10
Europe	1070.00	485.00	45
Oceana	200.00	37.00	19
Total	15090.00	2048.00	14

Source: Adapted from Laurie Burnham, Renewable energy, Washington, DC, Island Press, 1993, Table 3, p. 77.

energy production does not result in depletion of the energy-producing medium, water. Although water is not consumed by a hydroelectric facility, using water for electric power generation may limit its other uses, such as irrigation or maintaining flows required to sustain fish populations. Conversely, other water uses may restrict how and when a hydroelectric facility can generate power. These considerations, combined with seasonal and annual runoff variations, limit the ability of hydroelectric plants to operate continuously. Utilities have traditionally used hydroelectric facilities for satisfying increased power demands that occur only at certain times of the day or year, called peaking power, because hydroelectric generators can come on line quickly.

Hydroelectric power generation itself has few environmental impacts. Because no combustion is involved, hydroelectric projects have no significant atmospheric impacts. The environmental impacts typically associated with hydroelectric generation come from the construction and presence of dams, and not specifically from hydroelectric energy production. Dams have many significant environmental impacts, the relative importance of which varies from site to site. These impacts include: inundation of upstream habitat, displacing human, flora, and fauna communities; interruption of the natural hydrologic and sedimentation processes as the dam obstructs silt and nutrients that would otherwise flow downstream; increased erosion rates downstream of the dam; silt accumulation behind the dam, which eventually reduces the capacity of the reservoir; reduction in water quality due to reduced aeration; increased salt and mineral concentrations due to increased evaporation; elevated water temperature as a result of stagnation upstream and reduced flow downstream; destruction of fish migration routes and spawning areas; and loss of wild river recreational areas. The serious environmental consequences of pursuing a major hydroelectric project make it a very difficult energy choice. China is currently facing this decision as it proceeds with construction of the Three Gorges facility (see the box entitled *The Largest Hydroelectric Power Generator in the World*).

■ **Geothermal Energy.** Although geothermal energy cur-

rently meets much less than 1 percent of world and U.S. energy needs, it plays a significant role in certain regions. The Geysers in northern California, the world's largest geothermal development in the 1980s, recently produced as much as 8 percent of California's electricity. Geothermal energy is not likely to become globally significant in the near future; the number of suitable sites for its economic development is limited. However, it will play a significant role in geologically active regions such as the Philippines and El Salvador. In these countries, geothermal energy may eventually meet between 5 and 30 percent of each country's electric power needs. The first geothermal power plant began operation in Italy in 1913. As of 1990, approximately 6000 million watts of installed capacity worldwide were annually generating approximately a tenth of a quad of electricity. By the year 2000, estimates put worldwide generating capacity at approximately 15,000 megawatts. The primary end use for geothermal energy is electric power production; however, communities located near geothermal energy sources may use it for various domestic heating and drying purposes as well. The ancient Romans found that hot springs could be used not only for bathing but also to heat their public buildings. In Iceland today, over 85 percent of all residential buildings are heated geothermally.

Geothermal energy finds its origin deep in the Earth where the temperature is much higher than at the surface. In certain regions of the Earth, molten rock (magma) pushes up along faults and cracks in the crust, creating "hot spots" within 2 to 3 kilometers of the surface. The Earth's crust is thinnest at tectonic plate boundaries; the best geothermal sites exist along these boundaries (Figs. 6.6 and 6.7).

Although there are several types of geothermal resources, the most significant for energy production today are hydrothermal reservoirs. Heat for commercial energy production from other geothermal resources such as magma, hot dry rock, and geo-pressured reservoirs will not be economically recovered until costs are drastically reduced and significant technical advances are made. Hydrothermal reservoirs are crustal hot spots in which groundwater has accumulated. The high-temperature pressurized water in these reservoirs serves as a medium for removing thermal energy from the Earth's crust. Most hydrothermal reservoirs have moderate temperatures, in the 100 to 200°C range; however, some are very hot, containing pressurized water above 200°C. For a geothermal resource, the hotter the better. If an extremely hot hydrothermal field experiences relatively low pressures within the reservoir, the water may convert fully into steam before it reaches the surface. This is called a dry steam reservoir. The Geysers in northern California is this type of resource; steam is produced from wells drilled into the reservoir and transported to nearby power plants, where it is used to turn turbine generators to produce electricity. A wet steam reservoir has very high temperatures and pressures, and a portion of the water flashes to steam upon reaching the low pressures at the surface. In a liquid-dominated reservoir (one of lower temperature), hot water can be brought to the sur-

THE LARGEST HYDROELECTRIC POWER GENERATOR IN THE WORLD: THE PRC THREE GORGES PROJECT

The Chinese government has begun construction of the Three Gorges Dam, a structure that is 2 kilometers (1.2 miles) long and 170 meters (550 feet) high, located on the Yangtze River near Yichang in Hubei Province (Fig. 24.18). The first concrete was poured in 1995, and problems relating to sedimentation, solid wastes, and sewage have already emerged in construction. If completed, Three Gorges will be the world's largest hydroelectric facility, providing approximately 18 gigawatts (GW) of electric generating capacity (approximately the generating capacity currently in place in all of China) and will create a reservoir 590 kilometers (370 miles) long. The Chinese government justifies the dam on the basis of its nonenergy benefits alone; it will provide improved irrigation to the fertile river plain, further inland navigation on the river, and provide flood control for more than 5 million people living directly downstream. During the twentieth century alone, over 310,000 fatalities occurred on the Yangtze River from three major floods. Annual flood damage on the river is estimated to average over $100 million (U.S. dollars).

The Chinese government views the electric generating capacity that the dam will provide as not only critical to meeting the growing demand for electricity as the country industrializes but also critical to providing an alternative to coal, China's current dominant source of electricity. Each year the power plant is expected to produce over 80 billion kilowatts, providing power to eight provinces. The Chinese government claims that the Three Gorges project will increase agricultural and industrial productivity by approximately $70 billion per year.

The price tag of the project is expected to grow beyond the current estimate of $77 billion, and the World Bank has refused to fund the project for several reasons, including doubts about the project's economic viability. As Dan Beard, former commissioner of the U.S. Bureau of Reclamation, said, "Large dams are tremendously expensive. They always cost more than you thought and tie up huge sums of capital for many years." He adds, "There is no more visible symbol in the world of what we are trying to move away from than the Three Gorges Dam." The current trend in most of the world is toward smaller dams and more environmentally sensitive means of flood control, such as levees. The Chinese government is funding the project itself by taxing electricity sales, and many are concerned that this will exacerbate China's existing inflation problems. Technical problems have already affected the economics of the project. The Yangtze River carries approximately 500 billion cubic meters of silt each year, and the silt accumulation behind the dam is now anticipated to be a greater problem than originally envisioned. The project's engineers now admit that they will have to forgo 11 percent of the anticipated power output in order to allow more silt to flow through the dam and that several other dams will have to be built upstream to help reduce sedimentation problems. A lack of sewage treatment facilities in cities along what will become the shore of the reservoir may also create a severe water pollution problem.

Escalating costs form only part of the argument against the project. Construction of the dam requires one of the largest resettlement efforts in human history. It is estimated that at least 1.2 million people, many of whom are subsistence farmers, will be displaced by the project, and more than 20 cities will be partially inundated. The Three Gorges is China's equivalent of the Grand Canyon and is considered to be an area of spectacular natural beauty that has been a focus of art and civilization for centuries. Environmentalists argue that the dam will not only submerge thousands of years of human history but also threaten a number of endangered species such as the Siberian white crane and the river dolphin. Opposition to the dams is growing, especially outside of China. In 1992, the International Water Tribunal ruled that the dam should not be built until an impartial environmental assessment could be completed and the people who will be relocated have an opportunity to be heard. However, the tribunal has no enforcement power. All this ruling accomplished was a public relations problem that will probably succumb to the glaring need for electricity in central China.

Because of China's need for electricity and because so much money has already been poured into the project, it will be difficult to stop. The Chinese government views the dam as a national symbol of their ability to conquer nature on a scale never before achieved.

Figure 24.18. The site of China's Three Gorges Dam being constructed on the Yangtze River. (*Source:* Adapted from Stephen Magagnini, Taming the dragon, *The Amicus Journal*, Summer, 1993, p. 9–13).

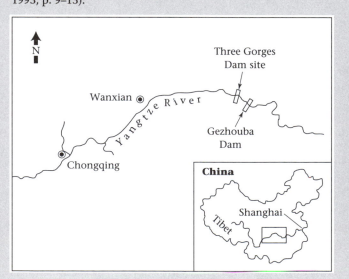

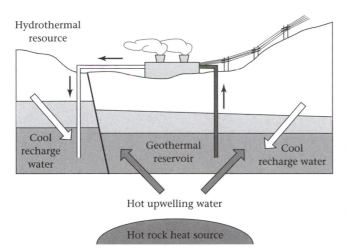

Hydrothermal resource

Cool recharge water

Geothermal reservoir

Cool recharge water

Hot upwelling water

Hot rock heat source

Figure 24.19. A high-temperature water geothermal system. The reservoir consists of porous rock and contains water that is heated by conduction from a magma heat source. Although the reservoir is surrounded by solid rock, hot water can escape to the surface as steam through fissures. This steam can be used to run turbines and produce electricity. (*Source:* Adapted from Earth Science Laboratory, *Geothermal energy*, University of Utah Research Institute, 1992.)

face and used to vaporize a low-boiling-point fluid in order to drive a turbine generator (Fig. 24.19).

The quality of a geothermal resources is determined by its proximity to a market. To date, the commercially viable, developed geothermal resources are located near hot spots and plate boundaries; however, the long-term potential for geothermal energy is globally widespread. As technology improves, lower-temperature geothermal resources will be developed. In time, certain geothermal applications may be economic almost anywhere on the globe.

Although most general publications and some experts in the field tout geothermal energy as a renewable resource, its renewability may be debated. Geothermal energy is truly renewable only as long as heat and hydrothermal fluid are replaced as quickly as they are withdrawn. Most geothermal power plants constructed to date appear to be removing hydrothermal fluid from the reservoir more rapidly than that fluid can naturally recharge from groundwater. As a result, electricity generation at these developments has tapered off in recent years, redefining the resource in these applications as depletable. In contrast, most small-scale developments of geothermal resources for direct, nonelectric use show little evidence of depletion. If properly managed, geothermal energy can be a renewable resource; however, this balance is not always consistent with economic considerations.

As with most other sources of energy, development involves tradeoffs important to the biosphere. The environmental consequences of exploration and drilling for geothermal resources are similar to those of oil: disruption of habitat at the well head, spills of drilling fluid, and releases of hydrogen sulfide or briny geothermal fluid. However, the environ-

mental impacts of energy production are quite different, including land use impacts, aesthetic impact of plants with prominent cooling towers and vapor plumes, noise, gaseous emissions, toxic elements such as mercury and arsenic, and waste disposal of brines and residual water. Geothermal resources are commonly located in areas of high seismic activity; consequently, there is an unusually high risk that earthquakes or volcanic activity may disrupt operations at the facilities. In some cases, the withdrawal of geothermal fluids may actually exacerbate seismic activity in the region of field development. Furthermore, subsidence and reduction in slope stability may occur near hydrothermal sites because geothermal facilities withdraw hydrothermal fluid that previously helped support the surface. Air pollution from hydrogen sulfide (H_2S) is typically the single largest concern at geothermal plants. Hydrogen sulfide emissions can be abated; however, the methods of doing so involve the transportation of toxic chemicals, creating a hazardous waste disposal problem. Other gaseous pollutants occur in such trace amounts that their environmental impact is insignificant. However, they are monitored carefully even though their concentrations rarely rise above background levels. The typical geothermal project emits only 5 percent as much carbon dioxide as a coal-fired plant of the same size.

■ **Wind Power.** The wind has provided energy for transportation (sailing ships), milling grain, and pumping water for thousands of years. Wind turbines were first used for electric power generation in Denmark almost 100 years ago. The introduction of dispatchable resources such as fossil fuels to remote rural areas led to the decline in the use of wind energy until the oil shocks of the 1970s spurred the United States and Denmark to invest in wind energy research and development. Today, wind energy meets an extremely small portion of the world's energy needs, primarily through electric power production from wind turbines. However, the utilization of wind power is currently growing faster than any other energy resource. The United States presently has over 50 percent of the world's wind power generating capacity; however, large wind power development incentives in European countries will likely lead to European dominance in wind power generation in the twenty-first century. Industry analysts estimate that worldwide installed capacity will more than double to over 6000 megawatts by the turn of the century and that over 60 percent of this capacity will be in the European Community. Globally, the wind industry is now growing at 20 percent per year (twice as much wind capacity was added in 1995 as in 1994). Some enthusiastic analysts even speculate that wind power could easily meet 20 percent of world electricity demand within the next five decades.

Wind turbines convert the wind's kinetic energy into mechanical energy, which is used to drive electric generators (Fig. 24.20). Most turbines have a horizontal axis mounted on a tower that rises 40 to 100 feet (12 to 30 meters) above the ground. Although some extremely large (> 1000 kilowatts) turbines have been built, the average turbine installed to date has approximately a 100-kilowatt generating capac-

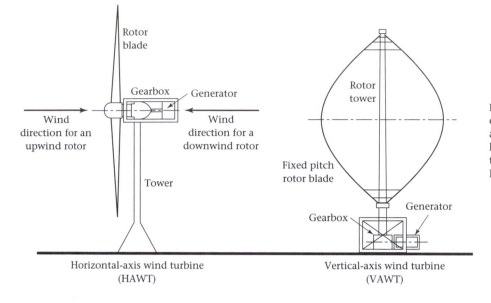

Figure 24.20. Horizontal- and vertical-axis wind turbines. Horizontal-axis turbines are more common; however, both designs have advantages. (*Source:* R. A. Hinrichs; *Energy*, Harcourt Brace Jovanovich, 1991.)

ity. The latest generation of turbines has ratings in the range of 200 to 600 kilowatts, which seems to be optimal for economies of scale. The electricity output of a turbine varies depending on the area swept by the rotor's blades and the cube of wind velocity. Because of the large dependence on the wind velocity, wind turbines must be properly sited for optimal performance.

The Sun's unequal warming of the Earth's surface creates wind (Chaps. 10 and 14). This warming creates large-scale atmospheric thermal gradients and smaller-scale temperature differentials between land and sea that drive the Earth's winds. The Earth's rotation and local topography also affect the direction and velocity of wind. Valleys and mountains can serve to amplify and channel winds to create highly predictable weather patterns that provide a good resource for reasonably reliable power production.

U.S. government studies suggest that the United States could meet 20 percent of its electric power demand with wind power produced on only 0.6 percent of its land. Although 0.6 of 1 percent of the United States constitutes a large land area (about the size of Vermont and New Hampshire combined), wind turbines occupy only 2 to 3 percent of the surface area, allowing them to coexist with other uses. In the United States, the most attractive undeveloped wind resources are located in the midwestern states, particularly Montana, Nebraska, and the Dakotas. Attractive wind resources are fairly well distributed worldwide The United States, the former Soviet Union, Africa, and, to a lesser extent, South America, Australia, southern Asia, and parts of Europe all have substantial wind resources.

The resource potential for wind is defined as those areas with wind speeds above the minimum necessary for economic production with current technology. This implies that as technology improves, more of the wind resource will become available to humankind. Wind resources are rated by

measuring the average wind speed 50 meters (164 feet) above the surface and then calculating the wind's generating capacity per square meter. At present, most installed turbines function at a minimum wind speed of approximately 22 kilometers per hour (14 miles per hour). Turbines currently being installed will operate at wind speeds as low as 16 kilometers per hour (10 miles per hour).

The production of energy from the wind has few environmental impacts. Wind energy production leaves over 95 percent of a wind energy site available for nonenergy purposes such as farming or cattle grazing, minimizing the land use impact. Some noise pollution is associated with the rotating turbines; however, this is noticeable only on-site. The visual impact that turbines have near a city often prompt "not in my backyard" (NIMBY) reactions and has made siting controversial in areas of scenic beauty, such as the Hawaiian Islands. The largest wind power installations to date, located in the mountain passes of California, also provide a home for a large number of avian species. Thus, in addition to being unsightly in the view of some people, the presence of these wind turbines has led to raptor deaths; however, studies conducted to date do not indicate that the long-term stability or size of local avian populations is threatened. The wind energy industry and several environmental groups are working to minimize these deaths. In the future, most of the new wind power installations will be built in areas where conflicts with avian species are less likely to occur. Overall, wind power is a very environmentally friendly resource, especially compared to the energy resource mix utilized today.

■ **Solar Electric Energy.** The Sun releases over 30 billion quads per day; however, only a small portion of this enormous amount of energy, a mere 5 million quads per day, reaches the Earth. This energy powers almost every system on the planet. It drives the winds, ocean currents, the hydrologic cycle, and plants. The Sun is the initial source of all our energy resources, except geothermal and nuclear resources. The amount of energy delivered to the planet on a daily basis equals over 5 million times world energy consumption. This abundance demonstrates the enormous potential for harnessing solar insolation directly instead of utilizing the small portion (0.02 percent) of it that becomes stored in biomass and the trace amount eventually sequestered in fossil fuels.

Locally, the Sun's radiative energy that reaches the Earth's surface is seasonal in strength, site-specific in its quality, and available only during daylight hours. Figure 24.21 shows the average amount of annual solar radiation in the United States in watts per square meter. The Southwest clearly has the best solar resources located in the United States. Worldwide, the best solar sites are in the Earth's equatorial regions.

Two technologies are used to convert the Sun's radiative energy into commercial electric energy: solar-thermal technologies and photovoltaic cells (PVs) (Fig. 24.22). Solar-thermal technologies use the Sun's energy to fuel a steam cycle that produces electric power, much like the steam-cycle generators used to produce electricity from fossil fuels. The major difference is that the energy that powers the cycle is attained by concentrating the Sun's rays instead of combusting a hydrocarbon. By reflecting and concentrating the Sun's radiant energy on a receiver containing a fluid, high-temperature steam capable of turning a turbine generator can be produced. The majority of the world's solar-thermal electric generation comes from a single cluster of facilities (350 megawatts total) near Barstow in southern California. Solar-thermal technologies are thought to be best suited to utility-scale generation because of cost considerations. Solar radiation, access to water, and the availability of fossil fuels for providing backup and supplemental heat to the generating system all determine the resource quality of solar-thermal technologies. In the United States, existing solar-thermal technology is estimated to be able to meet electric needs with power plants on 0.2 percent of U.S. lands (21,000 square kilometers). This is approximately the size of the Nevada nuclear test site or the state of Massachusetts.

Photovoltaic cells utilize the Sun's radiation directly, pro-ducing electricity with no intermediary conversions. These cells convert sunlight into electricity by taking advantage of the material properties of semiconductors such as silicon. The cells are "doped" (laced with impurities) so that there is an excess of electrons on one side of the cell and a lack of them on the other, creating an electric field. When light of sufficient energy hits the cell, it causes materials in the cell to release electrons, which move from one side to the other under the influence of the electric field, forming an electric current.

Photovoltaic cells create electricity with no noise, no moving parts, no emissions, and no water requirements. The scale of a photovoltaic cell application can be as small as a calculator power source or as large as a utility-scale power plant. Although sales of photovoltaic cells account for approximately 50 megawatts of new generating capacity per year, fewer than 20 megawatts are grid-connected for utility power production. Because of today's relatively high cost of electric generation by photovoltaic cells, most cells are used for distributed applications where the cost of building transmission lines is greater than that of purchasing the cells. With current technology, photovoltaic cells could meet the demand for electricity in the United States with a solar collector field occupying less than 0.5 percent of U.S. land (46,000 square kilometers, or half the size of Maine). In 1970, photovoltaic cells generated electricity at a cost of $60 per kilowatt hour, but, costs declined rapidly down to $1 per kilowatt hour in 1980 and then $0.30 per kilowatt hour by 1990. Enron Corporation recently announced a 100 megawatt photovoltaic cell project that would produce electricity at $0.055 per kilowatt hour. If costs continue to drop as they have for the past twenty-five years, photovoltaic cells will be used increasingly because of their wide range of applications. However, because photovoltaic cells are nondispatchable, their future will be highly dependent on energy storage technologies.

Solar electric technologies have only modest environmental impacts. Because the best solar-thermal sites are located in equatorial deserts where water is scarce, the water needed

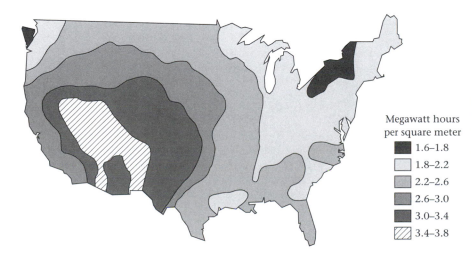

Megawatt hours
per square meter

▪ 1.6–1.8
☐ 1.8–2.2
▨ 2.2–2.6
▨ 2.6–3.0
▪ 3.0–3.4
▨ 3.4–3.8

Figure 24.21. The average annual solar radiation available in different locations in the United States. (*Source:* Adapted from Ken Zweibel, *Harnessing solar power*, Plenum Press, 1990.)

(A)

Figure 24.22. (A) Parabolic trough solar thermal collectors in the Mojave Desert and (B) an array of photovoltaic cells at a California generating station. (*Source:* Photograph courtesy of Marshall Thompson, Kramer Junction Company, 1994.)

to power the solar-thermal plants creates the biggest environmental concern. Shading and other microclimate changes resulting from photovoltaic cells or solar collectors may alter local habitat. The most significant negative impact of these cells is caused by the toxic chemicals used in their production. Fortunately, the long-standing semiconductor business in the United States and other countries has already set standards for manufacturing process controls that reduce these impacts. Other environmental impacts of solar energy production are associated with land use and aesthetics. It is likely that photovoltaic cells will be installed primarily on existing buildings or structures, minimizing costs and land use impacts.

■ **Biomass Energy.** Some people claim that plants are the ideal collectors of solar energy. However, plants store only 1 percent of incident insolation as carbohydrates, whereas modern solar (photovoltaic) cells can convert upward of 20 percent of the incident energy into electricity. Biomass consumption by humans for energy purposes is nearly impossible to assess because most of the global utilization is non-commercial, that is, not traded. Therefore, measurements of biomass consumption vary widely. Scientists estimate that biomass currently satisfies between 7 and 15 percent of world energy needs. The LDCs consume the majority of biomass energy (approximately 75 percent). In LDCs, biomass is the dominant energy source for approximately 3 billion people because of its availability and low cost. Typically, people who live in the LDCs convert biomass to heat for warmth and cooking by using primitive, low-efficiency technologies that have a negative impact on the environment. The amount of energy production from biomass is fairly significant in many countries and on a worldwide basis (Table 24.5); however,

(B)

TABLE 24.5. BIOMASS ENERGY USE IN SELECTED COUNTRIES, 1987

Country	Biomass Use (petajoules)	Share of Total Energy Consumption (percent)
United Kingdom	46	<1
United States	3482	4
Denmark	84	9
Thailand	206	20
Brazil	1604	25
China	9287	28
Costa Rica	31	32
Zimbabwe	143	40
India	8543	56
Indonesia	2655	65
Tanzania	925	97

Source: Christopher Flavin and Nicholas Lenssen, *Power surge: guide to the coming energy revolution*, New York, Norton, 1994.

biomass accounts for only approximately 1 to 4 percent of energy production in the United States.

The biomass energy resource encompasses a wide variety of sources, including:

Wood and Wood Wastes. Wood is the world's primary biomass fuel. It can be used for a wide variety of energy purposes including heating buildings and homes, cooking food, generating electric power, and producing heat for industrial processes. Wood and wood product companies use the largest amount of biomass energy in the United States and other industrialized countries because burning biomass not only generates heat and electricity but also serves as an economic means of waste disposal. Wood for energy production has traditionally been harvested from existing (wild) forests. In the future, wood energy resources may come from carefully managed plantations of fast-growing perennials such as cottonwood, silver maple, sweetgum, and sycamore trees.

Domestic and Farm Wastes. Utilizing biologic waste products in incinerators or digestors can be an extremely economic means of energy production. China, India, and Europe are the leaders in capturing "bio-gas" from the organic decay of human, animal, and agricultural waste using digestors. This bio-gas can then be used for electric power production or heating. Burning agricultural wastes can also be an economic means of power production. For example, the Hawaiian sugar industry burns sugar cane residues, and India burns rice husks to generate electricity.

Peat. Often referred to as "pre-coal," peat is partially decomposed plant matter that has accumulated in a watery environment. It produces approximately 6000 British thermal units per pound, or about half the energy of coal. Peat has been burned to produce heat for at least the past 2000 years, and it has been burned in power plants to produce

electricity in Europe since the early 1900s. Global use currently stands at approximately 30 million tons per year. The world has reserves of approximately 400 billion tons, the largest concentrations of which are in the former Soviet Union and Canada. The three largest producers and consumers of peat are the former Soviet Union, Finland (8 percent of all homes there are heated by peat), and Ireland (20 percent of electricity derives from peat).

Municipal Solid Wastes (MSW). Typically garbage is collected and then disposed of at landfills. However, this waste can be burned to generate electricity, produce steam, or produce gaseous and liquid fuels. Using MSW as an energy resource also greatly reduces the volume of waste dumped in landfills. Old landfills also generate an energy resource in the bio-gas they release as their contents decay. Unfortunately, MSW is a very dirty energy source, releasing many toxic gases when burned, and leaving behind ash rich in heavy metals.

Agricultural Crops (Annuals): Energy security concerns of the 1970s spurred ethanol production programs in the United States, Brazil, and other countries. These programs involved growing annual crops such as corn or sugar cane for the purpose of energy production through fermentation in much the same way as alcohol is produced. Annual crops are rarely beneficial sources of energy because the aggregate energy required to convert the crops to a useful product such as ethanol is more than the energy received from them. To date, subsidies and other national goals such as energy independence have supported the ethanol programs.

Aquatic Biomass. Aquatic plants or microalgae strains could be "digested" into liquid or gaseous fuels. To date, however, no aquatic energy farms have been established. Ocean biomass has an advantage over land-based production because it requires no irrigation, pesticides, or fertilizers. However, ocean-based biomass also competes with fishing, transportation, and recreational uses, and is susceptible to damage from waves and storms.

Biomass can be converted into energy services through four primary means: combustion, fermentation, anaerobic digestion, and gasification. Combustion is the direct burning of the fuel in the presence of oxygen, which releases energy when the chemical bonds within the material are broken. Fermentation converts sugars within the biomass into liquid fuels, which can then be combusted for energy. During anaerobic digestion, certain bacteria decompose biologic matter in the absence of oxygen, producing methane and carbon dioxide. A low-energy gas can also be formed from biomass by gasification, the process of heating biomass in the presence of oxygen.

Several fundamental factors restrict the use of biomass as an energy resource: weather patterns, water availability, soil quality, competition with other biomass uses (such as food, timber, or agricultural products), the low energy density of biomass, and the financial and energy costs of transportation to point of use. In the absence of subsidies, the use of biomass as an energy source in the United States and in other developed countries will most likely be competitive only if the biomass source is otherwise a useless or troublesome waste that is available near a site of needed energy production. As the costs of waste disposal increase, however, the relative value of biomass waste as a fuel will also increase.

The total recovery of crop, forest, and dung residues could currently provide approximately 10 percent of world energy needs. If the role of biomass as an energy resource is to grow, however, plantations will likely play an important role. Deforested and otherwise degraded lands hold the most promise as boimass plantation sites in developing countries because the plantation would also help decrease erosion rates. In industrialized countries, excess cropland could be used as plantations, reducing production surpluses. Land, water, and mineral resource constraints will be major factors limiting the size of the biomass resource.

Many argue that biomass is a renewable resource. However, biomass can be considered renewable only if the water, soil, and nutrients are managed in a sustainable manner and if the rate of biomass harvest does not exceed the rate of biomass growth. Biomass has two major advantages over other renewable resources: It is less site-specific, because biomass in some form can be grown almost anywhere, and it is dispatchable, because biomass can be stored until needed, much like fossil fuels. It is also different from the other renewable resources in that it involves capturing and transforming chemical energy, as opposed to mechanical or thermal energy. When biomass is grown in a sustainable manner, its production and use results in no net carbon dioxide emissions to the atmosphere, because the carbon dioxide released during conversion to energy services is recollected during regrowth through photosynthesis. Therefore, biomass may be superior to fossil fuels in terms of allaying global warming.

When burned, biomass emits less sulfur and fewer nitrous oxides than coal but produces more particulate emissions than any of the fossil fuels. The lower the combustion temperature of biomass, the lower the conversion efficiency and the greater the particulate production. Therefore, industrialized countries where air quality is a major concern have begun to restrict the use of fireplaces, which operate at very low temperatures. The low efficiencies of fuel wood stoves used in LCDs exacerbate deforestation and often leads to poor indoor air quality due to particulate emissions. This polluted air, in turn, causes respiratory problems – especially for women, who typically are the cooks in developing countries.

Deforestation is a serious environmental impact of biomass use that has regional and global significance. Deforestation results from harvesting trees, plants, and shrubs without replanting. The three primary activities that result in deforestation are clearing forests for agricultural purposes, gathering fuel wood for basic energy services (heating and cooking), and harvesting wood for its commercial value. Use of fuel wood to generate biomass energy accounts for approximately 50 percent of the total deforestation in recent history. Deforestation commonly leads to increased erosion because the land has lost the support provided by root structure

of the natural flora. Erosion strips nutrients and topsoil from the land, rendering once fertile land incapable of supporting vegetation and accelerating the accumulation of silt in downstream dams and rivers. Deforested and eroded acreage also reduces the carbon-retaining capacity of the land, resulting in net carbon emissions that contribute to global warming.

In the United States and many other countries, the government heavily subsidizes agriculture, distorting the true economic and environmental costs of using food crops for energy production. Regardless of cost, the energy received from biomass should be greater than the energy consumed in producing it. In most cases it is not optimal to grow grain crops for the purpose of ethanol production, because more energy is required to produce the ethanol than is received from it; consequently, ethanol production programs such as the one in the United States must be justified for other reasons. Ethanol production for energy use makes sense only when it uses agricultural wastes.

The production of biomass can also result in chemical pollution of the land and water. If not properly managed, pesticides, herbicides, and fertilizers accumulate in soil, groundwater, and rivers, causing serious harm to local species. Improper use of pesticides and herbicides can also hinder our ability to control weed and pest populations, because species become resistant to traditional chemical controls. Incineration of MSW can have severe environmental impacts when it releases hazardous chemicals into the atmosphere and creates ash that contains high concentrations of heavy metals. It can also compete with local recycling programs for waste materials. However, these facilities do reduce the amount of waste that is otherwise disposed of in landfills.

■ **Energy from the Ocean.** The ocean's tides, waves, currents, and thermal gradients all contain energy. To date, the contribution of these resources to meeting the energy needs of the human population has been insignificant, with human efforts to harness this energy largely limited to electric power generation. The only ocean energy technology that has been developed to any commercial extent is tidal energy.

Tidal energy facilities are very similar to hydroelectric plants in dams; however, the dam (in this context, a "barrage") is placed across the mouth of an estuary. Tidal facilities take advantage of the fact that the gravitational attraction of the Moon causes large diurnal (twice daily) oscillations in sea level, known as the tides. During high tide, the gates in the dam are left open to allow the estuary to fill. The barrage then holds the ocean water behind until sea level falls. Water is then released through turbines in the barrage, converting the potential energy of the water into electric energy. The process is repeated with each tidal cycle. Today's technology limits tidal energy production to areas that have at least fifteen feet of tidal range (Fig. 24.23). Most locations having adequate range are not feasible because of the huge capital costs and the severe environmental impacts on tidal systems. Potentially economically feasible sites are fairly well distributed globally. However, there are only two places in the world today where a significant amount of electricity is

Tidal energy

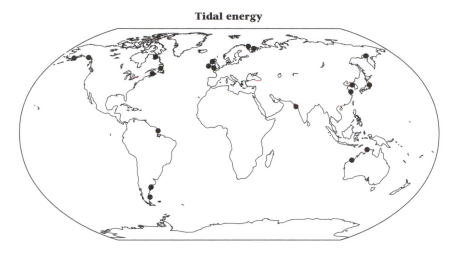

Wave energy

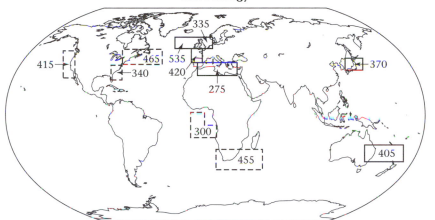

Ocean thermal energy

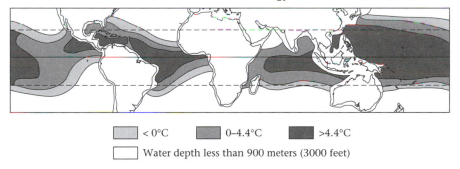

$\square$ < 0°C $\blacksquare$ 0–4.4°C $\blacksquare$ >4.4°C

$\square$ Water depth less than 900 meters (3000 feet)

Figure 24.23. Worldwide ocean energy resources: principal sites for tidal power development, annual wave energy in megawatt hours per meter from specific areas, and the average temperature differential between the surface and a depth of 900 meters for ocean thermal energy conversion (OTEC) utilization. (*Source:* Adapted from James E. Cavanagh, John H. Clarke, and Roger Price, *Renewable energy*, Island Press, 1993.)

generated by tapping the potential energy associated with changing tides. The La Rance River in France constitutes the largest project, with 240 megawatts of capacity, and the Annapolis River in Nova Scotia, Canada, has approximately 18 megawatts of capacity.

The ocean's energy can also be tapped to capture the energy contained in waves. Most attempts at capturing wave energy have been in association with breakwater projects. The most successful and economic applications thus far have involved a technology called oscillating water columns (OWCs). This technology uses wave energy to compress air in small vertical pipes, similar to pistons, forcing air through small turbine generators. This technology has largely been used in navigation buoys to provide power for lights. The areas best suited for wave energy development are those most influenced by the westerly winds (see Fig. 9.6). The west coasts of the United States, Japan, New Zealand, and Europe all have high potential for wave energy development.

A third form of ocean energy exists in the temperature differential between the warm surface water and the cool, deep water in tropical and subtropical regions. Ocean thermal energy conversion (OTEC) involves building an offshore or onshore power plant that either converts warm surface water to steam in a vacuum (open cycle) or uses it to evaporate ammonia at a very low pressure (closed cycle). The steam or ammonia vapor expands through a turbine to produce electricity. The subsurface cold water then condenses the steam or ammonia gas. Unfortunately, even under the most ideal conditions the conversion of the surface water's thermal energy to electricity is only 2.5 percent efficient; consequently, huge volumes of water must be circulated through the plant to produce an appreciable amount of electricity. Additionally, serious technical and cost obstacles have yet to be overcome. Given current technologic capabilities, OTEC potential is limited to tropical and subtropical regions with ocean temperature gradients greater than 20°C occurring within the first 900 meters (3000 feet). This region defines the resource base; however, cost, lack of experience, and technical challenges will make this resource base difficult to tap. Island communities with a need for fresh water will be the first to adopt this technology because the open cycle version of an OTEC plant doubles as a desalinization facility.

There are advantages and disadvantages to the utilization of ocean energy. Because ocean energy, like other renewable resources, could offset production of energy from fossil fuels, ocean energy production could reduce the emission of greenhouse gases and atmospheric contaminants. However, ocean technologies are not without detrimental environmental effects. Tidal energy projects have environmental impacts similar to those associated with large-scale hydroelectric projects. Both involve a dam or a barrage that blocks the flow of water, sediment, and marine life to form a reservoir with sufficient head. A typical barrage greatly reduces the inflow into the estuary, reducing the tidal range by half. Such an alteration can have significant effects on habitat and affect nearby areas in ways that are difficult to predict. On the positive side, barrages may protect vulnerable coastlines against surge tides. Wave energy devices of any significant scale may affect kelp growth, may get caught in the kelp itself, and will certainly have an effect on the coastal sedimentation patterns in much the same manner as breakwaters do. Wave energy devices would also need to be carefully sited so as not to interfere with the navigation of ocean vessels. The environmental impacts of OTEC are hard to assess, because no major systems have yet been built. The vast plumbing system required to produce electricity may cause artificial water circulation patterns to appear, dramatically altering the local ocean temperature gradient and the oxygen and mineral content of surrounding waters. These changes would have profound effects on local marine life and fishing industries.

OTEC systems would also result in the release of carbon dioxide to the atmosphere because the solubility of carbon dioxide in water decreases with increasing temperature; cold, carbon dioxide–saturated water brought to the surface from depth would release carbon dioxide as it warmed.

■ **Nuclear Fusion.** Although nuclear fusion is not a source of commercial energy production today, it is significant for several reasons. First, it has enormous potential as an energy resource. Second, industrialized nations have invested hundreds of millions of dollars into fusion research. Fusion produces a very large amount of energy from a very small amount of fuel; 1 gram of deuterium-tritium fuel can provide the energy equivalent of 2400 gallons (8 metric tons) of oil! The abundance of the fuel source for fusion energy production makes it an extremely exciting prospect; the necessary isotopes of hydrogen can be extracted from seawater. Millions of dollars have been spent because of the awesome promise of commercial fusion energy production and because its funding was tied to the Star Wars weapons program in the United States from the mid-1980s through the early 1990s.

Nuclear fusion has powered our universe since the beginning of time (the Big Bang theory; see Chap. 4). Earth's Sun, like other stars, is a natural fusion power plant. The energy we receive from the Sun originates from a fusion reaction that takes place within its core. Scientists have been working for almost forty years to imitate this reaction in the laboratory. Fusion combines two light nuclei to form a single heavier nucleus. The mass of the reactants is actually reduced during the reaction, and the lost mass is converted into energy. Several possible reactions have been hypothesized. The most commonly attempted fusion reaction in the laboratory involves fusing the nuclei of deuterium and tritium, two isotopes of hydrogen, to form helium and a neutron and releasing the difference in mass as energy (see Fig. 24.16). When a commercial reactor is developed, which is expected to occur by 2025 at the earliest, it will capture heat from the fusion process to power a conventional steam turbine generator, as found in most fossil and fission plants, to produce electricity.

In order for fusion to take place, the reactants must be confined. Three kinds of confinement are possible: gravitational (as in the Sun), magnetic, and inertial. Magnetic and inertial confinement are the methods used in research laboratories to confine a plasma of deuterium and tritium in order to achieve fusion. In the case of gravitational confinement, the pull of the Sun's gravity overcomes the tendency of hot nuclei to expand out into space. This results in the continuous fusing of nuclei and the release of enormous amounts of energy. In the laboratory, scientists have attempted to use magnetic fields to confine the fuel at very high temperatures long enough for it to fuse in a process called magnetic confinement. In the method called inertial confinement, scientists use powerful laser beams to compress the fuel to a very high density for a short time, commonly measured in billionths of a second.

Japan, Britain, the United States, and the former Soviet

Union have the most significant research fusion reactors in the world today. These reactors have created fusion reactions, however, thus far the energy required to produce the reaction is greater than the energy released by it. Currently, scientists are attempting to reach break-even, the point at which the fusion reaction produces as much energy as its creation consumes. The obvious goal is to move past break-even to ignition, the point at which the fusion reaction sustains itself by producing more energy than it consumes. Heat from this self-sustaining reaction could then provide energy for electricity production.

The negative environmental impacts associated with future fusion energy production are anticipated to be modest. Because the source of the fuel is seawater, few negative impacts will be associated with the fuel cycle. Deuterium represents 1 in every 6500 hydrogen atoms and is a resource that humans are not likely to exhaust for millennia. Chemically, deuterium does not behave differently from regular hydro-

gen; therefore, none of the ocean's systems will be significantly affected by its removal. Impacts associated with siting and building the power plants would be no worse than those for any other power plant existing today. The reactor vessel, which will become embrittled from neutron bombardment, will most likely need to be replaced during the life of the power plant. Disposing of these radioactive reactor vessels will be the primary environmental challenge. As with any other steam-cycle power plant that takes heat and transfers it to run a steam cycle, small-scale thermal pollution from waste heat will occur as well.

ENERGY EFFICIENCY

Consider the energy conversions between the chemical energy in fossil fuels to the radiative energy emitted from a light bulb in your home. This multistep process involves several energy transformations. Along the way, much of the original energy is lost; the overall efficiency of this process is the product of the efficiencies of each step. Although energy is conserved in any process, the efficiency of a conversion process is defined as the output of useful energy or work divided by the total energy input. The energy that does not go into useful work ends up in unusable energy forms such as low-temperature heat. For example, the efficiency of converting coal to electricity is 35 percent, the efficiency of transmission lines from the power plant to your house is approximately 90 percent, and the efficiency of converting electricity to light in an incandescent bulb is 5 percent. Therefore, the overall efficiency is

$35\% \times 90\% \times 5\% = 1.6\%!$

For every 100 tons of coal burned in a power plant, only 1.6 tons of coal equivalent energy is converted into lighting, the desired energy service. The other 98.4 percent of the energy primarily is converted to waste heat during all the conversion steps between the coal and the energy service it produces. The vast resource these energy losses represent can be harnessed by improving system efficiencies to meet rising energy needs. Table 24.6 shows the efficiencies of some other conversion devices and processes. The above examples make it clear that if any of the conversions could be made more efficient, the result would be the same energy service for less fuel. If less coal is required to produce the same illumination, then less coal needs to be mined, processed, transported, and combusted per unit of energy service delivered. Small downstream improvements can thus provide large environmental benefits.

Opportunities for energy efficiency can be categorized as generat-

TABLE 24.6. EFFICIENCIES OF ENERGY CONVERSION DEVICES AND SYSTEMS

Electric generators	70–99%
(mechanical → electrical)	
Electric motor	50–90%
(electrical → mechanical)	
Gas furnace	70–95%
(chemical → thermal)	
Fossil fuel power plant	30–40%
(chemical → thermal → mechanical → electrical)	
Nuclear power plant	30–35%
(nuclear → thermal → mechanical → electrical)	
Automobile engine	20–30%
(chemical → thermal → mechanical)	
Fluorescent lamp	20%
(electrical → light)	
Incandescent lamp	5%
(electrical → light)	
Solar cell	2–25%
(light → electrical)	

Source: Roger A. Hinrichs, *Energy*, Harcourt Brace Jovanovich, 1991, p. 40.

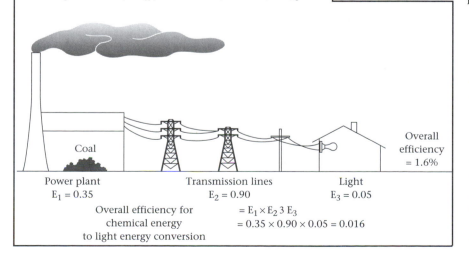

Power plant
$E_1 = 0.35$

Transmission lines
$E_2 = 0.90$

Light
$E_3 = 0.05$

Overall efficiency
= 1.6%

Overall efficiency for chemical energy to light energy conversion
$= E_1 \times E_2 \cdot E_3$
$= 0.35 \times 0.90 \times 0.05 = 0.016$

Coal

ing efficiency technologies (e.g., the improved conversion of coal to electricity), distribution efficiency technologies (e.g., energy transportation and load management systems), and end-use efficiency technologies (e.g., more efficient lights or motors). Generation technologies are gradually becoming more efficient. For example, high-efficiency combined cycle gas turbines convert natural gas into electricity by both combusting the gas in a turbine and using waste heat from the combustion process to run a simple steam cycle. By utilizing heat that is commonly wasted, combined cycle gas turbines now achieve efficiencies exceeding 50 percent, in contrast to the typical 35 percent efficient steam turbine power plant. Over 75 percent of all electricity today is generated at approximately this 35 percent efficiency level. Additionally, increases in efficiency can result in lower energy production capital costs by reducing the need for new generating capacity, encouraging the use of renewable technologies that can be installed on a small scale and may have marginally higher initial capital costs.

Distribution efficiencies can increase through the improved management of energy demand, also called load. This is especially true in the use of electricity. Utilities historically have charged their customers the same price for electricity regardless of when it is used. This level pricing structure has encouraged similar energy use patterns among most electricity users; consequently much more electricity is used during certain times of the day and year. For example, California may have "peak" loads during the summer afternoons when air conditioning energy requirements are greatest; these loads are three to five times the "base" load demand for power. Other areas such as Minnesota have their greatest peaks in the winter because of electric heating requirements. Utilities have had to build power plants for the sole purpose of meeting the peak load, even though that load may occur for only a few days per year. Power plants run most efficiently when used continuously, minimizing startup and shutdown losses. The higher the peak load, the higher the cost of the power needed to meet that demand. State agencies that regulate utilities now encourage several demand-side management schemes in order to flatten the peaks. A growing trend is the adoption of time-of-use billing practices that charge consumers more during peak demand times (similar to phone service rates), passing on the added cost of producing peak power to those who consume it. This billing scheme also reduces rates for those consumers who use electricity during off-peak hours. Utilities hope that the adoption of time-of-use billing will reduce the peak power demands, decreasing the need for additional power plants and allowing existing plants to operate at higher capacity and efficiency. Additional efficiency improvements may arise because higher energy prices during peak hours could encourage the use of more energy-efficient end-use technologies.

End-use technologies provide many opportunities for improving energy efficiency. For example, the United States could save approximately 25 percent of its electricity consumption by applying existing energy-saving technologies to

TABLE 24.7. EFFICACIES OF SELECTED LIGHTING TECHNOLOGIES	
Technology	Efficacy (lumens per watt)
Candle	0.2
Kerosene mantle	0.8
Incandescent	14.7
Halogen	17.0
Compact fluorescent	53.8

Source: Christopher Flavin and Nicholas Lenssen, *Power surge: guide to the coming energy revolution*, New York, Norton, 1994.

lighting devices and motors alone. In the United States, lighting consumes approximately 25 percent of electricity: 20 percent for lighting and 5 percent for lighting-related cooling. Estimates suggest that 50 to 90 percent of this energy could be saved through cost-effective changes in lighting and related technologies. Table 24.7 shows the *efficacy* measured in lumens per watt for different lighting technologies. Note the difference in efficacies between incandescents and fluorescents; fluorescent lights are four times as efficient! Motors, which run anything from pumps to elevators to industrial machines, consume 60 to 70 percent of industrial electricity and more than 50 percent of the overall electricity consumed in the United States. The energy efficiency of motors can be enhanced in many ways. Retrofitting motors with electronic, adjustable speed drives alone could reduce their electricity consumption by 20 percent. Significant opportunities also exist today for increased energy efficiency in homes and buildings through improved insulation, windows, appliances, shading, sensors, and controls.

Transportation represents another opportunity for energy efficiency. Consider the efficiency of converting crude oil to the energy service of transportation. Less than 15 percent of the energy in a gallon of gasoline is actually used to provide transportation. The remaining energy value in the fuel is lost during energy conversions. The government currently sets fuel efficiency standards for automobile at 27 miles per gallon, yet many prototype vehicles developed by manufacturers get better than 70 miles per gallon (up to 128 miles per gallon). Given that the United States imports approximately 50 percent of its oil and uses 60 percent of its oil for transportation, more efficient automobiles would have a powerful effect on everything from defense spending to the trade deficit – oil imports alone account for 10 percent of the U.S. trade deficit.

Energy efficiency has enormous potential benefits, offering the promise of the same energy services with less drain on our natural resources and environment. Despite the promise and economic sense of energy efficiency, it is not utilized to anywhere near its potential. Extensive policy work and government support are needed to create an environment in which an informed public adopts more efficient energy conversion technologies while keeping the potential environmental, resource depletion, and cost considerations in mind.

CONCLUSION: THE COMING ENERGY REVOLUTION

What will our energy future look like? The past may be a clue to the future. Think about what life was like when your parents were in college: a personal computer, let alone a laptop, was unimaginable for most; no voice mail, paging, faxes, e-mail, or portable or cellular phones. It would have been difficult for your parents then to imagine how you now accomplish tasks such as communication. Growing environmental concerns, global competition, and deregulation of long-time energy monopolies are setting the stage for a revolution in energy resource use as dramatic as the change from typewriter to laptop. The changes ahead will be much grander and more complex, including a fundamental shift in the energy resources humankind depends on and drastic improvements in energy efficiency. These changes will eventually take place with huge environmental benefits and no setback in lifestyle. The important questions are: How rapidly will we make these changes, and what can be done to encourage them other than crises?

Of fundamental importance is society's ability to look beyond the present and to have a common vision of the future that is consistent with the goals of sustainability. Many scientific experts, politicians, and policy makers believe that the energy future for the next generation will be one of "status quo plus," meaning that although other resources will gradually be utilized, fossil fuels will continue to dominate the energy sector. They believe that modest efficiency improvements will allow the fossil fuel supply to meet the needs of a growing industrialized population that will be 50 percent larger in 2020 than it is today. Others, including Christopher Flavin and Nicholas Lenssen at the Worldwatch Institute and Roger Rainbow and Georges DuPont-Roc in the Business Environment Group at Shell Oil Company believe that an energy revolution lies ahead: that a thirty-year continuation of the status quo is the least likely future.

Indeed, rapid change is already taking place, and many of the needed technologies exist today. Examples include:

1. *More efficient energy technologies:* From light bulbs to refrigerators, many new technologies are over 75 percent more efficient than the current standard; power plants built in the 1990s are 50 percent more efficient than those built a decade earlier.
2. *The shift to natural gas:* As the oil and coal industries mature, gas use is growing worldwide, continuing a trend toward cleaner, more versatile fuels; its use will also facilitate a trend toward smaller, more decentralized energy conversion and storage.
3. *Recent developments are making wind, solar, and other renewables more viable:* Costs are decreasing, and installations of wind turbines and photovoltaic cells are growing rapidly.
4. *The development of green pricing plans for electricity.* This strategy allows and encourages consumers to choose energy sources that cause minimal environmental impacts.
5. *Viable alternatives to the gasoline-powered internal combustion engine.* Lightweight, hybrid electric vehicles made of nonsteel, synthetic materials and powered by devices such as gas turbines, fuel cells, and flywheels will soon emerge from engineering laboratories around the world; these automobiles have three to four times the fuel efficiency, emit only 5 percent of the current standard for pollutants, and will enter the commercial market by the beginning of the twenty-first century.

These examples reflect the changes that are evident to date. One puzzle piece in the energy revolution involves the manner in which energy from the wind and Sun will be stored and transported. Many scientists point to hydrogen as a key to energy storage and transmission by the mid-twenty-first century. If the cost of electricity produced from renewable sources drops low enough, this electricity can be used to hydrolyze water to produce hydrogen, which can then be stored and used to produce electricity in fuel cells or other technologies when needed. Hydrogen is being studied and seriously considered as a transportation fuel as well.

The arrival of the revolution will depend on how quickly policy barriers and the vested interests of the long-established energy industry can be overcome. It will also be greatly influenced by the fact that the most important new energy technologies are relatively small devices that can be mass produced in factories – contrasting with the large mines, refineries, and power plants of the twentieth century. Moreover – much like today's consumer electronics industry – the economies of mass manufacturing are likely to reduce costs. There are signs that the corporate giants in the energy field know that the energy revolution is approaching, and they want a piece of it. For example, Bechtel is focusing corporate resources to study sites for renewable energy projects worldwide; also, a division of Enron, one of the largest natural gas companies in the world, has recently put in a compelling, low-cost bid for large-scale photovoltaic cells at the Nevada nuclear test site, which the U.S. government may convert to a solar enterprise zone. However, not all players in the current energy game are prepared to accept the changes that are already occurring. The threat of these changes has caused them to lobby tenaciously to preserve the energy policies that have favored a mix of fossil fuels over the past century.

As Flavin and Lenssen of Worldwatch point out, the numerous necessary policy changes fall into four categories:

1. *Reduce subsidies for fossil fuels, and raise taxes on fossil fuels to reflect environmental costs.* In 1991, the World Bank estimated that direct fossil fuel subsidies worldwide totaled approximately $220 billion per year, with China and the former Soviet Union receiving 75 percent of these subsidies. In 1989, the U.S. government provided over $36 billion in subsidies to energy industries, 60 percent going to

fossil fuels and 30 percent to nuclear resources. Ending subsidies is only half the task; all energy prices need to reflect the true social cost of the resource. Instituting energy taxes that accurately reflect the environmental damage associated with each energy resource is the best way to do this (see Chap. 27). The environmental costs will then be embodied in the price that consumers pay. Switzerland has adopted a modest carbon tax, the most commonly discussed option. Other options include strict environmental laws that have compliance costs and integrated resource planning by utilities that weighs the environmental costs in the planning process.

2. *Reform the energy research and development programs of governments.* Nuclear energy and fossil fuels have dominated government research funding. The twenty-three member countries of the International Energy Agency spent 85 percent of their $115 billion energy research budgets on these two resources between 1978 and 1991, and less than 6 and 9 percent, respectively, on energy efficiency and renewable resources.

3. *Ensure the expansion of commercial markets for technologies such as wind turbines, efficient motors, fuel cells, photovoltaic cells, and other innovations.* The creation of large markets by the government, or other groups who can place large multiyear orders for these technologies, will provide large gains in efficiency and reduce costs as growing markets provide incentives for competition and innovation to developers.

4. *Channel international energy assistance to developing nations.* Energy assistance programs that pushed developing nations down the fossil fuel path should be redirected to encourage those nations to focus on new renewable technologies instead. The destructive development path forged by the Western nations is not one they, or the rest of the world, can continue to follow.

Many of these policy changes will require a shift in the role of government, which in the past has been more centrally involved in the energy sector than in almost any other area of the civilian economy. Overall, the energy revolution will require greater reliance on the market and less government involvement. However, in a number of cases, governments will need to set the rules, focusing on ways to ensure that environmental costs are considered when economic decisions are made.

How quickly will the coming energy revolution take place? In 1882, Thomas Edison invented the light bulb; within twenty years, over 20 million bulbs were in use. In 1896, Henry Ford built his first automobile, and twenty years later there were 3.5 million vehicles on the road in the United States alone; five years later there were 23 million! Revolutionary technologic advances can change society seemingly overnight. As the barriers to the energy revolution break down, the way we live will change rapidly.

In an increasingly populated world, sustainable economic growth can be achieved only with the development of a new energy system. This system is imminent. The necessary changes in our use of energy resources will be accomplished because consumers demand that these changes take place. Consumers can make these demands only when they clearly understand the consequences of their actions and when the marketplace offers them choices that value sustainability of the Earth's systems. Ideally, the newly evolving information age and the continuing deregulation of markets will allow competitive forces to bring about the best decisions and technologies for the future.

QUESTIONS

1. Compare developing countries to industrialized countries in terms of the energy resources they use and the energy services they are provided. Given these differences, how do the environmental concerns vary from industrialized countries to the LDCs?

2. How significant an issue is resource depletion and possible resource exhaustion relative to the other concerns that fossil fuels pose? Why?

3. Nuclear power has produced energy that benefits consumers for a limited period of time. Because this resource was developed rapidly, with large subsidies and with poor understanding of private and social costs, future generations will have to pay a large share of the private and social costs (waste disposal, decommissioning, health impacts, and higher than expected operating costs) associated with energy produced to benefit a prior generation. What ethical issues does this raise for our energy consumption – past, present, and future? Large segments of our society believe that this situation poses few ethical issues. Can you justify their perspective?

4. Why have fossil fuels dominated the twentieth-century energy mix globally? How will the mix be different in the twenty-first century, and why?

5. Choose an example of an energy resource that plays a dominant role in your home community, and discuss how well its real, long-term costs to society are embodied in the price paid for the resource.

6. Among the depletable sources of energy, please rank the four sources in terms of their attractiveness as an energy resource in the twenty-first century and explain why. Do the same for renewable sources of energy. Of all these sources, which ones do you expect to play the most significant role in the twenty-first century energy mix?

FURTHER READING

British Petroleum. 1996. *BP statistical review of world energy.* London: British Petroleum Corporation Special Report.

Burnham, L. 1993. *Renewable energy,* Washington, DC: Island Press.

Energy Information Administration. 1995. *Annual energy outlook.* Washington, DC: U.S. Department of Energy, Annual Report.

Energy Information Administration. 1995. *Annual energy review.* Washington DC: U.S. Department of Energy, Annual Report.

Flavin, C., and Lenssen, N. 1994. *Power surge: guide to the coming energy revolution.* New York: Norton.

Golob, R., and Brus, E. 1993. *The almanac of renewable energy.* New York: Henry Holt.

Hinrichs, R. A. 1991. *Energy.* New York: Harcourt Brace Jovanovich.

Kahn, J. 1994. Dammed Yangtze. *The Wall Street Journal*, April 18, p. 1A.

Tyler, P. 1996. Cracks show early in China's big dam project. *The New York Times*, January 15, pp. D1–D3.

World Energy Council. 1993. *Energy for tommorrows's world.* London: St. Martin's Press.

Yergin, D. 1991. *The prize: the epic quest for oil, money and power.* New York: Simon and Schuster.

25 Natural Hazards: Prediction and Risk

W. G. ERNST

LIFE AND DEATH ON MOTHER EARTH

In 1970, 1985, and 1991, killer typhoons came ashore in Bangladesh – in all, claiming nearly 750,000 lives. Avalanches and volcano-induced mudflows caused nearly 50,000 fatalities in two recent Andean disasters: Peru (1970), and Colombia (1985). The great earthquake of 1976 killed more than 400,000 people in Tangshan, China. Risk is certainly a part of life, but natural disasters of this magnitude boggle the mind. Why and where do such catastrophies occur? More important, could they have been averted? This chapter addresses these questions.

Solar radiation impinging on the Earth drives the circulation of the atmosphere and the oceans. The dynamic flow patterns we know as weather and ocean currents are manifestations of the redistribution of thermal energy among the fluid envelopes of our planet. Hurricanes, typhoons, and cyclones are simply extreme examples of this energy transfer process. Because the Pacific Ocean is the world's largest solar energy sink, this basin is the site of the most devastating tropical storms and ocean-induced droughts. Coastal erosion and flooding also constitute a major hazard in low-lying portions of the region. Climatic hazards are well-understood phenomena; however, the underlying reasons accounting for geologic hazards are much less obvious. Hidden from view, the imperceptible mantle circulation, which is driven by buried heat, produces the crustal motions collectively known as plate tectonics. Such processes are responsible for the differential motions of lithospheric slabs. The boundaries of the plates are marked by intense seismicity, volcanic activity, and land slumpage. Climatic and geologic disasters cause great human suffering and financial loss, especially within the Pacific Basin and around its margins. If global temperatures continue to climb as a consequence of an intensifying greenhouse effect, the ferocity and number of tropical storms will also rise, coastal erosion and flooding will accelerate, and the vulnerability of the world's burgeoning tidewater communities to devastation by tsunamis will increase. Millions of lives will be endangered, not to mention the economies of the nations affected. What might be done to avert these potential disasters?

Sea-floor spreading, continental drift, and plate tectonics are inexorable manifestations of the transfer of thermal energy from the Earth's deep interior toward the surface, and severe cyclonic storms are a function of solar energy input to the tropics, followed by transfer between the hydrosphere and atmosphere. These processes will forever lie beyond the scope of humanity's ability to modify them; we can never hope to eliminate their accompanying natural catastrophes. However, understanding the basic causes of these phenomena will allow us to organize our lives and reinforce our structures in such ways as to minimize the dangers. At all scales, informed planning can avert potentially devastating loss of lives and shattered economies through clear recognition and assessment of hazards, appropriate land use, construction of safe facilities, and timely, accurate warning of an impending earthquake, volcanic eruption, landslide, tsunami, coastal flood, or hurricane-force wind. Reasonable risk assessment of site-specific hazards and the engineering of stable structures are within current economic and technical capabilities of the developed nations, but not of the Third World. In order to minimize the destruction and loss of life, we need to rectify this situation by providing quantitative hazard assessments for the entire planet, followed by the far more capital-intensive mitigation of hazards, including construction of disaster-resistant structures. Because the precise, timely forecasting of many types of impending natural disaster (especially geologic hazards) is not yet a reality anywhere on the globe, research needs to be carried out in order to understand and successfully predict such events.

The goals of natural hazard abatement include marked reduction of human suffering, minimization of economic loss, and preservation of environmental quality. To attain these ends, we must first clarify the reasons for natural catastrophes, then provide socioeconomic and engineering solutions for their mitigation. However, effectiveness requires political action, and in order to achieve this, we need to gain the attention, understanding, and support of decision makers worldwide. This can be realized only if the consequences of inaction are broadly perceived as totally unacceptable, and if predictions of natural disasters are so refined as to become both timely and accurate.

NATURAL DISASTERS ALONG THE PACIFIC RIM

The chronicling of natural catastrophes has marked human history since the dawn of civilization. Who among us has not heard of the eruption of Mount Pelée in 1902, which obliterated St. Pierre, Martinique (an entire city engulfed in an incandescent cloud of ash), or of Mount Vesuvius, which buried Pompeii and Herculaneum in 79 AD (cowering bodies entombed in ash, or felled as they tried to escape the deadly rain of cinders and noxious gases)? The course of history has been profoundly influenced by natural disasters. Consider the great sea storm that scattered and destroyed the Spanish Armada in 1588, and the so-called Kamikaze, or divine wind (a typhoon), which swamped the Mongol fleet invading Kyushu and Honshu in 1281. No area is immune from the effects of devastating storms; however, typhoons, coastal flooding, tsunamis ("tidal" waves), earthquakes, volcanic eruptions, landslides, and avalanches are especially severe in the Circum-Pacific and Alpine-Himalayan realms. Why should this be so? Given what we now know, can the effects of these hazards be avoided or, at least in large part, ameliorated? This chapter addresses such issues.

Study a topographic/bathymetric map of the Pacific Basin, as illustrated in Figure. 25.1, and you will immediately note two major features. (1) The margins of the basin are almost universally bounded by high mountain ranges – those of the Chilean and Peruvian Andes, Central America, western United States and Canada, southern Alaska, the Aleutians, Kamchatka, the Kuriles, Japan, the Marianas, Indonesia Tonga-Fiji, and New Zealand. (2) The adjacent off-shore regions are characterized by precipitous submarine slopes leading to astounding depths (7 to 10 kilometers) in the oceanic trenches. Young mountain belts such as these are the locus of the globe's most destructive geologic hazards, reflecting ongoing plate tectonic processes, as described in Chapter 6.

The Pacific Ocean is also the spawning ground for the most energetic, destructive cyclonic systems on the globe. Being the world's largest body of water, it receives an enormous input of solar energy, especially in the tropics; redistributions of this energy via currents in the hydrosphere and atmosphere, in turn, drive the Earth's most powerful storms. Typhoons, hurricanes, and coastal flooding, spectacular and extreme examples of the heat transfer process at work near the planetary surface, exact a terrible toll in destruction and

Figure 25.1. Bathymetric map of the Pacific Basin, showing major submarine spreading ridges, linear seamount chains, and marginal deeps, or trenches. (*Source:* Courtesy of B. C. Heezen and M. Tharp, *World ocean floor*, U.S. Navy 1977.)

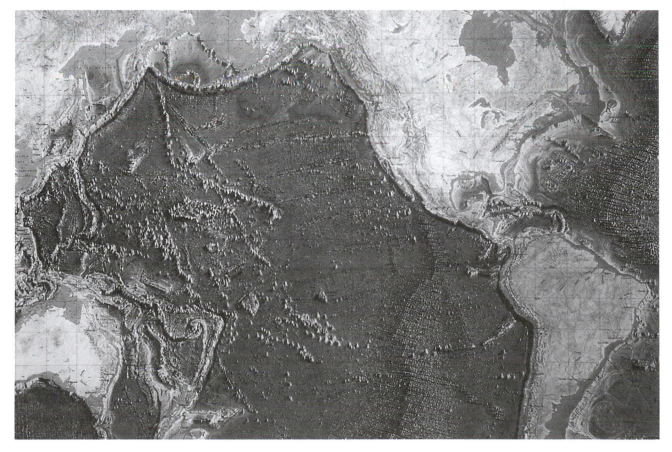

loss of life throughout the basin as well as along the Pacific Rim. Worldwide climatic disasters rival, and often exceed, geologic disasters in terms of the cataclysmic loss of life and property. The international impacts of the various types of natural hazards in terms of fatalities are shown in Table 25.1. Over the interval 1960–87, tropical storms were somewhat more devastating than earthquakes; however, mortality statistics are very closely correlated with specific events, such as the 1970 Bangladesh typhoon, and the earthquake of 1976 near Tangshan, China. Over the course of time, geologic and climatic disasters seem to be about equally deadly.

NATURAL DISASTERS AND HAZARD MITIGATION

Hazards are relatively sudden events (*disasters*) that endanger lives and property. The ultimate cataclysm for life throughout the solar system would be the explosion of our Sun as a supernova; nearly as devastating would be the terrestrial impact of a large asteroid, similar to the Chicxulub event (discussed in Chap. 3). Transfixing as such events would be, we confine our attention here to more familiar classes of events of lesser finality but higher probability of occurrence. Two principal types of hazard may be distinguished: *natural*, and *technologic*. Geologic catastrophes such as earthquakes, avalanches, volcanic eruptions, and seismic sea waves (tsunamis), and meteorologic-oceanographic disasters such as typhoons, droughts, floods, and lightning-strike forest fires, constitute some of the main types of natural hazard. Technologic, or human-induced, disasters reflect the unwise, or at least unlucky, impingement of human activities on the environment. Examples of technologic hazards include the collapse of seawalls, levees, bridges, and buildings; dam failure; chemical and radiation pollution; and war. Of course, the anthropogenic-induced gradual degradation of the environment and loss of biodiversity are calamitous as well, and may have more far-reaching impacts on the planetary web of life. For the purposes of this chapter, however, natural disasters are defined as characteristically sudden, brief, and, in some cases, unexpected events.

Historically, humans typically have failed to consider natural disasters in the appropriate physical and historic context. Rather, each is viewed as a unique event – typically, an "act of God." Following an earthquake, landslide, flood, or typhoon, structures are rebuilt by the survivors and life-styles are resumed, in many cases exactly as before the disaster; thus, the same vulnerability with regard to a future event is perpetuated. Moreover, as populations grow, their expansion into, and utilization of, hazard-prone, marginal environments increases; consequently, the overall risk is heightened.

Avoidance or alleviation of the adverse impacts of hazards necessitates their clear recognition, quantification of the processes involved, accurate assessment of associated risks, and avoidance or technologic mitigation. The first three steps require Earth systems scientific and engineering investigations; the fourth involves preventative procedures such as redesigning and reinforcing buildings, bridges, and dams, and constructing all-weather shelters and seawalls. Absolutely crucial to the fourth, sociopolitical process is widespread public understanding and appreciation of the problems of natural hazards. Once the potential danger is properly assessed, mitigation can take place through the implementation of engineering solutions, the optimization of wise land use, effective public education, and contingency planning for emergencies. However, effective implementation of our scientific knowledge and engineering capabilities requires widespread public understanding and support – something that is largely lacking at present, in the United States and around the world.

GEOLOGIC HAZARDS

EARTHQUAKE COUNTRY. We learned in Chapter 6 that earthquake hypocenters are concentrated along lithospheric plate boundaries, because these are the portions of the plates subject to differential movement (Fig. 6.11). Within the oceanic realm, a map showing epicentral locations of seismic activity is relatively simple for two reasons: (1) Ocean crust–capped lithosphere is exclusively young, geologically speaking – less than 100 to 200 million years old; and (2) beneath the oceanic lithosphere, the top of the asthenosphere lies at relatively shallow depths, within approximately 50 kilometers of the surface. Accordingly, earthquakes reflect the brittle behavior of the rather thin, homogeneous, ocean crust–capped plates.

In contrast, lithospheric slabs surmounted by complex continental crust may be as much as 200 to 400 kilometers thick. Because they are on the average much more ancient – up to 3.8 billion years old – the continents are riven by numerous discontinuities and zones of weakness that reflect a tortured history of repeated mountain building, and a diversity of rock types, anisotropies, and strengths. Furthermore, where ocean crust–capped lithosphere dives be-

TABLE 25.1. FATALITIES FROM NATURAL HAZARDS 1960–1987

Hazard Type	Deaths[a]	Largest Event and Date	Death Toll
Tropical cyclones	622,400	Eastern Pakistan (Bangladesh) 1970	500,000
Earthquakes	497,600	Tangshan, China 1976	250,000
Floods	36,400	Vietnam 1964	8,000
Avalanches, mudslides	30,000	Peru 1970	25,000
Volcanic eruptions	27,500	Colombia 1985	23,000
Tornadoes	4,500	Eastern Pakistan (Bangladesh) 1969	500
Snow, hail, wind storms	3,100	Bangladesh 1986 (hail)	300
Heat waves	1,000	Greece 1987	400

[a] Deaths rounded to the nearest hundred.

Source: E. Bryant, *Natural hazards*, Cambridge University Press, Table 2.2, p. 23.

TABLE 25.2. EARTHQUAKE INTENSITIES AND A COMPARISON OF THE RICHTER AND MODIFIED MERCALLI SAMPLES

Richter Magnitude[a]	Number Per Year	Modified Mercalli Intensity Scale[b]	Characteristic Effects of Shocks in Populated Areas
<3.4	800,000	I	Recorded only by seismographs
3.5–4.2	30,000	II and III	Felt by some people
4.3–4.8	4800	IV	Felt by many people
4.9–5.4	1400	V	Felt by everyone
5.5–6.1	500	VI and VII	Slight building damage
6.2–6.9	100	VIII and IX	Much building damage
7.0–7.3	15	X	Serious damage, bridges twisted, walls fractured
7.4–7.9	4	XI	Great damage, buildings collapse
>8.0	One every 5–10 yr	XII	Total damage, waves seen on ground surfaces, objects thrown in the air

[a] Richter numbers indicate the amplitude (in millimeters per 10000 of ground motion recorded by a standard seismometer sited 100 kilometers from the epicenter.

[b] Mercalli numbers reflect the amount of damage to human constructions and the degree to which ground motions are felt. These depend on magnitude of the earthquake, distance of the observer from the epicenter, and whether the latter is on solid ground or not.

Source: Bruce A. Bolt, *Earthquakes*, W. H. Freeman and Co., New York, 1993.

neath continents and island arcs along subduction zones (inclined shear surfaces), seismicity is situated at progressively greater depths as one proceeds inland. For these reasons, a map of the continents exhibits a much more complicated, broad, zonal distribution of earthquake epicentral distribution. Nonetheless, plate boundaries and subdomains are also defined, albeit less distinctly, by on-land earthquakes.

Large earthquakes are episodic, with intensities roughly proportional to the time of shaking and the length of the crustal segment that has been ruptured in the specific seismic event. The *Richter scale* quantitatively represents the amount of vibrational energy released, by referring the quake intensity to the response of a standard seismometer normalized for its distance from the hypocenter. The Richter scale in arbitrary units involves an energy increase of approximately

32 – not 10, as is commonly inferred – between successive numbers: thus an earthquake of magnitude 6 releases approximately 30 to 35 times as much seismic energy as a magnitude 5 earthquake, a 7 nearly 1000 times as much as a magnitude 5, and so forth. So-called giant earthquakes have magnitudes of approximately 7.5 or greater and, fortunately, occur much less frequently than small seismic events. Ranging from I to XII, the *modified Mercalli scale* relates earthquake intensity to its effect as felt in populous areas by humans and their constructs. A descriptive comparison of both methods of quantifying seismic intensity is presented in Table 25.2.

To illustrate the principles, we consider a prime example of earthquake country – California. As discussed previously, California represents a region where the North American margin has overridden the Pacific spreading system (the East Pacific Rise). The plate tectonic setting is shown diagrammatically in Figure 25.2. Although dynamic divergent and convergent plate junctions are situated near the northern and southern extremities of the state, most of the continental crust in California is fragmented by a system consisting of subparallel northwest-trending strike-slip faults. The San Andreas is one of the most active strands and certainly is the best known; however, as Figure 25.3 illustrates, it is only one of many major breaks in a state transected by innumerable major and minor faults. During California's brief recorded history, devastating quakes have occurred along the San Andreas fault in the vicinity of San Juan Bautista (1800), Cajon Pass (1857), San Francisco (1906), and Loma Prieta (1989). The most severe shaking recorded, however, took place near Lone Pine

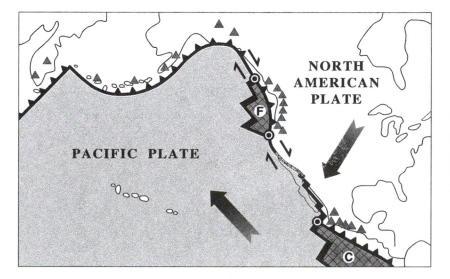

Figure 25.2. Plate boundaries and andesitic volcanic belts in western North America. The correlation of arc volcanism (open triangles) with zones of lithospheric underflow (barbed junctions) is clear. F = Farallon plate; C = Cocos plate. (*Source:* Adapted from R. V. Ingersol, Triple-junction instability as cause for late Cenozoic extension and fragmentation of the western United States, *Geology*, vol. 10, 1982.)

Figure 25.3. Fault map of California. New strands are found every year; undoubtably, many more major faults exist than those that have been recognized and mapped.

(1872) along the eastern Sierra Nevada–Owens Valley escarpment. Major temblors have struck throughout the state on other faults as well. Important seismic events include earthquakes at Santa Barbara (1812), Long Beach (1933), Bakersfield (1952), Hayward (1911 and 1984), Indio (1968), San Fernando Valley (1971 and 1994), Coalinga (1983), Whittier Narrows (1987), and Landers (1992). Obviously, the San Andreas, although important, is not the only earthquake-prone fault within California. Undoubtedly, many breaks capable of major shaking have not yet been recognized, let alone studied.

No area within California is more than a few tens of kilometers from the locus of historic seismic activity; hence, the entire state must be considered to be at risk. For better perspective, we also need to consider California within the broader framework of the Pacific Rim. Lessons learned in California can be applied throughout the Circum-Pacific, and vice versa. Ground shaking and loss of life and property have been constant dangers in the past, as evidenced by the devastation at Kobe, Japan (1995), Anchorage, Alaska (1964), San Fernando Valley, California (1971), and Mexico City (1985) (Figure 25.4). Because of the inexorable, continuing differential movements of lithospheric plates around the Pacific Rim, earthquakes assuredly will accompany us into the future.

The October 17, 1989, Loma Prieta earthquake, magnitude 7.1 on the Richter scale, provides an instructive example of the sort of earthquake hazard facing all dwellers along the Pacific Rim. At 5:04 P.M. that Tuesday afternoon, as much of

the local populace was settling back to watch a World Series baseball game about to begin in South San Francisco, the shaking began. Damage to freeways and buildings was extensive in coastal municipalities such as Santa Cruz, Oakland, and the Marina District of San Francisco, because vibration is intensified in weak, water-saturated muds and poorly consolidated landfill. (This phenomenon, known as liquefaction, is covered later in this chapter in the section dealing with mass wastage.) Landslides occurred in nearby mountains, reflecting slope instability. Sixty-three people were killed, 3757 were injured, and property damage totaled more than $6 billion. Considering the seismic energy released, the loss of life was minimal, partly because the epicenter was distant from major population centers; moreover, many people were at home rather than on the freeway or at work. Lower residential compared to industrial earthquake risks reflect the fact that flexible houses and apartments constructed principally of wood are less prone to catastrophic failure than are the larger unreinforced masonry and concrete buildings, freeways, and bridges. Perhaps more important, the duration of ground shaking was approximately twenty seconds, remarkably short for an event of this magnitude.

Most earthquake energy is released through the mechanism of rupture, initiating at one end of the fault segment undergoing rapid displacement, then propagating to the other end. In the Loma Prieta event, however, the initial break occurred near what became the middle of the active slip surface, and the rupture propagated both northwest and southeast along the San Andreas fault. Thus, vibrational energy was released in a little over half the time one would expect for an event of the measured intensity. Consequently, in most cases the unusually brief time interval of shaking was insufficient to allow vulnerable structures to commence swaying in synchronism with the arriving earthquake waves. Had the oscillation of human constructions achieved the natural seismic frequency as a result of longer shaking, many more buildings and bridges would have collapsed. Such fears were realized in the San Fernando Valley (Northridge) earthquake of 1994. In this temblor, the duration of intense vibration was more than thirty seconds. Although the magnitude was approximately 6.7 on the Richter scale, the epicentral location of this earthquake within a heavily populated area ensured severe damage, estimated as approaching, perhaps exceeding, $15 billion. Fortunately, relatively few lives were lost (thirty-three) because the event took place in the early morning hours while most people were asleep at home. Why

Figure 25.4 (opposite). Circum-Pacific earthquake damage due to loss of foundation stability. (A) Destruction in downtown Niigata, Japan, 1995; (B) devastated Government Hill School, Anchorage, Alaska, 1964; (C) on following page freeway bridge collapse, San Fernando Valley, California, 1971; and (D) total failure of a twenty-one-story office building, Mexico City, Mexico, 1985. (*Source:* Courtesy of National Geophysical Data Center. A: United States Geological Survey photograph. B, C, D: National Oceanic and Atmospheric Administration photographs.)

(A)

(B)

(C)

(D)

did this quake of somewhat lesser magnitude cause more extensive damage than did the Loma Prieta event? Also, was the Northridge event the main shock, or merely the precursor of an impending giant earthquake?

A minimum of six potentially dangerous strands of the San Andreas–Hayward fault system transect the San Francisco Bay area (Fig. 25.5). The U.S. Geological Survey estimates a 60 to 70 percent probability for a seismic event of similar intensity to that of the Loma Prieta earthquake occurring along one of these faults within the next three decades. As indicated on the map, other, earthquake-prone faults also lie within this region. Because of the proximity of these fault segments to major population centers, it is likely that such a temblor will cause at least an order of magnitude more casualties and property damage than the recent San Fernando Valley quake. In order to maximize the chance of survival in earthquake country, individuals and organizations need to plan for the inevitable "big one." Such planning makes especially good sense in light of the fact that the region is undoubtedly riven by more active faults than we have been able to detect thus far. Reinforcement of existing structures and the secure storage of emergency supplies are among the activities that should be undertaken prior to the onset of a major seismic event. The question is not whether such shaking will occur, but when! As of now, scientists have no answer to this question.

VOLCANISM IN THE PACIFIC NORTHWEST. Modern volcanic activity, predominantly of the andesitic-dacitic type (see Chap. 5, and Fig. 6.7), frames major segments of the Pacific Rim. Great stratovolcanoes stand along the margins of the basin, as in the Andes, High Cascades, Alaska-Aleutian chain, Kamchatka, northeastern Japan, North Island, New Zealand, and elsewhere along Circum-Pacific continental

Figure 25.5. Estimated likelihoods of earthquakes (Richter magnitude 7 or greater) related to specific faults in the San Francisco Bay region. Column heights are proportional to probabilities. Letters on columns indicate reliability of the forecast, on a scale of A to E, with A being most reliable, E the least. M = magnitude; P = probability. The San Gregorio, Calaveras, West Napa, Greenville, and Concord/Green Vally faults are excluded from this assessment of seismic hazards. (*Source:* Courtesy of United States Geological Survey.)

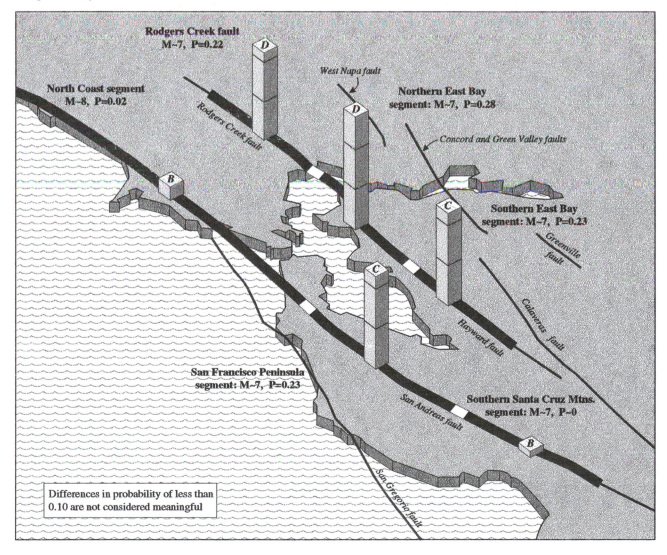

margins and island arcs. In all cases, the volcanism is sited on stable, nonsubducted plates directly above descending lithospheric slabs. Offshore trenches are the surface manifestations of these convergent plate junctions, with earthquake hypocenters outlining the landward-dipping seismic shear zones. Gaps in the volcanic chains are characterized by transform plate junctions (involving strike-slip lithospheric motion rather than subduction), typical of most of present-day California. Confirm for yourself the localization of andesitic volcanoes shown in Figure 25.2; all are sited landward from – and above – the convergent plate junctions, and are absent in the vicinity of transform plate boundaries.

Subduction is occurring beneath the Pacific Northwest, from southernmost British Columbia to northern California. Active volcanoes, reflecting this lithospheric underflow, constitute the High Cascades, principally situated in western Washington and Oregon. They are symmetric cones of great beauty, Mount Rainier and Mount Shasta being two famous examples; their relatively uneroded shapes demonstrate that the addition of igneous materials to the superstructures keeps pace with erosion. Most volcanoes of this chain have been regarded as dormant; however the eruptions of Mount Lassen in 1914–15 and Mount St. Helens in 1980 emphatically prove that the High Cascades remain very much alive. Scientists have noted a lack of seismicity beneath this arc, a puzzling fact because earthquakes should accompany lithospheric plate descent and the resultant igneous extrusion (and inferred intrusion). One disquieting explanation is that this lack could represent an ominous lull preceding renewed, intense earthquake and volcanic activity.

Every volcano possesses its own set of responses to the ascent of magma from below. Commonly, an underlying basement fracture system acts as the supply route, or conduit, for a rising column of melt. As molten rock enters the superstructure, a volcanic edifice tends to inflate slightly because melt is injected as dikes and sills along previously collapsed fissures and fractures, adding material to the whole, and thus enlarging it. Flank eruptions and satellite cones are constructed where such tabular igneous bodies reach the surface in a liquid state. Within the throat of an active volcano, the cooling rate may temporarily exceed the supply of new magma, allowing the molten material to congeal and gradually form a solid plug. Such an impermeable cap traps gases escaping from the cooling melt, resulting in volatile buildup, eventual rupture, and evisceration of the magma chamber as the gases rapidly expand outward and upward – entraining tiny glass shards of quenched, fragmented silicate liquid (ash) as well as larger molten blobs (bombs).

An example of volcanic eruption history is provided by a typical and well-documented andesitic volcano, Mount St. Helens. In early 1980, increased heat flow and steam discharge indicated heightened igneous activity near the summit of this volcano. Growth of a largely solid dacite plug and accumulation of underlying magma generated a prominent, oversteepened bulge at Goat Rocks, on the northern flank of the vent. On the morning of May 18, this igneous blister

failed in a landslide triggered by the upward movement of largely solidified melt in the conduit. Abrupt removal of some of the overburden through mass slumpage at Goat Rocks resulted in a pressure decrease, which, in turn, allowed the release of an almost horizontal blast of volatiles and entrained particulate matter. Relationships are detailed in Figure 25.6, where the sequence of events and a map view of the altered topography are presented. The blast initially approached sonic speeds, blowing down forests and wreaking havoc over a 400-square-kilometer area. Such a directed blast is an irresistible force, releasing enormous amounts of energy – as demonstrated by the devastated logging camp shown in Figure 25.7.

Depressurization of the magma body beneath Mount St. Helens was succeeded a few minutes later by a cataclysmic eruption of the contents of the volcanic neck, graphically illustrated in Figure 25.8. Much like the rapid expulsion of carbon dioxide from solution when the cap is removed from a warm, shaken bottle of soda, the release of expanding gases previously dissolved in the melt carried an ash cloud to a height of 24 kilometers – well into the troposphere. This towering column of gas-driven particulates and aerosols entered the jet stream of the upper atmosphere and moved rapidly east and southeast. Its dispersal is depicted in the satellite photograph of Figure 25.9, taken approximately eight hours following the onset of the eruption. Wind-blown ash falling near the orifice and immediately downwind accumulated as a thick blanket along the volcanic slopes. Melting snow and rainfall soaked this unconsolidated material, thus increasing its mass while weakening it; the resultant soggy material rushed downhill as mudflows, causing extensive damage.

In all, nearly 1 cubic kilometer of magma, including the top of the preexisting volcanic edifice, were deposited downwind and locally on the flanks of the cone. In a geologic sense, this was a relatively minor eruption, yet it snuffed out the lives of sixty people. Since the cratering event of 1980, renewed showering of ash at Mount St. Helens has resulted in the buildup of a small satellite cone within the summit caldera of the self-decapitated mountain. An even more spectacular example of this explosive beheading is exemplified by Crater Lake, Oregon, where prehistoric Mount Mazama "blew its top" approximately 6600 years ago. In 1883, the most famous event of this sort destroyed the island of Kra-

---

Figure 25.6 (opposite). The 1980 eruption of Mount. St. Helens. The congealed melt cap preventing upward movement of volatile-bearing magma (A) was relieved by a series of earthquakes and northward-descending landslides (B), which in turn gave rise to a directed blast of volcanic gas and entrained debris. This lateral blast moved northward at high speed, initially regardless of obstructions; then, as its energy dissipated, flow became channelized by the local topography (C). Subsequently, meltwater-soaked ash moved rapidly downslope as mudslides, especially westward in the Tootle River Valley. km = kilometer. (*Source:* Adapted from S. W. Kieffer, Geologic nozzles, *Reviews of Geophsyics*, vol. 27, 1989.)

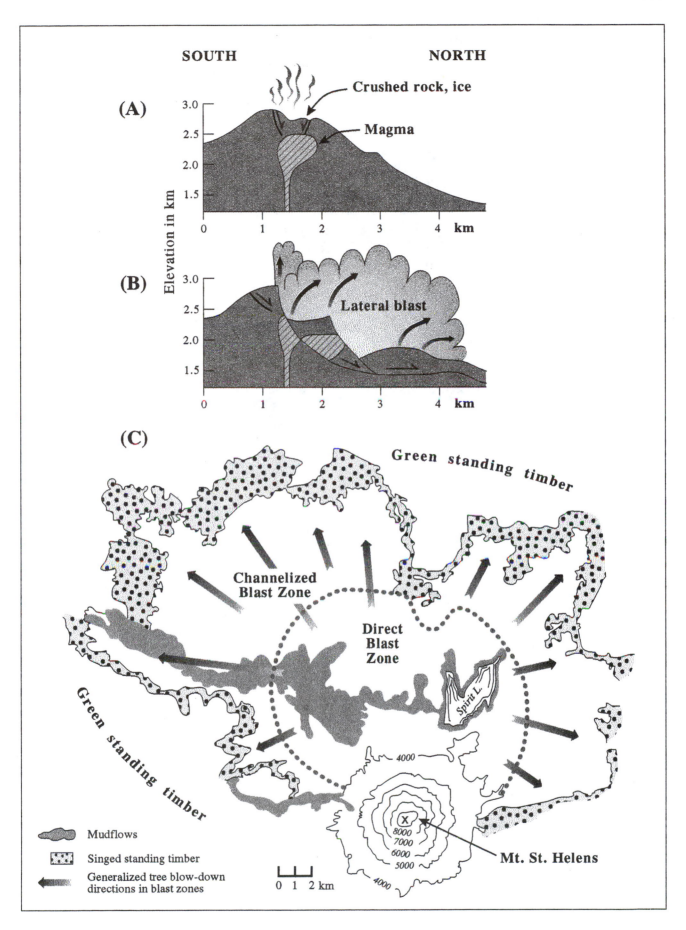

SOUTH   NORTH

(A)

Crushed rock, ice

Magma

Elevation in km

(B)

Lateral blast

(C)

Green standing timber

Channelized
Blast Zone

Direct
Blast
Zone

Spirit L.

Green standing timber

4000

8000
7000
6000
5000

4000

Mt. St. Helens

Mudflows

Singed standing timber

Generalized tree blow-down
directions in blast zones

0 1 2 km

Figure 25.7 (above). Wrecked logging camp on the north slope of Mount St. Helens. Trucks and stacked timber were strewn about like toys, reflecting the immense force of the directed volcanic blast. (*Source:* Courtesy of United States Geological Survey.)

Figure 25.8. May 18, 1980, eruption of Mount St. Helens, approximately five hours following the initial landslide at Goat Rocks, triggered the north-directed subhorizontal volcanic blast, succeeded in turn by the main-stage venting of ash shown here. (*Source:* Courtesy of United States Geological Survey.)

Figure 25.9. Satellite weather photograph of fine ash plume from Mount St. Helens, approximately eight hours after the eruption began. The volcanic dust is here shown being transported by the jet stream; some of the particulate matter rose as high as the tropopause. (*Source:* Courtesy of NASA.)

katoa, located between Sumatra and Java, killing more than 36,000 people; most deaths resulted from the associated tsunami. Volcanic particulates – including aerosols – were injected into the stratosphere and circled the Earth for more than two years, causing brilliant sunsets and a recognizable global lowering of temperatures. A similar climatic disturbance occurred in 1815, also known as the "year without a summer," following the eruption of Tambora, another Indonesian island volcano. The same worldwide cooling effect has been detected, but to a lesser extent, since the 1992 eruption of Mount Pinatubo in the Philippine Islands. Clearly, large eruptions can influence climate through the shielding of incoming solar radiation by volcanic dust and aerosols.

The events described above provide evidence that global climate may be influenced, if not controlled, by plate tectonic processes. The more rapid the sea-floor spreading, the greater the volcanic contribution of carbon dioxide, water vapor, and aerosols to the atmosphere, along both the basaltic mid-oceanic ridge systems and the largely andesitic island arcs and continental margins above subduction zones. The extent of this cause-and-effect relationship is not yet clear; however, in support of the hypothesized correlation, the Cretaceous Period was a time of unusual worldwide warmth, combined with especially rapid sea-floor spreading and inferred volcanism.

MASS WASTAGE. Land slumpage takes many forms; among them are rockfalls, rockslides, debris falls, debris slides, mudflows, avalanches, and landslides. All represent responses of unstable, oversteepened slopes to the unremitting pull of gravity. When the strengths of topographically elevated materials are exceeded, they move downslope by free fall or, more commonly, by bodily traction along the surface of the land; rates range from imperceptible creep to sudden avalanching. Seismic and/or volcanic events trigger some downslope movements; rainfall and water soaking cause others. Most take place simply because weakening of the Earth materials gradually occurs over the passage of time. Most of the entrained debris is composed of unsorted, chaotically mixed

Figure 25.8

Figure 25.9

(A)

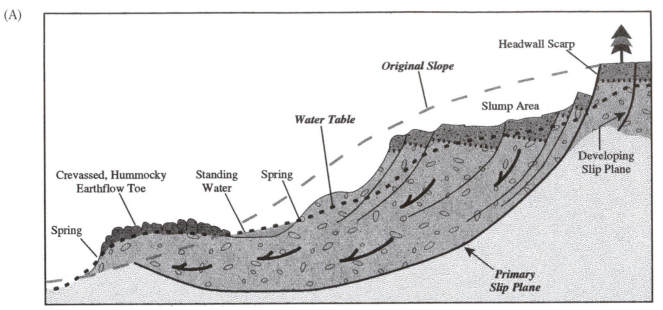

Figure 25.10. Schematic longitudinal section of (A) a small landslide, and (B) a perspective view of the same type of feature.

(B)

fragments and, depending on the manner of transport, ranges from coarse, meter-size blocks (rockfalls) to fine muds (mudslides), the average grain size of which is microscopic. Deposits are initially unconsolidated accumulations; the source area generally displays a pull-away scar or escarpment. A typical landslide is diagramed in Figure 25.10.

Swelling and settling also take place as soils respond, sponge-like, to changes in degree of hydration (that is, they expand and contract as a function of water content). Such processes can result in damage to lightly built structures, chiefly through tortional stress (warping). Of greater danger and considerably more concern is *liquefaction,* the property of water-saturated sediments and soils to loose coherence on shaking. Such pronounced weakening causes some Earth materials to quiver, on the application of stress, like a bowl of gelatin. Recent liquefaction events of a serious nature include those in Mexico City in 1985, and in San Francisco's Marina District in 1989, which occurred during the Mexico City and Loma Prieta earthquakes, respectively. Such violent shaking of the substrate causes the collapse of even sturdily constructed edifices.

The extreme form of natural hazard due to mass slumpage is embodied in debris and rock avalanches, landslides, and mudflows, as the following examples attest. A debris avalanche in the Peruvian Andes on May 31, 1970, raced down the Rio Shacsha Valley at speeds in excess of 300 kilometers per hour, destroying the towns of Ranrahirca and Yungay, killing more than 25,000 inhabitants. Continuing down the valley, it incorporated abundant water and mud and evolved into a rapidly moving mudflow that swept along the Rio Santa for a remarkable 160 kilometers before reaching the Pacific Ocean. A rocky, muddy wavefront 80 meters high was observed engulfing Yungay. Although the headwall of the Rio Shacsha Valley was not being monitored at the time, rockfalls and avalanches are frequent occurrences in this part of the High Andes.

A somewhat similar event took place in 1985 on the flanks of the active volcano Nevado de Ruiz, in Colombia. There, a minor summit eruption caused the widespread melting of glacial ice on the flanks of the mountain. As a result, ash-laden mudflows streamed down into the densely populated valley below, causing nearly 23,000 fatalities. In this case, scientific warnings were ignored by officials and the local people due to a breakdown in communications. Not all killer mudflows are so massive: On a far smaller scale, a destructive mudflow was associated with the eruption of Mount St. Helens (see Fig. 25.6C).

CLIMATIC HAZARDS

TROPICAL STORMS ALONG THE PACIFIC RIM. As presented in Table 25.1, tropical storms represent the deadliest natural disaster faced by the nations of the Pacific Basin and Rim – at least, the death toll over the interval 1960–87 supports such a conclusion. Of course, these statistics are very much a consequence of the irregular incidence of truly great calamities, such as the especially severe typhoons that hit Bangladesh in 1970 (500,000 fatalities), in 1985 (100,000 fatalities), and in 1991 (138,000 fatalities). Immense, equatorial-ocean-spawned cyclonic systems such as these routinely develop over a period of several days to one week (see Chaps. 10 and 14). With the advent of worldwide surveillance on an hourly pass basis by the National Oceanographic and Atmospheric Administration (NOAA) weather satellites, such severe storm systems impose far lesser danger to humanity than they did prior to the beginning of the space age. The technologies now exist (satellite monitoring, engineering constructs) to protect lives and property; where they are applied, climatic hazard reduction is a fact of life. Unfortunately, meteorologic disasters continue to take place with alarming regularity in the developing nations, chiefly reflecting their financial inability to implement preventative measures.

First it is important to understand the natural phenomenon. What terrestrial conditions cause the generation of intense tropical storms? Wind velocities in the atmosphere are a consequence of the flow of air from high-pressure regions. The actual transport direction is influenced by the Coriolis "force." This effect, in turn, is a function of the Earth's rotation (refer to Chap. 14), which tends to deflect winds clockwise in the Northern Hemisphere, and counterclockwise in the Southern Hemisphere. The stronger the pressure gradient, the higher the wind velocity, and the greater the curved path followed by the atmospheric flow; if the deflection is sufficiently marked, the regional air currents tend to assume a whirlpool-like form, or *vortex*. In addition, solar heating of the tropical oceans and attendant sea – atmosphere energy transfer causes moist, warm, buoyant air to rise rapidly. These two phenomena combine forces to create a low-pressure core in the swirling system, which effectively sucks replenishing surface winds into the vortex and substantially enhances its strength. The higher the temperature of the surface waters of the ocean, the more pronounced is this effect (why should this be so?). An extremely powerful vortex can thereby be

Figure 25.11. Areas of birth of tropical cyclones over the period 1952–71. Abbreviations, and average numbers of annual severe storms, are as follows: EA = eastern Asia (21); SP = southwestern Pacific (20); C = Caribbean (7); EP = eastern Pacific (6); BB = Bay of Bengal (6); M = Madagascar (6). A = Arabian Sea (2). and WA = western Australia (1). (*Source:* Adapted from E. Bryant, *Natural hazards*, Cambridge University Press, 1991.)

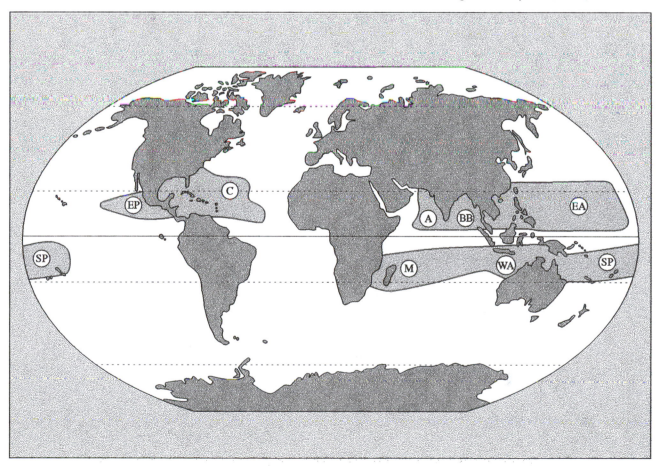

TABLE 25.3. WIND SPEED TERMINOLOGY

Wind Speed (Knots[a])	Wind Speed (km/hr)	International Description	U.S. Weather Bureau Description	Physical Aspect of the Sea
0	<1	Calm		
0–3	1–5	Light air	Light wind	Mirror surface to small wavelets
4–6	6–11	Light breeze		
7–10	12–19	Gentle breeze	Gentle–moderate	Large wavelets to small waves
11–15	20–28	Moderate breeze		
16–21	29–38	Fresh breeze	Fresh wind	Moderate waves, many whitecaps
22–27	39–49	Strong breeze	Strong wind	Large waves, many whitecaps, foamy sea
28–33	50–61	Moderate gale		
34–40	62–74	Fresh gale	Gale	High waves, foam streaks, and spray
41–48	75–88	Strong gale		
49–55	89–102	Whole gale	Whole gale	Very high waves, rolling sea, reduced visibility
56–63	103–117	Storm		
>63	>117	Hurricane	Hurricane	Sea white with spray and foam, low visibility

A knot is 1 nautical mile per hour or 1.15 statute miles, or 1.85 kilometers per hour

Source: N. Bowditch, *American Practical Navigator,* U.S. Navy Hydrographic Office, Publication 9, 1958.

produced. These features are known as hurricanes, typhoons, tornadoes, or cyclones, depending on the region in which they occur. The ocean regions spawning such violent tropical storms are illustrated in Figure 25.11. Note that, because these meteorological systems reflect the transference of near-equatorial solar heat to higher latitudes, global warming would increase their abundance and intensify their ferocity. Both the terminology of winds as a function of their velocities and the effect of wind speed on the ocean's surface are listed in Table 25.3.

CLIMATIC HAZARDS IN THE CONTERMINOUS UNITED STATES. In a developed and technologically advanced country such as the United States, monitoring and early warning of an impending tropical storm, combined with the potential for total evacuation of the populus, have vastly reduced the loss of life that might otherwise have resulted from intense climatic disturbances. As an example, the prediction and real-time tracking of Hurricane Hugo in 1989 by the U.S. National Weather Service is depicted in Figure 25.12. Within the United States, areas prone to coastal flooding, such as along the subsiding Eastern Seaboard and the Gulf of Mexico (the West Coast consists predominantly of an elevated, or emergent, coastline) are especially closely studied. At the same time, engineering projects have been undertaken to inhibit coastal erosion. Systematic cloud seeding and resultant precipitation are also employed to dissipate the energy of the most violent storms while such systems still lie well offshore. Finally, some local governmental agencies, having recognized the continuing threat of coastal flooding, have excluded development of land close to tidewater.

To place the danger from tropical cyclonic systems in perspective, the mortality in the lower forty-eight states over the period 1940–75 is illustrated in Figure 25.13. Deaths from lightning bolts exceed the tolls from other climatic hazards

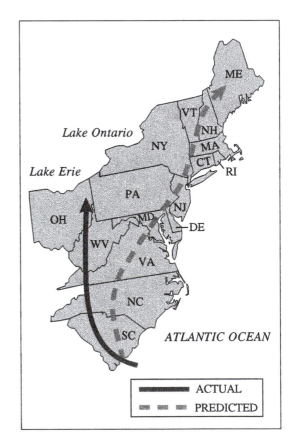

Figure 25.12. The track of Hurricane Hugo (September 22, 1989) was more westerly than expected because the storm came ashore faster and with greater intensity than had been forecast. Its strength gradually diminished as it moved inland. Prediction capabilities for meteorologic hazards have improved in recent decades; however, research and modernization of weather prediction facilities can still be effected, providing increased accuracy and lead time critical to decision makers responsible for implementing evacuation plans. (*Source:* Courtesy of National Oceanographic and Atmosphere Administration.)

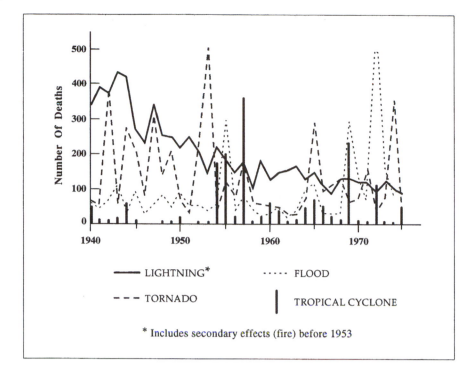

Figure 25.13. Number of deaths resulting from lightning (includes secondary effects (e.g., fire) prior to 1953), tornadoes, floods, and tropical cyclones in the United States, 1940–75. (*Source:* Adapted from J. R. Eagleman, *Severe and unusual weather*, Van Nostrand – Reinhold, New York, 1983.)

such as tornadoes, flooding, and tropical storms. In any case, all these natural disasters are associated with mortality rates ranking well below those of potentially catastrophic geologic hazards such as earthquakes, volcanic eruptions, landslides, and avalanches. The likelihood of flooding and lightning strikes is shown for the conterminous United States in Figure 25.14. A rough correlation between the annual frequency of thunderstorms, flooding, and forest fires can be surmised from this map – with forest fires and flash floods most prevalent in the Rocky Mountains, and regional inundation most common in the Deep South. Because of the relatively sophisticated level of weather forecasting in the United States, these regional and local hazards do not constitute as serious a threat to life and property as do other types of natural disaster.

Figure 25.14. Annual frequency of days of thunderstorm and other severe weather in the United States. (*Source:* Adapted from J. R. Eagleman, *Severe and unusual weather*, Van Nostrand–Reinhold, New York, 1983.)

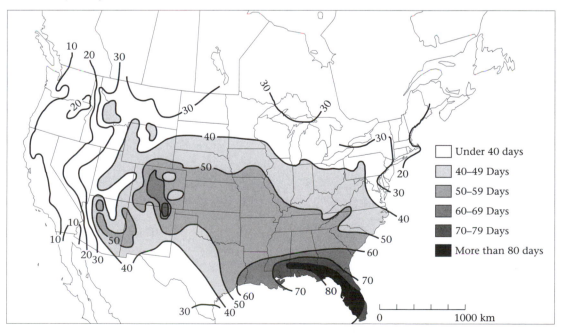

COASTAL FLOODING: BANGLADESH AS AN EXAMPLE.
For less developed nations, efforts to protect life and property from meteorologic hazards are beyond economic reach. Accordingly, even under the most favorable circumstances, when early warnings of an impending disaster have been received and acted upon, the population remains vulnerable to the vicissitudes of tropical cyclones, tsunamis, and seawater incursion. Similar to the Andean debris avalanche and mudflows, typhoons in Southeast Asia exact a terrible tribute in terms of lost lives and property. Up-to-the-minute monitoring, more efficient emergency planning, and better construction practices can make a significant difference, but only if the populus is well informed regarding both the dangers and effective survival strategies.

The tropical storm hazard exposure in Third World countries such as Bangladesh stands in stark contrast to the relatively mild exposure in the United States. The delta of the combined Ganges and Brahmaputra rivers (Fig. 25.15), an extraordinarily fertile region, represents a stupendous resource for its inhabitants. In the early twentieth century, when Dutch engineers designed dams and levees for new land, now a part of Bangladesh, the reclamation of low-standing marshland from the grasp of the sea was hailed as a technologic triumph for the local economy, which was based on subsistence agriculture. Due to intense overpopulation in the adjacent region, the drained land was rapidly occupied and intensively cultivated. Unfortunately, construction of the embankment system did not include protection from storm surges (which would have been prohibitively costly). Accordingly, a burgeoning population was placed in certain jeop-

ardy – as mentioned previously, tropical cyclones frequent the Bay of Bengal (Fig. 25.11).

NOAA satellites tracked an especially energetic typhoon three days prior to its arrival in the delta area in November 1970. (Figure 25.15 documents the path of this storm and the attendant swath of devastation.) During the night, a wall of water 15 meters high overran the delta, drowning approximately 500,000 people and 1 million head of cattle and other livestock. Although local authorities knew of the impending danger, they failed to issue warnings until too late. In addition to the awesome loss of life, rice paddies were obliterated or fouled with salt water, and heavy seas extensively eroded the reclaimed, low-lying agricultural land. Recovery occurred gradually; however, a mere 15 years later, a similar event claimed another 100,000 lives. And, during April 1991, yet another 138,000 people died when a deadly typhoon came ashore in this same area, inflicting economic losses estimated at $1.4 billion. Are such calamities inevitable?

Although reinforced, multistory, multipurpose typhoon shelters are now being constructed by the Bangladesh government, the delta and coastal lowland rice paddies remain largely unprotected from tropical storms because of a lack of shielding engineered structures such as dikes and levees. Because of chronic overpopulation, the vulnerable land is teeming with an influx of uneducated agricultural workers, most of whom have had no experience with severe tropical cyclones or coastal inun-

Figure 25.15. Track of the deadly typhoon of November 13, 1970. (A) The regional setting of the Bay of Bengal. (B) The resultant coastal flooding (stippled pattern) of the delta section of Bangdalesh and the path followed by the storm. (*Source:* Adapted from E. Bryant, *Natural hazards*, Cambridge University Press, 1991.)

(A)

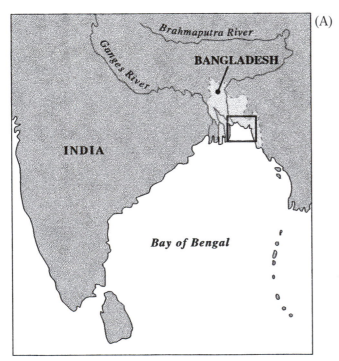

(B)

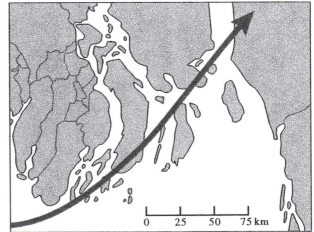

dation. Most important, public information systems remain woefully inadequate in terms of length of warning time prior to the arrival of the climatic hazard and efficacy in reaching the threatened populus. Moreover, the farmers lack the means for rapid transportation and evacuation from the danger zones, and most are reluctant to abandon their property because of the possibility of looting. Religious beliefs also decrease the mobility of women – and therefore families – during certain holy seasons. Hence, the danger of coastal flooding will remain a problem for Bangladesh for a variety of reasons, not the least of which is a fatalistic tendency to regard such intense storm systems as "acts of God" not amenable to human intervention.

NATURAL HAZARD PREDICTION

From a natural hazard mitigation standpoint, population centers should not be allowed to expand into dangerous regions; however, many do anyway because other considerations – including history, and climatic, geographic/topographic, socioeconomic, and political advantages – are thought to outweigh the perceived hazards. Taipei, Osaka, Kyoto, Tokyo, Vancouver, Seattle, San Francisco, Los Angeles, Mexico City, Lima, and Santiago are but a few of the world's major cities that are growing rapidly despite high risk from impending coastal flooding, tsunamis, earthquakes, volcanic eruptions, and/or land slippage. Long-term prediction is of value principally in order to upgrade building codes and to ensure the observation of stringent, safe construction practices in the future, and to retrofit older structures. Public awareness and an understanding of the problems associated with natural hazards is required for the process of realistic land-use planning to be effective. These and other municipalities have responded to the perceived dangers in varying ways and to differing degrees; how successfully they cope with their hazardous futures remains to be seen.

To reduce the loss of life and to lessen socioeconomic disruption, humankind must be able to predict impending disasters in a timely fashion. This applies to earthquakes, seismic sea waves, volcanic eruptions, landslides, avalanches, and related geologic hazards, as well as to other potential natural calamities, such as tropical cyclones, forest fires, droughts, and coastal flooding. Only by understanding the nature of the dangers can we take preventative action to ameliorate their effects, or to avoid them altogether. In pursuit of this understanding, several intergradational time scales are important – each associated with specific opportunities for saving lives and for damage mitigation.

Scientists are now able to provide accurate warning, on a time scale of hours to days, for tropical storms, river and coastal flooding, spread of forest fires, and tsunamis. Similar capabilities for short-term geologic hazard prediction would allow the evacuation of people and valuable mobile property from imperiled areas and unsafe human constructions; however, accurate forecasting requires the timely recognition of unambiguous premonitory phenomena. If prior emergency hazard plans have proved adequate through simulation testing, orderly evacuation should be possible but would not save stationary facilities at risk. Only realistic long-range planning and wise site choice can reduce potential damage to these structures. Unfortunately, the timely prediction of most geologic disasters is not yet feasible, and a far better understanding of such Earth processes is required in order for our forecasting to become reliable.

Long-term geologic hazard prediction – for example, on a time scale of years to centuries – provides the opportunity for planning land development and human site occupancy in an informed manner. Where areas prone to earthquakes, volcanic eruptions, floods, or avalanches are recognized, delineated, and understood scientifically, we can refrain from constructing critical facilities such as hydroelectric or nuclear power plants, petrochemical refineries, or chemical manufacturing plants in the vicinity. Schools, hospitals, and emergency control centers, including police and fire stations, should be sited in low-risk areas. Of course, in many cases, development was initiated long before the hazardous nature of the environment was fully appreciated, a fact that exacerbates both the extent of potential damage and length of recovery time.

TSUNAMIS AND CLIMATIC HAZARDS. With the advent of weather satellites, reliable short-term typhoon, hurricane, thunderstorm, and coastal sea-incursion warnings are a global reality today. Thus, timely predictions of climatic hazards are sharply reducing the mortality rates from such events in well-prepared regions. Moreover, the installation of worldwide seismometer and tide gauge networks have increased the advance knowledge of approaching tsunamis. Such warnings generally provide several hours' to a day's time for taking evasive action. Forest fires are well monitored from satellites as well as locally, and are combated effectively from both ground-level and airborne platforms. The most serious obstacle to saving lives threatened by imminent climatic disasters is inefficiency in the public communication system. Also, education and preparedness must be markedly improved.

GEOLOGIC HAZARDS. Although long-term prediction of eruptions, landslides, and earthquakes (e.g., over the next fifty years) is now a reality, great potential for loss of life and property remains. Why is this? Consider the fact that the likelihood of a disaster striking in anyone's lifetime is rather low; therefore, most populations are unwilling to provide the necessary resources for the research and development required to mitigate the effects of an inevitable natural calamity in the possibly far-distant future. Who wants to pay for protection they themselves may not need? Accurate short-term warning is not yet attainable because of the extremely complicated nature of the geologic hazards themselves. Moreover, predictions may result in severe social and economic disruptions that, if inaccurate, could generate loss of public confidence, along with a strong political backlash. Therefore, geologic hazard prediction studies are proceeding cautiously and carefully. What sorts of things are being done?

■ **Seismic Events.** Earthquake frequencies along a particular segment of an active fault may be approximated by studying and determining the ages of topographic features (see Chap. 7). For example, *escarpments* (long, low cliffs) are commonly produced by differential Earth movements across an active slip zone; degradation of these scarps by the agents of erosion proceeds at observable, measurable rates. Thus, if several different cliffs are present in varying stages of surficial modification, estimates of the earthquake recurrence interval can be made. Providing similar clues, *sag ponds* are ephemeral pull-apart features developed along fault traces immediately following a major seismic event. Where multiple layers of sedimentary bog deposits occur, the radiometric dating of preserved carbonized woody fragments may allow an estimation of the local earthquake episodic history.

One particularly useful observation is that stress accumulation along a particular fault is proportional to the regional slip rate. The length of the eventual rupture, fault offset (strain), and both duration and intensity of shaking during local stress release are proportional to the stress buildup.

Therefore, if we know how fast the latter accumulates, we can calculate the length of time between successive seismic events of a given magnitude and fault displacement. Suppose stress buildup occurs at a rate of 2 centimeters per year. What would be the average recurrence interval for earthquakes producing 5-meter offsets? Of course, a 250-year average frequency might be reflected in widely varying intervals between specific seismic events.

Active faults are being monitored quantitatively through a variety of techniques. Differential movements of crustal blocks, fluid and gas contents of fault-bounded rocks, fracture propagation mechanisms in different geologic media, and numerous other physical parameters are being measured in shallow bore holes, at the ground surface, in the laboratory, and by Earth-orbiting satellites. Certain portions of faults slip episodically but rarely, producing great releases of vibrational energy, whereas other portions undergo nearly constant creep. Both mechanisms are crustal responses to the continuous, inexorable, differential motions of the lithospheric plates. For example, seismic activity along the San Andreas fault on the San Juan Bautista–San Francisco segment during the twenty years leading up to, and the time immediately following, the Loma Prieta event is illustrated in Figure 25.16. Prior to the earthquake, the fault was "quiet"

Figure 25.16. Seismicity along the central section of the San Andreas fault, 1969–89. (A) Seismicity prior to the Loma Prieta earthquake. (B) The main event, plus aftershocks of the Loma Prieta earthquake. The size of the symbols increases with increasing earthquake magnitude. Rectangles in section (B) indicate the location of the Loma Prieta earthquake rupture inferred from geodetic data. (*Source:* Adapted from P. L. Ward and R. A. Page, The Loma Prieta earthquake of October 17, 1989, USGS Survey Pamphlet, January, 1990.)

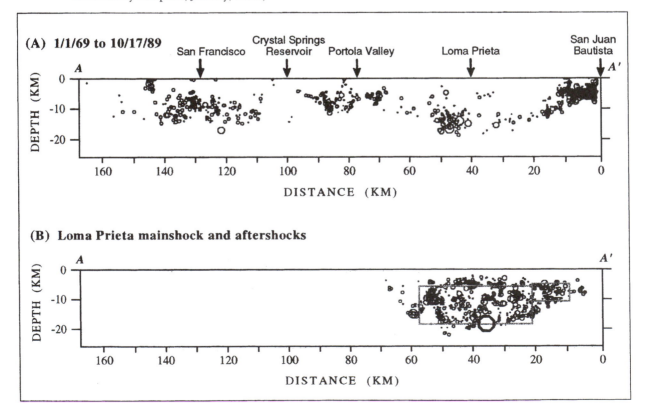

in the vicinity of the 1989 rupture; immediately following the Loma Prieta event, the *seismic gap* was filled in with a swarm of aftershocks.

If specific fault segments have been unusually inactive for a long time, or if a fault historically observed to generate an earthquake at periodic time intervals is due, the prediction of a future seismic event seems appropriate. Such a situation applies to the Parkfield area of central California. Since 1857, when records began to be maintained in this area, six moderate-intensity earthquakes have occurred, with recurrence intervals of approximately twenty-two to twenty-three years. The most recent seismic event took place in 1966. As shown in Figure 25.17A, the next quake was expected in 1988 but has not yet occurred. Stress accumulation is continuous at approximately 2.8 centimeters per year; Figure 25.17 B illustrates this deformation buildup as well as the amount of past slippage on the fault, averaging approximately 60 displacement per seismic event. Parkfield is heavily instrumented and under continuous monitoring by the U.S. Geological Survey because seismologists expect that a moderate-intensity quake will occur very soon on this portion of the San Andreas. If precursor phenomena (e.g., unusual ground tilts, abrupt changes in water level or gas release in wells, strain buildup in rocks, or cracking) are detected at Parkfield just prior to a seismic energy release event, such observations may be utilized in predicting other impending temblors. To date, the anticipated earthquake is long overdue.

■ **Volcanic Eruptions.** An increase in issuance of gases, hot spring activity, and/or measured heat flow near the summit of a volcanic vent, inflation of the volcanic edifice, ground tilting, and volcano-induced seismicity are all indicators suggesting the accumulation of magma beneath and in the conduit, and possible onset of an eruptive cycle. Instrumentation emplaced around Mount St. Helens in Washington state, and Kilauea volcano on the big island of Hawaii, provide case studies of a typical andesitic stratovolcano and a basaltic shield volcano, respectively. Which type do you suppose is more explosive (hence, dangerous), and why? Viscous andesitic/dacite melts contain considerable amounts of dissolved volatiles, whereas the hotter, more fluid basaltic magmas are quite low in gassy constituents.

■ **Mass Wasting.** Detailed surveying of ground-level, downhill movement of soil and bedrock, and abruptly fluctuating water levels and/or gas emissions in wells may provide warning of slow creep or impending onset of rapid landslide and avalanche activity. Instruments aptly termed *strainmeters* detect imperceptible movement of destabilized rock and soil. Such premonitory features need to be measured quantitatively and synthesized so that these data may be employed in the prediction of future mass downslope movements. Of course, many landslides and mudslides are induced by ground shaking in a seismic event, or by a volcanic eruption. Others reflect the decrease of frictional resistance to flow, and increase in density occasioned by aqueous seepage during rainfall and/or flooding.

CONCLUSIONS

In these and other examples of geologic hazards, environmental Earth scientists are trying to quantify the nature of the precursor indicators of an impending major natural disaster. They hope to provide accurate intermediate-and, especially, short-term warning. Progress toward the realization of these goals becomes ever more urgent as the Earth's human population burgeons, and as people occupy more marginal, hazardous environments in greater concentrations. Nowhere is this problem more acute than along the Pacific Rim, a region of spectacular sociopolitical, economic, and cultural growth. As the populations of these nations expand, they are exposed to special risks. The Circum-Pacific is a region of

Figure 25.17. Historic seismicity of the San Andreas fault near Parkfield, California. (A) Earthquake recurrence intervals are indicated by filled circles, with the nominal predicted date of the next event shown by a bull's eye. (B) Instantaneous slip during previous earthquakes (heavy downward-pointing arrows) and gradual strain buildup (inclined dashed lines). As of late 1998, the anticipated Parkfield earthquake is embarrassingly tardy, and is now nine years overdue! cm = centimeters. (*Source:* Data from United States Geological Survey.)

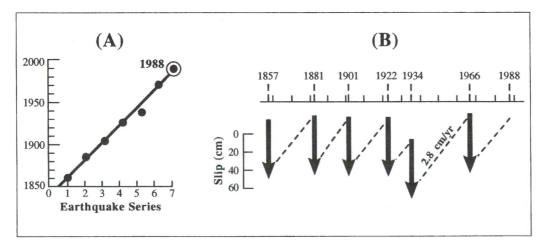

intense mountain-building activity; hence, geologic hazards pose a serious, ongoing threat – one that is heightened as marginal land utilization increases. The Pacific Rim also spawns the most energetic typhoons and hurricanes as well as seismic sea waves; hence, coastal flooding is intensified within and along the borders of this ocean basin.

The Circum-Pacific region is where the problems of natural hazard prediction and amelioration are most acute, and where practical solutions are most urgently needed. Better education, informed land use, enhanced public warning systems, and safe construction policies are all necessary in order to minimize the adverse effects of natural disasters, even assuming the increased ability to forecast accurately an impending catastrophe in terms of its time, site, and intensity. Such sociopolitical policies can be implemented, however, only as populations at risk come to understand the true cost of inaction.

QUESTIONS

1. Which of the natural disasters discussed in this chapter seems most easily mitigated, in terms of time? Money? Decreased threat to life? Ease of public awareness and willingness to plan for mitigation? Which appears to be most intractable, and why?

2. In a region with which you are familiar, identify three geologic and/or climatic hazards. Pick one and, as spokesperson for the local Natural Hazards Mitigation Board, indicate how you would proceed in order to alleviate its dangers to the populace. What obstacles might hinder the adoption of your recommendations by local decision makers?

3. Why is the Circum-Pacific region at significantly greater risk from natural disasters than most other areas – for example, the Atlantic coast? What special precautions would you take with regard to buying a house or starting up a business in some specific Pacific Rim location?

4. Describe and justify the emergency plans you would recommend for your family and for your community in order to survive a geologic catastrophe if you resided in one of the following cities: Hilo, Hawaii; Tokyo, Japan; Los Angeles, California; Dallas, Texas; Santiago, Chile; London, England; Mexico City, Mexico. Identify the hazards, their attendant risks, and both the likelihood and magnitude of potential damage.

5. Earthquake forecasts generally consist of four elements: likelihood, location, intensity, and timing of the seismic event. Compare the specific societal value to the citizens of San Francisco of the following predictions of an impending magnitude 8 seismic event on the Richter scale, centered beneath the Golden Gate Bridge: 10 percent probability, six hours; 67 percent probability, twenty to thirty years; 50 percent probability, two to three days.

FURTHER READINGS

Bryant, E. 1991. *Natural hazards.* New York: Cambridge University Press.

Eagleman, J. R. 1983. *Severe and unusual weather.* New York: Van Nostrand–Reinhold.

National Research Council. 1993. *Solid-earth sciences and society.* Washington DC:National Academy of Sciences.

Tilling, R. I. 1989. *Volcanic hazards and their mitigation: progress and problems. Reviews of Geophysics* 27: 237–269.

Wallace, R. E. (ed.) 1990. *The San Andreas fault system, California.* U.S. Geological Survey Professional Paper 1515, 283 pp.

Wright, T. L., and Pierson, T. C. 1992. *Living with volcanoes.* U.S. Geological Survey Circular 1073, 57 pp.

Figure 3.9. **(B)** Artist's conception of the instant of impact.

Figure 4.1. Hubble Space Telescope image of the core of galaxy M100, one of the more prominent members of the Virgo Cluster of galaxies. M100 is very similar to our own galaxy (the Milky Way) in its overall spiral arrangement of luminous matter (stars). (*Source:* Courtesy of NASA, 1995.)

Figure 4.9. Familiar Apollo 17 image of the Earth, with coverage extending from the Mediterranean Sea to the Antarctic ice cap. The blue planet, as it is sometimes called, contains H_2O in three physical states, solid, liquid, and gaseous (why is this?), dramatically illustrated in this photograph. (*Source:* Courtesy of NASA.)

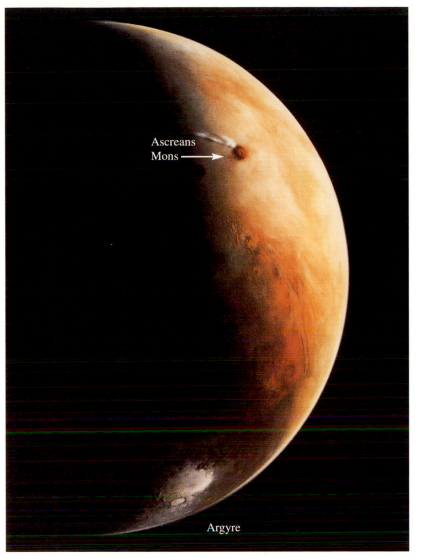

Ascreans
Mons →

Argyre

Figure 4.10. *Viking Orbiter 2* photographs of Mars. (A) At the top, with carbon dioxide ice on its western flank, is the giant Martian volcano, Ascreans Mons. The large, frost-encrusted, impact crater basin, Argyre, is located near the South Pole. (B) General view of cratered surface; the original shows the intense red color of Mars. (*Source:* Courtesy of NASA.)

Figure 4.14. Relief map of the surface of the solid Earth. Our planet is divided into two major physiographic provinces – continents and ocean basins. The ocean basins are full "to the brim" with seawater (in fact, overly so); the continental shelves are submerged portions of the high-standing continents and islands that lie below sea level. (*Source:* Courtesy of B. C. Heezen and M. Tharp, *World ocean floor,* U.S. Navy, 1977.)

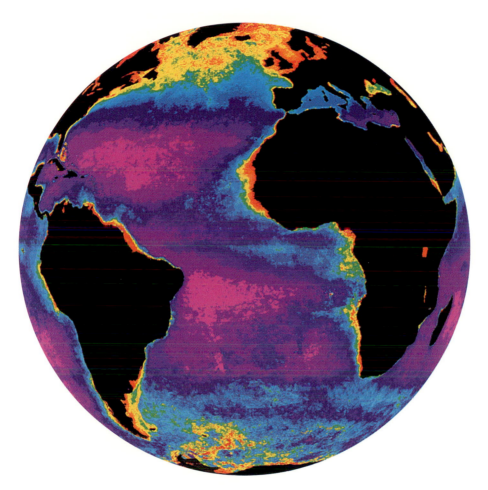

Figure 12.2. Distribution of phytoplankton in the ocean. Produced by scientists at the NASA Goddard Space Flight Center from data taken by the Coastal Zone Color Scanner on the *NIMBUS 7* satellite, the image is an ensemble of data from different seasons. Red, yellow: high concentrations; blue, purple: low concentrations. (*Source:* Courtesy of NASA.)

Figure 13.1, Part 1. The atmospheres of (A) Earth, (B) Mars, and (C) Venus differ immensely and determine the surface conditions and landscape. Only the atmosphere of Earth is greatly out of chemical equilibrium and is conducive to life. This disequilibrium is a signal of life that is probably detectable to any potential observers nearby in our galaxy. (*Source:* All photographs, with the exception of (A), courtesy of NASA. (A) courtesy of Catherine Flack.) Atmospheric pressures (proportions) are: Earth, 1 bar (N_2 = 77%; O_2 = 21%; Ar = 1%; H_2O ~ 1%); Mars, 0.007 bars (CO_2 = 95%; N_2 = 2.7%; Ar = 1.6%); Venus, 90 bars (CO_2 = 96%; N_2 = 3.5%).

(A)

(A)

Figure 13.1, Part 2. The atmospheres of (A) Earth, (B) Mars, and (C) Venus differ immensely and determine the surface conditions and landscape. Only the atmosphere of Earth is greatly out of chemical equilibrium and is conducive to life. This disequilibrium is a signal of life that is probably detectable to any potential observers nearby in our galaxy. (*Source:* All photographs, with the exception of (A), courtesy of NASA. (A) courtesy of Catherine Flack.) Atmospheric pressures (proportions) are: Earth, 1 bar (N_2 = 77%; O_2 = 21%; Ar = 1%; H_2O ~ 1%); Mars, 0.007 bars (CO_2 = 95%; N_2 = 2.7%; Ar = 1.6%); Venus, 90 bars (CO_2 = 96%; N_2 = 3.5%).

(B)

(B)

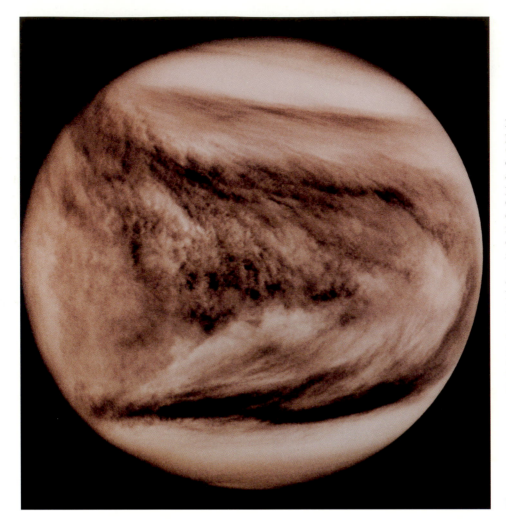

Figure 13.1, Part 3. The atmospheres of (A) Earth, (B) Mars, and (C) Venus differ immensely and determine the surface conditions and landscape. Only the atmosphere of Earth is greatly out of chemical equilibrium and is conducive to life. This disequilibrium is a signal of life that is probably detectable to any potential observers nearby in our galaxy. (*Source:* All photographs, with the exception of (A), courtesy of NASA. (A) courtesy of Catherine Flack.) Atmospheric pressures (proportions) are: Earth, 1 bar (N_2 = 77%; O_2 = 21%; Ar = 1%; H_2O ~ 1%); Mars, 0.007 bars (CO_2 = 95%; N_2 = 2.7%; Ar = 1.6%); Venus, 90 bars (CO_2 = 96%; N_2 = 3.5%).

(C)

(C)

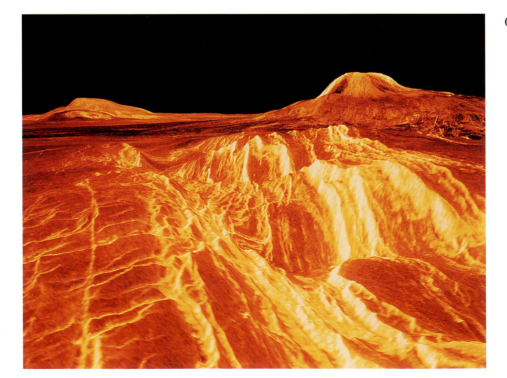

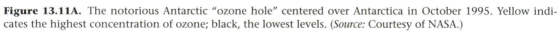

Figure 13.11A. The notorious Antarctic "ozone hole" centered over Antarctica in October 1995. Yellow indicates the highest concentration of ozone; black, the lowest levels. (*Source:* Courtesy of NASA.)

Figure 19.1. Aerial photograph of Lower Geyser Basin, Yellowstone National Park. Four consortia of microorganisms grow in the basin and effluent streams of Grand Prismatic Hot Spring. The distribution of each consortium is controlled primarily by temperature, and each has a distinctive color pattern that can be observed around the spring. (*Source:* Photograph by Julie K. Bartley.)

Figure 20.8. Part of the field component of the Jasper Ridge CO_2 Experiment. The foreground chambers (in the area with wildflowers) are on serpentine soil. The background chambers are on sandstone soil. Individual open-top chambers are approximately 0.65 meter in diameter and 1 meter tall.

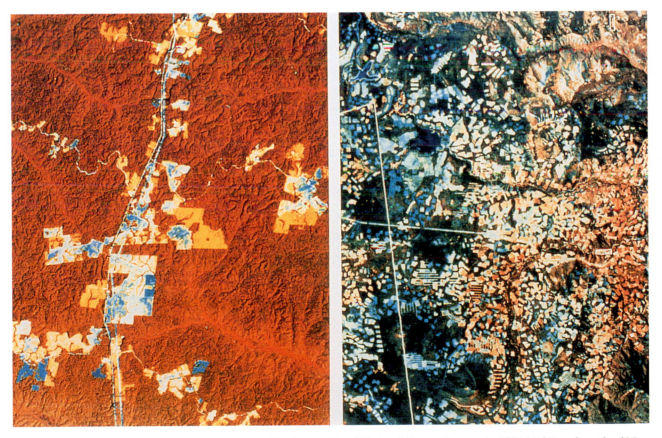

Figure 28.1. *LANDSAT* thematic mapper images of land use in Hood National Forest, Oregon, in 1991 (right) and north of Manaus Brazil in 1990 (left). The lightest areas represent recently cleared or harvested sites, gray areas represent regrowing forest, and the darkest areas represent intact forest. Each image covers approximately 1200 square kilometers. (*Source:* Courtesy of C. J. Tucker, NASA-Goddard Space Flight Center.)

SOCIETY, THE ENVIRONMENT, AND PUBLIC POLICY

26

Steps from Environmental Science to Effective Policy

LAWRENCE H. GOULDER

ENVIRONMENTAL PROTECTION REQUIRES MUCH MORE THAN SCIENTIFIC RECOGNITION OF ENVIRONMENTAL PROBLEMS

Previous chapters of this book have described important physical and biologic dimensions of Earth systems. Understanding these phenomena is an important precondition for making sensible policy to protect the environment. However, such understanding is only a first step: The road from *scientific recognition* to *effective policy* is long and involved.

Beyond recognizing the environmental problem, it is crucial to ascertain the costs involved in taking action to prevent or retard further environmental deterioration, and to compare these costs with the benefits associated with an improved or less degraded environment. Comparing costs and benefits is a difficult task requiring the expertise of philosophers, economists, policy analysts, and other social scientists.

The road to effective policy does not end with comparisons of benefits and costs, and the associated recommendations by policy analysts. A number of further hurdles must be overcome before sound environmental policies can gain political acceptability. One hurdle is limited public awareness of the nature of environmental problems. Another is "mobilization bias," which tends to limit the influence of environmental interests in the political process. A third is the (mistaken) view that individual responsibility is sufficient to protect the Earth – that public policies are not necessary. This chapter discusses each of these major obstacles on the road from descriptive science to effective policy.

The previous chapters of this text have introduced many "hard science" elements of Earth systems: geologic, atmospheric, oceanic, and biologic processes that interact to form the global Earth system. These chapters have described recent changes in these processes and the implications of these changes for the quality of the environment. Much of the "news" in these chapters is distressing; in many cases, the continuation of current trends implies substantial degradation of the natural environment.

As Earth scientists expose these various threats to Earth's systems, it is natural for us to think about how public policies might respond to these threats and help safeguard the environment.* However the path from the scientific recognition of an environmental problem to public policy change is typically long and involved. When natural or biologic scientists discover a new threat to the environment, these findings seldom translate immediately into changes in local, national, or global environmental policies. For example, discovery of a rapidly deteriorating ozone layer did not immediately lead to legislation requiring reductions in the use of ozone-depleting chemicals (see Chap. 3). Before natural or physical scientists' discoveries can induce policy changes, many other actors must enter the environmental arena, including economists, policy analysts, philosophers, journalists, lawyers, and politicians. This chapter describes the connections from environmental science to effective policy making. In so doing it provides a transition from previous chapters to the chapters that follow, which emphasize policy options.

FROM DESCRIPTIVE SCIENCE TO POLICY EVALUATION

WHY RECOGNIZING THE PROBLEM IS NOT ENOUGH: THE NEED TO CONSIDER TRADE-OFFS. Oftentimes physical and biologic scientists seem surprised when their scientific findings do not bring about immediate policy changes. Scientists frequently feel that their results should "speak for themselves," and that it should not be necessary to engage in policy analysis. The sobering fact of life is that scientific discovery of a problem is only the first step toward policy action.

---

* Of course, collective or public responses are not the only answer; there are also ways that each of us can respond to these problems as individuals. We return to the role of individual action below.

W. G. Ernst (ed.), *Earth Systems: Processes and Issues*. Printed in the United States of America. Copyright © 2000 Cambridge University Press. All rights reserved.

TABLE 26.1 A SAMPLING OF ENVIRONMENTAL PROBLEMS AND THE BENEFITS AND COSTS ASSOCIATED WITH POLICIES THAT ADDRESS THEM

Problem	Benefits from Addressing the Problem	Costs of Dealing with the Problem
Deterioration of the stratospheric ozone layer	Avoidance of increased skin cancers Avoidance of adverse impacts on wildlife	Need to employ higher-cost substitutes for ozone-depleting chemicals
Loss of biodiversity as a result of human encroachments on natural habitats	Future economic gains from commercial use of the medical and scientific contributions offered by biologic genetic resources Moral satisfaction associated with recognizing humans are protecting deserving species	Losses associated with prevention or slowdown of economic development
Hydrocarbon-related pollution from exhaust from gasoline-powered automobiles	Avoidance of damages to health from pollution Enjoyment of higher visibility	Need to utilize higher-cost alternatives to ordinary gasoline as fuel, or alternatives to the automobile Need to utilize expensive pollution control equipment

Consider the environmental problems listed in Table 26.1. The loss of biodiversity, the thinning of the stratospheric ozone layer, and the health impacts of hydrocarbon emissions from the automobile – all these problems require policy action. Nearly everyone thinks the world would be better if there were no automobile pollution, no loss of biodiversity, and no deterioration of the ozone. However, it is critical to go beyond recognition of the problems. Policy analysts also need *to consider the costs to society of dealing with these problems*.

Policy action seldom comes free. Virtually all environmental policy initiatives involve both benefits (stemming from a cleaner environment) and costs, or sacrifice. Consider, for example, a policy to prevent species loss. As indicated by the middle column of Table 26.1, the benefits might include the moral satisfactions associated with offering species a degree of protection we might think these species deserve, as well as economic gains if the protected species have current or future commercial value (see Chap. 31 for a closer examination of ethical issues associated with the preservation of nature). There are costs as well, as indicated by the right-hand column of the table. These include the sacrifices stemming from the reductions in economic development or in other activities that in the absence of a new policy to protect species, would lead to species loss. Some people think that the social benefits from new policies are worth the costs; others do not. Showing that these policies are worthwhile – and garnering the political support – requires attention to both the benefits and the costs involved. This type of scrutiny is not the usual business of natural and biologic scientists. However other analysts often concentrate on these issues of value. Economists, in particular, often aim to evaluate

the benefits and costs of various policy options. Philosophers often aim to assess the rights or protections that should be accorded to different living things. Legal scholars often examine the legal options for protecting certain species, and where certain policies might run into conflicts with existing law. Clearly, moving from descriptive science to public policy requires that the efforts of natural and biologic scientists be coupled with the work of economists, philosophers, and other social scientists.

If policy actions did not involve costs, then scientific findings would indeed "speak for themselves." Under these circumstances, whenever a threat to the environment was revealed, most people would easily be convinced that it made sense to take policy action to neutralize the threat. However, virtually *all* policy actions involve costs or sacrifice, as suggested by the right-hand column of Table 26.1. Making change is costly, whether it be preventing deterioration of the ozone layer, addressing potential climate change, or ameliorating water shortage problems in California. As a result, a crucial step in generating support for public action is an analysis of benefits and costs. If the benefits exceed the costs, this provides at least a partial justification for taking action.*

The issue of global climate change offers another example of the need to consider benefits and costs. Consider some policies to reduce or actually prevent global climate change. This would involve, among other things, reducing emissions of greenhouse gases that contribute to the greenhouse effect. Recognition of the possibility that calamitous climate change could occur is not enough. We must adduce costs as well, because taking action to prevent serious climate change appears to be quite costly.

Why is it costly? As Chapter 27 illustrates, finding substitutes for the fuels that emit the damaging gases is costly. A key element of climate change policy is to reduce carbon dioxide emissions; such emissions account for more than half of the radiative forcing from anthropogenic-produced greenhouse gases. One way to reduce carbon dioxide emissions is to reduce combustion of coal by electric power plants. This can be done by encouraging plants to switch

---

* As discussed below, there are limits to benefit/cost analysis. Benefits and costs cannot be measured with precision, and the assignment of benefits and costs involves value judgments. For these and other reasons, some people think that benefit/cost analysis does not provide a very good indication of whether a potential policy would be worthwhile.

from coal to other fuels such as oil or natural gas as a fuel source. However, this switch is costly, because in most cases the plants that now burn coal would find it more expensive to burn an alternative fuel.* Another option is to encourage the expansion of hydroelectric plants as an alternative to coal-fired (or other fossil fuel – based) plants. This is costly as well. Certain sources of hydropower are relatively available and cheap, but unfortunately, most of the cheap sources have already been tapped. New sources of hydropower tend to be at least as expensive – and usually more expensive – than existing coal-fired (or other fossil fuel–powered) plants.

Although one can debate the extent of the costs in particular cases, the reality is that reducing substantially the global level of coal combustion is a costly proposition. However, this fact does not make it unwise to do so; the costs may very well be worth bearing. The benefits from an active policy to reduce coal use in terms of reduced climate change may outweigh – perhaps even vastly outweigh – the costs. The key point is that justifying the idea that "public policy should reduce the rate of coal emissions" requires attention to both benefits and costs. This approach is essential to garnering public support for public policy action. As with prevention of species loss, reduction of global climate change is an area where economists and other social scientists, together with atmospheric scientists, climatologists, and others, can provide information to help make the case for policy action. As an example of such collaborative action, see the box entitled *Dealing with Stratospheric Ozone Depletion: Scientific Evidence, Economic Evidence, and Public Policy.*

LIMITATIONS OF BENEFIT/COST ANALYSIS. The previous discussion might imply that policy evaluation is fairly straightforward: Simply calculate the overall benefits and the overall costs, subtract the latter from the former, and – voilà – you have a measure of the policy's attractiveness!

This calculation is not that simple – for several reasons. First and foremost, the overall attractiveness of a policy depends on more than the overall net benefits. *How the benefits and costs are distributed* is another crucial consideration.

Policy analysts typically distinguish the concepts of *efficiency* and *equity*. A policy is deemed efficient if its aggregate benefits exceed its aggregate costs, that is, if there are positive aggregate net benefits. Efficiency is clearly no guarantee of equity. Before we make the leap from the fact statement, "The demand for California's water is growing faster than

> DEALING WITH STRATOSPHERIC OZONE DEPLETION: SCIENTIFIC EVIDENCE, ECONOMIC EVIDENCE, AND PUBLIC POLICY
>
> The case of stratospheric ozone depletion exemplifies the complementary contributions of physical scientists and economists on the road to public policy. The first critical steps on the issue of ozone depletion were taken almost exclusively by physical scientists, Mario Molina and Sherwood Rowland (see Chaps. 3 and 13). These scientists provided the theoretical basis and found the actual chemical evidence for the deterioration of the ozone layer. This information was then applied by health scientists and economists to obtain assessments of the benefits (in the form of avoided damages to health) from policies to reduce the rate of ozone depletion. In addition, economists used technologic information to generate estimates of the costs of policy action. The costs stem from the sacrifices that businesses must make to avoid the use of ozone-depleting chemicals; such sacrifices are experienced as costs in the form of lost profits and, to the extent that higher costs are passed on to consumers, higher consumer prices. Identifying the costs, and comparing them with the benefits, was an important step on the road to policymaking. Eventually, economists mounted considerable evidence indicating that many of the benefits from helping forestall the deterioration of the ozone layer could be realized at relatively low cost. This information helped win support for policies requiring reductions in the use of ozone-depleting chemicals.

supply," to the value statement "We should raise the prices of California's water to slow the growth of demand," most of us would want to consider who would benefit from higher water prices and who would lose. Higher water prices might offer net benefits to California – the gains to the environment and other benefits might exceed the costs from higher prices; however, the same policy might be considered very unattractive if predominantly poor households incur the higher prices. A common criticism of traditional benefit/cost analysis is that it takes no account of the distribution of benefits and costs. Clearly, it is important to supplement the findings of total benefits and total costs with information concerning how these costs are distributed. Beyond considering the distribution across income groups, one might also want to explore how benefits and costs are distributed geographically – across regions of a given country or across countries.† It is also important to consider the intergenera-

---

* Of course, in some cases plants may be forgoing a cheaper option, perhaps because they do not have good information about the other option. Economists tend to assume that circumstances of this sort – where cheaper options are left unexploited – are unusual. This assumption is caricatured in the following story. Two economists are walking down the street, and one notices a $20 bill on the sidewalk in front of the other's foot. This one says, "Hey, there's a twenty dollar bill on the sidewalk in front of you – why don't you pick it up?" The other exclaims, "There can't be a bill on the sidewalk: if there had been, someone would have picked it up already!"

† Much of the debate on acid rain policy that preceded the enactment of the 1990 Clean Air Act amendments reflected differences in the regional distribution of benefits and costs of such policy. The 1990 legislation required coal-fired electric power plants to reduce their emissions of sulfur dioxide, a key contributor to acid rain. Coal-producing states such as West Virginia vigorously opposed this legislation, fearing it would lead to considerable reduc-

tional distribution of benefits and costs. Many environmental policies involve sacrifices on the part of current generations, with benefits that accrue largely to future generations. Even if we know that the net benefits are positive (the benefits to future generations exceed the sacrifices made by current generations), we might be reluctant to compel one generation to make a sacrifice without any accompanying benefit. This issue looms large in the climate change debate: To what extent should the current generation be expected to make sacrifices (by curtailing greenhouse gas emissions) when associated benefits (avoided damages from climate change) will be enjoyed largely by other, future generations?

Benefit/cost analysis has another basic difficulty. In many cases, the ability of policy analysts to measure benefits and costs accurately is quite limited. What are the benefits from policies to curtail emissions of greenhouse gases? These benefits are the avoided damages to the environment that result from the prevention of or reduction in climate change. As previous chapters have made clear (see Chaps. 15 and 16), the uncertainties involved here are huge. To know the magnitudes of these benefits, one would need to know (1) the effect of given changes in greenhouse gas emissions on greenhouse gas concentrations, (2) the impact of given changes in greenhouse gas concentrations on climate around the globe (at each moment in time), (3) the impact of changes in climate on the environment (e.g., the extent of sea-level rise, agricultural productivity), and (4) the value of the environmental changes to society. Each link of this chain involves considerable uncertainty.

The presence of uncertainty in benefit/cost evaluations is unfortunate but should not hinder their implementation. The alternative to an imperfect evaluation and imprecise policy is no evaluation and policy at all. Arguably, it is better to make decisions on the basis of imperfect information than it is to make no decisions at all.

One fundamental objection to benefit/cost analysis is separate from, but closely related to, the uncertainty issue. Benefit/cost analysis translates the various effects of policies into dollars – dollar values for benefits and dollar values for costs. Economists play a principal role in making these translations. Some have argued that the value of many important amenities, including the environment (e.g., wilderness areas), cannot be measured in dollars. The issue here is not that the dollar values are uncertain but that it is *inappropriate* to express these values in monetary terms. Some claim that the very process of assigning dollar values to these amenities cheapens them. This is a fundamental criticism of benefit/cost analysis.

Two responses to this criticism are offered here (although

the issue deserves more detailed examination). First, some individuals who object to attaching dollar values to environmental amenities are implying that these amenities have infinite value. If one truly believes that every environmental amenity has infinite value, then every policy that preserves a bit more of the environment passes the efficiency test (assuming the costs are finite). This view does not disqualify benefit/cost analysis; rather, it argues that this analysis ought to justify virtually every environmental initiative. Some disagree with the imputation of infinite value to environmental amenities. The disagreement here is over the magnitude of the value of these amenities, not the appropriateness of benefit/cost analysis.

The second, perhaps more fundamental, criticism of benefit/cost analysis is that it is inappropriate to assign monetary values to the environment. Some critics argue that this process debases the environment, or vastly simplifies our appreciation for it. Clearly, a major simplification is involved when we try to express our appreciation for the environment in a dollar figure. At the same time, when we are faced with a decision to adopt or not to adopt an environmental policy, it helps to base this decision on information about monetary benefits and costs. Critics of this approach need to suggest other benefit/cost measures. Some argue that policy analysts should point out the environmental consequences of different policies and let decision makers or voters judge whether the policy is reasonable based on this information.

Clearly, assigning values to environmental amenities is not easy. In Chapter 31 we discuss in detail the philosophic and empirical issues involved in making such assignments.

Benefit/cost analysis is incomplete, imprecise, and controversial. It is incomplete in that its results need to be supplemented with other information, particularly, information on the distribution of benefits and costs. It is imprecise in that uncertainties about the relative magnitudes of benefits and costs are unavoidable. It is controversial in that some critics question its applicability to environmental policies. Despite its imperfections, benefit/cost analysis is useful, particularly when it is supplemented with information on other dimensions of policy impacts.

FROM POLICY EVALUATION TO EFFECTIVE POLICY

IS PUBLIC POLICY NECESSARY? The previous discussion suggests that *public* policy, or collective action, is critical to protecting the environment. An alternative viewpoint argues that the keys to "saving the Earth" lie in *individual* awareness and responsibility, not in public policies such as government regulations. This view asserts that we should focus more on cultivating our environmental awareness and on being more responsible in our roles as individual consumers and producers (corporate decision makers).

This alternative view has gained considerable attention. The idea is that a higher environmental consciousness will lead us to more responsible and environmentally benign be-

---

tions in demands for coal and associated unemployment in the coal industry. In contrast, states that were downwind from sulfur dioxide emissions (notably Pennsylvania, New York, and the New England states) favored this legislation, which was viewed as providing them with substantial benefits in the form of reduced damage from acid rain.

havior as consumers. Correspondingly, the "greening of corporate behavior" will ultimately lead corporate leaders to pursue business practices that are consistent with protecting the environment. Some adherents to these ideas believe that there is little need for public policy to protect the environment; rather, the solution will come from improved individual behavior.

■ **Responsibility in Our Roles as Consumers.** To what extent can greater individual responsibility substitute for public policy? First, consider your own activities as a consumer. Every purchasing decision you make affects the environment. Certainly there is much to be gained by orienting our purchases toward products whose manufacture or use involves less damage to the environment. Indeed, there are growing markets for so-called green products – products that are better for the environment. These products include less polluting (nonphosphate) laundry detergents and products whose packaging requires fewer resources. There is clear evidence that many people are are willing to buy these products, even when they cost more than their more environmentally damaging counterparts. However, the development of markets for green products will not take us far enough along the road to environmental protection.

Without complementary *public* policies, a heightened individual awareness and more responsible consumer decisions will not take us very far in the direction of protecting the environment. There are two main reasons for this. The first is temptation. In most circumstances, green products cost more than the "dirtier" substitute. The private production costs are typically greater for green products, which implies that their market prices (absent government intervention) will be higher as well. Although many consumers are able to resist the temptation to purchase the cheaper and dirtier product, many more are not.* The benefit to society from purchasing the greener product is the improvement of the environment. When an individual purchases this product, its benefit does not accrue directly to the purchaser but is spread among the entire population. The benefit to the individual purchaser associated with his or her purchase is negligible. In contrast, the added cost of purchasing the green product is quite perceptible to the individual purchaser. From this perspective, the resistance of many consumers to buying the higher-priced green product is understandable. For example, in many parts of the United States, disposable diapers create greater environmental damage (by requiring precious landfill space) than do cloth diapers.† Yet, because of their convenience and wide availability, many parents will be unable to

refrain from purchasing disposable diapers as long as they are relatively cheap in comparison with the cloth diaper alternative.

Even if a large share of the population were willing to spend more in order to "do the right thing" environmentally, there is a second problem that limits the extent to which a heightened environmental awareness can by itself assure a clean environment. This problem is information. For example, many of us are concerned about the prospect of global climate change. Suppose you want to do your part to reduce the prospects of such change by lowering your consumption of products whose manufacture involves the release of the greenhouse gas, carbon dioxide. For example, you might try to restrict your use of gasoline, because gasoline combustion releases carbon dioxide into the atmosphere. Virtually every product you buy involves, directly or indirectly, the combustion of carbon dioxide! A reasonable strategy might be to reduce your reliance on those products that release the largest amounts of carbon dioxide. However, it is extremely difficult to calculate these amounts for each good or service we use. Carbon dioxide is involved in every stage of the production process, from manufacture to distribution. In addition, it is generated in the production of the capital goods or equipment used to make the consumer goods we purchase. Even the most conscientious consumers do not have the necessary information to reprioritize their purchases in the environmentally appropriate way, given the complexities involved in the production of goods and services.

This problem can be solved by *public* policies that provide consumers with the information they need to put their environmental good intentions to work.

■ **Responsibility on the Part of Corporate Managers.** There is a lot of talk these days about a "new corporate environmentalism." Indeed, several books have suggested that an improved environmental consciousness on the part of corporate managers will cause the corporate world to make a fundamental change: Corporate decisions will soon be in keeping with the preservation of environmental quality. The new, green corporate ethic will ensure corporate environmental responsibility. Chemical manufacturers will adopt technologies that lead to less emission of airborne pollutants. Steel manufacturers will find cleaner ways to produce steel. This view implies that government rules are not necessary to protect the environment; the new corporate ethic will bring about such protection automatically.

For the most part, the "new corporate environmentalism" is nonsense. It may very well be that many of today's corporate managers have attained a heightened sensitivity to the environment. Many, perhaps most, of these managers are very fine people with high ethical standards. However, it is a fact of life that in a *laissez-faire* economy (that is, in an economy with no government interference in markets), the

---

* For households with especially low incomes, purchasing the lower-priced product may be closer to a necessity than an option. The word "temptation" is not applicable in this case.

† In some locations, cloth diapers involve greater overall resource cost (environmental cost, natural resource cost, plus other private cost). Laundering cloth diapers requires a considerable amount of water; in regions where water resources are particularly scarce, dis-

posable diapers may involve less overall resource cost than do cloth diapers.

competitive marketplace does not leave room for well-intentioned managers to "do the right thing" environmentally! Suppose you are a high-level manager of a corporation that produces steel. You and your competitors have relied on coal combustion to make the steel – a process that releases substantial emissions of pollutants (e.g., sulfur dioxide and carbon dioxide). You are sincerely concerned about the environment and want to change your production process so that it generates less pollution. Introducing the cleaner production process will involve higher costs than using the conventional process. This means that putting your good environmental intentions to work will raise your firm's costs relative to those of your competitors. If you insist on keeping your prices at their previous levels, your profits will fall and your stockholders will be very unhappy. If you raise your prices to cover the cost increase, your market share will fall, which also will upset your stockholders. In short, in the competitive marketplace, environmental priorities are likely to hurt your business (and perhaps cost you a job!). It is unrealistic to expect corporate managers, no matter how enlightened, to pursue environmentally sound strategies in view of these consequences.

What is the problem here? From society's point of view, the cleaner production process is actually *cheaper* considering the environmental benefits involved (and the savings in pollution-related health costs, cleanup costs, etc.). Yet, from the point of view of the competitive firm, the cleaner process is more expensive. This difference is at the heart of many environmental problems. There is a deviation between private cost and social cost: *What is socially less costly is privately more costly.* The difference between private cost and social cost also explains why we cannot rely on producers' good intentions alone to assure environmental quality. The capitalist system compels producers to select processes that involve the lowest *private* cost; however, in many cases, what is cheapest privately is not cheapest socially, when environmental effects are taken into account. Competitive producers have little choice other than to keep their private costs at a minimum. Stockholders demand reduced costs and increased profits.*

THE CRITICAL ROLE OF ENVIRONMENTAL PUBLIC POLICY: DEALING WITH EXTERNALITIES. Although a heightened environmental consciousness on the part of consumers

* In some cases, managers might choose to adopt a cleaner technology, even if it is more expensive. Firms might do this as a preemptive measure, in order to ward off government regulation. Suppose managers fear that the government will introduce stringent environmental regulations to reduce current pollution levels. In this case, managers might "voluntarily" adopt cleaner technologies to appease potential regulators and prevent an actual clamp-down. It may seem that managers are thereby opting not to minimize their costs. However, by reducing the likelihood of future regulations, the firm helps avoid the prospect of higher future costs associated with new regulations. Thus, the firm may lower its overall expected costs by introducing the new technology.

and producers is a very positive step, it cannot assure a clean environment. In the absence of government policies that change the market setting, consumers will continue to be tempted to "do the wrong thing." Even the most virtuous consumers who avoid the temptation of lower prices will lack the information necessary to put their good intentions to work. At the producer level, managers simply cannot afford to exercise their best environmental intentions in a competitive marketplace.

There is good news, however. Coupled with the appropriate public policies, consumers and producers can indeed pursue environmentally sound practices. What sorts of policies are needed? To answer this question, we need to consider a key concept: externality. An *externality* is a cost (or benefit) stemming from a market transaction that is not taken into account by the parties involved in that transaction. Table 26.2 provides some examples of externalities. Consider an electric utility that wants to purchase coal to drive its turbines. In deciding how much coal to purchase (or whether to use natural gas instead of coal), the utility is primarily concerned with the price it must pay for coal. In the absence of government regulation, the price of coal will reflect the *private cost* of extracting, processing, and delivering the coal. These private costs include payments to labor, for various materials inputs, as well as an established rate of profit. However, the private costs do not include a significant additional cost: the cost to the environment associated with the eventual combustion of coal. This cost is not taken into account by the purchaser of the coal, because it is not included in the private cost or the market price. In the absence of government regulation, this environmental cost is an externality associated with the use of coal.

Under these circumstances, there is a difference between the private and the social cost of using coal. The social cost is the private cost plus the external cost. Thus, in the presence of externalities of this sort, private cost (market price) falls short of the full social cost.

When market prices are less than the social cost, businesses and consumers tend to demand more of the goods involved than is best for society. (This issue is explored more fully in Chap. 27.) For example, utilities rely too heavily on coal – they convert too much coal into power, such that the social benefits from the last units of coal used fall short of the full social costs (including environmental costs). Similarly, consumers tend to purchase too many disposable (as opposed to cloth) for the sake of convenience, and because the prices of disposables are relatively low. Individuals tend to make too much use of gasoline-powered automobiles as opposed to other forms of transportation or consolidating automobile trips, because the cost of gasoline and of automobile use in general is below the social cost. Too many wetlands are being converted for agricultural use because farmers are able to purchase wetlands relatively cheaply, at a price that does not include the full social costs, including losses of important ecosystem functions (see Chap. 9).

Appropriate public policies can bring market prices closer

TABLE 26.2. EXTERNALITIES AND POLICIES TO DEAL WITH THEM

Activity	Externality	Potential Policy to Address the Externality
Burning of coal	Acid rain damage Local health damages Climate-change-related damages	Mandated scrubbers on coal-fired electric power plants Tradeable permits program for sulfur dioxide emissions Carbon tax
Use of disposable plastic diapers	Costs of lost landfill capacity	Tax on disposable diapers to incorporate landfill cost
Automobile use	Health costs from ground-level ozone Congestion (time) costs	Mandated catalytic converters "Tailpipe" emissions requirements Gasoline tax
Conversion of wetlands to agricultural use	Loss of ecosystem services (water filtration, breeding ground, habitat), with ultimate impact on humans	Tax on ecosystems conversion

ment such desires with public policies. In one sense, the need for public policies *underscores* the importance of a heightened environmental consciousness. Such policies will be adopted only if individual citizens are sufficiently environmentally aware to vote for them. Indeed, in a sense, heightened environmental consciousness is the most fundamental source of environmental protection. Such an awareness is critical not only to the advent of "green" consumers and producers, but also (and perhaps more important) to the existence of the "green" voters who ensure that vital public policies are adopted.

to social costs, which causes firms and consumers to purchase fewer goods that are harmful to the environment. The right-hand column of Table 26.2 indicates potential public policies that can be used to address the externalities associated with each commodity or activity. For example, a tax on purchases of disposable diapers could include the previously excluded costs of obtaining new landfill space. If the prices of disposable diapers rose to the full social cost, consumers would no longer be tempted to "do the wrong thing"; what is best for the environment (cloth diapers, in this example) would also be cheaper for the consumer. Similarly, if the government introduced a tax on coal combustion, then managers would have incentives to introduce cleaner technology for producing steel. The cleaner technology would now be less costly *privately* as well as socially! Alternatively, the government could require the use of the cleaner technology, in which case the cleaner technology is again privately less costly (assuming that the firm would have to pay large fines if it violated the government's requirement). Thus, public policies can enable consumers and producers to put their good intentions to work.

Public policies can be divided into two main categories: *price-based policies* (e.g., taxes on gasoline, diapers, or wetlands conversion) and *direct controls* (e.g., mandated scrubbers, required catalytic converters, and tailpipe emissions limits). Both types of policies help prices better approximate social cost. Price-based policies in the form of taxes help move prices toward social costs. Direct controls in the form of regulatory requirements force firms to alter their technologies in ways that account for externalities.

This discussion may appear to minimize the importance of an improved environmental consciousness, which is not our intention. Although greater environmental awareness and a desire to improve the environment on the part of consumers and producers are important, they are not sufficient to protect the environment: There is a need to comple-

WILL AN INFORMED PUBLIC SUPPORT USEFUL ENVIRONMENTAL POLICIES? The previous section of this chapter discussed the links from scientific recognition of an environmental problem to policy evaluation. Policy *evaluation* does not guarantee effective policy *making*. We now move from the domain of economics to that of political science.

Consider an environmental problem for which scientific, technologic, and economic knowledge are sufficient to identify a complete solution or at least an adequate improvement. Suppose we attempt to calculate, even with the uncertainties, some of the expected benefits and costs of a given environmental policy option. Suppose also that virtually all experts agreed that the policy in question was very likely to produce net benefits: The benefits would outweigh the costs. Further, suppose that this information about the beneficial aspects of the policy were made widely available to the public. Would this state of affairs assure sufficient political support to ensure adoption of the policy?

In many circumstances, the answer is no. The general recognition that a policy offers net benefits is no guarantee of political feasibility. In fact, for many potentially beneficial environmental policies, a systematic bias may militate against adoption. This is called *mobilization bias*. This bias, articulated originally by the famous political scientist Mancur Olson, may arise in various contexts; it is not limited to the area of environmental policy. Mobilization bias can be explained as follows. First, political awareness and political activity require effort and involve sacrifice. Keeping informed about political issues involves an expenditure of time and effort. Beyond that, political activity – voting, canvassing, lobbying, writing your congressional representative – also involves a sacrifice of time and effort. Often, we consider these efforts as worth making, in part because we think it is our duty as citizens and in part because we may think our efforts will produce political results that compensate for the sacrifices.

Olson noted that the expected return from a given sacrifice of effort or time is much larger in some cases than in others. In particular, he noted that the likelihood of becoming involved in the political process is closely related to whether the benefits or costs are highly concentrated among a small number or are widely dispersed across a large number. For example, suppose the nation is debating a policy that would reduce air pollution by taxing producers of fossil fuels. Suppose also that the benefits from cleaner air significantly outweigh the costs associated with reducing fossil fuel consumption. The "winners" from this policy would be the general public, that is, the people who breathe the air and who would experience pollution damage if the policy were introduced. The primary "losers" from this policy would be the producers of fossil fuels (and people who work in or own stock in the fossil fuel – producing industries). There is a basic asymmetry here. The "winners" in this example are large in number and relatively poorly organized, whereas the "losers" here are relatively few in number and highly organized.

This asymmetry is important, because it affects the likelihood of becoming involved politically. Because the potential losers are few in number and highly concentrated, each loser has a large stake in the outcome of this policy and a greater individual influence on the outcome. Under these circumstances it may be well worthwhile for the potential losers to incur the costs of assembling information about the policy, lobbying against its passage, and exerting whatever political influence is possible. In contrast, each of the potential winners has only a tiny stake in the outcome (although the overall "winnings" may be extremely large when added up across all the winners) and only an imperceptible individual influence on the outcome. Each individual "air breather" may feel that not enough is at stake personally to justify gathering the information and becoming politically involved. Political mobilization is more likely to occur among those who feel they have a larger individual impact on the political process; these are the groups or individuals for whom the impact of the policy is relatively highly concentrated.

Mobilization bias emerges: The "losers" in this example have a disproportionate political impact and are able to block passage of the proposed policy, even though the policy clearly offers greater overall benefits than costs. Mobilization bias also can hold in the case where the small, highly organized group constitutes the winners, in this case, however, the winners would actively campaign for passage of the proposed legislation.* See the box entitled *Mobilization Bias and the Clinton Energy Tax Proposal*.

Mobilization bias helps explain why special interest

MOBILIZATION BIAS AND THE CLINTON ENERGY TAX PROPOSAL

Mobilization bias helps explain the demise of the Clinton administration's proposed energy tax. In February 1993 the administration proposed an energy tax as part of its deficit reduction package. This tax was to be imposed on fossil fuels according to their energy (or BTU) content, with some adjustments to help maintain similar tax rates per dollar of fuel. Although most political observers initially seemed confident that the tax would be approved, the energy industries ultimately wielded substantial political clout and managed to rally a majority of legislators against the energy tax. The costs of the energy tax would be borne predominantly by a few industries. The environmental benefits, in contrast, would be distributed widely. The situation was therefore ripe for mobilization bias. Whether this bias was largely responsible for the defeat of the energy tax is difficult to determine conclusively. However, the effective and vocal organization of the opponents to the proposal, and the relatively lackluster organization of the proposal's supporters, certainly are consistent with this bias.

---

groups often wield disproportionate influence politically. To the extent that there is a significant difference in the breadth of the distributions of benefits and costs from an environmental policy, there is greater potential for mobilization bias. The political process may be stacked against environmental protection because the gains from such protection typically are much more widely dispersed across the population than are the costs.

Can mobilization bias be overcome, or must potentially beneficial environmental policies remain hobbled in the political arena? As long as environmental benefits are distributed more diffusely than economic costs, this bias will persist. However, some recent developments might mitigate its impact. In the past decade, membership in environmental advocacy groups has grown substantially. Buoyed partly by their larger membership, these groups have enjoyed greater political clout and have displayed a heightened awareness concerning the political process. Although the number of individual beneficiaries from reduced pollution is huge, the number of environmental groups is not. To the extent that these groups can act as proxies for the beneficiaries of environmental policies, they can overcome at least some of the mobilization bias. They can therefore exert a countervailing political influence relative to the "losers" from environmental protection.

Something akin to mobilization bias arises when benefits and costs are unevenly distributed across generations. Con-

---

* For example, consider the effects of farm price supports, according to which the government pays farmers not to grow certain crops in order to keep crop prices and profits up. Farmers benefit from these policies; consumers lose. Economists generally agree that these policies impose net costs to society: The costs to consumers are larger than the benefits to farmers. However, in this case, the

benefits are concentrated among a relatively small number, whereas the costs are dispersed widely. Here, mobilization bias works in favor of the farmers, who have been quite successful in maintaining these price supports, despite the inefficiencies.

sider the benefits and costs of policies to restrict emissions of greenhouse gases. The benefits from these policies would be enjoyed primarily by future generations, in the sense that these policies would allow future generations to avoid climate change–related damages. The costs of such restrictions, in contrast, are likely be felt as much by current generations as by future ones. In today's debates about greenhouse policy, current generations are overrepresented relative to future generations, because future generations have a political voice only to the extent that the present generation demonstrates altruistic concern for them. Thus, a type of political bias works against the introduction of policies whose benefits will be experienced in the distant future.

SUMMARY

The path from descriptive science to effective policy is long and complex. Biologic and physical scientists typically make the first part of the journey, providing key descriptive information about geologic, climatologic, atmospheric, or biologic processes. In order for policy analysis to proceed, this information needs to be translated into evaluations of the benefits derived from environmental policies. In addition, policy analysis needs to take into account the costs or sacrifices involved in policies to protect the environment. Here economists and other social scientists can supply their expertise. Economists, in particular, can help calculate not only the costs involved in protecting the delicate systems described by biologic and physical scientists, but also the benefits from such policies. This attention to efficiency impacts – the aggregate benefits and costs – is useful for policy evaluation. An overall policy assessment also requires attention to equity effects – the distribution of benefits and costs across income groups, across regions, and across generations.

Many have argued that individual responsibility, not government policy, provides the key to environmental protection. However, in the absence of environmentally oriented public policies, even the best-intentioned consumers and producers will have difficulties being effective in protecting the environment. As long as market prices poorly reflect social costs, the ability of consumers and producers to pursue environmentally sound practices will be fundamentally limited. Public policies can be effective to the extent to which they deal with externalities and thereby help prices reflect social costs. When prices better reflect social costs, consumers will have incentives to act in ways that take environmental impacts into account. Moreover, the profit-maximizing incentives of producers will become aligned with the goals of environmental protection. Thus, the public sector has a vital role in helping to provide the conditions that allow for socially responsible consumer and producer behavior.

Clear and cogent policy analysis does not assure the adoption of effective environmental policies. In order for potentially useful policies to gain political support, the public needs to be educated. However, even this is no guarantee of adoption. Often, mobilization bias prevents the adoption of policies that offer net benefits to society. Although the growth of environmental advocacy groups can help reduce mobilization bias, such bias is likely to persist. The effectiveness of future environmental policy depends critically on the quality of environmental education and the extent to which the general public expresses its concerns for the environment in the political process.

QUESTIONS

1. It has been suggested that policies to improve the environment (or to avoid further environmental degradation) almost always entail costs. Consider the policies to reduce automobile pollution. What are the costs of these policies? Can you imagine circumstances in which pollution could be reduced at zero cost?

2. Consider a policy that improves air quality in an urban area. How might policy analysts attempt to attach a value to the improvement in air quality? Can policy analysts avoid the need to translate the medical or health implications of the policy into dollar values?

3. "Benefit/cost analysis is not a good tool for policy analysis because it does not consider issues of fairness." Evaluate this claim.

4. "The environment is so precious, it should not be subjected to benefit/cost analysis." Assess this claim.

5. Former Environmental Protection Agency Chief William Reilly once asserted, "The power of the green consumer is much greater than that of the green voter." This chapter suggests otherwise. Which position do you agree with? Why?

6. What is mobilization bias? Does it undermine environmental protection efforts? If so, how can mobilization bias be overcome?

FURTHER READING

Daly, H. E., and Townsend, K. N. 1993. *Valuing the Earth.* Cambridge, MA: MIT Press.

Fraas, A. G. 1984. A benefit-cost analysis for environmental regulation. In *Environmental policy under Reagan's executive order*, ed. V. K. Smith. Chapel Hill, NC: University of North Carolina Press.

Goodstein, E. S. 1995. *Economics and the environment.* Englewood Cliffs, NJ: Prentice-Hall.

Kelman, S. 1981. Cost-benefit analysis: an ethical critique. *Regulation*, (January/February), pp. 33–40. (See also the collection of Replies to Kelman, in the March/April 1981 issue of *Regulation*.)

Olson, M. 1971. *The logic of collective action.* New York: Schocken Books.

Panayotou, T. 1993. *Green markets: the economics of sustainable development.* San Francisco: Institute for Contemporary Studies.

Ridley, M., and Low, B. S. 1993. Can selfishness save the environment? *The Atlantic Monthly* (September), pp. 76–86.

27 Confronting the Prospect of Global Climate Change:

Carbon Taxes and Other Domestic Policy Options

LAWRENCE H. GOULDER

PUBLIC POLICY AND GLOBAL WARMING IN THE AGE OF CARBON

Increased atmospheric concentrations of greenhouse gases raise the prospect of serious global climate change. How can public policies best address this possibility? To what extent should public policies curtail emissions of greenhouse gases? Which policies are most effective in achieving emissions reductions? To what degree should policies support the expansion of "carbon sinks"?

This chapter explores these issues, concentrating on unilat-eral, domestic policy options. We begin with the policy of a carbon tax, and then compare this policy with a number of alternatives.

The uncertainties surrounding the likely extent of climate change pose especially difficult challenges to policy makers. Should active policies be postponed until greater scientific evidence becomes available? Clearly the question of *when* to reduce greenhouse gas emissions is as important as the question of *how much* to reduce these emissions. We consider both of these questions in this chapter.

There is a real possibility that continued accumulation of greenhouse gases could bring about serious climate change (Chap. 16). Indeed, the Intergovernmental Panel on Climate Change (IPCC), a group of over 200 respected scientists from around the world, recently concluded that anthropogenic emissions of greenhouse gases have already induced changes in average global surface temperature. Many people are concerned about the prospect of further changes in temperature and precipitation, and of the associated changes in storm intensities, sea-level rise, and the distribution of climate belts. One wonders what can be done to avert such changes or at least reduce their magnitudes.

Under "business-as-usual" circumstances – that is, in the absence of significant changes in policies – anthropogenically caused global emissions of greenhouse gases are expected to rise over the next several decades as a result of expanding industrial output. In particular, the concentration of carbon dioxide is expected to double by the end of the twenty-first century, which raises the prospect of substantial climate change.

What can be done? How much, if at all, should greenhouse gas emissions be reduced? What is the best way to achieve such reductions? This chapter considers these issues. We concentrate on policies that could be undertaken unilaterally by the United States, leaving aside important issues of international policy coordination. We focus on policies

aimed at reducing the rate of buildup of carbon dioxide (CO_2) concentrations; by all accounts, carbon dioxide is the most significant of the greenhouse gases in terms of its contribution to global warming as a result of anthropogenic emissions. We begin by examining the carbon tax as a domestic policy option; later we compare this option with other policy alternatives.

There is a great deal of uncertainty concerning the potential seriousness of climate change. The scientific community is unsure as to how much a given increase in greenhouse gas contributions will raise global surface temperature. In addition, great uncertainty exists regarding the extent of damages – for example, from sea-level rise, changes in agricultural productivity, and increased storm intensities – associated with given temperature increases. These uncertainties make the problem of policy choice even more difficult, and have created considerable controversy.

In the United States, different presidential administrations have reacted to the uncertainties in different ways. The Bush administration held the view that undertaking costly policies to reduce emissions of greenhouse gases was premature, and that it would be better to wait until more scientific information was gathered. Consistent with this view, during the Bush presidency the United States avoided making any commitments to targets or timetables for reductions of greenhouse gas emissions. The Clinton administration has tended to embrace the view that taking action is now justified, despite the uncertainties, because dealing with climate change might be more costly if we wait two or three decades.

THE CARBON TAX OPTION

Table 27.1 displays some alternative options available to policy makers for retarding the rate of accumulation of carbon dioxide. Taxes on carbon based fuels diminish the demand for these fuels by raising their prices to users. Thus, they reduce emissions of carbon dioxide, which are closely tied to the amount of carbon in fuels. Direct controls on fuel use are an alternative to taxes. An example of such controls would be the requirement that electric utilities switch from coal to natural gas in power generation; natural gas releases significantly less carbon per unit of electricity generated.

As discussed in Chapter 19, concentrations of carbon dioxide can be reduced either by reducing the rate of emissions of carbon dioxide or by expanding the rate at which carbon dioxide is absorbed in "carbon sinks." Therefore, subsidies to afforestation, by expanding the supply of carbon sinks, is another way to reduce atmospheric carbon dioxide concentrations. The rate of growth of the human population contributes to the growth of industrial output and, hence, greenhouse gas emissions. To the extent that policies can reduce the growth of the world's population, they also slow the growth of emissions of greenhouse gases.

Population control itself does not involve an effort to abate carbon dioxide levels. Even this policy of "doing nothing" deserves close scrutiny; it is a viable alternative to the other policies listed. People who endorse this alternative tend to emphasize the current uncertainties about the severity of future global climate change and the possibility that sacrifices made today might later prove unnecessary.

We begin by examining the carbon tax option. This is a tax on fossil fuels – coal, oil, and natural gas – in proportion to carbon content. Carbon content typically determines the amount of carbon dioxide emitted into the atmosphere from the combustion of fossil fuels or the refined fuels derived from them. Hence, a carbon tax is essentially proportional to the carbon dioxide emissions from these fuels.*

In evaluating the carbon tax option, the following considerations seem especially significant:

- Efficiency: What are the aggregate net gains from the carbon tax?
- What are the impacts on prices and incomes?
- What are the distributional effects? That is, who wins and who loses?
- How effective is the carbon tax relative to the alternatives?

We address each of these questions in turn.

---

* An exception occurs when fuels (principally, petroleum) are used as feedstocks in the production of other goods such as plastics. In this case, the fuels are not burned and no carbon dioxide is released. In the United States, approximately 3 percent of petroleum is devoted to feedstock uses.

TABLE 27.1. GREENHOUSE POLICY ALTERNATIVES
Taxes on carbon-based fuels (carbon tax, BTU tax, or gasoline tax)
Direct controls on supplies or use of carbon-based fuels
Subsidies to carbon sinks
Population policies
No action now (wait for more scientific evidence)

THE EFFICIENCY RATIONALE. The notion of efficiency here is the economists' notion used in Chapter 26: It refers to the aggregate net benefits from a given policy. Efficiency considerations provide the fundamental rationale for a carbon tax. There may be other potential advantages of a carbon tax; however, it would be difficult to justify the tax if it didn't lead to an improvement in efficiency, that is, if the aggregate benefits did not exceed the aggregate costs.

The various damages stemming from global climate change are an important externality associated with the combustion of carbon-based fuels. This externality can help justify the introduction of a carbon tax and explain why this type of tax can lead to an efficiency improvement. As indicated in Chapter 26, in the presence of externalities the prices that emerge from unregulated markets typically do not reflect social costs. When firms or individuals purchase and burn fossil fuels, they pay only for labor, materials, and capital. They have no reason to consider the additional social cost from fuel burning: the economic and environmental damages caused by the release of carbon dioxide into the atmosphere. As a result, economic efficiency is not achieved. Fossil fuels are too inexpensive, demand for substitutes for such fuels is insufficient, and too much carbon dioxide is emitted. Because fossil fuel prices do not reflect their full social costs, incentives to develop alternative technologies that rely less on fossil fuels are diminished. In addition, all products and services dependent on fossil fuels end up being artificially cheap relative to other products and services.

A carbon tax can correct this inefficiency. "Correcting the inefficiency" means altering market conditions so that economic and environmental benefits exceed the costs associated with the higher fossil fuel prices brought about by the carbon tax. Figure 27.1 depicts the nature of the inefficiency and shows how it can be corrected. In the figure, the "D" curve is a market demand curve for coal, one of the carbon-based fuels of concern. The purchasers or demanders of coal include coal-fired electric power plants and steel mills. These demands, in turn, are a reflection of consumer demand for electricity and for consumer goods made of steel.

The economy's demands for coal are an inverse function of its price: The lower its price, the larger the quantity demanded. Hence, the demand curve slopes down and to the right. The MC curve represents the marginal private cost of bringing coal to the market. Underlying the MC curve are the costs of various private inputs used to produce coal: labor costs, materials costs, rental costs of machinery, and a normal return on invested capital. A portion of the social cost

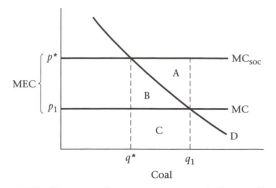

Figure 27.1. How a carbon tax can correct the inefficiency of coal use in the presence of externalities. Areas depicting benefits and cost are shown as A, B, C (see text); D is the demand curve. MC = marginal private cost, MC_{soc} = marginal social cost; MEC = marginal external cost; p = price; q = quantity.

associated with the use of fuels is not included as a private cost: The environmental cost in the form of damages related to global warming is an external cost. We can define the marginal external cost (MEC) as the external cost associated with each additional unit of coal supplied to the market. In this example, we concentrate on the global warming damages as the external cost, although other externalities are associated with coal use (e.g., the damages from sulfur dioxide emissions stemming from the combustion of coal, as well as local health effects associated with carbon monoxide and carbon dioxide emissions). The MC_{soc} curve is the marginal social cost of providing coal, that is, the marginal private cost plus the marginal external cost. The marginal social cost is the full cost to society of each additional unit of coal supplied. It is the sum of the marginal private cost and marginal external cost.

The failure of the private market stems from the fact that the marginal external cost is not taken into account when users demand coal. In the absence of regulation, the supply of coal is given by the marginal private cost curve, MC. Hence, in market equilibrium (i.e., where supply equals demand), the market price is p_1 and the quantity demanded is q_1. Why does this constitute a market failure? First, the demand curve represents the marginal benefit (the value of the additional output) attributable to the last unit of coal supplied to market. At q_1 this benefit falls short of the marginal social cost. In other words, the cost of producing the last unit of coal (including the external cost) exceeds the value of that last unit to users. The net contribution of that last unit of coal – the benefits to users minus the costs of supplying the coal (including environmental cost) – is negative! Coal producers do not consider the external cost when they decide how much coal to supply at different prices; from a social perspective, these prices are too inexpensive. Consequently, the economy tends to rely too heavily on coal. The quantity of coal used exceeds the level that would balance the benefits of coal use and the costs of burning coal.

Government policies can correct this market failure. They can incorporate the external cost in the market price of coal. One way to do this is to charge a tax equal to the marginal external cost of coal use. If this is done, the marginal private cost curve is no longer the original MC curve. Now it coincides with the marginal social cost curve! The government's tax is now an additional cost that private suppliers must take into account. If the tax on each unit of coal is equal to MEC, there will be a new equilibrium in the market, at price p^* and quantity q^*. The price of coal now equals its full social cost, p^*.

Introducing the tax leads to a better balance between the need to avoid greenhouse damages and the desire for cheap coal. At q^*, the marginal social cost of producing coal is just equal to the marginal benefit to society (given by the demand curve). If the quantity of coal produced were to exceed q^*, the incremental social cost would exceed the benefit. By introducing a tax on coal, the government causes producers and consumers to consider the full social cost, which in turn prevents excessive coal demand and production.

A tax on coal leads to an efficiency gain relative to the initial, unregulated situation. The tax causes production to fall from q_1 to q^*. The gross benefit from this reduction is the area A + B + C shown in Figure 27.1 – the area under the marginal social cost curve from q_1 to q^*. The gross cost (or lost benefit) from this reduction is the area B + C, the area under the demand (or marginal benefit) curve. Thus, the net gain from the government's policy is area A – the net gain from avoiding excessive reliance on coal.

Another way to view area A is as the net cost that would result if production of coal were to increase from q^* to q_1. As coal production expands from q^*, the marginal social cost curve is above the marginal benefit curve. Area A is the value of this excess cost as one moves from q^* to q_1.

This example of coal illustrates two principles that are central to the carbon tax. First, if the production or use of a marketed commodity involves an externality, the level of production of that commodity in an unregulated market will be inefficient. Second, by introducing taxes or other policies to make prices equal to social cost, the inefficiency can be eliminated.†

---

† Figure 27.1 illustrates a third principle as well: Generally it is not best, from an efficiency point of view, to avoid *all* use of coal. Suppose the quantity of coal used were reduced below q^*. Under these circumstances, an incremental unit of coal would have a value (given by the height of the demand curve) that exceeds the additional damage associated with that incremental unit (given by the height of the MC_{soc} curve). Thus, it would make sense to increase the use of coal. To put the matter another way: As the use of coal is reduced below q^*, each incremental reduction costs society more than the benefits in terms of avoided damages from greenhouse gases.

One can construct special cases where it makes sense to eliminate the use of coal: this would be the case where the demand curve is always below the MC_{soc} curve. For example, this will occur if there is a perfect substitute for coal in all its uses that is available at a price that is less than the social cost of using coal.

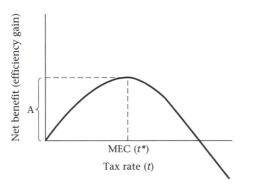

Figure 27.2. Net benefits from a carbon tax as a function of the tax rate. MEC = marginal external cost.

Thus far, we have discussed the efficiency gains that occur when a carbon (coal) tax is introduced at a level equal to the marginal external damages. Of course, it is possible to choose different rates for the carbon tax. Figure 27.2 indicates the relationship between the amount of the carbon tax and the gain in efficiency. If the goal is to maximize the efficiency gain, the carbon tax should be introduced at a rate t^*, where t^* is equal to the MEC. The value t^* is the most efficient carbon tax rate – it is equal to the marginal environmental or external cost associated with the use of carbon. If the tax is less than t^*, there is still a positive net benefit from the tax. In terms of Figure 27.1, this lower tax rate causes coal use to move from q_1 toward, but not all the way to, q^*. Hence only part of the potential efficiency gain (area A) is realized.† If the tax is higher than t^*, the efficiency gain is less than its maximal amount. When the tax exceeds t^*, it causes production to fall below q^*. Thus, production is less than the amount that best balances the benefits from coal use and the full social costs. Indeed, if the tax rate is far too high, the net benefits can become negative! In this case, the tax causes production to fall far short of q^*. Here the environmental benefits from reduced coal use fall far short of the costs in terms of higher costs of coal to its users. In terms of efficiency, it would have been better to have no tax at all than to have this very high tax!

If policy makers were certain about the levels of the environmental damage caused by carbon combustion, there would be little danger in setting the tax too high. Unfortunately, however, there is great uncertainty concerning these

damages. To pinpoint the ideal size of a carbon tax, one would need to know the environmental damages, today and in the future, from each marginal increment in fossil fuel (or carbon) use. Clearly, such knowledge is unavailable. The few estimates available today range from approximately $1 per ton of carbon to over $100 per ton! One might attempt to reduce the risk of too high a tax by endorsing a "small" carbon tax. Yet, how small is small enough? We return to this question in the section entitled "Comparison with Alternative Policies."

IMPACTS ON CURRENT PRICES AND INCOMES. Carbon taxes would push up the prices of the fuels on which they are levied. Table 27.2 indicates the likely impacts on the prices of these fuels from a carbon tax of $25 per ton. The $25 per ton tax rate is in the wide range of estimates for climate-related marginal external damages from the combustion of carbon-based fuels. This tax would raise the price of coal by over 60 percent, and raise prices of oil and natural gas by less than half that percentage. The differences reflect the fact that coal has the highest carbon content per dollar of fuel.

Higher fossil fuel prices translate into higher prices of consumer goods, especially goods with high "fuel content." Prices of gasoline, electricity, and delivered natural gas to residents would rise by 5 to 7 percent. The percentage increases for these goods are smaller than those for fossil fuels because fossil fuels represent only a portion of the cost of producing consumer goods.

Higher prices, in turn, translate into losses of real income to households. These losses are indicated by area B in Figure 27.1: When the price of coal rises, users of coal such as electric power plants lose some of the consumer benefit from cheap coal. These users must now pay higher prices for coal. Some users will find substitutes for higher-priced coal; however, these substitutes generally are higher in price than the original coal price. The higher prices of coal and of new substitutes cause the prices of electricity and other goods based on coal to increase, and this in turn results in a loss of

---

† Figure 27.1 shows the impacts of a carbon tax in the market for coal. An efficient carbon tax would apply to all sources of carbon, at a rate equal to the marginal external damages associated with each source. Thus, a carbon tax would have similar impacts in markets for other carbon-based fuels, notably crude oil and natural gas. The overall efficiency improvements from a carbon tax reflect the improvements in these markets as well as in the coal market. Correspondingly, when the tax is set at a rate differing from the marginal external damages, the shortfall in efficiency gains reflects the failure to exploit the full efficiency potential not only of the coal market but of other markets as well.

TABLE 27.2. PRICE IMPACTS OF $25 PER TON CARBON TAX

Fossil fuels	
Coal	60%
Oil	18%
Natural gas	22%
Consumer products	
Gasoline	5%
Delivered natural gas	7%
Electricity	6%

Note: These price impacts are based on results from the General Equilibrium Energy-Environment-Economy Model developed by Lawrence Goulder. Percentage changes are relative to the prevailing prices in 1995.

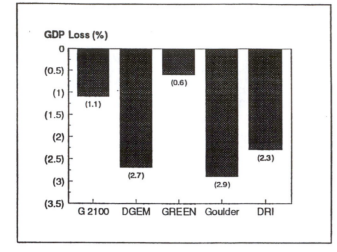

Figure 27.3. Loss of gross domestic product (GDP) per $100 of carbon tax. G2100 = Global 2100 Model developed by Alan Manne and Richard Richels; DGEM = Dynamic General Equilibrium Model developed by Dale Jorgenson and Peter Wilcoxen; GREEN = Greenhouse Policy Analysis Model developed by OECD, Goulder = General Equilibrium Energy-Environment-Economy Model developed by Lawrence Goulder; DRI = Mc-Graw-Hill/Data Resources Incorporated macroeconomic model. The models imposed the following carbon tax rates and measure the GDP losses in the following years: G2100 ($250/ton, 2010–2030), DGEM ($60/ton, 2020), GREEN ($160/ton, 2020); Goulder ($100/ton, 2020); DRI ($200/ton, 2010). (*Source:* Data from Figure 2–1 of Economic impacts of carbon taxes: detailed results, Electric Power Research Institute Final Report EPRI TR-104430-V2, November 1994.)

household real income. Of course, in the future a carbon tax can have a positive impact on incomes to the extent that it helps avoid damages to the economy that would otherwise result from global warming. In the present discussion, we focus on the current impacts on incomes. The beneficial effects from avoiding global warming will occur in the future; therefore, they are outside of our consideration.

Changes in gross domestic product (GDP) indicate the effects on incomes for the economy as a whole. GDP represents the total value of final goods and services produced in a given year, which roughly corresponds to the total income generated by the economy that year. Studies offer different predictions of the impacts of a carbon tax on GDP. Figure 27.3 displays the various impacts predicted by several economic models. The figure indicates the GDP impact per $100 of carbon tax.* These are results from hypothetical simulations in which a carbon tax is introduced at a given level in the year 1990 and maintained at that level indefinitely. The percentage of GDP losses is the percentage that the GDP falls in a given year relative to what it would have been in the same year if no carbon tax had been introduced. Most of the GDP losses are based on the impacts in the year 2010 – twenty years after the carbon tax is introduced.

* The magnitudes of the carbon taxes considered differ across the models, ranging from approximately $50 per ton to $120 per ton.

DISTRIBUTIONAL EFFECTS. The issue of efficiency concerns aggregate benefits and costs, and ignores how these benefits and costs are distributed among different groups. The issue of distribution – of who wins and who loses – is extremely important, however. Arguably, it is the issue of most intense concern in political discussions.

There are many dimensions of distributional impact. One is intergenerational distribution of benefits and costs. The benefits of a carbon tax and other climate-change policies are enjoyed primarily by future generations: the carbon tax primarily mitigates future damage to the world's climate. In contrast, the costs are borne by current generations to a significant extent. Even if the carbon tax remains in place indefinitely, future generations will bear a smaller tax burden to the extent that economies make the transition from conventional carbon-based fuels, which are subject to the tax, to alternative fuels, which do not incur the tax. This means that a carbon tax would tend to generate net losses to current generations and net gains to future generations.

Household income is another, vitally important, dimension of distributional impact. It is often claimed that a carbon tax would hurt low-income households more than higher-income households. This claim is supported by various studies. Table 27.3 shows the income distributional impacts predicted in a study employing the McGraw-Hill/Data Resources Incorporated Econometric Model of the United States. This study divides all U.S. households into five groups, or quintiles, with the richest 20 percent of households constituting the fifth quintile and the poorest 20 percent constituting the first. The impact on income distribution can be measured by comparing the percentage reductions in income that a carbon tax would impose on each household group or quintile. As indicated in the table, the poorest households experience the largest percentage reduction in income. Why is this so? The reduction reflects the fact that the poorest households tend to spend a larger fraction of their incomes on goods and services that have high energy (or fossil fuel) content, such as electricity, home heating, and transportation. A carbon tax would especially raise the prices of these goods. Because these goods constitute a larger share of poor households' budgets, lower-income families sustain a larger real income loss; that is, the basket of goods and services they can purchase shrinks by a greater proportion than for wealthier households.

Income distributional impacts can be tempered, to a degree. The revenues from a carbon tax could be used to finance cuts in income taxes for the poor. This could offset the adverse impacts on income distribution. Such a cut in income taxes does not undermine the purpose of the carbon tax, which is to discourage consumers from purchasing carbon-intensive goods and services. Even with the income tax cut, low-income households would have an incentive to substitute other goods and services for carbon-intensive ones, because these households would still perceive the cost of carbon-intensive goods and services as having increased relative to the other goods and services they purchase. How-

TABLE 27.3. INCOME DISTRIBUTIONAL IMPACTS OF A U.S. CARBON TAX

Income Quintile	Tax Impact in 2010 As Percentage of Income
1 (Poorest)	−0.45
2	−0.13
3	−0.07
4	−0.16
5 (Richest)	+0.08

Note: These results are based on simulations of a carbon tax introduced in 1990 at a rate of $15 per ton and growing at 5 percent per year in real terms.

Source: Results are taken from Figure 2a of Bruce Schillo, Linda Brannarelli, David Kelly, Steve Swanson, and Peter Wilcoxen, p. 241–266. The distributional impacts of a carbon tax, in Darius Gaskins and John Weyant, eds., *Reducing global carbon dioxide emissions: costs and policy options*, Energy Modeling Forum, Stanford University, pp. 241–266.

ever, income tax cuts might not entirely offset the income distributional problem. Many indigent families do not pay income taxes. It is difficult to reach such households through cuts in other taxes.

SOME INTERNATIONAL CONSIDERATIONS. Another key distributional dimension is the international distribution of benefits and costs. If undertaken unilaterally, a U.S. carbon tax would, to some extent, place domestic producers at a disadvantage relative to foreign competitors. Domestic energy companies are quick to remind the public and politicians of this fact. The carbon tax can be designed in a way that helps trim the adverse effects on competitiveness. Nearly all serious proposals of a U.S. carbon tax would include a levy on imported fossil fuels to match the tax on domestically produced fuels. This would help to ensure that domestic producers are on a par with to foreign suppliers in the U.S. market. In addition, most proposals would exempt exports of domestically produced fuels. With the exemption, domestic producers would experience no disadvantage relative to foreign competitors on the world market.

These provisions for imports and exports do not entirely undo the competitiveness problem. Other things equal, a unilateral carbon tax would raise the costs of production in the United States relative to the costs of production abroad for any firms that use fuels produced locally. For example, foreign producers of (energy-intensive) steel or aluminum, would have an advantage over U.S. producers.

There is no easy solution to this problem if the United States acts unilaterally in introducing the carbon tax. One might conceive of import taxes on all foreign products, with the tax proportional to the energy or fossil fuel content. However, it would be extremely difficult to calculate the energy content of all such products. In addition, a differentiated tax of this kind would be very costly to administer.

Many of these competitiveness problems would disappear if the carbon tax were implemented multilaterally (that is, by several nations at once) rather than unilaterally. If the United

States and all its trading partners introduced a carbon tax, then the economic playing field would be level. In addition, because foreign carbon taxes would cause the prices of foreign-made energy-intensive goods to rise, the price of energy-intensive goods produced in the United States would not be undercut.

Unfortunately, there are huge obstacles to achieving international agreements on multilateral carbon taxes. One key hurdle is the free-rider problem. For example, if country A reduces its emissions of carbon dioxide, other nations benefit; because carbon dioxide is relatively uniformly mixed throughout the globe, the impact of country A's action on the climate therefore occurs worldwide. This may tempt some countries to become "free riders," that is, to avoid making their own commitments to reduce emissions and instead simply reap the benefits of the abatement efforts undertaken by other countries.

A second key obstacle is that the impacts of higher greenhouse gas concentrations differ dramatically across countries and even within countries. Russia and Canada might benefit, overall, from an increase in average temperature. Therefore, they have weak incentives to undertake costly measures to reduce their carbon dioxide or other greenhouse gas emissions. Potentially, the countries that would benefit from reduced growth of carbon dioxide concentrations might have incentives to pay for the costs of reducing carbon dioxide emissions in countries that would not enjoy such benefits. However, reaching agreement on such international transfer payments is a very difficult enterprise.

A third problem is that there are serious disagreements over which countries are most responsible for the prospect of global climate change. The view put forward by the leaders in many developing country is that the industrialized nations are largely responsible for the problem, as a result of their historic contributions to carbon dioxide emissions. They argue that developing countries should not be required to make sacrifices to avoid future climate change. International agreements that involve only industrialized nations can significantly mitigate the global climate change problem; however, many of the problems of international competitiveness will remain if developing countries do not also undertake efforts to reduce carbon dioxide emissions.

COMPARISON WITH ALTERNATIVE POLICIES

A BTU TAX. One alternative to the carbon tax is a fossil fuel British thermal unit (BTU) tax. This tax on fossil fuels is based on the energy produced in burning the fuel, not on the carbon content (and carbon dioxide released). The relative merits of a carbon tax or a BTU tax depend on one's objectives. If the principal goal is to reduce carbon dioxide emissions, the carbon tax is superior because it targets more effectively the source of such emissions. This implies that a carbon tax will induce greater carbon dioxide reductions than a BTU tax that generates the same revenue.

The superiority of a carbon tax over a BTU tax is less clear if one's objective is to address a broad set of environmental problems associated with fossil fuel burning, not solely carbon dioxide emissions. Other externalities stemming from fossil fuel burning include the generation of sulfur dioxide and of nitrous oxide compounds. It is not clear whether the overall environmental damage from emissions of carbon dioxide, sulfur dioxide, and nitrous oxide is more closely associated with carbon or energy content; consequently, it is not obvious whether the carbon tax is preferable when one's environmental objectives are broad.*

AN INCREASE IN THE FEDERAL GASOLINE TAX. Because a carbon tax has a broader base than a gasoline tax, its impact is spread more evenly across industries. A gasoline tax would be less effective than a carbon tax in reducing carbon dioxide emissions per dollar of revenue raised, for the same reasons as discussed above in the context of a BTU tax.

The relative attractiveness of one policy compared with another again depends on one's objectives. A gasoline tax is an effective instrument for dealing with externality problems closely related to the burning of gasoline. An important example is ground-level ozone pollution from gasoline-powered vehicles. In principle, efficiency considerations allow room for both a carbon tax and a gasoline tax: The former would deal with environmental problems related to carbon dioxide, and the latter would address externalities stemming from the use of gasoline.

REGULATORY LIMITS ON THE USE OF FOSSIL FUELS. Thus far, we have concentrated on price-based approaches to handling environmental problems: carbon taxes and other taxes associated with fuel use. However, most environmental regulation in the United States takes the form of command-and-control regulation: mandated technologies or performance standards. Examples include required catalytic converters in automobiles and "tailpipe" emissions standards.

Would a command-and-control approach be most effective in dealing with carbon dioxide emissions? A potential policy of this type would consist of regulations limiting the use of fossil fuels by firms in industries that use these fuels directly. For example, electric power plants could be required to switch from coal to natural gas (which involves less carbon). Or steel manufacturers would be required to use less coal in the production of steel. Or automobile manufacturers would be required to achieve higher fleet averages for miles per gallon, thereby reducing demands for gasoline and the crude oil from which it is refined.

The price-based approach (fuels taxes) over direct controls has several potential advantages. First, relative to regulatory standards, fossil fuel taxes are inexpensive to administer. Each fossil fuel is demanded in numerous distinct industries. Economic efficiency would call for different standards for each of the industrial, commercial, and residential uses of each fuel. Given the thousands of distinct uses involved, determining, monitoring, and enforcing such standards would be very costly. In contrast, a carbon tax (or BTU tax) would require only three tax rates: one each for coal, oil, and natural gas.

A second, closely related advantage of the carbon tax is its flexibility. As more information becomes available concerning the environmental damage caused by carbon dioxide emissions, it makes sense to alter the policies in effect. It is much easier to adjust three tax rates than to make different adjustments to a huge number of standards.

Finally, in comparison with standards or other command-and-control measures, a carbon tax yields stronger incentives for technologic innovation. If a firm's use of a given fossil fuel happens to be below the standard for that industry, the standard exerts no further incentive for the firm to reduce its fuel use. Under a carbon tax, in contrast, firms reduce their tax obligations whenever they find means to lower the use of fossil fuels as inputs.

If the tax approach has so many advantages, why are direct controls favored as part of environmental policy? One reason is that the costs of standards tend to be less visible than those of taxes. Regulatory costs under a tax are quite visible: there is an explicit tax payment. Direct controls impose costs as well, by requiring firms to purchase equipment or inputs they otherwise would not have utilized. However, these costs are less obvious. The public also tends to trust direct regulation more than taxes. Direct controls seems to offer more assurance that the desired reductions in emissions will be achieved. However, this is more an issue of perception than fact: There is strong evidence that households and firms respond to tax-induced higher energy prices by reducing consumption of the higher-priced fuels or of the products whose prices are especially raised by fuels taxes. For example, many empirical studies indicate that for each percentage increase in the price of gasoline, the demand for gasoline falls between 0.5 and 1 percent in the long run.

SUBSIDIES TO AFFORESTATION. Climate damage depends on atmospheric concentrations of carbon dioxide, which depend on the rate of carbon dioxide emissions, and on the rate of absorption of carbon by "carbon sinks." Therefore, policies that increase the rate of absorption can help prevent climate damage much like policies that retard the rate of carbon emissions. Trees represent important carbon sinks; consequently, afforestation – the planting of new trees in places where trees have already been harvested or have never grown – is a potential policy worth considering.

---

* In comparison with a carbon tax that generates the same overall revenue, a BTU tax would tax coal less heavily, and natural gas more heavily. The BTU content of a ton of coal is approximately 3.8 times that of a barrel of oil and 21.3 times that of a unit (million cubic feet, or mcf) of natural gas. In contrast, the carbon content of a ton of coal is 4.6 times that of a barrel of oil and 36 times that of a unit of natural gas.

Although BTU taxes, direct controls, and gasoline taxes might "compete" with carbon taxes as policies to deal with emissions, there need not be any competition between a carbon tax and afforestation policies. It makes sense to pursue emissions reductions and afforestation simultaneously, because concentrations can be regulated more cheaply if both policies are implemented. In contrast, it does not make sense to employ both a carbon tax and a BTU tax as a greenhouse policy, because the carbon tax operates on the same entity – carbon dioxide emissions – and is always a less costly way of achieving emissions reductions than the BTU tax.*

This principle is illustrated in Figure 27.4. The horizontal axis indicates the percentage reduction in contributions to atmospheric carbon dioxide concentrations, either from reduced emissions or from expanded sinks. The vertical axis indicates the marginal or incremental costs of achieving these percentage reductions. Marginal costs are provided for carbon dioxide emissions reductions and for afforestation, and are based on values reported by William Nordhaus. For simplicity, these relationships ignore important issues of timing; in the real world, the costs would depend on when and how quickly the reductions are achieved. The marginal cost curves are upward sloping in keeping with the idea that achieving additional reductions requires increasingly more difficult sacrifices. For example, initial reductions in carbon dioxide emissions might involve relatively low marginal costs, because for some uses fairly cheap substitutes for carbon-based fuels can be found. However, additional reductions require that the use of carbon-based fuels is avoided in activities or processes in which relatively cheap substitutes for such fuels are not available. Similarly with afforestation: The first bits of afforestation can be achieved fairly inexpensively. However, additional afforestation requires use of land that might be less suitable for the purpose or that might cost more to obtain. In fact, the afforestation marginal cost curve here becomes vertical at concentration reduction of approximately 5 percent: This indicates that further contributions through afforestation cannot be achieved at any finite cost.

Suppose that the benefits from avoiding climate change justified a 5 percent reduction in net emissions. The cheapest way to achieve this overall reduction is to equate the marginal costs of such reductions under the two methods. Here afforestation would contribute approximately 20 percent of the reduction in net emissions, whereas carbon dioxide emissions reductions would contribute the remaining 80 percent. This is cheaper than attempting to achieve the 5 percent reduction in concentrations only by reducing emissions if emissions alone were cut back, the cost of the last increment

---

* This assumes that the goal of the policy is to reduce carbon dioxide emissions. As discussed earlier in this chapter, more than one tax policy can be effective if there is more than one policy objective. For example, it may make sense to have both a carbon tax and a gasoline tax if policy goals include both reducing carbon dioxide emissions and dealing with other, gasoline-related environmental problems.

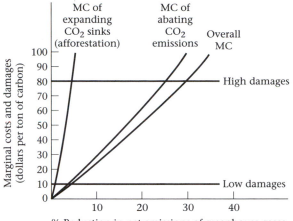

Figure 27.4. Achieving emission reduction at the lowest cost requires equality of marginal costs of reduction. CO_2 = carbon dioxide; MC = marginal costs.

would be approximately $14 per ton (as shown by the height of the marginal cost curve for carbon dioxide at the point indicating a 5 percent reduction in net emissions). The first unit of afforestation costs far less than this. If a bit of afforestation were applied instead of that last unit of carbon dioxide emissions, total costs of reducing carbon dioxide concentrations would be lower. The same reasoning applies whenever there is a difference in the marginal costs of reducing carbon dioxide concentrations across two methods. Overall costs are kept to a minimum when the incremental costs are the same across different approaches.

If benefits from avoiding global climate damage justified, for example, a 30 percent reduction in net emissions, the cheapest way to accomplish this would be for afforestation to contribute to approximately one-seventh of that amount, and for carbon dioxide emissions reductions to contribute the remainder. The marginal cost curves in Figure 27.4 suggest that the greater the overall amount of reduction in concentrations desired, the smaller the relative contribution by afforestation. This applies here because the costs from afforestation rise much more quickly than those from carbon dioxide emissions abatement.

POPULATION POLICY. The level of global emissions of carbon dioxide is the mathematic product of three factors: average carbon dioxide emissions per unit of industrial output, industrial output per person, and population. These three factors are the Ehrlich–Holdren I = PAT equation (where I is impact; P is population size; A is per capita consumption, or level of affluence; and T is a technologic factor, an index of ecologic damage) applied specifically to carbon dioxide emissions (see Chap. 22 for a discussion of this equation.) Clearly, carbon taxes and other measures aimed at fossil fuel production can help reduce the first factor. However, increases in the world's population ultimately contribute significantly to increased needs for industrial output and, through the second and third factors, to carbon dioxide emissions. Policies aiming to reduce population growth can usefully complement a carbon tax. Population policies can become highly controversial; however, this is not sufficient reason to avoid considering them seriously.

IS ACTION WARRANTED, GIVEN THE UNCERTAINTIES?

There is vast uncertainty about the extent to which continued accumulation of greenhouse gases would generate serious climate change and associated damages. In view of these uncertainties, it behooves us to consider one more policy option: postponing action on the climate front. Would it be better to wait for more scientific information before taking action? Those who suggest postponing action emphasize the possibility that we might take costly action today to reduce greenhouse gas emissions, only to discover, some decades later, that such action was unnecessary. However, if we do not take action today, it is possible that decades later, we will find that the failure to take action early was very costly.

A HEURISTIC MODEL OF DECISION MAKING UNDER UNCERTAINTY. Should aggressive policy action be deferred, to avoid the error of acting unnecessarily and thereby incurring unnecessary costs, or should it be accelerated, to avoid the greater costs of not acting soon enough? There is no simple answer to this question; however, the following heuristic model (which is a variant of the model developed by Alan Manne and Richard Richels) can shed considerable light on the issue. This model is meant to reveal qualitative aspects of the problem at hand. The model involves many simplifying assumptions; however, these simplifications are justified if they do not affect the qualitative results.

First, we divide time into two periods: the present and future. The "present" represents the interval from now to twenty to thirty years from now. The "future" represents the time from then on. Here, we assume that in the future (i.e., twenty to thirty years from now), much of the uncertainty concerning the extent of the damage caused by global climate change will be resolved.

Next, we specify the abatement costs of carbon emissions. For simplicity, we assume symmetric abatement cost functions, as shown in Figure 27.5. Suppose that the dimensions of each diagram are 1 by 1. Then the marginal costs of abatement functions have slopes of 1 and are given by $MC_1 = 1.0 \times a_1$ and $MC_2 = 1.0 \times a_2$. The amounts of carbon dioxide abatement are represented by the extent to which we move to the left in the diagram from point *1* (the intercept on the abcissa), which corresponds to zero abatement. The areas under the marginal cost curves – the shaded areas in each diagram – represent total abatement costs in each period associated with arbitrary specific abatement levels a_1 and a_2 (here a_2 equals $1 - a_1$).

Now we depict the uncertainty surrounding future global

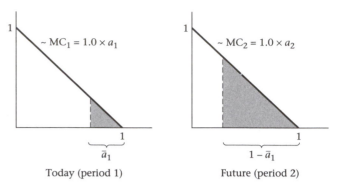

Figure 27.5. Costs of emissions abatement today and in the future. $\bar{a}_1$ is the abatement, MC_1 is the associated marginal private cost at stage 1; $\bar{a}_2$ is the abatement, MC_2 is the associated marginal private cost at stage 2.

warming as follows: There is a probability p that emissions would bring about global climate change serious enough to justify a *cumulative* emissions reduction of 1. Correspondingly, there is a probability $1 - p$ that, even with current and future emissions at the level of 1, no serious climate change damages will occur (and no need for emissions reductions). Call these two outcomes scenarios B and G (for "bad" and "good"). This is shown in Table 27.4

We assume that the uncertainty is resolved after period 1, that is, after the decision is made as to how much to abate emissions in period 1. Thus, we must make the decision concerning period 1 abatement in the presence of uncertainty. At the time of this decision, we do not know whether scenario B or G will occur – we have only probabilities to work with.

Suppose you are asked to determine the level of abatement today (if any) that minimizes the expected cost of abatement. The total expected cost (EC) of abatement is given by:

$$EC = 0.5 \times a_1^2 + (1 - p) \times 0 + p \times 0.5 \times (1 - a_1)^2$$

The first term is today's cost associated with abatement a_1. It is the area of the triangular shaded region in the left portion of Figure 27.5. The last two terms are the expected future costs. The middle term accounts for the probability $1-p$ that there will not be any need to undertake further abatement measures in the future: If scenario G applies, then there is no need for any further abatement. In fact, if scenario G applies, we will have learned that current abatement was not neces-

TABLE 27.4. ALTERNATIVE GLOBAL WARMING OUTCOMES

Scenario	Probability	Cumulative abatement needed to avert serious environmental damages
B	p	1
G	$1 - p$	0

sary either. The last term accounts for the probability p that scenario B applies, and the "remaining" abatement $1 - a_1$ will have to be undertaken so that cumulative abatement is 1. The greater the amount of abatement today (i.e., the greater is a_1), the smaller is the amount of abatement necessary in the future if the undesirable, or "bad," scenario applies.

What level of a_1 minimizes the total expected cost? Differentiating EC with respect to a_1 gives the following expression:

$$\partial EC/\partial a_1 = a_1 + p(1 - a_1)(-1)$$

The first-order condition for a minimum requires that this be set to zero. Rearranging gives

$$(1 + p)a_1 - p = 0, \quad \text{or}$$

$$a_1^* = p/(1 + p)$$

The above equation indicates that the cost-minimizing value of a_1 is a simple function of p, as indicated by Figure 27.6. Note that a_1^* increases with p: The greater the chances of serious environmental damage, the larger the amount of justifiable current abatement. Significantly, this equation indicates that some positive level of current abatement is justified for any (however small) positive value of p. Only if p is zero is there no rationale for current abatement! Why is some current abatement called for, even with the uncertainties?

It is thought that small amounts of current abatement cost relatively little but have large potential future payoffs. Figure 27.7 illustrates the cost of the "first unit" of current abatement. In the figure, the first unit of current abatement is g. The abatement cost associated with g is a (see Figs. 27.5 or 27.6). There is a future benefit to this current abatement. Suppose that the undesirable ("bad") scenario applies. By undertaking g today, somewhat less abatement will be needed in the future: Instead of having to abate by 1 in the future, society must abate by only $1 - g$. The shaded area b on the right hand diagram represents the future abatement cost that is avoided by undertaking abatement g today. This cost is quite large, because the avoided future abatement is the "last unit" of future abatement – the amount that is rendered unnecessary by undertaking the incremental abatement g today. Even if we reduce b to account for the fact that there is only a probability p that these savings will occur, the expected future benefit still exceeds g. The expected future

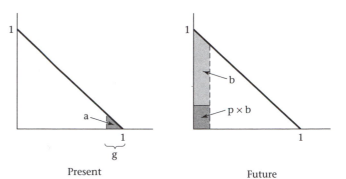

Present   Future

Figure 27.7. Why a small amount of current abatement pays for itself.

benefit associated with g is $p \times b$, this is the double-shaded region, which is still larger than the current cost a.

As Figure 27.7 indicates, when marginal costs of abatement are rising, instituting a small amount of current abatement can diminish the possibility of incurring exorbitant future marginal abatement costs. The expected saving in future abatement cost (indicated by the darkly shaded region) exceeds the current cost. Hence, it pays to undertake at least a small amount of current abatement, because doing so reduces expected future abatement costs by more than the cost of current abatement.

Undertaking prevention in the present is akin to purchasing insurance. Current prevention efforts have a cost (as do insurance premiums), which may turn out to be unnecessary (similarly, you may not ultimately file a claim – you may not become ill, your apartment may not be burglarized, etc.); however, it makes sense ex ante to pay for prevention now, just as it makes sense to buy insurance, in order to reduce expected future costs (which are incurred when some harm does in fact occur). In the case of global climate change, the reward for current prevention efforts is a lower expected value of future prevention costs that more than compensates for the cost of current prevention.

ARE THESE RESULTS GENERALLY APPLICABLE? Should we take these results as proof that we should be undertaking carbon abatement efforts today, despite the uncertainties? Is this strong evidence that the Bush administration was incorrect in arguing that current abatement efforts are premature? One might argue that the results are not general because the model is highly simplified. However, these results would apply even if the model were extended in a number of ways. It can be demonstrated that the same qualitative results would apply if the model were expanded to allow for (1) discounting of the costs of future abatement, and (2) a functional (negative) relationship between current abatement efforts and the marginal costs of future abatement.* Also, the same

Figure 27.6. Optimal current emissions abatement as a function of p. a_1 is the abatement, p is the probability that the bad scenario arises.

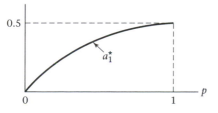

---

* Extension (1) would represent the marginal costs of abatement in period 2 by $1.0 \times a_2$ *δ, where δ is a discount factor less than

results apply in a more general model that explicitly incorporates the benefits of abatement as a function of the cumulative amount of abatement over the two periods.**

One simplification that could alter the qualitative results deserves mention. This model assumes there are no fixed costs or set-up costs associated with carrying out abatement. These are nonincremental costs that must be undertaken prior to the first small amount of abatement. An example would be the costs of setting up the bureaucratic infrastructure to administer and enforce a carbon tax. It can be shown that if these costs are sufficiently high, it is advisable to postpone abatement until the future, when the uncertainties are resolved.

IS THE EXPECTED-VALUE CRITERION APPROPRIATE? The problem described above was to minimize the expected value of necessary abatement costs. Some argue that this is not the most useful criterion for deciding upon a course of action, calling for an even more "conservative" approach. For example, consider the different outcomes under alternative abatement levels, as indicated in Table 27.5. Suppose the consensus probability of scenario B is 0.1. Then the level of period 1 abatement that minimizes expected total costs is $0.1/(1.1) = 1/11$, as indicated by the numbers in the right-hand column. Although this minimizes the *expected value* of total costs, it does not minimize the worst possible total cost. If scenario B arises, the actual total costs are 0.417 if a_1 is 1/11, whereas they are considerably smaller (0.250) if a_1 is 2. In fact, $a_1 = 2$ minimizes the worst possible cost.

Some argue that the decision criterion should be to minimize the worst possible cost – this is called the minimax criterion. Proponents of this criterion point out that, in this example, using the expected-value criterion leaves open the possibility that overall costs will be considerably higher than their minimized value. The minimax criterion is implicit in the "precautionary principle" advocated by some environmentalists, the notion that we should take whatever actions lead to the best outcomes in the worst possible scenario.

The minimax criterion reflects a higher degree of risk aversion than that underlying the expected-value criterion. In effect, the minimax criterion devotes complete

TABLE 27.5. ACTUAL AND EXPECTED COSTS OF CARBON DIOXIDE ABATEMENT AS A FUNCTION OF ABATEMENT UNDERTAKEN IN THE PRESENT

a_1	Actual total cost if scenario G materializes	Actual total cost if scenario B materializes	Expected value of total cost (EC)	
			$p = 0.5$	$p = 0.1$
0	0.000	0.500	0.250	0.050
1/11	0.004	0.417	0.212	**0.045**
1/4	0.031	0.312	0.172	0.059
1/3	0.056	0.278	**0.166**	0.078
1/2	0.125	**0.250**	0.187	0.137
1	0.500	0.500	0.500	0.500

attention to the worst scenario. One can also devise "compromise" criteria that imply a level of risk aversion between the "low" level implicit in the expected-value criterion and the "high" level implicit in the minimax criterion.

Which criterion is more valid? The answer is not easy, because it depends on our attitudes toward risk. People with less risk aversion argue that applying the minimax criterion would require society to undertake large expenditures "on average" (the expected value of costs is quite high) simply to ensure the best outcome in the worst scenario. (For example, if $p = 0.1$ the expected cost is 0.137 if the minimax criterion is used, as compared with only 0.045 if the expected-value criterion is used.) People with more risk aversion argue that applying the expected-value criterion would leave society vulnerable to excessively high costs if the worst scenario arises. There will never be unanimous agreement concerning the degree of risk aversion that should be included in the decision criteria used by policy makers. Which choice seems best to you, and why?

UNCERTAINTY AND CLIMATE CHANGE POLICY: GENERAL LESSONS. The answer to the question, "Should we abate today, despite the uncertainties?" is complex. When the costs of undertaking abatement are not fixed, it makes sense to initiate some abatement efforts today; this strategy offers a "hedge" against potentially prohibitive costs in the future. However, this hedging value can be undermined by high fixed costs. Ultimately, to answer this question we need to assign probabilities to different scientific outcomes, predict the marginal costs of abatement as a function of the levels of abatement, and recognize the magnitude of the fixed costs. In addition, we must make value judgments in choosing among alternative decision criteria (including minimizing the expected costs versus minimizing the costs of the worst possible outcome). Thus, complex conceptual, empirical, and moral issues are involved. No wonder the debate over climate change policy is so heated!

---

1. Extension (2) would represent the period 2 marginal costs of abatement by $\gamma * a_2$, where γ is less than 1. γ can be made a negative function of period abatement, a_1. Neither of these changes undoes the qualitative result that *some* period 1 abatement is required whenever p is positive.

** The original model implicitly assumes a discrete function for the environmental costs associated with emissions accumulation. These costs are implicitly infinite in scenario B if abatement is less than 1, and zero in that scenario if abatement equals 1. Environmental costs would depict the environmental costs as a function of cumulative emissions. The same qualitative results hold for the more general model.

CONCLUSIONS

This chapter has explored a number of options for dealing with the prospect of global climate change. The carbon tax is an especially effective policy instrument for this purpose. The tax causes the prices of carbon-based fuels to reflect their full social cost – including the damage related to climate change associated with the carbon dioxide emissions from, the combustion of these fuels. Increased prices, in turn, spur industry to find substitutes for these fuels and encourage consumers to shift their expenditures toward less carbon-intensive products. Because it deals with the externality associated with the use of carbon-based fuels, the tax can generate efficiency gains, that is, yield benefits (in the form of avoided damage to the environment) that exceed the economic costs associated with reducing the use of these fuels.

The carbon tax has several potential advantages relative to traditional, command-and-control approaches to reducing the use of carbon-based fuels. The tax is easier to administer and more flexible, and it provides continuous incentives for industry to develop technologic alternatives.

Goals of reducing the buildup of atmospheric carbon dioxide can be reached at lower cost if emissions reduction policies are pursued in tandem with policies to expand carbon sinks and with population policies. The relative emphasis on emissions abatement and afforestation policies depends on the marginal costs of these policies. Given concentrations of carbon dioxide are achieved at the lowest cost when the marginal costs of each of the approaches are the same.

The presence of uncertainty concerning the benefits of reducing carbon dioxide emissions is not sufficient reason to avoid taking costly abatement action today. If marginal costs of abatement are an increasing function of the amount of abatement, taking action today offers a hedge against potentially prohibitive future abatement costs. The hedging advantages justify current abatement efforts, except in cases where very large fixed (as opposed to marginal) costs are associated with abatement.

QUESTIONS

1. How can a carbon tax improve the efficiency of an economy?

2. Some people oppose carbon taxes because they would lead to higher prices of goods and services using fossil fuels. Is there a way to reduce the economy's use of carbon without causing price increases of this kind?

3. To what extent might direct limits on fossil fuel use be effective in reducing emissions of carbon dioxide? How effective might this option be in comparison with a carbon tax?

4. Why might a carbon tax have a regressive impact? Can policies be designed in order to prevent or significantly reduce such regressivity? If a certain amount of regressivity is unavoidable, should the carbon tax be rejected?

5. What are the difficulties in establishing international agreements to reduce carbon dioxide emissions? Is it appropriate to ask developing countries to join in such agreements? Why might the participation of such countries be important to the effectiveness of international efforts to reduce carbon dioxide emissions?

6. Why might it make sense to combine a carbon tax policy with a subsidy to afforestation?

7. ''There is much uncertainty about the likelihood of significant global climate change and about the associated impacts on the economy. Therefore it makes little sense to take costly action today to reduce greenhouse gas emissions.'' Evaluate this statement.

FURTHER READING

Barrett, S. 1993. International cooperation for environmental protection. In R. Dorfman and N. Dorfman, ed., *Economics of the environment: selected readings* (3rd edition). New York: Norton.

Daily, G., Ehrlich, P. R. Mooney, H. A. and Ehrlich, A. H. 1991. Greenhouse economics: learn before you leap. *Ecological Economics* 4. 1–10.

Manne, A., and Richels, R. 1992. Buying greenhouse insurance: the economic costs of CO_2 emissions limits. Cambridge, MA: MIT Press, pp. 67–75.

Morrisette, P. 1991. The Montreal Protocol: lessons for formulating policies for global warming. *Policy Studies Journal* 19(2): 152–61.

Nordhaus, W. D. 1990. Greenhouse economics: count before you leap. *The Economist* 20–4.

Nordhaus, W. D. 1991. The economics of the greenhouse effect. *Economic Journal* 101:920–37.

Nordhaus, W. D. 1993. Reflections on the economics of climate change. *Journal of Economic Perspectives* 7 (4):11–25.

Repetto, R., and Austin, D. 1997. *The costs of climate protection: a guide for the perplexed.* Washington, DC: World Resources Institute.

28 Land Use: Global Effects of Local Changes

EDWARD A. G. SCHUUR
PAMELA A. MATSON

HUMAN-INDUCED ALTERATION OF THE FACE OF THE EARTH

It is estimated that half of the ice-free terrestrial surface of the Earth has been transformed in some way by human activity. We have cleared or selectively harvested forests, converted grasslands to pasture or agricultural systems, drained wetlands, flooded uplands, and irrigated drylands. Much of this extensive transformation has occurred within the past 300 years, a relatively short period in the long march of human existence. For example, since 1700, land area devoted to crop production increased 466 percent to a current 1500 million hectares worldwide. If all of that agricultural land were contiguous, it would cover an area almost twice the size of the entire continental United States coast to coast with cropland. This magnitude and rate of change in land cover is unprecedented in the history of life on this planet.

Most of us have flown over or seen photographs of the major agricultural areas of the United States. Whether it is corn fields spread across Iowa, the orchards and vegetable fields of the San Joaquin Valley of California, or the vast cotton fields of Texas, the landscape is visually dominated by an endless patchwork of agricultural fields, obvious from 30,000 feet in the air. Similarly, patchworks of cut and regrowing forest, interspersed by small areas of old-growth forest, dominate the northwest and southeast of the United States. These patterns of land use are not restricted to North America; they are repeated on every continent settled by humans (Fig. 28.1). These human-dominated landscapes supply tremendous amounts of food and fiber to human populations locally and abroad, and thus far have kept pace with the ever-increasing demands of a growing world population. They also have left their mark on the Earth system as a whole.

We would all agree that most of these human-altered landscapes look different from natural lands. This is most obvious at the scale of a field or forest, where the altered system obviously supports plants and animals and fungi that are different from those supported by the unaltered one; not surprisingly, the processes that link plants, animals, soil, water, and air within the ecosystem also are altered. However, a bird's-eye view brings to mind a new set of questions about the effects of land use change on the Earth system *beyond* the forest or field. For example, what kinds of interactions take place among the natural and managed ecosystems on the landscape? Are the natural systems that surround agricultural areas indirectly affected by them? Can land use changes affect regional hydrologic patterns so that the effects of the change are felt far beyond the location of change? Is there an aggregate effect of all the world's land use changes on global biogeochemical processes or global climate or global biologic diversity?

These questions are relevant today as we unambiguously measure regional and global changes in the Earth system and seek to understand the sources of these changes. We are asking them today because the scale and rate of land transformation has become obvious, even to casual observers. Limits to the terrestrial biosphere are becoming visually apparent as individual plant and animal species and, indeed, entire ecosystem types are disappearing from the face of Earth, never to return. Perhaps more important, we are asking these questions now because we are able to do so: We have the technology (e.g., satellite remote sensing) to observe and quantify change at larger scales than previously possible, and over a range of time scales. Moreover, we now have the tools to look back into the Earth's past and describe present-day changes in the context of the longer-term patterns of natural change. It is clear that the global changes that we are measuring today are, in part, the cumulative result of land use changes over recent history. One goal for Earth system scientists is to develop approaches that limit the extent to which human needs require the alteration of the Earth's surface, and that minimize the effects of land use on the global system.

Figure 28.1. *LANDSAT* thematic mapper images of land use in Hood National Forest, Oregon, in 1991 (right) and north of Manaus Brazil in 1990 (left). The lightest areas represent recently cleared or harvested sites, gray areas represent regrowing forest, and the darkest areas represent intact forest. Each image covers approximately 1200 square kilometers. (*Source:* Courtesy of C. J. Tucker, NASA–Goddard Space Flight Center.) See color plate section for color version of this figure.

Land use change describes a discrete transformation that occurs at a particular site. Although such changes occur at a local scale, they can affect the entire Earth system in several ways. First, land use changes can lead directly to an alteration of the well-mixed, fluid envelopes of the Earth system: the atmosphere and the hydrosphere. Land use changes contribute to the increasing concentrations of carbon dioxide (CO_2) and other radiatively and chemically important gases in the atmosphere, which in turn play a role in determining climatic patterns. Second, although land use change occurs at discrete locations, it is extremely widespread. In total, land use changes must be considered a rapidly changing global phenomenon in their own right. This cumulative change has direct effects on climate by altering evapotranspiration rates (water evaporated by plants) and albedo (an index of reflected solar energy). It is a change that is having unprecedented effects on the diversity of Earth's biota. Land use causes massive fragmentation and even complete elimination of entire ecosystem types (e.g., tallgrass prairie and tropical deciduous forest), with concomitant loss of numerous species and populations (see Chap. 17 and 21).

In this chapter, we focus primarily on the ways in which changes in land use affect the fluid envelopes of the Earth system. It is our intention to illustrate that decisions made concerning individual pieces of land can affect the entire Earth system. Because the Earth's atmosphere is relatively well mixed and its constituents reasonably well quantified, and because some of the mechanisms that link atmospheric constituents to specific land uses are quite well understood, we emphasize the connection between land use change and atmospheric change. We describe how land use affects nitrogen and carbon cycling within ecosystems, which in turn controls the production of gases that are released to the atmosphere. Changes in these atmospheric gases are expected to change the heat balance of the Earth system and, ultimately, climatic patterns. Understanding the link between human land uses and global consequences will have important ramifications for future human transformation and use of Earth's terrestrial surface.

TYPES AND CAUSES OF LAND USE CHANGE

Land use takes many forms; humans use, at various levels of intensity, almost all types of terrestrial ecosystems. In our analysis we include land use changes that involve conversion from one type of land cover to another, as well as modification of a given land cover. *Conversion* refers to a complete change in the vegetation type – either conversion of a natural ecosystem type to a managed ecosystem or reversion of previously managed land. Typical examples of conversion include deforestation or conversion of grassland to agricultural use, reforestation of abandoned farmland, or conversion of any natural ecosystem to urban use. *Modification* refers to land use that involves alteration and management of the natural ecosystem, but not a complete change in the vegetation type. Typical examples of land modification include management of natural grasslands for livestock grazing, management of forests for timber, and intensification of agriculture through increased use of industrial fertilizers, irrigation, and pesticides. These categories are not necessarily always distinct from one another; for example, intensively farmed land may involve both conversion and modification.

Given the range of land uses, it can be difficult to pinpoint the forces that drive changes in land use. Although the fundamental cause of land use change is subject to intense debate, many scientists and sociologists agree that it is a combination of the rapidly increasing human population coupled with an equally rapid per capita increase in consumption of natural resources spurred by technologic advances. Figure 28.2 shows the exponential increase in human population and the parallel increase in energy consumption over the past 150 years, demonstrating how our population size is at a unique point in human history. It took all of human existence to reach the 1 billion population mark approximately 175 years ago. The current human population is now approximately 5.6 billion people; it has doubled since the middle of the twentieth century, and is projected to at least double again before possibly leveling off sometime in the twenty-first century.

Figure 28.2. Changes in human population and in energy use per capita over the past 150 years. KW = kilowatts (*Source*: Adapted from J. Holdren, Population and the energy problem, *Population and the Environment*, vol. 12, 1991.)

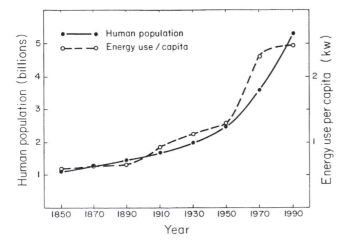

Such rapid population growth puts direct pressure on land in most regions of the world. For example, deforestation in some areas of Brazil and Indonesia results from resettlement of families out of overpopulated areas. These people may depend on subsistence farming for their livelihood, clearing forestland for agricultural use. Deforestation can also stem from the need for fuel wood as local energy source; this has occurred in Haiti and parts of Africa. Worldwide, much of the increase in agricultural land area and agricultural technology such as fertilizers, irrigation, biocides, and improved cultivars (the "Green Revolution") was driven ultimately by the need to feed more people (see Chap. 29). However, correlating the increase in the human population with land use changes does not always accurately identify the proximate causes of local land use patterns. In other words, on a regional level, the rate of land use change is not necessarily closely linked to the number of people who live there.

At local and regional scales, economic forces have a strong influence on land use change. Consequently, local land use change can be driven by policies and practices not directly connected to population growth. Resource extraction by international firms has caused extensive conversion of forests in sparsely populated tropical and temperate areas. This phenomenon is widespread; examples can be found in places such as Indonesia and British Columbia, Canada, where remote forest areas are currently subject to extensive timber extraction by both foreign and domestic firms. Domestic governmental policies such as price supports for export products, implemented in order to stimulate foreign exchange, are also critical forces in regional land use change.

Indirect forces may also have a strong influence on land use changes. These forces do not promote land use change by encouraging direct economic use of the land; however, they may influence land use changes for other reasons. Federal and state policies aimed at protection of borders or stimulation of colonization and development of a region can have the side effect of forcing land use change. An excellent example of this is the recently abolished policy that required clearing of the vegetation as a prerequisite to legal land ownership in some areas of the Brazilian Amazon. Colonists were required to convert their land prior to acquiring legal title, regardless of their original land use intent or characteristics of the land.

Given the range of forces driving land use changes, a global-scale understanding of why these changes are occurring and how they will change in the future is probably not possible. Instead, we must gain a detailed understanding of each region: its culture, its environment, the local politics and economics, and the interaction of the region with global politics and economics.

By now it is quite obvious that the phenomen of land use change is complex, encompassing a diverse range of change mechanisms, driven by an interacting set of individual, societal, and political forces, and distributed unevenly across Earth's land surface. Because of this complexity, it is not a simple task to evaluate the consequences of land use change for the Earth system. In the following sections, we describe some of the better understood ecologic and physical consequences of various land use changes, with an emphasis on global system effects. Global system effects such as increasing levels of atmospheric carbon dioxide are among the best documented; however, they are not the only consequences of land use change. We are only beginning to document and understand some effects, such as those associated with the loss of biologic diversity; other effects may not be understood or even detected at this point. At the same time, it is important to remember that land use changes are only one component in an array of causes of global change.

LAND USE CHANGES AND BIOGEOCHEMISTRY

DEFORESTATION. Forests cover approximately 30 percent of the terrestrial land surface, or approximately three times the total agricultural land area. This definition of forest encompasses a wide range of vegetation types including boreal, temperate deciduous, evergreen rain forest, tropical deciduous, and open woodland (open tree canopy interspersed with grasses or shrubs). Significant amounts of forested land of each type have been converted to agricultural systems throughout human history. For example, large portions of the European and the Indian subcontinents were prehistorically blanketed by forests; today (and for the past 5 to 10

centuries) they are extensively agricultural. Likewise, North America was once contiguously wooded from the Atlantic Seaboard to the Mississippi. This forest originally occupied 170 million hectares of land area; however, it was rapidly and extensively deforested by European settlers to a low point of a mere 10 million hectares of fragmented forest by the early twentieth century (see Fig. 22.5). Global land area covered by forest has declined 15 percent since pre-agricultural times as a result of anthropogenic conversion. This estimate, however, is a conservative measure of the human impact on forests because it *does not* include forests temporarily cleared for purposes such as timber harvesting. This estimate also masks the dynamic and heterogeneous nature of forest clearing around the globe; areas that were untouched in the past are now undergoing extremely rapid rates of deforestation, whereas some regions that were extensively deforested in past centuries are now undergoing reversion to forest cover.

Today, conversion of forested land to pasture or agriculture is one of the dominant land use changes in the humid tropical regions of Earth (Fig. 28.3). Annual land-clearing estimates are uncertain, in part because clearing rates are determined on a country-by-country basis, employing methods of estimation of varying sophistication. Some countries estimate clearing rates as a constant proportion of human population, whereas other countries employ satellite remote

Figure 28.3. Percentage changes in forested and woodland areas globally, 1974–84. (*Source:* B. L. Turner, *The Earth as transformed by human action: global and regional changes in the biosphere over the past 300 years*, Cambridge University Press, 1990.)

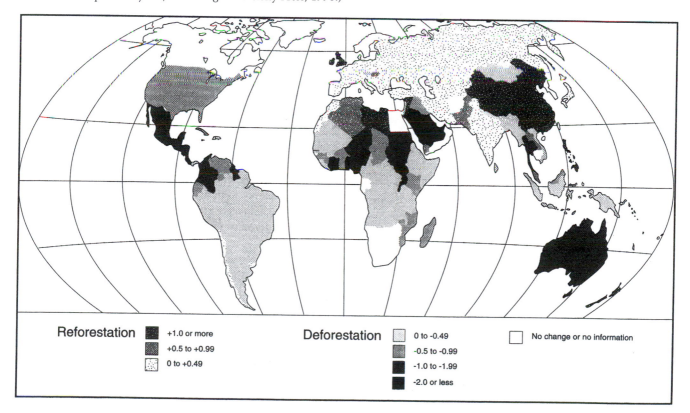

Reforestation			Deforestation			
■	+1.0 or more		▦	0 to -0.49	□	No change or no information
▨	+0.5 to +0.99		▦	-0.5 to -0.99		
▦	0 to +0.49		■	-1.0 to -1.99		
			■	-2.0 or less		

sensing to measure change directly in forest cover over time. One intermediate estimate suggests that 7.5 million hectares of primary tropical forest (forest that has never been cleared) is cleared annually, either permanently or in the process of shifting cultivation (see the section on "Shifting Agriculture"). Another 3.7 million hectares of primary forest are logged for commercial use. Finally, approximately 14.5 million hectares of secondary forest (forests that have regrown after earlier clearing) are cleared each year. At this rate of clearing, most primary tropical forests in such regions as Southeast Asia, East and West Africa, and Central America may disappear in the next several decades. However, these regional patterns are far from homogeneous, and primary forests are likely to persist longer in places such as Amazonia and Central Africa.

In order to calculate the effect of deforestation on global bigeochemistry, we can examine the physical and biologic changes that occur in the ecosystem as forests are cleared. As a forest is cleared, the vegetation is cut and left to dry in place until it can be burned. Combustion of dead plant material releases carbon, nitrogen, and sulfur as a suite of gases that go directly into the atmosphere. This is a loss of biologically important elements from the ecosystem that will affect the regrowth of vegetation and can directly alter future ecosystem dynamics. Following removal of the vegetation cover, the microclimate (temperature and moisture conditions) at the soil surface is drastically altered. Because vegetation no longer shades the soil surface, soil temperatures increase. Similarly, lack of evapotranspiration (plant uptake and evaporation of water) causes soil moisture to increase. These temporary warm, wet conditions lead to an increase in the decomposition rate – the rate at which organic matter in the soil is broken down and utilized as an energy source by microorganisms. In the process of decomposition, carbon dioxide is released to the atmosphere; increased rates of decomposition transfer more carbon dioxide from where it was stored in the soil to the atmosphere. Along with the direct increases in carbon dioxide fluxes out of the ecosystem through biomass burning and increased rates of decomposition, elimination of plant uptake of carbon dioxide (through the process of photosynthesis), even for a short period, shifts the balance from net carbon dioxide uptake to carbon dioxide loss, causing the deforested area to become a net source of carbon dioxide to the atmosphere. This means that the ecosystem is giving off more carbon dioxide to the atmosphere than the plants are taking up, and that it is a *source* of carbon to the atmosphere, instead of a *sink* for carbon.

Although it is relatively easy to estimate whether a given area of land is a net source of carbon dioxide to the atmosphere at any given time, calculation of the degree to which deforestation is a net source at a global scale is much more difficult. Global estimates attempt to construct annual accounts of total land area deforested, the type of conversions (conversion to pasture results in lower carbon losses than does conversion to permanent agriculture), the type of forest and soil present prior to clearing, the pools of carbon stored in vegetation and soils prior to clearing, and the amount and form of carbon lost in biomass burning. Global estimates also take into consideration the extent to which reforestation is a net sink (more carbon dioxide is taken up than lost), which partially offsets carbon transfer from deforestation (see the following section, which describes reforestation). Given all these necessary pieces of information, many of which are local or regional measurements, it is not surprising that there is significant uncertainty in the global estimate of carbon dioxide lost from deforestation. Most researchers agree that deforestation annually contributes a net flux of carbon dioxide to the atmosphere of approximately 0.9 petagram (1Pg is equal to 1×10^{15} grams) of carbon. Although it may be difficult to imagine how much carbon this is, we can compare this figure to the amount released by fossil fuel combustion. Fossil fuel combustion contributes carbon dioxide at an annual level of 5 to 6 Pg of carbon, more than four times the average annual estimate of release from deforestation. Although significant in its own right, deforestation is only one human source contributing to the observed atmospheric increase of carbon dioxide (see Chap. 19).

Deforestation also has an important impact on nitrogen cycling, often leading to losses of nitrogen from the cleared ecosystem, either in gaseous form or in solution (Fig. 28.4). In the process of decomposition, mineral nutrients such as nitrogen and phosphorus are transferred (*mineralized*) from organic matter to inorganic forms that are readily available for plant uptake or, potentially, loss from the system. Ammonium (NH_4^+), the first inorganic nitrogen product of decomposition, can be used by plants, immobilized (taken up and stored in biomass) by microorganisms, or oxidized to nitrate (NO_3^-) by a specialized group of bacteria in the process of *nitrification*. Once nitrate is produced, it can be taken up by plants, consumed by microorganisms, volatilized to several different gaseous forms in the process of denitrification, or leached by water movement out of the soil into groundwater or adjacent surface waters.

The same nitrogen cycling processes outlined in Figure 28.4 take place in both undisturbed and deforested ecosystems; however, the loss pathways of nitrogen are more significant following disturbance. This is because: (1) increased decomposition rates following land clearing or other disturbance increase the rate of nitrogen mineralization; and (2) plant uptake of nitrogen is reduced, at least temporarily. These two factors cause inorganic nitrogen to accumulate in the soil; leached into groundwater, typically in the form of the nitrate ion; or become lost in gaseous forms such as nitrous oxide (N_2O) (Fig. 28.3; see also Chap. 19). Numerous studies from many regions of the world have shown that removal of vegetation leads to increased nitrate leaching and elevated nitrate concentrations in surface waters. If concentrations in drinking water are high enough, nitrate can be detrimental to human health, especially the health of infants. Nitrate leaching also has consequences for the natural downstream riparian (streamside), lake, and estuarine ecosystems; it can lead to eutrophication (nutrient enrichment) of

Undisturbed forest

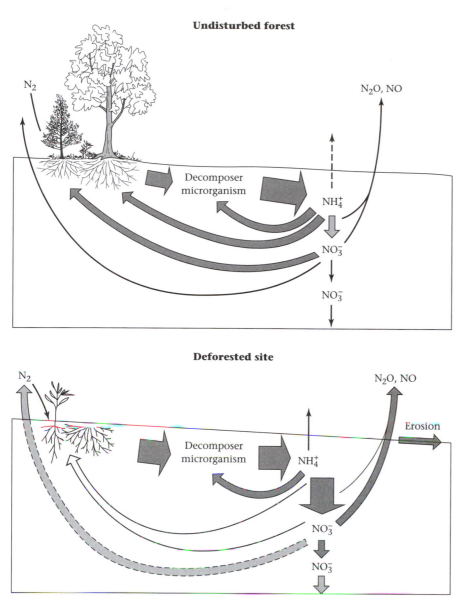

Deforested site

Figure 28.4. Nitrogen cycle processes in an undisturbed forest versus a deforested site. Following clearing, decomposition and release of inorganic nitrogen increase, plant uptake decreases, and the proportion of ammonium converted to nitrate (NO_3) (NH_4) increases. As a consequence, greater emissions of nitrogen trace gases (N_2 = nitrous oxide; NO = nitric oxide; N_2 = di-nitrogen) and greater nitrate leaching are possible (as shown by the width of the arrows). (*Source:* Adapted from B. Bolin and R. B. Cook, J. Wiley, *Major biogeochemical cycles and their interactions*, Chichester, 1983.)

some aquatic systems. Similarly, studies of tropical deforestation and subsequent conversion to pasture have shown that emissions of nitrous oxide often increase significantly following clearing and conversion, and in some cases remain elevated for ten years after clearing. Such increases may contribute to the observed atmospheric increase of nitrous oxide, a greenhouse gas. These losses of nitrogen from terrestrial ecosystems to water systems and to the atmosphere also can be critical to the terrestrial ecosystem itself; losses of nitrogen cause a reduction in the availability of nitrogen to plants, and nitrogen is an essential element for plant growth.

Deforestation can also have direct effects on regional and global climate. Vegetation cover controls soil water recharge by protecting the soil surface from erosion and by contributing the organic matter that maintains soil structure and drainage. Vegetation is also the main link for water transfer to the atmosphere through evapotranspiration. These factors help determine the amount of water cycled between ecosystems and the atmosphere. Moreover, vegetation cover influences the amount of incoming solar radiation that is absorbed – low plant canopy cover and bare soil reflect more radiation (higher albedo) than do dense plant canopies. Changes in energy and water exchange resulting from forest clearing have the potential to change climate and rainfall patterns, at least at a regional scale. Climate model simulations of the Amazon Basin suggest that complete conversion of forest to grassland could lead to increased surface temperatures, decreased total rainfall, and lengthening of the dry season (see Chap. 16 for a description of climate models). It is likely that such changes would limit rain forest regeneration in the region.

Other effects of deforestation with potential global significance are less well understood than are changes in biogeochemistry and climate. Vegetation cover protects the soil surface from impaction of precipitation, thereby reducing the potential for erosion to occur. Following removal of vegetation, erosion by water and wind leads to losses of soil organic matter and essential nutrients associated with the soil particles. Erosional losses together with the gaseous and leaching losses of nutrients (mentioned earlier) can lead to site degradation and decreased productivity (the ability of plants to fix carbon and grow). Site degradation can inhibit regrowth of pre-clearing vegetation, shifting the balance to plant species tolerant of nutrient-poor environments. Additionally, deforestation is a dominant cause of loss of biologic species with both degradation of the soil system and fragmentation of habitat as contributing factors.

REFORESTATION. Conversion of forested ecosystems to agricultural land is not a unidirectional process. Abandoned agricultural land can regrow to forest either through natural succession (the process by which ecosystems develop successive series of plant populations) or through active tree planting, called *afforestation*. Although deforestation is the dominant change in many regions of the world, reforestation is increasingly apparent in regions such as the eastern United States, Europe, China, and the former Soviet Union, especially on relatively unproductive or abandoned agricultural lands. Areas that are currently undergoing reforestation of some type were largely deforested by human activity in the past. One example of extensive reforestation occurred in the eastern United States between 1910 and 1980. Abandoned farmland reverted to forest on approximately 23.9 million hectares, almost 20 percent of the original land cleared.

The rate of which reforestation occurs in any given site, and the composition of the forest that results, depend on many site-specific factors that are strongly influenced by the history of land use. Conditions on abandoned agricultural land are substantially different from conditions immediately following forest clearing or natural disturbance. The new array of conditions influences the rates and trajectories of successional processes and silvicultural (trees planted and managed for timber) forest regeneration. Long-term and intensive agricultural practices can result in compaction of soil, depletion of the seed bank (supply of seed in the soil) of native species, depletion of soil organic matter, alteration of soil structure and drainage capability, reduction of soil water holding capacity, reduction in nutrient availability, and introduction of new weed species into the system. Natural reforestation under these conditions may proceed very slowly or not at all. In contrast, natural regeneration and/or afforestation in recently cleared areas tends to be comparatively rapid.

Similarly, grazing intensity and accompanying land management practices influence potential revegetation. Studies in the tropics revealed that eight years following abandonment, pastures grazed with moderate intensity returned to successional forest and had accumulated 12 percent of the biomass of a mature forest. However, land that was intensively grazed and managed accumulated only 1.5 percent of the mature forest biomass in the same time period. The regenerating vegetation in the latter site was dominated by grasses and other herbaceous species that restricted the growth of woody species and the biomass accumulation associated with them. In this study, the time period following abandonment was relatively short, accounting for the small biomass accumulation as compared to a mature forest; however, if the documented trend continues, the ecosystems that develop on the different sites may be radically different in terms of their species composition as well as their productive potential.

At a global scale, reforestation has the potential to influence the carbon cycle and atmospheric concentrations of carbon dioxide. Actively growing forests typically fix and store more carbon in vegetation annually than do mature forests, and young forests tend to be carbon sinks (they take up more carbon from the atmosphere than they release). The significance of this can be seen in the temperate zone, where carbon accumulated by reforestation approximately offsets current annual carbon dioxide releases from temperate zone deforestation. Despite the fact that reforestation removes carbon from the atmosphere, it cannot be thought of as an absolute, long-term solution to the problem of increasing atmospheric carbon dioxide concentrations. In order for reforestation to control atmospheric carbon dioxide levels, forests would need to cover nearly all land currently being used for agriculture as well as for forestland. Moreover, carbon would be sequestered by these regrowing forests only as long as they remained relatively young and actively growing, or as long their wood products were not combusted or decomposed. Therefore, reforestation can be considered only as a viable delay tactic; thus, it may contribute to the variety of approaches aimed toward control of carbon dioxide in the atmosphere.

BIOMASS BURNING. Fire is the predominant tool used to convert forests to agricultural uses and is also used widely to manage pastures and savanna systems for grazing and to maintain grassland productivity. The term *biomass burning* encompasses these practices as well as the use of wood as cooking fuel or heating, and the removal of postharvest agricultural debris by burning. Biomass burning is not itself a specific land use; however, it is a critical mechanism of land use change that in itself has significant effects on the global system. Fire represents a means by which biogeochemical elements are transferred directly from terrestrial pools to atmospheric pools.

Combustion of biomass releases a variety of gases that affect the atmosphere in numerous ways; these gases reflect the elemental concentrations in the vegetation as well as the intensity of the fire. Because approximately 50 percent of dry biomass consists of carbon, the predominant gases released are carbon compounds in various stages of oxidation: carbon dioxide (CO_2), methane (CH_4), carbon monoxide (CO), as well as a number of nonmethane hydrocarbons in much smaller amounts. The atmospheric role of these gases varies.

For example, carbon dioxide is a greenhouse gas, as is methane. Carbon monoxide and nonmethane hydrocarbons are not infrared absorbers, but they are involved in critical chemical reactions in the troposphere, including those that lead to the production of tropospheric ozone and other atmospheric pollutants (see Chap. 13).

Nitrogen is released as a variety of compounds also in various states of oxidation: molecular di-nitrogen (N_2), nitric oxide (NO), nitrogen dioxide (NO_2), ammonia (NH_3), and nitrous oxide (N_2O). The proportional release of these forms also depends on the intensity of the burn. Di-nitrogen commonly accounts for approximately 33 to 50 percent of the transformed nitrogen; nitric oxide and nitrogen dioxide (together known as NO_x) on average contribute approximately 12 percent of the nitrogen emissions, and nitrous oxide typically is a much less significant emission. As with the carbon gases, the atmospheric roles of these gases vary considerably. Nitrous oxide is a greenhouse gas and contributes to global warming; and nitrogen dioxide and nitric oxide play critical roles in atmospheric chemistry and the production of atmospheric pollutants such as tropospheric ozone.

In addition to these carbon and nitrogen gases, many sulfur-containing gases are released in fire. Moreover, soot and aerosol particles that consist of organic matter, elemental carbon, and a variety of trace species of carbon, nitrogen, and sulfur compounds are also released. These compounds are produced in varying amounts, depending on the composition of the biomass that is burned and the nature of the fire. Some of these compounds can have important regional and global effects. Satellite and aircraft data clearly show the transport of gases and aerosols in biomass-burning plumes from one region to another (Fig. 28.5). Aerosols act as cloud condensation nuclei, providing a condensation surface for water vapor in the atmosphere. Computer models have suggested that increased cloud cover due to an increase in aerosols from biomass burning could actually cool Earth by reflecting solar radiation. Aerosols associated with sulfur releases from coal combustion are likely to have an effect of similar or greater magnitude than biomass burning. Processes such as those associated with aerosols may in part counteract warming due to greenhouse gases and illustrate the difficulties of developing unambiguous predictions of global climate change.

SHIFTING AGRICULTURE. *Shifting agriculture, slash-and-burn,* and *swidden agriculture* are several terms used to describe a type of cultivation that has been practiced for thousands of years. Shifting agriculture is currently employed most extensively in tropical areas of the world; however, in the past it played an important role in the forest clearing of Europe and eastern North America. Typically, small areas of forest are cleared of most trees and burned to release organically bound nutrients. Soil is left untilled and planted in complex patterns (as opposed to the cultivated row-crop style) of mixed plant species and successive plantings and harvests. As fertility decreases and insect and plant pests encroach (often within 3 to 5 years), the agricultural plots are abandoned and allowed to regrow to forest; at the same time, other forest plots are cleared (Fig. 28.6). If the fallow periods during which plots are allowed to regrow to forest are sufficient, productivity and fertility of the land can be sustained, perhaps indefinitely.

Under ideal shifting agriculture, the soil system is left largely intact and loss of soil organic matter is transient; levels in the tropics are restored in as little as three years following abandonment to natural succession during the fallow period. Essential nutrients such as nitrogen reaccumulate in the regrowing forest through biologic fixation, rainfall deposition, and uptake of nutrients from deep in the soil profile. Although forest regeneration to the original state may take much longer, fertility may approach precultivation levels of nitrogen and phosphorus within several decades. As

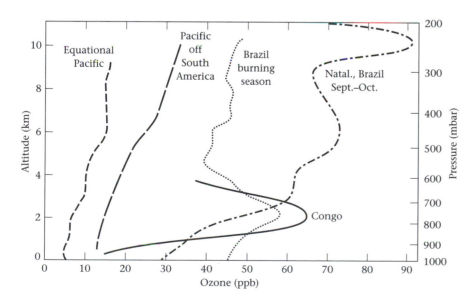

Figure 28.5. Concentrations of tropospheric ozone with altitude in the atmosphere. The highest concentrations are evident in tropical areas during the dry season, when biomass burning occurs. The bulges in ozone at low altitudes in the Congo and Brazil occur as a result of the development of plumes of smokes and gases. km = kilometers, mbar = millibars; ppb = parts per billions. (*Source:* Adapted from P. J. Crutzen and M. O. Andreae, "Impact of biomass burning in the tropics: impact on atmospheric chemistry and biogeochemical cycles." *Science*, vol. 250, 1990, pp. 1669–1678; figure adapted from J. Fishman and J. C. Larsen, Distribution of total ozone and stratospheric ozone in the tropics: implications for the distribution of stropospheric ozone. *Journal of Geophysical Research*, vol. 92, 1987, pp. 6627–6634.)

Figure 28.6. Aerial photograph of shifting cultivation in the Brazilian Amazon (*Source:* Photograph by E. A. G. Schuur and P. A. Matson.)

a result of human population levels that have allowed for sufficient fallow periods and judicious selection of land for cultivation, shifting agriculture has existed sustainably for thousands of years. Averaged over time, such land-clearing systems have relatively little effect on global biogeochemical systems.

Today, traditional shifting agriculture in many tropical regions is moving away from this ideal as land and population pressures become more intense. As fallow periods are shortened or eliminated entirely, nutrient and organic matter losses resulting from agriculture cannot be recouped, and the systems become degraded. Similarly, the long-term productivity and fertility of shifting cultivation systems can be difficult to maintain on marginal lands, such as those with steep slopes, soils prone to erosion, or highly infertile soils. For example, cultivation of lands prone to erosion occurred fifty years ago in the Bragantina Zone in eastern Amazonia; the results – 35,000 square kilometers of unproductive scrubland – persist today. This type of land conversion decreases the productivity of the successive ecosystem and can represent substantial net losses of carbon and other elements to the atmospheric pool.

INTENSIVE AGRICULTURE. The majority of agricultural land use involves tillage as a part of the management system

– that is, the plowing of soil on a regular basis (see the box entitled *No-Till Agriculture*). In the past several decades, potential agricultural yield per unit of farmland has increased substantially, primarily through the development and use of high-yielding crop varieties combined with industrially produced fertilizers, pesticides, herbicides, and irrigation (Chap. 29). This transition in agricultural practices represents a major shift in the way humans have traditionally practiced agriculture. The combination of tillage, new cultivars, and intensive inputs into agricultural systems has been largely responsible for keeping food production in step with the rapid human population growth of the past several decades. However, the practice of intensive agriculture carries significant consequences for the global as well as local and regional environments.

Globally, organic matter stored in the soil profile represents a larger carbon pool than the carbon stored in the biota and the atmosphere combined. Tillage disrupts the physical structure of the soil and exposes organic matter that is normally physically protected by its aggregation with soil minerals, thereby making it available for microbial decomposition. This physical change, along with the alteration of soil microclimate, results in increased rates of decomposition and increased potential for erosion. Thus, carbon stored as soil organic matter decreases when subjected to tillage. Of the

NO-TILL AGRICULTURE

Agriculture is the most extensive and dramatic alteration of natural ecosystems practiced by humans. Because of the widespread nature of agriculture, specific farming practices can have global consequences. Decisions in farming practices, by necessity, are typically based on short-term economic return to the farmer and not on long-range global impacts. Nonetheless, management practices can help mitigate the trends that we are documenting in this chapter. For example, the development of *no-tillage* agriculture in the United States and other regions of the world has the potential to reduce the amount of carbon transferred from the soil to the atmospheric pool under current intensive agricultural practices.

Agriculture takes many forms throughout the world, depending on the particular culture and environment. One of the most common practices, which unites a diversity of farming styles, is the plowing of the soil to prepare for planting. Whether this plowing is achieved through the use of hand-held tools (e.g., in the Peruvian highlands), the bullock (e.g., in some areas of India), or the gasoline-powered tractor (e.g., in the United States), they share this common link. Plowing (or tillage) prepares the soil for the planting of crops by disrupting weed plants that have colonized the fields, by breaking up and incorporating any plant residues left on the surface, and by breaking up the soil structure and the plant roots that keep the soil in place. This creates an ideal place for crop plants to establish. However, it also leaves the topsoil vulnerable to erosion by wind and rain, and it increases decomposition rates of soil organic matter, with concomitant losses of carbon dioxide to the atmosphere as well as increased mineralization and potential loss of nitrogen compounds.

For centuries, farmers have been addressing the impact of organic matter loss on crop yields. Until recently, farmers all over the world spread manure, human wastes, and waste plant material on their fields to maintain fertility and crop production. Similarly, for thousands of years Asian agriculture relied on periodic dredging of irrigation canals to recover eroded topsoil. These additions of carbon-containing residues balanced the increase in decomposition losses of soil organic matter that accompanied tillage; the approach was practiced by agrarian cultures worldwide.

In the North American and European nations, and now less developed nations, farming practices have deviated from the pattern described above. On the North American continent, the arriving European colonists found ''limitless'' acres of land for cultivation. As soils became depleted through cropping, new soils were cleared of native vegetation and old soils were abandoned to forest or prairie regrowth. Plowing of virgin soil allowed crop production without the necessity of adding organic matter because the native soil was initially very fertile. A second reason for the farming change was the widespread availability and use of industrially produced nitrogen. Farmers could rely on chemical nitrogen to fertilize crop plants, rather than mineralized nitrogen released from organic matter additions. This new development, occurring only in the past half century, has made the consequences of the loss of topsoil less obvious.

Over time, it has become clear that the consequences of the loss of topsoil associated with farming changes cannot be alleviated by the addition of nitrogen alone. Scientists have discovered that in addition to being the source of biologically available nitrogen, soil organic matter performs other functions in the soil profile. It provides structure to the topsoil, increasing the water-holding capacity and aeration of the soil. Also, because soil organic matter retains added nitrogen, fertilizer can remain available to crops and not leach into the groundwater. These scientific discoveries paralleled the experiences of farmers, who saw crop yields decline even as fertilizer use increased. This knowledge resulted in the development of management practices for industrial agriculture that would help retain soil organic matter in the agricultural fields in order to maintain crop yields.

Agricultural research on soil conservation practices spans a wide range, from improved plowing techniques and hardware to the breeding of perennial grain crops that could require less frequent tillage. One truly effective solution for industrialized agriculture is *no-till agriculture*. As the name suggests, this farming technique reduces the loss of soil organic matter by eliminating tillage entirely. This technique became possible with the development of modern chemical herbicides to control weed growth, which replaced plowing for the same purpose. No-till differs from conventional agriculture in that crop seed is planted directly into the unplowed soil surface, still containing root remnants from last year's crop. In this way, soil structure is maintained, soil organic matter is conserved, and crop yields are high.

No-till agriculture has gained popularity with farmers in some regions because of the associated economic practicalities. With the elimination of tillage, less tractor work is required to prepare the field for planting, resulting in reduced fuel demand. The tradeoff for this practice is apparent: an increased dependence on chemical herbicides for weed control. The short-term economic viability of this farming practice is balanced by the cost of herbicides versus the cost of tractor fuel. Longer-term effects of this practice are more difficult to quantify in dollar terms, however, the benefits of reducing carbon loss include maintaining future crop yields, balanced by unknown costs incurred by increased herbicide use.

Although not the direct target of soil conservation measures, no-till has the unintended effect of reducing the flux of soil carbon to the atmosphere. We can calculate the increase in the soil carbon pool that results when farmers use soil conservation measures as compared to current agricultural methods. In this calculation, we are combining the effects of no-till with several related low-till soil conservation practices such as *ridge-till* and *mulch-till*. Currently, soil conservation practices in the United States (27 percent of total cropland) conserve an estimated 255 teragrams of carbon in the soil profile as compared to the equivalent land area under conventional tillage. Additionally, if increases in the extent of soil conservation continue as they have since their introduc-

continued

NO-TILL AGRICULTURE (*continued*)

tion in the 1970s, by the year 2010, 76 percent of cropland in the United States will be managed by some type of soil conservation practice (Fig 28.7). This will result in an additional 377 teragrams of carbon sequestered in the soil profile, effectively removed from the atmospheric pool. To place these figures in a global perspective, the total carbon conserved over thirty years using the second estimate is equivalent to 0.9 percent of the carbon released by projected fossil fuel consumption in the United States over the same time period.

These projections demonstrate that soil conservation measures can help mitigate carbon dioxide increases in the atmosphere. However, although the effects of soil conservation measures are globally significant, they are dwarfed by the carbon dioxide levels released by fossil fuel consumption, especially in high fossil fuel–consumption nations such as the United States. Therefore, if we want to mitigate carbon dioxide increases in the atmosphere, regulating our use of fossil fuels is a more efficient target for change than is altering the pool of carbon in the soil. Nonetheless, it is encouraging that land use decisions that improve long-term viability of agriculture and also help reduce global atmospheric change are tenable. Such dual benefits should be factored into cost/benefit analyses made for land use changes in the future.

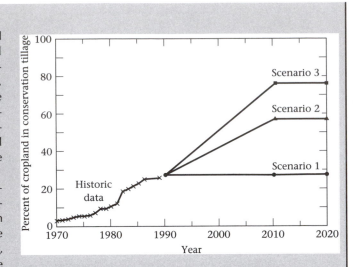

Figure 28.7. Historic and projected conservation tillage usage to the year 2020 for current levels of conservation tillage (scenario 1, 27 percent of cropland), and increases to 57 percent (scenario 2) and 76 percent (scenario 3). (*Source:* Adapted from J. S. Kern and M. G. Johnson, *Soil Science Society of America Journal*, Vol. 57, 1993, pp. 200–210.)

original carbon content of the soil profile, 30 to 50 percent is lost in permanently cultivated agricultural systems, with the highest levels of loss concentrated in the surface soil horizon. These soil carbon losses commonly occur within twenty years in temperate ecosystems, and within five years in tropical systems, depending on the temperature and moisture levels controlling microbial decomposition rates. The soil reaches a new carbon equilibrium based on the particular soil characteristics of that area, management practices, and carbon inputs from the cultivated plant species (Fig. 28.8). In addition to altering the productive potential of the soil, these carbon losses represent a net transfer of carbon in the form of carbon dioxide from the soil pool to the atmosphere.

Modern intensive agriculture also plays a significant role in the biogeochemical cycle of nitrogen. Intensive agriculture is dependent on the use of synthetic nitrogen fertilizer, which became widely abundant after 1945 (see the box entitled *Alterations of the Global Nitrogen Cycle*). Because grain is harvested and removed from the fields annually, agriculture depends on the continual input of nitrogen to maintain yields. Relatively recently, industrial nitrogen fertilizers have largely supplanted organic nitrogen applied as animal manure or supplied by nitrogen fixation by organisms (such as leguminous plants) as a nitrogen source to crops. The increasing use of chemical nitrogen fertilizers has consequences for both water and air quality, on local and global scales.

Regardless of the form in which it is applied (e.g., ammonium, urea, nitrate, organic manures), nitrogen fertilizers can undergo a series of microbial processes that result in the potential loss of nitrogen from the system. The potential fates of urea, an industrially produced form of organic nitrogen that represents approximately 50 percent of the fertilizer used worldwide, are shown in Figure 28.11. Urea is first hydrolyzed to ammonium in a chemical reaction. Ammonium can then undergo plant uptake, microbial immobilization, or oxidization to nitrate (termed *nitrification*); during nitrification, nitric oxide can be emitted. Once nitrate is produced, it

Figure 28.8. Loss of carbon and nutrients from soil following land clearing and permanent conversion to agriculture. The time at which carbon and nitrogen reach approximately 50 percent of original mass (denoted by arrow) occurs within approximately twenty years in many temperate sites, but may be reached within five years in tropical sites.

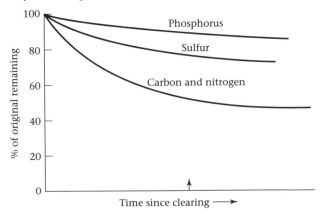

ALTERATIONS OF THE GLOBAL NITROGEN CYCLE

The global nitrogen cycle consists of a large, well-mixed, and relatively nonreactive pool of di-nitrogen in the atmosphere and smaller amounts of a number of nitrogen compounds that are stored in and cycle among atmosphere, water, soil, and biologic systems. Biologic processes mediate many of the transfers between these pools, because nitrogen is a critical element for all living organisms. Nitrogen is a building block for protein synthesis and often is the single most limiting element to growth of individual organisms, as well as to productivity of entire ecosystems.

Human activity has changed the global nitrogen cycle to an extraordinary extent. Under background conditions, involving no human influence, biologic nitrogen fixation is the predominant way in which nitrogen is transferred from the large, and largely unavailable, di-nitrogen pool in the atmosphere into the biosphere. Nitrogen-fixing microorganisms are able to fix di-nitrogen into usable forms that are incorporated into their biomass and the mass of plants associated with them. As they die and are decomposed, that nitrogen becomes available to other components of the biogeochemical cycle. Biologic nitrogen fixation in terrestrial ecosystems worldwide has been estimated at approximately 100 teragrams (1 teragram is equal to 1×10^{12} grams) of nitrogen per year, and nitrogen fixation in marine ecosystems is 5 to 20 teragrams per year. The other pathway by which nitrogen travels from the nonreactive di-nitrogen pool in the atmosphere to the biosphere is through lightning fixation: Nitrogen is converted to oxides of nitrogen during the high-energy reactions that lightning fuels. This pathway accounts for approximately 10 teragrams per year. In contrast to these natural fluxes of nitrogen, industrial nitrogen fixation for nitrogen fertilizer now accounts for more than 80 teragrams of nitrogen per year. An additional 25 teragrams of nitrogen are fixed by internal combustion engines and released as oxides of nitrogen. Finally, approximately teragrams 30 of nitrogen are fixed in biologic nitrogen fixation of crop plants such as soybeans. Together, these anthropogenic sources of nitrogen now fix more nitrogen globally each year than do the natural sources! Perhaps most amazing is how rapidly this phenomenon has arisen (Fig 28.9). Furthermore, land use activities are over whelmingly important in causing these changes.

What are the consequences of these enormous changes? Because nitrogen is so critical to biologic systems and is typically limited in its supply, this huge increase in biologically available forms can be expected to affect biologic systems at all levels of organization, including individual organisms, plant populations, plant communities (assemblages of populations), and entire ecosystems (the soil-plant-microorganism-animal-atmosphere-water system). Numerous experimental studies demonstrate that added nitrogen leads to a dominance of one or a few nitrogen-responsive plants, thus shifting species composition and species diversity in plant communities. Additions of nitrogen also affect animals that consume plants (termed *consumers*) as well as the parasites and predators of those consumers. Furthermore, when plant levels of nitrogen are changed, microorganisms that decompose those plant materials may also be affected. It is

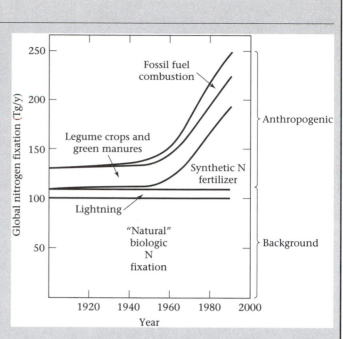

Figure 28.9. The total global amount of nitrogen (N) "fixed" into available forms (1 teragram = 10^{12} gram) each year over the twentieth century. Tg/y = teragrams per year.

easy to see how the alternation of nitrogen inputs can create change throughout an ecosystem and, indeed, throughout entire regions and the world.

Additions of nitrogen to terrestrial ecosystems can also cause changes in the outputs from ecosystems, including nitrogen leaching to groundwater and surface water (see the sections "Deforestation" and "Intensive Agriculture") and gas emissions. Field-scale experimental studies have shown that fertilizer inputs result in increased emissions of both nitrous oxide, the greenhouse gas, and nitric oxide, the chemically active gas involved in regional air pollution. For nitrous oxide, global-scale analyses are possible because it is a long-lived, nonreactive gas whose atmospheric accumulation is readily apparent. Globally, increased anthropogenic nitrogen fixation and input into terrestrial systems has occurred concurrently with the increase in atmospheric nitrous 16oxide (Fig 28.10), suggesting the possibility that nitrous oxide is responding to the land use–related increase in anthropogenic nitrogen fixation. However, global nitrogen fixation has been altered at least two-fold by human activity, whereas detailed analyses of the global budget of nitrous oxide suggest that it has changed by only approximately 30 percent. To date, the increases in nitrous oxide leaving ecosystems are much less than the increases in nitrogen entering ecosystems worldwide as a result of human activity. Where is the remaining nitrogen? Will we detect it in increasingly high nitrous oxide fluxes in the future? Or is it being lost through other pathways, including: the volatilization of nitric oxide, which leads to the formation of tropospheric ozone; the volatilization of ammonium with other atmospheric chemical impacts; the volatilization of di-nitrogen, with no atmospheric impact; or the leaching of nitrate to aquatic systems, with potential effects on water quality? Or is the nitrogen being retained in terrestrial ecosystems, perhaps in interaction with the uptake of the ever-increasing concentrations of atmospheric carbon dioxide? The answers to these questions are critical as we attempt to understand current global changes, predict changes, and develop prevention or mitigation strategies.

Because nitrogen fertilization and nitrogen fixation by crops are dominant sources of change in the global nitrogen cycle, anthropogenic alteration of nitrogen can be expected to continue to increase well into the twenty-first century. Fertilizer use is predicted to reach 120 teragrams annually by 2025, given current trends and practices. However, changes in management practices – such as subsurface placement of fertilizer, applying fertilizer in small amounts at frequent intevals, and carefully timing fertilizer application to plant demand – hold promise for minimizing losses of nitrogen from agricultural systems to the atmosphere. Such precise nitrogen management in intensive agriculture may be a means to maintain high crop yields, reduce the requirement for anthropogenic nitrogen fixation, and minimize local and global atmospheric and aquatic effects.

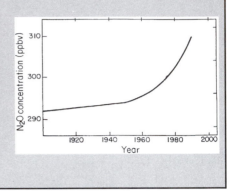

Figure 28.10. The change in atmospheric concentration of nitrous oxide (N_2O) over the twentieth century. The increase mirrors the increase in anthropogenic nitrogen fixation shown in Figure 28.9. ppbv = parts per billion by volume.

can be taken up by plants or microorganisms, volatilized to several different gaseous forms, or leached out of the soil by water movement.

Approximately 50 percent of the applied fertilizer nitrogen typically is taken up by crop plants as ammonium or nitrate and used for growth and reproduction, as intended by the farmer. Nitrate leaching is one common fate of at least a portion of the remainder. Numerous local and regional studies have measured elevated nitrate concentrations in systems adjacent to intensive agricultural areas; as noted earlier, such increases in nitrate have consequences both for human health and for the functioning of aquatic ecosystems (see the section entitled "Deforestation").

The use of nitrogen fertilizers also influences atmospheric processes through the emissions of a number of trace gases. The various nitrogen gases and their atmospheric roles are listed in Table 28.1. All these gases are produced by physical or biologic processes in the soil. Ammonia (NH_3) is often

Table 28.1 The Atmospheric roles of nitrogen gases

Nitric oxide	Regulates hydroxyl radical
	Regulates tropospheric ozone
	Precursor of nitric acid
Ammonia	Neutralizes precipitation
	Transports "available" nitrogen
Nitrous oxide	Greenhouse gas
	Regulates stratospheric ozone
Di-nitrogen	No effects on atmospheric processes

volatilized from surface soils soon after application of urea or ammonium-based fertilizers. Nitrous oxide and nitric oxide are produced in the biologic processes of nitrification and denitrification, and di-nitrogen gas is produced in the process of denitrification. Numerous studies have reported large increases in the emissions of all these gases following fertilization. Figure 28.12 illustrates the magnitude and rapidity of elevated nitrous oxide and nitric oxide emissions following fertilization in tropical sugar cane systems.

When nitrogen trace gases (nitrogen gases not including di-nitrogen) enter the atmospheric pool, they play a number of critical roles in radiative and chemical processes (Table 28.1). Some of the products of these reactions can be deposited to terrestrial systems downwind. For example, nitric oxide reacts to form nitrogen dioxide and can be taken up in that form by plant leaves. Nitric oxide and ammonia can also go into solution and be washed out of the atmosphere in rain, ending up in ecosystems some distance from where the fertilizer was applied.

The total amount as well as the particular forms of nitrogen gases lost from fertilized agricultural systems depends on soil properties such as organic matter levels, pH, moisture, and texture, as well as management practices such as the type and placement of fertilizer and the nitrogen requirements of

Figure 28.11. The potential fates of urea fertilizer in the nitrogen cycle in agricultural sites. N = nitrogen, N_2 = di-nitrogen; NO_3^- = nitrate; N_2O = nitrous oxide, NH_4^+ = ammonium; NO = nitric oxide; NH_3 = ammonia.

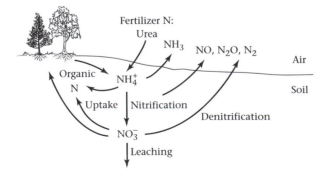

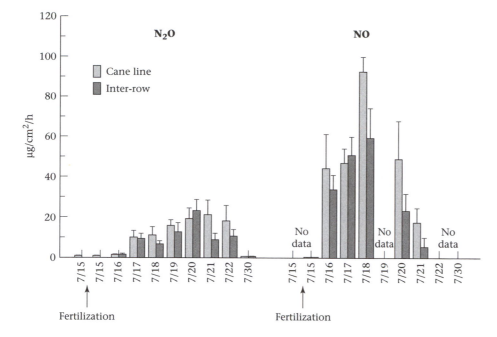

Figure 28.12. Nitrogen trace gas response to urea fertilization in Hawaiian sugar cane systems. Fluxes of both nitrous and nitric oxide increase rapidly following fertilization, and return to background levels within 2 to 3 weeks postfertilization. $\mu g/cm^2/h$ = milligrams per square centimeter per hour; NO = nitric oxide; N_2O = nitrous oxide.

the crop at the time of application. For example, application of nitrogen fertilizer on the soil surface results in higher losses of nitric oxide and ammonia than when it is applied with subsurface fertilization. This suggests that specific management of agricultural systems could be directed at reducing emissions of some globally significant gases, simultaneously reducing losses of fertilizer nitrogen from the field and thus minimizing the cost to the farmer (see Chap. 29).

At the global scale, intensive agriculture appears to have significant cumulative effects on the greenhouse gases carbon dioxide and nitrous oxide (and methane – see the following section, "Paddy Agriculture"), and thus is implicated in global warming. The conversion of a natural ecosystem to agricultural fields decreases carbon storage in soils and represents a net transfer of carbon to the atmosphere. Global estimates suggest a total net release of 275 petagrams of carbon from soils since the beginning of settled agriculture with a current annual flux of 0.85 petagram. Nitrous oxide emissions from fertilized agricultural soils are estimated to range from 0.03 to 3 teragrams (1 teragram is equal to 1×10^{12} grams) of nitrogen annually. Although the uncertainties are great, agriculture explains a significant, perhaps dominant portion of the current atmospheric increase in nitrous oxide (estimated at approximately 3.5 teragrams).

PADDY AGRICULTURE. Rice feeds 50 percent of the world population, with 90 percent of the production located in Asia. The most intensive rice cultivation takes place in periodically flooded fields conducive to rice plant growth (Fig. 28.13). Although this type of agriculture can certainly be termed "intensive," it differs from the upland agriculture described in the previous section in several ways. Because the soils are typically flooded, soil organic matter and soil carbon

levels remain higher than in upland agricultural systems, in large part because anaerobic decomposition is much less efficient than that in aerated conditions. Emissions of carbon dioxide are quite small from these systems. However, the flooded conditions create a perfect habitat for a unique group of microorganisms, *methanogens*, that produce methane during decomposition of organic matter under anaerobic (no oxygen) conditions. Because of these conditions, paddy agricultural systems are critical sources of atmospheric methane. As with carbon dioxide and nitrous oxide, methane is a greenhouse gas whose atmospheric concentrations are rapidly increasing.

As might be expected, methane is also produced in many natural wetlands – globally, wetlands are the single most important natural source of the gas. Today, rice cultivation has nearly matched wetlands in total annual flux of methane to the atmosphere, followed in importance by cattle production. Demand for rice is projected to increase by 65 percent in the next thirty years. If this increased demand is met by an increase in the areal extent of paddy land or in the number of rice crops grown per area each year, methane emissions can be expected to continue to increase. However, a number of recent studies have suggested that water management in paddies – for example, occasional drainage of water from paddies – can reduce methane fluxes over a crop cycle. Management practices in paddy lands that explicitly attempt to lower methane flux could significantly reduce the global environmental consequences of rice agriculture.

CONCLUSIONS

In this chapter we have illustrated how land use changes affect biologic and physical processes within ecosystems.

Figure 28.13. Rice paddy ecosystems in the Phillipines. Flooding provides optimum conditions for methanogenesis. (*Source:* Photograph courtesy of R. Naylor.)

These individual changes are occurring on such a broad scale that the consequences are influencing the global biogeochemical cycles of the Earth system, with ultimate effects on the composition of the atmosphere and on climate. Deforestation contributes approximately 1 petagram of carbon as carbon dioxide to the atmosphere each year; intensive agriculture contributes another 0.85 petagram; together, they contribute approximately 25 percent of the anthropogenic source of atmospheric carbon dioxide. Deforestation and intensive agriculture are the key sources of the increase in atmospheric nitrous oxide, potentially accounting for nearly all of the increase. Similarly, paddy agriculture and cattle sources can account for most of the annual increase in atmospheric methane.

It is clear that land use change is one of the dominant drivers of global atmospheric change and thus, of climate change. Unfortunately, land use change may also be one of the more intractable of our global problems. Although the fossil fuel source of carbon dioxide and the industrial sources of chlorofluorocarbons are enormously difficult to deal with, they are largely tied to the standard of living and their solutions lie in the realm of engineering alternatives. In contrast, land use changes are tied closely to the production of food and fiber for our growing human population. These are pressures that will certainly increase at least into the twenty-first century. If we are to have the ability to predict and mitigate future atmospheric change, it will be necessary to understand which specific land use approaches are most critical in influencing global change. Perhaps more important, we will need to understand more about the human social, economic, and policy drivers of land use changes worldwide so that we can develop viable alternatives that sustain the human population while reducing if not eliminating environmental impacts.

QUESTIONS

1. As the head of agricultural development for a nongovernmental organization, you know that food production must increase in order to meet the demands of a growing population. Do you implement policy supporting *extensive* agriculture, which utilizes fewer inputs but requires conversion of large areas of land, or *intensive* agriculture, which relies on high levels of fertilizers, water, and pesticides to produce the same amount of food on a much smaller area of land? Discuss the pros and cons of these two different strategies.

2. On a map of the United States, identify the major crop-producing regions and the kinds of products that are grown there (including nonfood crops). Can you identify the types of natural ecosystems that occupied these areas prior to agricultural conversion? How do these agricultural areas differ in terms of the types of crops they produce and the inputs that they require for agricultural production? Is there any link between the natural ecosystem type and the current agricultural systems? Is this relationship similar for food-producing agricultural systems and non-food-producing agricultural systems?

3. Consider a forested area that undergoes conversion to row crop agriculture. Draw a simple box model of the carbon pools of these two systems, showing the components representing major pools of carbon in the forests and the agricultural field. Can you identify pathways of carbon loss from the systems during conversion? What is the fate of this released carbon?

4. Identify the biogeochemical pool that is the source of nitrogen in fertilizer production. Draw a simple box model of the nitrogen cycle, showing the production of nitrogenous fertilizer and its use in an agricultural system. Complete the cycle by including the fate of nitrogen not taken up by crop plants.

FURTHER READING

Aber, J. D., and Mellilo, J. *Terrestrial ecosystems.* Philadelphia: W. B. Saunders, 1991.

Human dominated ecosystems: a special report. 1997. *Science* 277 (25 July): 485–525.

King, F. H. 1927. *Farmers of forty centuries; or, Permanent agriculture in China, Korea and Japan,* ed. J. P. Bruce. New York: Harcourt Brace.

Turner, B. L. (ed.) 1990. *The Earth as transformed by human action: global and regional changes in the biosphere over the past 300 years.* New York: Cambridge University Press with Clark University.

Vitousek, P. M., and Matson, P. A. 1993. Agriculture, the global nitrogen cycle, and trace gas flux. In *The biogeochemistry of global change: radiatively active trace gases,* R. S. Oremland, ed. New York: Chapman and Hall, pp. 193–208.

29 Agriculture and Global Change

ROSAMOND L. NAYLOR

THE GREEN REVOLUTION

The onward march of the Green Revolution in wheat, rice, and maize is an ecological and economic imperative in population-rich but land-hungry countries like Bangladesh, China, and India. If the Revolution is allowed to falter, the poverty of small-farm families will persist, since they will have very little marketable surplus. . . . Nor will it be possible to prevent further expansion of cultivated area at the expense of forests and soils vulnerable to erosion or other forms of destruction.

M. S. Swaminathan

When Paul Ehrlich wrote *The Population Bomb* in 1968, the world's population was 3.5 billion – far above, he argued, the Earth's carrying capacity for humans. The argument was supported by the fact that over one-third of the world's population was chronically undernourished, and agriculture was pushing the limits of arable land. Severe droughts in South Asia in the mid-1960s threatened famines of disastrous proportions, and almost one-half of China's population lived in chronic hunger.

Since that time, daily calorie intake per capita has risen steadily in most regions of the world. Real cereal prices in international markets have declined by over 50 percent, and chronic undernutrition has been reduced to one-fifth of the world's population. Cereal production in the developing world more than doubled during this period, which enhanced food availabilities as well as rural incomes and employment in many regions. What was at the root of this transition?

On the supply side of the global food balance sheet, the Green Revolution has been responsible for transforming agriculture in developing countries, particularly throughout Asia. The idea behind the Green Revolution, initiated in the 1960s with the help of the Rockefeller Foundation and the Ford Foundation, was to transfer and disseminate modern agricultural technologies to poor countries through the creation of a series of international agricultural research centers that were crop-specific. The revolution has been characterized most generally by the introduction of high-yielding, nutrient-responsive seed varieties of rice, wheat, and corn. These varieties have three main attributes: (1) they have reduced maturity periods in comparison with traditional seed varieties, so that crops can be planted two or three times a year on a single plot of land given sufficient water inputs; (2) they have been bred for resistance to a wide range of diseases, pests, and climatic variables, such as drought, flooding, and frost; and (3) they have a short, thick-statured architecture (semidwarf) that enables the plants to take up more nitrogen than traditional varieties and supply it to the grain as opposed to the plant biomass. Because water is needed to move nutrients from the soil into the plant, Green Revolution seed varieties have been most successful in irrigated areas.

The result of the Green Revolution has been a dramatic increase in land productivity. When grown under conditions of reasonable rainfall or irrigation, the modern seed varieties produced yields double or triple the traditional varieties used in developing countries. As a result, yield increases accounted for over 90 percent of the total gains in output in the developing world between 1970 and 1990. Achieving high rates of production growth through yield increases as opposed to area expansion has helped to preserve many natural ecosystems that otherwise would have been converted to cropland during this period. However, the Green Revolution has contributed significantly to land use change through the intensification of farm inputs. In particular, it has led to large increases in chemical fertilizer and pesticide use, which have damaged the environment and altered the Earth's interactive systems.

The significant questions concerning agriculture and global change today are: Can the yield growth experienced during the past two decades be sustained far into future? If so, at what cost to the global environment? What would be the toll on the Earth's systems if crop yields did not continue to increase? These questions are important because, for example, growth in rice yields in some of the most productive regions of Asia is already beginning to level off, and the major rice-wheat system in South Asia that supplies much of the region's cereal needs is proving to be unsustainable. In most parts of Africa, the Green Revolution has not even taken hold. Given likely population growth and food demand in the future – and where the largest populations of humans will be located – it is quite possible that the chronic hunger and famine that threatened several continents in the late 1960s will return early in the twenty-first century.

W. G. Ernst (ed.), *Earth Systems: Processes and Issues.* Printed in the United States of America. Copyright © 2000 Cambridge University Press. All rights reserved.

Previous chapters have clearly shown that the Earth's dynamic environment is changing at an unprecedented rate. To address this problem, policies are focusing on the influence of industrialization and toxic wastes, fossil fuel combustion and emissions of carbon dioxide, air pollution, and the contamination of water supplies. Human agricultural activity has wrought profound effects as well; its impact on the Earth's systems through changes in land use are wide ranging (Chap. 28). For example, clearing land for cultivation generates fluxes of radiative (nonabsorbing) gases such as nitrogen oxide (NO_x) and carbon dioxide (CO_2) to the atmosphere; it alters climate by disrupting hydrologic cycles and changing the albedo (reflectivity) of land surfaces; and it reduces biodiversity through habitat destruction. In addition, settled agriculture represents an important avenue for exotic species invasion. Although intensifying agricultural production reduces the need to clear land, it alters biogeochemical cycles and causes changes in air and water chemistry through high inputs and the accumulation of chemical fertilizers and pesticides.

Global population growth (see Chap. 22) will continue to place pressure on existing agricultural systems and will continue to destroy natural ecosystems in the process of new cultivation. The need to increase food production will affect intensive, high-productivity areas that sustain rural and urban areas, as well as marginal agricultural areas in fragile ecosystems, such as rain forests, upland systems, and coastal areas, that provide subsistence for a growing number of poor people. These pressures imply that agricultural production will play a more important role in global environmental change in the future. At the same time, agricultural productivity will remain vulnerable to changes in the Earth's systems, such as climate change (Chaps. 15 and 16) and elevated levels of atmospheric carbon dioxide (Chap. 19).

This chapter focuses specifically on the role of agriculture in global environmental change. It begins with a description of the world food economy, then shows how both consumption and production of food can lead to changes in the Earth's systems. The discussion is set in the context of agricultural policy, which strongly influences demand and supply in the agricultural sector. The environmental impacts of farming practices are as severe, or perhaps more severe, than in any other sector in the global economy. Understanding the interaction between government policy and private decision-making is essential for understanding the role of agriculture in global change; it is also critical for assessing global food balances and the ability of countries, households, and individuals to meet their food needs in the future.

THE WORLD FOOD ECONOMY

The geographic distribution of nations, classified according to their economies, is illustrated in Figure 29.1. Consider the following differences in agricultural trends between developed and developing countries. In 1945, the United States had almost 6 million farms. By the mid-1980s, this number had declined to approximately 2.2 million. In contrast, at the time India became independent in 1947, the country had approximately 50 million farms. By the early 1990s, this number had risen to approximately 100 million. Today, India has over 3000 people per square kilometer of land, and over half of the country's land is being used for crop production. The United States has one-tenth as many people per square kilometer, and only 20 percent of its land is devoted to agriculture.

The percentage of gross national product (GNP) in rich countries that comes from agriculture ranges from 8 percent in New Zealand down to only 2 percent in Germany, the United Kingdom, and the United States. In these countries, employment in the services sector exceeds that in manufacturing, agriculture, and all other natural resource–based enterprises combined. In low-income countries, by contrast, over 30 percent of GNP comes from agriculture (Table 29.1). Even more significant, over half of the labor force in low-income countries depends on agriculture for their livelihood, and the vast majority of export earnings comes from the primary (agriculture and natural resource–based) sector. This reliance places great pressure on the natural ecosystems of poor countries as they try to expand their national incomes or pay off their foreign debts.

Table 29.1 shows that the decline in agriculture associated with economic growth occurs over time as well as across countries. In almost all countries for which data are provided, the agricultural sector played a smaller role in gross domestic product and total employment in 1990 than in 1965. In the few countries where this phenomenon did not occur, economic growth was low or negative, and exports of primary products continued to account for virtually all export revenues.

Despite the relatively small role that agriculture plays in the economies of affluent countries, per capita food availabilities are much higher in these countries than in low-and middle-income countries. The high-income market economies of North America, Europe, Australia, and New Zealand are able to supply their populations with approximately 3300 calories per capita per day. In contrast, sub-Saharan African countries supply less than 2200 calories per capita per day. Moreover, the United States, Europe, and Oceania typically run a trade surplus in agriculture, whereas many poor countries in Africa and Asia run an agricultural trade deficit. The world food economy can thus be characterized as a closed system globally, but an open system at the national level. It is a complex system, with significant regional differences in production, consumption, and trade.

Large differences exist, for example, in the types of crops that are grown throughout the world. These differences are dictated primarily by agroclimatic conditions, such as sunlight, temperature, rainfall, soil type, and topography; however, they are also influenced by each country's agricultural policy, access to technology, infrastructure, tradition, and relative abundance of arable land and agricultural labor. In most agricultural areas, more than one crop is produced an-

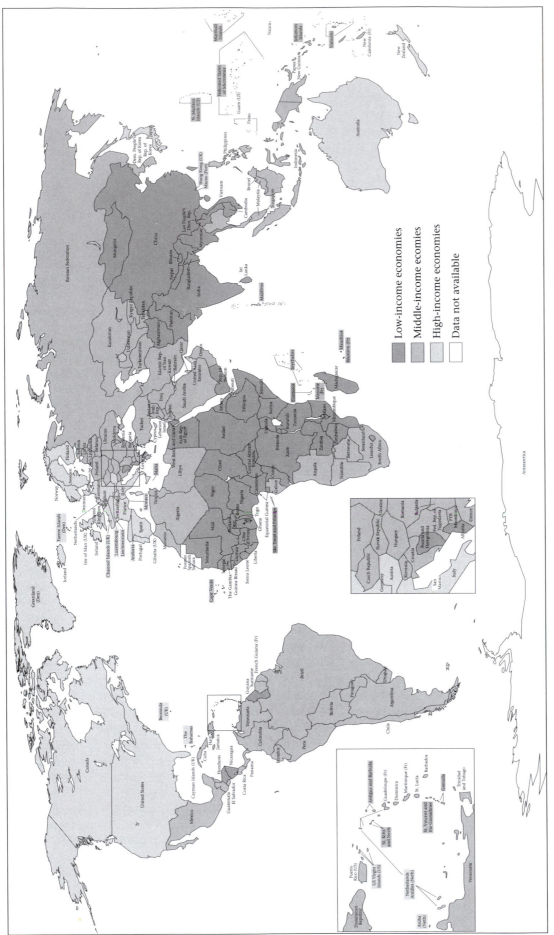

Figure 29.1. Groups of economies, classified by income groups. Low-income economies are those with a gross national product per capita of $725 or less in 1994; middle-income, $726–$8955; high-income, $8956 or more. Six middle-income economies – American Samoa (U.S.), Fiji, French Polynesia (France), Kiribati, Tonga, and Western Samoa – and Tuvalu, for which income data are not available, are not shown here because of space constraints. (*Source:* Adapted from World Bank; World Development Report, 1996. Copyright © 1996 by The International Bank for Reconstruction and Development/The World Bank. Used by permission of Oxford University Press, Inc.)

TABLE 29.1. THE DECLINING ROLE OF AGRICULTURE OVER TIME AND SPACE

Economies[a]	Agricultural Production as a Percentage of GDP, %		Employment in Agriculture as a Percentage of Total Employment, %		Exports of Primary Products as a Percentage of Total Exports, %	
	1965	1990	1965	1990	1965	1990
Low income						
Burma (Myanmar)	35	48	64	47	99	97
China	39	27	81	68	35	26
India	47	31	73	66	51	27
Indonesia	56	22	71	48	99	64
Sri Lankja	28	26	56	52	99	53
Ethiopia	58	41	86	80	99	97
Ghana	44	48	61	59	99	99
Kenya	35	28	86	81	90	89
Nigeria	53	36	72	65	97	99
Tanzania	46	59	92	56	87	89
Middle income						
Bolivia	23	24	54	47	96	96
Colombia	30	17	45	27	93	74
Costa Rica	24	16	47	25	84	74
Thailand	35	12	82	64	97	36
Senegal	25	21	83	81	97	78
Zimbabwe	18	13	79	65	71	64
Industrial market						
Canada	6	3	10	3	63	37
Japan	9	3	26	7	9	2
Spain	15	6	34	11	60	24
United States	3	2	5	2	35	22

[a] Definitions of low-, middle-, and industrial market economies are based on annual GNP per capita. In 1994, The World Bank ranked the groups as follows: low income at $675 per capita or less; middle income at $676 to $8355 per capita; and high income at $8356 per capita and above.

nually and even seasonally. In fact, most farmers in the world produce more than a single crop, and many grow more than one crop on a given plot of land at one time. In the aggregate, however, several regional patterns are discernible. Over 90 percent of the world's rice is produced and consumed in Asia, where monsoon weather patterns are compatible with flooded rice agriculture; virtually all wheat is grown in the temperate zone, where the climate permits the control of certain plant diseases; millet and sorghum are found primarily in dryland agricultural regions of the world, where seasonal rainfall is variable; and fruits and vegetables are most abundant in Mediterranean and tropical climates, where the temperature remains above freezing.

Cereals (also referred to as grains) form the backbone of the world food economy. Grains are consumed directly by humans; they are also used to support livestock production that, in turn, is consumed by humans in the form of meat, dairy products, and eggs. Approximately 50 percent of the world's total arable cropland is used for cereal production. Cereals account for roughly one-half of the total calories consumed globally, and one-half of the world population's protein intake (Table 29.2).

The grain sector is composed of a number of different crops, including wheat, rice, corn, sorghum, millet, barley, rye, and oats. Of these crops, rice and wheat are by far the most important, followed by corn, which is used predominantly as a livestock feed. Table 29.2 shows that rice and wheat each account for approximately 40 percent of annual grain consumption globally. Rice accounts for approximately 50 percent of the calories and 30 percent of the protein consumed directly through cereals on a global basis. Wheat accounts for just over 30 percent of the calories consumed directly through cereals, although wheat consumption is now rising more rapidly than rice consumption with income growth and urbanization in many parts of the developing world. As a result, wheat is expected to surpass rice as the primary staple food crop in the twenty-first century. Wheat is widely traded and consumed throughout the world; by contrast, rice remains principally an Asian cereal crop (Table 29.3).

In addition to the regional variation that exists in crop production and consumption, countries differ greatly in the amount of arable land that is available to feed their populations (Table 29.4). For example, in China, arable land per capita is less than one-tenth of a hectare. In contrast, in Australia and New Zealand, arable land per capita is more

TABLE 29.2. WORLD FOOD CONSUMPTION PER CAPITA, 1997

Commodity	Kilograms/year	Calories/day	Grams of protein/day
Cereals, total[a]	160.3	1384.0	33.5
Rice	58.7	588.0	10.9
Wheat	72.2	558.0	16.5
Corn	17.6	144.0	3.4
Pulses	6.0	56.0	3.5
Starchy roots	61.9	142.0	2.1
Oil crops	6.8	51.0	2.5
Vegetable oils	10.0	239.0	—
Vegetables	82.5	57.0	3.1
Fruits	67.5	78.0	0.9
Meat	35.9	198.0	12.1
Seafood	15.9	28.0	4.4
Sugar	22.2	197.0	—
Other	—	352.0	11.8
Total food	—	2782.0	73.9

[a] Direct grain consumption, not consumption through livestock.

Source: 1997 Food Balance Sheet (United Nations, Food and Agricultural Organization).

TABLE 29.3. CEREAL CONSUMPTION BY REGION, 1997 (KILOGRAMS PER CAPITA PER YEAR)

Region	Rice	Wheat	Corn	Total Cereal[a]
Africa	17.4	46.2	41.5	144.3
Asia	88.0	71.5	11.6	178.9
Europe	3.9	110.4	5.8	133.0
North and Central America	10.0	71.0	40.4	126.7
Oceania	15.0	64.8	3.5	88.9
South America	29.8	55.2	23.3	110.3
World	58.7	72.2	17.6	160.3

[a] Direct grain consumption, not consumption through livestock.

Source: 1997 Food Balance Sheet (United Nations, Food and Agricultural Organization).

TABLE 29.4. ARABLE CROPLAND BY REGION, 1994 (MILLION HECTARES[a])

Region	Total Land Area	Arable Cropland	Arable Land per Capita (Hectares)
Africa	2963	190	0.27
North America	1838	233	0.79
Central America	256	41	0.34
South America	1753	115	0.37
Asia	3085	622	0.18
Europe	2260	316	0.43
Oceania	849	52	1.84
Developed	5462	667	0.57
Developing	7586	799	0.18
World	13389	1239	0.22

[a] Hectares are the metric measure for land area; 1 hectare equals 2.47 acres, or 10,000 square meters.

Source: World Resources Institute, World Resources 1998–99.

than 2 hectares. The amount of arable land per capita in each region helps to explain why food production is very labor-intensive in some regions, and very land-intensive in others. In Asia, it is common to find many people in the rice fields at any given time preparing the land, planting, weeding, and harvesting. In California rice fields, by contrast, it is more common to see mechanical equipment in the fields being used for land preparation and harvesting, and crop dusting aircraft overhead being used for planting, fertilizing, and applying pesticides. Indeed, one can drive for miles in the California rice delta during the growing season and not see a single person in the fields.

It is clear from this brief description that agriculture is by no means homogeneous throughout the world. Large variation exists among regions, in terms of both the crops that are grown and the resources that are utilized. Moreover, regional consumption patterns differ in terms of the types of cereals, the mix of cereals and noncereals, and the level of calories consumed in the diet. Understanding the patterns of consumption and production in the world food economy is important for understanding how agriculture and global change are related. What are the mechanisms by which agricultural consumers and producers influence global environmental change? How will agriculture itself be affected by Earth system changes?

THE DEMAND FOR AGRICULTURAL PRODUCTS

The human population places direct pressure on the Earth's evolving systems through its demand for agricultural and food products. The most important determinants of agricultural demand in the aggregate are population and income growth. At a more disaggregated level, tastes and cultural dietary habits strongly influence food demand.

POPULATION GROWTH. Population growth plays a particularly large role in determining food demand in developing countries, where four-fifths of the world's population resides (see Chap. 22). The developing world's population share is expected to increase in the future; fertility rates in most low- and middle-income countries, including some of the most populous countries such as India, Indonesia, Brazil, Mexico, Pakistan, and Nigeria, range from 2.5 to 6.0 (i.e., the average number of children born to women in the child-bearing cohort), and show few signs of declining to replacement levels in the near term. Over 90 percent of the people added to the planet at the turn of the twenty-first century will be born in developing countries.

In rural areas of the developing world, population growth places increasing demands on high-productivity agricultural areas as well as marginal areas where crop production can be expanded. Population pressure is reflected in expanding agricultural settlements along the floodplains of Bangladesh, on the steep hillsides of the Philippines, in the drier areas of Kenya, in the rain forests of Costa Rica, and along the coast in the mangrove swamps of Indonesia. Urban population growth, which is also most rapid in developing countries, places higher demands on intensive agricultural systems that are capable of producing a surplus for the cities. Moreover, as cities expand geographically to accommodate growing populations, they compete with the agricultural sector for land and water – resources that are critical for productivity growth. Given the limited amount of land suitable for agricultural production at reasonable costs, productivity growth is absolutely necessary to maintain even the present level of per capita food availabilities in the future.

INCOME GROWTH. As economies develop, income growth becomes a more important determinant of aggregate food demand than population growth. An empirical regularity that is observed over time and across countries is *Engel's law*, which states that people spend a smaller share of their total incomes on food as their incomes rise. This law helps to explain why agriculture becomes a less significant factor in the GNP as economies expand (as shown in Fig 29.2). Engel's law refers to expenditures on food, not the physical quantities of food demanded. As shown in Figure 29.2 absolute expenditures will continue to rise well after calorie intake has stabilized. This discrepancy is compensated for by an increase in food quality through changes in diet composition, which is reflected in the consumption of higher-valued food products.

Another empirical regularity of food demand – one that has more direct significance for the Earth's systems – is *Bennett's law*. Bennett's law states that the proportion of calories derived from starchy staples (grains, roots, and tubers) declines with income growth as diets become more diversified to income higher-valued foods (Fig. 29.2). An important implication of Bennett's law is that the total amount of grains consumed directly by individuals falls with higher income levels, whereas the amount of grains consumed indirectly

Figure 29.2. Changing nature (quality, quantity, and amount of starchy foodstuffs) of food consumption with income growth. (*Source:* Adapted from C. P. Timmer, W. P. Falcon, and S. R. Pearson, *Food policy analysis*, Johns Hopkins University Press, 1983.)

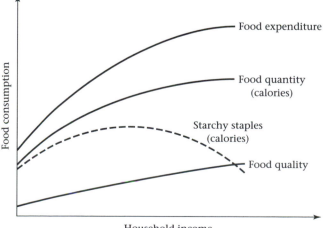

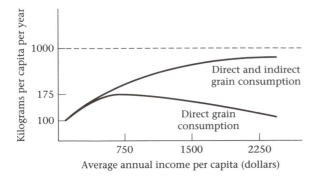

Figure 29.3. Movement from grain-to meat-based diets with income growth. This figure illustrates the principle that is real per capita income increases, the intake of grain first increases, then declines as the intake of meat increases quite substantially.

through livestock increases (Figure 29.3). The amount of grain needed to derive the same caloric intake from livestock as from grains directly (commonly referred to as the *energy requirement*) is approximately 2:1 for poultry, 5:1 for pork, and 8:1 for grain-fed beef. These ratios imply that income growth and increased livestock consumption place proportionately larger pressures on agroecosystems than does population growth.

The composition of food demand is determined to a great extent by relative prices. Figure 29.4 shows that real prices for staple grains fell according to trend during the second half of the twentieth century. This trend, begun earlier, has occurred despite increased demand from population and income growth over the period. Declining grain prices make livestock production and consumption more affordable, and thus facilitate the transformation of high-income growth

countries into livestock economies. Several countries in Asia have already experienced rapid income growth and increased livestock consumption. Given the high population levels in these countries, particularly in China, this transition could have dramatic effects on the Earth's systems through pressures to intensify grain production and to clear land for livestock production globally (see the box entitled *Human Appropriations of Net Primary Product*).

Will the natural resource costs of intensive grain and livestock production be reflected in higher market prices in the long run? The answer to this question remains open for debate. If prices *do* rise, a reversal in consumption patterns toward more directly consumed grains is likely. In the context of global carrying capacity, resource limitations to agricultural production will thus be reflected first in the quality of diets, and later in the quantity of food consumed. It is conceivable that, with additional technical change, the Earth's systems could support a global human population of 12 to 15 billion; however, this population would most likely have to consume a diet that, on average, contained more vegetable products and fewer animal products than the diet of the current population.

AGRICULTURAL PRODUCTION

The decline in real grain prices shown in Figure 29.4 has resulted from strong productivity growth in cereals in many regions of the world. Increased yields, as opposed to area expansion, has been responsible for virtually all of this growth in recent decades. Most of the good arable land globally has now been developed for cultivation. Few opportunities exist to expand future production into new fertile areas,

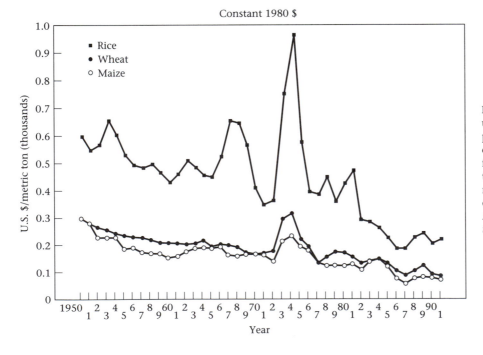

Figure 29.4. World cereal prices. The unusually sharp rise in commodity prices in 1973–4 was caused by a severe drought in Asia and the escalation of fossil fuel prices that disrupted the fertilizer market. Filled squares represent rice; open circles represent maize; filled circles represent wheat. (*Source:* Adapted from World Bank, *Commodity trade and price trends*, 1993.)

particularly in light of the increasing constraints of water availability in most areas of the world.

Despite dramatic gains in global grain production since the late 1960s, significant variation in agricultural productivity has occurred among countries and even across continents. The variation in growth is a function of agroclimatic conditions, investments in irrigation, access to technology and farm inputs (including high-yielding seed varieties and fertilizers), and policies that affect producer incentives. These factors determine the potential productivity for a given crop; they also influence decisions on what crops to produce, which inputs to use, how many crops to grow each year, and the seasonality of crop production.

The decisions that farmers make in both rich and poor countries are based on a complex set of valuations (implicitly or explicitly calculated) that assess the relative gains and technical possibilities of growing different crops and the relative costs of inputs for alternative cropping patterns and farming practices. In general, the individual farmer's objective is to maximize expected profits (including the implicit value of products for home consumption) over time, given a set of resource and technologic constraints on production. Many farmers, particularly in poor countries, also seek to minimize risk or variation in expected returns. They do so, for example, by diversifying their crops within a season or over the course of a year.

TECHNICAL CHANGE AND INPUT USE. In their efforts to maximize expected profits, farmers generally adopt new agricultural technologies that either: (1) enable substitution of lower-cost inputs in agricultural production: or (2) increase output in proportion to resource costs. For example, farmers might switch from hand weeding to the use of herbicides when real wages rise rapidly. Even if real wages do not increase relative to the price of herbicides, farmers might opt for chemical weed management if the chemicals are more effective at controlling weed growth. In the first case, the new technology helps to lower costs; in the second case, it helps to increase output. Both these options enhance farmers' ability to increase their net return.

The introduction of new technologies in response to changes in relative resource costs and profitability is widely referred to as *induced innovation*. (This concept is more fully developed by Yujiro Hayami and Vernon Ruttan, in Chapter 4 of their book *Agricultural Development: An International Perspectives*.) The process of induced innovation has been largely responsible for the adoption of Green Revolution seed technologies in the developing world. Innovation has been induced, in this case, by the rising cost of land relative to labor, which has resulted from rapid population growth on a limited area of land. The Green Revolution is characterized by its success in raising output per unit of land, both through yield growth and through higher cropping intensities (more crops grown annually on a piece of land). Induced innovation in agriculture has also occurred in developed countries such as the United States, Canada, and Australia. In these countries, high labor costs relative to land costs induced the adoption of tractors for land preparation, planting, and harvesting, and the use of airplanes to apply agricultural chemicals, such as pesticides and fertilizers, and to disseminate seeds.

Technologic change in agriculture is, indeed, one of the most important – and one of the most contentious – issues in the global change debate. What does technologic change in agriculture really mean? How will it affect the Earth's systems? Will it ensure the ability of the human population to feed itself in the future? Can it be managed appropriately in terms of human equity and the environment?

DEVELOPING AN ANALYTIC FRAMEWORK. In order to answer these questions, it is important to understand the interactions between technologic change, agricultural inputs (such as fertilizers), and productivity growth. Within the context of the Green Revolution, technologic change with respect to wheat and rice provides an excellent starting point

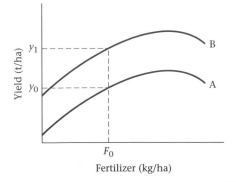

Figure 29.5. Increase in crop yields in response to added fertilizer, with modern varieties (curve B) compared to traditional varieties (curve A) of irrigated rice and wheat. kg/ha = kilogram per hectare, t/ha = metric tons per hectare; F_0 = a given level of fertilizer use. Y_0 = the yield employing traditional crops, Y_1 = the enhanced yield employing modern technologies.

for analysis. Consider first a very simple model of rice or wheat production in an irrigated ecosystem. In this case, the new technology is the modern high-yielding seed variety (HYV), and the critical input for yield improvement is nitrogen fertilizer.

Figure 29.5 charts crop yields against nitrogen fertilizer use, holding all other inputs constant. Line A represents the initial relationship between yields and fertilizer with traditional varieties. With the introduction of HYVs, which are more responsive to added nitrogen, the line shifts up to B. That is, for a given level of fertilizer use, F_0, yields increase from Y_0 to Y_1 with the new technology. The process is similar to introducing a new fuel-efficient car; for a given amount of gasoline, the new car can travel farther than the old fuel-inefficient one.

How do farmers choose how much fertilizer to apply? Does it make sense for them to apply just enough fertilizer to maximize plant growth? The answer is *maybe*. It depends on the cost of fertilizer, and whether the gains in productivity are sufficient to cover this cost. Farmers should apply just

enough fertilizer so that the last unit of fertilizer applied pays for itself. This practice, referred to as *maximizing profits*, is illustrated in Figure 29.6. Price$_1$ represents the ratio between the price of fertilizer and the crop price. This price ratio depends on the markets for fertilizer and wheat or rice, and on government policy in the form of fertilizer subsidies or crop price supports. In principle, farmers should produce at the tangent point between the price line and the fertilizer response function, where the gain in revenue just equals the increased cost of additional fertilizer. That is, they should operate at point A for traditional varieties and at point B for HYVs. These points may or may not represent the point at which plant growth is maximized, because the objective of farmers is to maximize expected profits, which include costs as well as revenues.

When Green Revolution seed technologies were introduced in the late 1960s, many farmers thought that they were too risky, and many farmers did not have the financial resources to buy fertilizer early in the season. In order to stimulate productivity growth through the adoption of HYVs and nitrogen fertilizers, policy makers in many countries chose to lower the price of fertilizer relative to the crop price through a subsidy (price$_2$). Farmers responded by applying more fertilizer, which resulted in higher yields. In other words, they moved up the fertilizer response curve. As they continued to apply more fertilizer (reaching, for example, point D), however, the marginal gains from additional fertilizer use in terms of yields declined.

In Figure 29.7, the economic optimum, assuming no policy distortions in the crop or fertilizer price, is defined by point B, where the revenue gains from additional fertilizer use in relation to the input costs switch from being positive to negative. If farmers do not understand the benefits of fertilizer use, they may operate at point A. Would farmers ever rationally operate at point C or beyond? If they did so, what would happen to their profits? What would happen to crop yields?

Answers to these questions assume *rational* behavior by

Figure 29.6. Physical and economic responses of yields to added fertilizer for modern and traditional varieties of irrigated rice and wheat. kg/ha = kilogram per hectare; t/ha = metric tons per hectare. For discussion of A, B, and C, and prices 1 and 2, see text.

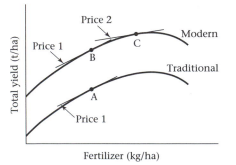

Figure 29.7. Fertilizer response time for modern irrigated rice and wheat. For discussion of loci A, B, and C, see text.

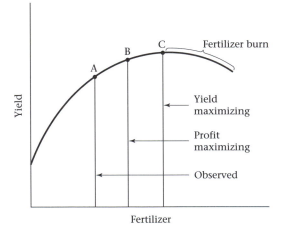

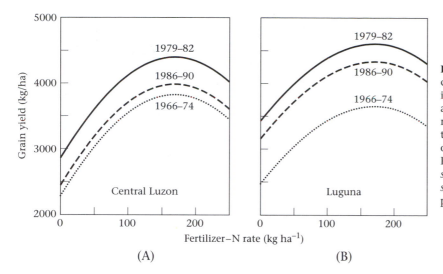

Figure 29.8. Changes in fertilizer response curves for modern rice variations over time in two Southeast Asian sites, Central Luzon and Laguna. (*Source:* Adapted from K. Cassman P. Pinglai, Extrapolating trends for long-term experiments to farmers' fields: the case of irrigated rice systems in Asia, in V. Barnett, R. Payne, and R. Steiner (eds.), *Agricultural sustainability in economic, environmental and statistical terms*, John Wiley and Sons, Ltd., in press.)

farmers. Farmers are generally rational, but they do not always have sufficient information to do what is best on their farms. For example, in very intensively farmed rice systems in the Philippine regions of Central Luzon and the Laguna, continuous cropping and high levels of nutrient inputs have altered the soil chemistry and have caused the marginal gains from fertilizer inputs to decline (Fig. 29.8). Under these circumstances, the decline has resulted from downward shifts in the entire fertilizer-response functions rather than a movement along the response functions. That is, for all levels of fertilizer use, yields of modern varieties in 1986–90 were lower than in 1979–82, but still higher than yields of traditional varieties in 1966–74. Farmers without knowledge of the changing ecologic conditions have continued to add more fertilizers, even though further intensification of their land may lower yields in the long run.

Most farmers respond relatively quickly to changes in agroecologic conditions, however, and either diversify their cropping patterns or farm less intensively. This is particularly true if the changes in farm management practices increase their yields in the short run. The response is not always uniform among all farmers, however, and it is not always matched by changes in government policy. For example, some governments maintain high subsidies on nitrogen fertilizers, even when farmers are using more nitrogen than is optimal. Because of this inefficient use of fertilizers, nitrogen uptake by the plants may be suboptimal, and nitrogen losses to the environment may be quite large.

PRIVATE VERSUS SOCIAL BENEFITS AND COSTS. A wide body of empirical evidence shows that farmers carefully evaluate prices of inputs and outputs in their choice of cropping patterns, farming practices, and technology. The prices that farmers use in making decisions are private prices; that is, prices they observe in the market. In general, these prices do not account for the true social value of inputs and outputs when *environmental externalities* are considered. For example,

the price of pesticides in the market does not incorporate the potential negative health and environment effects associated with their use. These effects include damages to farm workers' skin, eyes, and respiratory systems; pollution of surrounding surface water and groundwater supplies; and damages to the health and habitat of wildlife that are exposed to the chemicals. The price of fertilizers also does not account for the negative environmental externalities reflected in leaching and groundwater contamination, nitrogen emissions to the atmosphere, and alternations in local and regional biogeochemical cycles. Finally, the price of some outputs, such as maize, may not accurately reflect the impact of production on soil erosion – at least, not in the short term.

In principle, government policy could correct for the difference in private and social costs by taxing harmful use of pesticides and fertilizers and crop production that degrades the soil or water resources. In practice, the opposite usually occurs; governments often subsidize chemical inputs and support crop prices in order to stimulate production. A market value is placed on the flow of agricultural outputs and purchased inputs; however, no value is placed on the depreciation of natural resource stocks that support the production process.

Due to the absence of markets for environmental goods and services, private decisionmaking at the farm level may lead to input use and output production that are not consistent with the long-run preservation of productive resources and balance in the Earth's systems. Even when environmental damages affect productivity on the farm, producers do not always adopt more environmentally sound practices. What are the reasons for this paradoxical behavior?

The first is that some farmers strongly discount the future in their calculations of long-run payoffs from agricultural investments and production. Behavior characterized by a high *rate of time preference* is particularly widespread in poor countries, where the short-run imperative of hunger alleviation overshadows the long-run objective of resource sustainability. Unfortunately, there is no long-term dimension to hunger; starving people must eat immediately or they simply will not survive.

Some land tenure arrangements also encourage myopic thinking by farmers. Farmers who rent land or who produce in common property areas (especially with short leases) may have little incentive to preserve the land for long-run productivity growth. Without some guarantee of future ownership or land use rights, they often opt for farm management practices that lead to increased output in the short run, but a reduction in productivity in the long run.

Another factor leading to nonsustainable production decisions is *imperfect information*. Farmers may not be aware of the magnitude of environmental damages – either on or off the farm – caused by their production practices. For example, they may not know the extent of nitrogen losses from fertilizer applications, or the impact of these losses on crop productivity and the functioning of surrounding ecosystems. Without reasonably good knowledge of the magnitude and process by which environmental damages occur, farmers are not likely to change their behavior. Part of the problem lies in the type of information conveyed to farmers through agricultural extension services. Farmers receive farm management information that is based more on agronomic principles (characterized by single-plot, single-season analysis) than on ecologic principles (characterized by whole-system, multiperiod analysis). They may also receive much of their technical information from agrochemical companies that have a vested interest in promoting a particular product, irrespective of its longer-run health and environmental consequences.

Overall, it is clear that agricultural intensification has significant benefits and costs on a global scale. Yield growth has been responsible for most of the gains in global agricultural production in recent history, and has thus helped to alleviate problems of chronic hunger and poverty in many parts of the world. Intensification has also eased pressures to clear new land for cultivation, especially in more fragile tropical ecosystems. On the other hand, continuous cropping with high inputs of fertilizers and, in some cases, pesticides has altered water and atmospheric chemistry in ways that contribute substantially to global environmental change (see the box entitled, *Pathways of Nitrogen Loss from Fertilization*). Moreover, the need to raise agricultural productivity in order to meet population-and income-driven demand in the future will call for even higher input levels in agriculture. These trends suggest that agricultural production will continue to alter the Earth's systems dramatically unless new farm management strategies are pursued – strategies that are both profitable and environmentally sustainable.

THE ROLE OF POLICY

Despite apparent resource limitations in agriculture, the world has not yet reached a point of absolute food scarcity. Global grain production continues to increase, and interna-

---

PATHWAYS OF NITROGEN LOSS FROM FERTILIZATION

Rice and wheat are the most important staples in the human diet globally. They also represent the most significant crops in terms of grain yield, area covered, and nitrogen fertilizer applied. On average, only 40 to 60 percent of applied nitrogen is taken up by the rice and wheat plants. The remainder is lost to the environment through nitrate leaching, which affects regional surface and groundwater quality; ammonia volatilization and subsequent deposition, which affects the functioning of regional ecosystems; and nitrification and denitrification, which result in emissions of nitrous oxide, nitric oxide, and dinitrogen to the atmosphere.

The atmospheric effects of nitrogen fertilization are potentially enormous. Nitrous oxide has an atmospheric lifetime of approximately 150 years, and it traps outgoing radiation with an effectiveness that is 200 times greater than that of carbon dioxide and 10 times greater than that of methane. Therefore, even relatively small amounts of nitrous oxide emitted from fertilization could have a significant impact on global atmospheric processes and climate change. Furthermore, other nitrogen oxides emitted from fertilization combine with hydrocarbons and sunlight to enhance the production of tropospheric ozone. Relatively little is known about the environmental effects of high levels of nitrogen fertilizer use and loss at the farm, regional, or global levels – particularly in tropical ecosystems – yet these effects may have serious consequences for the long-run productivity of intensive rice and wheat systems, as well as for atmosphere and groundwater chemistry. For a more extensive discussion of these effects, see Peter Vitousek and Pamela Matson, "Agriculture, the Global Nitrogen Cycle, and Trace Gas Flux," in R. Oremland (ed.), *The Biogeochemistry of Global Change* (Chapman and Hall, 1993), New York, pp. 193–203.

---

tional prices for grain continue to fall in real (inflation-corrected) terms. More than a billion people, however, still live in chronic hunger. What explains the discrepancy between global food surpluses and regional food shortages?

Government policies play a significant role in determining the incidence and levels of poverty and hunger that exist throughout the world. Policies affect food supplies. For example, price supports and export subsidies in the 1980s and early 1990s have encouraged excess grain production in the United States and Europe, and protection against rice imports in Japan has allowed that country to remain self-sufficient in rice production, albeit at an extremely high cost (rice prices in Japan in the early 1990s were ten times higher than the price of rice in the international market). Policies also influence food demand through their impact on local and regional income and employment opportunities. The latter is particularly important for understanding the world food situation, inasmuch as chronic hunger is caused primarily by the lack of income needed to purchase food, not by physical shortages of food.

The agricultural sector has several characteristics that make it unique in a policy context. First of all, it consists of a very large and disaggregated group of producers, and the size of farming operations is often small, especially in developing countries where the average farm size is typically less than 1 hectare. As a consequence, food and agricultural policies must be implemented with broad instruments – such as taxes, subsidies, interest rates, and exchange rates – in order to be effective. Furthermore, many households in developing countries are both producers and consumers of food, which makes the design of agricultural policy particularly challenging. For example, price supports that provide incentives to producers may hurt poor consumers, and price subsidies for consumers may jeopardize the process of agricultural development by lowering incentives to producers. (An extensive discussion of these characteristics and agricultural policy more broadly can be found in C. P. Timmer, W. P. Falcon, S. R. Pearson, *Food Policy Analysis*, 1983.)

The agricultural sector is also characterized by a relatively high degree of risk and uncertainty. The most uncertain variables in farmers' production decisions are weather and prices; the former is controlled almost entirely by the Earth's systems, and the latter is dictated primarily by government policy. Plant breeding and genetic improvements have been successful, to some extent, in adapting cultivars to variable weather conditions and other risks, such as disease and pest infestations. Varietal improvements have been particularly important for agricultural productivity growth in poor countries, where there is virtually no formal insurance against a bad crop.

Nonetheless, technology cannot eliminate all the risks that these farmers face. Farmers throughout the world live with some degree of uncertainty as the Earth's climate changes, and as the chemistry of the Earth's atmosphere, soils, and water are altered. Farmers in high-latitude regions of Siberia and Canada may actually benefit from a warmer climate, whereas farmers – and the population as a whole – in Bangladesh may suffer. Indeed, in Bangladesh, where one-third of the arable land already floods to an average of 5 feet annually, a rise in the base sea level of 30 centimeters as a result of global warming would be devastating. Moreover, climate change could magnify the intensity of cyclones in the region, which would further jeopardize the stability of human settlements and agricultural production (see chap. 25).

Agricultural technology will continue play a significant role in the future, however. Technologic change that increases crop tolerance to temperature extremes, drought, and flooding could help to mitigate the impact of climate change on agricultural production in many parts of the world. Unless major surprises occur in the Earth's interactive systems (Chap. 16), climate change may be gradual enough to allow most farmers to adapt over time.

It is much more difficult for farmers to adapt to sudden swings in government policy. For example, the elimination of price supports for corn in Mexico as a result of the North American Free Trade Agreement (NAFTA) and changes in domestic agricultural policies in the mid-1990s are likely to reduce the livelihood of many poor Mexican farmers. In the short run, these farmers will be compensated partially by income supports by the government to help them adapt. A more difficult situation is faced by farmers in Africa, where the political and economic environment is highly unstable, or in countries such as Cambodia, where prolonged civil war and political unrest eliminate most chances for successful agricultural development. Farmers in the United States are also vulnerable to sudden changes in policy; for example, when interest rates jumped in response to new monetary policies in the late 1970s and early 1980s, many farmers with large agricultural loans outstanding went out of business.

Understanding agricultural policy is therefore extremely important for predicting regional food surpluses and shortages in the future. It is also important for assessing how the agricultural sector in different regions might respond to changes in the Earth's systems. Moreover, understanding the policy environment is critical for predicting the future state of hunger and poverty throughout the world. Producing more food is not a sufficient condition for ending hunger; more food is of no help to those who are too poor to buy it.

Two broad categories of policies affect agricultural development: microeconomic and macroeconomic (including trade policies). Microeconomic policies in the agricultural sector include taxes, subsidies, and price supports. This set of policies directly affects cropping patterns and farming practices by altering prices of inputs and outputs. By influencing production decisions, these policies help determine the level of farm income and output, food prices, and employment in agriculture as well. For example, several governments in Southeast Asia raised the guaranteed price of rice relative to soybeans and maize in the 1970s in order to encourage the adoption of high-yielding rice varieties; this policy generally had a positive impact on rural employment, because the production of rice requires more labor than the other crops. In the 1980s, however, some governments implicitly subsidized herbicides through macroeconomic policies, which reduced job opportunities because the amount of labor required for hand weeding fell.

Investments in irrigation and marketing infrastructure, which are an integral part of a country's overall development strategy, similarly affect rural income and employment. These policies determine the economic and technologic possibilities for different cropping patterns and farming practices within each ecosystem. For instance, the introduction of new irrigation systems in South Asia during the 1970s and 1980s allowed farmers to begin growing rice two or three times a year instead of once, which enhanced rural incomes and employment. Moreover, rising supplies of rice kept rice prices relatively low, on average, for consumers. Given that many of the rural poor are net purchasers of food, development policies that lead to low-cost rather than high-cost agriculture are especially effective at alleviating poverty and hunger in both urban and rural areas.

Macroeconomic and trade policies, although not necessarily directed to the agricultural sector, often have larger impacts on rural income and employment than do agricultural price and investment policies. The most widely used policy instruments in this category are trade restrictions, such as tariffs and quotas, and adjustments in interest rates and exchange rates. Manipulating these policy levers may alter the cost of credit to the farmer, change the relative value of domestically produced products to imports, and shift the relative value of tradable inputs (e.g., machinery) to nontradable inputs (e.g., labor and land) in the farm economy. By altering relative prices in the agricultural economy, macroeconomic and trade policies can have a significant impact on technical change and the levels of rural poverty and food consumption. For example, the maintenance of an overvalued exchange rate in several countries of West Africa during the 1970s and 1980s made food imports cheap in comparison to domestic agricultural products, and therefore lowered the demand for local food products, which was detrimental to rural incomes and employment. Unfortunately, macroeconomic policies and agricultural sector policies are almost always handled by different ministries and are often ineffectively coordinated.

Rural poverty tends to be highest in countries where microeconomic, macroeconomic, and trade policies discriminate against the agricultural sector. The extent to which poverty persists throughout the world has serious implications for the future of the Earth's systems. Poor people have few options to preserve ecosystems for long-run productivity and environmental benefits if their own survival is threatened in the short run. Moreover, a lack of income and employment opportunities in one region typically leads to migration to other regions, either within or across national borders. For example, in many Central and South American countries, inequalities in farm size have encouraged the movement of low-income people from settled agricultural areas into rain forests and cities. Similarly, migration in several African countries has been aggravated by the absence of secure property rights in rural areas that leads to short-run exploitation of land and water resources. Migration has now become a dominant demographic trend worldwide; it poses a threat to natural ecosystems and opens up a new area of international concern centered on environmental security. It is becoming increasingly clear that poverty, along with population growth, are among the most toxic agents to the global environment. (see the box entitled *Sustainable Development Through Poverty Alleviation*).

Given these pressures on the Earth's systems, the obvious question arises as to whether it will be possible to achieve the degree of agricultural productivity growth that is needed to feed the world and reduce global hunger in the future. To date, technologic change in agriculture has allowed food supply to grow faster than food demand on a global basis. Despite improvements in technology, however, agricultural production in the short run remains vulnerable to weather and other biotic disturbances. When the long-run uncertain-

---

SUSTAINABLE DEVELOPMENT THROUGH POVERTY ALLEVIATION

The production of food and other basic commodities in developing countries is closely linked with three interrelated problems: poverty, environmental decline, and population growth. This triad of problems defines the development challenges facing the world today. This dynamic set of interrelationships is particularly critical in rural areas of developing countries.

Poverty is the pivotal element of the triad. It limits the opportunities for protecting and enhancing the environment, because poor people have no option but to mine the natural resource base in order to survive. Poverty hinders efforts to reduce population growth because, for poor people, children represent additional sources of income, labor, and social security.

The way forward is through economic development aimed at broad-based poverty alleviation. Historic experience shows that raising the productivity of rural economies is the key to this process. Because poor people spend up to 80 percent of their income on food, success in reducing poverty depends critically on rising real wages and on declining real prices for food, and these in turn hinge on increasing agricultural productivity. Policies and programs to generate agricultural technologies that increase productivity while conserving natural resources are thus central to the entire development process. Source: Consultative Group on International Agricultural Research, "Food and Resources for the Future," 1993.

---

ties of global environmental change – including changes in atmospheric chemistry, climate, hydrologic cycles, biogeochemical cycles, and genetic diversity – are factored into the supply equation, the outlook for the global food economy becomes even more complex and uncertain.

Whether the world can grow enough food to feed a human population of 10 billion or more remains an open question. An even larger and more contentious question remains: Can enough food be produced *and* enough income and jobs be generated so that the poor can have access to this food? Finally, one of the greatest challenges of the twenty-first century will be to produce the food and provide access to it in ways that do not destroy the environment.

QUESTIONS

1. In what ways are the objectives of hunger alleviation and sustainability of agroecosystems compatible? In what ways are they in conflict?

2. What are the pathways in which agriculture contributes to global change? How has the contribution changed over time, and how do you think it will change by the end of the twenty-first century? In answering this question, what are your assumptions about population growth, income growth, and technologic change?

3. How can agricultural practices be modified to reduce their impact on the Earth's systems? What types of policies might encourage this modification?

4. In what ways will the ongoing transition in the Earth's systems affect hunger and poverty throughout the world in the twenty-first century? Based on your answer, do you think that it will be possible to feed a global population of 10 billion or more? Why or why not?

FURTHER READING

Consultative Group on International Agricultural Research. 1993 ''Food and Resources for the Future,'' National Academy of Sciences, Washington, DC

Edwards, C. A., Lal, R., Madden, P., Miller, R. H., and House, G. (eds.) 1990. *Sustainable agricultural systems*. Delray Beach, FL:. St. Lucie Press.

Ehrlich, Paul E. 1968. *The population bomb*. New York: Ballantine.

Hayami, Y., and Ruttan, V. 1985. *Agricultural development: an international perspective*. Baltimore, MD: Johns Hopkins University Press.

Kaiser, H. M., and Drennen, T. E. (eds.) 1993. *Agricultural dimensions of global climate change*. Delray Beach, FL: St. Lucie Press.

Timmer, C. P., Falcon, W. P., and Pearson, S. R. 1983. *Food policy analysis*. Baltimore, MD: Johns Hopkins University Press.

Vitousek, P., and Matson, P. 1993. Agriculture, the global nitrogen cycle, and trace gas flux. In R. Oremland, ed. *The biogeochemistry of global change*. New York: Chapman and Hall, pp. 193–203.

30 Water Allocation and Protection: A United States Case Study

BARTON H. THOMPSON, JR.

DYING OF THIRST ON THE BLUE PLANET

Many world regions are running short of their most essential natural resource – water. Water is indispensable to life and to all forms of economic development. Irrigation has been central to the world's efforts to feed its soaring population. Although many regions can produce crops simply with rain, some of the world's most fertile lands require supplemental irrigation. Dependable water supplies also permit farmers to maximize their crop yield. Although only 16 percent of global cropland is irrigated, this land currently produces 36 percent of the world's food.

As the global population continues to expand, competition mounts for the limited supply of fresh water. In regions that are already short of water, growing urban populations and industrial development threaten to take water away from farming regions that must somehow feed additional people. Partly due to water shortages, both irrigated land and worldwide grain production have fallen on a per capita basis since the mid-1980s, following several decades of growth. As countries squeeze their available water supplies, periodic droughts also take a greater toll on their already thirsty populations and economies.

Water-short regions have typically utilized two strategies to meet needs. First, many regions have built dams, reservoirs, and aqueducts to store and import the runoff of distant watersheds. To date, more than 40,000 large dams (defined as dams more than 15 meters high) and millions of smaller dams have been built worldwide. Second, regions have tapped the vast quantities of water amassed over the millennia in underground aquifers.

Neither strategy, however, provides a sustainable solution. Silt builds up and diminishes the storage capacity of most reservoirs. Because nations typically develop the best dam sites first, new water projects grow ever more technologically complex and expensive. Moreover, time has revealed the large environmental price tag of water projects: the extinction of fish species, desertification of former wetlands, and harmful concentration of salt and other contaminants in depleted waterways. For all these reasons, new dam construction has dropped over 70 percent since its peak in the 1950s and 1960s.

The second strategy is not sustainable because excessive groundwater withdrawals are depleting thousands of aquifers. As the aquifers dry up (or steps are taken to save them from overuse), those who are using the groundwater will need to find alternative sources, reduce their water use, or both. In the meantime, water tables are dropping, overlying land is subsiding, and rapid pumping rates are drawing salt water and other contaminants into many aquifers.

Two questions dominate water policy at the the beginning of the twenty-first century. Can the world's nations reverse the dewatering of surface waterways and groundwater aquifers that has accompanied traditional solutions to water shortages? Can they find alternative means of meeting the water needs of their growing populations? The answers to both questions depend on nations' abilities to stimulate conservation, better allocate water among competing users, and employ new, more environmentally benign means of expanding water supplies.

This chapter analyzes the issues that countries confront in trying to meet growing water demand on a sustainable and environmentally benign basis. We begin by surveying the tools countries can employ to shape their populations' use of water resources. We then examine in detail how water policy has evolved in the United States, the consequences of that policy, and possible reforms.

Spread over a vast and climatically varied terrain, the United States provides a useful case study for water policy. Water has long been scarce in the nation's arid Southwest, and regional shortages are now developing elsewhere. Having followed the traditional solutions of dams and groundwater mining, the United States is now awakening to the environmental, economic, and social consequences. Although the nation is making some progress toward modifying its water policies, reform has been slow and remains

W. G. Ernst (ed.), *Earth Systems: Processes and Issues*. Printed in the United States of America. Copyright © 2000 Cambridge University Press. All rights reserved.

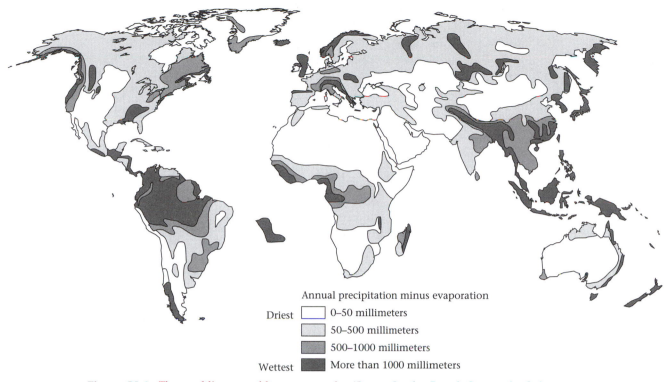

Annual precipitation minus evaporation

Driest ☐ 0–50 millimeters

  ▨ 50–500 millimeters

  ▨ 500–1000 millimeters

Wettest ▰ More than 1000 millimeters

Figure 30.1. The world's renewable water supply. (*Source:* Sandra Postel, *Last oasis: facing water scarcity,* copyright 1992 by Worldwatch Institute. Reprinted by kind permission of W. W. Norton and Company, Inc. 1997; based on *Atlas of world water balance,* State Hydrological Institute of Russia, UNESCO, 1977.)

uncertain. Will the United States meet this challenge successfully? What lessons in water allocation and protection does the United States experience provide to other nations?

POLICY CONSTRAINTS AND TOOLS

In trying to meet the water needs of their populations, governmental and private officials throughout the world assume that some factors are beyond their control. For example, as shown in Figure 30.1, nature has bequeathed a vast imbalance in water supply among regions. Approximately 15 percent of the world's population lives in arid areas with average rainfall of less than 30 centimeters (12 inches) per year. With such meager precipitation, crops die without artificial irrigation and water is always an issue. Even normally wet regions frequently experience multiyear droughts. From 1989 to 1990, precipitation levels in European countries dropped by as much as 60 percent.

Unwilling to concede that precipitation levels are totally beyond human control, some nations have seeded rain clouds and investigated other means of weather modification. To date, the results of these experiments have been inconclusive. Although some appear to have produced brief showers, results have been neither predictable nor reproducible. The experiments have also generated social and environmental concerns. Weather modification, if successful, promises not only to increase rainfall in one region, but also

to reduce precipitation elsewhere and perhaps produce unknown changes in the global weather system.

Precipitation levels represent only part of the water equation confronting policy makers. Equally important is population growth, which, paired with precipitation levels, determines per capita water availability. Countries begin to encounter serious water shortages when yearly water availability falls below 5000 cubic meters per person. When this figure falls below 1000, water scarcity endangers agricultural and economic production and threatens the natural environment. Water experts label these countries "water-scarce." As shown in Table 30.1, approximately twenty countries are currently "water-scarce." Experts believe that by the year 2050 another forty countries or more will join the list, encompassing almost half the world's population.

Continued population growth increases water needs at the same time that it decreases per capita water availability. Indeed, because of increasing living standards, global water use has been growing faster than world population. To make matters worse, most nations with already low per capita supplies also experience rapid population growth. As Table 30.1 indicates, experts predict that, on average, the population of water-scarce countries will double in less than thirty years, halving current per capita water availability unless new sources are identified.

Water officials indirectly influence population growth and development by regulating water use. Some regions purpose-

TABLE 30.1. WATER-SCARCE COUNTRIES, 1995

	Renewable Water Supplies[a] (cubic meters per capita)	Population Doubling	
		Population (millions)	Time (years)
Africa			
Algeria	528	27.9	27
Burundi	563	6.4	21
Cape Verdi	654[b]	0.4	21
Djibouti	19[b]	0.4	24
Egypt	923	62.9	28
Libya	111	5.4	23
Rwanda	792	8.0	20
Tunisia	443	8.9	33
Asia			
Singapore	211	2.8	51
Middle East			
Bahrain	184[b]	0.5	29
Israel	382	5.6	45
Jordan	314	5.4	20
Kuwait	103	1.5	23
Oman	892	2.1	20
Qatar	103[b]	0.5	28
Saudi Arabia	254	17.9	20
Yemen	359	14.5	20
Other			
Barbados	195[b]	0.3	102
Malta	85[b]	0.4	92

[a] Except where noted, freshwater data generally include water flowing in from neighboring countries.

[b] Freshwater data do not include water flowing in from neighboring countries.

Source: Except where otherwise noted, water and population data are from *World Resources 1996–97: A Joint Publication by the World Resources Institute,* the United Nations Environment Programme, the United Nations Development Programme, and the World Bank, Oxford University Press, 1996. Population doubling time is from Sandra Postel, *Last oasis: facing water scarcity,* Norton, 1997.

fully have used water policy to limit growth. For example, several California communities have tried to restrict growth by declaring moratoria on new water supplies. Despite the pressure that population growth and economic development can place on water supplies, however, most water officials have shied away from a direct role in setting population or development policies. In the view of most water officials, their job is to develop policies that can meet expected population growth and economic demands, not to shape population growth and development to fit water availability.

Water officials can use a variety of tools in trying to meet their populations' demands within their regional water budget. Statutory restrictions, regulatory programs, and other legal rules constitute the principal means of shaping water use in most countries. Informal local norms, however, can also be an important policy tool. For example, Tucson, Arizona, has reduced peak water demand during summer months through a voluntary "Beat the Peak" campaign that promotes the civic virtues of water conservation.

The price of water is another critical policy tool. Although we need a minimum amount of water for survival, most water use is discretionary and responsive to price. Economic studies of water use in the United States suggest that a 10 percent increase in the price of water will produce anywhere from a 2 to 14 percent decrease in consumption, depending on such variables as the region, type of user, and time of year. Price changes are most effective over the long run as farmers switch to less water-intensive crops or install new irrigation systems, domestic users relandscape and adopt low-flow plumbing, and industries modify their manufacturing processes.

To resolve anticipated shortages, water officials have turned frequently to engineering projects and technologic innovation. Most major cities in the world import at least part of their water supply through elaborate systems of dams, reservoirs, and aqueducts. As large-scale water projects have run into greater fiscal, environmental, and technologic obstacles, officials have also turned their attention to an array of other engineering options designed either to increase supply, such as desalination and reclamation, or to reduce demand, such as drip irrigation and high-precision sprinkler systems.

A CASE STUDY OF UNITED STATES WATER POLICY

THE UNITED STATES WATERSCAPE. Similar to the world at large, the United States has a considerable imbalance in water supplies. As shown in Figure 30.2, the mountain ranges of the Pacific Coast and the 100th meridian represent important divides. North of San Francisco, the coastal ranges create a "rain shadow effect," ensuring considerable precipitation for the Pacific Northwest. Historic storm patterns have also guaranteed significant rainfall to most of the nation east of the 100th meridian. By contrast, the western flatlands are parched, leading early cartographers to label the area as "The Great American Desert." Even small distances can make dramatic differences. For example, the Puget Sound area of Washington receives over 40 inches of rain in an average year; only 100 miles east, over the Cascades Mountains, precipitation drops to less than 9 inches.

Differences in precipitation are reflected in river variations. The four largest rivers in the United States – all in the East – are the Mississippi, Columbia, Ohio, and St. Lawrence. The Colorado River, which dominates the Southwest's geography, ranks only twenty-fifth nationally and carries approximately 3 percent of the water found in the Mississippi.

Imbalances in supply are joined by tremendous variations in the timing of precipitation. The Northeast boasts relatively year-round precipitation and thus continuous river flows. By contrast, other regions regularly receive only a few months of precipitation, sometimes accompanied by flash flooding, followed by many months of drought during which rivers often dry up entirely. Approximately 80 percent of the meager rainfall near Tucson, Arizona, occurs during the summer. Virtually all of the Pacific Coast's precipitation is wedged into the period from November to April.

As early as the nineteenth century, some policy makers and scientists advocated a strategy of ecologic adaptation to these differences in natural water supply. In the view of John Wesley Powell, who headed the U.S. Geological Survey from 1881 to 1894, the West should be divided into hundreds of "hydrographic basins," or watersheds, with development reflecting each watershed's natural characteristics.

Similar to virtually all nations, however, the United States has built cities, planted crops, and developed industries with little regard for these natural variations. Six of the nation's ten largest cities, and all ten of its fastest-growing cities, rise out of the arid West. In California, the nation's most populous state, only 3 percent of the natural water supply, but over 50 percent of the population, is located in the southernmost third of the state.

The dramatic growth of irrigated agriculture in the western United States is a prime example of the mismatch between water supply and economic development. Table 30.2

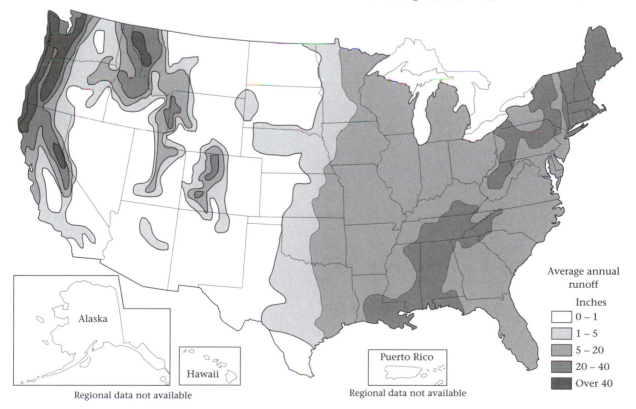

Figure 30.2. Average annual runoff in the United States. (*Source: The nation's water resources*, The First National Assessment of the Water Resources Council, Washington, DC, 1968.)

Alaska

Regional data not available

Hawaii

Puerto Rico

Regional data not available

Average annual runoff

Inches

- 0 – 1
- 1 – 5
- 5 – 20
- 20 – 40
- Over 40

TABLE 30.2. FRESHWATER USE AND CONSUMPTION IN THE UNITED STATES
BY REGION, 1990

Consumed Region	Withdrawals (millions of gallons per day)	Consumption (millions of gallons per day)	Percent Consumed of Withdrawals
Eastern United States	170,300	19,900	12
Western United States	151,000	73,300	48
Total	321,300	93,200	29

Source: Estimated Use of Water in the United States in 1990 (U.S. Geological Survey Circular 1081).

contrasts the amount of water withdrawn in the East and West, as well as the amount of water irretrievably "consumed" in the two halves of the country. Typically, eastern farmers do not need to irrigate their crops because of the large rainfall. As a result, thermoelectric power generators, which are quite low-percentage "consumers" (as shown in Fig. 30.3), use the vast majority of water in the East. By contrast, irrigating farmers, who are high-percentage consumers, withdraw approximately 80 percent of the water in the western United States.

Unwillingness to match development with local water supplies, however, is not unique to the arid West. Many other regions of the United States, particularly along the eastern seaboard, have also knowingly outgrown their local water supplies. New York City receives an average annual rainfall of 44 inches; however, its water demands long ago surpassed its natural water supply.

For much of the twentieth century, cities and agricultural communities that faced local water constraints simply turned to distant watersheds for supplemental supplies. New York meets the needs of its 7 million residents by importing over 1.5 billion gallons of water per day from watersheds up to 125 miles away. When Los Angeles began running out of water early in the twentieth century, it reached out to the Owens Valley 250 miles northeast (setting off a sometimes violent battle over water between valley farmers and the city); today, the Los Angeles region also draws large quantities of water from the Colorado River, some 200 miles to the east, and from northern California rivers over 450 miles away. Most major cities in the United States, including Boston, Denver, Phoenix, and San Francisco, have adopted similar import strategies.

The national government's major water distribution program dwarfs even these efforts. In 1902, Congress initiated a "reclamation program" to encourage small family farming in arid regions of the West by importing and distributing needed irrigation water. Over the past 90 years, the national reclamation program has built hundreds of separate reclamation projects, involving over 300 dams and 7000 miles of canals. Although originally designed to promote small farms, the program today provides water to over 20 percent of all irrigated land in the West, much of it held by large corporate farms, as well as to 20 million domestic users.

SURFACE WATER POLICY

A Brief Primer on Water Institutions. A complex web of institutions shapes water policy in the United States. States have principal authority over how much water can be diverted from their rivers, for what uses, and by whom. The states have turned to two principal doctrines to allocate surface water among competing entities and individuals. States east of the 100th meridian, following the lead of England and a number of its other former colonies, use the riparian doctrine. Only water users who own "riparian" land (i.e., land through which or tangential to which a waterway flows) can withdraw water from a surface waterway. Riparian landowners are limited to a "reasonable" quantity of water and typically can use the water only on their riparian land. Cities commonly obtain water supplies for their residents, many of whom are not riparian, by condemning the water through their power of "eminent domain."

Because the riparian doctrine would pose major problems for the arid West, where land is seldom riparian to a waterway, western states allocate water by the prior appropriation system. Anyone, riparian or nonriparian, is entitled to "appropriate," or divert, water from a river or lake for use even scores of miles away as long as the water is not already being used by someone else and is put to a "reasonable and beneficial" use. The West thus allocates water on a first-come, first-served basis.

Eastern and western states also administer their water systems in very different ways, reflecting differences in water availability. In the West, state water agencies (often known as "state engineers" or "water boards") oversee who uses the region's scarce water supplies and police against unlawful diversions. Anyone wishing to divert water from a stream must apply for an "appropriation permit" from the state agency, which will check to ensure that unappropriated water is available in the stream and that the applicant plans to use the water for a reasonable and beneficial purpose. Eastern states, by contrast, historically have not needed to regulate water use as carefully. Disputes over water use remain rare in many Eastern states; when arguments arise, typically they are resolved in the court. As water demand has grown, however, some eastern states also have set up permit systems to track water use and resolve disputes.

Although states control who can divert water from a river

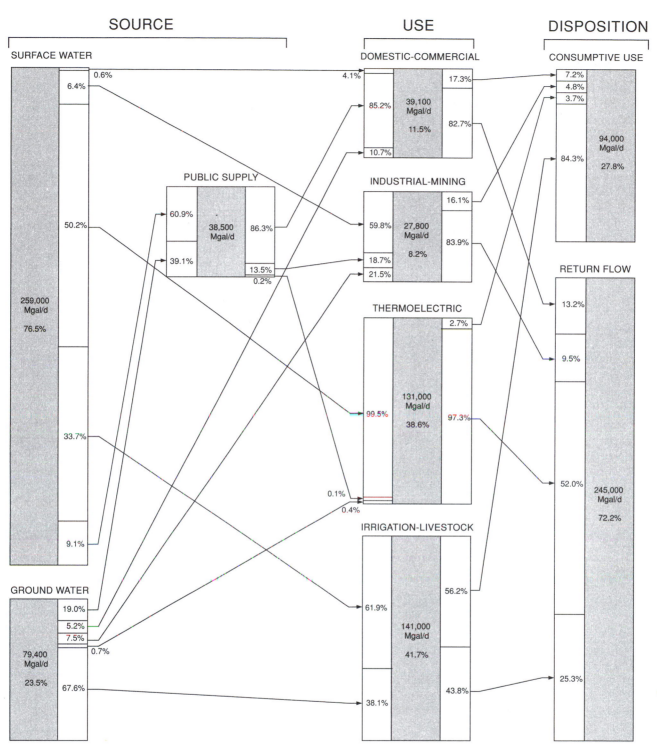

Figure 30.3. Source, use, and disposition of freshwater in the USA, 1990. Mgal/d = millions of gallons per day. (*Source*: Water Resources Council, *The nation's water resources*, Washington, DC, 1968.)

or lake, local governmental agencies and private companies distribute most water to the ultimate users. Diverting, transporting, and (when necessary) purifying water can be expensive. Therefore, most domestic users secure their water from cities, counties, municipal agencies, or privately owned "public utilities," which can spread the costs across many

users. Farmers typically receive irrigation water from private "mutual water companies" or governmental "irrigation districts."

The national government plays the role of the nation's chief engineer. The United States Army Corps of Engineers has modified thousands of waterways to improve navigation and reduce floods. The national Bureau of Reclamation has built hundreds of massive reclamation projects that, as described earlier, service more than 4 million hectares of agricultural land and 20 million domestic users.

Today, the national government is assuming an even more critical role: that of environmental regulator. As the environmental consequences of current water practices become better understood, national environmental statutes such as the Clean Water Act and the Endangered Species Act increasingly shape water policy.

Waste and Conservation. The need for more effective water conservation is one of the most important water issues confronting the United States. Far from constraining waste, traditional United States water policy has encouraged excessive levels of water use. The nation has long promoted new residential, agricultural, and industrial development and has perceived water as key to that growth. To encourage new development, national, state, and local governments have fostered a policy of cheap water – undercutting financial incentives to conserve water. Although state law technically outlaws gross water waste, states have never vigorously looked for and proscribed overuse.

■ **Water Pricing.** Few, if any, water users pay the true cost of their water. There are several reasons. First, in awarding rights to use water, states do not charge for the value of the water itself. Water is a valuable resource. When cities or farmers divert water from a stream, there is an "opportunity cost": instream flow is reduced at the expense of fish, wildlife, and the environment generally; other potential consumers, moreover, cannot use the water for domestic or commercial purposes. Yet cities and farmers pay nothing for the water they divert (other than a small permit fee, perhaps). As a result, water users and distributors divert water even when some or all of the water would be economically more valuable left in the stream or used by someone else. (The national government's practice of underpricing timber and mineral deposits also is criticized for encouraging uneconomically high use. Only water and air, however, are given away free.)

Second, water users do not generally pay for the environmental damage incurred in storing and distributing water. Many of the dams, reservoirs, pipelines, and aqueducts used to store, divert, and transport water have destroyed or injured valuable ecosystems. For example, in the 1920s, San Francisco flooded the Hetch Hetchy Valley – which naturalist John Muir likened in grandeur to its neighbor Yosemite – as a storage site for supplemental water supplies. Water users should, but do not, pay the cost to society of these and other environmental impacts.

The prices that cities, irrigation districts, and other water distributors charge for providing water to the ultimate consumer also promote excessive water use. A number of water distributors charge customers a flat rate regardless of how much they consume. Even distributors that charge by usage spread the cost of expensive new supply projects across all consumers (which often results in only a minimal rate in-crease) rather than charging new water users for the true cost of their water.

The government, in addition, often subsidizes the costs of collecting, storing, transporting, purifying, and distributing water. Many cities use property tax revenues to reduce their water rates. The U.S. Bureau of Reclamation permits farmers fifty or more years to pay back, *interest free*, the costs of the irrigation projects that it builds for the farmers. National reclamation subsidies have been massive. Some farmers pay 20 percent or less of the full cost incurred by the national government in developing and distributing the water. According to the Department of the Interior, reclamation subsidies through 1986 totaled approximately $10 billion.

■ **State Regulation.** State law under both the riparian and prior appropriation systems theoretically prohibits the use of "unreasonable" quantities of water. State regulation of water waste, however, has been relatively ineffective for several reasons. First, state agencies and courts have been reluctant to second-guess water users and distributors on their water needs. When a city has decided to build a new water project, water agencies and courts have not wanted to substitute their judgment for that of elected officials. Nor have water agencies or courts considered themselves capable of making independent economic and technologic decisions on what conservation measures would be feasible for individual water users. To avoid making such decisions, agencies and courts normally have looked to community custom to determine what degree of conservation is practicable. If a farmer is using no more water than her neighbors, agencies and courts generally have not intervened even though all the farmers in the area might have significant conservation opportunities.

Second, states have not regularly monitored water use. In the arid West, water uses are reviewed automatically only when a water user first applies for an appropriation permit. (Typically, even this review is perfunctory because few agencies have sufficient numbers of personnel necessary to identify and ferret out possible waste. In order to identify at least gross levels of waste, some states set maximum agricultural "water duties" that prohibit the application of more than a specified quantity of water to each irrigated acre.) After the permit is granted, agencies and courts typically reevaluate a use only if someone complains that the use is wasteful. In the East, where only a minority of states have adopted permit systems, most water uses are never automatically reviewed.

■ **Consequences.** Absent effective regulatory constraints, water subsidies have stimulated the diversion of large quantities of water for uses that do not justify the cost. The national reclamation program has been a particular source of waste. A 1989 study of eighteen irrigation districts receiving national reclamation water estimated that the financial costs of providing water to eleven of the districts (not including the opportunity cost of the water or any environmental costs) outweighed the economic benefits to the farmers receiving the water. In one case, the reclamation project cost over $1500 per acre served, whereas the project's value was only approximately $200 per acre.

National reclamation subsidies also encourage irrigation of crops in the West that could be produced less expensively elsewhere in the United States. By subsidizing western irrigation water, the United States often displaces economically more efficient agriculture from the Midwest or South. As an ironic aside, many Western farmers long used national reclamation water to grow so-called surplus crops, such as wheat, cotton, and rice, that the United States paid Midwestern and Southern farmers elsewhere millions of dollars *not* to grow! In the early 1990s, almost half of all lands irrigated with national reclamation water grew surplus crops.

Water subsidies also encourage water users to divert and use *more* water than is socially optimal. Although estimates of the exact amount of such waste vary substantially, most analysts agree that there is significant opportunity for conservation. Mohamed El-Ashry of the World Bank estimates that the United States wastes 50 percent of the water it withdraws (slightly less troubling than the 65 to 70 percent of water he believes is wasted worldwide).

Irrigating farmers consume only a fraction of the water that they divert from the nation's waterways. Only a percentage of the water withdrawn by a typical irrigation system reaches the farm (the remainder is lost to evaporation and leakage); crops, in turn, consume on average only slightly more than half of the water that reaches the farm. Although most of the unused water flows back to a waterway or into an underground aquifer where it is reusable by others, significant quantities are lost to use through contamination, evaporation, or seepage into unaccessible aquifers. Farmers can economically reduce their water use through various technologic improvements, including lining canals, carefully monitoring irrigation needs by computer, and switching from furrow irrigation systems to more efficient systems such as drip irrigation or precision sprinklers. (See the box entitled *Confronting Water Scarcity in Israel*.)

Waste is not unique to agriculture. Urban water use could be cut dramatically and economically through relatively simple conservation measures. In older cities, the repair of leaking water mains can save up to a third of the public water supply. Water-saving appliances, such as low-flow shower heads, can effectively cut indoor water use by up to 20 percent. "Xeriscaping" of yards with drought-tolerant plants can reduce outdoor water use, which represents the largest fraction of domestic water use, by a third.

■ **Current Reforms.** Increased recognition of the need for water conservation has led to some U.S. reforms. A number of states, for example, have adopted planning laws requiring cities, irrigation districts, and other water distributors to develop conservation plans. Few of these laws, however, mandate actual conservation measures or require real water savings.

Some state water agencies also have begun to police water use more carefully. For example, California agencies and courts have taken to task a handful of the state's more egregious water squanderers. In the most famous instance, California's water board found that the Imperial Irrigation District (IID), a large agricultural district near the California – Mexico border, was wasting approximately 100,000 acre feet of water each year (enough water to service a city of over 500,000 people) by transporting water in unlined canals, delivering excessive quantities of water to farms, and failing to install runoff recovery systems. After the board ordered IID to eliminate the waste, the Metropolitan Water District of Southern California (which supplies water to Los Angeles, San Diego, and many other southern California areas) agreed to pay for the necessary conservation measures in return for the right to use the conserved water for at least thirty-five years. State agencies and courts, however, generally remain reluctant to mandate conservation.

What appears to be "waste," moreover, sometimes is not. Agricultural water that is currently lost because of inefficient irrigation practices goes somewhere. If the water evaporates or is otherwise lost to others' use, the water indeed is "wasted" and conservation can increase the water supply. Yet, as previously noted, most "wasted" water returns to a waterway and is used by others. The vast majority of IID's excess water flowed into the nearby Salton Sea and was lost to human use because of the sea's high salinity. According to farmers across the border in Mexicali, however, some of the water from IID's unlined canals seeped into underground aquifers that the Mexican farmers used for their own irrigation. If so, California's efforts to eliminate "waste" by lining the canals would harm the Mexican farmers by lowering their water table.

The past decade has also seen pricing reform. At the local level, a number of cities and other water suppliers with flat-rate pricing systems have begun to meter water use. A growing number of water suppliers are moving to "inclining tier" prices in which the price per unit of water increases as individual use increases. Many municipal water agencies have decreased their subsidization of water rates.

Since 1982, the U.S. Congress also has passed a number of bills reforming prices under the national reclamation program. Most of the bills have raised water rates that farmers must pay the national government, although the rates remain significantly less than the true cost of the water. Congress has also required some irrigation districts to use tiered prices.

Pricing reform has not been easy. Over time, water users make investment decisions based on existing water prices. Farmers purchase additional acreage or new farm equipment; domestic residents plant lawns and foliage. When water agencies threaten to eliminate subsidies and raise prices, water users object that the government is belatedly and unfairly changing policy and undercutting these investments. Beyond these equity objections, most politicians and bureaucrats are loathe to take the politically unpopular step of rais-

CONFRONTING WATER SCARCITY IN ISRAEL

Nations confronted by scarce water supplies often have stretched supplies through conservation or other sustainable means, providing a model for future water policy elsewhere in the world. By anyone's definition, Israel is a water-scarce country. In 1995, its annual renewable water supply was only 380 cubic meters per capita. The comparable number in the United States is approximately 10,000 cubic meters (a thirtyfold differential); average annual consumption in the United States is over 2300 cubic meters. Israel receives virtually no rain for six months each year, and rainfall varies considerably from year to year around Israel's already low mean. Yet from 1973 to 1988, Israel's population doubled, and irrigated acreage expanded by over 25 percent.

Israel has met the needs of an increasing population and agricultural sector largely through conservation. Israel pioneered drip irrigation in the 1960s and today uses the technique on over 50 percent of its irrigated land (the highest percentage in any nation other than Cyprus, where drip irrigation is used on over 70 percent of irrigated land). Holes in plastic hoses deliver the correct amount of water to each plant, eliminating potential loss through evaporation, percolation, or runoff. Drip irrigation can be anywhere from 20 to 50 percent more efficient than standard sprinklers and is dramatically more efficient than the open-ditch flood irrigation systems still used in much of the world. Israeli farmers have increased their water effi-

ciency another 10 to 30 percent by linking their irrigation systems to computers that detect leaks, that shut off faulty lines, and that precisely adjust water application amounts for wind speed, air temperature, and soil moisture. Researchers are examining various farming techniques that could further reduce water consumption. For example, by covering exposed soil with polyethylene, farmers could reduce evaporation and increase the water efficiency of plants by raising root-zone temperatures.

Israel also is making extensive use of recycled water. Israel has achieved a high quality standard, suitable even for domestic use, by taking treated effluent and letting it percolate into underground aquifers from which it is later pumped back up; the soil provides natural purification as the water passes through it. Because agriculture does not demand the same high level of quality as domestic use, the Dan Region Project also transports treated effluent from Tel Aviv and other northern cities south to farms in the Negev Desert – freeing up fresh water for expanding domestic needs. Many parts of the country use dual distribution systems, with farms receiving untreated storm water and some effluent, while homes receive high-quality drinking water. Waste water currently constitutes 30 percent of Israel's agricultural water (with the percentage projected to grow to 80 percent by 2025).

Israel also has begun to look for other ways of expanding its limited natural

water supply. For example, the country uses available capacity in its groundwater aquifers to store water. Israel artificially recharges the aquifers with excess water during wet seasons and then pumps the water back up during dry periods. In 1997, Israel fired up its first major desalination plant to purify water from the Red Sea for use in the growing coastal town of Eilat. (Approximately 60 percent of the world's current desalination capacity is found in the neighboring Persian Gulf region. Saudi Arabia, with over twenty plants and 30 percent of the global capacity, continues to lead the way.) Israeli farmers are also experimenting with a variety of crops that are more tolerant of salt.

Despite these various efforts, Israel could do more to promote water conservation and ensure a sustainable environment. Because of Israel's completion of the National Water Carrier, a system of aqueducts, which taps water from the Sea of Galilee for use as far south as the Negev Desert, virtually no water now flows into the southern Jordan River. Israel also continues to use over 15 percent more water than its renewable supply by overdrafting its underground aquifers. The overdrafting has permitted salt water to invade Israel's coastal aquifer, one of its two principal groundwater aquifers. Israel encourages excessive water use in some sectors and areas by subsidizing water deliveries to stimulate agricultural production and the settlement of new territories.

ing prices. Consequently, most water rates remain substantially below actual water costs.

Depletion of Waterways. The United States also has done a poor job of protecting its rivers from depletion, particularly in the West where water demand long ago outstripped most local supplies. Until recently, state water agencies did not weigh the value of instream flows in deciding whether to permit diversions. Under the prior appropriation doctrine, moreover, people interested in preserving instream flow for fishing, environmental, or other purposes were not permitted to "appropriate" water themselves for use in the stream. Western states wanted to encourage development and rewarded only those who planned to divert water for offstream consumption.

As a result, many western waterways are drained of water for some or all of the year. Water users reduce other waterways to mere trickles. Little of the Colorado River, which dominates the southwestern waterscape, reaches its mouth in the Gulf of California; in dry years, it peters out in a saline pond 15 miles from its natural destination.

The principal costs of depleted waterways are environmental. Low water flows, along with deteriorated quality, are currently threatening over one-third of all fish species in the United States. Depletion of waterways has led to the total extinction of some fish species such as the Truckee River cutthroat trout. Wildlife has also been threatened; low water flows in California's Trinity River have reduced the black-tailed deer population by thousands.

As scientists have learned more about the harmful effects of depleted streamflows, policy makers have begun to enact measures to protect existing instream flows. In one of the earliest reforms, most western states authorized their water agencies to consider environmental, aesthetic, and recreational impacts before permitting new appropriations from a river. Several western states have also authorized their environmental or wildlife agencies to reserve instream flows, preventing others from appropriating and diverting the water for offstream consumption. A number of states, and the national government, have adopted scenic river legislation proscribing or limiting new appropriations from rivers designated as having special aesthetic or recreational value.

National environmental legislation now provides the greatest protection for instream flows. Under section 404 of the Clean Water Act, no one can excavate or fill a waterway without obtaining a permit from the U.S. Army Corps of Engineers. Although section 404 grew out of a century-old statute designed to protect navigability, it now plays a number of other important roles. Section 404 is the principal national tool for regulating the filling and development of wetlands. Moreover, because almost all major water diversions require some excavation or filling of a waterway, section 404 also serves as an important constraint on new water projects.

Before issuing a permit in connection with a planned water diversion, the Army Corps of Engineers must determine that the water project will not have "an unacceptable adverse effect on municipal water supplies, shellfish beds and fishery areas (including spawning and breeding areas), wildlife, or recreational areas." The corps must follow guidelines developed by the national Environmental Protection Agency (EPA), which can veto any permit that it believes should not be issued. Under the U.S. Endangered Species Act, moreover, the corps cannot issue a permit if the planned water diversion would "jeopardize the continued existence" of an endangered species or "result in the destruction or adverse modification" of the species' critical habitat.

Section 404, along with the Endangered Species Act, has killed a number of recent water projects and given environmentalists a powerful weapon for opposing many more. The demise of the planned Two Forks Dam in Colorado is illustrative. Denver proposed to build the $1 billion project on a stretch of the South Platte River to meet the city's future needs. Even though studies suggested that conservation would be a cheaper means of meeting Denver's future water needs, nothing could derail the project until it ran into section 404. The Army Corps of Engineers issued a permit for the project; however, the EPA vetoed it after finding that the project would inundate a prize trout stream, eliminate a major recreation area lying within Pike National Forest, and destroy valuable wildlife.

These measures, powerful as they may prove to be in avoiding further depletion of our waterways, unfortunately do nothing to restore the thousands of already depleted waterways. Here progress has been slower because restoration requires water users to reduce their current diversions. Proposals to reduce existing diversions directly threaten current water users and generate stronger political opposition than do efforts to limit new diversions. Because the U.S. Constitution prohibits the government from "taking" private property without paying compensation, current water users also argue that the government cannot unilaterally order reductions in existing diversions unless they are willing to pay for the water.

Some progress has been made. A few state courts have reduced water diversions through innovative application of long-standing legal doctrines. The law contains a fair amount of flexibility and, spurred by new needs, courts can often remold or reinterpret the law to meet the needs. An example is the "public trust" doctrine. Courts have long held that states, upon formation, receive title to lands beneath navigable waterways in "trust" for the public. Although states can sometimes convey these lands to private owners, the public trust still applies in most cases and limits what the private owners can do with the lands. In *National Audubon Society v. Superior Court* (see the box entitled *Salt Lakes*), the California Supreme Court held that the public trust doctrine also extends to the water itself and imposes on the state "a duty of continuing supervision" over use of the water. If at any time the state concludes that the environmental impact of a diversion outweighs its value, the state must order the water user to cut back or entirely stop the diversion.

Environmental statutes have also forced some reductions in current diversions. Diversions can violate national and state water quality laws by increasing the concentration of existing contaminants or encouraging saltwater intrusion into coastal estuaries. Diversions can also violate national and state endangered species laws by threatening the continued existence of native fish species. California's two major rivers, the San Joaquin and Sacramento, meet to form an immense delta that flows out to San Francisco Bay and the Pacific Ocean. Years of major water diversions have caused salt water to move up into the delta; as a result, both the delta smelt and certain runs of chinook salmon have declined to the point that they are now listed as threatened species. Despite years of government foot dragging, environmental groups in the 1990s were able to use the Clean Water Act and Endangered Species Act to force at least temporary cutbacks in these diversions.

Meeting Growing Urban Needs. With water supplies already pressed in many regions, meeting the demands of rapidly growing urban and suburban populations looms as another critical issue. Traditionally, cities adopted "supply solutions" to water shortages. Taking the level of demand as unchangeable, cities built large projects, such as San Francisco's Hetch Hetchy reservoir, to import water from distant watersheds. As discussed earlier, most cities now are addressing prospective shortfalls also from a demand side. Although far more could be done to reduce urban demand, particularly through the pricing of water, however, many cities will still need to find sources of additional water.

SALT LAKES

Lakes are classified in two categories: freshwater lakes, and salt or saline lakes. Freshwater lakes, such as the Great Lakes of North America, drain to the sea and contain less than 3 grams of salt per liter. In contrast, salt lakes are sinks with no outlet to the sea; as a result, salt accumulates. The salinity of salt lakes almost always exceeds 3 grams per liter and can sometimes surpass 300 grams per liter. (For comparison, ocean water contains approximately 35 grams per liter.) Many of the world's most famous lakes, including the world's largest (the Caspian Sea) and its lowest (the Dead Sea), are salt lakes. Salt lakes together account for almost half of the total volume of all inland surface waters.

Salt lakes are far more sensitive to both natural and human-induced change than are freshwater lakes. In salt lakes, inflows are delicately balanced with outputs from evaporation and seepage into underground aquifers. If water diversions reduce inflows into salt lakes, the lakes will shrink and grow even saltier – threatening the local ecosystem. Most of the world's principal salt lakes, including the Dead Sea and Lake Corangamite (the largest permanent lake in Australia), currently are shrinking.

The best-known example is the Aral Sea in Russian Central Asia, which was once the fourth largest inland lake in the world, approximately the size of Lake Huron. Two north-flowing rivers, the Amu Darya and Syr Darya, historically drained into the Aral Sea. Until the latter part of the twentieth century, the Aral Sea, also known as the "blue sea," covered 64,000 square kilometers and was home to thriving fishing, trapping, and lumber industries. In the early 1950s, fishermen harvested 50,000 tons of fish per year, which were canned in coastal cities.

However, the former Soviet Union recognized the potential for producing massive quantities of cotton on lands neighboring the Aral Sea. Obsessed with becoming cotton-independent, the Soviet Union made it the "patriotic duty" of Central Asians to grow cotton and, by 1937, was exporting cotton. To grow the cotton, however, the Soviet Union diverted most of the flows of the Amu Darya and Syr Darya. Water planners recognized that the diversions would lead to the shrinkage of the Aral Sea, but they believed that the environmental consequences would be minor.

The water planners were wrong. The Aral Sea has shrunk to slightly more than one-third its original size, splitting into two lakes with a large desert chasm in between. As deltas and islands that once supported hundreds of species have disappeared, 135 wildlife species and 50 percent of all bird species have also disappeared. The lake's salt content has tripled to approximately 30 grams per liter, and 20 of 24 indigenous fish species have become extinct. Commercial fishing ended in 1981, and many former fishing villages are now marooned in the middle of a desert. Winds pick up the salts and other particulates from the exposed lake bed, contributing to increased incidence of human diseases such as emphysema and other respiratory ailments. The shrinking of the Aral Sea also has had an impact on the local climate. Evaporation from the sea used to deflect cold northern winds in the winter and provide rain and shade in the summer. Today, the average temperature in the winter is approximately 2°C colder, and the average summer temperature comparably warmer, than before.

Saving the Aral Sea will be difficult. Conservation measures, such as lining irrigation canals, installing drainage systems, and improving irrigation methods, could substantially reduce water use. However, local officials argue that, given the weak local economy and growing population, any water saved should go toward growing more food rather than into the Aral Sea. Even if the conserved water was dedicated to the Aral Sea, the lake probably would continue to shrink. Current restoration efforts thus are focused on stabilizing and making small improvements to the local environment. For example, the United Nations is funding the creation of "green belts" that will shield particularly sensitive areas from the Aral dust. In the meantime, the Aral Sea continues to shrink and is expected to stabilize at slightly less than one-third of its current size.

Mono Lake provides a valuable contrast. The second largest lake in California, Mono Lake sits on the eastern side of the Sierra Nevada escarpment near Yosemite National Park and has sometimes been called the "Dead Sea of California." Mark Twain, who wrote of Mono Lake in *Roughing It*, observed that "Half a dozen little mountain brooks [actually five] flow into Mono Lake, but not a stream of any kind flows out of it. It neither rises nor falls, apparently, and what it does with its surplus water is a dark and bloody mystery." The lake's most notable feature is its "tufa towers," limestone-like stalagmites that form over freshwater springs in the lake bed.

In 1941, California granted Los Angeles the right to divert water from the streams that feed Mono Lake and transport the water 250 miles to meet the needs of the growing population of Los Angeles. By 1982, the lake level had dropped by 14 meters (or 45 feet), its surface area had shrunk by one-third, and its salinity had nearly doubled from 48 to 90 grams per liter. As the lake shrank, one of its two principal islands became a peninsula, exposing roosting gulls to coyotes and other predators. Trout fisheries in the feeder streams all but disappeared.

A coalition of environmentalists and sports fishermen sued to reduce the diversions of water to Los Angeles, arguing that these diversions violated the public trust doctrine and various environmental laws. The California Supreme Court ruled in favor of the coalition in 1983. Acting to implement the court's decision, the California State Water Resources Control Board in 1994 prohibited Los Angeles from diverting any water from the feeder streams until Mono Lake rises 17 feet. Although the board initially believed that 20 years might elapse before this goal could be achieved, high amounts of rainfall raised the lake level 6 feet by the middle of 1997. Los Angeles also will ultimately spend more than $10 million trying to restore the habitat of the lake and its environs.

A number of cities are experimenting with technologic solutions such as desalination and recycling. By the end of the twentieth century, local governments in California had built more than a dozen desalination plants. However, the plants are small and largely experimental. Despite potential promise, desalination of seawater remains expensive because of the large amounts of power needed to drive the process. Water from the United States' largest plant in Santa Barbara, California, is four times the average cost of the city's other water supplies. Desalination also presents a number of environmental concerns, including disposal of saline waste, air pollution from the needed energy production, and destruction of coastal vistas by large desalination plants and power facilities.

A less expensive option is to recycle waste water. St. Petersburg, Florida, has become the first city in the world to recycle all its water. The city distributes fresh water for drinking and household use, and makes recycled water available (at only 70 percent the fresh-water prices) for irrigation of parks, golf courses, and lawns, and for a variety of other outdoor uses. A growing number of other cities, including Los Angeles, Denver, Tucson, and El Paso, plan to use recycled water to meet up to 20 percent of their local needs by early in the twenty-first century. The future of water recycling will depend on the ability of cities to reduce the cost of water even further (the cost is already competitive with major new water projects) and to reduce psychological barriers to the use of recycled water. The greatest promise for recycled water lies in using it for agriculture, which could free up fresh water for domestic use (see the box entitled "Confronting Water Scarcity in Israel").

Many western cities regard transfers of water from agricultural to urban areas as the best solution to their water needs. Because agriculture currently accounts for almost 85 percent of all water consumption in the United States (see Fig. 30.3) and an even higher percentage in the West, even small reductions in agricultural water use can provide sizable increases for urban areas. For example, if Arizona farmers reduced their current water consumption by only 5 percent, the freed-up water could support the domestic needs of 1.5 million people (over half the current population of the state). As discussed earlier, economic studies suggest that far greater levels of agricultural conservation are cost-justified.

Given the opportunities for trade, one would expect farmers to conserve and sell water to cities and other urban water suppliers. However, water law in the western United States historically has not favored market transactions in water. Because states did not charge water users to divert water, water officials did not believe that users should be able to profit by selling the water to others. Water officials also feared that people might "speculate" in water by appropriating far more than needed, with an eye toward marketing it. Ten western states banned water transfers.

Although all ten states now have repealed the bans, the law still presents significant obstacles to market transfers. Ironically, the antiwaste provisions of western water law pose a hurdle to the conservation and sale of irrigation water. The very act of conserving water suggests that the farmer may have been wasting and therefore may not have been entitled to the saved water. According to some courts, farmers and other water users cannot use or sell conserved water; instead, the conserved water reverts to the state and is available for new appropriation.

Irrigation districts and other agricultural water distributors pose perhaps the greatest barrier to water transfers. Most farmers do not own their water rights; they receive water from irrigation districts or other organizations. Farmers therefore cannot generally transfer water rights to cities or other urban agencies without the approval and cooperation of their irrigation district or supplier. Most irrigation districts, however, oppose such transfers. Where water can be freed up, district managers prefer to keep the water within the district where it can be used to satisfy the demands of farmers who need water or to support local growth.

A number of western states, as well as the national government, have enacted legislation to overcome these barriers. For example, California has explicitly authorized the sale or lease of conserved water. Oregon permits water users to submit "conservation plans"; if the plans are approved by the state water agency, users can then sell 75 percent of the conserved water (the remainder must return to the river as a form of environmental tax). In an attempt to circumvent the opposition of irrigation districts to water transfers, Congress authorized farmers in the Central Valley of California to sell limited quantities of national reclamation water without the approval of their local districts.

The prospect of active water markets has generated considerable fear within rural agricultural communities. These communities fear that cities and other urban water agencies will pay farmers not only to conserve water but also to fallow their fields. Water is economically far more valuable in urban use than on many farms; with active water markets, farmers may often find it more profitable to close up shop and sell their water than to continue farming. Many policy analysts believe that this is how the market should operate. Water should not remain permanently tied to marginally profitable farms when it could be used more valuably elsewhere.

The rural communities at risk, however, argue that water transfers will impose economic, public, environmental, and social costs on local residents that the nation should not require them to bear. Farm closings can generate unemployment and reduce business in support industries. Closings can also decrease local tax revenues, requiring a curtailment of public services. In some areas, the fallowing of fields can lead to dust storms and local desertification. Many communities fear that farm closings will have a snowball effect. The problems created by initial closings will encourage other farms to close and businesses to leave, increasing the local impacts and leading to the termination of yet more farms and businesses.

Many states try to protect agricultural communities against such harm. For example, states often authorize water boards to prohibit transfers that might injure local econo-mies or otherwise damage the "public interest." Such restric-tions protect agricultural communities, but limit the ability of the marketplace to help meet shifting demands.

GROUNDWATER POLICY. Groundwater use has grown dra-matically in the United States over the past fifty years and today constitutes approximately 25 percent of all freshwater use. Irrigated agriculture is the largest consumer of ground-water, using approximately 70 percent of the total. A number of factors have helped produce growth in groundwater use. The advent of cheap rural electricity and new technology has made groundwater economically competitive with surface water supplies. Dwindling surface supplies also have forced growing cities and agricultural belts to search for water un-derground (this scenario is also occurring globally (see the box entitled *Global Groundwater Depletion*).

Groundwater aquifers (Fig. 30.4) contain two sources of water: *annual recharge* (the water that seeps into the aquifer each year from overlying precipitation, waterways, or irriga-tion) and *fossil water* (the water that has seeped into the aquifer over thousands of year without being withdrawn). When a region has outgrown its renewable water supply, it can continue to meet its water demand by "mining" fossil water in local aquifers. As shown in Figure 30.5, many regions of the United States are doing exactly that. In the West, agricultural users mine or "overdraft" nearly 25 per-cent of all aquifers. Nationally, recharge averages approxi-mately 60 billion gallons per day, an encouraging figure until one realizes that water users are extracting approximately 75 billion gallons per day.

Mining aquifers, however, is at best only a temporary so-lution to unmet water demands. The fossil water will ulti-mately run out, leaving the areas that have grown reliant on it with the need to either find alternative supplies or dramat-ically reduce their water use. The Ogallala (or "High Plains") Aquifer, which underlies large portions of Kansas, Nebraska, Texas, and four other states, is probably the largest under-ground body of fresh water in the world (see Fig. 30.4). Yet it has virtually no recharge, partially due to the low precipita-tion in the region and primarily due to an impervious layer that forms a watertight cover over most of the aquifer. Local residents turned to the Ogallala several decades ago to sup-port agricultural and residential growth and now are quickly draining it dry. With water levels falling 3 feet every year in some areas, most of the aquifer will be effectively depleted by early in the twenty-first century, leaving an estimated 5 million acres of farmland without water.

Even before an aquifer runs dry, groundwater mining can impose severe economic and environmental costs. As an aq-uifer is mined, its water table drops. Water users must pump the groundwater greater distances, increasing both pump and electricity costs. (Water users seldom deplete an aquifer of all its water. Long before this occurs, water tables drop to a level where pumping is no longer economical.) Lower water tables already have taken a sizable toll. Although total groundwater use has increased in the United States since 1975, spiraling pumping costs have led to an actual decline in the quantities

Figure 30.4. Major aquifers of the United States. (*Source:* William Ashworth, *Nor any drop to drink,* Summit Books, 1982.)

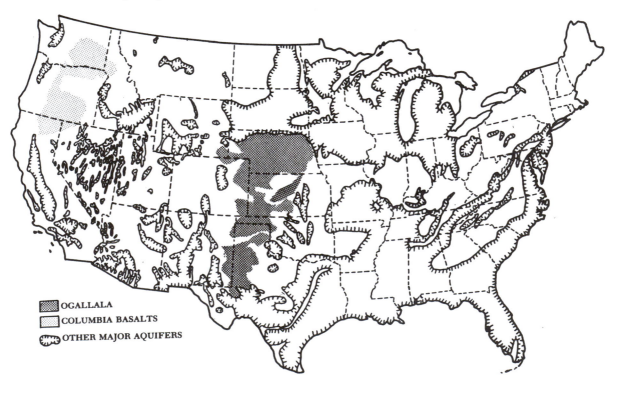

OGALLALA
COLUMBIA BASALTS
OTHER MAJOR AQUIFERS

GLOBAL GROUNDWATER DEPLETION

Overdrafting of groundwater aquifers is a global problem. Mexico City is depleting the groundwater aquifer that lies beneath it twice as fast as nature is replenishing it. As a result, the land on which the city is built is subsiding at the rate of approximately 6 centimeters annually (or approximately 1 foot every five years). Some parts of the city are dropping at rates up to three times this decline. In Mexico City's historic center, the ground has dropped 9 meters (almost 30 feet) in the past 100 years, leaving churches and palaces leaning precariously.

China also is seriously overdrafting its groundwater. As a result of local pumping, China's most populous city, Shanghai, has been sinking by approximately 10 millimeters annually (or 1 foot every thirty years) since the mid-1980s. Subsidence has been more severe in other Chinese coastal cities, causing pipes to break and houses, bridges, and other structures to collapse; coastal subsidence also has threatened increased flooding from local rivers. With that nation's capital also overdrafting its local aquifer, the land underlying Beijing has sunk more than 60 centimeters (or approximately 2 and 1/2 feet) since the middle of the twentieth century and is continuing to sink at a rate of up to 2 centimeters per year (approximately 1 foot every fifteen years). Groundwater tables are dropping by approximately 1 meter every year, increasing the cost of pumping water for surface use. Northern China's second largest city, Tianjin, has sustained serious building damage from subsidence rates that are over four times those of Beijing.

One can cite examples of groundwater overdrafting virtually everywhere in the world that major population or agricultural centers are found. Tokyo is sinking. Water tables are dropping rapidly in important agricultural regions of India, forcing the abandonment of thousands of wells and threatening the regions' long-term sustainability. Overpumping in the western Indian state of Gujarat has led to saltwater intrusion and contamination of local drinking water supplies. The Middle East faces particularly dire groundwater problems. Hydrologists believe that Bahrain will exhaust its groundwater aquifer early in the twenty-first century. At the current rate of pumpage, Saudi Arabia's groundwater supply is likely to last from 25 to 100 years, with most estimates at around 50 years.

Dropping groundwater tables, surface subsidence, and even saltwater intrusion have not moved many governments to act. Most of the major options, from importing alternative water supplies to regulating groundwater withdrawals, will either raise water costs or require significant conservation, which will be unpopular among local populations. Faced with the choice of political criticism or the status quo, many officials choose the status quo.

Of the options for addressing these problems, most countries choose to find alternative sources of supply. Some nations import water from distant watersheds for direct distribution to the population or for artificially recharging the aquifer. By itself, importation of new water supplies is only a short-term solution. Mexico City now supplies approximately one-third of its water by importing water over 150 kilometers (or 90 miles) from distant provinces. Yet, as the city has continued to grow by half a million people per year, groundwater overdrafting has continued apace. Moreover, water importation projects, often damage the environment, replacing one set of water problems with another.

A number of countries, including China, have begun to restrict the amount of water that can be withdrawn from overpumped aquifers. Unless wells are registered and metered, however, restrictions cannot be enforced. Also, regulation of individual aquifers often simply increases pumping from neighboring aquifers. In 1997, Chinese officials reported that new Shanghai restrictions had led local industries to obtain water from neighboring provinces where restrictions were fewer.

Long-term solutions to groundwater overdrafting will require areas with dropping water tables not only to limit groundwater pumping, but also to conserve water or otherwise stretch their limited supplies. Mexico City finally has begun to encourage conservation by raising water rates, replacing conventional toilets with low-flow alternatives, and implementing an education campaign. A number of countries with severe groundwater problems, including China and Saudi Arabia, have begun to treat effluent both as a replacement for groundwater and to recharge depleted aquifers.

Even these measures ultimately will be ineffective unless population growth is brought under control. As mentioned, Mexico City's population is expanding by more than half a million people every year. The population of Saudi Arabia is likely to double in twenty years. Even with conservation and recycling, high population growth rates inevitably will increase water demand and place pressure on local groundwater resources.

---

used by farmers, who often cannot afford the same costs as municipal and industrial users.

If an aquifer consists of poorly consolidated materials (such as sand or gravel), overlying land can subside. The Shipping Channel area of Houston, Texas, dropped 6 inches per year in the late 1960s and early 1970s (resulting in millions of dollars in property damage). Overlying flora that rely on a high water table perish, leading to desertification of the surface. Natural springs dry up, threatening species reliant on their flow. Finally, dropping water pressure levels lead to destructive groundwater contamination. As discussed in Chapter 9, mining of coastal aquifers frequently leads to saltwater intrusion from the ocean. In other areas, groundwater mining has accelerated the spread of existing contaminants or drawn in polluted drainage water and other impurities.

None of this means that groundwater mining should be

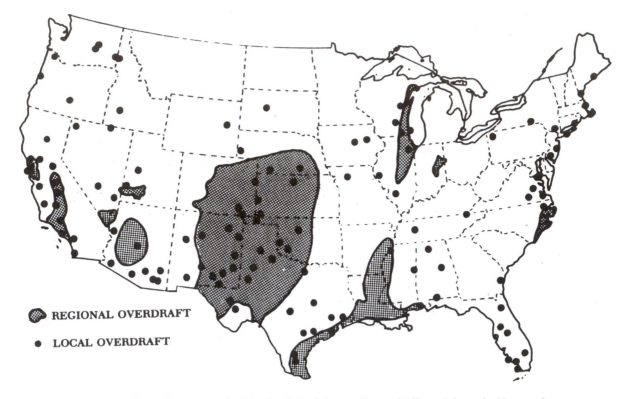

Figure 30.5. Groundwater overdraft in the United States. (*Source:* William Ashworth, *Nor any drop to drink*, Summit Books, 1982.)

totally outlawed. Fossil water is a valuable economic resource, and the benefits from a limited degree of mining can outweigh the associated costs. Groundwater mining must be carefully regulated, however, to ensure net benefits.

Most states long tolerated groundwater mining, no matter what the costs. Eastern states originally employed a rule of "absolute ownership" that entitled landowners to withdraw as much groundwater as they wanted from underlying aquifers. This set the stage for a tragedy of the commons in which individual landowners saw little reason to limit their withdrawals. Landowners did not experience the damage their withdrawals caused others by, for example, lowering the water table. Moreover, no individual landowner could eliminate an overdraft; if one landowner voluntarily stopped pumping, others would still continue.

As groundwater mining has become a more serious problem, a growing number of states have taken at least some steps to regulate pumping. Yet reform has been slow. Except when they are faced with critical and immediate problems such as saltwater intrusion, farmers and cities generally have been unwilling to give up any of the groundwater on which they have grown reliant. Even in the midst of a crisis, disagreements over how to cure existing groundwater problems and who should bear the burden of groundwater reduction have derailed reform efforts or led to ineffective compromises.

Arizona is a case in point. By the late 1960s, Arizona was

mining its aquifers at almost twice the recharge rates. Initially the state hoped that the Central Arizona Project (CAP), when completed in the 1980s, would cure the problem by importing enough Colorado River water to displace excessive groundwater withdrawals. Fearing that CAP water might instead simply feed new growth, the national government refused to finish constructing the CAP unless the state solved its groundwater problem. Although a state water commission informally estimated that Arizona could cure its problem simply by buying and retiring a few hundred thousand acres of agricultural land in key areas, no one promoted this approach. Instead, after lengthy negotiations led by the then-Governor Bruce Babbitt, Arizona's major water interests agreed in 1980 to a compromise that satisfied the national government but has yet to eliminate the state's overdraft. The Groundwater Management Act of 1980 pledges to eliminate Arizona's most severe water mining problems by 2025; however, current progress suggests that the state will not achieve even this extended goal. Instead of imposing firm aggregate limits on pumping, the act relies on a variety of conservation standards that, although they have resulted in pumping reductions, appear too ineffectual to eliminate overdraft completely.

Environmental laws might again prove a saving grace. Texas, the only state that continues to use the rule of "absolute ownership," is experiencing very serious groundwater depletion. Reform efforts largely have failed. Yet when falling water levels in the Edwards Aquifer near San Antonio threatened endangered species in the aquifer and in waterways fed by the aquifer, a national district court ordered the state and local users to reduce water withdrawals. Texas responded by forming the Edwards Aquifer Authority to regulate groundwater pumping and restrict total withdrawals.

QUESTIONS

1. How do the issues in this chapter relate to water policy in your own community? Do some research to answer the following questions: From whom do you receive your water, and where does the water originate? How, if at all, are you charged for the water? Are any efforts made to encourage you to conserve water? If so, have the efforts been successful? What, if any, water problems confront your community (e.g., groundwater overdrafting or surface water depletion)? If you were in charge of water policy in your community, what would be your goals? What policies would you adopt to achieve those goals?

2. What are the relative advantages and disadvantages of the riparian doctrine and the prior appropriation doctrine as methods of allocating the water in surface waterways? Which doctrine is likely to lead to the most valuable economic use of the available water? Is one doctrine preferable from an environmental standpoint? Why do the western United States employ the prior appropriation doctrine and the eastern United States use the riparian doctrine? In what other ways could a state or nation allocate water rights? Are any of these methods preferable to the prior appropriation and riparian doctrines?

3. Debate the following resolution: water should be freely transferable among users through unrestricted water markets. Specifically, address the issue of whether farmers should be free to transfer the water they use to distant growing cities, even over the opposition of the local community or an irrigation district from which they are receiving their water.

4. Should groundwater overdrafting ever be permitted, given the attendant problems? What tools could a government use to address groundwater overdrafting? What are the comparative advantages and disadvantages of the tools?

5. Governments can try to eliminate wasteful water practices through regulation (e.g., prohibiting waste), education, and economic incentives (e.g., higher water rates). How effective do you believe each of these approaches is likely to be? What obstacles does a government face in trying to use each approach?

6. Efforts to restore depleted waterways will inevitably require existing water users to reduce the amount of water they have historically withdrawn from the waterway and used. How would you respond to the argument that it is unfair to force existing users to cut back, given the investments they have made (e.g., in equipment or land cultivation) based on previously existing water policies? Should users be compensated for any forced water reduction? How would you overcome the likely opposition of existing water users to forced curtailments?

7. Many water analysts believe that the only long-term solution to the world's water problems is to develop a new "water ethic." What should such an ethic say? How can an advocate of this water ethic try to promote the ethic among members of society? How successful is the advocate likely to be?

FURTHER READING

Anderson, T. L., and Hill, P. J. 1997. *Water marketing: the next generation.* Lanham, MD: Rowman & Littlefield.

Postel, S. 1997. *Last oasis: facing water scarcity.* New York: Norton.

Reisner, M. 1993. *Cadillac desert* (revised and updated). New York: Penguin Books.

Worster, D. 1985. *Rivers of empire.* New York: Pantheon Books.

31 Valuing Nature

LAWRENCE H. GOULDER
DONALD KENNEDY

HOW CAN THE WORTH OF THE ENVIRONMENT BE ASSESSED – AND BY WHOM?

Societies are often in the position of making choices among different ends they might pursue. Should the vacant corner lot be used for public parking, or reserved for housing? Should limited public financial resources be used for education programs or for welfare payments to the indigent elderly? Decisions of this kind, which require choosing between or among competing social "goods," are always difficult. They challenge us because they bring into play considerations of *efficiency* (what are the overall benefits in relation to the overall costs?) and *equity* (how are those benefits and costs distributed, and how does that distribution accord with society's concepts of fairness?). These questions become even more challenging when it is difficult to attach precise value to one or another of the outcomes, as is often the case when we find ourselves choosing among environmental values.

Consider a proposal to dam a wild river in order to produce a set of multiple claimed benefits: downstream protection from flooding for those who live near the river; a supply of seasonally reliable irrigation water for farmers; and hydroelectric power for distant urban consumers. The benefits of the project can be calculated, though not without some controversy. The power contributed by the dam has a value, as does irrigation water and the avoided costs of flood damage. However, calculating the costs of damming the river presents a new kind of problem. The construction costs associated with building the dam are easy enough. Yet, river rafters, trout fishermen, hikers, and bird watchers – all users of the river in its natural state – point out that not all the costs have been counted. They claim that the project will ruin a natural environment of great value. When contests develop between such claims and those of farmers or other workers who will benefit from the dam, the outcome is likely to be controversial, as in the celebrated struggle over the spotted owl in the old-growth forests of the U.S. Northwest, where it was claimed that the issue was "jobs versus environment."

Despite the power of the economic growth argument, it is plain that the environment has for many people a special kind of value. Our aim in this chapter is to consider what factors underlie these attachments, and how we might go about measuring their strength. "Valuing nature" means exactly that: the assessments of worth that people make of undisturbed (or, at least, relatively undisturbed) ecosystems.

When we talk about "environmentalism," we are in fact referring to attitudes people have about their surroundings. On a basic level, people want the place where they live to be free of threats to their health – for example, no toxic substances in the groundwater, and no pollutants in the air. However, as controversies such as the one over logging old-growth forests in the U.S. Northwest suggest, people are also deeply concerned about conserving something called "nature."

Often, we are forced to choose between making use of the land and leaving it as it is. How can we make such choices intelligently? Are there reasonably precise ways of attaching value to "nature," so that we can compare the benefits of conversion with those of conservation? Valuation is not difficult when we are talking about things that are bought and sold in markets, such as concrete, or construction labor, or even irrigation water. Nature, however, is not traded in the same way, and the indicators of its value are less obvious.

In approaching the problem of valuing nature, we pose three questions:

1. What, precisely, do we mean by "nature?" Are there different versions of what we are interested in saving? If so, how can we decide among them?
2. What is the fundamental source or basis of value; that is, what "interests" or "rights" are at stake, and whose interests and rights should take precedence? This question challenges us to define carefully what we mean by "value," and exactly what it is that we are valuing. It also brings us head to head with some fundamental philosophic issues.

W. G. Ernst (ed.), *Earth Systems: Processes and Issues*. Printed in the United States of America. Copyright © 2000 Cambridge University Press. All rights reserved.

3. How can nature's value be measured empirically? The answer to this question will presume particular answers to the first two – that is, we must base any methods for quantitative valuation on our decisions about exactly what it is that we care about, and about what constitutes a legitimate basis for value.

WHAT ARE WE VALUING?

We begin by examining what we mean by the word "environment," or as it is more popularly called, "nature." The box entitled *To Root or not to Root – Feral Pigs in Paradise* provides an illustration of the very different definitions people have of nature, and suggests the difficulties those differences may present to policy makers.

For most conservationists in industrialized nations, the term "nature" typically means the entire ecosystem under discussion – all of it. The subject matter of this book, of course, is exactly that: the oceans, the atmosphere, the physical landscape (landforms, soil, climate), and the assemblage of living things that inhabit it. However, when people are asked to express more carefully the precise entity in which they are interested, they emphasize the biosphere. For example, in the case of the damming of a scenic river, the water and its flow may be important – especially to kayakers. However, it is the surrounding forest, the deer, the birds, the trout in the river, and the mayflies on which they feed that people emphasize when they talk in nontechnical ways about environments they value. Often, they express special fondness for particular species: the enormous scale of the conservation efforts expended on the African rhinoceros, the Siberian tiger, or the bald eagle attest to the capacity of a single kind of organism to capture the active support of the human spirit, both for individual animals and for the entire species. Yet, even as these costly and heroic efforts are under way, we are

---

TO ROOT OR NOT TO ROOT – FERAL PIGS IN PARADISE

A lively controversy has arisen over policies being pursued by the Nature Conservancy, a nonprofit environmental organization in the United States dedicated to the purchase and preservation of lands containing unique or threatened ecosystems. Among its holdings are preserves on the Hawaiian Islands, especially Molokai and Maui, that include forests of native vegetation with several endangered species of birds. Similar to other groups of isolated oceanic islands, Hawaii has experienced unique evolutionary pressures that have led to the development of many endemic species – organisms found there but nowhere else.

The Nature Conservancy's preserves on Hawaii are heavily threatened by feral exotic animals, especially pigs. Pigs were introduced into the islands with the arrival of Polynesian peoples approximately 1200 years ago. These animals quickly became established in the wild and were later hybridized with larger European hogs, which were introduced following Captain Cook's "discovery" of the islands in the late eighteenth century. Large numbers of feral pigs now roam the forest, severely damaging native plant populations. A large hog can thoroughly root up 1000 square feet of ground in a day, destroying most of the ground-level vegetation.

As on other islands, a control strategy was put in place to limit the pig population on the preserves. The terrain is difficult, and the only workable method is to set snares. In one of the areas on Maui, snares captured approximately 100 animals in the first year, but by the third year the number had been reduced to 3 – and forest recovery was already evident.

Although the method of using snares is effective, two quite different groups have strongly objected to it. Some native Hawaiian hunters regard pig hunting as an important part of their culture. For the current generation of hunters, the pigs appear to be perfectly "natural" elements of the ecosystem, and hunting them with packs of dogs is both economically and culturally functional. In certain cases, the Hawaiian hunters have objected to eradication programs on the grounds that their preferred use of the preserve is being interfered with. The second group, actively represented by the organization called People for the Ethical Treatment of Animals (PETA), is opposed to the snaring on the grounds that it causes unacceptable animal suffering. The strict "animal rights" position holds that humans have no right, for any purpose, to take the life of another living creature. In practice, the concerns of PETA and other organizations have focused strongly on mammals, especially those that (for reasons of taxonomic closeness or other attributes) elicit particularly strong human sympathies. In opposing the Nature Conservancy's program, they have organized economic boycotts and demonstrations against corporate supporters of the organization, disrupted meetings of the Conservancy, and located and removed snares from Conservancy preserves on Molokai. In the latter activity, they have had active support from some groups of hunters – a strange alliance, in view of the convictions of animal rights advocates concerning hunting.

In this strange controversy, three views of "nature" are clearly distinguishable. The Nature Conservancy views the ecosystem as a whole, and is especially concerned to "save all the parts." Moreover, it takes the view that exotic organisms – late arrivals that are present because of human intervention – may be differentiated from the historic, coevolved components. For the hunters, many of whom are fond of the habitat, "nature" is the home for a particular species that is important to their economic livelihood and meaningful to their culture. The animal rights advocates focus strictly on the individual organism; their concern for it is displaced not only from the ecosystem as a whole, but from other individuals of the same species.

losing entire species of less prominent, less "charismatic" organisms at an accelerating rate.

Although we value nature at several different levels, a powerful scientific rationale, as well as an emotional attachment, applies to the value of ecosystems in their entirety. The whole is greater than the sum of its parts, that is, in valuing ecosystems as collections of plant and animal species, we do more than add up the values of the individual species. Because of the interactive nature of the system, something extra, emergent, gives value to the assemblage. One practical reason exists for viewing ecosystems this way: To the extent that there is uncertainty about the economic value of particular species, preserving collections of them constitutes a kind of "hedge," a protection against unexpected loss. (A given species may have germ that can be employed in improving crops, or may provide "services" important to the global environment, but these potential contributions may be highly uncertain.) In this analogy the ecosystem is likened to a mixture of investment assets; it can be thought of as a diversified "nature portfolio." On a less practical but perhaps more meaningful level, we view nature as harmonious. Part of what we seek from wild environments is a sense of higher order, a feeling that the things we are viewing in some sense belong together.

The intrinsic value of the totality of nature is a somewhat diffuse notion, difficult to defend in precise form. The evolutionary biologist E. O. Wilson has coined the term "biophilia" to describe what he believes to be an innate human need for contact and continuity with nature. However, we must ask: What kind of nature? Is our recognition of nature-ness dependent on the completeness of the ecosystem we are looking at? Is our appreciation of it a function of the length of time the process of coevolution has had to increase the number of elements and fit them together seamlessly? Much of what contemporary western men and women think of as "nature" comprises ecosystems that have lost many of their parts. Thoreau, at Walden Pond in 1856, mourned the loss of wolves, mountain lions, and the "charismatic megafauna," asking: "Is it not a maimed and imperfect nature that I am conversant with?" In northern California, two hundred years ago, tule elk, pronghorn antelope, and grizzly bears roamed across the landscape where Stanford University is now located. And only 10,000 years before that, the Paleolithic hunters who crossed the Bering Straits land bridge found camels, dire wolves, mammoths, and saber-tooth tigers. They hunted them down and drove many species to extinction.

What we call "nature" is already altered and incomplete because of human impact – yet we still care for it. Nor can the length of the coevolutionary process be a critical criterion. For example, the Hawaiian forests are relatively young, yet extraordinarily beautiful and appealing.

We cannot completely define the appeal of nature to people – let alone how that appeal might be quantified in some valuation exercise. Surely there is some sort of higher human resonance with natural systems; however, it is not easy to specify criteria for our "love of nature." We only know that there is a value that is exceedingly difficult to quantify, yet meaningful enough often to inspire real economic commitment!

THE BASIS OF VALUE

From these reflections it becomes clear that "nature" refers not only to individuals and species but also to ecosystems as a whole. Thus, valuing nature is a matter of valuing living things on different scales – as individual organisms, as species, and as ecosystems. In the end, we must define the basis for that assessment. If we are choosing between leaving a natural environment as it is or altering it for some alternative purpose, we need to have some rationale for valuing the original state.

Because the question of how to value nature is a philosophic one, different analysts offer different answers. One answer – the one preferred by many philosophers and ecologists and by the vast majority of economists – is that living things possess value insofar as they provide satisfaction to human beings. Clearly, this answer is anthropocentric: It asserts that the ultimate determinant of a given species' value is how much it contributes to *human* satisfaction or *human* welfare! We refer to this as the *utilitarian view*, because it focuses on utility or satisfactions. The notion of "satisfaction" should be interpreted very broadly here, to encompass both our most mundane enjoyments of other forms of nature (e.g., our consumption of plants or animals as food) and our more lofty appreciations (e.g., our appreciation of the beauty of an eagle).

The utilitarian approach does not necessarily imply ruthless exploitation of other species; indeed, it is consistent with our fervent protection of the well-being and survival of other living things. After all, we may feel that the protection of certain forms of nature is important to our satisfaction or well-being, and therefore be led to place a high value on these forms. Thus, the utilitarian approach does not rule out our going to great lengths to protect and maintain other living things. However, it does assert that we should assign value (and therefore protect other forms of life) *only insofar as we take satisfaction from doing so*.

In effect, the utilitarian view puts humankind at the top of a species hierarchy, assigning values to other, "lower" organisms based on their contributions to human satisfactions. Not all philosophers put humanity in this top position. For example, in the *Republic*, Plato argued that the value of every species, including humans, is based on how well that species performs the function that the Creator assigned to it. According to this view, the Creator (not humankind) dictates the potential value of all things, and the actual value of each thing depends on the extent to which it realizes that potential. Few modern philosophers adhere to the notions that the functions of each living thing are fixed and preordained, and that value depends entirely on how well these particular functions are performed.

However, not all philosophers endorse the utilitarian approach, either. An alternative view, which, like the Platonic view tends to grant humans a less exalted position among species, is the *intrinsic rights view*, according to which species have intrinsic rights to exist and prosper, independent of whether human beings derive satisfaction from them or not. Many "animal rights" advocates appeal to a particular intrinsic right. In the book *Animal Liberation*, ethicist Peter Singer argues that nonhuman animals have the basic right to be spared suffering deliberately caused by humans. According to this view, the slaughtering of livestock, even if it provides food for humans, violates this right and therefore is immoral. This argument is grounded in the notion that, like humans, other animals are sentient creatures, capable of experiencing pleasure and pain, and that there is something fundamentally wrong about causing pain to any creature.

Some nonwestern belief systems incorporate elements of both utilitarian and intrinsic rights viewpoints. For example, the Yup'ik people of the Bering Sea coast derive a significant fraction of their summer food supply from harvesting the eggs and adults of nesting eider ducks and geese. Because the birds, like other living things, are believed to have souls and consciousness, they are regarded as deserving of respect. At the same time, the Yup'ik also believe that the birds exist for human use consumption and that if they are not hunted when offered for that purpose, they will become scarce. Accordingly, the Yup'ik explain the dearth of spectacled eiders, an endangered species, by saying: "They disappeared because we don't hunt them no more."*

The contrasting notions of utilitarian values versus intrinsic rights produce different judgments as to the appropriate social decision in given circumstances. Suppose that constructing a certain dam would destroy the last remaining habitat of a particular endangered species – a not uncommon situation. Suppose, further, that this species cannot be moved to another area or habitat. Thus, building the dam implies the elimination of the species. For the adherent of the utilitarian view, the question of whether the dam should be built rests on whether the benefits from constructing the dam (the value created) are greater than the costs from such construction (the value forgone). Included among the costs would be the loss of satisfactions that the endangered species would confer upon humans. The adherent of the utilitarian approach would want information on the magnitude of this lost satisfaction, add this information to other cost information, and compare these costs with the benefits. The benefits also relate to satisfactions. These are the satisfactions that constructing the dam offers through reduced costs of electric power, reduced costs of providing agricultural output as a result of more accessible water for irrigation, and perhaps recreational opportunities at a lake produced by damming the river. See the box entitled *Dam Building and the Endangered Species Act*.

In contrast, for the supporter of the intrinsic rights view, this is not a matter of comparing benefits and costs. The intrinsic rights advocate would maintain that this species has a right to exist regardless of how much human beings care about or derive satisfaction from its existence. Therefore, under no circumstances should the dam be built (assuming that its construction would eliminate this species). To put the matter another way: The intrinsic rights advocate would argue that there is an *infinite* cost associated with losing this species. Hence, it is impossible to produce benefits larger than these costs: The project will never satisfy the appropriate cost/benefit test!

In many cases, defenders of other species invoke both utilitarian and intrinsic rights arguments. Consider the controversies over commercial whaling. Over the past two centuries, such whaling has resulted in drastic decreases in the numbers of several of the larger species. International agreements to regulate whaling were initiated around 1930; however, they established only modest limitations on the world catch. Following World War II there was a massive rebuilding of whaling fleets, and the International Whaling Commission was formed. Although the commission established quotas, whale populations continued to decline until the early 1980s. By that time, the commission had added many member countries, and it then instituted a moratorium on commercial whaling. The ban protected a number of seriously threatened species and also benefited populations of the minke whale, a small species whose numbers in the northeast Atlantic had grown to over 85,000 animals by 1993. The Scientific Committee of the Whaling Commission recommended a revised management plan that could have led to a commercial quota for minke whales; however, the commission voted it down.

Norway, which had reserved its rights under the previous moratorium decision, voluntarily abstained from whaling

---

* Interview with a man from Scammon Bay, by Erica Zavaletta, 1994.

DAM BUILDING AND THE ENDANGERED SPECIES ACT

Perhaps the most celebrated case of this kind of conflict in the United States occurred in the late 1970s, as a result of the Tennessee Valley Authority's proposal to build the Tellico Dam. Conservation groups argued that the dam's construction would cause the extinction of a species of fish, the snail darter, by flooding its habitat in the Little Tennessee River – the only place in which the fish was known to exist. Citing provisions of the Endangered Species Act, the Environmental Protection Agency opposed construction of the dam. The two agencies failed to resolve the conflict, and Congress ultimately allowed construction of the dam. This case resulted in an amendment to the act that provided for a high-level interagency committee to settle interagency disputes wherever the Endangered Species Act applies to government projects or public lands. This group, informally called the "God squad," was called in to resolve the logging controversy in old-growth forests in the Pacific Northwest.

until 1992, when it proposed a small quota for scientific and consumption purposes. The Norwegians assert that this catch is necessary for the local economic health of coastal villages, although it is insignificant in terms of their national economy. Norway's resumption of whaling produced an international outcry. In the U.S. Congress, some members called for economic sanctions against Norway. Greenpeace and other environmental organizations published advertisements advocating a boycott of all Norwegian products by American consumers.

Beyond the disagreements concerning scientific issues – for example, the size of whale populations, the numbers of whales actually harvested – the whaling controversy reveals different philosophic bases for valuing particular species. Some defenders who accept as legitimate the practice of harvesting whales also make a utilitarian argument that continuing this rate of harvest will be very costly to humans because it will lead to extinction. Other defenders challenge whale hunting, claiming that whales, by virtue of their relatively high intelligence or "charismatic" nature, are of special value to humans. This claim owes much to a claimed relationship – taxonomic, or perhaps communicative – between whales and people. Still other defenders appeal to intrinsic rights; however, in this rigorous form of the argument it ought to make no difference whether the particular whale under consideration is the last member of its species or one in a population of one hundred thousand. Thus, the preservation of whale populations raises strong and even inconsistent feelings and arguments, perhaps in part because of the degree to which we humans identify with them. Other species and ecosystems do not fare so well, or enjoy such popular support.

The whaling case illustrates how difficult it is to find consistency within the various versions of the intrinsic rights approach. Do all species have a right to be preserved, or only the most advanced or human-like? Does the right to survival apply to species (permitting limited harvesting) or to individuals (permitting none)?

Assessment of the relative merits of the utilitarian and intrinsic rights approaches depends on how broadly these rights are assigned, and also on some complex philosophic issues. Without fully examining the latter here, we will confess our own preference for the utilitarian approach. Far from rejecting the laudable human feelings that motivate many of those who hold the intrinsic rights view, we believe that the utilitarian approach fully incorporates them. This approach takes the broadest view of human satisfactions as we learn in the following section of this chapter.

TYPES OF VALUE

The two approaches we have been discussing may argue for different decisions about how or how much to protect nature. We now investigate this issue further. We begin by outlining the variety of ways in which nature confers human satisfactions. We then consider where attention to these satisfactions might fail to generate sufficient value to appease an advocate of the intrinsic rights approach.

USE VALUES. Nature offers satisfactions in various ways, and we can categorize values accordingly. *Use values* refer to our enjoyment of various forms of nature by actually using them. Sometimes, use values are reflected (to a degree) in market prices. For example, the satisfactions that we may derive from eating fish provide a basis for establishing the demand for fish and for the prices of fish in various markets. However, in many cases important use values have no obvious reflection in market prices. For example, many of us enjoy trout fishing in wilderness areas. Although such enjoyment of nature is very real and the value may be quite substantial, there are no obvious market prices here – we cannot obtain sales data for a single commodity called "recreational fishing." The absence of market prices makes measuring these values more challenging; however, measurement methods do exist for them, as we see in the section entitled "Nonuse Values."

We can classify use values as *direct* and *indirect*. When we eat apples, we obtain direct use value. In contrast, plankton offers indirect use value: It provides nutrients for other living things in the food chain, which in turn feed humans. In doing so, it forms an integral part of the ecosystem that allows fish to thrive. Hence, it contributes use value indirectly: It helps produce other forms of nature that we enjoy directly.

We can also distinguish between *consumptive* and *nonconsumptive* use values. When we eat duck for dinner, the use value is consumptive: The duck is "used up" in the process of yielding value. In contrast, when people engage in the activity of bird-watching, the pleasure of this activity does not (one hopes!) compromise the viability of the birds being observed. This is a case of a nonconsumptive use value.

Ecologists sometimes rebuke economists concerning these notions of value. In theory, economists take account of all the use values mentioned above; however, in practice, they tend to focus less on values that are more difficult to measure, thus downplaying important natural values. Usually, it is quite difficult to assess the values afforded by "simple" natural organisms that provide ecosystems services, because the contributions to human satisfactions are very indirect. Although most economists would acknowledge that these simple organisms do confer value, they tend to neglect these values in typical cost/benefit analyses. A hopeful sign is that many economists are now working closely with ecologists to gain a better sense of how ecosystems function and thereby obtain better assessments of indirect use values.

NONUSE VALUES. Certain values are largely removed from the use of nature. These are termed nonuse values. Perhaps the most important of these is *existence value* (or passive use value). This is the satisfaction one obtains from the mere recognition of the existence of some entity. For example, a New Jersey resident who has never seen the Grand Canyon and who never intends to visit it can derive satisfaction simply from knowing that it exists. As another example, many people experienced a *loss* of satisfaction or well-being upon learning of the ecologic damage that resulted from the 1989 *Exxon Valdez* oil spill. The spill caused a loss of existence value. One can experience a loss of existence value even if the environmental-damage involved is temporary. Of course, the more severe or more long-lasting the damage, the greater the loss of existence value.

Various natural amenities can confer existence value in addition to the other values we have discussed. In the case of the *Exxon Valdez* spill, many of the costs stemmed from the lost values from commercial fishing and the lost recreational values. The losses of existence values are in addition to these costs.

Tremendous controversy surrounds the subject of existence values. The debate is not about the validity of the concept (nearly all analysts accept it) but about the magnitude of such values in particular situations. Existence value is extraordinarily difficult to measure, and estimates of such values generate controversy. One of the most heated debates surrounding the *Exxon Valdez* oil spill – a debate that continues to this day – concerns the magnitude of the lost-existence values. Exxon argued that the estimates of lost-existence values had no scientific basis and that these values were relatively small. In contrast, environmentalists supported estimates indicating substantial losses of existence values. The issue of the magnitude involved were important because they had a bearing on the size of the fines the company would need to pay to compensate for the damages it created from the spill.

Another nonuse value is *option value*. This value arises from situations in which there is uncertainty about the future use of a natural resource. We convey the notion here informally by way of an example. Suppose you are not sure whether you will be interested in visiting a certain lake next year. You figure that if you go to the lake next year, your satisfaction from doing so is worth 20. If you do not go to the lake, you will derive no satisfaction from the lake. Suppose you believe there is a 50–50 chance that you will decide to go to the lake next year. What is the value to you of this lake? It is tempting to say 10: $0.5 \times 20 + 0.5 \times 0$. In fact, however, many people would be willing to pay more than 10 to guarantee the option of going to the lake next year. The *difference* between this willingness to pay and 10 is the option value.

INTRINSIC RIGHTS RECONSIDERED

Now that we have examined the various types of value that stem from a utilitarian approach, we are in a better position to assess what this approach might omit. We might try to imagine some species that would receive little or no value

under the utilitarian approach – yet would be valued under the intrinsic rights approach.

The species we have in mind must produce no use value – neither directly nor indirectly. Accordingly, this organism must be something we do not enjoy eating (no consumptive use values) and it must be something we do not enjoy observing (no nonconsumptive use values). In addition, the organism must not serve any position ecosystem function (no indirect use value). To maintain the consistency of the example, assume there is no uncertainty about any of this – we are convinced that human tastes will not change and that no positive ecosystem function will develop. This covers the use value aspect. To complete the picture, the organism must also have zero existence value – humans must not enjoy contemplating it.

Such an organism would receive no value from the utilitarian approach. Is there any real-world organism that fits this picture? Perhaps some species of cockroach would come close. Whether it *exactly* fits the picture is not important. The key point is that such a creature would be given virtually no value in a cost/benefit analysis; if we were considering a development project that threatened its existence, this threat would not cause us (as utilitarians) to refrain from undertaking the project. As long as there are some benefits from the project and no "significant" species were put at risk, we would not prevent the loss of this particular species. In contrast, a supporter of intrinsic rights might argue that this species has the right to existence, even if there are no use or nonuse values associated with it.

Two comments on this example may be helpful. First, one can easily detect contradictions here. In particular, the appeal to intrinsic rights in the case of the cockroach might be regarded as proof of the cockroach's existence value – which we had tried to rule out. To the extent that arguments for intrinsic rights require existence value, the line between the utilitarian and intrinsic rights approaches is blurred. Second, the differences between the two approaches may be relatively unimportant in practice. Arguably, in many cases involving the protection of species, one will arrive at similar decisions under the utilitarian and intrinsic rights approaches. The cockroach example, which leads to different decisions, is likely to be an unusual one in the real world. To bring this issue into even sharper focus, see the box entitled *Do Some Species Have Negative Value? Eradication and the Smallpox Virus*.

THE MEASUREMENT OF VALUE

It is one thing to define and classify the bases of value; measuring those values is another. The previous discussion indicated that much of the controversy over the valuation of nature centers on the validity of the attempts to measure them. This section presents some important measurement methods. Over the years, considerable progress has been made in developing such methods. However, the science is far from perfect. Controversies persist.

DO SOME SPECIES HAVE NEGATIVE VALUE? ERADICATION AND THE SMALLPOX VIRUS

Similar arguments have come to life recently in the context of the smallpox virus. This virus, which has killed, blinded, and disfigured millions of humans since its emergence more than 10,000 years ago, has largely been eradicated. However, what appear to be the last remaining stocks of the virus are stored in freezers at the Center for Disease Control Prevention in Atlanta and at the Research Institute for Viral Preparation in Moscow. Recently, there has been much debate about whether to destroy these last samples. After considerable argumentation, the World Health Organization recommended unanimously in September 1994 that these samples be destroyed on June 30, 1995. Here is a virus with a strongly negative use value – it creates human misery, not satisfaction. Yet some would argue that preserving at least one sample may prove to be useful for future scientists who wish to study the virus. (One could also invoke the intrinsic rights notion to justify preventing the complete loss of this element of nature, although few people have thought to apply this notion to viruses!) This argument has been countered by scientists who claim that most of the useful scientific information can be recorded and does not require the actual survival of the virus. Still others indicate that science is not that far away from being able to re-create extinct living things, particularly relatively simple forms of life such as viruses, so that destroying the current stock of the virus does not preclude its future examination.

This controversy raises some intriguing questions. Suppose that scientists eventually obtain the ability, dramatized in the movie *Jurassic Park*, to re-create extinct species. How would this alter our attitude toward extinction? If, in principle, every species can be brought back, need we worry so much about the (possibly only temporary) loss of a species? If species have intrinsic rights to existence, do these rights proscribe the "temporary" elimination of the species or only some permanent form of extinction? We hasten to add that primary genetic material from the extinct species must exist for replication of the organism. This is manifestly not the situation for most forms of past life; furthermore, in contrast to the movie, scientists are far from capable of re-creating organisms even when complete genetic material is available.

VALUES AND WILLINGNESS TO PAY. A key concept that underlies virtually all the measurement methods is *willingness* to pay. Nearly every empirical approach assumes that the value of a given natural amenity is revealed by the amount that people are willing to pay or sacrifice in order to enjoy it. Willingness to pay is regarded as an indicator of the extent of satisfaction that the amenity provides. Hence, these measurement methods implicitly invoke the utilitarian principle: they appeal to human satisfactions, as reflected in the willingness of human consumers to purchase those satisfactions.

More specifically, economists regard the prices that people are willing to pay as the *marginal* value, that is, the value they place on the *last unit* purchased. For example, consider what you would be willing to pay for tap water in a given month. You might well be willing to pay a huge sum for the privilege of consuming the first 10 cubic feet, because doing without them would deprive you of even the most fundamental uses of water for that month. You would be reluctant to miss the occasional shower, access to drinking water, and so on. The *next* 10 cubic feet would probably not be worth quite as much to you. They would allow you an extra shower or two a month, and additional opportunities to fill a glass from the faucet; however, these would not be as critical to you (or to people with whom you associate!) as the first 10 cubic feet. Extending this line of reasoning, we can see that the marginal willingness to pay – the amount one is willing to pay for each successive increment – will fall steadily.

Figure 31.1 displays a willingness-to-pay schedule. The first cubic foot is shown to be worth a great deal more than the 50th, which in turn is worth much more than the 100th. In reality, of course, households do not have to purchase each unit of water at its marginal value. If they did, they would be charged larger amounts for the first increments than for later ones. (Such a pricing schedule would present serious metering difficulties.) Instead, households typically pay a given price per unit of water, regardless of how much they consume. There are exceptions. In some cases, the unit rate is a certain amount for up to a certain quantity of water, then another amount for consumption in excess of this quantity. This is a case in which two prices are charged; however, the charge is not based on willingness to pay for each unit. That would involve a multitude of prices.

Suppose the willingness-to-pay schedule shown in Figure 31.1 reflects your own preferences, and also suppose that the price is $0.02 per cubic foot. On the basis of the schedule, you would be willing to purchase up to 400 cubic feet of water in a given month. You would rather not use more than that, because each cubic foot used beyond the 400th has a value less than its cost to you of $0.02. Thus, in a given

Figure 31.1. Willingness-to-pay schedule for household water use.

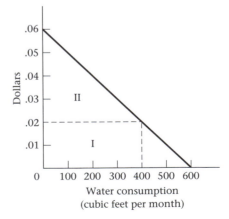

month, with the willingness-to-pay schedule shown here, you would limit your water consumption to 400 cubic feet per month, at a total cost of $8.00.

Area I in Figure 31.1 represents the cost to you ($8.00) of the water you consume. What is the *total value* to you of this water? It is not only the $8.00 you paid for it – it is worth much more than that to you! In fact, it is worth the *sum* of the amounts that each unit is worth. On the diagram, this sum is the area under the willingness-to-pay schedule for the first 400 cubic feet of water – areas I *and* II. *The value of the water to you (areas I and II) exceeds its cost to you (area I).* The difference between its value and its cost is called *consumer surplus* (area II on the diagram). What makes this surplus possible? It is that the price you pay per unit corresponds to the incremental value of the *last* unit of water you consume; on all "previous" units the incremental value exceeds the price, and thereby contributes to the surplus.

ASCERTAINING VALUES DIRECTLY FROM MARKET TRANSACTIONS. The notions of willingness to pay and consumer surplus are central to the measurement of the value of nature. To the extent that we can calculate the willingness-to-pay schedules that apply to human "consumers" of nature, we are in a position to obtain measures of its value.

Unfortunately, obtaining these willingness-to-pay schedules is not always easy. The simplest cases occur when the natural amenity in question is transacted directly in a market. A few years ago, the Mediterranean fruit fly threatened to destroy a significant portion of California's apricot crop. What is the value of the crop? What would be the loss if it were eliminated? Because apricots are directly purchased and sold in markets, statisticians can obtain information indicating the willingness to pay for them as a function of the quantity provided. Suppose the usual retail price of apricots is $5.00 per pound. If the willingness-to-pay schedule for California apricots were as depicted in Figure 31.2, the consumer surplus per year (aggregated across all consumers)

from this product would be $6250 million (area II). This number represents the direct use value of the apricot crop. If the fruit fly eliminated the crop in a given year, the loss of direct use value would be $6250 million. Some would claim that under these circumstances, society should be willing to pay up to this amount to avoid the loss of the crop. To the extent that the crop confers other values (such as indirect consumptive use values or nonconsumptive use values), the $6250 million figure would understate the cost from losing the crop.

INDIRECT METHODS BASED ON ACTUAL BEHAVIOR OR PRICES

■ **The Travel Cost Method.** In the previous example, value could be ascertained relatively easily. Things are more difficult when there are no markets that directly involve the natural amenity in question. Fortunately, in these situations it is sometimes possible to employ indirect methods to assess values.

The travel *cost method* is a widely used indirect approach. The method has been applied to evaluate the values offered by parks, lakes, and rivers or, equivalently, the cost that results from the loss of these elements of nature. Such natural amenities are, of course, not directly bought or sold in markets; usually, prices are not even charged for their use. In those instances when use prices *are* charged (e.g., through the use of entry fees), the prices are unlikely to be good indicators of (marginal) value. That is because the users of these resources actually "pay" more than the entry fees to use them. For example, the cost of the family visit to Yosemite National Park is much greater than the $15 per day use fee. The travel cost method recognizes that fact by adding to the entry fee (if any) both the transportation cost and time cost forgone in visiting a particular site, in order to ascertain the overall travel cost. This method regards the overall travel cost as a measure of the marginal willingness to pay by a visitor to the park; this is considered to be the same as the marginal value of the park to the visitor. The underlying assumption is that people will continue to visit the park until the value of the last visit (i.e., the marginal value) is just equal to the travel cost.

If every visitor paid the same entry fee and incurred the same transportation and time costs, then each visitor would effectively pay the same "price" to visit the resource in question. Suppose that at a price of $35, there are 20,000 visits to a given park each year. This would not be enough information to ascertain the value of the park; it would supply us with only one point on a willingness-to-pay schedule. Fortunately, there is often a great deal of variation in the overall prices or travel costs incurred by visitors. If we assume visitors have the same preferences, then any differences in numbers of visits to the park will be entirely due to differences in travel costs. (This assumption obviously requires a departure from reality. However, it can be shown that the travel cost method provides a reasonably good approximation of value

Figure 31.2. Willingness-to-pay schedule for California apricots.

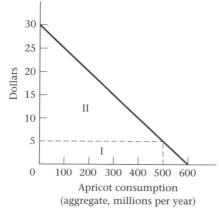

if the average preferences for visiting the park are very similar across the various localities from which people initiate their trips to the park.) Under these circumstances, we can use the different travel costs to assemble a willingness-to-pay schedule for the park. Based on this schedule, we can calculate the value of the park.

The following example offers an illustrative calculation. Suppose the park is visited by residents of three nearby towns: A, B, and C. (Suppose that other towns are so far from the park that travel costs are high, causing residents to prefer to visit other, closer parks.) Suppose we have the following information about costs and visits to the park:

Town (1)	Number of Residents (in thousands) (2)	Visits Per Year (in thousands) (3)	Overall Travel Cost Per Visit (average) (4)
A	10	20	$40
B	30	30	$60
C	15	45	$20

Travel costs are highest for residents of town B and lowest for residents of town C. This might reflect differences in proximity to the park, or differences in the terrain, road quality, and so forth, on roads accessing the park from a given town. By dividing the numbers of visits (column 3) by the numbers of residents (column 2), we obtain, for each town, the visits per resident. Visits per resident are a function of the overall travel costs. This relationship is conveyed in the aggregate willingness-to-pay schedule (or "demand curve") for the park, as illustrated in Figure 31.3. Strictly speaking, a demand curve is somewhat different from a willingness-to-pay schedule because of "income effects."

The travel cost method allows us to infer the value of the park based on this cost and demand information. What is the value of the park? According to this method, it is the consumer surplus enjoyed by its users. For each user in Figure 31.3, the consumer surplus is given by the area between the demand curve and the horizontal line indicating the costs of visiting the park. Thus, it is area I (on average) per resident from town B, areas I plus II per resident from town A, and areas I plus II plus III per resident from town C. To calculate the total surplus provided by the park (per year), we multiply for each town the surplus per resident times the number of residents, and add the results:

Town	Surplus Per Resident (in dollars)	Aggregate Surplus (in thousands)
A	40 (areas I + II)	400
B	10 (area I)	300
C	90 (areas I + II + III)	1350

Thus, the total surplus to all users of the park is $2.05 million. (To calculate the total value over the "lifetime" of

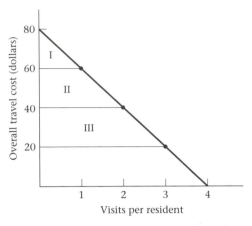

Figure 31.3. Aggregate willingness-to-pay schedule for Yosemite National Park.

the park, we would need to project the annual values every year based on estimates of changes in population and, perhaps, tastes for visiting the park. We would then calculate the present value of the stream of these surpluses.) This value also represents the annual loss associated with the destruction of the park. So, on a strict cost/benefit basis, we could not justify sacrificing this park for other purposes unless the annual benefits associated with the alternative project exceeded this value, on average. Strictly speaking, we should compare the present discounted value of the benefits from the alternative project with the present discounted value of the lost surplus associated with destroying the park. Calculating the future values and costs is not easy, because it requires us to make projections about future travel costs and tastes (the extent to which future generations are likely to enjoy visits).

■ **The Hedonic Method.** Another indirect approach is the *hedonic method*. This widely used approach infers the values of nature or environmental improvements by observing changes in prices of fixed factors such as real estate. This approach assumes that, other things being equal, the better the quality of the surrounding environment, the more people are willing to pay to live in that environment. Accordingly, this greater willingness to pay will be reflected in housing prices. This approach uses information from one market (housing) to infer values for environmental amenities that are not directly exchanged in markets.

Economists have employed this approach to assess the environmental damage from chemical (PCB) contamination in New Bedford Harbor in Massachusetts. A 1992 study by Mendelsohn, Hellerstein, Huguenin, Unsworth, and Brazee distinguished several types of single-family homes and examined the changes through time in the sales prices of each type. Using statistical regression, they examined how sales prices were influenced by the significant increase in PCB contamination that began in 1982. The statistical approach attempts to take into account the changes in other variables (e.g., interest rates, average incomes) that also contribute to changes in house prices. The authors estimated that PCB contamination tended to reduce the prices of nearby homes by between $7000 and $10,000. Summing up these costs for all the affected homes (whether actually sold or not) gives an estimate of the aggregate loss in the harbor's value as a result of PCB contamination. The central estimate obtained in this study for aggregate damages was $35.9 million.

■ **Indirect Methods Based on Hypothetical Behavior.** The methods previously described have all used actual behavior as direct or indirect indicators of value. However, in some situations, we do not have a "behavioral trail" from which we can infer value. This is the case when we attempt to measure existence value.

In the film *Grand Canyon*, the Grand Canyon itself plays an important symbolic role for a family living in Los Angeles, none of whose members have ever visited the park itself. It would be very difficult to measure the satisfaction gained by the father from knowing that the Grand Canyon exists. Nonetheless, the existence value is real for him, and comes through clearly in the film; its loss would be tragic for the man and his family. How can we measure this kind of value? The only available method, it seems, is to construct hypothetical situations, and then ask people how they would behave in those situations. In response to survey questions, people are thus asked to make a quantitative assessment of the "existence value" of a given natural amenity.

The contingent valuation (CV) method is a survey technique used to infer existence values. For example, it might appear that the most appropriate question to ask the father in the film (*Grand Canyon*): "If the Grand Canyon were to be destroyed, what amount of money would you need to receive in order to feel compensated for its loss?" In fact, however, practitioners of the CV technique typically do not ask the question in this form, because it invites the surveyed individuals to place no bounds on the required compensation. CV practitioners instead would ask the question in the following form: "How much would you be willing to pay to prevent the loss of the Grand Canyon?"

The CV approach is highly controversial. Its detractors argue that people's responses to hypothetical situations, such as the possible loss of the Grand Canyon, are very unreliable indicators of their true preferences (true willingness to pay). The critics claim that the information obtained from such surveys is so misleading that it would be better not to use the results at all. They argue that the circumstances about which people are asked to comment are usually fairly unfamiliar, with the result that peoples' responses are very much influenced by the way the information is presented or selected by the interviewer. The proponents of CV counter that this method is the only possible technique to employ in cases where there is no "behavioral trail"; they maintain that the CV approach, while imperfect, provides a useful gauge of

people's preferences. We now examine one case of CV in practice, in order to consider the strengths and weaknesses of this controversial method.

Under the Oil Pollution Act of 1990, the National Oceanographic and Atmospheric Administration (NOAA) is responsible for assessing damages to natural resources resulting from major oil spills. One important aspect of oil spill damage, associated with the destruction of natural habitat and marine life, is lost existence value. This was a key issue following the 1989 *Exxon Valdez* oil spill, which resulted in extensive degradation of intertidal areas in Prince William Sound. Estimates of the lost existence value from this spill, based on CV surveys, ranged from $0.5 to 3.0 billion. Many researchers – including a large number of prominent economists hired by Exxon! – argued that the numbers obtained from the CV studies offered no useful information. Many urged the NOAA to discontinue the use of CV in carrying out its damage assessments. This prompted the agency to commission a panel of experts to assess the validity of the CV method. The panel's report offered a partial defense of CV, indicating that, with careful safeguards, the approach could provide useful information. At the same time, the report indicated that many CV studies did not include important tests that would help reduce the risk of survey bias and improve the reliability of the information. (Some examples of natural resource evaluations are presented in Table 31.1.)

■ **A Real-World Valuation Study.** Most methods for measuring nature's values have been known for decades. However, until the early 1980s, most cost/benefit analyses tended to focus on *nonenvironmental* benefits and costs of given projects. For projects that produced significant net costs in the environmental dimension, the failure to consider environmental effects implied a bias against nature.

The pattern is changing. An important precedent was established during the controversy in the early 1980s as to whether to construct a new hydroelectric dam on the Tuolumne River in California. The Tuolumne flows westward 158 miles from Yosemite National Park to the San Joaquin Valley. By 1970, most of the river had been dammed; however, there remained a 30-mile stretch of "wild" river that passed by the town of Wards Ferry. In the mid-1970s, the city and county of San Francisco and two irrigation districts proposed to dam the Tuolumne near Wards Ferry in order to provide additional electric power and water for agriculture. Proponents of the project argued that the new dam was needed to meet the additional electricity needs of the state. They also indicated that although the dam would eliminate some recreational opportunities of the river such as trout fishing and whitewater rafting, it would create some new ones, including opportunities for flat-water fishing and recreational fishing along a new lake produced by the dam. Environmentalists countered that the dam was unnecessary to meet energy needs and that it would destroy the river's last wild stretch, eliminate the beauty of Tuolumne Canyon, destroy prime

---

TABLE 31.1. NATURAL RESOURCE VALUES AND MEASUREMENT METHODS: SOME EXAMPLES

Natural Amenity	Type of Value	Possible Measurement Method	
Commercial fishing in Monterey Bay	Consumptive use	Direct inference from market prices (willingness to pay)	direct method
Recreational fishing in Monterey Bay	Consumptive use	Travel cost	indirect method
Sightseeing along the Amazon	Nonconsumptive use	Travel cost	
Contemplating the grandeur of the Amazon	Nonuse (existence value)	Contingent valuation	
Unpolluted air	Nonconsumptive use[a]	Hedonic	

[a] Whether to call individuals' enjoyment of clean air a consumptive or nonconsumptive use is somewhat arbitrary. One could regard it as nonconsumptive, because the act of breathing does not diminish the quantity of air. At the same time, breathing alters the air (replacing oxygen with carbon dioxide), which could be regarded as an act of consuming air of a given chemical mix.

---

spawning beds for trout and other fishes, and threaten the southern bald eagle, an endangered species.

An original feasibility study provided by a local consulting firm indicated that the benefits of the project in terms of providing cheaper electric power would outweigh its costs. Many interested citizens and politicians regarded the outcome of the feasibility study as a solid justification for undertaking the project. However, many environmental groups, including the Environmental Defense Fund (EDF), were dissatisfied with the limited scope of the feasibility study and argued that attention to environmental impacts could tilt the cost/benefit balance the other way. EDF soon decided to develop its own, broader cost/benefit study. The EDF study contained many of the estimates from the original feasibility study, and supplemented these numbers with benefits and costs to the environment.

An outline of the EDF cost/benefit assessment is provided in Table 31.2. The benefits from the project can be classified as "internal" (reflected in ordinary market transactions) and "external" (not so reflected). The principal internal benefit from the project is cheaper power: the figure of $146.7 million is the *difference* between the cost of providing power with the proposed plant and the cost if the plant is not built. An additional benefit was the provision of water that could be used for agricultural irrigation. The study assumed that each acre-foot of water was worth $105 to farmers (the value

---

TABLE 31.2. BENEFITS AND COSTS FROM PROPOSED TUOLUMNE DAM PROJECT (AVERAGED OVER LIFETIME OF PROJECT, IN 1994 DOLLARS)

Benefits	Value ($millions)	Measurement Method
Internal		
Annualized energy cost savings	146.7	Direct
Annualized electricity capacity cost savings	37.5	Direct
Increased water yield	3.4	Direct
External		
Recreational benefits: flatwater fishing and boating	0.3	Travel cost
Total benefits	187.9	
Costs		
Internal		
Annualized construction and maintenance	134.2	Direct
Forgone profits to local enterprises	1.2	Direct
External		
Lost consumer surplus (forgone value of recreational fishing, whitewater rafting)	78.9	Travel cost
Total costs	214.2	

is assumed to represent the increase in net profits to farmers for each acre-foot of additional water provided). The external benefits were the values of flat-water boating and reservoir fishing to potential users. These benefits were derived from estimates of willingness-to-pay by the U.S. Water Resource Council: $20 per user per day for flat-water boating and $30 per user per day for recreational (flat-water) fishing. The willingness-to-pay numbers were obtained using the travel cost method.

The principal internal costs of the project were the construction and maintenance costs; other internal costs included the forgone profits to existing enterprises such as river-rafting outfits. The external costs were meant to capture the lost surplus to potential recreational users of the river – specifically, river rafters and trout fishermen. (There is rich opportunity for political controversy even within the universe of fishermen. Those who fish for bass in flat water would favor the dam, whereas trout fishermen who prefer moving water – and have more elegant tastes in other respects as well – would oppose it.) In Table 31.2, the internal benefits of the project exceed the internal costs. However, the overall (internal plus external) benefits fall short of the *overall* costs. Taking account of the external benefits and costs is critical to the bottom line.

The EDF analysis, one of the earliest cost/benefit studies to incorporate environmental impacts, appeared to play a significant role in the political debate over the Tuolumne Dam proposal. In September 1984, a year after the EDF study was released, Congress passed the California Wilderness Act, which granted "wild and scenic" status to the stretch of river that would have been affected by the proposed dam, thereby proscribing construction of the dam.

Although the EDF study expanded the horizons of cost/benefit analysis, its scope remained limited in may ways. The study made no attempt to measure the potential costs from constructing the dam in terms of lost existence value. Nor did it attempt to measure the costs that would result from destruction to ecosystems associated with the damming of the Tuolumne. In this sense, the analysis appears to understate the overall costs from the project. On the other hand, the EDF may have employed unrealistically high estimates for the costs in terms of lost recreational values. The study assumed very high costs in terms of lost "option values" associated with the project.

■ **Summary of Measurement Approaches.** The methods for measuring the values of nature vary considerably. All these methods associate values with the willingness of people to pay for environmental amenities; hence, they all invoke the utilitarian approach to value described previously. The difficulties of measurement become greater as we move from direct methods to indirect ones. Among the indirect methods, those based on hypothetical situations (as opposed to revealed behavior) are the most difficult and controversial of all. Over time, we can expect some improvements in all these techniques; however, controversies are likely to remain.

One further aspect of these measurement approaches deserves mention. Such methods obtain a social willingness-to-pay schedule by adding up the willingness-to-pay schedule of individuals. However, because people differ in their attitudes toward nature, they differ in terms of their willingness to pay for preserving nature. Consequently, how can we calculate the overall value of the natural amenity in question? For example, if "enlightened" naturalists are willing to pay $500 each to save Mono Lake, while many others are willing to pay only $50, which "willingness to pay" should we use in establishing the total value of the lake for society? Should we simply add up the willingness to pay of the tree-hugger and tree-cutter, even though these values differ, or should we apply to each person the willingness to pay of the "more enlightened" members of our society (whoever they may be)? A strict utilitarian would argue for the former solution. In practice, most economic researchers apply the utilitarian method, giving each person the same voice in the calculus of aggregate willingness to pay. However, many environmentalists would insist that some people are better informed about nature or occupy a higher "moral" stance for other reasons, and that their valuations deserve more weight than the valuations of other people. It is possible to embrace the utilitarian view that the basis of value is human satisfactions while repudiating the notion that each person's satisfactions are equally deserving of attention when it comes to calculating an overall social value. Indeed, many scholars would contend that the business of moral philosophy is to make clear why some satisfactions or attitudes are superior to others. Although we consider this a significant problem, we cannot solve it here.

CONCLUSION

The problem of valuing nature begins with the question of what exactly is being valued. Are we valuing individuals, as the animal rights advocates would have it? Are we valuing species, as the activities of those who would invest heavily to save the California condor would suggest? Or are we valuing the concept of nature as an ecosystem, such as the "wilderness" that writers from John Muir and Henry David Thoreau to Wallace Stegner have described so movingly? Human beings ascribe values to nature at all these levels – although for most people in the western world, the concept of nature as entire or not-quite-entire ecosystems may be especially important.

Attaching value to nature is difficult; it depends on whether one believes in an intrinsic right of other life forms or adopts instead the utilitarian position, holding that nature has value in (and only in) that it provides human satisfactions. Taking the utilitarian view as the most workable one, what types of value are there? Some are use values: Either by consuming some asset or simply by walking around and experiencing it, we demonstrate value by our decision to enjoy nature. Sometimes these decisions, if they involve purchasing access in markets, are easily measurable as willingness to pay. Sometimes they must be assessed indirectly, by hedonic measurement of the changes in price of some related thing – like the reduction in sale prices of homes near noisy airports. Perhaps the most difficult measurement problems are encountered with existence, or nonuse values. Such difficulties arise, for example when we try to evaluate what it would mean to lose a natural landmark that we never intend to view directly, but appreciate merely by contemplating its existence.

A striking conclusion from this examination of valuing nature is that it is precisely those things that seem most important that are the most difficult to measure. For example, virgin tropical rain forest is being lost at between 1 and 2 percent per year worldwide – a frightening rate to those concerned with the preservation of biodiversity. The level of alarm among many persons who have never seen a rain forest is high; yet the important use and nonuse values of the rain forest are very difficult to measure.

Thus, ascertaining the value of nature in terms of willingness to pay poses many difficult (if not impossible!) challenges. Even if we are able to obtain accurate estimates of individuals' willingness to pay to preserve nature, we face another problem. Different people have different attitudes toward nature, and thus differ in terms of their willingness to pay to protect nature. How do we address these differences? Should residents of New York be allowed to claim a valuation right on forests in Ecuador merely on the basis of existence value? If so, what principles of social justice would we use in deciding how much their votes are worth compared with the votes of Quichua natives in Ecuador?

To this list of difficulties to ponder, we add one more. In the quantitative approaches we have set out in this chapter, willingness to pay has been a defining principle. It rests on the assumption that human desires will always be reflected in their willingness to pay for it. Yet does the willingness-to-pay principle encompass all the recesses of our relationship to the natural world?

The special value many people attach to nature was expressed beautifully by the late Wallace Stegner:

What is such a resource worth? Anything it costs. If we never hike it or step into its shade, if we only drive by occasionally and see the textures of green mountainside change under the wind and sun, or the fog move soft as feathers down the gulches, or the last sunset on the continent redden the sky beyond the ridge, we have our money's worth. We have been too efficient at destruction; we have left our souls too little space to breathe in. Every green natural space we save saves a fragment of our sanity and gives us a little more hope that we have a future.

The passage refers to the San Francisco Peninsula. However, Stegner could as well have been speaking of the special love many people have for their favorite natural places. Although he speaks in terms of economic value, it is clear that he is referring to something more. Is he describing the resonance with the rest of life that is meaningful in an essentially spiri-

tual sense – a sense so removed from material choice that we could never measure it in that way? The biologist in us wonders; even so the economist.

QUESTIONS

1. What are the differences between utilitarian and intrinsic rights approaches to decisionmaking with regard to the preservation of endangered species? Under what circumstances might the two approaches lead to different decisions? Which approach seems more defensible to you, and why?

2. What system of ethics is implicit in cost/benefit analysis? Explain.

3. What types of values are most likely to be ascertainable from market transactions, and what types need by be measured by indirect methods?

4. What is the difference between use and nonuse values? Are all nonuse values existence values?

5. Under what circumstances might our marginal "willingness to pay" for a resource that is essential to life be relatively low?

6. Consider the following list of activities associated with nature. What principal types of values are involved in each, and what approaches might be taken to measure them?

- Pack trips in the Grand Canyon
- Duck-hunting in the Central Valley of California
- Rock-climbing in Yosemite National Park
- Appreciation of the beauty of Yosemite National Park
- Spring bird-watching at Big Bend National Park in Texas

7. Your city can no longer support the maintenance of a small (150-acre) urban park. A number of citizens, primarily those with residences near the park, have claimed that the park is of great value. As city manager, you are charged by the City Council to assess that claim. How might you proceed?

FURTHER READING

Bell, F., and Leeworthy, V. 1990. Recreational demand by tourists for saltwater beach days. *J. Environ. Econom. Man.*, 18(3): 189–205.

Boyle, K. J., and Bishop R. C. 1987. Valuing wildlife in benefit-cost analysis: a case study involving endangered species. *Water Res. Res.* 23: 942–950.

DesJardins, J. R. 1993. *Environmental ethics: an introduction to environmental philosophy.* Belmont, CA: Wadsworth Publishing.

Freeman, A. M. 1993. *The measurement of environmental and resource values: theory and methods.* Washington, DC: Resources for the Future.

Hirschleifer, J. 1988. *Price theory and applications.* 2nd edition. Englewood Cliffs, NJ: Prentice-Hall.

Kant, I. 1963. Duties to animals and spirits. In *Lectures on ethics*, L. Infield, New York: Harper and Row.

Mendelsohn, R., Hellerstein, D., Huguenin, M., Unsworth, R., and Brazee, R. 1992. Measuring hazardous waste damages with panel models. *J. Environ. Econom. Man.*, 22: 259–271.

Mitchell, R. C. and Carson R. T. 1989. *Using surveys to value public goods: the contingent valuation method*. Washington, DC: Resources for the future.

Portney, P. R. 1994. Contingent valuation: why economists should care. *J. Econom. Persp.*, Fall.

Singer, P. 1990. *Animal liberation*. 2nd edition. New York: Random House.

Watson, R. A. 1983. A critique of anti-anthropocentric biocentrism. *Environ. Ethics*, 5: 245–256.

Wilson, E. O. 1994. *Biophilia*. Cambridge, MA: Harvard University Press.

32 The Life Support System – Toward Earth Sense

W. S. FYFE

HUMAN VISION AND THE FATE OF THE TERRESTRIAL BIOSPHERE

On Earth we are facing a new situation for our species, a situation in which we know that we must control our numbers, the genetic drive to reproduce. We must use "sense" as the *Oxford English Dictionary* defines it: "practical wisdom, judgment, common sense." We are increasingly aware of the complexity of the interactive systems that support life on this planet. The system, Sun-atmosphere-hydrosphere-geosphere-biosphere, controls our environment; however, at present, the dominant agent forcing change is human activity, and the effects of such activity increase with population increases. We must understand the life support system. We must be aware of the maximum human population that can survive well on Earth given present technologies. All humans on Earth must be educated about the fundamental aspects of our life support system.

At present, a great human tragedy is being enacted. We have vast knowledge to enable us to live well; however, as our numbers increase, the gap between those who benefit from knowledge and those who do not, grows ever wider. Perhaps the greatest question facing all those involved in science and technology today is this: Can we produce the food, fiber, materials, and energy that 10 billion humans will require in the twenty-first century to live well, without destruction of the life support system?

In the recent report of the New Delhi Population Summit, (see Graham-Smith 1994) M. Atiyah, president of the Royal Society of London, stated, "Most of the problems we face are ultimately consequences of the progress of science, so we must acknowledge a collective responsibility, we cannot postpone action on grounds of ignorance." In the same volume, C. Tickell states, "It would be nice to think that the solutions to some of our present problems could be drawn from past experience, but in this case the past is a poor guide to the future. Our current situation is unique." (p. 373).

James E. Lovelock's Gaia hypothesis involved the intriguing idea that life is vital in regulating the "comfortable" environment. It is remarkable that Earth's environment has been, in general, so comfortable. We now know that oceans have existed for almost all of recorded history, and that the hydrosphere has never completely frozen or boiled away (Chap. 3). What has been the role of our massive hydrosphere and the very special properties of water (the thermodynamics of heating, freezing, vaporization)? Its ability to absorb and release energy is remarkable. We also know that, for most of Earth history, microorganisms have dominated life. Recently, it has been reported that these microorganisms may have been quite complicated and included early forms capable of photosynthesis. This seems plausible when one looks at ancient rocks and sediments. Although there is evidence for changes in the planet's convective style and the intensity of convection,

ancient rocks and minerals are all rather familiar, suggesting that surface geophysiology has been relatively constant.

For the future, new factors must be considered in the way we live on Earth. A very clever species is flooding the planet, and reducing biodiversity at a rate quite equal to many major natural catastrophes of the past. We have technologies that are capable of vast destruction. As British geologist Sir Charles Lyell noted in 1872,

In reference to the extinction of species, it is important to bear in mind that, when a region is stocked with as great a variety of animals and plants as its productive powers will enable it to support, the addition of any new species to the permanent numerical increase of one previous established, must always be attended either by the local extermination or the numerical decrease of some other species.

There may undoubtedly be considerable fluctuations from year to year, and the equilibrium may be again restored without

---

W. G. Ernst (ed.), *Earth Systems: Processes and Issues.* Printed in the United States of America. Copyright © 2000 Cambridge University Press. All rights reserved.

any permanent alteration; for, in particular seasons, a greater supply of heat, humidity, or other causes, may augment the total quantity of vegetable produce, in which case all the animals subsisting on vegetable food, and others which prey on them, may multiply without any one species giving way: but whilst the aggregate quantity of vegetable produce remains unaltered, the progressive increase of one animal or plant implies the decline of another.

Nevertheless, in the long run, there will be a tendency in the higher and more perfect organisms, to survive and multiply, at at the expense of the lower, with most of whom they will never come into competition, but at the expense of those which are most nearly allied to them.

I would also like to consider the question of whether or not humanity can, in fact, fundamentally make the planet "uncomfortable" for most or all species and if, or if not, there are limits to our sustainable population. At a 1987 conference on resources and world development (see D. J. McLaren and B. J. Skinner, editors), D. Pimentel stated, "World resources and technology can support an abundance of humans, e.g., 10–15 billion humans living at or near poverty, or support approximately one billion humans with a relatively high standard of living." Given the present human population numbers, we must address such questions.

In 1991, at a similar meeting (see C. Barlow, editor) Dawkins stated, "We are survival machines – robot vehicles blindly programmed to preserve the selfish molecules known as genes. This is a truth which still fills me with astonishment. Though I have known if for years, I never seem to get fully used to it. One of my hopes is that I may have some success in astonishing others." Whether or not this statement is true, it is certainly worth thinking about!

THE PRESENT REALITY

We live in the age of observation of natural systems on all scales – from galaxies, to our genetic code, to atomic nuclei – and we live in the age of information systems (qualified by a comment from the *Economist*: "Information is not knowledge"). Currently, we live in a world with nearly 6 billion human beings. The human population is increasing by approximately 100 million per year (Chap. 22); approximately 50 percent of the current human population is under twenty years of age. Unless there is a catastrophe on a scale almost (but not quite) beyond imagination, the human population will exceed 10 billion by the year 2100; some have estimated the number at over 20 billion.

Another reality, unique in the development of any previous organism, is that humanity now has the tools and techniques to destroy ourselves (and take many other species with us). Most of the population increase today is occurring in Third World, developing countries. Many developing countries might be better described as overpopulated and overdeveloped; China and India developed culturally and economically long before Europe.

Our present world is divided into those who do not worry about food for tomorrow, education, health services, owning a car, or flying on a jet plane . . . and those for whom many of these things are almost beyond the realm of dreams. It is thought that the major cause of premature human death on this planet is lack of adequate nutrition. I would place the fraction of those who live with a high degree of basic security (access to our great knowledge base, access to opportunity), at about one-fourth of the total world population. There is far too much poverty, disease, illiteracy, malnutrition, and violence. Every day, over 40,000 children below the age of five die, and the major causes are rooted in malnutrition and pollution of water. More than 100 animal and plant species disappear from the Earth every single day.

In 1992, the Earth Summit took place in Rio de Janeiro, Brazil. This meeting of many world political leaders demonstrated an increasing awareness that problems exist in our systems of development and that such systems are presenting a serious threat to the most basic components of the life support system. New words are appearing in international discussions: "sustainable development," "environmental protection," "preserving biodiversity," "holistic planning and management" . . . even "no-growth, zero-growth, negative-growth economics." Although the *words* are prolific, however, the actions are not. It is interesting to reflect on the "actions" that followed the Stockholm declaration of 1972: "To bear a solemn responsibility to protect and improve the environment for present and future generations."

THE BASIC COMPONENTS OF THE LIFE SUPPORT SYSTEMS

Our modern life support system, our carrying capacity based on evolving technologies, is related to several basic components:

- Climate, which never was and never will be constant (Chap. 15)
- The chemistry and physics of the atmosphere (Chaps. 13 and 14)
- Energy, first from the Sun, which fluctuates on a number of time scales and from our technologic resources today, primarily from nonrenewable fossil carbon (Chap. 24)
- The availability of usable water (Chaps. 9 and 30)
- Mineral-rock resources from the Earth's crust – advanced societies use approximately 20 tons of such resources per person per year (Chap. 23)
- Soil, the thin surface layer of the crust that supports much of the photosynthetic biosphere (Chap. 8)

With a few exceptions, these resources are common to most living species.

CLIMATE. Over recent decades, the new data from ice cores, ocean and lake sediments, and the like have increasingly shown that our climate fluctuates on many time scales, and on many geographic scales. We are now aware that factors influencing climate include the following: fluctuations in the Sun itself, the Sun–Earth orbit, ocean–wind current transport of energy on the planet, the total planetary albedo (including distribution of vegetation, clouds, ice, deserts), global topography, and volcanic events (which may cause dramatic short-term fluctuations). It is a fundamental law of all convecting systems that as the temperature rises, the system becomes more chaotic. Many of these processes represent highly non-linear phenomena that are difficult to predict (as with the recent record floods in North America). One thing is now certain: At this time, at least 1 billion people are severely malnourished. In order to increase food security, there should be a continuous massive food surplus – at least a year's supply in any nation – and a distribution system to supply those in need. A recent report from the Worldwatch Institute stated that over 1 billion people have an iodine deficiency.

ATMOSPHERE. We live in the atmosphere. In order for humanity to survive, this atmosphere must have a very limited range of composition, with the major components of nitrogen and oxygen in a ratio of approximately 4:1. For many minor components – for example, carbon dioxide and water – there is probably a wide range of tolerable variability, because humans are a water-rich species. However, humans could not tolerate a major shift in oxygen for long: Too much, and we would burn up, too little, and we would suffocate. Our hemoglobin system is tuned to a steady level of oxygen.

Humans also cannot tolerate major changes in the level of sulfur, ozone, dust, and other atmospheric constituents. Recent data from regions of Eastern Europe clearly show the deleterious influences of atmospheric loading by industrial chemicals. Human reaction to such changes progresses from massive increases in problems related to allergies to dramatic damage to the immune system and decreased life expectancy. The state of our atmosphere is coupled to the present systems of photosynthetic organisms, and to the global carbon cycle in general. We must never forget that our present atmosphere is not in equilibrium with our planet, because its iron system – with ferric and ferrous iron and a metallic core – at equilibrium would produce an atmosphere with essentially no oxygen. This recognition led to the Gaia concept, and we must not ignore the fact that our climate, our temperature, is a function of a few trace gases: CO_2, CH_4, H_2O. The Gaia hypothesis suggests that the environment is "regulated" at a comfortable state for life because of the behavior of living organisms. Although there is little doubt that the word "regulate" is too strong, it is clear that the total biosphere plays a vital role in maintaining environmental systems.

Recent data have also shown clearly that massive dust input from volcanism can severely reduce ozone levels in the stratosphere (Chap. 25). It is also clear that periods of abnormal continental or submarine volcanism could even perturb the carbon dioxide and oxygen levels of the atmosphere. This may have been particularly significant in causing fluctuations on the early planet, when volcanism and meteorite bombardment must have been more intense.

ENERGY FOR HUMANITY. Growth of the human population exploded with the scientific-technologic advances of the past centuries. One of the most critical technologies has been the development of "fossil" energy resources.

There is no question that, to a large extent, our standard of living is related to the availability of energy (Chap. 24). Energy resources are fundamental to our systems of education, health, nutrition, recreation, and communication. At the present time, most of the world's energy is derived from oil (40 percent), followed by coal (30 percent), and then natural gas (20 percent). Thus, fossil carbon supplies approximately 90 percent of the world's energy. Once a significant energy source, hydropower now supplies only 7 percent of global energy. It is the by-product waste gas from burning fossil carbon, carbon dioxide, that is one of the most significant causes of changing world climate, temperature, and rainfall distribution. Present trends in the use of fossil carbon fuels are not sustainable, and the potential atmospheric pollution will almost certainly have intolerable consequences.

In most parts of the world, the demand for energy is rising, even where the population is stable. Currently, in many developing and less developed nations, more energy is needed, and in vast quantities. If this energy were supplied, even transiently, from fossil carbon, particularly coal, the results could be a global disaster. Here, there is only one caveat. If a cost-effective technology could be found that would eliminate carbon dioxide injection into the atmosphere (e.g., acid components), the fuels still might be usable. This solution is being discussed by a few; however, no such technology is in sight at this time. Even given such technology, however, costs will be much higher.

Not so long ago (100 years or more), in much of the world, biomass (wood) was the dominant source of energy for a population of 1 billion. Population growth, and the associated deforestation, have removed this source as a global option for the present. However, for a few nations with low population density, excess farmland, excess food production, sound soil conservation, and robust soils, biomass fuels (wood, ethanol, methanol) are viable options. Biomass fuels use solar energy and recycle carbon dioxide. However, let there be no illusions. The total annual production of biomass carbon fixation, from all cultivated land is approximately 0.4×10^{10} metric tons of carbon. If this total biomass were burned, the energy production would be 1.6×10^{20} joules. Given that people use approximately 4×10^{20} joules per year, the product from all cultivated land, if dedicated to fuel alone, would last us approximately five months. The carbon stored on land (90 percent in trees) is approximately 8×10^{11} metric tons of carbon, equivalent to 20×10^{21} joules. This would last five years. It is obvious that natural products of photosynthesis cannot satisfy the energy needs of 10 billion humans (Chap. 19).

Future potential for hydropower is limited to a few areas. North American and Europe have already developed approximately 50 percent of this potential; however, in parts of Asia, South America, and Africa, 10 percent or less is currently used. It has been estimated that, in these countries, potential capacity exceeds 100,000 megawatts (or approximately 100 major power plants). Water and the environment of rivers are increasingly needed for food production and irrigation. Today, almost half the reliable runoff from the continents is manipulated by humans (Chaps. 28 and 29). For a population of over 10 billion, the use of water will require planning with a wisdom never demonstrated before. In addition, the source regions of the water typically are located far from the regions of energy needs (e.g., Brazil), or even in neighboring nations (e.g., Canada–United States). The ecologic impact of hydopower development (downstream and upstream), and the potential for geohazards, induced seismic activity, salinization, and the like require the most careful analysis (Fig. 32.1). Also, when major river systems no longer flow freely to the oceans, the causes of observed changes in stream flow, sediment load, and solution chemistry are poorly understood. Change in light water fluxes to the oceans can change ocean current patterns and, hence, influence local and global climate regimes. When a resource that is dependent on precipitation is developed, changing climate patterns must be understood. When one considers the future hydropower needs of agriculture, industry, urban centers as the human population doubles, one must contemplate that many or most of the world's great rivers will be diverted. Runoff from the continents will be replaced by evaporation. What will be the impacts – local and global – on climate? Those who model future climates (Chap. 16) must consider this factor.

There is simply no doubt that the ultimate, sustainable, energy resource for humanity is the Sun. Today, humans use approximately 4×10^{20} joules per year. Every day, the amount of solar energy that the Earth receives is 1.5×10^{22} joules. If we could store a few hours of solar energy per year, we could provide the energy now used by humans! There would be no change in the overall radiation balance of the Earth. Solar energy clearly must be our future energy resource – with the exception of a few special geographic locations (near the North and South poles). To provide all our energy

Figure 32.1. The vast lake behind the Tucurui dam on the Tocantins River in the Amazon. The lake and drowned forest covers hundreds of square kilometers. (*Source:* Photograph by W. S. Fyfe.)

needs, we would need to collect and store all this energy on approximately only 0.01 percent of the surface of Earth (much of it in deserts). There are other applications of solar energy, as with wind and tidal systems.

In addition to the solar flux, in many regions of Earth – particularly near the great plate boundaries – igneous activity constitutes a replenishable source of power. Such environment are characterize by volcanism and high heat flow. Thus, geothermal energy is another resource of considerable potential (Chap. 24).

The Earth possesses vast sources of renewable, sustainable energy. If there are limiting factors on the human population, it is clear that the limit is not energy itself.

WATER. If world human population reaches 10 billion and if these 10 billion are to have adequate nutrition and supplies of clean water, it will be necessary to manage the global water supply with great care and attention. When we come to consider the water resource question, we see the potential for real limits or potential environmental disasters.

Globally, precipitation is approximately 525,100 cubic kilometers (1 cubic kilometer equals 10^{15} grams or 10^9 metric tons). However, much of this precipitation falls over the oceans (78 percent). On land, more than 50 percent evaporates. The reliable runoff available for human use is approximately 14,000 cubic kilometers (much more occurs but only during floods). The reliable amount of usable water is estimated to be approximately 9000 cubic kilometers. Currently, human beings manipulate approximately 3500 cubic kilometers, almost 40 percent. After we use water, its chemistry and biology are changed (Chap. 9). It is interesting to note that each Canadian uses 5000 liters of water each day. If this usage rate were applied to 10 billion humans, the amount would be $5 \times 10^3 \times 3.65 \times 10^2$ liters per year (or 18,000 cubic kilometers), about twice the reliable supply!

For any given region, the water supply can be quantified. Consider the "island" model (e.g., a river may flow from one region to another, but who can guarantee that this inflow – through flow – will be maintained?). In an island model, the water inventory involves the rainfall, the groundwater penetration, the evaporation or evapotranspiration, and the runoff to the oceans. One of the most important parameters is the reliability, or variations, of the rainfall. Over what time period does one average the availability of the total supply? The numbers to be used will, in part, reflect the technologies that may be available, or possible, to store water over long periods of time. Certain realities must be taken into account in the preparation of a plan of management for the island water supply:

- The amount of runoff must be sufficient to prevent salinization – runoff removes salts and other wastes from the land surface.
- The evaporation processes depend on the nature of the island's surface cover. This cover will vary, depending on the vegetation, urban developments, transport systems, and so forth.

- The potentially useful groundwater resources depend on the geology, and the deep porosity, permeability, and chemical nature of the rocks, which will influence the chemistry of the water and its suitability for various purposes – agricultural and industrial.
- Surface storage removes land area, and can increase evaporation.
- Subsurface storage is sustainable only if recharge exceeds withdrawal, and if the cycling does not reduce long-term porosity and permeability.

It is remarkable that in many regions today, such basic data are not considered in the management of water use. Groundwaters are being "mined" in many regions (Chap. 30).

In the near future, it is likely that massive technologies will be developed for the wholesale recycling and reuse of water, particularly in urban and industrial systems. Given an appropriate climate, urban-industrial runoff could be redistilled by abundant solar energy – a sort of "spaceship Earth" technology (imagine a human colony on Mars!). With the use of appropriate rocks, the distilled water could be easily remineralized to the required levels of mineral content for human nutrition. For industrial uses, modern filtration systems (inorganic or biologic) can allow reuse and reduction of waste discharge.

A great difficulty (with potential for major problems) involves the agricultural energy uses of major systems of runoff. A large river is dammed for such purposes. The runoff to the oceans now fluctuates over large values, depending on fluctuations in rainfall. Downstream, the aquatic biodiversity and biomass are seriously perturbed. At sites of ocean discharge, the nutritional supply to the ocean biomass is heavily impacted, and satellite images show us clearly that the marine biomass is concentrated near continental margins.

Perhaps most problematic of all, ocean current systems may be altered. Approximately 10,000 years ago, the great Gulf Stream current, which transports energy to the Atlantic Arctic regions, was perturbed. The North went into a little ice age, the Younger Dryas event. For a period of time, the Mississippi River system was diverted into the Saint Lawrence system. This placed a vast flow of light continental water onto the surface of the North Atlantic, which modified the Gulf Stream and the northern energy transport. Very rapidly, the North Arctic froze. This model warns us that if a large continental outflow is changed, the patterns of ocean currents, ocean mixing, may change, affecting local and global climate. Such phenomena must be considered when there are plans to modify major components of continental runoff! I am sure that most water engineers have never heard of the Younger Dryas cold event.

Recent world data are simply mind-boggling. For example, in 1900, 8750 kilometers of river had been modified for navigation; in 1980, 498,000 kilometers. In 1950, the world had 5270 large dams; in 1985, 36,562. Will any rivers run to the oceans naturally by 2050?

MINERAL RESOURCES. We are surrounded by modified materials, derived primarily from the top kilometer or so of the Earth's crust. We live with concrete, glass, bitumen, ceramics, steel, copper, aluminum, stone, zinc. . . . Our transport machines, our computers, contain components from half the elements in the periodic table of the elements. Advanced societies use approximately 20 metric tons of rock-derived materials per person per year. For a population of 10 billion living at an advanced quality of life, the consumption rate would be 2×10^{14} kilograms of rock per year, or almost 100 cubic kilometers per year. This quantity exceeds the volume of all the volcanism on the planet, on land and submarine, by an order of magnitude. Human activities now constitute a major component of the processes that modify the planet's surface (Fyfe, 1981). I am always impressed by the general ignorance of people concerning "Where does it come from?" A standard gold wedding ring weighs approximately 10 grams. Many of the gold ores mined today contain approximately 3 grams of gold per metric ton. In order to produce a wedding ring, approximately 4 metric tons of rock was processed (ground up, leached with cyanide, and so forth) and dumped. Often, the gold ores contain elements such as sulfur and arsenic at high concentrations, causing a large impact on local water systems. It is time that humans were better informed about their resource base (see Fig. 32.2).

Can we supply these materials for all humankind? If we have the necessary energy to drive the machines and transport the materials, the answer is probably yes. However, it is also clear that for the less common materials – such as copper and zinc – with rigorous attention to recycling, the current rates of environmental modification can be reduced. The new trend to design a product with recycling in mind must become the rule for the future. The savings in total use of materials and energy, and in the total production of wastes, makes recycling an economic necessity of the future.

A major component of the materials we use today is derived from wood. It is clear that the present styles of the use of wood products for housing, paper, packaging, energy, and the like cannot continue. The environmental impact of the removal of forests is too serious to allow present and past careless harvesting to continue. Wood will not be a major material for the 10 billion humans of the twenty-first century. It is not necessary, given the potential of modern systems and materials.

Figure 32.2. The famous gold mine, Serra Pelada, in the Amazon region of Brazil. The total environmental impact is vast. (*Source:* Photograph by W. S. Fyfe.)

SOIL. Soil, water, and climate are at the core of the human food supply. Today, at least 1 billion humans suffer from malnutrition. In nations such as Egypt and India, close to 70 percent of the population suffers from classic anemia. Data in a 1992 paper by B. Moldan indicate the critical nature of the present problem. He shows that, at present, human activity manipulates 39 percent of the net primary bioproduction of the planet. Much of this "use" involves waste. What will the situation be when we once again double population?

Soil is formed by the reactions of rock with air, water, and living organisms (Chap. 8). Soil science is perhaps one of the most underdeveloped subjects of modern Earth science. Although science requires the exact description of a system, necessary descriptions of soil on the microscale have not been possible previously, except in rare cases. The reactive components of soil consist of very small particles. Modern tools are required for these descriptions: geochemical analysis, structure analysis (e.g., high-resolution electron microscopy), and, in particular, surface analysis and particulate analysis. These tools were not available for normal descriptive procedures until very recently.

Soil consists of many components. Debris from the parent rock sets the stage, in large part, for the future chemistry, mineralogy, and biology of soil. Airborne dust may also be a significant component. New minerals and amorphous materials form by weathering; typically, these minerals are hydrated, and commonly oxidized. A multitude of fine-grained particulate material is present. As well, soil is the home for a host of microorganisms, macroorganisms, and their debris, which are involved in key processes of nitrogen fixation, carbon fixation, and mineralization (Chaps. 28 and 29). Soil bacteria are excellent ion exchange phases. Soil provides the physical support system for the roots of plants, which are often finely adapted to local conditions. Soil contains gas and liquid phases that are present in its pore spaces and, as surface films, on inorganic and biologic materials. Soil physics (thickness, porosity, permeability, rigidity, viscosity, strength, cohesion, compressibility, albedo) reflects all the above.

Soil is a dynamic system. Its biology varies with seasonal changes; its liquid and gas phases vary on many time scales. Wetting and drying are reflected in transient solution and precipitation processes, primarily on surfaces, of and in thin films of fine-grained materials. The physics, chemistry, and biology of soil is not steady-state. Soil is a very complex system. It is dynamic on all microscales, as it become wet or dry or warm or cold, and the biology changes in response to changes in the physics and chemistry. The in-place interactions of all the pieces, living and dead, make this system most difficult to describe.

Although soil is complex, one defining property or quality is significant for the biosphere (and humankind). This property or quality is based on the following question: Given a certain climate (e.g., temperature, rainfall, light, wind speed), what is the capacity of a soil to support photosynthetic organisms, to support carbon fixation, to support the production of food and fiber on a truly sustainable basis? This ca-pacity determines the soil and plant modulation of the atmosphere, by means of reactions of the soil with oxygen, carbon dioxide, and moisture, for example. For humankind, and particularly for economists, other key questions arise (see Fig. 32.3): Is the soil changing? How is it changing? How rapidly? Is it becoming more, or less, productive?

Today, one of the most dramatic changes occurring on a global scale is related to soil thickness. Erosion is the primary cause of soil change; however, there are other causes of change, such as irrigation and salting, nutrient reduction because of overcropping, compaction, and human-induced anaerobic conditions. The relation between soil thickness and food and fiber productivity can be highly nonlinear (Chap. 8).

The introduction of many herbicides, pesticides, new plant species, and industrial pollutants such as lead must surely change the total biologic–geologic interactions. The new agro-technologies, which appeared to be successful over past decades, are now showing massive fatigue, even failure. Many people are also worried about the new genetic engineering, the move to monocultures, and their almost certain instability and vulnerability to fluctuations within the biosphere and the lithosphere.

How do we estimate the local and regional health of soil? I would like to suggest that there are simple ways to obtain a useful estimate of the health of soils on a local and regional basis. All organisms, from bacteria to trees, require a wide range of macronutrients and micronutrients. Key macronutrients are carried in the rivers of the world (Chap. 12). Most of the water in these rivers passes through surface soil and rock. The levels of the macronutrients and the total dissolved inorganic material (TDS) speak eloquently about the state of the soils that the rivers drain. The Mississippi and Ganges river systems pass through younger soils that are loaded with rock-forming minerals; the Amazon and Negro systems of Brazil flow through old, deeply weathered laterite terrains. The water in the latter systems indicates a low capacity to support intense bioproductivity, probably the reason this fragile region has not been heavily populated by the human species, historically.

The great fertile regions of the world – for example, Hawaii, southern Ontario, and the Nile Valley – contain young soils with massive mineral debris. Plate boundaries are fertile, volcanic terrains are fertile, and regions that have been covered by sediments from diverse source rocks as a result of river, ice, or wind erosion are fertile. In contrast, the global agricultural-disaster regions have not had a sediment shock or an igneous mountain-building shock, for millions of years. The macronutrients and micronutrients in these regions have passed to the oceans. In fact, in some of these regions, the primary nutrient input today may be from the airborne particulates in dust and rain.

According to the Worldwatch Institute, topsoil is being lost at a global rate of 0.7 percent per year, a rate that must be accelerating. If this is true, we are heading for a food catastrophe! How long does it take to form a layer of fertile

Figure 32.3. A scene in an Atlas mountain valley of Morocco. The consequence of too many people and animals is evident. Among the most important: The soil has been removed. (*Source:* Photograph by W. S. Fyfe.)

soil 10 centimeters thick? ("Fertile" is defined in terms of the capacity of soil to produce food and fiber.) The answer to this question is found in nature itself. For example, in Hawaii, we can watch well-dated volcanic flows evolve into a cattle ranch or a forest. In northern Canada, where clean rock was exposed after the ice retreated 10,000 years ago, there is often little or no soil. A range of 1000 to 10,000 years is necessary for the formation of topsoil; therefore, this is essentially a nonrenewable resource, in terms of human time scales.

Conversely, rocks can be formed as a result of the erosion of soil (e.g., deltaic sedimentary rocks), and the formation of soil from these rocks may be much faster. However, the conversion of these rocks into fertile soil requires time for the creation of porosity and other physical properties that will allow the development of diverse biologic support systems and that will create water-holding systems. How fast these processes occur depends critically on the history of sediment compaction.

For sustainable carbon fixation (i.e., agriculture), simple laws must apply. The rate of soil loss must not exceed the rate of soil formation. The nutrients taken out by crops must be balanced by input for the entire spectrum of essential elements. The ability of the soil to hold water must not be diminished because of agricultural practices. How often do we have the data to quantify such laws for large or small regions?

As mentioned above, global loss of topsoil is currently running at 0.7 percent per year. Although this figure may be challenged, any figure even close to this is cause for alarm. Further, the consequences of topsoil loss go far beyond the problems of the global food supply. Today, atmospheric carbon dioxide levels are rising faster than ever. If soil is failing, its carbon dioxide fixation processes will decline as well; this combination of problems has the potential to accelerate greenhouse warming.

As human beings invade the marginal areas of the globe, the problem of soil erosion can only become worse. We must quantify soil erosion more accurately, and we can. Global-scale observations are now possible, using satellite imagery, and even schoolchildren can play a local role in quantitative monitoring of erosion in their region. Erosion can be controlled now, with intelligence and with awareness!

The systems to control erosion are well known. A few very simple, obvious examples apply. Steep slopes should not be used for intensive farming (consider Switzerland and Japan). Clear-cutting and burning, commonly practiced in Canada

Figure 32.4. As a result of clear-cutting and burning in the great forests of Vancouver Island, Canada, the soil is now vulnerable to extreme erosion. (*Source:* Photograph by W. S. Fyfe.)

even on steep slopes, must be outlawed! Steep slopes requires forests for their protection (Fig. 32.4). Even gentle slopes, which have fragile soils, heavy rainfall, and rain-impact erosion, require continuity of ground cover and no use of herbicides or monoculture agriculture. The same considerations apply to areas that have frequent high winds. In my own home country of New Zealand, where I was born on a farm on the Canterbury Plains, wind erosion in the 1930s was severe. As a result of using new crops that emphasize grass cover, and using small fields and prolific planting of belts of trees, the problem no longer exists. In fact, erosion in some of these regions may be lower than before human existence, when natural fires could sweep across huge areas. I recently visited the Federal Republic of Germany. One can only be impressed by the productivity of their careful agriculture, small fields, crop rotation, organic fertilizers, and trees in profusion. Perhaps the most spectacular cases of the nonlinear results of soil erosion can be seen in regions of former tropical rain forest. In these areas where evapotranspiration dominates the pre-human water cycle, the cutting of protective forests on steep topography can lead to catastrophic soil loss within a very short time. In fact, reasonable productivity may go to zero productivity within a period of years, or even days.

CONCLUSION

As we move toward a human population of 10 billion, we can no longer afford to be careless with our planet. There will be vast development on Earth. We now have extensive knowledge, but vast unknowns remain. We do not have adequate climate models, and we have only recently recognized the complexity of climate forcing. We have an urgent need for extreme models that will consider:

- The climate of Earth without life – the ultimate test of Gaia
- The climate of Earth with vastly reduced biodiversity (e.g., no forests)
- The climate of Earth when no rivers flow to the oceans and all continental runoff is used on land, and evaporates

Perhaps we need such models to shock us into action. We need to study Mars. I am sure it once had life (microorganisms and hot springs). Did Gaia fail? Is there hope for a population of 10 billion? I think it is possible that we can provide energy and mineral resources. However, my concerns are with soil, biodiversity of biomass, and water, and the natural fluctuations. Will we see famines on ever-increasing scales with all the social instabilities we now see around the planet? I think we should all read again the wonderful essay written in 1969 by C. P. Snow, which states: "At best this will mean local famines to begin with. At worst, the local famines will spread into a sea of hunger. . . . Peace, food, no more people than the earth can take. That is the cause" (p. 43).

There is hope, however, and the data are hard. Where there is a high standard of education, universal literacy, and gender equality, populations are declining. Our objective must be literacy, numeracy, science for all. All who live on this planet must understand life support, and be eco-geo-literate. There must be Earth Sense. Let us hope that we can outsmart our selfish genes. We require educated citizens to plan our future. As King and Schneider stressed in a 1991 report by the Council of the Club of Rome, our systems have failed.

Earth scientists must leave their cocoons, and must join with other specialists to solve world problems. There *is* a global crisis. As we enter the twenty-first century, we must develop new technologies to maintain and improve our quality of life, and preserve the remarkable, the necessary, biodiversity of Earth. New energy, waste, agricultural, resource, and water technologies are urgently needed. We must never forget that we are a small part of a complex biosphere living in the atmosphere, and our resources come from the hydrosphere and the top kilometer or so of Earth. Those of us fortunate enough to live in a world of opportunity must accept our responsibility.

QUESTIONS

1. Account for the habitability of our planet in terms of its place in the solar system and the evolution of life.

2. What are the strong points of the Gaia hypothesis? In what ways is it a less persuasive accounting for the conditions on the Earth?

3. Given a mid-twenty-first-century human population of 10 billion, what levels of existence do you envision for: the industrialized nations? the developing nations?

4. Justify your best estimation of the steady-state human carrying capacity of the planet.

FURTHER READING

Dawkins, R. 1991. A new world view, in C. Barlow (ed.), *From Gaia to selfish genes.* Cambridge, MA: MIT Press, 195–213.

Ehrlich, P. R., Ehrlich, A. H., and Daily, G. C. 1993. Food scarcity, population and environment. *Pop. Devel. Rev.,* 19:1–32.

Fyfe, W. S. 1981. The environmental crisis: quantifying geosphere interactions. *Science,* 213:105–110.

Graham-Smith, F. (ed.) 1994. *Population—The complex reality.* The Royal Society, London, 404 pp.

King, A., and Schneider, B. 1991. *The first global revolution,* a report by the Council of the Club of Rome. London: Simon and Schuster.

Molden, B. 1992. Geochemistry of the environment, in A. Cendrero, G. Luttig, and F. C. Wolff (eds.), *Planning the use of the Earth's surface.* Berlin: Springer-Verlag, 237–258.

Pimentel, D. 1987. Technology and natural resources, in D. J. McLaren and B. J. Skinner (eds.), *Resources and world development.* New York: Wiley, 679–695.

Snow, C. P. 1969. *The state of siege.* New York: Scribner.

SUMMARY

33 Synthesis of Earth Systems and Global Change

W. G. ERNST

LIFE AND DEATH IN THE GALAXY

As this book goes to press, several startling, but perhaps inevitable, scientific discoveries have burst upon the scene. (1) Astronomers have confirmed the presence of planetary bodies orbiting several nearby stars. The existence of planets in other solar systems has proven what cosmologists have long suspected: there is nothing unique about our little corner of the galaxy, and planetary bodies are probably as abundant as the stars themselves, perhaps more so. When a powerful new generation of telescopes comes into service within the next decade, the detection of additional planets is confidently anticipated. (2) Among the meteorites that have fallen on the Antarctic ice cap in geologically recent times, several have been recognized as derived from Mars (blasted off the surface of that planet by the impacts of incoming asteroids). The evidence presented by chemical and isotopic analyses of one of these meteorites – in which certain hydrocarbons and microscopic traces of minerals that seemingly could have been made by living organisms – suggests the possibility of primitive life on the surface of the red planet approximately 3.6 billion years ago. (3) Here on Earth, chemical and isotopic data from glacial ice cores, pollen and coral records, shallow-marine sediments, and desert rock varnish have shown that rapid temperature fluctuations characterized the Pleistocene and Holocene climates – both atmospheric and oceanic. These observations indicate that biologic activity is sufficiently hardy and resilient to withstand moderate climatic variability; moreover, it probably flourishes throughout the universe where planetary conditions are hospitable. However, no matter how robust the life force is, we need to understand quantitatively how the complicated, interactive Earth system functions in order to safeguard and perpetuate it.

Humankind's central place in the universe, and even the unique habitability of the third planet within our solar system, are being reevaluated, as is the past rate of environmental change. Among the terrestrial biota, humans are especially capable of (and prone to) modifying portions of the ecologic system. As our numbers grow and our technical capacities increase, this ability can only be magnified. Global change is inexorable, and is now due at least in part to anthropologic influence. How will our current practices influence the future carrying capacity of our thus-far benign planetary home? How long do we have to "get it right"?

As the world's current population reaches 6 billion people, on the way to an "equilibrium" value of 10 billion by approximately 2050 (assuming the most optimistic scenario regarding minimal population growth), our ever-increasing impact on the Earth system soon must become understood quantitatively if the biosphere is to survive in anything resembling its present form. History provides few indications that we will achieve such an appreciation of our interrelationship with the environment. Rather, it demonstrates that humans typically exhaust their resource base (examples include the Mayan civilization of Yucatan, the Anisazi Cliff Dwellers of the American Southwest, and the currently malnourished inhabitants of the sub-Sahara, Madagascar, and Haiti). The profligate consumption of nonrenewable and slowly renewable resources – especially topsoil, fresh water, and fossil fuels – constitutes an example of the human proclivity for living for today and letting tomorrow take care of itself. However, borrowing from the future carrying capacity of the planet will exacerbate the situation as rising demands for materials goods and services from Third World peoples (approximately 85 percent of the global population, and growing) are registered.

It seems likely that humanity will be obliged to reach a sustainable level of terrestrial development – no further deterioration of the Earth's integrated ecosystem – by about the middle of the twenty-first century. In other words, within the next two generations, we surely must "do or die." The alternative seems to be a retreat to a level of subsistence compatible with an ever more severely damaged environment, which would involve Earth systems that are massively

W. G. Ernst (ed.), *Earth Systems: Processes and Issues*. Printed in the United States of America. Copyright © 2000 Cambridge University Press. All rights reserved.

out of kilter, and an evolution toward a new dynamic equilibrium among biosphere, atmosphere, hydrosphere, and solid Earth. What have we learned in this book that might help us to cope more successfully with the future?

THE EARTH AS A SYSTEM

We need to study the Earth as a whole in order to understand the reality of climate change. The greenhouse effect is an incontrovertible fact, and an anthropogenic influence is no longer seriously debated by knowledgeable scientists. The question is simply one of the magnitude of our influence, not of the existence of the phenomenon. Studying the planet as an interconnected, integrated system allows us to comprehend more fully the evolving global interactions of smaller-scale effects, such as urbanization, volcanic eruptions, coal burning, and deforestation. The perceived nature of the terrestrial processes are a function of the degree of magnification at which they are examined; we are obliged to investigate the Earth systems at spatial and temporal scales ranging from the atomic (the angstrom level) to the universal (billions of light-years).

It is clear that human activities induce ecological changes and habitat fragmentation, and an accelerating, intensive decrease in biodiversity. Economic progress often comes at the expense of the overall health of the environment. How important is this, and in what ways should we modify our behavior? A quantitative knowledge of the Earth systems will help us to decide; however cost/benefit analyses inevitably require an evaluation of intrinsic goods (e.g., the quality of life, the existence of wilderness areas, clean air and water) that defy conventional, economic measures of worth (see Chap. 31). The daunting challenge of anthropogenic environmental degradation is the verity that future generations and innocent neighbors (including all life forms, not only humans) will bear the consequences of actions taken by specific groups of people today.

Earth systems science is actually twenty-first century geography: It encompasses the study of the environmental physical and life sciences and engineering, coupled with an analysis of human constructs and political and economic policies. It employs space-age technologies to identify, measure, and manage diverse global data bases that serve as a framework and foundation for a coherent discipline.

Because of their small size and spatial isolation, islands constitute microcosms – test beds – of the larger Earth systems. Kauai is a readily understandable example. Basaltic in origin and only 5 million years old, this mid-Pacific member of the Hawaiian chain lies distant from all continental environments. Consequently, although this ideal habitat is abundantly nourished through precipitation (supplied by the northeast trade winds) that soak into its richly fertile volcanic soil, Kauai has been colonized by plants and animals only as a result of a series of accidental introductions (e.g., storm-driven birds and insects, creatures riding on floating debris). A unique indigenous ecosystem thus evolved through chance arrival and subsequent adaptive radiation. However, with the coming of Polynesians and later, Europeans, the late-stage anthropogenic introduction of alien species (and disease vectors), pervasive and invasive agricultural practices, and hunting have vastly altered the web of life on Kauai. Environmental change is accelerating on this island paradise.

North America represents an infinitely more diverse "island universe." Geologically far more complex than a small basaltic island, the continent's legions of interpenetrating ecologic niches have been changing over at least the past 3.8 billion years, since the time of the first self-replicating, non-nucleated cells on Earth. These gradual changes reflect the evolution of the hydrosphere and atmosphere, as well as the accretion and rifting of continental crust due to plate tectonic processes. Climatic zones have migrated as a consequence of the combined effects of continental drift, variable overlap by shallow seas, and mountain building. Native Americans arrived via the Bering Straits land bridge at least 10,000 (probably closer to 18,000) years ago, and gradually diffused throughout the continent. Sophisticated civilizations flourished, especially in Latin America. Because of relatively small populations and limited technologies, their aggregate modification of the environment, while locally intense, was modest. All that changed irrevocably, however, with the arrival of Europeans, especially following the end of the eighteenth century. Despite the vast area of the continent, the pervasive anthropogenic impact on North America has been comparable to that on the relatively diminutive habitats characteristic of the Hawaiian Islands.

Several distinct but intergradational time scales attended the formation and subsequent evolution of our planet and various Earth systems processes: universe evolution spans 10 to 15 billion years; geologic history spans millions to 4.5 billion years; biologic evolution spans thousands to millions of years; sociopolitical transitions spans months to hundreds of years; and cataclysmic events spans fractions of a second to days. The evolution of individual plant and animal species spans thousands to millions of generations, commensurate with, or more rapidly than, geologic alteration of the landscape, but infinitely slower than the devastation wrought by a destructive earthquake or the cataclysmic arrival of an asteroid from outer space. Compared to biologic evolution, human times scales are yet shorter, and rates of change are clearly accelerating: The transition from a primitive food-gathering stage to a culture typified by agricultural settlement proceeded imperceptibly, far more slowly than the subsequent rise of city-states; these, in turn, formed more slowly than the rate of global industrialization and ongoing conversion to an electronic information society. What we may conclude from the imperfectly recorded history of the Universe is that processes in progress today also operated in the distant past, and that these evolutionary processes are – and always were – governed by the immutable laws of physics and chemistry. However, the age of consciousness has witnessed a remarkable acceleration in the rate of environmental change, both locally and globally.

THE GEOSPHERE

The Earth formed along with the rest of the solar system during the collapse of a locally dense region of intergalactic gases. This event took place 4.5 to 4.6 billion years ago, long after the Big Bang approximately 10 billion years earlier. Most of the mass of the solar system, chiefly hydrogen, was gravitationally concentrated in the central object; the enormous pressures and temperatures attending such a condensation initiated the fusion of hydrogen in what was becoming a new star – the Sun. Preservation of angular momentum requires that the massive central star rotate more rapidly than the primordial gassy nebula; consequently, minor clumps of matter possessing sufficient radial velocities were sprayed into orbits defining a common plane at right angles to the rotation axis of the Sun, the so-called plane of the ecliptic.

The hospitable environment for life that characterizes the Earth resulted from the accretion of planetesimals to a particular size at an appropriate distance from the central star such that abundant liquid H_2O was stabilized, resulting in the hydrosphere. The physicochemical environment that can support biologic activity apparently has rather narrow limits. Equable, mild terrestrial conditions involving a somewhat saline aqueous medium in equilibrium with an initially strongly reduced, but increasingly oxygenated, atmosphere must have been ideal from almost the beginning, or else we would not be here to describe terrestrial biologic evolution. Since its formation, the Earth has experienced "the Goldilocks conditions" – not too hot, not too cold, but just right.

The planets in our solar system exhibit systematic and progressive variations in their chemical constitutions and surface features as a function of their orbital distances from the Sun. Some planetary satellites – such as the Earth's Moon, the four Galilean satellites of Jupiter, Saturn's Titan, Neptune's Triton, and the largest asteroids – are bigger than, or approach the sizes of, Pluto and Mercury, the smallest planets. Where we draw the line between planet and satellite is somewhat arbitrary. Pluto has an elliptic orbit that crosses that of the eighth planet, Neptune. The innermost planet, Mercury, is silicate- and, especially, metal-rich, whereas the next three, Venus, the Earth and Moon, and Mars, are stony. The outer planets represent progressively more gassy giants, and finally, even farther out, proportionately more icy bodies. The Oort cloud, at the gravitational limit of our solar system, is the sparsely populated home of icy comets.

Each planet is unique in its external appearance and, evidently, in its internal constitution. Each planetary interior is hotter than its surface because of trapped heat. The most efficient mechanism for removal of large quantities of buried energy is by mass circulation, with cooling taking place at the planetary surface. Larger bodies contain more heat, and because they possess larger mass-to-surface ratios, they cool more slowly; hence, they remain hot and convective for longer periods of time. The Earth is the largest of the terrestrial planets; its proportionally larger store of trapped heat accounts for both its vigorous internal circulation and plate tectonics.

Although it has a composition similar to both Venus and Mars, the Earth is obviously very different from them surficially. This is a reflection not only of relative size and internal heat, but also of the position of our planet relative to the Sun. Solid, liquid, and gaseous forms of H_2O are stable at and near the Earth's surface; carbon dioxide is precipitated from the terrestrial oceans and largely sequestered as limestone and dolomite. In contrast, the proximity of Venus to the Sun is responsible for its much higher surface temperatures and lack of liquid H_2O oceans. Accordingly, the carbon dioxide complement of Venus occurs as a gas, and resides in the dense venusian greenhouse atmosphere. In contrast, Mars is so distant from the Sun that on its surface, H_2O frost is presently confined to seasonal polar caps (which actually consist primarily of frozen carbon dioxide); the carbon dioxide on Mars is thus trapped in icy solid layers. We have much cause to cherish the environmental conditions that characterize planet Earth.

With the exception of sunlight, all the resources supporting life are derived from the planet itself. Natural hazards are also manifestations of the dynamically interactive Earth systems. Humanity's attempt to live comfortably and harmoniously with nature depends on recognition, understanding, and protection of the interconnected systems that operate on our terrestrial home. What are these local to worldwide systems?

The thermally driven chemical evolution of the Earth has produced a planet surrounded by a tenuous global atmosphere consisting of 78 percent nitrogen, 21 percent oxygen, 1 percent argon, and traces of rare gas above a thin, rocky rind or crust. Approximately one-third of the planet is surfaced by a continental crust some 20 to 60 kilometers thick, possessing a composition roughly equal to that of granite; the oceanic crust is only 4 to 6 kilometers thick and closely corresponds to basalt in composition. All of the latter and the margins of the former are covered by the world's oceans. Beneath the rocky exterior, the solid Earth passes downward into magnesium silicate– and oxide-rich shells of progressively increasing density; together, these concentric layers constitute the mantle, are similar in chemistry to stony meteorites, and represent 83 percent of the planetary volume. Below the mantle, the core of the Earth is made of iron-nickel metal, and is similar chemically to iron meteorites; the core constitutes approximately 16 percent of the Earth by volume, and consists of a molten shell in its outer portions and an iron-nickel solid sphere at the center. Rocks and their constituent minerals represent the materials from which the solid Earth and its various layers are constructed.

A rock is a naturally occurring multigranular aggregate of one or more minerals and/or amorphous (noncrystalline) solids. We recognize three main rock-forming processes, and therefore three principal classes of Earth materials. Primary additions to the crust are supplied as molten magma, which cools and solidifies to either glass or an aggregate of minerals: such rocks are termed igneous. Sedimentary rocks consist of mechanically accumulated fragments of preexisting particles, as well as chemical precipitations from water. Metamorphic rocks include all those rocks whose original minerals and/or textures have been altered markedly by changes in temperature and pressure and/or by deformation; typically, metamorphism takes place at considerable depth within the crust. Soils are the weathering products of rocks. They represent the chemical breakdown and mechanical disaggregation of preexisting Earth materials; most contain abundant decomposing organic material as well. Virtually all land plants take their nutrients from soils; inasmuch as terrestrial animals ultimately are dependent on the botanic cover for sustenance, soils are crucial for the nonmarine portion of the web of life.

The Earth's surface topography, unique in our solar system, consists of ancient, high-standing continents and geologically youthful, low-riding ocean basins. The terrestrial crust is constantly being reworked; a dynamic equilibrium has been established between the competing agents of erosion and deposition (an external process) and crust formation (an internal process). Driven by deep-seated thermal convection, imperceptible but inexorable differential movements characterize the outermost skin of the planet. It is this mobility that accounts for the growth and persistence of continents and ocean basins, although the former are being planed down by erosion, and the latter are being filled through sedimentary deposition. Planetary dynamism, energized by solar radiation and the escape of internal heat, is responsible for oceanic/atmospheric circulation and, through mantle flow, sea-floor spreading and the drifting of continents. (Solar energy absorbance and transfer mechanisms generate the terrestrial climate and its variations, as well as droughts, cyclonic storms, and coastal flooding.) Plate tectonic activity is the ultimate cause of many geologic hazards, including earthquakes and tsunamis, volcanic eruptions, and landslides; differential plate motions also concentrate fossil energy, mineral resources, and other Earth substances that sustain humanity and, in fact, all living things. Less obviously, drifting continents interact with the oceanic and atmospheric circulation patterns, causing global climate change over time, and thus influencing the course of biologic evolution. A wobbling planetary spin axis also contributes to climatic cycles.

Lithospheric plate boundaries are of three end member types: divergent, convergent, and transform. These are topographically represented, respectively, by spreading ridges (e.g., the Mid-Atlantic ridge), oceanic trenches and island/continent arcs (e.g., the Chile-Peru trench and Andes chain), and strike-slip faults (e.g., the San Andreas fault). Plate margins are the sites of generation of new oceanic crust, its destruction, the production of continental crust, and the displacement of exotic terranes. Such boundaries are also the locus of most natural hazards as well as the production of Earth resources. Plate interiors are relatively benign, geologically less exciting places, although the central portions of continents are subjected to harsh climatic extremes.

Fluvial landforms are geomorphic features produced by running water. Such types of topography (alluvial fans, floodplains, deltas) provide clues regarding the nature of ancient terrestrial climates. Inasmuch as landforms are constructed and modified with the passage of time – either being obscured by younger deposits, or being removed by erosion – geologically youthful landforms are the most helpful in deciphering past climates. Fluvial geomorphology can be fully understood only through the quantitative application of the concepts of hydrology. The solar energy–driven hydrologic cycle is the endless process (we hope) whereby evaporated ocean water is transferred along with latent heat to the various temporary storage reservoirs on land. On the land, H_2O is sequestered in biomass, continental ice caps and glaciers, groundwater, the unsaturated (vadose) zone, surface bodies, and streamflow. Geomorphic features attending this hydrologic cycle depend as well on weathering and erosion. The result of these combined processes is a continuously evolving landscape in dynamic equilibrium with the agents of both construction and destruction.

Anthropomorphic impacts on the naturally produced topography are a function of the increasingly widespread and intrusive practices of mining, farming, grazing, logging, and urbanization, among others. Change is obvious. Better land management will require that developers and conservationists recognize and utilize the hydrologic principles underlying the formation of fluvial landforms.

The geosphere, hydrosphere, atmosphere, and biosphere come together and interact in a thin, superficial film at the Earth's surface known as soil. Consisting mainly of weathered bedrock, but also containing important amounts of organic materials and minor volatile components extracted from the more fluid envelopes of the planet, soils provide the nutrient base for all terrestrial vegetation. Because land animals ultimately are dependent on botanic cover for sustenance, virtually all nonmarine life exists because of the presence of soils. The world's crops are grown from soils; hence, humans are critically dependent on this thin, ephemeral, fertile layer.

Soil is produced from solid rock by chemical reaction with atmospheric gas components, water (solution and/or precipitation), and organic acids, and by mechanical disaggregation attending heating-cooling, mass slumpage, and wind, glacial, or stream abrasion. A typical soil profile, from the surface downward, consists of: zone A – a humus-rich layer, leached of iron, aluminum, and manganese oxides; zone B – a clay- and iron and manganese oxide – rich layer; and zone C – progressively fresher, partly disaggregated rock. The type of soil formed reflects the mineralogic and chemical composition of the parent rock, the attending climate, and the duration of the weathering process. Most soils require thousands of years to develop; however, accelerated human activities result in the progressive erosive loss of topsoil, buildup of contaminants (salts, insecticides, herbicides), and leaching of important nutrients on a much shorter time scale. Among the wide range of Earth materials, soils are perhaps the most vulnerable to degradation through the diverse activities of humans.

Because the mineralogies of ancient soils are a reflection of the paleoclimate, their presence, and that of sedimentary strata containing transported soil materials, provide methods for evaluating the Earth's past surface conditions. Results are independent of, but generally compatible with, temperature data obtained by isotopic analysis of atmospheric gases trapped in ice cores or amber, as well as by other geochemical techniques.

THE HYDROSPHERE

Seawater covers approximately 70 percent of the Earth's surface; its abundance is undoubtedly the most distinctive aspect of our planet, which we may appropriately call the blue planet or the watery planet. Most of the Earth's H_2O is present as seawater (97 percent); views from space clearly show the polar ice caps (2 percent), but not the subequal amount of water stored underground in porous rocks (1 percent). Water residing in lakes, rivers, the atmosphere, and the biosphere, while absolutely crucial for life, is volumetrically negligible. Volatiles present in the hydrosphere and atmosphere have been derived from the degassing of the planetary interior. Degassing may be continuing on a small scale today; however, to a first approximation, the H_2O sequestered in the hydrosphere and cryosphere (ice caps) is a fixed quantity. It is, however, involved in a dynamic equilibrium, with solar radiation absorbed in the equatorial oceans causing net evaporation and heat absorbance, and poleward and landward transport of H_2O, followed by net precipitation and a release of heat. Some of the condensed water seeps into the ground, recharging groundwater reservoirs, and some flows down the land surface and back to the seas. This thermally driven hydrologic cycle generates bodily currents in both oceans and atmosphere; these currents transport the H_2O and its contained heat to higher latitudes, and to the continents, where the water is condensed as precipitation and ultimately returns to the global hydrosphere.

The length of time water is present in any specific reservoir is a function of the geologic environment, and accordingly is enormously variable. However, the mean residence time (MRT) for a particle of water can be approximated as follows: the MRT for atmospheric moisture is 10 days; for seawater, 1000 to 3000 years; and for groundwater, 10,000 years. We can quickly detect the effects of atmospheric pollution; however, we have progressively greater difficulties assessing anthropogenic influences on the global oceans and the subterranean water that occur in porous rocks. The impact of human activity on bodies of water small enough to show the effects of uncontrolled industrialization and agribusiness (e.g., Lake Erie) are disquieting; however, increased awareness and regard for the environment can result in a partial to virtually complete restoration (again, Lake Erie).

Radiant energy from the Sun powers the surficial circulation of both hydrosphere and atmosphere, in the process bodily carrying heat poleward from the tropics, and transferring moisture from the oceans to the land surface via the atmosphere. This is the life-giving hydrologic cycle, characterized by both advection (mass and energy transport) and convection (differential density–propelled motion). The thermally driven motions of the liquid and gaseous fluids enveloping the Earth define enormous cells, the circular motions and dimensions of which are controlled by the rotation and size of our planet, a fact that can be demonstrated by comparing the six main atmospheric belts on Earth with the plethora of circulation zones surrounding Jupiter. Ocean surface currents and gyres are a function of the prevailing wind shear and, like the dynamic patterns of atmospheric motion, are a reflection of the Coriolis effect. Of course, the sizes and dispositions of the ocean basins (regions of extremely low albedo) and the polar ice caps and/or sea ice (regions of very high albedo) influence the nature of the near-surface ocean currents and airflow as well. Local upwelling of nutrient-rich deep-sea water is a complex function of the above factors, and is largely responsible for the biomass proliferation that sustains the marine web of life.

In general, the hydrosphere is well stratified, mixing/overturning on an approximately 1000 to 3000-year cycle. The layering is manifested in terms of contrasts in water temperature, salinity, and density – the latter two quantities increasing gradually downward. The temperature of near-surface ocean water approaches the mean ambient value in the approximately first 100 meters (i.e., to the base of wave and current activity), then declines from about 10 to 3.5°C at a depth of approximately 1000 meters; in the deep sea, the water temperature is essentially constant (liquid H_2O achieves its maximum density at 3.5°C).

The surface flow of ocean currents is clearly related to poleward transfer of solar heat preferentially absorbed by the tropical seas, with gyres related to both planetary rotation and locations of the continents. Deep-sea flow is far less obvious, but involves immense circulating cells in all the major oceans. Although contrasts exist among the deep-sea current regimes in the Atlantic, Pacific, Indian, and Southern oceans, differential motions are a consequence of contrasting densities of the various water masses. Relatively dense water sinks, whereas relatively light water buoyantly rises. The density of H_2O, in turn, is a reflection of both temperature and dissolved constituents: the colder and saltier the H_2O, the greater its density. Another process adding to the complexity of deep-sea circulation and overturn is the phenomenon of caballing, whereby the mixture of two water masses of contrasting temperatures and/or salinities results in a water density that is greater than the weighted mean of the two initial densities.

The freezing and evaporation of saline water produces nearly pure H_2O ice and water vapor, respectively; in either case, the residual aqueous liquid left behind thereby becomes enriched in dissolved salts. Thus, in polar regions where sea ice forms, and in low-latitude seas of restricted circulation, highly saline, dense waters are produced. In the Arctic and Southern oceans, the resultant brines are also cold, hence doubly dense, whereas in the eastern Mediterranean, Red Sea, and Persian Gulf environments, the salty water is warm. Accordingly, westward outflow of deep-sea water through the Straits of Gibralter produces an intermediate-level water mass in the North Atlantic (salty but warm). The coldest, densest current of all, of course, is the Antarctic Bottom Water. Overturn of these oceanic cells is slow by comparison with sea surface gyres – on the order of 1000 years, as indicated by oxygen and nutrient contents.

Our blue planet is distinguished from its solar system neighbors by the presence of a globe-encircling hydrosphere. Life apparently sprang from seawater, and the hydrologic cycle is a necessary requisite for sustenance of the biosphere. Circulating ocean currents play an enormous role in the transfer of solar energy from equatorial to high latitudes, moderating the climate of the Earth in the process, and ensuring its habitability. The oceans are salty (approximately 3.5 percent by weight) and seem to have been of approximately constant salinity over the sweep of geologic time. Moreover, although the concentrations of the elements in seawater vary slightly, their proportions are constant. How and why is this so?

Rapid mixing is part of the answer. Fresh waters draining the continents, and aqueous volatiles expelled from submarine and subaerial volcanoes and hot springs, contain minor amounts of dissolved substances; reaction of seawater with hot spreading centers also adds material in solution. The continuous evaporation of H_2O leaves these constituents behind in the residual seawater. However, with the exception of marine embayments with restricted circulation, nearly constant salinity is maintained through removal of dissolved salts, calcium and magnesium carbonates, silica, and other constituents through inorganic precipitation, biogenic growth of shell material, reduction-oxidation (redox) reactions, and adsorbtion on, or incorporation in, settling clay minerals.

The residence time of various chemical species in the ocean is a function of the rate of supply versus the areally adjusted rate of removal, and in nonanthropogenic cases, a dynamic steady state has been achieved. Chemical oceanographers monitor the circulation of the surface ocean and deep-ocean water masses by employing radiogenic and chemical tracers, and have calibrated the rate of circulation and overturn of water bodies, as well as the residence times for the various dissolved constituents. Defense- and industry-related disturbances of marine chemistry – chiefly the introduction of toxic metals, insecticides, and carbon dioxide, which is the end product of fossil fuel burning and deforestation – are altering the dynamic chemical/isotopic balance of the global hydrosphere and atmosphere. How these perturbations will play out in the future is yet to be quantified.

Geologically speaking, however, we know that higher carbon dioxide levels track with warmer paleoclimates.

THE ATMOSPHERE

The Earth's atmosphere is unique in the solar system on at least two counts: It has a peculiar, biogenically mediated composition, 78 percent nitrogen, 21 percent oxygen, and 1 percent argon; and it is far removed from chemical equilibrium with the condensed portions of the planet, being too oxidized and nitrogen-rich. Consisting principally of simple diatomic molecules and monatomic noble gases, the terrestrial atmosphere does not strongly absorb visible sunlight, and therefore is of remarkable clarity. Water vapor and other less abundant light absorbers such as carbon dioxide, methane, and nitrous oxide, as well as traces of particulate matter (aersols, dust, and soot), provide clouds and obscuring haze.

Variation in density of the atmosphere is a function of its total mass and of the intrinsic properties of temperature and pressure. As altitude increases above the Earth's surface, all these properties fall off: in the absence of moisture, the dry lapse rate is approximately 10°C per kilometer, whereas in humid, cloudy regions, it is closer to 6°C per kilometer. Differential densities signify contrasting buoyancies of adjacent air masses, resulting in mixing of the lower atmosphere, or troposphere. The overlying stratosphere is sited between 8 to 16 kilometers and approximately 55 kilometers; its top is heated by solar energy; consequently, cooler air parcels below become neutrally buoyant and flatten out in this zone. Upward, the stratosphere grades into the mesosphere, which in turn grades upward to the thermosphere at altitudes exceeding 85 to 100 kilometers. These latter two regions are characterized by increasingly energetic ultraviolet (UV) radiation, ionizing the very tenuous atmosphere, and in aggregate producing the ionosphere (where the solar wind interacts with the Earth's magnetic field to produce the aurora). The "top" of the atmosphere is an ill-defined boundary between approximately 100 and 400 kilometers, where the extremely thin atmosphere passes by degrees into the interplanetary medium of outer space.

Stratospheric ozone (O_3) is produced and destroyed by interaction with UV radiation from the Sun. Under steady-state conditions, most of these energetic rays are absorbed in the stratosphere and do not reach the terrestrial surface. Because such radiation is harmful to life, the sustained presence of ozone in the stratosphere acts as a shielding layer and increases planetary habitability. Ironically, ozone is itself a poisonous gas and attacks organisms; produced in minor amounts in automobile exhaust, the ground-level ozone problem had been effectively eliminated through modern pollution controls. However, ozone in the stratosphere has declined substantially over the past 15 years, and a continent-sized hole has opened over Antarctica. This phenomenon reflects the upward diffusion of human-generated free radicals – chemicals such as nitric oxide and chlorofluorocarbons – into the strtatosphere where, on reaction with UV light, they decompose and strip oxygen from ozone, converting it to oxygen. Fortunately, the Montreal Protocol of 1989 produced an agreement among the industrialized nations that has curtailed the production of such chemical species; over the long term, the ozone content of the stratosphere should ultimately regain its previous balance.

Circulation of the atmosphere is a consequence of pressure differences of neighboring parcels of air, which are in turn functions of the ratio of solar energy absorbed to terrestrial energy radiated back out to space. Motions of the gravity-driven air masses are also affected by rotation of the planet. With the spinning Earth as a reference frame, the Coriolis "force" deflects flowing air masses to the right in the Northern Hemisphere and to the left in the Southern Hemisphere; the magnitude of turning is directly proportional to the velocity of the moving stream of air and its latitude.

The greenhouse effect is the phenomenon whereby polyatomic gas molecules – for example, H_2O (water vapor), CO_2 (carbon dioxide), CH_4 (methane), N_2O (nitrous oxide) – absorb incident solar radiation, heat the atmosphere, and reemit energy in all directions. The Earth is in a steady state between the incoming solar energy and that returned to space – otherwise, the planet would either warm or cool over time, depending on whether absorbed energy exceeded or was less than that emitted back to space. Greenhouse gases provide an important thermal blanketing effect: In the absence of our atmosphere, the Earth would become the frigid "white planet," with oceans frozen solid. In fact, because outgoing radiation is an exponential function of the effective surface temperature of the planet, whereas incoming solar energy is inversely related to the distance from the Sun and is virtually constant, a strong negative feedback mechanism exists that helps to prevent runaway heating. On the other hand, increasing amounts of H_2O, CO_2, NO_x, and CH_4 in the atmosphere provide a positive feedback whereby increased temperature of the Earth's surface would increase the volatilization of the hydrosphere and heighten greenhouse warming.

With regard to climate change, a major complicating factor is provided by cloud cover. Clouds both reflect incident solar radiation back to space, and insulate the planetary surface from radiative loss. How increased global temperature will influence the proportions and nature of the cloud cover is not yet well understood; moreover, whether changes in cloudiness would help ameliorate or exacerbate global climate change is also unknown. In any event, anthropogenic modification of the atmosphere's composition is underway – whether we like it or not, we are running the experiment in real time.

For the well-being of humanity and, in fact, the biosphere in general, it is crucial that Earth systems scientists be able to quantitatively assess the causes and effects of climate and to predict accurately impending local-to-global trends. Economists possess 20–20 hindsight when it comes to analyzing the performance of the New York Stock Exchange. However, investors could markedly improve their personal financial situations if their brokers were able to predict near-term, future stock performance. Similarly, with regard to the Earth's surface conditions, scientists can readily account for prehistoric climatic variations but are much less certain concerning future trends. Is global warming in progress, or are we about to enter a new Ice Age? Despite a great deal of research, the answer to this question is still unclear. However, we cannot even hope to understand future changes if we are unaware of the paleoclimatic history of the Earth.

It is comforting to realize that liquid water has characterized the surface of our planet throughout the entire geologic record – the past 3.9 billion years. Evidently, the terrestrial environment has not ever departed from the 0 to 100°C temperature range required for the stability of liquid H_2O. However, within this range, gradual, seemingly nonsystematic variations in surface conditions were the rule; Precambrian time was, on average, warmer than the more temperate Phanerozoic. Glaciation occurred episodically, at the end of Archean time, in the middle and near the end of the Proterozoic Era, during Permian time, and in the Neogene (the Miocene, Pliocene, and Pleistocene epochs). Great intervals of ubiquitous warmth occurred throughout the Precambrian, during several periods in the Paleozoic, during the Cretaceous Period, and even as short-lived events within the Pleistocene Epoch. Geochemical and isotopic data support the correlation of increased atmospheric carbon dioxide and methane levels with warm, humid times, and low concentrations of these greenhouse gas components in the air during dry, cold events. None of the evolutionary changes, however, occurred as rapidly as the anthropogenic buildup of such greenhouse gases coincident with industrialization, involving the massive combustion of fossil fuels and destruction of the world's rain forests. The current speed with which carbon dioxide, methane, nitrous oxide gases, chlorofluorocarbons, and other atmospheric species are increasing may result in surprising and unexpected disturbances of the Earth systems – only the passage of time will tell.

Accurate prediction of global climate change, as distinct from local weather patterns, is a complex function of numerous forcing factors. These latter consist of positive and negative feedback mechanisms that, in concert, have resulted in periods of long-term thermal stability, and in progressive warming or cooling of the climate within rather narrow temperature limits. Some forcing factors are external, residing outside the system, and not influenced by the planet, whereas others are internal, reflecting the system's response to a perturbation. Early in Earth history, the impacts of incoming asteroids, the decay of initially abundant radioactive isotopes, and the inward migration of the iron-nickel core all released heat within the surficial to deep portions of the Earth. However, by far the greatest contribution to the planetary heat budget and surface conditions always was – and still is – radiant energy supplied by the Sun.

Solar radiation impinging on the planetary surface is a function of the constancy of the energy source, the Earth's latitude, the ellipticity of the Earth's orbit, the inclination of its spin axis, and the transparency/opacity of the interplanetary medium. The terrestrial albedo (reflectivity) is related to the proportion of the Earth's surface covered by water, land, and ice, as well as the extent and thickness of cloud cover. The efficiency of energy absorbtion is itself a function of these factors, as well as the atmospheric concentration of polyatomic (greenhouse) gas molecules and particulate matter such as aerosols and volcanic dust. Terrestrial heat transfer is accomplished by advection and convection of the ocean currents and atmospheric circulation. These dynamic fluid flows are ever-changing because of the evolving nature of the surface distribution of the Earth's continents and ocean basins – itself a reflection of plate tectonic processes. Currents in the oceans and the atmosphere effectively transport thermal energy poleward from the tropics, and provide a net transfer of both heat and fresh water from the oceans to the land masses.

Because of the complexity of the natural climate system, the influence of anthropogenic factors must be evaluated quantitatively. Unfortunately, the global climate is a function of a large number of intimately interactive components, some of which are as yet poorly understood. What is clear is that human additions to the atmosphere, such as increased greenhouse gas emissions, aerosols, and dust, overall contribute to elevated temperatures. The sign of the perturbation is well known; only the magnitude of the effect is debated. Moreover, such human-induced changes are set in a moving baseline of natural climatic fluctuations, as recorded in the terrestrial rock record over nearly 4 billion years, making it difficult to ascertain what is anthropogenic in origin and what is not. What we do know is that these anthropogenic modifications are occurring at an accelerated rate, far faster than the stately geologic pace of past climate changes.

THE BIOSPHERE

Biodiversity is a measure of the total range of genetic material present in an area, whether viewed at the local or global scale. The concept encompasses organic variation in terms of numbers of distinct species, population sizes, and richness of organismic communities. Terrestrial biodiversity is a measure of the health and viability of life on this planet. We recognize that habitats and the occupying organisms are in dynamic equilibrium, with the overall environment in flux as a result of both natural and anthropogenic changes; the former take place imperceptibly, under long-term, nearly geologic rates, whereas the latter occur rapidly, during brief sociopolitical time intervals.

Organic evolution is the gradual change in plant and animal life resulting in the transitional emergence of new types of life, and the dying out of the old. This progression is an aggregate consequence of the rates of speciation, extinction, and radiation (colonization). Speciation occurs when populations of progenitor organisms fail to interbreed as a result of some form of isolation (e.g., geographic, ecologic, temporal, behavioral), resulting in environmentally induced population contrasts. Over a sufficiently long time interval, this process gradually produces distinct life forms that cannot interbreed because of fundamental genetic differences. Extinction is a reflection of unequal capacities for adaptation to a changing environment, resulting in unequal competition and the failure of the declining species to perpetuate their genetic material in the reproductive pool. Rabbits are successful competitors not because of superior intellect, fuzziness, or even fleetness of foot, but because they produce at least as many offspring that reach reproductive age as their parents, instead of prematurely falling prey to cougars and coyotes. Radiation (colonization) is the gradual diffusion of species into hospitable ecologic environments. Evolution is the sum total biologic reflection of speciation plus colonization minus extinction. The processes are most readily comprehended in the study of a geologically youthful, simple island community because the biodiversity of such a site is severely limited, as is the opportunity for colonization – thus removing many variables. Although marine and continental biodiversities are far more complex, these communities appear to conform to the same evolutionary principles.

The genetic characteristics of a species may gradually evolve over many generations as a reflection of environmental pressures. Such adaptation results in enhanced survivability of the species within its total environment or niche (i.e., the physical surroundings, as well as interaction with the rest of the biosphere). Competitive exclusion is the phenomenon whereby two organisms possessing very similar competitive needs/profiles cannot simultaneously occupy the exact same niche: Because the more efficient species will outcompete the other for food and in reproduction, the less well-adapted species ultimately will be eliminated. In order for niche overlap to occur, other aspects of the environment must not be shared. Alternatively, small but increasing differences between two types of closely similar organisms may be accentuated in cases where the contrasts contribute to the long-term viability of both species.

The food web consists of several levels. Primary producers utilize solar radiation or chemical energy in order to generate their biomass. Herbivores graze on primary producers, primary carnivores eat herbivores, and secondary carnivores may dine on primary carnivores. Omnivores are not so choosy, and consume a range of dietary types that may include primary producers, herbivores, and/or carnivores. Scavengers process dead and decaying biomass. The mass conversion of plants into herbivores and prey into predator is approximately 10 percent; hence, the biomass of primary producers is no less than ten times that of herbivore body mass. Similarly, the total biomass of primary carnivores must not exceed approximately 10 percent that of herbivores – the food pyramid has a very broad base and low slope indeed! Biologic communities consist of a range of species dependent on the food chain. The larger the community in terms of numbers of species and/or numbers of individuals, the greater the stability during times of environmental stress. Where changing conditions disfavor some groups of organisms, other species invade their niches, and the total biomass may remain nearly constant. When the ecologic structure is modified or destroyed by a natural event or human activity, a succession of communities replaces the earlier configuration over time, until a stable climax community is produced. The climax community itself is a function of the nature of the original disturbance, reflected in the surviving pool of colonizing species.

Drifting continents and climate change over the course of geologic time have influenced the extent and direction of biologic evolution through environmental pressures, geographic isolation, and the introduction of species. Most such natural changes have taken place slowly enough that entire ecologic zones could diffuse more or less as a unit. However, because of the substantial habitat fragmentation that has resulted from human activities, rapid migration and/or adaptation now may not be possible for many species. As a consequence, disassembly of well-integrated biologic communities is probable, with attendant extinction cascades. Biodiversity is declining, and this process will surely accelerate as humanity further influences the solid Earth, the hydrosphere, and the atmosphere. Awareness of the accelerating and destabilizing loss of biologic diversity is the first step toward remedial action to restore the interactive Earth systems to a healthier configuration.

Water is absolutely essential for terrestrial and, obviously, marine forms of life; almost everywhere that liquid H_2O is present, biologic activity accompanies it. Three broad groups of organisms are recognized: Eucarya, Bacteria, and Archaea. Collectively, members of the first group are called eukaryotes, and have nucleated cells; the latter two groups constitute the prokaryotes, and are characterized by nonnucleated cells. Among the diverse types of life, autotrophs produce organic matter from inorganic carbon, whereas heterotrophs consume organic substance as an energy source in order to manufacture their cellular material. Individual members of both types function as aerobic or anaerobic organisms, depending on whether or not they utilize oxygen in their life processes. Interdependence of dissimilar organisms, or symbiosis, complicates and enhances the spectrum of biologic activity.

Microorganisms are essential players in the process of biogeochemical cycling (i.e., the paths followed by chemical species during their organic uptake, sequestration, and elimination); the chief elements incorporated in living cells are carbon, hydrogen, oxygen, nitrogen, sulfur, and phosphorus. We tend to think of macroorganisms when we consider biogeochemical cycling. However, although animals and plants (especially the latter) are important in the dynamics of carbon dioxide emission and fixation, microorganisms, chiefly prokaryotes, are responsible for almost all aspects of biospheric recycling of carbon, hydrogen, oxygen, nitrogen, sulfur, and phosphorus. Stratified sulfide ore bodies, coal, and petroleum (oil and natural gas) are but three of the economically significant resources whose origins are inextricably linked to microorganisms. Biogeochemical cycling, primarily by the prokaryotes, has profoundly influenced the course of

Earth history, causing a reduction in atmospheric ammonia, methane, and carbon dioxide, and an increase in the concentration of oxygen over geologic time. These and other biogeochemical trends are recorded in sedimentary rocks in the contained remains of the microbes themselves (impressions, shells, biomarkers – biogenic chemical species), and in the fractionations of stable isotopes of the light elements.

We are living in the Carbon Age. The Industrial Revolution required an abundant, inexpensive source of energy, and fossil fuels (hydrocarbons) have effectively served this purpose worldwide for nearly two centuries. An inevitable consequence of the burning of coal, petroleum, and natural gas is the production of atmospheric carbon dioxide. Global deforestation represents a substantial secondary source of such carbon dioxide emissions. In addition to human-induced exhaust to the atmosphere, carbon dioxide is increasingly dissolved in seawater, and another fraction is fixed by growing plants; it is disquieting to realize that a significant amount of anthropogenically produced carbon dioxide is as yet unaccounted for. This so-called missing carbon is being stored in a terrestrial reservoir (possibly in the deep oceans); as this excess carbon dioxide continues to accumulate and eventually equilibrates with the atmosphere, accelerated climate change may take place.

How does the well-documented increasing carbon dioxide content of the atmosphere affect terrestrial ecosystems, particularly land plant productivity? The problem is complex, and the answers are not clear; however, some tentative conclusions can be reached based on experiments utilizing well-monitored ecosystems under controlled, increased partial pressures of carbon dioxide. Although separate plant species respond differently, all exhibit a degree of heightened growth rate, sequestration of carbon, and H_2O retention. Modifying factors include elevated temperatures caused by the greenhouse effect, changing organic contents of the soil as a function of temperature, and changing patterns of precipitation and moisture retention, in turn affecting the extent of grazing by heterotrophs. What seems certain is that ecologic modification will occur in response to the steadily increasing carbon dioxide content of the atmosphere. Whether agricultural crops will be favored or disfavored by the evolving conditions is not at all clear; however, dislocations are never without social and economic costs.

Ours is a dynamic planet, with inexorable changes proceeding at various speeds throughout the solid Earth (infinitely slowly), the hydrosphere, the biosphere, and the atmosphere (extremely rapidly). These are interactive, interdependent systems, and none is more sensitive to the total environment than is the web of life. Individual species and the biosphere in aggregate can approach, but not exceed, the carrying capacity of available ecologic niches. The globe's ability to support life must reflect evolving planetary conditions, and increasingly – especially since the rise of civilization and, most particularly, the Industrial Revolution – the impact of humanity on the environment has been substantial, pervasive, and growing both in the nature of the alterations and the speed at which these changes are taking place.

Anthropogenically induced changes in populations of organisms and climate will require surviving species to migrate, causing a gradual decomposition and reconstitution of the diverse biologic communities. Because of the increasing fragmentation of ecologic niches caused by human activities such as urbanization, manufacturing, commercial fishing, deforestation, mining, and farming, the dispersal of genetically varied organisms as an evolutionary attempt to optimize habitability will be impeded. Accordingly, trends already under way will be exacerbated, resulting in further decline in terrestrial biodiversity. The crucial point here is that although the web of life is in a state of constant flux, the changes now in progress are occurring at unprecedented rates.

RESOURCE USE AND ENVIRONMENTAL TECHNOLOGY

The number of individuals of a species is a composite function of the rates of birth, death, and migration. For humanity (ignoring migration), three gradational stages of population development are recognized: (1) a high birth rate proportional to a high death rate; (2) a high birth rate far exceeding a low death rate; and (3), a low birth rate proportional to a low death rate. Stages 1 and 3 are typified by roughly constant numbers of individuals; the total number of individuals in stage 3 massively exceeds that of stage 1, whereas stage 2 is characterized by exponential population growth. Most industrialized nations have achieved the so-called demographic transition to stage 3; however, the Third World is still largely in the throes of stage 2. The global population, currently 5.6 billion, is therefore growing rapidly – a net increase of approximately 90 million per year, the equivalent of the population of Mexico – and is expected to reach 10 or more billion within the twenty-first century.

This burgeoning global population generates an environmental stress in some ways comparable to the terminal Cretaceous event responsible for the widespread extinction manifested in the biosphere, including the dying out of the dinosaurs and many other forms of life. The bolide impact 65 million years ago was virtually instantaneous; on a geologic time scale, the modern anthropogenic influence on the environment is scarcely less abrupt, and equally profound. The alteration of ecosystems has been pervasive and intense, as mirrored by the well-documented and accelerating decrease in terrestrial biodiversity. As the habitability of the planet changes, we are also modifying the ultimate carrying capacity of the Earth: nonrenewable natural resources, such as mineral deposits and fossil fuels, are being extracted at an increasing rate; even many renewables (e.g., groundwater, topsoil, forests, marine fisheries) are being utilized far more rapidly than they can recover, hence are being depleted – some, perhaps beyond recovery.

Although population growth is the principal cause of the severe degradation of the natural environment through farflung, intrusive human activities, the level of consumption is a critical factor to be reckoned with. Greater extraction of natural resources causes a more substantial impact on the environment; hence, the industrialized western nations (including some Pacific Rim countries) have a far larger negative influence per capita than do the peoples of developing nations.

Mineral resources are the naturally occurring Earth materials used by humans for industrial, agricultural, and related developmental purposes. Conventionally, fossil fuels are assigned to a separate category. Mineral resources include crustal minerals, rocks, and unconsolidated surficial materials (e.g., iron ore, diamonds, building stone, gravel); they are of economic value, and concentrated substances are called ore, in areas where they can be extracted, beneficiated (refined), and sold at a profit. A resource is the world's crustal inventory of the specific commodity in question, whereas a reserve is an economically profitable ore deposit of the substance. In order for a mineral resource to possess value, its concentration or mode of occurrence must significantly exceed the global average; for example, we mine iron from banded iron formations in locations where its concentration is approximately 40 to 45 percent or more (the Earth's crust averages 5.8 percent iron). Geologic processes such as transportation in (and precipitation from), a geothermal aqueous solution, weathering and dissolution, evaporation of brines, deposition from a volcanic vent or hot spring, reaction between magma and carbonate-rich wallrocks, for example, result in the unusual accumulation of sought-after minerals. Because the richest and easiest to mine, near-surface deposits are worked first, we are now engaged in the development of lower-concentration deposits, typically located at greater crustal depths. In most cases, the minerals that are extracted contain large amounts of the desired commodity, for example copper in a sulfide mineral. Where concentrations decline below a specific threshold (the so-called mineralogic barrier), the sought-after element may be present as a minor constituent dispersed in the rock-forming minerals, and not as a separate, element-rich phase. In this case, the amount of effort required to concentrate the element is often prohibitive because of the energy costs in the refining process.

The extraction of mineral resources and processing of the concentrates, even for high-grade ores, commonly results in massive alteration of the environment. Although the areal dimensions of many mines are negligible, the hydrologic (surface runoff and groundwater) and atmospheric effluents can be quite deleterious to the environment because of the chemistry of the ore; for example, the smelting of sulfides produces vast amounts of locally or regionally dispersed hydrogen sulfide, and sulfur dioxide; these chemicals travel downwind as well as downstream, producing acid and toxic heavy-metal precipitation/fallout. Even where the Earth materials are not toxic, habitats may be profoundly changed; for example, dredge gold mining of river gravels irrevocably alters the nature of stream channels, and open pit mining of copper or iron ore bodies and coal seams leaves a massive topographic depression, as do large stone quarries.

Modern civilization relies on a vast array of mineral commodities, and the natural resource base is an important factor influencing the gross national product (GNP). Emerging nations are now increasingly utilizing their mineral resources in an attempt to raise their standard of living, although a commensurate degree of environmental degradation is inevitable. If a burgeoning global population is to achieve a better quality of life, it must make considerably greater efforts at recycling the diminishing complement of unextracted resources, and it must substitute relatively abundant materials for substances in ever-decreasing supply. As yet unrecognized types of mineral resources (unconventional deposits) must be discovered as well. Equally important, remediation of the environment will require that a larger fraction of the GNP of every nation be devoted to cleanup and that more environmentally benign mining and beneficiation practices be implemented.

Cheap, readily available energy is the hallmark of civilization – or, at least, the currency of a technologically advanced society. Since the Industrial Revolution began nearly two centuries ago, the world's technologically advanced nations have relied upon coal (now 27 percent), oil (40 percent), and natural gas (25 percent), to energize their economies. The less developed countries (80–85 percent of the global population and growing) employ animal power and biomass burning for most energy-consuming tasks. Fossil fuel energy deposits are derived from economic concentrations of decaying organic matter; however, the rate of natural production is far exceeded by the anthropogenic rate of consumption. Accordingly, with an expanding, industrializing global population, the Carbon Age will probably come to a close within the next century or two, because we will have exhausted the carbon resource base. In the meantime, atmospheric concentrations of carbon dioxide, the inevitable combustion product of carbon-based fuels, will continue their upward trend – and, forced by this phenomenon, global climate change will continue. The Earth's unexpended coal reserves are much greater than its remaining oil and gas deposits. In terms of pollution this is unfortunate, because the amount of carbon dioxide, sulfur dioxide, nitrous oxide, and toxic trace metals released to the environment per unit of derived energy is far greater for coal than for liquid petroleum or for natural gas (the cleanest of the hydrocarbons). Methods to clean up coal are now available but add to the energy cost, and none will moderate the liberation of carbon dioxide into the air during coal burning. Climate change seems unavoidable as long as civilization relies on carbon-based energy.

Alternative energy sources include very inefficient and dirty – but inexpensive – biomass burning, along with a host of high-tech power systems: nuclear fission, nuclear fusion, hydroelectric, geothermal, wind, solar-voltaic, solar-thermal, ocean-tidal, ocean-thermal, and chemical (hydrogen oxidation). With the exception of nuclear fission, all these systems represent renewable energy sources in the sense that the rate of human consumption presently is equal to or less than the rate of natural generation. However, hydroelectric and geo-thermal power tend to dwindle with time at any specific installation as a result of reservoir sedimentation and thermal drawdown, respectively. Perhaps the most exciting, and also practical, increase in energy availability lies in the realm of enhanced efficiency of existing systems. Order-of-magnitude increases in end-use power can be effected by employing known methods and technologies for increasing efficiency. We have the capabilities now to enhance the efficiency of production, transmission, and utilization of energy. In the long run, of course, the world population will be obliged to come to equilibrium with total energy usage at a sustainable, constant rate.

Natural hazards pose a growing threat to humanity around the world, and are especially severe in the vicinity of lithospheric plate junctions and along tropical coastlines. Stable continental interiors are distant from regions of active mountain building; consequently, the inhabitants there have little to fear from volcanic eruptions, earthquakes, and landslides. Temperature extremes characterize such environments; however with the exception of small-scale but intense cyclonic storms and episodic flooding, natural calamities are infrequent. In contrast, many continental margins and island arcs, located along convergent or transform plate boundaries, are the sites of ongoing crustal accretion and deformation; they are typified by unstable, seismically active volcanogenic crust subject to mass slumpage and avalanching. Such margins are also vulnerable to devastating tsunamis and to ocean-spawned tropical storms generated by heat transfer to the atmosphere from warm seawater. Most of the world's people inhabit the near-shore regions of the land; there, the population is rapidly growing. Accordingly, coastal flooding and the threat of rise in sea level as a consequence of global warming are additional impending dangers.

We can minimize loss of life, destruction of property, and environmental devastation through sensible civil organization, realistic development practices, and accurate, efficient warning systems. However, we must first arrive at a better scientific understanding of the various natural hazards themselves and their ultimate causes, identify unambiguous premonitory phenomena, and then educate society and our leaders so that we can take effective action to mitigate – and, in favorable cases, to eliminate – the adverse impacts of natural disasters.

SOCIETY, THE ENVIRONMENT, AND PUBLIC POLICY

Comprehensive scientific analysis of an environmental problem is a necessary prerequisite to making effective policy; however, achievement of the former does not automatically ensure the latter. Before discussing and implementing any program, a cost/benefit study must be undertaken, and the different environmental policy options considered in such a way as to maximize the benefit and minimize the cost to the populous involved. Finally, effective action can be initiated only if (1) the public has been made aware of the facts, (2) the environment can be substantially improved as a result of the proposed action relative to other options, and (3) a common policy is adopted by both private and public sector participants. Facts (the sciences) describe the world as it is, and are objective; values (judgments) define how it should be, and are subjective. Effective policy must be based on scientific facts, and the proposed actions need to be in conformity with the value judgments of those empowered to ensure and support its implementation.

What are the major stumbling blocks to the formulation and acceptance of a sound environmental policy? The problem must be widely acknowledged and the facts well understood. This is the arena in which scientists and engineers interface most comfortably with policy makers and the general public; however, thus far the results of their interaction have been relatively ineffective. Any cost/benefit analysis identifies "winners" (those who would benefit from institution of a proposed policy) and "losers" (those disfavored by the new policy). Every contested action has an associated cost – otherwise, implementation would be universally advantageous, and there would be no hindrance to action. However, when even a small sector of society is severely disadvantaged, it will struggle fiercely to prevent implementation of the policy. Conversely, when winners individually gain only slightly (even though in the aggregate, the gain is substantially more than the avoided loss), they tend to be apathetic in their support of the proposed action. This is especially the case where environmental degradation is regional or transnational; for example, a small number of developers might be severely disadvantaged by the imposition of regulations preventing further loss of endangered species, whereas the vast majority of humanity probably would enjoy only a small gain in terms of maintaining the health of the biosphere. This phenomenon is termed mobilization bias. Many view protection of the environment as a personal responsibility – and, of course, it is. However, in the absence of coherent, clear public policy, some individuals and institutions may fail to exercise their responsibilities in order to receive specific benefits or advantages, or may undertake conflicting courses of action. Although voluntary efforts represent a positive step, they do not ensure universal compliance with recognized environmental goals. Because of the accelerating and pervasive impact of human activities on terrestrial habitability, these various issues must be satisfactorily resolved if public policy is to be successfully implemented.

When everyone in general is responsible for the preservation of a valuable asset, but individuals or groups gain temporary advantage through detrimental, destructive consumption, the asset eventually will be destroyed. This is the so-called tragedy of the commons. International issues such as air quality, pollution in deep-sea, near-shore, and river waters, ocean fisheries management, preservation of wilderness regions, and protection of wildlife fall into this category. Some have clearly assignable economic values, whereas others represent intrinsic goods, and are less readily quantifiable. Given the complex nature of this political and socioeco-

nomic problem, what can an individual nation effect in terms of public policy actions to lessen the adverse impact of human activity on the global environment? With regard to the buildup of atmospheric greenhouse gases, specifically carbon dioxide, several policy options are available, among them: Refrain from any action (business as usual); institute a tax on carbon-based fuels; introduce controls on carbon dioxide emissions; institute human population controls; and carry out extensive reforestation.

How may we choose from among these (and possibly other) options? Business as usual would be the most sensible policy if atmospheric scientists could assure us that the measured accumulation of carbon dioxide in the air will not trigger global climate change; unfortunately for us, the opposite is the reality of the situation. According to most observers, global warming is already in progress. The greenhouse effect is quite well understood – only the magnitude and rate of the worldwide warming is debated. A carbon tax would be particularly effective in achieving the desired climatic stabilization inasmuch as it would provide an incentive for conservation of fossil fuel usage while providing funds to ameliorate the environment – all without the complex bureaucratic infrastructure required by command-and-control measures such as gasoline rationing) Combined with the institution of carbon fixation systems (e.g., reforestation) and the curbing of population growth (creating the attendant, commensurate demand for fossil fuel energy), such a policy would provide a mechanism for achieving the greatest possible reduction in carbon dioxide emissions at the lowest possible cost. It would also provide us with some insurance if global warming has a more significant negative effect than is currently anticipated. The downside is that any nation unilaterally expending capital in excess of that necessary for its own well-being could realize a competitive disadvantage in the world marketplace. For this reason, it is important to assess accurately the magnitude and national impact of the greenhouse effect – a calculation not yet possible with current science and technology.

How has local land management influenced the global Earth system? Humans have, to a greater or lesser extent, modified the environment over approximately half of the ice-free land surface in order to produce food, fiber, shelter, energy, and the raw materials of commerce. Human activities include draining wetlands, irrigating dry areas, deforestation, mineral resource extraction, and urbanization – the last particularly along seacoasts and inland waterways, but also including encroachment of cities onto prime agricultural land. The magnitude of the alteration reflects the growing human population; the area under cultivation has increased almost fivefold during the past three centuries. Land use intensification and change will undoubtedly continue into the future, causing fragmentation of natural ecosystems, soil erosion, pollution of land, air, and water, mass extinctions of species, desertification, and increased greenhouse gas emissions.

In addition to the increase in atmospheric carbon dioxide caused by logging and burning trees, deforestation results in a net loss of nitrogen and organic matter from the soil. The resultant albedo increase influences solar energy absorption, heat transfer, and evaporation and runoff of H_2O. As a consequence, the soil becomes impoverished in nutrients, is less able to retain moisture, and therefore, is not as fertile. In some extreme but common cases, desertification is a consequence. In others, recovery due to natural reforestation or managed afforestation can be severely retarded because of this induced infertility. Intensive agriculture and paddy agriculture tend to deplete the soil in organic carbon while enriching it in nitrogen and pesticides/herbicides; where irrigation is practiced, increased soil salinity or waterlogging may result. Overall, deforestation and intensive agriculture contribute approximately 25 percent of the yearly anthropogenic carbon dioxide to the atmosphere; these activities also constitute the main source of the global increase in nitrous oxide levels. Paddy agriculture and cattle raising account for most of the annual increase in methane. Power generation, manufacturing, and transportation are responsible for most of the carbon dioxide added to the atmosphere; therefore, their use could conceivably be diminished through technology and modified human behavior. However, a growing world population will require sustenance. Because inherent momentum will swell the human population for generations even if we successfully curb the rate of reproduction, more widespread, intensified land use must be anticipated as we begin the twenty-first century.

Currently, civilization utilizes nearly 40 percent of the global net primary production of foodstuffs from land ecosystems, leaving less than two-thirds of this terrestrial food productivity to sustain the rest of the nonmarine biosphere. Direct human consumption, livestock feed, and conversion of the land from original habitats to modified agricultural and nonagricultural activities are responsible for this remarkable, if relatively recent, utilization pattern. Where personal income rises, people spend a smaller proportion of their money on food but switch to higher-quality diets, typically richer in meat. The amount of grain required to obtain the same energy intake from livestock as from cereals is approximately 2:1 for poultry and 8:1 for grain-fed beef. As impressive as these ratios are, the amount of water needed to produce meat versus vegetable matter is even more compelling. Thus, increasing affluence exacerbates the need for grain production to a level far above that required for basic subsistence. Humans are now consuming approximately 40 percent of the global net primary production. What will be the impact on the web of life when the world population has doubled – as it will in the next fifty years?

Since the late 1960s, the so-called Green Revolution has resulted in dramatic increases in global crop yields of cereals (rice, wheat, and corn), with little concomitant expansion in the total area of land under cultivation. This feat has significantly reduced world hunger, and was accomplished through the introduction of chemical fertilizers, pesticides, herbicides, and genetically engineered grains. Among the latter are new strains of crops characterized by a brief growing season, rapid maturation, short, stout architecture, ease of uptake of water, and hardiness (resistance to physical damage, temperature changes, crop diseases, and pests). The global decrease in chronic undernourishment has helped to accelerate population growth. The question now is: Can the spectacular gains in food productivity over the past thirty years be duplicated in the first decade of the twenty-first century? If not, famine will return to Asia and South America, and it will persist in Africa. Government policies (subsidies) typically are attempts to increase agricultural output. However, food producers are responsive to personal cost/benefit analyses, and not to environmental externalities, such as health and remediation costs associated with regional and transnational pollution. Accordingly, environmentally benign farm policies must take into account global tradeoffs, not simply food production; these policies will be difficult to implement because farms and agricultural expenses are geographically contained within national boundaries, whereas their effects have global consequences.

Life on Earth sprang from the sea, and water is the most indispensable of all biologic resources. Civilizations are created assuming the adequacy (or, at least, ready availability) of a sufficient water supply for their future. In arid regions of the planet, water clearly represents the ultimate limit to growth. In order to increase the carrying capacity of the land, the human population has constructed dams, storage reservoirs, and aqueduct systems, and has withdrawn a large amount of groundwater. However, the waters stored underground and on the surface are a strict function of precipitation on the land – the hydrologic cycle – and we must eventually achieve a steady-state configuration regarding its usage. Given that only a finite amount of water will be available on a long-term basis, per capita consumption can be decreased through regulation, pricing, and technologic fixes.

In the United States, federal and state governmental adjudication of water rights assures the equitable distribution of this resource, depending on land ownership and/or prior usage, and consonate with the need for compliance with both Clean Water and Endangered Species acts. The goal is beneficial usage for society as a whole. To the extent that the government designs, constructs, and maintains mammoth water diversion and delivery facilities, it possesses adequate control over both allocation and consumption. Currently, those who propose new water projects must also demonstrated that those projects will not degrade existing bodies of water (lakes and rivers) in terms of water quality, wildlife habitat, recreational opportunities, and scenic beauty. Even for on-line water diversion systems, where severe and irreparable damage is being caused by excessive withdrawal of water, reallocations are being authorized in such a way as to minimize adverse environmental impacts.

In the past, the price of water in the United States has been heavily subsidized by the federal government, in an effort to spur growth. An unfortunate side effect of this policy has been inefficient, wasteful consumption because historically, water has been so inexpensive. As demand ultimately outstrips water supply as a result of a growing population and a finite amount of available water, conservation will become crucial. This can be accomplished largely through voluntary changes in usage if water prices rise toward the full, realistic (loaded) costs of providing the resource. Loaded costs include not only those of building and maintaining water projects, but also those of attendant environmental degradation. Although industrial and residential water usage could be reduced considerably, agribusiness is the major consumer of water in the United States (83 percent of total usage) as well as worldwide. Fortunately, there is great opportunity for reducing overall consumption through technologic improvements such as lining aqueducts and canals, transitioning to precision sprinklers or to drip-system irrigation, and sealing water conduits. Moreover, the recycling of agricultural runoff appears to represent the most promising way of extending the capacity of existing water systems to meet the growing needs both in the city and on the farm. Desalination is another option, but a rather expensive one because of the energy required and the transportation costs. In any case, humans will be obliged to come to an equilibrium with water supply in the foreseeable future.

Humans value the environment. The question is, how much? What are we willing to do – or to sacrifice – in order to preserve our immediate natural surroundings? How about protection of the global environment? Individuals have markedly different answers to these questions, and none are readily quantifiable. This is a reflection of the fact that nature is not traded in the marketplace; there is no objective dollar value assignable to a wild river or a species of bird. Clean air, because it is utilized by humanity, can be priced – at least hypothetically – after a comprehensive cost/benefit analysis of scenarios involving various airborne pollutants and their effects on terrestrial ecosystems.

How do we assess the value of nature? Conservationists cherish the entire, interactive biosphere, and the physical environment in which it exists. Publically popular activities, in contrast, include the defense and protection of various "charismatic" species such as whales, harp seals, bald eagles, and the like; however, the average person attaches little or no sympathy or value to the existence of skunks, flies, and mosquitos. To the devout environmentalist, of course, all terrestrial organisms are equally worth preserving; some would even argue that all species have an intrinsic right to existence without human interference. As scientifically compelling justification for this view, it is pointed out that, in aggregate, the biosphere represents the genetic portfolio of life and responds to natural and anthropogenic changes in the environment by evolutionary adaptation or extinction. As biodiversity is reduced, the options for future organic evolution become correspondingly restricted.

Two philosophies may thus be defined, if not quantified, concerning the evaluation of nature: intrinsic, and utilitar-ian. Accepting the former view, any and all aspects of the natural environment possess values greater than alternatives involving anthropogenic change. Employing the latter view, we can recognize various kinds of value: (1) use values include direct benefits (hiking in a redwood forest) and indirect benefits (saving the spotted owl), as well as consumptive benefits (quail hunting) and nonconsumptive benefits (bird-watching); (2) nonuse values include existence values (the Grand Canyon National Park exists), and optional use values (you could visit the Grand Canyon). The objective measurement of worth is exceedingly difficult to quantify, and is conventionally regarded as a willingness-to-pay value (how much is it worth to you not to dam a scenic river?). Clearly, the natural environment has been extensively altered by past and current human activities, and this process will continue into the future. The problem is one of proportion, balancing conflicting values, and subjective evaluations of perceived costs and benefits – a tremendously difficult but increasingly imperative task. Throughout history we have made, and in the future will continue to make, decisions that profoundly affect the complex Earth systems. Henceforth, such choices and tradeoffs will need to involve value judgments considered within the global framework.

A FINAL WORD

This book attempts to provide an integrated overview of the diverse processes that constitute our planet's interactive systems. We have tried to describe some of the incredible array of both positive and negative feedback mechanisms that constitute the dynamic, evolving equilibrium among the Earth's living organisms, its watery envelope, its atmosphere, and its not-so-solid geologic substrate. We have also laid out some of the daunting sociopolitical and economic issues facing humanity. If many of the relationships seem obscure, the mechanisms puzzling, the data fragmentary, and the conclusions tentative, this is at least in part a reflection of our present limited understanding. However, to the extent that this book gives you a fuller appreciation for the complicated and intricate interconnectedness that characterizes our planet, we will have at least partly succeeded. This book provides you with a springboard for deeper examination of Earth systems topics and issues, to spark your enthusiasm for further scientific and socioeconomic inquiry, and it teaches you the basic language and concepts of the many environmental subdisciplines for the purposes of effective information integration and communication. If it inspires you to contribute actively to ameliorating the damage we humans have caused to the physical and biologic environment, we will have succeeded completely. Your generation must accomplish far more than ours: You must leave the world a healthier place than you found it. The time is already late.

Glossary

abundant resource Natural resource whose concentration in the available parts of the Earth is very high; accordingly, deposits of economic value must be very rich in the resource in question, and significantly above the average terrestrial concentration. Silicon, aluminum, and iron are examples of abundant resources.

accretionary wedge Largely sedimentary units (but including fragments of oceanic crust) scraped off, or decoupled from, the downgoing lithospheric plate near a convergent junction, and laminated against the stable, nonsubducted plate. Also called a subduction complex.

acid deposition Impact of acidic components (nitrate and sulfate) from the atmosphere on terrestrial surfaces, either as dry deposited aerosols and particles or as acid precipitation (e.g., rain, sleet, snow).

acid rain Gas species such as carbon dioxide (CO_2) and sulfate(SO_3) dissolve in water and form acids through reactions of the sort $CO_2 + H_2O = H_2CO_3$ (carbonic acid), and $SO_3 + H_2O = H_2SO_4$ (sulfuric acid). When these and other noxious gases (e.g., nitrites, or NO_2) accumulate in the atmosphere due to the combustion of fossil fuels by industrial, vehicular, and home usage, rainfall becomes acid, polluting lakes and rivers.

adiabatic lapse rate The expression for the fall (or lapse) of temperature with increasing height that occurs when air parcels simply trade gravitational energy for heat energy, without any additional energy coming into or leaving the air parcels (adiabatic conditions).

adsorption The adherence of gaseous or aqueous species (molecules, ions) on surfaces of solids. Where internal interfaces are involved, the process is termed absorbtion.

advective heat transfer Movement of energy by bodily transport.

aerobe An organism capable of utilizing molecular oxygen (O_2) during metabolism.

aerobic photosynthesis Type of photosynthesis carried out by cyanobacteria and photoautotrophic (cells that manufacture biomass using light as an energy source); eukaryotes (nucleated cells) in which hydrogen from water is used to reduce carbon dioxide, and during which molecular oxygen is released.

aerosol An ultrafine suspension of liquid droplets and/or solid particles in a gas, such as the atmosphere.

age structure Refers to the number of people in various age categories in a population.

agglomerate Volcanic ash deposit containing volcanic bombs (see **ejecta**).

air mass A volume of air with reasonably uniform temperature and humidity, generally described by the nature of the land or ocean region determining its character: Maritime tropical airmasses are warm and moist, compared to mean conditions throughout the troposphere; arctic continental air masses are frigidly cold and dry.

albedo The percentage of incident sunlight bounced back, hence, the reflectivity of a substance.

albite, $NaAlSi_3O_8$. *See* **feldspar**.

algal reef A biogenic carbonate bank (i.e., a bioherm) formed by calcium carbonate ($CaCO_3$) secreted by colonial algae.

alkali feldspar A feldspar solid solution series, $(Na,K) AlSi_3O_8$, ranging in composition from albite ($NaAlSi_3O_8$) to orthoclase ($KAlSi_3O_8$).

alluvium Unconsolidated detrital material deposited by a stream or river. Alluvial deposits include midchannel bars, point bars, floodplain deposits, alluvial fans, and fluvial terraces.

almandine, $Fe_3Al_2Si_3O_{12}$. *See* **garnet**.

alpha (α) particle A radiation product that is actually a helium atom (two protons, two neutrons, two electrons).

amorphous Any solid or liquid that lacks an ordered atomic arrangement (crystal structure).

amphibole Any of a class of hydrous double-chain silicates typified by $(Si_4O_{11})^{-6}$ groups; common examples include anthophyllite ($Mg_7Si_8O_{22}(OH)_2$), tremolite ($Ca_2Mg_5Si_8O_{22}(OH)_2$), and hornblende ($NaCa_2(Mg,Fe^{+2})_4AlSi_6Al_2O_{22}(OH)_2$).

amphibolite A lineated, medium-grade metamorphic rock characterized by intermediate calcium-sodium plagioclase, hornblende, and, typically, epidote and/or garnet. (*See* Chap. 5 for descriptions and classification.)

anaerobe An organism that does not require molecular oxygen for metabolism.

anaerobic photosynthesis Type of photosynthesis carried out by photosynthetic bacteria in which hydrogen from a source other than water is used to reduce carbon dioxide. Molecular oxygen is not released.

andesite A volcanic rock – typically, lava – intermediate in chemical composition between basalt and rhyolite. (*See* Chap. 5 and Table 5.1 for descriptions and classification.)

anhedral Adjective denoting a mineral grain bounded by irregular surfaces, not crystal faces.

anion A negatively charged ion ($Z < e$), so called because in an electrolyte solution, negatively charged ions migrate toward the positive electrical terminal, or anode.

anion complex The covalent bonding of several anions about a central cation, which yields an overall negative charge to the structural subunit. Examples include carbonate (CO_3^{-2}), sulfate (SO_4^{-2}), and nitrate (NO_3^{-}).

anomaly A disparity from the anticipated relationship. In Earth science, most commonly used for geophysical departures from a projected field gradient, e.g., gravity, magnetic, seismic transmission velocity, etc.

anorthite, CaAl₂Si₂O₈. *See* **feldspar.**

anorthosite Nearly monomineralic felsic igneous rock consisting chiefly of calcic plagioclase. (*See* Chap. 5 for descriptions and classification.)

anoxic Reducing conditions, depleted in oxygen.

anthophyllite *See* **amphibole.**

anthracite *See* **coal.**

anthropogenic Refers to the influences of humanity (on nature).

anticline An arch-like fold in layered rocks in which the beds dip downward away from a central high.

anticlinorium A great, linear fold or upwarp, consisting of subparallel, dominant anticlines and subsidiary synclines. Basically a large anticline with superimposed minor synclines (folding at two different scales).

anticyclone Large, cylindrical circulations of air moving opposite to the Earth's rotation (hence, "anti-"). They move clockwise in the Northern Hemisphere; generally their center is a region of higher than normal pressure. The Bermuda Anticyclone, or Bermuda High, extends thousands of kilometers out to sea from the southeastern United States.

antimatter Negative substance – when combined with an equal amount of matter, produces energy but not mass. Antimatter was entirely consumed in the Big Bang.

apparent polar wander paths Presuming that a segment of the Earth's crust is fixed in its geographic location relative to the spin axis, the remnant magnetism of rocks of different ages yields an apparent wandering of the Earth's magnetic poles. We now know that it is the crust itself, not the Earth's magnetic field, that wanders. The magnetic field wobbles to some extent, and changes polarity episodically; however, the field is essentially a dipole coincident with the Earth's rotation axis.

aquifer A porous geologic unit that contains sufficient saturated permeable material to yield significant quantities of water to wells and springs.

arc–trench gap The geographic region between seaward oceanic deep and landward volcanic arc. It is the locus of the fore-arc basin.

Archaea A group of prokaryotic (nonnucleated) microorganisms whose members include all methanogenic bacteria, some halobacteria, some sulfur-utilizing bacteria, and some heat-loving (thermophilic) bacteria. Also called *Archaebacteria*.

Archeozoic Eon (or Archean) The time interval between about 3.8 and 2.5 billion years ago.

artesian (or confined) aquifer Where water completely fills an aquifer that is overlain by a confining bed, the aquifer is said to be confined. Such aquifers are known as confined aquifers or artesian aquifers; they are free-flowing at the surface of the Earth.

assimilation Any metabolic process that consumes a substance for incorporation into biomolecules.

asteroid Subplanetary mass of condensed matter in our solar system. Most are the size of meteorites (up to a few meters in diameter), and all are less than 1000 kilometers across. Most are confined to orbits between Jupiter and Mars – the so-called asteroid belt.

asthenosphere The soft, plastic portion of the upper mantle of a planet lying directly beneath the lithosphere. On Earth – the only planet for which such a weak, ductile zone has been recognized with certainty – the asthenosphere gradually passes downward into the mantle transition zone at depths of 220 to 400 kilometers.

atmosphere The gassy envelope surrounding the solid and/or liquid outer surface of a planetary body.

atoll An island characterized by a concentric, fringing coral reef.

atom The fundamental chemical particle that in sufficient numbers, displays all the characteristics of the element. An atom consists of a positively charged nucleus (protons plus neutrons) and an enveloping, negatively charged electron cloud.

atomic group Vertical column in the Periodic Table of the Elements (*see* Table 3.1). Each neutral atom in a particular group has the same number of valence electrons, but different numbers of electron shells.

atomic nucleus Central, massive portion of an atom, consisting of protons and neutrons.

atomic number Known also as Z, it is the number of protons of an element, and defines its chemical nature.

atomic series Horizontal row in the Periodic Table of the Elements (*see* Table 3.1). Each neutral atom in a particular series has the same number of electron shells, but a different number of valence electrons.

atomic weight The sum of the masses of the protons and neutrons (and electrons) of an element. For example, the common isotope of oxygen, ^{16}O, has eight neutrons and eight protons, which equals an atomic mass of 16.

attenuation Reduction in amplitude of vibrational energy or, in another context, thinning of solid material. Attenuation of seismic wave energy means reduced amplitude at the same frequency; attenuation of continental crust refers to the stretching or necking down of the continent by pulling it apart.

autotroph An organism that uses inorganic carbon (carbon dioxide or bicarbonate) as its major source of carbon.

axis of *n*-fold symmetry A rotation of $360°/n$ about an *n*-fold axis of symmetry results in self-coincidence. For example, a common pencil is a hexagonal prism, and possesses a sixfold axis of symmetry parallel to its length; ignoring the lettering on one side of the pencil, a rotation of 60° results in an aspect identical to that prior to rotation.

azimuth The angular direction from north measured on or parallel to the surface of the Earth. An azimuth of N60E is a direction 60° clockwise (i.e., east) of north.

bacteria A domain of prokaryotic (nonnucleated) microorganisms with a diverse array of metabolisms. Includes most familiar prokaryotic forms (e.g., *Escherichia coli, Salmonella, Bacillus*). Also called Eubacteria.

banded iron formation (or BIF) Well-stratified sedimentary layers of considerable thickness and great lateral extent. BIFs consist of chert and iron oxide, precipitated chemically in a stable shallow sea far from a source of clastic debris. Most such deposits are Early Proterozoic in age of accumulation. Economic iron ore deposits generally are secondary; that is, they form subsequent to sedimentation, due to the preferential dissolution of silica over time by vast quantities of circulating groundwater or hydrothermal solution. (*See* Chap. 5 for descriptions and classification.)

bankfull discharge The discharge corresponding to a water level at the crest of the stream bank.

basalt A dark, ferromagnesian lava typical of the upper part of the oceanic crust. Basaltic flows also occur in continental crust. (*See* Chap. 5 and Table 5.1 for descriptions and classification.)

base flow   The relatively flat portion of a stream hydrograph between storms; in small watersheds, the base flow is primarily due to discharge from the aquifer into the stream.

base level   The level below which a land surface cannot be reduced by running water.

bauxite   A rock composed principally of various amorphous or finely crystalline oxides and hydroxides of aluminum. A product of extreme chemical weathering of aluminum-rich rocks, and an important economic source of aluminum.

beneficiation   Concentration of an ore by one or more of several mechanical and/or chemical processes.

beta (β) emission   The process whereby a neutron emits an electron and is transformed into a proton.

Big Bang   The time of origin of the expanding universe, approximately 13 billion years ago. Matter and antimatter (whatever that was) were concentrated in a tiny clump characterized by incredibly high temperatures and pressures. Rapid expansion then ensued. What things looked like prior to the Big Bang is anybody's guess.

bimodal topography   Two contrasting elevation levels – on Earth, the high-standing continental platforms versus the low-lying oceanic basins.

binding energy   The subatomic forces that hold the atomic particles together in the nucleus.

biochemical sediment   Sedimentary accumulation of organically precipitated material.

bioclastic sediment   Sedimentary accumulation of shell material produced by organisms.

biodiversity   The variety of genes, species, and populations of plants, animals, and microorganisms, and the ecosystems they comprise. These include both the communities of organisms within particular habitats and the physical conditions under which they live.

biochemical cycling   The recycling of elements by organisms, in the context of geologic processes.

biogeography   Study of the geographic dispersion of organisms on planet Earth.

biomarker   An organic compound found in sediments that is diagnostic of a particular type of metabolism or group of organisms.

biosphere   The portion of the atmosphere, ocean, and crust that is inhabited by living organisms.

biome   The largest unit of terrestrial communities that is easily recognizable, e.g., tundra, temperate forest, tropical rain forest, desert, and temperate grassland.

biotite K $(M_a,Fe)_3Si_3AlO_{10}(OH)_2$.   *See* **mica**.

birth rate (or crude birth rate)   The annual number of births per 1000 individuals in a population.

bituminous coal   *See* **coal**.

block-fault mountains   Mountains that originated not from folding and crumpling of the constituent rocks, but by fault movements. Some blocks are elevated, others depressed as differential creep takes place at depth.

blueschist   A lineated, low-grade, high-pressure metamorphic rock characterized by glaucophane and/or lawsonite, and hydrous ferromagnesian silicates. (*See* Chap. 5 for descriptions and classification.)

body waves   Vibrational energy propagated through a body – such as seismic waves passing though the Earth.

bog   A peat-accumulating wetland that has no significant inflows or outflows and supports acidophilic ("acid-loving") mosses.

bonding   The forces that attract atoms to one another. They are of four principal types: ionic, covalent, metallic, and van der Waals.

borderland   A land mass lying off the margin of the continent. In actuality, most borderlands were adjacent continental masses that drifted against or away from the reference continent during sea-floor spreading.

Bowen's Reaction Series   *See* **reaction series**.

breccia   Rocks of a broad range of surface or near-surface origins, all of which are characterized by angular fragments. Without modifiers, the word commonly refers to sedimentary gravels and conglomerates that contain sharp, angular clasts. Igneous and metamorphic breccias consist of preexisting rocks that have been brittly deformed and broken into angular pieces due to internal flow, meteorite impact, crustal deformation (faulting), and the like.

bridging oxygen   In a silicate structure, an oxygen that is bonded to two silicon-oxygen tetrahedra. Tetrahedrally bound oxygens not shared between two tetrahedra are termed nonbridging oxygens.

British Thermal Unit (BTU)   Common unit of energy. 1 BTU equals 1055 joules.

C_3   A biochemical pathway of plant photosynthesis in which the initial carbon dioxide fixation product is a three-carbon compound. C_3 is also used to describe plant species that utilize this photosynthetic pathway. In C_3 photosynthesis, ambient concentrations of oxygen cause some of the carbon dioxide that is fixed to be respired via a supplemental form of respiration termed *photorespiration*, which reduces the net amount of carbon dioxide fixation.

C_4   A biochemical pathway of photosynthesis in which carbon dioxide is initially fixed in a four-carbon compound and later released and refixed in separate cells via the same pathway used by C_3 plants. The initial fixation pathway and spatial separation of the second fixation prevent photorespiration in C_4 plants.

calcite   The most common carbonate, with the composition $CaCO_3$.

calcite compensation depth (CCD)   The depth in the ocean below which seawater is undersaturated in calcium carbonate due to increased pressures and low temperatures; calcite dissolves and therefore carbonate strata are not deposited. At present, the CCD lies approximately 4 kilometers below sea level.

caliche   Accumulation of calcium carbonate in soils due to capillary rise from a moist substrate. Particularly common in arid regions, where it takes on a variety of nodular forms.

CAM   (Crassulacean Acid Metabolism). This is a photosynthetic pathway that is relatively common in desert succulents. Plants with this pathway fix CO_2 into a C_4 (a molecule containing 4 carbon atoms) organic acid during the night. During the day, they release the CO_2 from the C_4 acid and capture it for incorporation through the normal pathway of photosynthesis. This pathway dramatically decreases water loss.

camouflaged element: A trace element substituting to a very minor extent for a common element in a particular structural site in a common rock-forming mineral, such as Rb for K in orthoclase, Ag for Pb in galena.

cap rock: Strata, impermeable to fluid flow, that overlie a petroleum-bearing reservoir rock, and thus prevent the oil and gas from migrating away.

capillary force   An attractive force between pore water and host rocks. As a result of this force, water clings to the surfaces of rock particles and rises in small-diameter pores against the downward pull of gravity.

capillary fringe   The lowest part of the unsaturated zone. Greater water content results from capillary forces drawing water upward from the saturated zone.

carbohydrate   A class of chemical compounds based on building blocks of simple sugars. The basic chemical formula of a carbohydrate is $(CH_2O)_n$; the n subscript means that the basic formula is repeated many times. For simple sugars such as glucose, n is 6. For sucrose (cane sugar), n is 12. For starch and other complex carbohydrates, many simple sugars are linked together, and n can be in the thousands. Carbohydrates are the initial products of photosynthesis and the raw material for all the compounds in living organisms.

carbon dioxide   A colorless, odorless gas with the chemical formula CO_2. Carbon dioxide is one of the end products of the complete metabolism or the complete combustion of organic compounds, including plant and animal biomass and fossil fuel. Carbon dioxide is also the raw material for photosynthesis, the plant pathway through which light energy is captured in chemical bonds. Carbon dioxide is an important greenhouse gas, with a current atmospheric concentration of approximately 360 parts per million, or 0.036 percent.

carbonaceous chondrite   A relatively rare type of chondritic meteorite (a stony meteorite) containing carbon and H_2O.

carbonate rock   A common sedimentary rock such as limestone and dolomite composed of carbonate minerals.

carrying capacity   The maximum population size that can be supported by a given environment and its resources without reducing that region's ability to support future generations.

cataclastic metamorphic rocks   Metamorphic rocks that develop from preexisting rocks through the process of differential shear, fracturing, and mechanical milling down of the original grains without significant chemical reconstruction (recrystallization).

cation   A positively charged ion ($Z < e$); so called because, in an electrolyte solution, positively charged ions migrate toward the negative electrical terminal, or cathode.

Cenozoic Era   The time interval between approximately 65 million years ago and the present.

central eruptive-type volcano   Central cone, constructed of viscous lava and ash. Eruptions characteristically are violent.

chalk   Very fine-grained bioclastic limestone. (*See* Chap. 5 for descriptions and classification.)

chemical sediment   Accumulation of inorganically precipitated material.

chemical weathering   The process of weathering by which chemical reactions (hydrolysis, hydration, oxidation, carbonation, ion exchange, and solution) transform rocks and minerals into new chemical combinations that are stable under conditions prevailing at or near the Earth's surface.

chemoautotrophy   An autotrophic metabolism that relies on chemical energy to drive the synthesis of organic matter.

chert   Sedimentary rock consisting of biochemically or chemically precipitated very fine-grained silica. (*See* Chap. 5 for descriptions and classification.)

chilled contact   The rapidly cooled or quenched margin of an igneous body.

chlorite   A hydrous sheet silicate typified by $(Si_3AlO_{10})^{-5}$ groups; a common example is clinochlore, $Mg_5Al(Si_3AlO_{10}(OH_8)$.

chlorofluorocarbon (CFC)   Any of a group of halogen-bearing hydrocarbons, such as $CFCl_3$ and CF_2Cl_2. Inert in the lower atmosphere, it breaks down in the stratosphere in the presence of strong sunlight and is mediated on the surfaces of ice crystals. The chlorine then forms compounds with ozone, thus accounting for the "ozone hole" positioned over Antarctica during the austral spring and early summer.

chondrite   A stony meteorite containing abundant blebs or spheroids of magnesium-rich silicates; the droplets are former molten condensates, called chondrules.

chondrule   A spherical, formerly molten droplet of silicate material in a meteorite. Generally, formerly molten chondrules consist of intergrown prismatic sprays of magnesium-rich silicates such as olivine and/or enstatite. A meteorite that contains chondrules is termed a chondrite.

clastic (detrital) sediment   Accumulation of particulate matter or clasts (also called detritus, and detrital sediments).

clay   An unconsolidated mixture of one or more clay minerals, very fine grained.

clay mineral   Any of a class of fine-grained hydrous sheet silicates typified by $(Si_4O_{10})^{-4}$ groups; common examples include kaolinite $(Al_4Si_4O_{10}(OH)_8)$ and montmorillonite $(Na, Ca)_{0-1}(Al,Mg)_2(Si_4O_{10})(OH)_2 \cdot H_2O)$.

claystone   A massive, clay-rich variety of mudstone; a rock consisting of very fine (clay-size) detritus.

cleavage, mineral   Planar fractures in a mineral that are parallel to planes of atoms in the three-dimensional atomic arrangement (or crystal structure); that is, the cleavage surfaces are parallel to possible crystal faces.

cleavage, rock   Planar fracture surfaces in a metamorphic rock that are parallel to the orientation of the constituent minerals. Fracture cleavage consists of discrete planar breaks separated by intervening, undeformed rock. Flow cleavage is a three-dimensional fabric of the rock and is pervasive, penetrative.

clinochlore, $Mg_5AlSi_3AlO_{10}(OH)_8$.   *See* **chlorite**.

coevolution   Evolutionary changes in parallel, often due to interdependency.

coal   The accumulation of vegetal matter, chiefly trees, in a swampy environment, later compressed and heated by burial in a sedimentary basin, now consisting primarily of carbon. Increasing temperatures and pressures during burial result in the progressive devolatilization series peat, lignite, bituminous coal, anthracite.

coal rank   Refers to the gradual loss of volatile constituents hydrogen gas, water vapor, sodium, hydrogen sulfide, carbon dioxide, and sulfur dioxide, leaving behind more nearly pure carbon at progressively higher ranks.

color index   Volume percentage of ferromagnesian (dark-colored) minerals in an igneous rock.

columnar joints   Planar fractures that develop at right angles to the rapidly cooling surface of lavas and intrusive igneous melts injected at very shallow crustal levels. Seen on the cooling surface itself, the fractures (or joints) form polygonal sets, typically hexagonal in plan view.

comet   A relatively small planetary body that is surrounded by a gaseous halo; its tail points away from the Sun due to the pressure of the solar wind. Comets are composed chiefly of ices of the volatile elements, and have been described as "dirty snowballs."

commercial energy source   Energy resources used for services that are traditionally traded in the marketplace or industrially produced.

composite igneous body   A complex consisting of two or more distinct magma injections of different compositions.

compound   An electrostatically neutral chemical species consisting of definite proportions of two or more elements (e.g., H_2O, Na-$AlSi_3O_8$).

compressional mountain belt   A deformed belt exhibiting folded shortened sections of rock.

compressional seismic waves   So-called P waves, transmitted through matter by a push–pull mechanism (compression–rarefaction), similar to the propagation of energy along a row of touching balls when the first is struck with a hammer. The waves are called "P" for "primary" because these are the fastest waves and are the first arrivals received by a seismograph station.

compressive bend   Constraining bend along an active strike-slip and/or transform fault boundary. Commonly the site of a topographic high.

conchoidal fracture   Any curviplanar fracture, such as that characterizing the glassy lava obsidian, or the mineral quartz.

concordant igneous body   A tabular intrusion, the margins of which are parallel to the layering of the wall rocks.

conductive heat transfer   The migration of thermal energy through a continuous medium, driven by movement of heat down a temperature gradient.

cone of depression   A cone-shaped decrease in the head (water table for an unconfined aquifer) in the area surrounding a pumping well.

conglomerate and gravel   Clastic sedimentary rock and unconsolidated detritus, respectively, the particles of which are coarser grained than sandstone. (*See* Chap. 5 for descriptions and classification.)

congruent dissolution   Dissolution of a substance such that all its components are reduced to ions in solution and no solid secondary phase is formed.

conservative plate boundary   A near-vertical fault or fracture system (the most common examples are oceanic fracture zones) along which differentially moving lithospheric plates are being neither created nor consumed. They are the locus of transform faults.

contact metamorphic rocks   Metamorphic rocks that develop from preexisting rocks through the process of recrystallization localized around a hot spot, such as an igneous pluton. The progressive zonation of recrystallization in the country rocks is called an aureole.

contact metamorphism   The heating and consequent baking (recrystallization) of country rocks surrounding a hot mass, such as a molten igneous body.

continental drift   The differential motion of continents, first elaborately described and detailed by Alfred Wegner.

continental rise   The broad inclined submarine surface from the base of the continental slope (water depths of approximately 3 kilometers) to the deep, abyssal ocean basin (water depths of approximately 5 kilometers).

continental shelf   Portion of continental crust covered by shallow seas (water typically less than 200 meters deep).

continental slope   The transition zone between the base of the continental shelf (water depths of approximately 200 meters) and the continental rise (water depths of approximately 3 kilometers).

continental transform basin   A basin formed by differential motion along a transform fault transecting a portion of the continental crust. Where the depression is produced at a releasing bend of a transform, it is called a pull-apart basin.

convection   Mass circulation of a flowing medium in a gaseous, liquid, or plastic solid state. Thermal convection is a consequence of heating a medium at depth too fast to achieve steady-state energy transfer through conduction, and at too low a temperature for radiative transfer to be important. As material at depth is heated, it expands, becomes buoyant, and therefore rises toward a free surface; cool upper layers tend to be displaced laterally, and ultimately sink into the hotter substrate due to their greater density. Thus, a gravity-driven circulation pattern – convection – is established.

convective heat transfer   Migration of energy through gravitative instability and overturn of the medium.

convergent continental margin (active margin)   Also called Pacific-type margin. Formed in the vicinity of a subduction zone and on the landward side. Characteristic of island arcs as well as continental margins. Trench sediments are scraped off and underplated to the nonsubducted plate during convergence, in the process becoming severely deformed (folded, and thrust-faulted); in contrast, sediments of the landward fore-arc basin (i.e., arc–trench gap) are virtually undeformed.

convergent plate boundary   A linear feature along which a preexisting lithospheric plate is being destroyed by sinking into the mantle beneath another nonsubducted plate (the downturn is marked by an ocean trench). Also known as a consumptive or destructive plate boundary.

coordination number   The number of nearest-neighbor anions surrounding a central cation (or vice versa) in a crystal structure.

coordination polyhedron   The figure formed by connecting the centers of all anions surrounding a central cation. Polyhedra with three, four, six, eight, and twelve sides (triangle, tetrahedron, cube or hexahedron, octahedron, and dodecahedron, respectively) are common; others, much less so.

coquina   A sedimentary deposit consisting dominantly of macroscopic shells and shelly debris, compacted and cemented into rock.

core   The inner, central sphere of a planetary body. On Earth, the core extends from depths of approximately 2900 kilometers to the center at 6378 kilometers; its composition appears to be comparable to that of iron meteorites.

Coriolis effect   An *apparent* deflection of a freely moving object or fluid to the right in the Northern Hemisphere and to the left in the Southern Hemisphere, caused by the rotation of the Earth.

Coriolis force   A name given by Earth-bound observers to an ever-present tendency for persistent motions in the Northern Hemisphere to veer off to the right, with the veering effect proportional to the speed of the motion. It is referred to as a "force" because it seems to cause a rightward acceleration. Viewed from space, or any nonrotating frame, the effect is explainable in terms of the motion of the ground-based observer, as the observer's reference frame moves. Related to, but distinct from, the "centrifugal force."

country rock (wall rock)   Preexisting rock into which a molten or partly molten igneous body is intruded.

covalent bonding   Outer-orbital electrons are shared between neighboring atoms, whose electron clouds thereby interpenetrate one another. Discrete, strong bonds are formed by this mutual electronic interpenetration.

craton (shield)   A continental nucleus, generally of Precambrian age (older than approximately 570 million years). Such nucleii are also termed Precambrian shields. Cratons constitute the ancient basement of a continent. Many consist of granite + greenstone belts and/or high-grade metamorphic gneiss complexes, or both.

creep Virtually continuous differential movement of contiguous portions of a solid, either as discrete slip along a fault or as distributed shear within the body.

Cretaceous Period Abbreviated K, the time interval between approximately 144 and 65 million years ago.

crust The outer, compositionally distinct solid rind of a planetary body. On Earth, it consists of two types: oceanic crust approximately 5 kilometers thick, and continental crust approximately 40 ± 20 kilometers thick.

crystal A mineral grain bounded by planar growth surfaces (crystal faces) that bear a simple geometric relationship to the ordered atomic arrangement – or crystal structure.

crystal defect A discontinuity in the otherwise perfect, three-dimensional repeat periodicity of a crystal structure. Included are point defects, such as the omission of a cation or anion, line, edge and surface irregularities where an extra row or plane of atoms is present or absent, and stacking mistakes.

crystal faces Planar growth terminations to a mineral grain that bear a simple geometric relationship to the ordered atomic arrangement. Crystal faces are typically structural planes of high atomic concentration.

crystal form All faces of the same type and shape, which share a common geometric relationship to the crystal structure. An example is the eight equilateral triangle faces of an octahedron.

crystal structure The three-dimensional, systematic (ordered) atomic arrangement of a crystalline solid. Periodic, identical repeats of the local atomic arrangement occur in any direction; however, in general the structure is different in different directions. Although the analogy is imperfect, crystal structure may be likened to three-dimensional wallpaper.

crystalline Any solid that possesses a systematic, ordered atomic arrangement – or crystal structure.

cumulates Layers of crystals that, due to density contrasts with the melt, sink to the bottom of the magma chamber (dunites, peridotites, pyroxenites) or, being lighter, buoyantly float to the top (anorthosites).

cyanobacteria A group of photoautotrophic prokaryotes (Bacteria) that conduct aerobic photosynthesis.

cyclic sea salts Input of salts derived from the oceans to land as either dry deposition or precipitation.

cyclone Cylindrical circulations of air moving in the same sense as the Earth's rotation. These move counterclockwise in the Northern Hemisphere. Cyclones are generally more compact than anticyclones. The word "cyclone" is used particularly to refer to relatively compact circulations such as winter storms or hurricanes. Cyclones in the Bay of Bengal are examples of the same feature called a "hurricane" in the West Atlantic and a "typhoon" in the Far East; all are fully developed, destructive examples of what are generally called "tropical cyclones."

cyclonic storm A storm system, usually in mid-latitudes or polar latitudes, with well-developed warm and cold fronts in the classic pattern of Figure 14.5, formed around a moving low-pressure center or cyclone. Also called "winter storms," because they are particularly strong only in that season.

D" layer The boundary between the Earth's outer core and the base of the mantle, hypothesized to be characterized by strong compositional and thermal gradients.

dacite Extrusive igneous rock compositionally intermediate between andesite and rhyolite. (See Chap. 5 and Table 5.1 for descriptions and classification.)

Darcy's Law An empirical relationship between flow and hydraulic gradient (change in head over a given distance).

dark matter In the universe, luminous matter is readily detected. However, motions of some galactic bodies suggest gravitative attraction by nonluminous (or dark) matter, hitherto completely unknown. As much as 90 percent of the mass of the universe may consist of dark matter.

death rate (or crude death rate) The annual number of deaths per 1000 individuals in a population.

deep-focus earthquake Vibrational energy release (hypocenter) at depths of 200 to 700 kilometers.

demographic momentum The continued growth of a population for a period of time even after birth rates have been reduced (as a result of the number of young women already in the population who have not yet reached their reproductive years).

demographic transition Theory that refers to the historic change in industrialized regions from (1) high death rates and high birth rates; to (2) low death rates and high birth rates; to (3) low death rates and low birth rates and population stability.

demography The study of the dynamics, distribution, and age structure of populations, especially human populations.

dentrification The metabolic conversion of nitrate (NO_3) to nitrous oxide (N_2O) or nitrogen gas (N_2)

density Mass per unit volume, commonly given as grams per cubic centimeter (essentially, the same as specific gravity).

depletable resources Resources that do not replenish at a rate sufficient for reasonable human consumption. Use of the resource depletes the stock.

depleted mantle Composition of mantle after low-melting (crustal) constituents have been preferentially removed through partial fusion, melt separation, and ascent of buoyant magmas.

desertification The process of land being converted to desert primarily as a result of deforestation, overgrazing, and erosion due to poor land management.

detritus Loose sedimentary particulate matter that accumulates to form detrital or clastic rocks.

developed countries (or industrialized countries) Those countries with higher levels of per capita income, industrialization, and modernization (e.g., North America, Europe, Japan, Australia, New Zealand, and the former Soviet Union).

developing countries (or less developed countries) Typically, those countries with per capita income below $4000 (U.S.).

devolatilization The driving off of gaseous constituents, particularly H_2O, CH_4, CO_2, CO, H_2S, SO_2, F_2, and Cl_2.

diabase Very shallow-level intrusion of mafic magma. Textures are intermediate between compositionally equivalent fine-grained basalt and coarse-grained gabbro. (*See* Chap. 5 and Table 5.1 for descriptions and classification.)

diagenesis The chemical and physical changes in a sediment following deposition during burial and prior to weathering. Lithification and very low-grade metamorphism are the principal processes.

diamond A mineral formed exclusively from elemental carbon; its structure consists of an interlocking framework of relatively densely packed atoms covalently bonded to four nearest-neighbor carbon atoms.

digital elevation model (DEM) An ordered array of numbers that represents the spatial distribution of elevations above some arbitrary datum in a landscape; a subset of digital terrain models (DTMs).

digital terrain model (DTM) An ordered array of numbers that represents the spatial distribution of terrain attributes.

dike A tabular igneous intrusive body discordant (not parallel) to the layering of the wall rocks.

dimictic Lakes that overturn twice per year; typical of temperate lakes.

diopside, $CaMgSi_2O_6$. *See* **Pyroxene**.

discharge Water leaving an aquifer (or lake or stream).

discordant igneous body A tabular intrusion, the margins of which transect and cut the layering of the wall rocks.

dissimilation Any metabolic process that consumes a substance to obtain energy for metabolism.

divergent plate boundary A linear feature along which new lithosphere is forming and being transported away at right angles to the plate boundary (the latter is marked by an oceanic ridge). Also known as a constructive or accretionary plate boundary.

dolomite A fine-grained, chemically precipitated sedimentary rock formed by the settling out of carbonate on the sea floor or a lake bottom. Some dolomite may represent a primary precipitate; however, many such rocks evidently have been formed by the later chemical replacement of primary calcite. (*See* Chap. 5 for descriptions and classification.) Also a mineral, $CaMg(CO_3)_2$.

domain The largest taxonomic division of living organisms. Three domains encompass all life on Earth: Archaea, Bacteria, and Eucarya.

doppler effect Apparent change in frequency of light and/or sound radiation as a consequence of the differential motion of emitter and observer (detector).

double-chain silicate A group of minerals characterized by a pair of silicon-oxygen tetrahedral chains polymerized to infinity along a single direction, and cross-bonded to the neighboring chain by every other tetrahedron. The unit repeat of the chain in $(Si_4O_{11})^{-6}$, and each tetrahedron, on average, possesses 2.5 bridging oxygens and 1.5 nonbridging oxygens. (Of course, only half of each bridging oxygen can be assigned to a specific tetrahedron.) Some of the silicon may be replaced by aluminum.

dunite A nearly monomineralic ultramafic rock consisting chiefly of olivine. (*See* Chap. 5 for descriptions and classification.)

dynamo A machine or construct of nature capable of generating an electromagnetic field. The Earth's core is thought to act as a dynamo: by virtue of convection in the outer, liquid core, the surrounding magnetic field is produced.

dynamothermal metamorphism Regional metamorphism typified by accompanying deformation.

Earth systems science Study of the planet as an integrated, interactive set of systems (atmosphere, hydrosphere, biosphere, solid Earth).

earthquake focus The site at depth (hypocenter) of seismic energy release, usually along a fault surface.

earthquake The release of vibrational energy (and heat) through rupture of a segment of the lithosphere along a fault surface.

eclogite High-pressure metamorphic rock characterized by intermediate sodium-calcium pyroxene solid solution and garnet. (*See* Chap. 5 for descriptions and classification.)

ecologic succession The vertical separation of microorganisms in a water column and sediment according to the availability of light, oxygen, and other nutrient and energy sources.

ecosystem services The essential services supporting humanity that are provided by ecosystems. These include controlling the proportions of gases in the atmosphere, regulating the hydrologic cycle, generating and maintaining soils, disposing of wastes, and recycling nutrients.

ecosystem The collection of living things in an area and the physical environment with which they interact. Ecosystems can be defined and studied on a range of scales, from microscopic soil particles to the entire globe.

efficacy The level of illumination a lighting implement can provide per unit of power input.

ejecta Particulate matter thrown from a volcano. Depending on grain size, it is termed ash (very fine-grained), lapilli (sand-sized), and larger blocks (angular) or bombs (rounded).

electron A negatively charged atomic particle that is confined to an orbital zone – or cloud – surrounding the nucleus. It is approximately 1/1837 the mass of a proton or neutron.

electron capture The subatomic process whereby a proton and an electron combine to produce a neutron.

electron orbitals The disposition of electrons about a central nucleus. Referred to as an electron cloud, discrete electron shells and suborbitals represent differing energy levels and numbers of electrons statistically contained therein. *See* electron shells.

electron shells The electron orbitals that define the energy levels and statistical complement of electrons in the cloud surrounding the central nucleus. In the Bohr atom, these shells are designated K, L, M, N, and can accommodate up to two, eight, eighteen, and thirty-two electrons, respectively.

electron subshells Subdivisions of the electron shells, whose orbits are, for the most part, elliptical, not circular. Subshells are termed s, p, d, and f, accommodating up to two, six, ten, and fourteen electrons, respectively.

electronegativity A quantitative measure of the ability of an atom to attract electrons. Atomic species with high electronegativities become anions, whereas those possessing low electronegativities become cations.

end member Where compositional variation is possible for a mineral series, end members mark the theoretical chemical limits. For example, in alkali feldspar, $(Na, K)AlSi_3O_8$, compositional variation may be described in terms of two end members, $NaAlSi_3O_8$ and $KAlSi_3O_8$; such compositions are represented by real minerals in nature, albite and orthoclase, respectively.

energy The ability to do work.

energy currencies Energy units that can be moved and traded on the market (e.g., gallons of gasoline and kilowatts of electricity).

enstatite, $Ma_2Si_2O_6$. *See* pyroxene.

epeiric seas Marine waters contained on continental shelfs and platforms, such as the Grand Banks and the Gulf of Mexico. Water depths typically are less then 200 meters.

epicenter   The surface location directly over an earthquake energy release at depth (i.e., focus, or hypocenter).

epidote   A hydrous silicate series containing paired silicon-oxygen tetrahedra, and typified therefore by $(Si_2O_7)^{-6}$ groups; the most common type has the formula $Ca_2Fe^{+3}Al_2O(OH)(Si_2O_7)(SiO_4)$.

equilibrium   The minimum energy configuration, in either a mechanical or chemical system.

erosion   The process by which materials of the Earth's crust are loosened, dissolved, or worn away and moved from one place to another on the Earth's surface.

estimated reserve   An educated guess at the amount of reserve for a particular natural resource (see reserve).

Eucarya   The domain of organisms with nucleated cells (eukaryotes). This domain includes all protists, fungi, plants, and animals.

eugeoclinal sediments   Poorly sorted, poorly bedded, typically very thick, somewhat deformed, dominantly clastic sections laid down along a convergent margin in the vicinity of a trench. Chaotic sediments include submarine debris flows and tectonically disrupted strata. Many such deposits accumulate in and near oceanic trenches.

euhedral   Adjective describing a mineral grain bounded chiefly by crystal faces.

eucaryote   An organism possessing a cell nucleus. It reproduces by either of two mechanisms: cell division (mitosis); or cell combination and genetic material rearrangement followed by division (meiosis).

evaporite deposit   Chemically precipitated salt beds resulting from the evaporation of briny seawater.

evapotranspiration   The passage of water vapor from the Earth's surface into the atmosphere by the combined processes of evaporation and plant transpiration.

excited state of an atom   One or more inner-orbital electrons may, on absorbing electromagnetic energy, be displaced to, and reside in, an outer orbital, or may be missing completely.

exfiltration   The removal of water from the soil at the ground surface (e.g., evapotransiration).

extrusive igneous rock   The surficial accumulation of magma-derived material. Extrusives include lava flows, ash falls, ash flows, etc.

facultative aerobe   An organism capable of aerobic respiration but also able to grow in the absence of molecular oxygen.

fault   Any fracture in a solid geologic body across which differential motion has occurred.

fault breccia   Materials composing the walls of a fault zone that have been broken into angular fragments during differential movement (generally confined to the upper crust).

fault gouge   Unconsolidated, fine-grained, milled-down material along a shallow crustal fault zone.

fayalite, Fe_2SiO_4.   *See* olivine.

feedback mechanisms   (Positive and negative): positive feedback enhances the original effect; negative feedback counteracts it.

feldspar   Any of a class of framework silicates typified by $((Si,Al)O_4)$ groups; common end members are orthoclase ($KAlSi_3O_8$), albite ($NaAlSi_3O_8$), and anorthite ($CaAl_2Si_2O_8$).

felsic igneous rock   Magmatic rock units rich in the felsic constituents – silicon, sodium, and potassium, and generally poor in the ferromagnesian constituents – iron, magnesium, and calcium.

fen   A peat-accumulating wetland that receives some drainage.

fermentation   A metabolic process carried out in the absence of oxygen that converts organic compounds into carbon dioxide and a small organic molecule such as ethanol or lactate.

fertile mantle   Mantle that still contains crust-forming elements. Also commonly called primitive or undepleted mantle.

fertility   The actual number of offspring produced by a female.

first-cycle sediment   A clastic sediment, the grains of which were derived primarily from a preexisting igneous or metamorphic rock, that exhibits only limited effects of sedimentary differentiation.

fission   The process of splitting the nucleus of an atom.

fissure-type volcano   A volcano that erupts quiescently along a fracture system; such volcanoes have low, broad profiles, and are built up chiefly of highly fluid lava.

flux   A measure of flow through a given area per unit time (dimensions of velocity).

foliation   The preferred orientation of platy mineral grains in a metamorphic rock that imparts to it a layered structure and ready cleavage (foliation).

forcing   Process whereby resultant state is required.

fore-arc basin (marginal basin)   A structural trough situated outboard (seaward) from a volcanic continental margin or island arc, and inboard (landward) from an oceanic trench complex. The site of the fore-arc basin is sometimes referred to as the arc-trench gap portion of a convergent plate junction.

forsterite, MG_2SiO_4.   *See* olivine.

fossil fuels   Energy resources originating from the remains of ancient plants and animals, including oil, coal, natural gas, and others.

foundering of crust   The sinking of a solid, high-density crust through an underlying, less dense, molten, or partly molten substrate.

fracture, mineral   Any breakage surface that does not coincide with a plane of atoms in the three-dimensional atomic arrangement (or crystal structure) of a mineral grain.

fracture, rock   Any planar breakage surface in a rock.

framework silicate   A group of minerals characterized by cross-linkage of silicon-oxygen tetrahedra to infinity in three directions. The unit repeat of the framework is SiO_2, and each tetrahedron has four bridging oxygens. (Of course, only half of each bridging oxygen can be assigned to a specific tetrahedron.) For framework minerals other than the SiO_2 polymorphs, a portion of the silicon is replaced by aluminum.

free radical   A molecule or atom without a normal complement of bonds, that is, with unpaired electrons. They are electrically charged species. Free radicals are typically very reactive because they seek to establish a normal bonding structure. C1, C1O, NO, and NO_2 are all radicals.

freeboard   That portion of a material rising above the surface of the sea. As continents attain increasing freeboard, the forces of erosion tend to intensify – wearing down the crust toward sea level.

fusion   The process of combining or fusing the nuclei of two atoms.

fusion power   Energy liberated by the combination of two hydrogen molecules to make one helium molecule: $2 H_2 \rightarrow He + energy$. Called hydrogen burning, this is the energy system that fuels the Sun and, as far as we know, all other stars.

gabbro   A dark, ferromagnesian rock, cooled at depth from magma of basaltic composition, typical of the lower portion of the oceanic crust. Gabbroic intrusions also occur in continental crust. (*See* Chap. 5 and Table 5.1 for descriptions and classification.)

garnet   A silicate series containing isolated silicon-oxygen tetrahedra, and typified therefore by $(SiO_4)^4$ groups; a common example is almandine $(Fe_3^{+2}Al_2Si_3O_{12})$.

geochronology   Quantitative study of the age of formation of Earth materials utilizing isotope ratios in mineral and rock systems as "atomic clocks." Time of origin is determined by measuring isotopic proportions and knowing rates of radioactive decay.

geographic information system (GIS)   An integrating information technology that can include aspects of surface culture, demographics, economics, geography, surveying, mapping, cartography, photography, remote sensing, landscape architecture, and computer science.

geomagnetic time scale   A measure of the geologic past normal and reversed magnetic fields as a function of time.

geothermal energy   The flow of heat within the Earth toward the cooling outer surface of the lithosphere.

geothermal gradient (geotherm)   The increase of temperature within the Earth as a function of depth, commonly given in units of °C per kilometer.

geyser   Episodic or periodic conversion into steam of surface water and/or groundwater draining into a constricted fissure in hot rock. The wall rock, of course, is hotter than the boiling point of H_2O.

gigaton 10^{15}g. This is the same as 1 petagram or 1 billion metric tons. As a reference point, a coal hopper railroad car commonly carries 50 to 70 tons. A long coal train is 100 cars, or 10^4 tons. One gigaton is the coal carried by 100,000 coal trains, each 100 cars long!

gigawatt (gw)   A common unit of power equal to 10^9 watts (10^6 kilowatts).

global change   Long-term systematic change in planetary conditions, some of which may be permanently anthropogenic in nature.

global climate model (GCM)   A model of the Earth's climate, arrived at by computational simulation.

glossopteris flora   A distinctive, latest Paleozoic group of fossil plants geographically confined to the Gondwana supercontinent.

glycolysis   The metabolic process that converts sugars into pyruvate, generating energy.

gneiss   A foliated metamorphic rock characterized by mineralogically heterogeneous but parallel layering. The foliation is called gneissosity. (*See* Chap. 5 for descriptions and classification.)

Gondwana   The late Paleozoic assembly of South America, Africa, India, Australia, and Antarctica, at times separated from the other great continental mass (Laurasia) by a narrow seaway (Tethys).

granite + greenstone belt   A linear downwarped zone of upper crustal, low-grade metamorphosed besalts and interlayered volcanogenic sediments, intruded around the margins by more buoyant granitic plutons. Such assemblages are nearly confined to the Archean Era.

granite   A highly silicic, alkali-rich rock, cooled at depth from magma approximating rhyolite in composition, typical of the middle and deeper parts of the continental crust. (*See* Chap. 5 and Table 5.1 for descriptions and classification.)

granodiorite   An igneous intrusive rock of intermediate composition, typical of subvolcanic portions of convergent continental mar-

gins and island arcs. (*See* Chap. 5 and Table 5.1 for descriptions and classification.)

great circle   Any circle described by a plane that passes through the center of a sphere. On the Earth, the equator and lines of longitude are examples of great circles and arcs of great circles, respectively.

greenhouse effect   The natural or anthropogenic heating of the atmosphere and hydrosphere due to the gradual buildup of carbon dioxide and related gas species. Because these gas molecules absorb incoming solar radiation, which in turn is reemitted at different (lower) frequencies and trapped by the atmosphere, global warming results.

greenhouse gas   Gases species that increase the heat retention of the Earth's atmosphere. Water vapor, carbon dioxide, and methane are common natural gas species, whereas chlorofluorocarbons are synthetic greenhouse gas molecules.

greenschist   A foliated low-grade metamorphic rock characterized by albite, chlorite, epidote, and actinolite. (*See* Chap. 5 for descriptions and classification.)

greenstone   A term describing a green-colored, non-foliated, low-grade metamorphic rock, the protoliths of which are chiefly, but not exclusively, ferromagnesian igneous rocks such as basalt and diabase. (*See* Chap. 5 for descriptions and classification.)

greenstone belt   A long, narrow deformed zone consisting of feebly recrystallized basalts and interlayered volcanogenic sedimentary rocks. The rocks are green because of an abundance of the hydrous, ferrous aluminosilicate, chlorite. Other important green minerals are actinolite and epidote.

ground state of an atom   The lowest energy configuration of the atom in which the electrons occupy orbitals closest to the nucleus.

groundwater   Water residing beneath the surface of the solid Earth in the interstices of porous rocks and soil. Its upper surface (the boundary with unsaturated, porous rock and soil) is known as the water table.

guyot   A flat-topped submarine mountain. These bathymetric features are produced by eruption, which results in construction of a conical edifice, subsequently planed off by postvolcanic erosion approximately to sea level, followed by regional subsidence of the entire structure and subjacent sea bottom (thought to be due to gradual cooling of the lithosphere).

Hadean time   The interval between the formation of the Earth and the beginning of the Archean Eon, from approximately 4.5 to 3.8 billion years ago.

half-life   The time required for half the amount present to be converted from reactant to product assemblage. Most commonly employed to describe the rate of decay of a radioactive isotope.

hardness   A substance's resistance to abrasion. Harder minerals readily scratch softer minerals, but not vice versa. Mohs' scale of hardness for minerals of increasing relative hardness is: talc, gypsum, calcite, fluorite, apatite, feldspar, quartz, topaz, corundum, and diamond.

heat conduction   Transmission of thermal energy by propagation of atomic vibrational energy through a medium – most effective in condensed (solid and liquid) systems, much less so in gases.

hectare   Metric unit of area that is equivalent to 100 meters × 100 meters, or 2.2 acres.

hemipelagic sediment   Mixed very fine clastic and chemically–biochemically precipitated debris drifting downward and deposited as a thin blanket on the deep ocean floor. Consists principally of silica and/or carbonate + manganese nodules.

hermatypic   A type of coral that contains photosymbionts called zooxanthellae.

heterotroph   An organism that uses organic matter as its main source of carbon.

high-grade metamorphic gneiss complex   A broad region of deep- or mid-crustal, intensely recrystallized, interlayered mafic and felsic rocks, which may represent the deep roots of mountain ranges, island arcs, and/or granite + greenstone belts.

high-grade ore   A rich, highly concentrated ore deposit. Even small-volume occurrences can be worked profitably.

Holocene   Geologic time since the end of the Pleistocene.

hornblende   *See* amphibole.

hornfels   A type of contact metamorphic rock characterized by a lack of strain; such rocks do not exhibit aligned mineral grains, foliation, or cleavage, and instead possess a homogeneous, mosaic texture.

hot spot   A rising column, or plume, of hot – and therefore buoyant – mantle or crustal material.

Hubbert curve   Behavioristic model designed to estimate the supply of depletable resources.

hydrocarbon   Any of a family of organically produced hydrogen–carbon compounds. Coal is solid, oil is liquid, and natural gas is gaseous hydrocarbon.

hydrograph   Graph of the rate of water runoff plotted against time for a point on a stream channel or hillslope.

hydrologic cycle   The complex circulation of water between the oceans, atmosphere, polar ice caps, surface waters (lakes and rivers), and underground waters (soil water and groundwater).

hydrolysis   The reaction of water with cations to form (OH)-rich complexes, and with anions to form hydrogen-rich groups. A common type of chemical weathering reaction, typically involving silicate minerals.

hydrophyte   A "water-loving" plant adapted to wetland conditions. Typical examples include cattail and bulrush.

hydrosphere   The liquid sheath, or layer, situated between solid inner portions and gaseous outer portions of a planetary body. Within our solar system, a globe-encircling ocean is confined to the Earth.

hydrothermal fluid   Hot aqueous fluid of uncertain origin that causes metamorphic alteration and the formation of a wide range of economic mineral deposits. The process of chemical change in the host rocks is known as metasomatism.

hydrothermal solution   Hot aqueous fluid derived at depth from a local or regional heat source (typically a congealing magma, or heated metamorphic rock) that moves into the surrounding rocks. As H_2O-rich fluids cool, dissolved substances (e.g., quartz, carbonates, metal-bearing sulfides) precipitate out, forming veins and disseminated mineralization.

hypocenter   The actual location – or focus – of an earthquake at depth.

igneous fractionation (differentiation)   Partitioning of elements between melt and crystals, followed by separation of the liquid, results in changing composition of late-stage melts. Other fractionation mechanisms for igneous systems include liquid immiscibility, thermally driven diffusion, and assimilation of wall rocks.

igneous rock   Any rock formed from the cooling of a molten (or partly molten) solution of silicates ± carbonates ± oxides, i.e.,

magma. Two broad categories of igneous rocks are recognized, depending on where in the crust they have solidified: (1) surficial (e.g., lavas, ash) = extrusive; (2) deep (e.g., batholiths, plutons, dikes) = intrusive.

imbrication   In structural geology, the interlayering of diverse rock, generally through juxtaposition by subparallel faults.

impact breccia   Pulverized angular rock fragments produced by the collision of a planetary body with a meteorite, asteroid, or comet.

incipient melting   A very low degree of partial fusion of a preexisting solid.

inclination (dip)   The angle measured in a vertical plane between the horizontal and a magnetic field line (the north-seeking end of a compass), or between the horizontal and any other linear or planer feature, such as layering in a sedimentary rock.

inclusion   A fragment of country rock engulfed by magma.

incongruent dissolution   Dissolution of a mineral in which a secondary solid weathering product is formed.

index mineral   A phase that indicates the intensity – or grade – of metamorphism. Typical index minerals include zeolites (e.g., laumontite), chlorite, biotite, garnet, staurolite, cordierite, jadeite, kyanite, sillimanite, and andalusite.

individual tetrahedron $(SiO_4)^{-4}$ tetrahedral groups that are not linked to other silicon-oxygen tetrahedra, but instead share their oxygens with other, higher coordination cations.

infall   Refers to the melting of dispersed iron metal in the primitive Earth, and migration downward of the dense, coalescing metallic droplets to produce the iron (+ nickel) core.

infiltration   The entry of water, made available at the surface, into the soil.

inner core   The inner, spherical portion of the Earth's iron-nickel core; seismic evidence indicates it is solid.

interflow   Shallow subsurface flow of water only through the soil zones.

intermediate-focus earthquake   Vibrational energy release (hypocenter) at depths of 50 to 200 kilometers.

intermittent resource   Energy resources that are not always available.

interstitial solid solution   A solid compositional series in which chemical variation is accomplished by the addition of another atom or ion to a previously unoccupied structural site. An example is steel: Here, the addition of small amounts of carbon interstitial to the much larger-scale packing of iron atoms converts iron into steel.

intrusive igneous rock   Magma-derived material that congeals at depth within the Earth. Intrusives include batholiths, stocks, sills, and dikes.

inversion, or temperate inversion   A situation in which warmer air lies above colder air, e.g., when the land surface is cooling at night or in winter. Warming of air above, e.g. when ozone absorbs light energy, can also cause inversions. Inversions tend to be the exception in the troposphere; however, they are characteristic of the stratosphere.

ion   An atom charged by virtue of its loss or gain of electrons. In the neutral atom, the number of protons, Z, equals the number of electrons, e; however, for ions, $Z \neq e$.

ionic bonding   Attractive atomic forces resulting from unlike neighboring ionic charges. Electrons are donated or acquired so that the ions can achieve the noble gas electronic configuration. General cohesiveness is a result of the mutual attraction of unlike changes; however, discrete bonds are not recognized.

iron meteorite   Any meteorite composed chiefly of iron (+ nickel) metal alloy.

irrigation district   A local governmental agency that holds water rights and distributes the water to farmers and other users within its geographic borders.

isograd   A line on a map, or a planar surface within the Earth's crust, that locates the first appearance with advancing metamorphic grade (intensity of recrystallization) of a particular index mineral.

isostasy   The gravitative adjustment of segments of the Earth's crust and uppermost mantle whereby all vertical columns down to a depth of compensation have the same mass per unit cross section (surface area). An analog example is icebergs floating in the sea.

isotopes   Where the number of neutrons associated with a fixed number of protons for an element is variable, the different atomic species are characterized by different masses, and are referred to as distinct isotopes. For example, the heavy isotope of oxygen, ^{18}O, has ten neutrons and eight protons, which equals an atomic mass of 18, whereas ^{16}O has eight neutrons and eight protons.

K-T boundary   The time 65 million years ago when the (older) Cretaceous Period gave way to the (younger) Tertiary Period.

kaolinite, $Ae_4Si_4O_{10}(OH)_8$.   *See* **clay mineral.**

kilowatt   Common unit of power equivalent to 1.34 horsepower.

kilowatthour   Common unit of energy; work done by applying 1 kilowatt for 1 hour.

kimberlite   A rare variety of igneous rock derived from the mantle beneath old cratons, and consisting chiefly of hydrous minerals such as serpentine, but also typically containing calcium and/or magnesium carbonates. Kimberlites rise from the mantle in a fluidized condition by virtue of the presence of abundant associated volatile constituents, especially H_2O and carbon dioxide. Kimberlites are well-known because some contain trace amounts of diamond.

kimberlite pipe   An upward-enlarging conduit filled with hydrated mantle materials (kimberlite), a fluidized igneous rock derived from the upper mantle; such H_2O-and CO_2-rich complexes ascend into the continental crust in a relatively cool state due to the expansion of the associated volatiles.

komatiite   Ultramafic lava carrying major amounts of olivine; the bulk composition of the lava is similar to that of mantle peridotite. Komatiites are found almost exclusively in Archean crust. (*See* Chap. 5 and Table 5.1 for descriptions and classification.)

land bridge   Continuous land or a string of islands that span an ocean basin and link otherwise mutually isolated continents. The Isthmus of Panama and Indonesia are examples of real land bridges; however, many hypothetical land bridges, evoked by paleontologists to explain faunal and floral similarities in now geographically dispersed regions, are no longer required because of the discovery and acceptance of continental drift.

laterite   The conversion of preexisting aluminous or ferruginous rocks to bauxite (aluminum ore deposit) or iron ore through the preferential leaching away of all other constituents by tropical weathering and groundwater circulation. A product of extreme chemical weathering.

Laurasia   The late Paleozoic assembly of North America and Eurasia; at times this supercontinent was separated from the other great continental mass (Gondwana) by a narrow seaway called Tethys.

light year   The distance traveled by electromagnetic radiation through a vacuum in one Earth year.

lignite   *See* **coal.**

limestone   A fine-grained, chemically precipitated sedimentary rock formed by the settling out of calcite on the sea floor or a lake bottom. (*See* Chap. 5 for descriptions and classification.)

lineation   A preferred orientation of prismatic mineral grains in a metamorphic rock that imparts to it a linear (pencil-like) structure.

lithification   The process of compaction, dewatering, and cementation that converts an unconsolidated sediment into a sedimentary rock.

lithosphere   The outer rind of a planet that behaves as an integral, relatively rigid, brittle unit. The Earth's lithosphere consists of crust + uppermost mantle. It extends to depths of 50 to 200 or more kilometers and is divided into several enormous segments, or plates, which move relative to one another.

lithospheric plate   A relatively rigid, coherent slab of the Earth's crust and uppermost mantle – 50 to 200 or more kilometers thick, and hundreds of thousands of kilometers in lateral or surface dimensions – which moves differentially with respect to other plates.

lithostatic pressure   The pressure per unit cross section generated by an overlying column of rock. In low-density sedimentary terranes, the lithostatic pressure increases downward by approximately 1 kilobar per 4 kilometers, whereas in the upper mantle, a kilobar increment in lithostatic pressure results from slightly in excess of a 3-kilometer increment in depth.

low-angle thrust fault   A planar break (dipping less than 15 to 20°) in which the overlying, hanging-wall block has moved along the fault upward relative to the underlying footwall.

low-grade ore   A mineral deposit lean enough to be near the break-even point for profitability. To be worked, such deposits generally must be very large before economics of scale are achieved.

lower mantle   The poorly understood, deep portion of the Earth's mantle, located at depths between approximately 670 and 2900 kilometers.

lunar highlands   Elevated, light-colored areas of the heavily pock-marked (cratered) surface of the Moon, characterized by ancient anorthosite breccias.

lunar maria   Topographically subdued, basinal, dark-colored areas of the slightly pockmarked (cratered) surface of the Moon, typified by basaltic lava flows younger than the highlands.

mafic igneous rock   Magmatic rock rich in ferromagnesian constituents iron, magnesium, and calcium, and generally poor in silica and alkalis.

magma   A molten (or partly molten) solution of silicates + carbonates that, when congealed, forms a rock (*see* **igneous rock**). Magma extruded onto the Earth's surface is called lava.

magma ocean   Refers to the theorized existence of a near-surface molten silicate layer on the primitive Earth and/or Moon.

magnetic anomaly   The magnetic field strength and orientation that differs from the regional field, in either intensity or direction, or both. A local magnetic intensity exceeding the present regional field is termed a positive magnetic anomaly.

magnetic field lines   The direction of a magnetic force field. On the Earth, the field line coincides with the direction in which a magnetic compass needle points.

magnetic reversal   The phenomenon whereby the polarity of the Earth's magnetic field reverses orientation by 180°.

mantle   The shell of a planetary body directly beneath the crust, and extending to the outer boundary of the core. On Earth, the mantle extends from beneath the Moho (the base of the crust) to depths of approximately 2900 kilometers (the top of the core); its chemical composition appears to be comparable to that of stony meteorites.

marble   A coarse-grained carbonate-rich metamorphic rock, the precursor of which was generally limestone or dolomite (*see* Chap. 5 for descriptions and classification).

marsh   An area frequently or continually inundated, nontidal wetland characterized by emergent vegetation adapted to saturated soil conditions.

mass extinction   The abrupt termination of many kinds of plant and/or animal life, brought about by rapid change of the environment.

matter   Substance, or detectable mass.

maturity of sediments   Repeated cycles of sedimentary processing – weathering, erosion, transportation, deposition – are required to produce diverse, chemically concentrated deposits; consequently, first-cycle sediments are compositionally unfractionated (i.e., immature), whereas multicycle sediments are strongly differentiated compositionally (i.e., mature).

mean residence time (MRT) (of water)   The time that an average water molecule can be expected to spend within a reservoir before moving on to another reservoir.

meandering channel   A river pattern consisting of a series of curves.

meggawatts (MW)   A common unit of power equal to 10^6 watts (10^3 kilowatts).

melange   Chaotically deformed sedimentary rock. Two types are common: (1) submarine slump deposits, or olistostromes; and (2) strata folded and sheared in a subduction zone by deformation, or tectonic melanges.

mesosphere   The "middle" region of the atmosphere, actually from 55 to 85 kilometers, lying between the stratosphere and the thermosphere.

Mesozoic Era   The time interval between approximately 245 and 65 million years ago.

metallic bonding   Weak cohesive force characteristic of metals, whereby positively charged metal ions are bonded together by a loosely configured, negatively charged swarm of electrons, or electron "gas," which permeates the material. High mobility of the electrons is responsible for the great electrical and thermal conductivities of metals.

metamorphic aureole   Zonal development of newly generated minerals in wall rocks around a hot spot (typically, an igneous pluton).

metamorphic grade   The relative intensity of recrystallization, correlated with physical conditions: the higher the grade, the higher the inferred pressure and/or temperature that attended the metamorphism.

metamorphic mineral facies   An assemblage of minerals – defined in rocks of basaltic composition – that reflect the relative intensity of metamorphic recrystallization.

metamorphic rock   Any rock whose preexisting phases and/or textures have been modified by recrystallization and/or deformation – usually taking place at depths in the Earth. Two chief types are recognized: (1) contact metamorphic rocks, disposed about a thermal anomaly such as an igneous intrusion; and (2) regional, or dynamothermal metamorphic rocks, typically deformed and recrystallized over a broad region, and in many cases unrelated to igneous activity.

metasomatism   A metamorphic process whereby the bulk chemical composition of an original rock is altered, typically by exchange with an infiltrating hydrothermal solution charged with alkalis and silica.

meteor   Condensed matter of subplanetary dimensions within the solar system that has entered the Earth's atmosphere. Frictional dissipation causes most meteors to burn up as they pass through this gaseous medium at high velocity.

meteorite   Any meteor that has survived entry into the atmosphere and has landed on the surface of the Earth. Meteorites are of two general types: metallic (iron-nickel alloy); and stony (magnesium-rich silicates, principally olivine and enstatite).

methanogenesis   The biologic production of methane. This metabolic process is carried out by certain members of Archaea.

methanogenic bacteria   Prokaryotes, all of which are members of Archaea, capable of producing methane metabolically.

methanthropy   The metabolic oxidation of methane, producing biomolecules and carbon dioxide. This process is carried out by certain members of Bacteria.

mica   Any of a class of hydrous sheet silicates typified by $(Si_3AlO_{10})^{-5}$ groups; common examples include muscovite, $KAl_2Si_3AlO_{10}(OH)_2$, and biotite, $K(Mg,Fe^{+2})_3Si_3AlO_{10}(OH)_2$.

microcosm   An experimental situation designed to duplicate many aspects of ecosystem structure and function, but in a context where the scale is manageable and where key aspects of the ecosystem can be manipulated. Microcosms are more or less test-tube ecosystems that make it possible to test hypotheses that would be much more difficult to approach in nature. Most microcosm experiments involve many individuals of a number of species and a reasonably natural soil. Commonly, inputs and outputs of water, nutrients, and energy are carefully controlled.

migmatite   A rock characterized by solid but plastic, refractory ferromagnesian layers, and by felsic layers that crystallized from a melt. Many migmatites form by ultrametamorphism from the in situ partial fusion of a preexisting rock.

migration   The movement of individuals into (immigration) and out of (emigration) a given geographic area.

mineral   A naturally occurring, inorganically grown solid that has a fixed chemistry, or limited range in composition between end members, and a three-dimensional atomic order, or structure.

mineralogic barrier   Where the concentration of a sought-after element is below its saturation limit (i.e., the mineralogic barrier) in common rock-forming minerals, the element is broadly dispersed throughout these phases (i.e., it is camouflaged) and is not easily concentrated. When the abundance of the element in question increases, these minerals become saturated, and at yet higher concentrations a separate phase containing large amounts of the sought-after element forms. The latter may be of economic significance.

mineraloid   Any naturally occurring solid or liquid that lacks an ordered atomic arrangement – or crystal structure. Such substances are said to be amorphous.

minerotrophic   Groundwater flowing into a wetland from adjacent areas of mineral soils and aquifers.

miogeoclinal sediments   Well-bedded, usually thick sections laid down along a passive (Atlantic-type) margin or in the fore arc of an active (Pacific-type) margin on the continental shelf, slope, and rise.

Mohorovicic discontinuity (moho) Also known as the M discontinuity. The horizontal or subhorizontal boundary between overlying crust and underlying mantle. It represents an abrupt change in physical properties such as density and composition of the materials; it may be vertically gradational over 1 or 2 kilometers, especially under the continents.

Mohs' scale of hardness *See* **hardness**.

molecule Combinations of two or more atoms bonded together, e.g., H_2, O_2, CH_4, S_8, H_2O. These are all gas or liquid molecules. Molecules, as such, do not exist in minerals because of their three-dimensional bonding. For example, one cannot recognize an NaCl molecule in the structure of halite (common table salt).

moment of inertia Defined as the product of the mass and the square of an effective radius, which is related to the distribution of matter in a rotating body.

montmorillonite *See* **clay mineral**.

mudstone A fine-grained, nonlaminated, massive, clay-rich clastic sedimentary rock formed by the accumulation of tiny mineral flakes and grains, commonly in a marine basin or lake. (*See* Chap. 5 for descriptions and classification).

multicycle sediment A clastic sediment, the grains of which have repeatedly undergone cycles of weathering, transportation, abrasion, and deposition. Such rocks display pronounced effects of sedimentary differentiation.

muscovite A white mica (layer silicate), possessing the formula $KAl_2Si_3AlO_{10}(OH)_2$.

mycorrhiza Fungi that establish symbiotic relations with plants, aiding the plants by providing increased nutrient and/or water uptake in exchange for carbohydrates produced through photosynthesis. Mycorrhizas associated with some plant species, especially trees, form dense sheaths around the plant roots. Mycorrhizas associated with herbaceous plants penetrate the plant roots but extend their hyphae far into the surrounding soil.

mylonite A very fine-grained metamorphic rock, milled down by shearing (the process is called cataclasis) but not recrystallized (chemically reorganized). Mylonites are characteristic of fault and shear zones.

natural increase Refers to the difference between the birth rate and the death rate. The rate of natural increase is commonly expressed as the average increase or decrease per 100 individuals per year.

natural resource Naturally occurring material, including energy, useful for supporting life. A resource is considered to be the total planetary inventory of the material or energy source in question, whether recoverable or not.

negative magnetic anomaly A local magnetic field less than the regional field strength.

neutron An atomic nuclear particle possessing unit mass and lacking an electrical charge. Differing numbers of neutrons for an element (fixed number of protons) yield differing masses (i.e., isotopes) of the element.

nitrification The metabolic conversion of ammonia to nitrite and nitrate.

noble gas configuration Atoms, ions, and molecules typified by completely filled or completely empty outer electron (valence) shells.

nondispatchable resource Energy resource that cannot be immediately called upon to produce energy.

nonbridging oxygen In a silicate structure, an oxygen ion bonded to one silicon and one or more nonsilicon cations.

nonrenewable resource A resource of finite abundance that is not being continuously generated – such as gold, iron, and silica – or is being generated at a rate that is infinitesimal compared to the rate of consumption – such as the fossil fuels oil, gas, and coal.

normal fault A high vertical-angle fault in which the upper block, or hanging wall, has moved down relative to the lower, footwall block.

normal magnetic polarity The Earth's magnetic field is similar to that of the present, whereby the north-seeking end of a compass needle points north.

nuclear fission The spallation of heavy (high atomic number) elements such as uranium, thorium, and plutonium into lighter, lower Z (atomic number) daughter products, liberating energy.

nuclear fusion The combination, or "burning," of light (low atomic number) elements such as hydrogen and helium to produce heavier, higher Z (atomic number) elements plus energy.

nucleus The membrane-bound organelle in eukaryotes that contains a cell's genetic material.

nuée ardente Glowing avalanche cloud of volcanic ash (*see* **ejecta**).

obligate aerobe An organism whose metabolism requires oxygen.

obligate anaerobe An organism that cannot use molecular oxygen in its metabolism, and is adversely affected by the presence of oxygen.

oceanic deep (trench) An especially deep linear trough in an ocean basin (water depths characteristically are 6 to 8 kilometers or more). These negative bathymetric features mark the locus of downturn of convergent lithospheric plates as the latter descend into the deeper mantle.

oceanic fracture zone Planar features at right angles to oceanic ridges that are marked by submarine escarpments. These zones are caused by differential motion of lithospheric plates (*see* **transform fault**).

oceanic ridge (rise) Linear submarine volcanic mountain chains, almost totally submerged, that mark the near-surface expression of rising limbs of a pair of mantle convection cells.

oceanic rise (oceanic ridge) Within the oceanic crust, a submarine mountain chain of great length – the bathymetric reflection of upwelling convective currents in the asthenosphere. Near-surface cooling produces the pair of capping lithospheric plates that are continuously conducted away from the oceanic ridge or linear spreading center.

oceanic trench (oceanic deep) Linear submarine depression well below the general level of the sea floor – the bathymetric expression of a downturning ocean crust–capped lithospheric plate that is sinking into the deeper mantle.

oil pool A concentration of petroleum within the pore space (interstices) of a reservoir rock, typically a porous sandstone, or a fractured or cavernous limestone. Being of very low density, gas occupies the highest portion of a stratal reservoir. The condensed liquid petroleum lies beneath it, but is less dense than water; consequently, the oil overlies and floats upon the yet denser water.

oil shale A muddy, stratified sediment in which the original organic matter has neither thermally matured (has not been heated and distilled) nor migrated away.

olivine A silicate series containing isolated silicon-oxygen) tetrahedra, and typified therefore by $(Si_4)^{-4}$ groups; examples of the olivine solid solutions series include the end members forsterite (Mg_2SiO_4) and fayalite ($Fe_2^{+2}SiO_4$).

ombrotrophic Wetland waters originating from direct rainfall without passing through soils and aquifers.

omission solid solution A solid solution series in which chemical variation is achieved by the systematic absence of a small proportion of the constituent atoms – either cations or anions. An example is pyrrhotite, which has the nominal formula FeS; a small iron deficiency in natural pyrrhotites is accommodated by oxidation of minor Fe^{+2} to Fe^{+3}, or a comparable reduction of S^{-2}.

Oort cloud The region outside the gravitational pull of the Sun occupied sparsely by comets and other icy condensed matter.

orbital theory Celestial mechanics, the process whereby the motions of heavenly bodies are determined by their masses, distances, and velocities.

ordering In mineralogy, a term referring to the confinement of atoms to specific sites in a crystal structure. As temperature increases, a less systematic, less ordered, more nearly random distribution of the atoms in the various structural sites takes place.

ore (mineral) deposit A natural resource sufficiently concentrated above the terrestrial average composition so that it can be profitably extracted.

orogenic belt A mountain belt in terms of geologic structure, which may be, but is not necessarily, reflected in the topography.

orogeny The process of producing structural (not necessarily topographic) mountains. This complex of structures is thought to result from compressive and tangential forces accompanying differential plate tectonic motions, especially along and near convergent plate boundaries and on-land transform lithospheric plate junctions.

orthoclase, $KAlSi_3O_8$. *See* **feldspar.**

outer core The outer shell of the Earth's iron-nickel core, thought to be molten. Circulation (convection) within this liquid may be responsible for the Earth's magnetic field.

outgassing Loss of volatiles from an extrusive or near-surface magma, as well as from deep crustal and upper mantle sections sited at profound depths, because of high temperatures.

overturned fold (nappe or recumbent fold) A fold in which both limbs are inclined (i.e., dip) in the same direction.

oxbow lake A crescent-shaped lake formed in an abandoned river bend by a meander cutoff.

ozone Tri-atomic oxygen molecule, O_3.

ozone layer The tenuous layer in the lower stratosphere (elevation 15 to 30 kilometers) in which ozone is relatively plentiful due to its natural production from oxygen by the action of ultraviolet solar radiation.

paired metamorphic belts Contemporaneously produced metamorphic zones, an oceanward high-pressure blueschist belt, and a continentward high-temperature recrystallized rocks + andesitic/granitic igneous belt. The narrow, seaward, linear metamorphic terrane is thought to have formed in a subduction zone, whereas the broad landward belt is regarded as produced in the wall rocks of a magmatic arc.

paired tetrahedra The linking of two $(SiO_4)^{-4}$ tetrahedra to produce a $(Si_2O_7)^{-6}$ complex. One oxygen, a so-called bridging oxygen, is shared by the two tetrahedra.

paleoclimate The climate of the geologic past, evidence of which is contained in – or inferred from – the rock record.

paleolatitude The angular distance in degrees from the equator for a specific portion of the Earth's crust at the time of consideration in the geologic past.

paleomagnetism The magnetic field of the geologic past, evidence of which is contained in the rock record as remnant magnetism.

paleosol A preserved ancient soil.

Paleozoic Era The time interval between approximately 570 and 245 million years ago.

palustrine wetlands A term applied to all nontidal wetlands, excluding those wetlands that are adjacent to or part of large river channels or the edges of large lakes.

Pangea The late Paleozoic assembly of all major continental masses into a single supercontinent. It included both Gondwana and Laurasia.

partial fusion Partial liquification of a preexisting rock, whereby the lowest-melting, most fusible constituents are concentrated in the melt.

peat Thick accumulations of slightly to moderately decomposed organic materials, under wet conditions. Peats often underlie wetlands, and dried peat is used as a relatively low-cost fuel in some regions, particularly Ireland.

pegmatite A very coarse-grained felsic, silicic, intrusive igneous rock comparable in chemistry and mineralogy to granite. Commonly occurs as irregular or tabular bodies in and adjacent to granite.

peridotite A very magnesian, low-silica rock rich in ferromagnesian constituents characteristic of the upper mantle, similar in chemistry to stony meteorites. It is an ultramafic rock containing essential olivine ± pyroxenes. (*See* Chap. 5 and Table 5.1 for descriptions and classification.)

permeability A measure of the interconnectedness of pores of a soil or a rock.

petroleum Liquid and/or gaseous hydrocarbons produced dominantly by the distillation of microscopic aquatic plant and animal life, typically in a marine environment, less commonly in freshwater lakes.

pH A chemical term, defined as the negative logarithm of the hydrogen-ion activity (concentration). In the pH scale, a solution with a pH of 7 is neutral, whereas solutions with pH > 7 are alkaline and with pH < 7 are acidic. These definitions derive from the equilibrium dissociation of H_2O at ambient conditions, whereby the dissociation constant K is 10^{-14} = (concentration H^+ × (concentration OH^-).

Phanerozoic Eon Geologic time interval from the end of Precambrian time, 570 million years ago, to the present.

phase A specific state of aggregation, such as gas, liquid, or solid. Solid phases include minerals, glasses, and related substances. Rocks consist of one or more phases.

photoautotroph An organism that uses light energy and inorganic carbon to synthesize biomolecules and to obtain energy.

photosynthesis The combination of CO_2 and H_2O in the green leaves of plants catalyzed by sunlight, resulting in the production of organic compounds and free oxygen. Through photosynthesis, some of the Sun's energy is captured as chemical energy, in a form that can be stored and used to drive the chemistry of life. Photosynthesis is the primary sustenance for life on Earth, the source of all of the organic compounds required by plants and therefore the source of the plant food required by animals.

photosynthetic bacteria Prokaryotes, members of Bacteria, that conduct anaerobic photosynthesis.

physical (mechanical) weathering The process of weathering by which frost action, salt-crystal growth, absorption of water, temperature and/or pressure fluctuations, and other physical or mechanical processes break down a rock or mineral to fragments, involving no chemical change.

physiographic features The topography or surface aspect of the dry land. Underwater analogs are referred to as bathymetric or hydrographic features.

pillow lava Ellipsoidal blobs of basaltic lava, extruded and chilled under water.

placer A deposit of high-density clastic grains in stream or beach gravel, characterized by settling out under conditions of high flow rate of the carrier water.

plagioclase A feldspar solid solution series ranging in composition from albite, $NaAlSi_3O_8$, to anorthite, $CaAl_2Si_2O_8$.

plane of symmetry A reflection across a mirror of symmetry results in self-coincidence. For instance, an automobile and a dog each possess a vertical, longitudinal plane of symmetry – at least, as far as their external morphologies are concerned.

planetesimal Any of a group of relatively small, condensed matter planetary bodies, including asteroids, meteorites, and comets.

plasma An incandescent ultrahigh-temperature ionized gas, such as occurs in the Sun.

platelet stage The Archeozoic Era in Earth history typified by the rapid circulation of numerous, small mantle convection cells. Crust produced consisted of small, mafic island arcs, swept together in circumoceanic areas by the sea-floor spreading process to considerable thickness, and invaded by granitic melts derived by partial fusion of their lowest portions during subduction.

platform (epeiric) sediments Thin-bedded, areally continuous, well-sorted and chemically differentiated, multicycle sandstones, shales and carbonate strata laid down on the stable craton on shallow-water continental shelves (i.e., epeiric seas).

Pleistocene Epoch The time interval between approximately 2 million and 10,000 years ago, characterized by continental glaciation.

plug A solid body of igneous rock that has cooled and congealed, typically in a volcanic conduit. Many plugs are thought to represent the throats of volcanoes.

plume In plate-tectonic theory, plumes are rising columns of hot mantle material (*see* **hot spot**).

pluton An intrusive igneous body of irregular or unknown shape.

point-bar deposits Convex bank deposits that are gradually extended streamward.

polarization of an ion Distortion of the cloud of electrons surrounding an atomic nucleus from a highly symmetric, spherical, to more oblate or prolate (pumpkin-like or football-like) shape, due to the attractive forces of a neighboring atomic nucleus.

polymerization The linking together of atomic, molecular, or ionic substructures (or polyhedra) to form a larger structural array. Pyroxenes contain infinite silicon-oxygen chains, and are said to be polymerized in one direction. In contrast, silicon-oxygen tetrahedra in quartz are cross-linked in three directions; hence, the structure is described as a three-dimensional framework (polymerization is in three directions).

polymorphs Minerals having the same composition but different crystal structures. Well-known examples include the polymorphs of carbon, diamond, and graphite, and the polymorphs of silicon dioxide, α-quartz, β-quartz, tridymite, cristobalite, coesite, and stishovite. Because of their contrasting three-dimensional atomic arrangements, polymorphs have different physical properties such as hardness, density, melting point, refractive index, etc.

pore fluid Fluid occupying the soil or rock pore space.

porosity The fraction of the volume of voids (filled with air and/or water) in a bulk volume of rock or sediment.

porphyry copper deposit Widely disseminated, low-grade copper sulfide mineral-bearing shallow-level portions of granitoid rocks. Such occurrences are especially typical of the western Cordillera of North and South America, and of Indonesia.

positive magnetic anomaly Local magnetic field that exceeds the regional field strength.

power The rate at which energy can be produced or consumed.

Precambrian shield Ancient nucleus of a continent – or craton.

Precambrian time The combined Hadean, Archean, and Proterozoic eons, or the time interval from the formation of the Earth approximately 4.5 billion years ago to 570 million years ago (the beginning of the Cambrian Period).

precursor phenomena Physical changes that occur prior to a geologic event such as an earthquake, landslide, or volcanic eruption. Clear recognition of unambiguous premonitory phenomena would allow prior action to be taken to reduce the loss of life and reduction of damage during a geologic catastrophe.

premonitory (precursor) events Natural geologic features that, due to abrupt or unusual changes, signal the onset of a sudden event such as an earthquake, volcanic eruption, or landslide.

preplate stage The Hadean time interval in Earth history (postaccretion but pre-Archean time) typified by a terrestrial magma ocean, followed by turbulent or poorly organized mantle convective overturn. No crust survived this stage.

primary rocks Original additions to the crust of igneous material derived from the mantle.

primitive mantle Original composition of mantle (stony meteorite) prior to partial melting and removal of liquids and volatiles to form the Earth's crust, hydrosphere, and atmosphere.

prior appropriation doctrine A legal doctrine that awards water on a first-come, first-served basis to anyone who "appropriates" the water by diverting it and putting it to a "reasonable and beneficial" use.

procaryote An organism that lacks a nucleus, which must reproduce by cell division or binary fission. Prokaryotes may belong to Archaea or to Bacteria.

Proterozoic Eon The time interval between approximately 2.5 billion and 570 million years ago.

protolith The original rock prior to a change in its present configuration.

proton An atomic nuclear particle possessing unit mass and a single positive charge. The number of protons, Z, in an atom define the atomic number characteristic of that element, as well as its chemical behavior.

proven reserve Demonstrated, quantitative evaluation of the amount of a reserve that is known to exist and is recoverable with the current technology at the current price for a particular natural resource (*see* **reserve**). Commonly, proven reserves are documented through drilling or excavation.

pull-apart   Releasing bend along a strike-slip (transform) fault boundary. Typically the site of a topographic basin.

pumice   Glassy, low-density lava dominated by bubbles (expanding and escaped gas).

pyroclastic rock   Rock formed dominantly by the accumulation of solid and liquid particles (called ejecta) expelled from a volcano.

pyroxene granulite   A high-grade metamorphic rock characterized by intermediate calcium-sodium or calcic plagioclase, one or two pyroxenes, cordierite at low pressures, and garnet at high pressures. (See Chap. 5 for descriptions and classification.)

pyroxene   Any of a class of single-chain silicates typified by $(SiO_3)^{-2}$ groups; common examples include enstatite $(MgSiO_3)$ and diopside $(CaMgSi_2O_6)$.

pyroxenite   Ultramafic rock consisting of either orthopyroxene (hypersthene) or clinopyroxene (augite) or both.

quad   Large unit of energy equaling 1 quadrillion BTUs.

quantum   A discrete amount of energy, used in reference to the electron orbital levels surrounding a central nucleus of an atom.

quartz   A framework silicate with the formula SiO_2.

radiative heat transfer   Transmission of thermal energy by wave propagation. It is most important at high (incandescent) temperatures.

radioactive   An element of high atomic number that spontaneously is converted to other, lower atomic number elements ± atomic particles at a fixed rate proportional to its abundance, liberating large amounts of energy in the process.

radiometric age   The quantitative age of a mineral or rock as determined by measurement of an originally incorporated radioactive element(s), and the newly produced radiogenic daughter product(s).

radius ratio   In crystal structures, the fraction defined by the quotient: cation radius / anion radius. For ionic structures, the distance between neighboring cation and anion centers is the sum of the respective radii.

reaction series   Also known as Bowen's reaction series, the sequence of crystallization of minerals from a cooling magma, the composition of which is changing from mafic to felsic (*see* Chap. 5).

recharge   The entry into the saturated zone of water made available at the surface, together with the associated flow away from the water table within the saturated zone; basically, the net addition of the water entering an aquifer (or lake or stream).

recovery factor   The percent of a resource that can be recovered economically from an existing field.

recumbent fold (nappe)   A tightly appressed fold that has been overturned. Both limbs are inclined (dip) in the same direction.

reflected wave   Vibrational energy bounced back from a planar discontinuity at a path angle equal to the angle of transmission incidence.

refracted wave   Vibrational energy is propagated through a planar discontinuity into a new medium at an angle different from the angle of transmission incidence (angle *i*). The bending, or refraction angle (angle *r*) is related to the transmission velocities, V_i and V_r, and the angle of incidence by the equation: $V_i \sin i = V_r \sin r$.

regional metamorphic rocks   Metamorphic rocks that develop from preexisting rocks through the process of recrystallization on a regional scale, unrelated to local igneous phenomena such as pluton emplacement.

regional metamorphism   The recrystallization and deformation of rocks over a broad region due to the imposition of stress and markedly altered physical conditions, such as temperature and pressure.

reinfiltration   The infiltration of excess precipitation into more permeable media, downslope from the point of its original contact with the soil surface, after running off.

remnant magnetism   Alignment of magnetic domains in the constituent mineraloids and/or minerals of a rock coincident with the Earth's magnetic field at the time of formation of the rock.

renewable resource   A resource that is being continuously generated, such as groundwater, biomass, and some forms of energy, including tidal, hydroelectric, wind, geothermal, and solar power. With this kind of resource, human use has no bearing upon its future availability, or the rate of renewal exceeds that of reasonable use.

repeat periodicity   The distance in any chosen direction within a crystal structure required to produce an identical atomic arrangement.

reserve   That portion of a natural resource, discovered or undiscovered, that can be profitably extracted, using present-day, economically feasible methods.

reserves to production rato   The ratio of reserves to annual production rate. This represents the length in time that a depletable resource can be exploited at current rates assuming no new supplies are found.

reservoir rock   A porous rock whose interstices are occupied by a fluid of interest, such as oil and/or gas, or steam.

reservoir species   A species that can hold potential radical species together, at least temporarily. For example $ClONO_2$ can hold both ClO and NO_2 radicals together; however, its bonds can be split by light or by reactions on solid ice surfaces.

resource base   The ultimate amount of a resource that is believed to be present.

respiration   A biochemical process through which all living organisms metabolize organic compounds. Respiration releases much of the energy stored in chemical bonds, making it available to drive everything an organism does, from growth and biosynthesis to locomotion and thermoregulation. For organisms exposed to normal atmospheric levels of oxygen, the major end products of respiration are carbon dioxide and water. Thus, photosynthesis removes carbon dioxide from the atmosphere, and respiration returns carbon dioxide to the atmosphere.

retorting   The process of heating, thermally maturing or distilling, and extracting a substance, such as hydrocarbons originally residing in oil shale.

reverse fault   A high vertical-angle fault in which the upper block, or hanging wall, has moved up relative to the lower, footwall block.

reversed magnetic polarity   A situation in which the Earth's magnetic field is the opposite of that of the present. During a period of reversed magnetic polarity, the north-seeking end of a compass needle would point south.

rhyolite   A highly silicic, alkali-rich lava typical of the continental crust. (See Chap. 5 and Table 5.1 for descriptions and classification.)

Richter scale   Scale of quantitative measurement for vibrational energy released in an earthquake. The magnitude of the event is proportional to the linear deflection of the recording pen of a standard seismometer normalized for a specific distance from the hypocenter. Intensity increases approximately thirty-two times between successive numbers; hence, a 7.3-magnitude quake is approximately ten times more energetic than a 7.0-magnitude quake.

rifted continental margin (passive margin)   Also called an Atlantic-type margin. Formed as a result of the breakup of continental crust by the sea-floor spreading process. Although the margin initiated as a divergent lithospheric plate junction, continued spreading removes the passive margin from the vicinity of the ridge axis and divergent plate boundary. Sediments are laid down on the thermally subsiding continental shelf, slope, and rise as a prism of undeformed strata.

rifting, drifting, and suturing stage   The Phanerozoic time in Earth history, typified by very slow, laminar flow of enormous mantle convection cells, and by the modern cycle of plate tectonics.

riparian doctrine   A state legal doctrine that awards a "reasonable" share of water from a natural waterway to anyone who owns land along (i.e., "riparian" to) the waterway.

rock cycle   The continuous or episodic reworking of preexisting crustal and uppermost mantle materials during the geochemical evolution of the lithosphere. The cycle includes igneous, metamorphic, and sedimentary processes.

rock   A naturally occurring aggregate of grains of one or more minerals and/or mineraloids, having a common origin. In order to be designated by the term "rock," such aggregations should be present in units large enough to represent an important part of the Earth.

roof pendant: A large expanse of country rock invaded and surrounded by a plutonic igneous body.

rubisco   One of the key enzymes in photosynthesis. Rubisco is short for Ribulose-1,5-bisphosphate carboxylase/oxygenase. Rubisco is the enzyme responsible for the initial fixation of carbon dioxide into organic compounds in most plants, algae, and photosynthetic bacteria. In plants that use the C_4 or CAM (crassulacean acid metabolism) photosynthesis pathway, a different enzyme, called PEP carboxylase, catalyses the initial fixation of carbon dioxide into an organic compound; however, the carbon dioxide is later removed from this compound and refixed by Rubisco. Rubisco is one of the world's most abundant proteins, constituting 15 to 30 percent of the total protein in most leaves.

saltwater intrusion   Movement of salt water into a freshwater aquifer; often caused by excessive pumping of freshwater from wells.

sandstone   A medium-grained (grain size between 1/16 and 2 millimeters) clastic sedimentary rock formed by the accumulation of granular fragments of rocks and minerals, generally in a marine basin or lake, but much less commonly as river or dune deposits. (See Chap. 5 for descriptions and classification.)

scarce resource   A natural resource whose concentration in the reachable parts of the Earth is quite low. Therefore, deposits of economic value may be very low in concentration for the resource in question, but nevertheless significantly above the average terrestrial concentration. Gold is a good example of a scarce resource.

schist   A foliated metamorphic rock characterized by homogeneous concentrations of minerals throughout. The foliation is called schistosity. (See Chap. 5 for descriptions and classification.)

sea mount   A submerged mountain within an ocean basin.

sea-floor spreading   The process whereby new ocean crust – capped lithosphere is generated in the vicinity of an oceanic ridge. The latter marks the site of a rising column or sheet of hot, buoyant (convecting), but largely solid mantle material.

secondary rocks   Reworked crustal lithologic units, including some igneous rocks, all sedimentary rocks, and all metamorphic rocks.

sedimentary breccia   A rock that consists chiefly of angular conglomerate pebbles.

sedimentary differentiation   The chemical fractionation of protoliths by sedimentary processes into product rocks greatly enriched in certain elements, e.g., SiO_2, $CaCO_3$, MnO, Fe_2O_3, or Al_2O_3.

sedimentary rock   Any rock formed at or near the surface of the Earth through the settling of material. Two broad classes are recognized: (1) materials chemically and organically precipitated from aqueous solutions (principally seawater); (2) discrete grains (or clasts) of particulate matter (detritus), brought to a site of deposition by running water, moving ice, wind, or downhill slumpage.

seismic waves   Vibrational energy mechanically transmitted through the Earth (a variety of sound waves).

seismograph   An instrument that, at a particular location, records and amplifies the vibrations of the Earth in specific directions.

shale   Clastic sedimentary rock possessing a grain size less then 1/16 millimeter – typically, less than 1/256 millimeter. (See Chap. 5 for descriptions and classification.)

shallow-focus earthquake   Vibrational energy release (hypocenter) at depths of 0 to 50 kilometers.

shear   differential motion between two masses of matter. For solids undergoing shear, deformation – or distortion of the material – may be elastic (recoverable) or plastic (permanent), or failure by rupture (faulting) may ensue.

shear strength   The internal resistance offered to shear stress, measured by the maximum shear stress that can be sustained without failure.

shear waves, seismic   So-called S waves, transmitted by an undulatory vibration transverse to the propagation direction, similar to the movement of energy along an oscillating rope or string with fixed ends. The waves are called "S" for "secondary," because they travel at lesser speeds than do primary waves and hence are received later by a seismograph station.

sheet silicate   A group of minerals characterized by a layer of silicon-oxygen tetrahedral six-member rings polymerized to infinity in two directions. The unit repeat of the sheet is $(Si_4O_{10})^{-4}$, and each tetrahedron has three bridging oxygens and one nonbridging oxygen. (Of course, only half of each bridging oxygen can be assigned to a specific tetrahedron.) Up to one-fourth of the tetrahedral silicon is replaced by aluminum in certain minerals.

shield volcano   A volcanic edifice characterized by a broad, low outline, formed primarily through the extrusion of very fluid lava. Mauna Loa is a good example. Like Mauna Loa (and Mauna Kea), most shield volcanoes are basaltic.

sial   Continental crust of the Earth, relatively rich in SiO_2, Al_2O_3, K_2O, Na_2O, volatiles, and radioactive elements.

silicic igneous rock   Magmatic rock rich in silica, generally also sodium and potassium, and poor in ferromagnesian constituents such as iron, magnesium, and calcium.

sill   A tabular igneous intrusive concordant (parallel) to the layering of the wall rocks.

silstone   A slightly finer-grained equivalent of sandstone. (See Chap. 5 for descriptions and classification.)

sima   Oceanic crust of the Earth, relatively rich in CaO, FeO, and MgO and poor in alkalis, volatiles, and SiO_2.

single-chain silicate   A group of minerals characterized by a chain of silicon-oxygen tetrahedra polymerized to infinity along a single direction. The unit repeat of the chain is $(SiO_3)^{-2}$, and each tetrahedron possesses two bridging and two nonbridging oxygens. (Of course, only half of each bridging oxygen can be assigned to a specific tetrahedron.)

site-specific resource A resource that is by nature confined to a particular location (e.g., hydroelectric power).

slate A fine-grained metamorphic rock, the precursor of which was typically a mudstone. Numerous parallel fractures impart a cleavage to such rocks. (See Chap. 5 for descriptions and classification.)

small circle Any circle described by a plane that passes through a sphere but does not include the center of the body. On the Earth, lines of latitude are examples of small circles.

smelting (roasting) Heating an ore to high temperatures in order to drive off volatiles such as sulfur.

soil The product of a weathering "front" passing through bedrock; the material consisting of both organic and inorganic detritus and secondary products of weathering that accumulate as a weathering front moves through bedrock.

soil horizon A layer of soil that is distinguished from adjacent layers by physical properties including structure, color, or texture, or by chemical properties including organic matter or acidity. Soil horizons are subhorizontal and do not reflect sedimentation events; instead, they reflect the migration of materials within a weathering profile.

soil profile Zonation from surficial dirt downward to fresh rock: zone A = organic-rich, unconsolidated; zone B = Al_2O_3 and Fe_2O_3-rich; zone C = fresh bedrock. (See Chap. 8 for a more complete description.)

soil skeleton Soil matrix.

solar insolation The level of solar radiation (given in watts per square meter) reaching an area of the Earth's surface.

solar nebula Primordial cloud of gas, condensation of which resulted in formation of the Sun and encircling planetary bodies, meteorites, and comets of the peripheral Oort cloud. In the aggregate, these condensates constitute our solar system.

solar wind (solar flux) The emission of energetic particles (hot electrons and protons, minor He^{+2} ions, plasma) as well as electromagnetic radiation from the Sun. The solar wind is responsible for the fact that comet tails point directly away from the Sun, regardless of their trajectory through the solar system.

solid solution Chemical variation of a mineral series among several (two or more) end member compositions. A good example is the plagioclase solid solution series, specific mineral examples of which range from albite ($NaAlSi_3O_8$) to anorthite ($CaAl_2Si_2O_8$).

sorptivity Parameter representing the surface tension dominated component of infiltration.

source rock Organic-rich rock from which oil or gas originates.

specific gravity Mass of an object divided by the mass of an equivalent volume of water (at 4°C).

spreading axis (spreading center, oceanic ridge, divergent plate boundary) The locus where ocean crust–capped lithospheric plates move orthogonally apart due to convection in the asthenosphere.

spreading center A linear zone along which new ocean crust–capped lithosphere forms – commonly called a divergent plate boundary. This zone is marked by an oceanic ridge.

sterane A biomarker compound, derived from cholesterol or other sterol, that indicates that eukaryotes contributed to the organic matter preserved in a sediment.

stomatal conductance A parameter that characterizes the degree of opening of the stomates, the pores in leaves that regulate the inward diffusion of carbon dioxide and the outward diffusion of water vapor. Stomatal conductance is physiologically regulated to maintain an acceptable compromise between carbon dioxide uptake (greater uptake is a benefit) and water loss (greater loss is a liability). Because nature has never developed a membrane that allows passage of carbon dioxide but not water vapor, plants must continuously form a linkage between carbon dioxide uptake and water loss.

stony meteorite Any meteorite composed chiefly of silicate materials – typically, magnesium-rich minerals such as olivine and/or enstatite.

strain Change in shape or volume of a body as a result of stress. The response of a substance – in the geologic case, a rock – to differential forces. Strain includes brittle failure (fracture), elastic distortion, and ductile deformation.

strata Sedimentary layers of rock, formed parallel to the original floor of the basin of deposition.

strategic mineral A resource that is essential for the efficient running of a nation, and that must be imported from abroad.

stratigraphic trap Lateral changes in the petroleum-bearing reservoir rock that cause the reservoir volume to decrease to zero thickness, thus pinching out and confining the oil-and gas-bearing horizon.

stratigraphy The study of layered sedimentary (and volcanogenic sedimentary) rocks.

stratosphere The layer of the atmosphere overlying the tropopause, and extending to an altitude of approximately 50 kilometers above the surface of the Earth. This region contains the main ozone layer, in which ozone's absorbtion of light energy causes warmer air to overlie cooler air. The motion of air in the stratosphere is restricted to very thin layers, or strata.

stratovolcano A symmetric central cone, formed primarily through the venting of viscous lava and voluminous ash, typically andesitic or dacitic.

stress Differential forces, or pressures, applied to a substance – in the geologic case, a rock.

strike-slip fault A fault characterized by differential horizontal motion parallel to the length of the fracture. Most large strike-slip faults are actually continental transforms.

stromatolites Layered rocks, formed from the lithification of laminated microbial communities. The organisms are primitive, carbonate-secreting algae that build colonial mats and lagoonal incrustations and mounds in marine water.

structural trap Folds and faults in which impermeable cap rocks overlie, or are juxtaposed against, a petroleum-bearing reservoir rock, preventing migration and escape of the oil and gas.

subduction The process whereby lithosphere is destroyed by descending into the deeper mantle in the vicinity of an oceanic trench. Typically, the sinking slab is old lithosphere capped by oceanic crust. As it sinks into the mantle, it gradually heats up and loses its distinctive plate-like character.

subjacent pluton (batholith, stock, cupola) Downward-enlarging igneous body of irregular shape, at least partly discordant. Large ones (> 300 square kilometers) are called batholiths; successively smaller exposed bodies are termed stocks, plugs, and cupolas.

subsilicic igneous rock   Magmatic rock poor in silica and alkalis, generally rich in ferromagnesian constituents such as iron, magnesium, and calcium.

subsistence agriculture   Farming at a level that provides only enough food to meet the needs of farm families.

substitution solid solution   A solid solution series in which chemical variation is described as replacement of one (or more) ion(s) by another of similar properties. This is the most common type of solid solution in minerals. For example, the hypersthenes are intermediate $Fe^{+2} + Mg$ orthopyroxenes whose compositions may be derived from the magnesian end member enstatite ($MgSiO_3$) by the substitution of ferrous iron for magnesium. The chosen example involves cation exchange; however, anion replacement (e.g., F^- for OH^-) is also possible in some minerals.

sulfate reduction   The metabolic process that converts sulfate to sulfide, oxidizing organic material in the process.

sulfuretum   An environment in which the various metabolic transformations of the sulfur cycle take place.

supercratonal stage   The Proterozoic Era in Earth history typified by slow, laminar flow of middle-sized mantle convection cells. Supercontinental assemblies represent the accretion of Archean microcontinental and island-arc fragments, plus continental crust generation.

superposition, law of   For stratified rocks deposited at the Earth's surface, older units are located near the base of a layered section, with the oldest at the base; progressively younger units occur farther upsection because they were laid down later.

surface water   Liquid water above the ground surface. Commonly refers to lakes and steams, not to oceans.

surface waves   Vibrational energy propagated along the surface of a body – such as the Earth.

sustainable development   Growth and development of the anthrosphere (humanity) in dynamic equilibrium with the natural environment (i.e., producing no long-term depletion in natural resources and ecosystem services).

swamp   Nontidal wetland dominated by trees and shrubs.

symbiosis   The intimate union or very close association of two dissimilar organisms.

symmetry   That property of an object, such as a crystal, whereby a translation, rotation, or reflection across a mirror plane results in self-coincidence (an identical aspect).

syncline   A hammock-shaped fold in layered rocks in which the beds are inclined (dip downward) toward the central low.

synclinorium   A great, linear downwarp consisting of subparallel, subsidiary anticlines and dominant synclines. Basically, a large syncline with superimposed minor anticlines (folding at two different scales).

talc   A hydrous sheet silicate typified by $(Si_4O_{10})^{-4}$ groups, and possessing the formula $Mg_3Si_4O_{10}(OH)_2$.

talus   The clastic debris that accumulates at the base of a steep slope – often building up a cone of material toward the source area.

tar sand   A petroleum reservoir rock in which the liquid hydrocarbons have been degraded by near-surface oxidation and devolatilization to viscous, semisolid asphalt or tar.

tension   Condition in which the hydrostatic pressure of soil water is less than atmospheric pressure.

terrane   A fault-bounded block of the Earth's crust, the lithologic units of which share common or related histories, but which in aggregate contrasts strongly with adjacent blocks. Because of their lack of genetic relationships, terranes appear to be far-traveled or exotic compared to neighboring, juxtaposed terranes.

Tertiary Period   Abbreviated T, the time interval between approximately 65 and 2 million years ago.

tetrahedral chain   Infinite polymerization of silicon-oxygen tetrahedra in one dimension. Gives rise to $(SiO_3)^{-2}$ (single-chain) and, through cross-linking, $(Si_4O_{11})^{-6}$ (double-chain) silicate mineral groups.

tetrahedral framework   Infinite polymerization of silicon-oxygen tetrahedra in three dimensions. Gives rise to SiO_2-type silicates. Some aluminum may substitute for tetrahedral silicon.

tetrahedral rings   Three-, four-, or six-member rings, having formulas for the structural subunits of $(Si_3O_9)^{-6}$, $(Si_4O_{12})^{-8}$, and $(Si_6O_{18})^{-12}$, respectively. In such configurations, each tetrahedron contains two bridging oxygens (each shared with another tetrahedron) and two nonbridging oxygens.

tetrahedral sheet   Infinite polymerization of silicon-oxygen tetrahedra in two dimensions. Gives rise to $(Si_4O_{10})^{-4}$ sheet silicates. Some aluminum may substitute for tetrahedral silicon.

therapsid reptiles   Distinctive, latest Paleozoic mammal-like reptiles geographically confined to the Gondwana supercontinent.

thermal gradient   Any change of temperature with distance. The increase of temperature with depth within the Earth is called the geothermal gradient, and is commonly given in units of degrees Celsius per kilometer.

thermal maturation   The distillation process that converts low-grade organic matter to liquid and gaseous hydrocarbons during heating by driving off volatile constituents.

thrust fault   A low-vertical angle fault in which the upper block, or hanging wall, has moved, relatively, up and over the lower, footwall block.

transform fault   A special type of fault, the active portion of which links segments of divergent and/or convergent lithospheric plate boundaries. Also known as a strike-slip fault or a conservative plate boundary.

transition zone   The part of the Earth's mantle sited between the upper and lower mantle, at depths of approximately 400 to 670 kilometers. Here, physical properties change rapidly but continuously with depth.

transpiration   Water loss through evaporation from plant surfaces. Under normal circumstances, the cell walls inside a leaf are saturated with water. As this water evaporates, it diffuses outward through the stomates and is replaced by additional water moved from the soil through the plant's vascular system. Transpiration increases with increasing stomatal conductance and with decreasing atmospheric. Because evaporating water consumes a large amount of energy, transpiration is a key process for dissipating solar energy absorbed by plants.

tremolite, $Ca_aMg_5Si_8O_{22}(OH)_2$.   *See* **amphibole**.

trench basin   The linear bathymetric low associated with a downturning ocean crust–capped lithospheric plate. The trench marks the surface expressing of the subduction zone, or inclined convergent plate junction.

trench complex The rocks and rocky debris that accumulate adjacent to or within the confines of an oceanic trench. Because of convergent plate motion, material originating some distance away from the trench may ultimately be brought into this site, decoupled from the downgoing, subducting plate, and accreted to the stable, nonsubducted slab.

triple junction The surface point common to three lithospheric plates. Several types are common, including ridge-ridge-ridge, and ridge-trench-transform. Such junctions may evolve rapidly with time as a result of differential plate motions.

tropopause The layer of the atmosphere separating the higher stratosphere and the lower troposphere.

troposphere The layer of the atmosphere underlying the tropopause, and extending from the surface of the Earth to an altitude of approximately 16 kilometers near the equator and approximately 10 kilometers near the poles. The troposphere is characterized by vigorous overturning motions.

tuff A pyroclastic deposit consisting chiefly of volcanic ash (*see* ejecta).

twinned crystals Composite crystals consisting of two (or more) portions that have identical geometries but contrasting orientations. Simple twins consist of two parts, whereas multiple twins consist of many individual crystal segments. Twinning accounts for the striations observed on plagioclase.

ultramafic rock Richer in the ferromagnesian constituents, – iron, magnesium, and, in some cases, calcium – and poorer in alkalis and silica than are mafic rocks. "Ultramafic" commonly is used nearly synonymously with the terms "peridotite" and "mantle material."

ultrametamorphism Partial melting of metamorphic rock to produce layered refractory solid and more fusible molten lithologic materials – in the aggregate, such banded units are termed migmatites.

ultraviolet-b (UV-B) Radiation at the shortest wavelengths of the solar spectrum received at the Earth's surface, 280 to 315 nanometers. The amount of UV-B reaching the Earth's surface increases as the stratospheric concentration of ozone decreases. The waxy surface of the outer layer of leaves filters most UV radiation, partially protecting plants from the damaging effects of UV radiation on living tissues.

uniformitarianism, hypothesis of The theory that the types of geologic processes operating today attended the Earth in the distant geologic past, reflecting current modes of geologic activity as well as the immutability of the laws of physics and chemistry.

unit cell The atomic repeat periodicity along the three principal crystallographic axes of a crystalline material. Mineral grains and crystals are edifices constructed by the combination of unit cells in exactly the same orientation, neither leaving gaps nor allowing overlaps.

upper mantle The outer portion of the Earth's mantle extending from the base of the crust to depths of approximately 400 ± 50 kilometers.

valence electron(s) Outer-orbital electron(s) that may be removed to form a positively charged ion (cation), or added to form a negatively charged ion (anion); both types of stable ion (+ and −) possess the noble gas electronic configuration.

van der Waals bonding Extremely weak or "residual" attractive forces resulting from the nonspherical distribution of positive and negative charges between bonded atoms.

vein deposit Passing through fractures in rocks, hot waters encounter changing physical and/or chemical conditions. As they cool or chemically react with their surroundings, those hydrothermal solutions precipitate minerals. Whether the resultant mineralogic aggregate is economic or not, the deposit is termed a vein.

viscosity The relative difficulty with which a substance flows; the opposite is fluidity.

viscous drag The entrainment of buoyant material on a sinking, relatively dense slab because of structural coherence and integrity of the entire unit. Low-density sediments are carried down a subduction zone piggyback on the descending ocean–crust capped plate as a result of viscous drag.

volatile phase A gas, consisting of one or more highly volatile constituents, e.g., H_2O, CO_2, H_2S.

volcanic breccia A rock that consists chiefly of angular fragments of lava, pumice, conduit walls, etc.

volcanic neck A pipe-like or fissure-like conduit filled with igneous rock, thought to represent the throat of a volcano.

volcanic/plutonic arc An island arc or continental margin characterized by voluminous volcanic rocks and their deep-seated, intrusive equivalents. Andesite and dacite are the characteristic – but not exclusive – rock types.

water table The surface in an unconfined water body at which the hydrostatic pressure is equal to atmospheric.

watershed A drainage basin.

wave base The ocean depth (typically, a few meters) below which wave action, and therefore erosion, does not exist.

weathering The complex physical and chemical processes resulting in the breakdown and decay of rocks and minerals and the formation of soils under near-surface geologic conditions by air, water, and biologic products (*see* Chap. 8).

wetland Transitional marshlands between terrestrial and aquatic ecosystems. Here the water table is usually at or near the surface, or the land is actually submerged beneath shallow water.

zeolite Any of a group of hydrous framework silicates typified by low densities and relatively open crystal structures.

zeolite-facies rock A very low-grade metamorphic rock characterized by zeolite minerals such as laumontite, alkali feldspars, quartz, clay minerals, and hydrous ferromagnesian phases.

zero population growth The cessation of population growth (technically, achieving a stationary population) that occurs when the rate of natural increase is zero; that is, when birth and death rates are equal.

zircon An accessory mineral possessing the formula $ZrSiO_4$.

zodiacal disk (plane of the ecliptic) The plane normal to the Sun's rotation axis that contains the planets, satellites, asteroids, and almost all the nonstellar matter within our solar system as far out as the Oort cloud.

General Index (see also *Glossary*)

Geographical Index

Significant Events and Figures Index